Atomic Weights *continued*

Element	Symbol	Atomic Number	Atomic Weight
Mercury	Hg	80	200.59
Molybdenum	Mo	42	95.94
Neodymium	Nd	60	144.24
Neon	Ne	10	20.179
Neptunium	Np	93	237.0482
Nickel	Ni	28	58.70
Niobium	Nb	41	92.9064
Nitrogen	N	7	14.0067
Nobelium	No	102	(259)
Osmium	Os	76	190.2
Oxygen	O	8	15.9994
Palladium	Pd	46	106.4
Phosphorus	P	15	30.97376
Platinum	Pt	78	195.09
Plutonium	Pu	94	(244)
Polonium	Po	84	(209)
Potassium	K	19	39.0983
Praseodymium	Pr	59	140.9077
Promethium	Pm	61	(145)
Protactinium	Pa	91	231.0359
Radium	Ra	88	226.0254
Radon	Rn	86	(222)
Rhenium	Re	75	186.2
Rhodium	Rh	45	102.9055
Rubidium	Rb	37	85.4678
Ruthenium	Ru	44	101.07
Samarium	Sm	62	150.4
Scandium	Sc	21	44.9559
Selenium	Se	34	78.96
Silicon	Si	14	28.055
Silver	Ag	47	107.868
Sodium	Na	11	22.98977
Strontium	Sr	38	87.62
Sulfur	S	16	32.06
Tantalum	Ta	73	180.9479
Technetium	Tc	43	(97)
Tellurium	Te	52	127.60
Terbium	Tb	65	158.9254
Thallium	Tl	81	204.37
Thorium	Th	90	232.0381
Thulium	Tm	69	168.9342
Tin	Sn	50	118.69
Titanium	Ti	22	47.90
Tungsten	W	74	183.85
Uranium	U	92	238.029
Vanadium	V	23	50.9415
Xenon	Xe	54	131.30
Ytterbium	Yb	70	173.04
Yttrium	Y	39	88.9059
Zinc	Zn	30	65.38
Zirconium	Zr	40	91.22

Numbers in parentheses are mass numbers of most stable or most common isotope.
Atomic weights shown above were used to calculate formula weights given on back cover.

ANALYTICAL
CHEMISTRY
Principles

ANALYTICAL
CHEMISTRY
Principles

John H. Kennedy
University of California, Santa Barbara

HARCOURT BRACE JOVANOVICH, PUBLISHERS

San Diego   New York   Chicago   Atlanta   Washington, D.C.

London   Sydney   Toronto

*To my wife, Victoria,
and our prayer group
without whose support this
project would not have
reached fruition.*

Copyrights and Acknowledgments

24 Table 2-3 adapted with permission from G. H. Morrison (1971). Permission to reprint is granted by the American Chemical Society. Copyright 1971 by the American Chemical Society. **34** Table 2-7 taken from I. M. Kolthoff and P. J. Elving, eds., *Treatise on Analytical Chemistry*, 2nd ed., pt. 1, vol. 1, p. 267. (New York: John Wiley & Sons, Inc., 1978). Copyright 1978 by John Wiley & Sons, Inc. Reprinted by permission of the publisher. **37** Table 2-8 adapted with permission from R. B. Dean and W. J. Dixon (1951). Permission to reprint granted by the American Chemical Society. Copyright 1951 by the American Chemical Society. **51** Example 2-6 from J. J. Delfino and D. B. Easty, "Interlaboratory Study of the Determination of Polychlorinated Biphenyls in a Paper Mill Effluent," from *Analytical Chemistry* (1979). Permission to reprint granted by the American Chemical Society. Copyright 1979 by the American Chemical Society. **121** Table 5-1 adapted with permission from J. Kielland (1937). Permission to reprint granted by the American Chemical Society. Copyright 1937 by the American Chemical Society. **125** Table 5-2 taken from data in H. Stephen and T. Stephen, eds., *Solubilities of Inorganic and Organic Compounds*, vol. 1 (New York: Pergamon, Inc., 1963). Reprinted by permission of Pergamon Press, Inc. **125** Table 5-3 taken from I. M. Kolthoff and E. B. Sandell, *Textbook of Quantitative Inorganic Analysis*, 3rd ed. (New York: Macmillan Publishing Company, 1952). Reprinted by permission of Macmillan Publishing Company. **251** Table 9-3 taken from data from E. Bishop, ed., *Indicators* (Elmsford, N.Y.: Pergamon Press, Inc., 1972). Reprinted by permission of Pergamon Press, Inc. **300** Table 10-2 data taken from D. G. Peters, J. M. Hayes, and G. M. Hieftje, *Chemical Separations and Measurements* (Philadelphia: W. B. Saunders Co., 1974). Reprinted by permission of W. B. Saunders Company, Inc. **306** Table 10-3 information excerpted from G. Schwarzenbach and H. Flaschka, *Complexometric Titrations* (H. M. N. H. Irving, trans.) 2nd ed., (London: Methuen & Co. Ltd., 1969).

Copyrights and Acknowledgments on page 737 constitute a continuation of the copyright page.

Preface

As the title implies, *Analytical Chemistry: Principles* attempts to describe many of the theoretical principles upon which modern analytical chemistry rests, as well as some of the practical aspects that must be considered in the development of an analytical procedure. My primary objective in writing this text has been to continue the time-honored traditions and to include modern applications (for example, environmental and clinical analysis). This approach gives students the opportunity to see that analytical chemistry can be interesting as well as exacting. The laboratory manual, *Analytical Chemistry: Practice*, also reflects this approach.

The concept of a "total analysis process," from sampling to evaluation of results, is stressed throughout the text. The objective is to prevent students from coming away from the course thinking analytical chemistry is merely performing a quantitative measurement — getting a number.

Library projects listed in Chapter One require students to go through the entire analytical process for a particular analytical problem. Such a project may be assigned near the end of the course in place of a traditional problem-solving examination.

The material is divided into three equally weighted divisions. The first covers traditional classic principles, including data handling and gravimetric and volumetric methods. The second addresses the more modern analytical methods, commonly thought of as instrumental techniques. The parts in this division are of equal importance: Spectrochemical Methods, Electroanalytical Methods, and Analytical Separations. The third division is the separate laboratory manual, *Analytical Chemistry: Practice*, which presents experimental procedures. The experiments complement the lecture material; the procedures do not attempt to give a complete theoretical background.

At the end of each chapter are many problems, whose scope reflects the intertwining of theoretical principles and applications presented in the chapter. In addition, there are problems relevant to previous material, such as error analysis or comparing two analytical methods.

The appendixes include some mathematical developments that are more advanced than the main body of the text and, in any event, would be distracting if presented there. Some simple computer programs are included to solve analytical problems, such as titration curves, presented in the test. BASIC language is used because it is so widely adopted in desktop computers often available to undergraduate students. Even when computers are not available, students can see the flow of logic in solving a problem. This should help them learn problem-solving techniques.

The theoretical material fits comfortably into a two-term course covering Parts One to Three in the first term and Parts Four to Six in the second term. There are more than enough experimental procedures from which to choose for a two-term course. However, the author has used the text material to teach a one-term course by choosing the main themes in each chapter and omitting some of the detailed developments.

I want to thank many people who helped in the preparation of this text. Professor Harold MacNair of the Virginia Polytechnic Institute reviewed the entire first draft, and his suggestions were of great value. Helpful comments were also received from Professor Richard Linton of the University of North Carolina, Chapel Hill, who reviewed, in detail, the chapters on spectrochemical methods of analysis. Other reviewers who made helpful suggestions were: C. B. Boss, North Carolina State University, Raleigh; Dale Chatfield, San Diego State University; Geoffrey Coleman, University of Georgia; Glen Dryhurst, University of Oklahoma; Quintus Fernando, University of Arizona, Tucson; Robert Michel, University of Connecticut, Storrs; John C. Wright, University of Wisconsin, Madison.

Many thanks are due to my graduate teaching assistants, Susan Stuber, John Stickney, and Ann Schuler, who helped check many of the problem solutions, and to my students in analytical chemistry, who were often guinea pigs. Thanks go to the many persons and companies who gave permission to use their illustrations, experimental procedures, and data. The painstaking work required to prepare the manuscript for publication, particularly by David Esner and Barbara Girard, is gratefully acknowledged. Finally, I thank my wife, Victoria, for the many long hours spent typing the manuscript and making the necessary sacrifices to keep the project on schedule.

John H. Kennedy

Contents

ANALYTICAL
CHEMISTRY
Principles

BASIC CONCEPTS

PART ONE

Analysis as a Process

1

Many people find the word "chemistry" sufficiently forbidding, but when it is modified by the term "analytical," the resulting phrase may seem even more cold and exacting. Even the dictionary definition of "analysis"—a separating or breaking up of any whole into its parts so as to find out their nature—suggests something destructive. On the other hand, the definition of "synthesis"—putting together of parts or elements so as to form a whole—appears to be more constructive. However, the analytical process is a constructive attempt by chemists to determine the composition and to understand the nature of materials.

Another distorted view of analytical chemistry is that it consists merely of measuring some property—obtaining a number. This highly important step is called *quantitation*. However, this one step may not be the most critical one for completing a successful analysis. For our purposes, we will assume that the total analysis process consists of the following five basic steps:

1. Sampling—obtaining a *representative* sample.

2. Method—choosing an appropriate analytical method.

3. Separation—separating the substance being analyzed (the *analyte*) from any interfering substance.

4. Quantitation—performing a quantitative measurement.

5. Evaluation—evaluating the results.

1-1 SAMPLING

Before the analysis can begin, the analytical chemist must define the problem, and this task may not be as easy as it sounds. In the controlled environment of an analytical chemistry laboratory course, carefully prepared samples are handed out and students are instructed to follow a specified method to determine the amount of some element the sample contains. Outside the classroom, the problem is seldom so straightforward.

"Does Company *A*'s product infringe Company *B*'s patents?" "Does the latest shipment of steel bars meet specifications?" "Does the amount of detergent discharged into the river by Company *C* violate Federal and state environmental protection laws?" "What is the value of the iron ore in the barge that just arrived at the pier?" To answer such questions, an analytical chemist must decide what analyses are required and how to obtain representative samples. The term *representative* implies that the sample chosen for analysis will reflect the entire body from which it came. Consider the iron ore in the barge. How does an analytical chemist choose an ore sample, weighing a few grams at most, that will contain the same percentage of iron as the several tons of iron ore in the barge?

The iron ore problem deals with *inhomogeneous* samples—those with a composition that is not uniform. However, sometimes the samples are homogeneous, but the system itself fluctuates, as is the case in most environmental analyses. For example, to measure the waste water effluent leaving Company *C*, it would be useless to base your analysis simply on samples gathered when the factory was shut down or operating at reduced capacity. Even more difficult would be deciding where and when to take a sample of air that would be representative of a large area such as the Los Angeles Basin. In this case, samples would have to be taken periodically or, better yet, monitored continuously from many sampling stations at various elevations. Even then, it might still be impossible to determine definitively whether the air quality in Los Angeles is now better or worse than it was 20 years ago.

Another problem that analytical chemists confront when attempting to obtain a representative sample is how to transport the sample from the field to the laboratory without altering the sample. For example, the atmospheric concentration of a pollutant, such as sulfur dioxide (SO_2) or ozone (O_3), may have a steady-state value. At a *steady state*, the material is being produced and transformed into some other form at the same rate. For example, SO_2 is slowly oxidized in air to sulfur trioxide (SO_3). But when an air sample is removed from the environment, the production reaction may no longer occur although the oxidation reaction continues. By the time the analyst gets the sample back to the laboratory, a significant amount of the pollutant may have disappeared, thereby yielding a meaningless analytical result.

Transportation of samples may lead to other problems, especially when trace quantities of some element are involved. If a water sample is collected in a glass bottle, impurities in the glass may "add" to the total pollutants in the water. If the water sample is collected in a plastic bottle, some pollutants may be absorbed by the plastic, thereby lowering the concentration of pollutants in the water.

In short, sampling—the first step in the total analysis process—is the basis for all the other steps. If it is not done carefully, the entire analysis will be invalidated.

1-2 METHOD

No single "best" way to analyze for a particular constituent exists. The "best" method is the one most suited to the circumstances of the analysis. For example, a highly accurate classical method for determining chloride content involves precipitating silver chloride (AgCl) and weighing the precipitate. Thousands of students have carried out this analysis over the years, usually obtaining results well within 0.5% of the true value. However, for determining the chloride content of drinking water, the precipitation of AgCl would be an inappropriate method. The inherent solubility of AgCl, an insignificant factor for samples containing large amounts of chloride, would lead to such low results for samples containing trace amounts of chloride that an analyst might easily conclude that the sample contained no chloride at all because the AgCl solubility product was not exceeded and no precipitate was formed. Conversely, a suitable method for measuring trace amounts of chloride in drinking water might not be sufficiently accurate for samples containing larger concentrations.

The analytical chemist must decide which of the numerous analytical methods available will be appropriate for the problem at hand. Factors that the analyst may have to consider include accuracy, sensitivity, selectivity, speed, cost, and legality.

Accuracy The analytical chemist normally prefers to use the most accurate method available. However, other criteria, some of which follow, may preclude the use of the most accurate method. The analyst must assess the required level of accuracy and eliminate from consideration only those methods that do not meet this requirement.

Sensitivity The term *sensitivity* is frequently used to denote the minimum quantity of detectable analyte.[1] Sensitivity is especially important in trace analysis. The method used must allow the chemist to detect and determine the desired constituent (the *analyte*) at the level it is likely to be encountered. For example, the concentration of fluoride in fluoridated drinking water is approximately one part per million (5×10^{-5} M), that of ozone in a polluted atmosphere may be a few tenths of a part per million, whereas the maximum acceptable amount of phenol in drinking water is one part per billion according to Federal law.

Selectivity In some samples, constituents other than the analyte may interfere with the analysis. The interference may add to the measured signal (causing a positive error) or may decrease the measured signal (causing a negative error). For example, in the method for determining ozone content by measuring iodide oxidation to iodine, nitrogen dioxide (NO_2) gives a positive interference because it, too, oxidizes the iodide ion, whereas SO_2 gives a negative interference because it reduces iodine.

If the nonanalyte constituents present in the sample are known, the analyst may be able to find or develop a method that will not be sensitive to those constituents. In a truly specific method, no concentration of any substance other than the analyte would exhibit an interference. The best that can usually be attained is a highly selective method in which few substances interfere and then only when present at much higher concentrations than that of the analyte. For

---

[1] *Sensitivity* is also used to denote the amount of measurement signal observed for a unit amount of analyte. In this text, sensitivity will generally have this second definition. The two meanings are related in the sense that the higher the signal observed for a given amount of analyte, the lower the minimum detectable values will usually be.

example, when the glass electrode is used for determining acidity (the hydrogen ion concentration), the only significant interference is from sodium ion when it is present at concentrations 10^{12} times greater than the hydrogen ion concentration.

Speed   Although some methods are accurate, sensitive, and selective, they are still inappropriate because they are too slow. For example, if your method for monitoring highly toxic beryllium dust in a machine shop involved collection of a sample over several hours, dissolution, precipitation of a beryllium compound, filtration, ignition, and weighing of the product, everyone in the shop might be dead by the time the results were available! But the criterion need not involve personal danger. For example, a slow analysis may endanger company profits. A quality-control analyst monitoring a high-speed production line cannot afford to use a slow method of analysis. By the time the analyst discovers that the product is inferior, another million just like it may have been produced.

Cost   After considering all these factors, the analytical chemist may recommend a method of analysis requiring an instrument that costs $100,000. Although the chemist might have to argue the case that there is no less expensive method available, high-cost equipment often saves considerable labor time per sample. Clearly, saving a few dollars of labor costs per analysis would not be a reasonable return on investment when only a few analyses are to be conducted. However, when several thousand analyses are contemplated, the return could be substantial.

In a few situations, the analyst can use a method of choice no matter what the cost. When *Apollo* astronauts brought back moon rocks to be analyzed, the cost of fetching the samples far exceeded the cost of any analytical method. It would have been pointless to use an inferior analytical method only to save a few dollars.

Legality   Today the analytical chemist must be aware of how well an analytical method will stand up in court. There may be an "accepted" method for the analyte in question. For example, the Environmental Protection Agency (EPA) publishes the methods accepted for measuring various air and water pollutants. An analyst who uses some other method must be prepared to prove that this method is equal to the accepted method. In many situations, because this burden of proof may be too time-consuming, the analyst should use the accepted method.

1-3 SEPARATION

In choosing an appropriate analytical method, the analyst will seek highly selective methods. However, even a method that is not highly selective may still meet all the other requirements for an appropriate method. In such cases, a separation step must be taken before a quantitative measurement (*quantitation*) can be made. The separation step can take many forms: The interfering substances may be separated from the analyte by precipitation or extraction into another solvent, or the analyte may be removed from the interfering substances by one of these techniques. For some substances, the separation step may be quite specific. For example, because of its ability to dissolve hydrogen selectively, palladium metal is used to remove H_2 from a gas stream. Another example is the use of distillation to remove highly volatile mercury from an ore sample after reduction to the elemental state. Several general-use techniques can even be used to separate all the components of a mixture. Several chromatographic techniques (see Chapter 19) are widely used.

Although it is not strictly "separation," the technique of *masking* can be used in the separation step to render some interfering substances innocuous. For example, in the determination of fluoride ion in water some metal ions, such as iron and aluminum, interfere because they form complexes with fluoride ion. These metal ions can be "masked" by adding a substance, such as sodium citrate, that forms stronger complexes with the metal ions, thereby preventing the formation of fluoride complexes.

The separation step may be so critical to the success of the analysis that the analytical method will be designated by the separation method instead of the more traditional quantitation step. In fact, after a mixture has been separated into its components, the quantitation step may be nonspecific—that is, it may merely indicate the amount of any component after completion of the separation step.

1-4 QUANTITATION

Numbers would be meaningless if the three previous steps were not taken correctly. *Quantitation* is the taking of a quantitative measurement functionally related to the analyte. However, because all steps of the total analysis process are interrelated, analytical chemists frequently use the term to mean the total analysis process itself. Remember that this functional relationship may take many forms; only on rare occasion is the quantitative measurement numerically *equal to* the mass of the analyte. More often the *measurement*, or *signal*, S is directly proportional to the analyte A:

$$S = kA \qquad\qquad (1-1)$$

The proportionality constant k is a measure of the sensitivity of the method—that is, the larger k is, the greater S will be for a given level of analyte A. However, in principle, any functional relationship between analyte and measurement signal can be used, even if it is only an empirical relationship determined by measurements of known samples. Moreover, some functional relationships between measurement signal and analyte—such as those for transmitted light or electrochemical potential—are known from theory but are not linear.

Analytical methods are normally classified by the measurement signal employed. However, as mentioned earlier, some separation methods are so critical to the analysis that the method becomes known by the separation method. Several examples of both methods of classification are listed in Table 1-1.

TABLE 1-1 Classification of Analytical Methods

Measurement signal	Analytical method
by measurement signal	
mass	gravimetric analysis
volume	volumetric analysis
any electrical quantity	electroanalysis
potential	potentiometry
voltage and current	voltammetry
current	amperometry
coulombs	coulometry
electromagnetic radiation absorbed or emitted	spectroscopy
heat or temperature	thermal analysis
by separation step	
electrical resistance	gas chromatography (thermal conductivity detector)
electrical current increase	gas chromatography (flame ionization detector)
electrical current decrease	gas chromatography (electron capture detector)
electromagnetic radiation absorbed	liquid chromatography (ultraviolet light detector)
electrical conductivity increase	ion chromatography

1-5 EVALUATION

After the first four steps of the total analysis process have been successfully completed, the analytical chemist still must evaluate the results. If a reported value is 56.38% calcium, how confident is the analyst that the true value is not greater than 56.4, 56.5, or 57.0% calcium? Has the analytical question initially posed been answered? Although quantitative measurements may have been required, the answer to the original question may not involve a number. Several such questions were suggested in the discussion involving sampling.

Statistical analysis is an important evaluative tool for the analytical chemist. Unfortunately, even if the statistical treatment is beyond reproach—which would be highly unlikely because statistics can be used to prove or disprove most anything—the final answer will be couched in probability terms. The analyst may say, "I am 95% certain the value lies between 56.28% and 56.48% calcium" or "I am 99% certain Company *A*'s product infringes upon Company *B*'s patents." However, other than a personal feeling about the validity of the methods used, the analytical chemist is never 100% sure of any analysis.

Statistical analysis may be used to determine the precision required of the analytical technique and the number of samples that must be analyzed to answer the question to within specified limits of uncertainty. For example, suppose you must determine the percentage of silver in an alloy to $\pm 0.02\%$ (with 95% confidence), but the best available analytical method has an inherent uncertainty of $\pm 0.05\%$. At first glance, it may seem impossible to make a determination to $\pm 0.02\%$. However, as will be demonstrated in Section 2-5, it can be done by analyzing at least 25 samples.

1-6 SUMMARY

The most illustrative way to summarize the total analysis process is to show how it can be used to tackle a real analytical problem. Suppose you must establish a method for the monitoring and control of your city's fluoridated water supply. How would you accomplish each of the five steps in the total analysis process?

1. **Sampling**   To ensure that the fluoride level is not drifting, it will be necessary to collect samples frequently. Therefore, a continuous monitoring system should be initiated. Samples should be gathered far enough away from the point of fluoridation to ensure that adequate mixing has taken place but close enough to minimize the delay between fluoridation and analysis. If samples are to be taken to the laboratory, polyethylene rather than glass containers should be used because fluoride ion reacts with glass and could lead to artificially low results.

2. **Method**   Few analytical methods are available for the determination of fluoride ion at the low concentrations in drinking water ($\sim 10^{-5}\ M$). One of the best methods involves an ion-selective electrode with a potential related to the fluoride ion concentration (Section 15-3). This method is also well suited to continuous monitoring, provided that any interfering substances can be separated or masked.

3. **Separation**   The electrode is highly selective; only hydroxide ions (OH^-) will add to the measurement signal, thereby causing a positive error. If the pH of the water is below 8, little interference will be contributed by OH^-. Therefore, the pH of the water should also be monitored. (This is a normal procedure. Moreover, a water supply would rarely have a pH above 7 because of the natural presence of carbon dioxide.) A more serious concern would be the possible presence of trivalent ions, such as Fe(III) or Al(III), which form complexes with fluoride ion and thereby decrease the concentration of free fluoride ion. These interferences may preclude direct measurement of fluoride in the water. In such an event, a flow-sampling system would be designed to draw samples of water and add an agent that forms strong complexes with these interfering ions (a masking technique).

4. **Quantitation**   The electrochemical potential of the fluoride ion-selective electrode would be measured in comparison to a reference electrode potential.

Because the potential is also temperature dependent, a constant-temperature environment must be used or correction made for temperature changes. The basic relationship between electrochemical potential and fluoride ion concentration (neglecting nonideality of ions in solution, which will be discussed in Chapter 4) would then be calculated (using techniques that will be discussed in Chapter 15). A calibration chart, as shown in Figure 1-1, can be prepared for water samples containing known amounts of fluoride ion. The fluoride level in the water supply can be determined from the measurement of potential and the calibration chart.

5. **Evaluation**   The necessity for keeping accurate records of any analysis is self-evident. In the extended example of analysis for fluoride level, good records would provide the basic evidence in possible court cases claiming that fluoride concentrations reached unsafe levels at some point in time. This is one reason why continuous monitoring would be preferred.

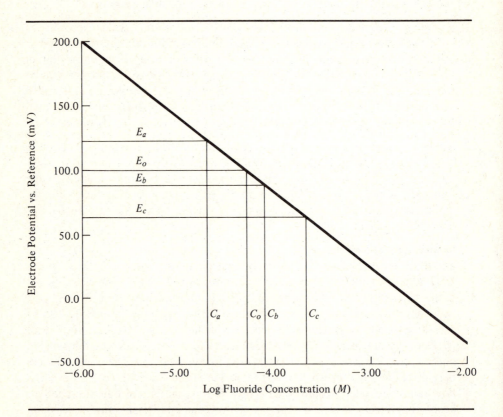

FIGURE 1-1   Fluoride ion concentration in fluoridated water.

Another function of the analysis would be for process control. As shown in Figure 1-1, if the potential rises above E_a, additional fluoride should be added because C_a is significantly lower than the recommended fluoridation level, C_0. If the potential drops below E_b, fluoride addition must be decreased or stopped entirely because C_b is significantly higher than the recommended fluoridation level. At the alarm level, E_c, more drastic action must be taken because the fluoride concentration, C_c, has reached a dangerous level. Naturally, to ensure the validity of the control function, it might also be necessary to monitor the water supply *before* it is fluoridated. The rate at which fluoride would be added would depend on any incoming levels of fluoride ion from natural sources.

The extended example of fluoride in the water supply demonstrates that many factors must be considered before a meaningful analysis can be made and that the results of such analysis may be positively or negatively affected by the validity of any step in the total system.

1-7   ORGANIZATION OF TEXT

It would not be prudent to attempt such an analysis or any kind of laboratory work without knowing basic calculations and methods by which to evaluate your results. Therefore, this book begins with step 5—evaluation (Chapter 2) and a review of stoichiometric (Chapter 3) and equilibrium-constant (Chapter 4) calculations. Then various methods of analysis suggested in Table 1-1 will be explored (Chapters 5 to 17) followed by separation methods (Chapters 18 and 19). A separate volume treats laboratory and sampling procedures and provides concrete examples of the various theoretical methods discussed in this text. Thus all steps in the total analysis process will be covered with the exception of specific ways to choose an appropriate method for a particular analysis. This step depends on the particular analytical problem under consideration. Some examples will be described throughout the text, and the library projects listed as problems at the end of this chapter can be used to familiarize yourself with the methods analytical chemists use to accomplish all five steps in the total analysis process.

SUPPLEMENTARY READING

General References Describing Analytical Methods

Beilby, A.L., "Comparison of Analytical Methods," *Journal of Chemical Education*, 49:679 (1972).

Cleaning Our Environment: A Chemical Perspective, chap. 2. Washington: American Chemical Society, 1978.

Kratochvil, B., and J.K. Taylor, "Sampling for Chemical Analysis," *Analytical Chemistry*, 53:924A (1981).

Official Methods of Analysis, 13th ed. Washington: Association of Official Analytical Chemists, 1980.

Standard Methods for the Examination of Water and Wastewater, 15th ed. American Public Health Association (1981).

Solving Analytical Problems

"Analysis of Orange Juice" (focus feature article), *Analytical Chemistry, 52*:1269A (1980).

Bentz, A.P., "Who Spilled the Oil," *Analytical Chemistry, 50*:655A (1978).

Budinger, P.A., T.L. Drenski, A.W. Varnes, and J.R. Mooney, "The Case of the Great Yellow Cake Caper," *Analytical Chemistry, 52*:942A (1980).

Hall, L.H., and R.F. Hirsch, "Detection of Curare in the Jascalevich Murder Trial," *Analytical Chemistry, 51*:812A (1979).

Hills, J.P., "Legal Decisions and Opinions in Pollution Cases," *Environmental Science and Technology, 10*:234 (1976).

McCrone, W.C., "Authenticity of a Medieval Document," *Analytical Chemistry, 48*:676A (1976).

Morrison, G.H., "Evaluation of Lunar Elemental Analyses," *Analytical Chemistry, 43*:22A (June, 1971).

Reynolds, H.L., H.P. Eiduson, J.R. Weatherwax, and D.D. Dechert, "Chemistry for Consumers," *Analytical Chemistry, 44*:22A (November, 1972).

PROBLEMS

Note: The following analytical problems are intended as possible library projects to be undertaken near the end of the term, after the student is familiar with various analytical methods.

After graduation, you get a job in an analytical lab. Your first assignment is one of the following analytical problems. Describe in detail an analytical procedure to accomplish this assignment, keeping in mind the total analysis process. For example, in choosing your method, you might compare different possible techniques, demonstrating why your method is the most appropriate for this problem.

1-1. Determine the level of nitrate ion in the city's water supply.

1-2. Determine the level of SO_2 being evolved from your company's smokestacks.

1-3. Determine the level of NO_x in the exhaust of your company's cars.

1-4. Help an environmentalist group by determining the DDT content of body fat in sea gulls.

1-5. Determine the lead content in soil near a freeway.

1-6. Help establish the value of a bauxite ore by determining the aluminum content.

1-7. Determine the lead content in pottery that is being considered for import to this country.

1-8. Determine the level of trace amounts of copper in powdered milk.

1-9. Establish a public screening test of cholesterol levels in blood.

1-10. Establish a continuous monitoring system to test for hydrocarbons in the air of your company's solvent-storage warehouse.

1-11. Establish evidence for a district attorney's office by comparing trace-element contents in hair samples.

1-12. Help the district attorney's office to verify that an oil spot on the victim's driveway matches oil from the suspect's automobile.

1-13. Establish a vinyl chloride monitoring system for your company's plant.

1-14. Help a local museum verify the authenticity of an old map supposedly drawn in A.D. 1450 by analyzing ink used in drawing it.

1-15. Determine the level of chromium in lunar rocks.

1-16. Determine whether or not the cardboard used in packaging your company's cereal contains PCB's.

1-17. Verify that an excavated ivory plaque dates from the ninth century B.C.

1-18. Demonstrate that cloth fibers found under a victim's fingernails were from a suspect's shirt.

1-19. Assist "Nader's Raiders" by determining the concentration of cyanide in a sewer line downstream from a plating factory.

1-20. Determine the level of phosphates in the city's sewer system.

Evaluation of Analytical Results

2

This chapter deals with typical results from analytical determinations and will present methods of evaluating their reliability and communicating the results to others. Remember that, unless results are handled and communicated accurately, the many hours spent in the laboratory obtaining them will be wasted. Moreover, analytical chemists do not live in a vacuum: Usually they are asked to solve problems by people who need accurate, understandable answers. Even university professors of analytical chemistry who have chosen their own research problems must present their results clearly and accurately in published chemical literature.

2-1 DATA REDUCTION

One of the most critical steps in communication of scientific results is the reduction of the accumulated data to an amount that can be transmitted to and efficiently assimilated by another person without the loss of important information. Information may be lost if the scientist omits important results or the factors that led to them. Similarly, the scientist should not generalize beyond what is justified by the results.

Most analyses are conducted several times on the same sample (*replicate measurements*) for reasons discussed later in this chapter. For now, suffice it to say that even the person who performed the analysis should have some doubts about an analytical result that was not reproducible. Suppose an analyst seeking to determine the amount of iron in a mineral sample carries out four replicate measurements resulting in the measurements given in Table 2-1. At this point, the analyst could tell the person requesting the analysis, "The mineral sample contains 56.12, 56.45,

TABLE 2-1   Iron Analysis Results

Trial	Sample weight (mg)	Iron content (mmol)	Iron weight (mg)	Fe (%)
1	417.7	4.197	234.4	56.12
2	486.1	4.913	274.4	56.45
3	513.4	5.166	288.5	56.20
4	451.5	4.657	260.1	57.60

56.20, or 57.60% iron." However, the person paying for the analysis probably wants an answer, not a multiple-choice selection. Therefore, it will be not only more concise but more understandable to report the single value that is most consistent with the analytical results.

Mean Value   Assuming that all results are distributed around some true central value, the analyst ought to report the best estimate of this value. One method of estimating the true central value is to postulate that all observed values are distributed on both sides with equal probability. In that case, the central value is simply the average, or **mean**, value. It is calculated by dividing the totaled observed values, x_i, by the number of observations N. Expressed mathematically, this familiar equation is

$$\text{Mean} = \bar{X} = \frac{\sum\limits_{i=1}^{N} x_i}{N}$$

The mean result for the iron analysis listed in Table 2-1 is 56.59% Fe.

With a small set of measurements, as is usually the case for analytical results, the distribution assumption can be misleading. In Table 2-1, there are 3 values between, say, 56.1 and 56.5%, but there is no very low value to counterbalance the very high value of 57.60%, which skews the distribution so significantly that the mean is higher than three of the four observations. This skewing effect is more evident if we calculate the average income of ten people, one of whom earns $1 million per year. If the other nine each earn $20,000 per year, the mean value is $118,000 per year. This value has been so severely distorted by the one high value that it reflects little of the nine other values. In fact, if the other nine earned nothing, the mean value would still be $100,000 per year. If we want to know the average income for these particular ten people, the mean value of $118,000 is correct. However, if we want to know the average income for U.S. citizens, we should be concerned that not one person out of ten earns $1 million per year. That is, our sample inadvertently included an outlying value and therefore is not representative of the whole population.

Median Value   A different method of estimating the true central value assumes that the observed values are distributed on both sides of the true central value with equal probability but that not every high value is counterbalanced by an equally low value. The high and low values (*outlying values*) are counted, but not weighted according to their magnitude. When the values are ordered numerically, the center observed value is called the **median**. For an odd number of observations, the median is the middle value. For an even number of observations, the median is the average of the two middle values. The median for the results listed in Table 2-1 is 56.32%. Note that the median is within the group of three values between 56.1% and 56.5%.

The median *focuses on the middle values and gives little weight to the outlying values, whereas the* mean *can be strongly affected by outlying values.* For the ten wage earners, the median income of $20,000 per year would more properly reflect what the average person earned in the group. For this reason, median values are often used in reporting such statistics. *For a small set of observations the median is a better estimate of the true central value than the mean.* As a practical guide, if there is little difference between the mean and median, either may be reported. But if there is a significant difference, it is prudent to report the median for a small set of observations.

2-2   ACCURACY VERSUS PRECISION

ACCURACY

Accuracy and precision have different meanings to chemists. **Accuracy** is a measure of how close the experimental results are to the "true," or accepted, value. Although the true value is usually unknown, an accepted reference value is usually available, to which all measured results can be compared. It is an accepted true value.

Accuracy is normally reported in terms of error. **Error** is the difference between the true and measured value. Suppose that the accepted value for the mineral samples given in Table 2-1 is 56.39% Fe. In this case, the error is $+0.20\%$ if the mean value is reported, but -0.07% if the median value is reported. We shall frequently use **percentage error**—the error divided by the accepted value (multiplied by 100 for percentage)—because it is dimensionless and allows easier comparison of samples or methods. Unfortunately, as was the case for the iron analysis, results are also commonly expressed in percentage terms, which can produce some confusion. The difference between error and percentage error is clearer if we consider an analysis in which the results are expressed in nonpercentage units. For example, for the weight of ascorbic acid in a vitamin pill, the reported median value might be 245 mg per tablet with the true value at 249 mg per tablet. The error would then be -4 mg per tablet, whereas the percentage error $[(-4/249) \times 100]$ would be -1.6%.

Once a true value is available, the error for each replicate measurement can also be reported. For the iron analysis, suppose the accepted (true) value is 56.39% Fe; therefore, the errors are: trial 1, -0.27% Fe; trial 2, $+0.06\%$ Fe; trial 3, -0.19% Fe; trial 4, $+1.21\%$ Fe. Trial 2 was the most accurate determination—an even more accurate measurement than the mean or the median.

PRECISION

Precision is a measure of how close the experimental values are to each other (rather than to a true value). As such, it is a measure of the reproducibility of the analysis. Because the analytical chemist usually has no true value with which to compare test results, accuracy (error) cannot be determined. However, precision of the results can be ascertained through the use of replicate measurements. A measure of precision usually implies a measure of minimum probable error. In other words, the accuracy of results that vary by several percent from trial to trial is unlikely to be greater than that percentage.

Range   There are several methods for reporting precision. The most widely used method in statistical analysis is called the *standard deviation*, which is discussed later in this chapter. However, a quick and surprisingly useful method is to quote the range. The **range** is the difference between the highest and lowest values. For the data in Table 2-1, the range is 1.48% Fe (57.60% − 56.12%).

Average Deviation   Precision can also be expressed as average deviation. The *deviation* is the difference between an observed value and the central value (mean or median). The sum of all deviations can then be divided by the number of observations to yield the **average deviation**. No algebraic sign is used (absolute deviations) because, if the sum were carried out with algebraic sign, the average deviation (from the mean) would equal zero and would not reflect the amount of scatter in the results. In Table 2-9, the average deviation will be calculated and compared with the standard deviation for the iron analysis results.

SIGNIFICANT FIGURES

Significant figures represent an order-of-magnitude estimate of precision and, by inference, the best accuracy we may expect for the reported value. Without additional information (such as the average deviation), the reader of the report should assume that the last quoted digit has an uncertainty of ± 1. For the iron analysis, we could report median values of 56, 56.3, or 56.32%. The reader would assume that the precision for each of these reported values is $56 \pm 1\%$, $56.3 \pm 0.1\%$, or $56.32 \pm 0.01\%$ Fe. Which would be most appropriate? The mean, the median, and three of the four reported values are between 56 and 57% Fe. There is little doubt about the first two digits (unless trial 4 is the only correct value). However, we should have considerable doubt about a report of 56.3% Fe. In fact, although the values were scattered around it, none of the trials gave this value. The third significant figure involves uncertainty. If no precision data were quoted, a report of 56.32% Fe would be misleading because the reader would assume the average deviation was approximately 0.01%.

   Table 2-2 shows the number of significant figures for several numbers and the uncertainty a reader could assume if no accuracy or precision information were

TABLE 2-2   Significant Figures

Number	Scientific notation	Significant figures	Assumed uncertainty (%)
56.3	5.63×10^1	3	$\frac{1}{563} \cong 0.2$
56.32	5.632×10^1	4	$\frac{1}{5632} \cong 0.02$
0.00247	2.47×10^{-3}	3	$\frac{1}{247} \cong 0.40$
85,000.	8.5×10^4	2	$\frac{1}{85} \cong 1.2$
85,000.	8.50×10^4	3	$\frac{1}{850} \cong 0.12$
85,000.	8.500×10^4	4	$\frac{1}{8500} \cong 0.012$
85,000.	8.5000×10^4	5	$\frac{1}{85,000} \cong 0.0012$
99	9.9×10^1	2	$\frac{1}{99} \cong 1.0$
101	1.01×10^2	3	$\frac{1}{101} \cong 1.0$
$\log_e{}^* x = 3.25$	—	2	$\frac{0.01}{1} = 1.0$
$\log_{10}{}^* x = 3.25$	—	2	$\frac{0.01}{1} \times 2.3 = 2.3$
$\log_{10} x = 56.32$	—	2	$\frac{0.01}{1} \times 2.3 = 2.3$

*In this chapter, natural logarithms will be symbolized by $\log_e$, and common logarithms by $\log_{10}$. In the remainder of the book, these symbols will be replaced, respectively, by the conventional ln and log.

available. Note that zeros used to fix the decimal point are not usually significant figures. The use of scientific notation eliminates much confusion concerning significant figures. Also note that, when a logarithm quantity is expressed, figures to the left of the decimal point (characteristic) indicate the exponent and are, therefore, not significant. When the mantissa of the logarithm (figures to the right of the decimal point) contains two digits (for example, 3.25) the assumed uncertainty is ± 1 part per 100 for natural logarithm but is increased by 2.3 (conversion factor between natural and base 10 logarithms) when using base 10 logarithms, as shown in Table 2-2. Also note that logs have much greater uncertainties—56.32 has an assumed uncertainty of $\sim 0.02\%$ whereas the uncertainty in a number whose common logarithm is reported as 56.32 has an assumed uncertainty of 2.3% (100 times greater).

For reporting the correct number of significant figures—for example, in multiplication and division operations—the answer should have the same number of significant figures as the number in the calculation having the *fewest* significant figures. It will have the greatest assumed uncertainty and, therefore, will determine the uncertainty in the answer. However, some common sense is needed when an examination of the assumed uncertainty is required. As can be seen in Table 2-2, 99 has two significant figures, and 101 has three; yet both have an assumed uncertainty of about 1%. Therefore, for the problem $101/99 = 1.02$, the answer should be given to three significant figures, whereas for the problem $101/102 = 0.99$, the answer should be given to two significant figures because in both cases the numbers involved have assumed uncertainties of about 1%. For addition or subtraction, the operation can be carried to the last digit for which information is known for each of the numbers. Typical problems are given in Example 2-1.

EXAMPLE 2-1   Handling Significant Figures in Arithmetic Operations

Addition and Subtraction   Carry out operations to the last digit for which information is known for each of the numbers.

$$
\begin{array}{r}
3761.2 \\
1.0|063 \\
47.3|1 \\
+\quad 0.0|002 \\
\hline
3809.5
\end{array}
\qquad
\begin{array}{r}
49.6 \\
63.8 \\
+25.1 \\
\hline
138.5
\end{array}
\quad \text{(note the gain in significant figures)}
$$

$$
\begin{array}{r}
721.34\,1 \\
-\quad 29.63 \\
\hline
691.71
\end{array}
\qquad
\begin{array}{r}
25.4463 \\
-25.2161 \\
\hline
0.2302
\end{array}
\quad \text{(note the loss in significant figures)}
$$

Multiplication and Division   Retain the same number of significant figures as the number in the calculation having the fewest significant figures, *but* your answer should reflect approximately the same assumed uncertainty as the number having the fewest significant figures.

$$
\frac{46.32 \times 0.173}{51.743} = 0.155
\qquad
\frac{3{,}472 \times 61{,}842}{0.01621} = 1.325 \times 10^{10}
$$

$$
\frac{101}{99} = 1.02
\qquad
\frac{101}{102} = 0.99
$$

2-3 DETERMINATE ERRORS

When an analyst conducts a procedure incorrectly or reads a meter incorrectly these mistakes are called *determinate errors*. In principle, determinate errors can be determined and corrected. Someone watching the analyst could determine the error made and correct it (sometimes only by throwing out the result and starting over). The analyst may spot the error by conducting a second determination on the sample—another reason for replicate measurements. Often, however, determinate errors are subtle and not readily discovered. Three kinds of determinate errors are method, instrumental, and personal.

METHOD ERRORS

Analytical methods based on some chemical reactions usually involve simplifying assumptions, such as 100% completion of the reaction. For example, in most gravimetric procedures, a compound is precipitated from solution and the chemist assumes that the analyte has been completely removed from solution. However, no compound is totally insoluble. Some small amount of the substance to be analyzed will remain in solution. This is a determinate error because the solubility can be measured, even though the error is often so insignificant that it is neglected. A related but more serious error can occur when the chemical reaction does not reach chemical equilibrium, thereby invalidating all calculations based on equilibrium properties. In such a case, a detailed study of the reaction kinetics may uncover (*determine*) the error.

A determinate method error may arise when a color change is used for an endpoint signal in a volumetric analysis. An excess amount of reagent may be required to cause the color change that indicates completion of the chemical reaction between analyte and reagent. If this excess amount of reagent is not determined, the analytical results will have a positive error. Such analytical methods usually call for a "blank titration" to establish how much reagent is needed to cause the color change when no analyte is present.

It is very difficult to determine a method error when the sample to be analyzed contains interfering substances. For example, in the determination of iron content in a steel sample by oxidizing Fe(II) with $KMnO_4$ reagent, any other constituent in the steel that also reacts with $KMnO_4$ will interfere, causing a positive error. This is a determinate error; if the entire composition of the sample were known, we could separate the interfering constituents prior to the quantitation step. However, because complete knowledge of the sample is not always available, especially for environmental analyses, this source of error may be determinate in principle only.

INSTRUMENTAL ERRORS

Instruments have been a major force in the tremendous advances in analytical chemistry during the last 30 years. However, they are also "black boxes" that can hide determinate errors. One of the most prevalent determinate errors is caused by analytical instruments that are "out of calibration"—that is, on which the numbers on the meters or dials are no longer accurate measurements. For example, a pH meter must be calibrated using a buffer solution of known pH, say, adjusting the meter to read $pH = 4.00$ when a buffer of pH 4.00 is measured. Analytical instruments also include all measuring devices such as balances and volumetric glassware. Burets and pipets are now manufactured of such high quality that they rarely need to be calibrated. From experience or the manufacturer's literature, the analytical chemist knows whether the various instruments should be calibrated on a monthly, daily, or hourly basis.

However, even calibrating the instruments for each sample does not necessarily preclude determinate errors. Calibration could change as the analyst moved the electrodes from the standard buffer solution to the sample to be analyzed. The calibration of glassware may change if the temperature of a solution to be measured differs from the temperature at which calibration was made. Moreover, glassware that has been heated will expand, but it may not contract to the original volume when cooled.

PERSONAL ERRORS

It is easy to blame the method or the instruments used in an analysis. However, the analyst may also be at fault. A physical impairment, for example, may lead to determinate errors. An analytical chemist who is color-blind might not be able to see color changes accurately; one who wears bifocals may have problems reading the liquid level in a buret.

Carelessness or lack of knowledge concerning the analytical procedure may lead to large determinate errors. When an analyst uses an analytical method for the first time, the results often improve with each succeeding trial; this improvement represents the analyst's personal "learning curve." Therefore, experienced chemists often practice a new method using "known" samples before attempting to solve an analytical problem involving unknowns.

The learning-curve syndrome may also play a part in the determinate error of *personal bias*. Although many analytical methods require that the analyte must be in a certain concentration range for a successful analysis, the analytical chemist should have no preconceived notions about what the results *should* be. For example, suppose an analyst is titrating 10-mL portions (*aliquots*) of a sample. The first aliquot requires 33.89 mL of reagent to reach a color-change endpoint. Now the analyst titrates a second 10-mL portion and again sees a color change at 33.89 mL. But did the solution actually change color or was the analyst biased because of the first trial? This can be an unfortunate determinate error. At best, only the first titration was meaningful, and the succeeding ones supported only the analyst's preconceived ideas. At worst, if a learning curve is involved, the first determination will be the least accurate—in which case, the analyst should be biased *against* this result, not toward it.

DETECTION OF DETERMINATE ERRORS

The term *determinate* implies that the analytical chemist can determine the source of error and, with that information, correct it. To detect such errors, chemists have learned to use blanks, standard samples, independent analysis, and variations in sample size.

Blanks   Blanks are used as zero points in many analytical procedures and are especially common in instrumental methods such as spectrophotometry (see Chapter 13) as well as titrations. The amount of reagent needed to make the indicator change color can be determined by using a *blank* titration. The blank solution should have the same background color and volumes as the sample solution and, of course, must not contain any substance that will react with the reagent.

Standard Samples   Another method for detecting determinate errors is the use of *standard samples*. Using prepared samples that contain known amounts of the analyte and similar amounts of other constituents, the analyst can verify that the analytical method chosen gives results with the required accuracy. This aspect is so important that commercial samples (*controls*) are available with the same constituents that will be present in the "unknown" samples to be analyzed to ensure that other constituents do not interfere with the analysis.

Independent Analysis   If samples of known composition are unavailable, samples may be sent to another lab for an *independent analysis*. If possible, the other lab should use an alternate accepted method of analysis; but even if the same method must be used, factors such as instrumental and personal errors will be different. Agreement between the two labs implies that both carried out the analysis with no determinate errors. For extremely important analyses, samples may be sent to several labs. Table 2-3 shows the analysis of moon rocks performed at several labs using different analytical methods.

Sample Size   A variation in sample size can also prove useful in detecting determinate errors, if a constant error exists as opposed to a proportional error. A **constant error** is one in which the error is constant with change in sample size, whereas a **proportional error** is one in which the *percentage error* is constant. As the sample size increases, any constant error will have a smaller relative effect and the observed percentage error will decrease.

For example, suppose a sample that is being analyzed for calcium content is dissolved in 200 mL of H_2O that contains some calcium as an impurity. The results will show a positive error due to the calcium already in the water. Because the amount of calcium introduced from the water is constant, as the sample size increases, the percentage error will decrease. A set of possible results is shown in Figure 2-1 (curve *A*).

On the other hand, when a proportional error is present, no change in percentage error will be observed as the sample weight increases. In the calcium analysis, the existence of a magnesium interference in the sample would be an example. If the sample size doubles, the amount of magnesium doubles and the percentage error will remain constant, as shown in Figure 2-1 (line *B*).

When sample size is varied with standard samples, the analyst can sometimes identify the source of a determinate error by noting whether it is a constant or proportional error.

After all determinate errors have presumably been accounted for, individual trials will still scatter around the true value. Even if there are no determinate errors, each analytical result contains some *indeterminate error*. In the next section, we will assume that no determinate errors are present and develop methods for handling indeterminate errors using statistical analysis.

TABLE 2-3   Analysis of Lunar Rocks

Element	Method	Reported values*
chromium (%)	spark source mass spectrometry	0.28 0.7 (rejected)
	neutron activation analysis	0.227 0.243 0.248 0.26 0.28 0.28 0.30
	X-ray fluorescence	0.208 0.29
	chemical isolation—gas chromatography	0.270
	emission spectroscopy	0.28 0.303
lanthanum (ppm)	spark source mass spectrometry	32 36 82 (rejected)
	neutron activation analysis	22 26.8 32.1 33 33 33.5 36 36.8 38
	isotope dilution—mass spectrometry	30.1
	X-ray fluorescence	29 33 43
	emission spectroscopy	40

SOURCE: Results are for sample 12070 brought back by the *Apollo 12* mission. (Adapted from G.H. Morrison, "Evaluation of Lunar Elemental Analyses," *Analytical Chemistry*, *43*:22A (June, 1971).

*When more than one value appears for a given technique they represent values reported by different laboratories using that technique. Two values shown were so far outlying that they were rejected. (See Section 2-6.)

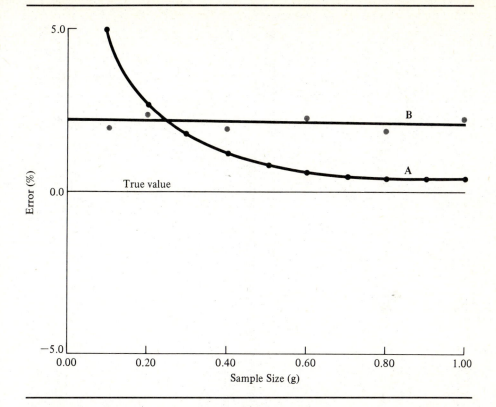

FIGURE 2-1 Constant and proportional errors. (*A*) The constant determinate error is due to calcium impurity in the water used for analysis; the percentage error decreases as the sample size increases. (*B*) The proportional determinate error is due to magnesium interference in the calcium sample; the percentage error remains essentially constant as the sample size increases.

2-4 INDETERMINATE ERRORS

Indeterminate errors represent random fluctuations in procedures and measuring devices (including the human observer) that are beyond the control of the analyst. They occur when measuring devices are pushed to their maximum capability. Consider the thermometer shown in Figure 2-2. If you asked 100 observers what the temperature is to the nearest degree, there would be no fluctuation in the result. Everyone would probably agree that the temperature is between 98.4° and 98.5° (remember, no determinate error is present — the thermometer is accurate). But is this the limit for the measuring device? Whenever information is available, it should not be neglected; in this case, we may estimate the temperature to ±0.01°. Now if we ask 100 observers to read the thermometer to ±0.01° there will be considerable fluctuation, which will be an indeterminate error. The uncertainty (that is, the

FIGURE 2-2 A thermometer as a measuring device. Uncertainty in reading this thermometer is introduced when readings to $\pm 0.01°$ are attempted.

average deviation) may be 0.03°, with some observers reporting values as low as 98.43° and others as high as 98.49°, but it is customary to read devices to tenths of a division anyway.

In an analytical procedure, random positive or negative fluctuations may occur at each measurement made. As you might guess, a negative fluctuation at one step will tend to balance a positive fluctuation at another, so that the most probable result will be the true central value. However, there is a finite probability that the fluctuations will not completely compensate for each other, so that results will be scattered around the central value. The probability also decreases for observing values farther away from the central value.

A more graphic example of this random fluctuation distribution involves flipping a coin 10 times. Table 2-4 lists all the possible combinations of heads and tails; Figure 2-3 is a graphic representation of these combinations. To obtain 10 heads,

TABLE 2-4   All Possible Combinations and Permutations for Flipping 10 Coins

Combination type	Possible permutations
10 heads, 0 tail	1
9 heads, 1 tail	10
8 heads, 2 tails	45
7 heads, 3 tails	120
6 heads, 4 tails	210
5 heads, 5 tails	252
4 heads, 6 tails	210
3 heads, 7 tails	120
2 heads, 8 tails	45
1 head, 9 tails	10
0 head, 10 tails	1
total	$\overline{1024}\ (2^{10})$

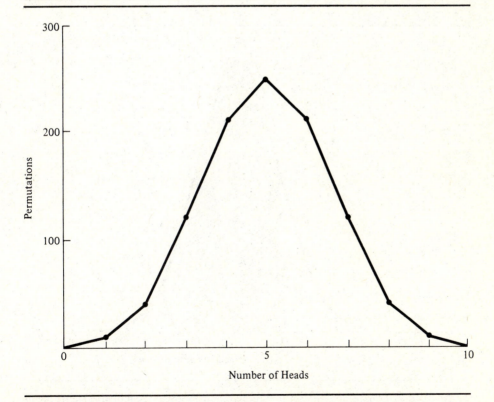

FIGURE 2-3 Distribution of coin-flip observations. Number of permutations will be proportional to frequency of observation for large number of trials in which each trial consists of 10 coin-flips.

every flip must come up heads. With 9 heads and 1 tail, however, the one tail could be any one of the 10 flips; therefore, there are 10 permutations leading to this net result. For 8 heads and 2 tails, the two tails could be flips 1 and 2, 1 and 3, 1 and 4, ... 2 and 3, 2 and 4, ...9 and 10, leading to 45 permutations. The most probable specific combination is 5 heads and 5 tails, but combinations near the central value have a high frequency, too. In fact, there is less probability that the final outcome will be a 5 to 5 split than there is that it will be 6 to 4 (6 heads to 4 tails, with 210 permutations, or 6 tails to 4 heads, with 210 permutations, for a total of 420 permutations).

When a very large number of fluctuations is possible, the distribution becomes a smooth mathematical function called the *normal error curve*, as shown in Figure 2-4. Although it is not obvious that the finite number of fluctuations occurring in an analytical procedure can be described in terms of a normal error curve, the 10 coin-flip distribution is similar in shape, even though it is not a continuous function. This type of distribution has also been observed when a large number of replicate determinations have been made on the same sample.

The equation for the normal error curve is

$$y = \frac{e^{-(x-\mu)^2/2\sigma^2}}{\sigma\sqrt{2\pi}}$$

(2-1)

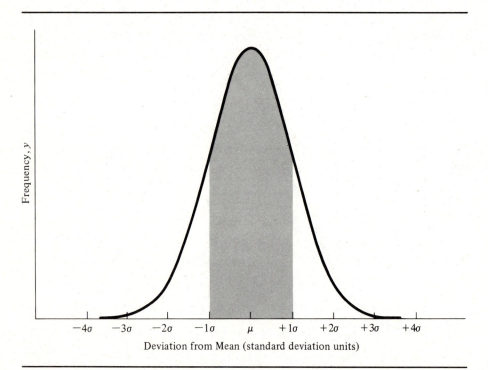

FIGURE 2-4   The normal error curve. Shaded area represents all observations within one deviation, σ, of the true mean, μ.

Here y is the frequency of occurrence for each value of x, μ is the central value and the true mean for an infinite number of measurements, and σ is the standard deviation, which determines the width of the bell-shaped normal error curve. The frequency depends on how far x is from the mean and drops off rapidly as the distance increases. The difference $(x - \mu)$ is the deviation from the mean which was noted earlier as a measure of precision. For the normal error curve, the important function is $(x - \mu)^2$, and the average of the deviations squared, σ^2, is called the *variance*. Provided the true mean, μ, is known, the variance can be calculated as follows:

$$\sigma^2 = \frac{\sum_{i=1}^{N} (x_i - \mu)^2}{N} \tag{2-2}$$

The true standard deviation, σ, is a convenient measurement because it has the same dimensions as x. It is calculated simply as the square root of the variance:

$$\sigma = \sqrt{\frac{\sum_{i=1}^{N} (x_i - \mu)^2}{N}} \tag{2-3}$$

Example 2-2 shows how σ and σ^2 can be calculated when μ is known. In a multistep process (n steps) in which each of the steps is independent, such as the coin-flip example, each step may have its own variance. The overall variance for the total process will then be the sum of the individual variances, as shown in Equation 2-4:

$$\sigma_{\text{process}}^2 = \sigma_1^2 + \sigma_2^2 + \sigma_3^2 + \cdots + \sigma_n^2 \tag{2-4}$$

Only the variances are additive, *not* the standard deviations. This additive property is useful for estimating the expected precision for a multistep analytical method.

The properties of the normal error curve are illustrated in Figure 2-4. The most frequently observed value is μ; 68% of all observed values are within one standard deviation of μ (shaded area). Table 2-5 shows the percentage of observed values for various intervals around the true mean. If the equation for the normal error curve were accurate for small sets of data commonly encountered in the analytical lab, of three replicate measurements on a sample, two would be within one standard deviation of the mean and one would be outside this interval. However, if 20 trials were run, only one would be expected to lie outside two standard deviations from the mean.

To use the normal error curve, we must know the true mean—that is, the mean obtained from analyzing a sample an infinite number of times. Although the true mean doesn't always demand an infinite number of measurements, our measurement must include every member of the total population. For example, the true mean for the height of students in a class could be obtained by measuring each person in the class (a finite population). However, if that is too time-consuming, we may measure only a *sample* of that total population—10 students out of a class of 100. This is the situation encountered in analysis. Although we could run numerous

EXAMPLE 2-2

Using the iron analysis results given in Table 2-1, calculate the variance, σ^2, and standard deviation, σ, *if* the true value, μ, is known to be 56.39% Fe.

Solution

Trial	Fe %	$|x - \mu|$	$(x - \mu)^2$
1	56.12	0.27	0.0729
2	56.45	0.06	0.0036
3	56.20	0.19	0.0361
4	57.60	1.21	1.4641
total			1.5767

$$\sigma^2 = \frac{1.5767}{4} = 0.394$$

$\sigma = \sqrt{0.394} = 0.63\%$ Fe (σ should be reported to same decimal place as the original data)

TABLE 2-5   Percentage of Observed Values for Intervals around the True Mean

Observed values*	Interval	Standard deviations, z[†]
50.	$\mu \pm 0.67\sigma$	0.67
68.	$\mu \pm 1.00\sigma$	1.00
80.	$\mu \pm 1.29\sigma$	1.29
90.	$\mu \pm 1.64\sigma$	1.64
95.	$\mu \pm 1.96\sigma$	1.96[‡](~ 2)
98.	$\mu \pm 2.33\sigma$	2.33
99.	$\mu \pm 2.58\sigma$	2.58
99.7	$\mu \pm 3.00\sigma$	3.00
99.9	$\mu \pm 3.29\sigma$	3.29

*This percentage is also the confidence level for observing a value within this interval.
[†]z will be used to denote the number of standard deviations away from the mean (central value). This can be expressed mathematically as $z = (x - \mu)/\sigma$.
[‡]The 95% confidence interval is frequently used in analytical chemistry; for convenience, the approximate value of $z = 2$ is used. In other words, $\sim 95\%$ of all observations will be within ± 2 standard deviations of the true mean.

determinations, we actually run a small finite number, a sample from the total population of possible determinations.

At this point let us differentiate statistics and probability. With *statistics*, we attempt to predict the total population from knowledge of a sample. With *probability*, we attempt to predict a sample from knowledge of the total population. With chemical analysis, we use statistical analysis to predict the true mean value from the results of our sample.

Remember there are two common methods for estimating the central value or true mean—the mean and the median. Therefore, we can approximate the central value as follows:

$$\mu \cong \bar{X} \quad \text{(mean or median)}$$

According to Equation 2-3, we can also calculate a standard deviation by replacing μ with $\bar{X}$. However, the data used to calculate the standard deviation are also used to calculate the central value. Given the values of $(x_i - \bar{X})^2$ for the first $N - 1$ points, we can calculate $(x_N - \bar{X})^2$. Therefore, the last point (or any one of the points) is not independent of the others; we have lost one *degree of freedom*. With this in mind, we can approximate the true standard deviation, σ, with a standard deviation, s, determined from the measurements. When $\bar{X}$ is used (because μ is unknown), the standard deviation is calculated as follows:

$$\sigma \cong s = \sqrt{\frac{\sum\limits_{i=1}^{N} (x_i - \bar{X})^2}{N - 1}} \tag{2-5}$$

The $N - 1$ term represents the degrees of freedom. We can also approximate the variance as s^2.

Now we can calculate the standard deviation, variance, and the normal error curve that would be consistent with the data in Table 2-1 (page 16). The calculations and results are given in Table 2-6. In Figure 2-5, curve A is the normal error curve predicted from these data. Of the four trial values shown, only one is more than one standard deviation from the median (used here as the best estimate of the central value).

In Table 2-6, the calculation of the frequency values, y, is standard for all normal error curves—that is, the y value for $\bar{X}$ will always be $0.399/s$, y for $\bar{X} \pm s$ is always $0.242/s$, and y for $\bar{X} \pm 2s$ is always $0.054/s$. These values can be used to construct a normal error curve for any experimental data.

2-5   CONFIDENCE INTERVAL

Although you are now ready to report the iron content in the sample analyzed, you cannot be too confident that the sample contains 56.32% Fe. As an estimate of the true central value, μ, the low precision (high scatter) of the experimental results

TABLE 2-6   Iron Analysis Statistical Analysis

Trial	Fe (%)	$|x_i - \bar{X}|$	$(x_i - \bar{X})^2$
1	56.12	0.20	0.040
2	56.45	0.13	0.017
3	56.20	0.12	0.014
4	57.60	1.28	1.638
sum	226.37	1.73	1.709

$\bar{X}$ (mean): $\dfrac{226.37}{4} = 56.59\%$ Fe

$\bar{X}$ (median): $\dfrac{56.20 + 56.45}{2} = 56.32\%$ Fe

The following calculations are based on $\bar{X}$ (median):

variance: $s^2 = \dfrac{\sum\limits_{i=1}^{N}(x_i - \bar{X})^2}{(N-1)} = \dfrac{1.709}{3} = 0.57$

standard deviation: $s = \sqrt{0.57} = 0.75\%$ Fe

Calculations for normal error curve are: $y \cong \dfrac{e^{-(x-\bar{X})^2/2s^2}}{s\sqrt{2\pi}}$

Chosen x values	x	y values
$\bar{X}$	56.32	$(e^{0}/\sqrt{2\pi})s^{-1} = 0.399s^{-1} = 0.53$
$\bar{X} + s$	55.57, 57.07	$(e^{-1/2}/\sqrt{2\pi})s^{-1} = 0.242s^{-1} = 0.32$
$\bar{X} + 2s$	54.82, 57.82	$(e^{-2}/\sqrt{2\pi})s^{-1} = 0.054s^{-1} = 0.07$

95% confidence interval:

confidence interval $= \bar{X} \pm \dfrac{ts}{\sqrt{N}} = 56.32 \pm \dfrac{3.18 \times 0.75}{\sqrt{4}} = 56.32 \pm 1.19$

$= 55.13\text{--}57.51\%$ Fe

indicates the true value may easily be somewhat removed from 56.32% Fe. Consider the following argument. Assuming a normal error curve distribution, if we analyze the sample once, there is

$$68\% \text{ probability} \quad x = \mu \pm \sigma$$

$$95\% \text{ probability} \quad x = \mu \pm 2\sigma$$

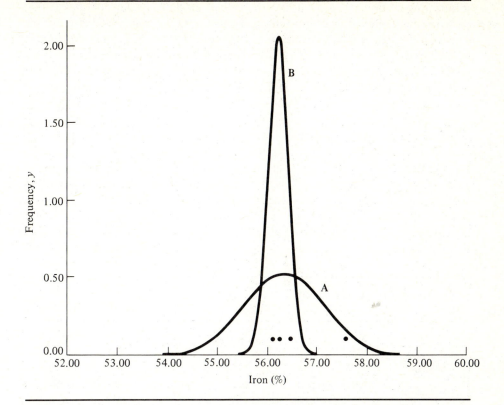

FIGURE 2-5 Normal error curve. (*A*) Calculated from iron analysis results given in Table 2-6. The dots are trial values. (*B*) Calculated from iron analysis results given in Table 2-9.

If so, then there is

$$68\% \text{ probability} \qquad \mu = x \pm \sigma$$

$$95\% \text{ probability} \qquad \mu = x \pm 2\sigma$$

Because we usually want to be at least 95% confident about our results, we will consider only the 95% confidence interval. For the results in Table 2-6, the standard deviation was 0.75% Fe. *If* that were a good approximation of σ, using trial 1, we could state with 95% confidence that the iron content is 56.12 ± 1.50, or 54.62–57.62% Fe. However, to calculate the confidence interval we should use the data available from all four trials.

The true mean, remember, is the mean from an infinite number of determinations (assuming no determinate errors are present). Therefore, as the number of actual determinations increases, the central value estimate (mean or median) should move closer to the true mean. The central value estimate approaches the true mean in a meandering motion—somewhat like the path of a drunken sailor finding his way back to ship. In fact, this motion has given us the term *random-walk calculation*. There is a 50% probability that each determination made will be higher than the true value and a 50% probability it will be lower than the true value. Under these conditions, the central value estimate moves toward the true mean at the rate equal to the square root of the number of determinations, $\sqrt{N}$.

Therefore, using the central value estimate, $\bar{X}$, the 95% confidence interval is $\bar{X} \pm 2\sigma/\sqrt{N}$. When N approaches infinity, $\mu = \bar{X}$. Now the only problem in calculating the confidence interval is to determine σ, the *true* standard deviation. Our estimate, s, turns out to be too small because it was calculated using $\bar{X}$ from a small subset of the total possible determinations. Instead of revising s, it is customary to increase z (see Table 2-5) to allow for this uncertainty and to continue using s as the standard deviation. The new value of z is given the symbol of t and will depend on the number of determinations.[1] It is usually tabulated in terms of degrees of freedom (which equals $N - 1$ for N determinations on a single sample). Values of t are given in Table 2-7.

TABLE 2-7   Values of t

Degrees of freedom	t Factor for confidence interval (%)					
	50	80	90	95	99	99.8
1	1.00	3.08	6.31	12.7	63.7	318.
2	0.82	1.89	2.92	4.30	9.92	22.3
3	0.76	1.64	2.35	3.18	5.84	10.2
4	0.74	1.53	2.13	2.78	4.60	7.17
5	0.73	1.48	2.02	2.57	4.03	5.89
6	0.72	1.44	1.94	2.45	3.71	5.21
7	0.71	1.42	1.90	2.36	3.50	4.78
8	0.71	1.40	1.86	2.31	3.36	4.50
9	0.70	1.38	1.83	2.26	3.25	4.30
10	0.70	1.37	1.81	2.23	3.17	4.14
12	0.70	1.36	1.78	2.18	3.06	3.93
15	0.69	1.34	1.75	2.13	2.95	3.73
20	0.69	1.32	1.72	2.09	2.84	3.55
30	0.68	1.31	1.70	2.04	2.75	3.38
60	0.68	1.30	1.67	2.00	2.66	3.23
∞	0.67	1.29	1.64	1.96	2.58	3.09

SOURCE: From I.M. Kolthoff and P.J. Elving, eds., *Treatise on Analytical Chemistry*, 2nd ed., pt. 1, vol. 1, p. 267 (New York: John Wiley & Sons, Inc., 1978).

---

[1]The *t* function is known as Student's *t* because it was first described by W.S. Gosset, who published under the pen name "Student."

Returning again to our data for the iron analysis, we can now calculate the 95% confidence interval as follows:

$$95\% \text{ confidence interval} = \bar{X} \pm \frac{ts}{\sqrt{N}} \qquad (2\text{-}6)$$

using the values of 56.32% Fe (median as $\bar{X}$), $s = 0.75$, $N = 4$, and (from Table 2-7) $t = 3.18$ (4 trials = 3 degrees of freedom). As given in Table 2-6, the 95% confidence interval is 55.13–57.51% Fe.

NUMBER OF TRIALS TO REACH DESIRED CONFIDENCE LIMIT

An inherent indeterminate error in all analytical procedures will determine the lowest standard deviation the analyst can hope to achieve; any sloppiness in the analysis will lessen the precision. Equation 2-6 implies that no matter how poor the precision, the confidence limits ($\pm ts/\sqrt{N}$) can be decreased to any desired value if N is sufficiently large. However, the square root function also means that for even a small improvement in confidence limits, a large increase in the number of trials may be necessary. Example 2-3 illustrates this problem.

EXAMPLE 2-3

As the instructor of an analytical chemistry course, you have just prepared a new batch of calcium samples. (Last year's class had a great deal of trouble with the previous ones.) If your laboratory technique is such that your standard deviation is 0.12% CaO (only slightly better than the students') and you wish to be at least 95% confident that the true value is within 0.04% CaO of your analyzed value of 41.37% CaO, how many trials should you run?

Solution

$$\mu = \bar{X} \pm \frac{ts}{\sqrt{N}}$$

Then the desired uncertainty is $\mu - \bar{X} = \pm 0.04$. Therefore, $0.04 = ts/\sqrt{N}$ and $\sqrt{N} = ts/0.04$. If N is reasonably large, t approaches the limit of 2 for the 95% confidence level. Therefore, $\sqrt{N} = 2 \times 0.12/0.04 = 6$ and $N = 36$. (*Note*: The assumption made here is reasonable: $t < 2.04$)

2-6 REJECTION OF DATA

Sometimes even the best of us make mistakes, and sometimes a very reliable instrument gives an erroneous result. Certainly we should eliminate such sports from our collected data, but on what grounds are we allowed to drop a result?

Three methods should be considered:

1. Observe the results objectively.

2. Retain the result, but minimize its effect by using the median.

3. Use a statistical test to justify rejection of data.

Observation, if it is made without personal bias, is the most reliable method for rejecting a result. If trial 4 (Table 2-1) in our iron analysis produced a green reaction while all the other trials produced yellow, we could conclude that trial 4 had led to an erroneous result. However, "without personal bias" is an important qualifier. Suppose *one* of the trials turned green but which one wasn't noted before the results were calculated. Now it is not being completely objective when trial 4 is dropped on the grounds of observation—perhaps trial 3 was the one that turned green. The numerical results have biased the analyst. Before the results are calculated, there is a 25% probability that the analyst would conclude that trial 4 turned green. After the results are calculated, there is a 100% probability that this conclusion will be reached.

Of course, if your observations have been methodically entered into your lab notebook as they are observed, when the results show an outlying value, you may have an objective observation to justify dropping this particular value.

In Table 2-6 we saw how the median can decrease the effect of an outlying value. But note the difference between outright rejection of data and using the median. Rejection of trial 4 is the same as never having run this trial. The mean value would be 56.26% Fe, the median would be 56.20% Fe, and the number of observations would be 3. By using the median, we admit to the existence of the outlying result, but we minimize its effect on our estimate of the true central value.

The third method by which experimental data can be rejected involves the use of a statistical test. The ideal test would retain those values that rightfully belong to the set, but eliminate values that indicated spurious results. Although no such ideal test exists, some statistical tests can indicate with a reasonable probability that a particular value should be retained or dropped. The Q test is one such test.

Use of the **Q test** enables the chemist to examine how far the *questionable* result lies from the rest of the set. In Figure 2-6, this test is applied to the iron analysis data. The distance, a, to the nearest neighbor in the set is compared to the entire range, b, by calculating the quotient as follows:

$$Q_{expt} = \frac{a}{b} = \frac{57.60 - 56.45}{57.60 - 56.12} = \frac{1.15}{1.48} = 0.78$$

The outlying result becomes more questionable and less likely to be retained as Q_{expt} nears a maximum value of 1. Therefore, we need some Q value beyond which we are justified in dropping the outlying result. This critical value, Q_{crit}, depends on the desired confidence level (we can never be 100% certain that the outlying value should be dropped) and the total number of observations. It is customary to accept a 90% confidence level for the decision to reject data; Table 2-8 lists Q_{crit} as a function of the total number of observations. For the iron analysis, Q_{crit} is 0.76, slightly less

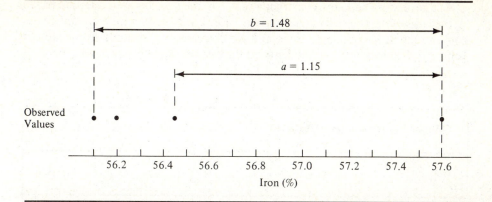

FIGURE 2-6 The Q test. The value of *a* is the difference between the outlying result (57.60% Fe in this example) and its nearest neighbor (56.45% Fe), whereas *b* is the total range of observed values. $Q_{expt} = a/b$.

TABLE 2-8   Q Test for Rejecting an Observation

Number of observations	Q_{crit} (90% confidence level)
2	—
3	0.94
4	0.76
5	0.64
6	0.56
7	0.51
8	0.47
9	0.44
10	0.41

SOURCE: From R.B. Dean and W.J. Dixon, "Simplified Statistics for Small Numbers of Observations," *Analytical Chemistry, 23*:636 (1951).

than Q_{expt}. Therefore, we may drop 57.60% Fe on the basis of the Q test. Note how stringent this test is—although a value of, say, 57.45% Fe might have seemed too high, it would have led to a Q_{expt} of 0.75 and could not be dropped according to the Q test. This is why the first two methods for handling outlying data (observation and use of median) are so important. Given the small sets of data commonly encountered in the analytical laboratory, statistical analysis will not allow the rejection of many observations because a very real (albeit small) finite probability exists that the outlying result rightfully belongs to the set.

As a review, we can now construct a report of the iron analysis, rejecting trial 4 on the basis of the Q test. This is shown in Table 2-9. Note that the standard deviation and the 95% confidence interval have decreased significantly from those in Table 2-6. We are now 95% confident that the true value is less than 56.67% Fe, even

---

TABLE 2-9   Iron Analysis Revisited

Trial	Fe (%)	$|x - \bar{X}|$	$(x - \bar{X})^2$
1	56.12	0.08	0.0064
2	56.45	0.25	0.0625
3	56.20	0.00	0.0000
4	(57.60)	(Rejected on basis of Q test)	
sum	168.77	0.33	0.0689

mean: $\dfrac{168.77}{3} = 56.26\%$ Fe

median: 56.20% Fe

best estimate of true central value: $\bar{X} = 56.20\%$ Fe

(Because the mean and median are quite close, the mean could be reported with equal validity.)

average deviation: $d = \dfrac{0.33}{3} = 0.11\%$ Fe

relative average deviation: $d_r = \dfrac{0.11}{56.20} \times 100\ (\%) = 0.20\%$

variance: $s^2 = \dfrac{0.0689}{3 - 1} = 0.0344$

standard deviation: $s = \sqrt{0.0344} = 0.19\%$ Fe

relative standard deviation: $s_r = \dfrac{0.19}{56.20} \times 100\ (\%) = 0.34\%$

95% confidence interval: $56.20 \pm \dfrac{4.3 \times 0.19}{\sqrt{3}} = 55.73 - 56.67\%$ Fe

calculation for normal error curve:

	x	y
$\bar{X}$	56.20	$0.399/0.19 = 2.10$
$\bar{X} \pm s$	56.01, 56.39	$0.242/0.19 = 1.27$
$\bar{X} \pm 2s$	55.82, 56.58	$0.054/0.19 = 0.28$

though trial 4 indicated a considerably higher percentage. The Q test enabled us to state with 90% assurance that the value for trial 4 contained some *determinate* error and could be rejected from further consideration.

The new normal error curve (B) is plotted behind the old one in Figure 2-5. Note that it has become much sharper with higher frequency values (y). Although the comparison of the old and new curve may suggest that presenting the results of an analysis as a predicted normal error curve is not very exact, remember that the "true" value is usually another normal error curve because it is an accepted value based on measurement.

Because they are now so close in value, both the mean and the median could be used as a reasonable estimate of the true mean (central value). In Table 2-9, we chose to use the median in order to emphasize the fact that, for a small set of observations, the median is normally a better estimate of the true mean than is the mean.

Before we leave the subject of the Q test, be forewarned that the Q test may suggest the elimination of data which, by common sense, should be retained. If two out of three observations fortuitously happen to be identical, the Q test will suggest that the odd observation should be dropped no matter what its value, because Q_{expt} will equal one. With small sets of data, $N-1$ observations frequently form a close cluster with little or no range. In such a case, the chemist should question whether or not this subgroup exhibits a precision which is unexpectedly high. If the precision of the entire group of N observations is reasonably consistent with the inherent precision of the method, then the outlying result should not be dropped simply because the Q test indicates that it should be.

In order to use the Q test with confidence, then we need to develop some technique for estimating the precision (the standard deviation) of an analytical method. This involves the concept of error propagation. First we shall consider the propagation of determinate, then indeterminate, errors. Then we shall develop a method for estimating the inherent precision of an analytical method.

2-7 PROPAGATION OF DETERMINATE AND INDETERMINATE ERRORS

PROPAGATION OF DETERMINATE ERRORS

The measured analyte value, A, often depends on several measurement signals (S_1, S_2, S_3, and so on). Determinate errors in any of these measurements will affect the final result. The nature of the effect depends on the mathematical relationship between the measurement signals. Addition and subtraction form one type of operation, whereas multiplication and division form another.

For addition and subtraction, the error in the final result can be determined as shown in Equation 2-7, in which the analyte value is assumed to depend on the sum of signals S_1 and S_2 and on the difference between their sum and signal S_3:

$$A = (S_1 + S_2) - S_3 \qquad \textbf{(2-7)}$$

The error in A will be ΔA and the error in each of the signals will be ΔS—that is,

$$A + \Delta A = (S_1 + \Delta S_1) + (S_2 + \Delta S_2) - (S_3 + \Delta S_3) \tag{2-8}$$

Subtraction of Equation 2-7 from Equation 2-8 leads to:

$$\Delta A = (\Delta S_1 + \Delta S_2) - \Delta S_3 \tag{2-9}$$

In other words, the absolute error in A simply depends on the absolute errors of the measurements needed to determine A.

   The situation is a little more complicated when multiplication or division is required to relate the analyte value to the measurement signals, but the same approach is used. For multiplication, the basic formula is as follows:

$$A = S_1 \times S_2 \tag{2-10}$$

$$A + \Delta A = (S_1 + \Delta S_1)(S_2 + \Delta S_2) \tag{2-11}$$
$$= S_1 S_2 + S_1\,\Delta S_2 + S_2\,\Delta S_1 + \Delta S_1\,\Delta S_2$$

Subtracting Equation 2-10 from Equation 2-11 yields

$$\Delta A = S_1\,\Delta S_2 + S_2\,\Delta S_1 + \Delta S_1\,\Delta S_2 \tag{2-12}$$

The last term is the product of two small errors (if our analysis has any merit) and therefore should be much smaller than $S_1\,\Delta S_2$ or $S_2\,\Delta S_1$ and may be dropped:

$$\Delta A \cong S_1\,\Delta S_2 + S_2\,\Delta S_1 \tag{2-13}$$

Although Equation 2-13 may seem more complicated than Equation 2-9, it can be put in a more useful form by dividing both sides by A (which also equals $S_1 S_2$):

$$\frac{\Delta A}{A} = \frac{S_1\,\Delta S_2}{S_1 S_2} + \frac{S_2\,\Delta S_1}{S_1 S_2} = \frac{\Delta S_2}{S_2} + \frac{\Delta S_1}{S_1} \tag{2-14}$$

$\Delta A/A$ is the *relative* error in A and is equal to the relative error in S_1 and S_2.
   Division is handled in the same way:

$$A = \frac{S_1}{S_2} \tag{2-15}$$

$$A + \Delta A = \frac{S_1 + \Delta S_1}{S_2 + \Delta S_2} \tag{2-16}$$

Therefore,

$$(A + \Delta A)(S_2 + \Delta S_2) = S_1 + \Delta S_1 \tag{2-17}$$
$$AS_2 + S_2\,\Delta A + A\,\Delta S_2 + \Delta A\,\Delta S_2 = S_1 + \Delta S_1 \tag{2-18}$$

However,

$$AS_2 = S_1 \quad \text{and} \quad \Delta A\, \Delta S_2 \ll S_2\, \Delta A \quad \text{and} \quad A\, \Delta S_2$$

Therefore:

$$S_2\, \Delta A + A\, \Delta S_2 = \Delta S_1 \tag{2-19}$$

By dividing Equation 2-19 by S_1 (remembering that $A = S_1/S_2$) we obtain

$$\frac{S_2\, \Delta A}{S_1} + \frac{A\, \Delta S_2}{S_1} = \frac{\Delta S_1}{S_1}$$

$$\frac{\Delta A}{A} + \frac{\Delta S_2}{S_2} = \frac{\Delta S_1}{S_1} \quad \text{or} \quad \frac{\Delta A}{A} = \frac{\Delta S_1}{S_1} - \frac{\Delta S_2}{S_2} \tag{2-20}$$

Again, relative errors are involved; the negative sign shows that errors of the same sign (both positive or both negative) will tend to compensate for the division operation. In multiplication, errors of the same sign will compound the error in A.

In Example 2-4, these principles are applied to the iron analysis. Similar logarithmic, exponential, or trigonometric approaches can be made when the functional relationship between analyte and measurement signal is more complicated.

PROPAGATION OF INDETERMINATE ERRORS

Although determinate errors are not usually known, knowing what indeterminate errors have been made in the various measurements should lead us to the expected indeterminate error for the overall process. However, because we don't know the sign on an indeterminate error, the second error has a 50:50 chance of compensating the first error. A random mix of errors is accomplished through use of the variance.

As mentioned in Equation 2-4, the variance for the process will equal the sum of the variances for the individual steps:

$$s_{\text{process}}^2 = s_1^2 + s_2^2 + s_3^2 + \cdots + s_n^2 \tag{2-21}$$

Therefore, the predicted standard deviation for the entire process will be

$$s_{\text{process}} = \sqrt{s_{\text{process}}^2} = \sqrt{\sum_i s_i^2}$$

The same rules apply with regard to functional relationships between the measurement signals. For addition and subtraction, absolute variances are summed; for multiplication and division, relative variances are summed.

EXAMPLE 2-4

One method for accomplishing the iron analysis involves addition of a reagent solution of known concentration (N) and measuring the volume (mL) needed to carry out a chemical reaction with the iron analyte. The analyte value, $A(\% \text{ Fe})$, can then be calculated using the following equation:

$$\% \text{ Fe} = \frac{\text{Concentration of reagent} \times \text{Volume of reagent} \times \text{Atomic weight of Fe} \times 100(\%)}{\text{Sample weight}}$$

Given the following measurement signals and determinate errors (shown in parenthesis), calculate the error in the iron analysis.
Concentration of reagent: $0.1281N$ ($+0.0004$)
Volume of reagent:

$$\text{final} = 32.84 \text{ mL } (-0.09)$$

$$\text{initial} = \;\;0.02 \text{ mL } (-0.02)$$

Sample weight:

$$\text{sample} + \text{container} = 25.6473 \text{ g} (-0.0004)$$

$$\text{container} = 25.2296 \text{ g} (-0.0002)$$

Atomic weight of Fe: 55.847 (-0.001)

Solution

First the absolute error associated with any sums or differences is calculated; then all errors are converted to a relative basis (percentage error) to calculate the error in the iron percentage calculation.

Volume of reagent:

$$32.84 \text{ mL } (-0.09)$$

$$\underline{0.02 \text{ mL } (0.02)}$$

Net volume of reagent: 32.82 mL

Error in volume:

$$-0.09 - (-0.02) = -0.07 \text{ mL}$$

Percentage error:

$$\frac{-0.07}{32.82} \times 100 \,(\%) = -0.21\%$$

EXAMPLE 2.4 (cont.)

Sample weight:

$$25.6473 \text{ g} (-0.0004)$$

$$\underline{25.2296 \text{ g} (-0.0002)}$$

Net sample weight: 0.4177 g = 417.7 mg (consistent with mL)

Error in sample weight: $-0.0004 - (-0.0002) = -0.0002$ g
Percentage error:

$$\frac{-0.0002}{0.4177} \times 100 \, (\%) = -0.05\%$$

Concentration of reagent percentage error:

$$\frac{+0.0004}{0.1281} \times 100 \, (\%) = +0.31\%$$

Atomic weight of iron percentage error:

$$\frac{-0.001}{55.847} \times 100 \, (\%) = -0.002\%$$

Given the equation for calculating % Fe, we find

$$\% \text{ Fe} = \frac{0.1281 \times 32.82 \times 55.847 \times 100 \, (\%)}{417.7}$$

$$= 56.21\%$$

Percentage error in % Fe:

$$(+0.31) + (-0.21) + (-0.002) - (-0.05) = +0.148\% \cong +0.15\%$$

Absolute error in % Fe:

$$+0.15 \times 56.21/100 = +0.08\% \text{ Fe}$$

True value:

$$56.21 - 0.08 = 56.13\% \text{ Fe}$$

2-8   ESTIMATION OF PRECISION INHERENT TO ANALYTICAL METHOD

We can now estimate a standard deviation for the iron analysis and compare it with the experimental standard deviation. We will assume that the iron analysis involved a volumetric method in which a standard reagent was added to the analyte until some endpoint signal was reached, as described in Example 2-4.

The first measurement is the weighing of the iron sample, which involves some degree of uncertainty.

$$\text{Sample + Container} \quad 25.6473 \text{ g} \pm 0.0002 \text{ g}^{*}$$
$$\text{Container} \quad \underline{25.2296 \text{ g} \pm 0.0002 \text{ g}}$$
$$\text{Sample} \quad 0.4177 \text{ g} \pm 0.0003 \text{ g} = 0.072\%$$

In the subtraction of the weight of the container from the weight of the container and sample absolute uncertainty is propagated.[2] The overall uncertainty is ± 0.0003 g because, as shown in Equation 2-21, only the variances are additive. Therefore, the individual uncertainties must be squared and summed; then the square root is taken as follows:

$$\text{uncertainty in mass} = \pm \sqrt{(0.0002)^2 + (0.0002)^2} \cong \pm 0.0003 \text{ g} \qquad \textbf{(2-22)}$$

After the sample has been weighed, a volume of standard reagent such as potassium permanganate ($KMnO_4$) is added. This addition is called a *titration*. As was the case for weighing the sample, this measurement necessitates two volume readings, each of which involves some degree of uncertainty. An initial reading must be made as the standard reagent is added from a buret, even if the buret is adjusted to read zero mL. For example, as shown in Figure 2-7,

$$\text{Final reading:} \quad 32.84 \text{ mL} \pm 0.03 \text{ mL}$$
$$\text{Initial reading:} \quad \underline{0.02 \text{ mL} \pm 0.03 \text{ mL}}$$
$$\text{Volume of reagent:} \quad 32.82 \text{ mL} \pm 0.04 \text{ mL} = 0.12\%$$

The overall uncertainty in volume was calculated in the same way as the sample weight uncertainty in Equation 2-22.

The very nature of such a volumetric procedure may contribute additional significant uncertainty to the analytical process. Although there is no uncertainty associated with the initial addition of reagent, there is usually some uncertainty

---

[2]The ± 0.0002 g estimated uncertainty will depend on the balance, its location, and the skill of the user. It can be measured experimentally by having several members of the class weigh a small container and calculating the standard deviation. It will probably be 0.1 to 0.2 mg for the typical single-pan balances found in the undergraduate analytical laboratory. Each weighing has a relative uncertainty of approximately 2 parts per 250,000, but the weight of the sample has a considerably greater uncertainty of 3 parts per 4,177 (equivalent to 180 parts per 250,000). The weighings have six significant figures, whereas the sample weight only has four. Taking the difference between two numbers close in magnitude leads to loss in precision (see Example 2-1) and should be avoided when designing experiments. Fortunately, analytical balances are so accurate that weighing by difference can be used with high reliability.

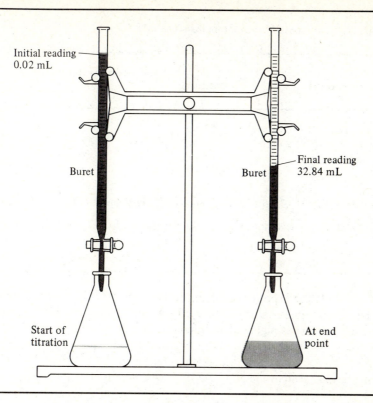

Initial reading 0.02 mL

Buret

Buret

Final reading 32.84 mL

Start of titration

At end point

FIGURE 2-7 Volume readings required for a titration. The buret on the left shows an initial volume reading of 0.02 mL $KMnO_4$. The buret on the right shows final volume reading of 32.84 mL $KMnO_4$ when the endpoint has been reached.

concerning when the reaction has been completed so that the final reading may be made. Some type of endpoint signal, such as a color change in the solution, is required. A typical case would involve an uncertainty of 1 to 2 drops of reagent (0.03 to 0.06 mL). Because students may be observing the endpoint signal for the first time, we will assume an endpoint uncertainty of ±0.06 mL.

Still other uncertainties, such as those related to drying the sample and transfers from one container to another, may be involved in this process. For simplicity, we will exclude them from this calculation. However, we will expect that our estimated uncertainty will be somewhat low because some uncertainties have not been considered.

We will now use *relative* uncertainties to calculate the overall uncertainty of the process because the calculation for percent iron involves multiplication and division as given in Example 2-3. If different reagents were used on successive trials, any uncertainty in the reagent's concentration would also be important in calculating the expected precision. However, because an analyst usually uses the same reagent solution for each trial, reagent uncertainty would not affect the precision observed for replicate measurements on a given sample.

TABLE 2-10   Estimated Precision of Iron Analysis

Measurement	Uncertainty	Relative uncertainty (%)	%²
sample weight: 417.7 mg	±0.3 mg	0.072	0.0052
volume reading: 32.82 mL	±0.04 mL	0.122	0.0149
endpoint: 32.82 mL	±0.06 mL	0.183	0.0334
total			0.0535

Therefore, $s^2_{process} = 0.0535\%^2$ and s_r (relative standard deviation) $= 0.23\%$.

predicted standard deviation: $\bar{X} = 56.20\%$ Fe; $s = 0.23 \times 10^{-2} \times 56.20 = 0.13\%$ Fe

observed standard deviation: $s = 0.19\%$ Fe (from Table 2-9)

In Table 2-10, the expected standard deviation for the iron analysis is given as 0.13% Fe. The observed value was 0.19% Fe which is in reasonable agreement but, as expected, somewhat higher. The standard deviation calculated in Table 2-6— before trial 4 was rejected—was 0.75% Fe, considerably higher than expected.

Now we can use the Q test with decision-making based on expected precision. If the iron analysis results were 56.15, 56.19, 56.20, and 56.41%, then

$$Q_{expt} = \frac{56.41 - 56.20}{56.41 - 56.15} = \frac{0.21}{0.26} = 0.81 > Q_{crit} = 0.76$$

Therefore, the 56.41% value could be dropped. However, when 56.41% is dropped from the set, the standard deviation turns out to be 0.03% Fe (0.05% relative). This value is considerably lower than the 0.23% predicted from the measurement uncertainties. When the 56.41% value is retained, the standard deviation is 0.12% Fe (0.22% relative), which is almost as expected. Therefore, the outlying result should be retained.

2-9 ACCURACY OF METHOD

A similar procedure can be used to estimate the probable error for the analysis. The probable error for each step would be estimated in the same way as was the estimated precision, provided that no determinate errors are anticipated. However, some methods are inherently more reproducible than accurate. (For example, the use of uncalibrated pipets suggests the presence of a determinate error, since calibration would *determine* the error.) The estimated accuracy will normally be reasonably close (within a factor of 2) to the estimated precision for most analytical procedures. However, for trace analysis or environmental analyses, standards needed for calibration may have large probable errors, even though the measurement signals are quite precise.

2-10 FITTING DATA TO A STRAIGHT LINE

In Chapter 1, we noted that many analyses involve a measurement signal that is directly proportional to the amount of the analyte (Equation 1-1). Occasionally theoretical considerations can provide a proportionality constant that is highly accurate, but in many analyses the proportionality constant must be determined by analyzing known samples. For example, polarography, a common electroanalytical method discussed in Chapter 17, involves the measurement of an electrical current, I, which is directly proportional to the analyte concentration, C:

$$I = kC \qquad\qquad (2\text{-}23)$$

Solutions of known concentration are analyzed, resulting in data of the type shown in Figure 2-8. There may also be some small blank current, I_0, and we may wish to

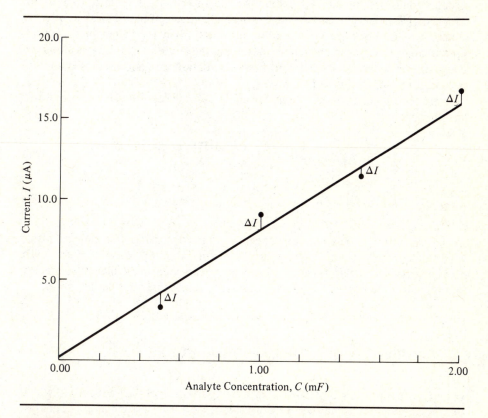

FIGURE 2-8 Best line fit. In this polarographic analysis to determine the lead content of gasoline, calibration points are marked with a dot. Sum of $(\Delta I)^2$ values has been minimized.

determine how well the data fit a straight line of the form:

$$I = kC + I_0 \qquad\qquad\qquad (2\text{-}24)$$

where k is the slope of the line and I_0 is the current value when $C = 0$ (the y-intercept).

For situations of this kind, the "best" straight line is obtained using the *method of least squares*. This method is based on the principle that the most probable value of a quantity is obtained from a set of measurements by choosing the value that minimizes the sum of the squares of the deviations of these measurements. The reason for minimizing the squares goes back to the normal error curve, Equation 2-1, which shows that the probability (frequency) increases as $(x - \mu)^2$ decreases. Therefore, the most probable distribution is the one that minimizes the square of the deviations.

For the best fit line in Figure 2-8, it will be necessary to determine which value of k and I_0 minimizes the sum of $(\Delta I)^2$ values. This is a straightforward problem using calculus and is presented in Appendix 1-A. The uncertainty is assumed to be along the y-axis, (electrical current). In other words, it is assumed that the values of C for our known solutions are known more accurately than the values of I. (A least squares analysis can also be made by assuming the largest uncertainty is along the x-axis or even in both variables, but these calculations are not often used and are beyond the scope of this text.)

From the equations developed in Appendix 1-A, the best value for the slope and intercept are:

$$\text{Slope} = k = \frac{N \sum C_i I_i - (\sum C_i)(\sum I_i)}{N \sum C_i^2 - (\sum C_i)^2} \qquad\qquad (2\text{-}25)$$

$$\text{Intercept} = I_0 = \frac{(\sum I_i)(\sum C_i^2) - (\sum C_i I_i)(\sum C_i)}{N \sum C_i^2 - (\sum C_i)^2} \qquad\qquad (2\text{-}26)$$

In the past, such calculations have been so time-consuming that many analysts would determine the best line "by eye." However, desktop calculators and even many hand-held calculators can now carry out a least squares calculation so quickly that there is no reason why the most probable straight line can't be calculated.

2-11 USE OF SINGLE TAIL OF NORMAL ERROR CURVE

In many situations, the central value is less important than the frequency at which certain outlying results may occur on one side of the normal error curve. The analyst must determine that no more than a small percentage of the samples will be larger or smaller than some predetermined value. As shown in Example 2-5, it may be worthwhile to improve precision by decreasing the standard deviation so that the mean value is as close as possible to the predetermined value.

EXAMPLE 2-5

The octane number of gasoline increases with the addition of tetraethyllead, but the limit on lead additives is 0.50 g per gallon. If no more than 0.5% of gasoline samples may exceed 0.50 g per gallon and the standard deviation for lead content in Company A's gasoline is 0.05 g per gallon, what is the average amount which may be used in the gasoline? If Company B maintains tighter control on its processing so that $\sigma = 0.01$ g per gallon, how much lead can be added on the average?

Solution

Table 2-5 shows that 99% of the samples will be within 2.58σ of the mean. This means that 0.5% will be $< \bar{X} - 2.58\sigma$ and 0.5% will be $> \bar{X} + 2.58\sigma$. Figure 2-9 shows that the average amount of lead in the gasoline can only be $0.50 - (2.58 \times 0.05) = 0.37$ g per gallon.

As Figure 2-9 shows, for Company B, the average amount of lead can now be $0.50 - (2.58 \times 0.01) = 0.47$ g per gallon.

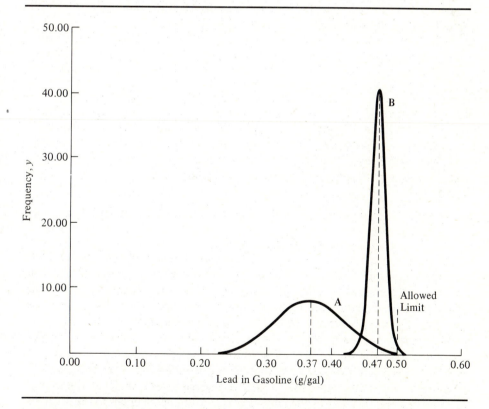

FIGURE 2-9 Lead in gasoline. Normal error curves for Company A and Company B. The limit of 0.5 g per gallon is not to be exceeded by 0.5% of gasoline samples.

2-12   HYPOTHESIS TESTING USING STATISTICAL ANALYSIS

Many real-world analysis problems require the chemist to arrive at a numerical result *and* a qualitative conclusion. To reach such a conclusion, the chemist often poses an hypothesis and uses statistical analysis of the data to determine the probability that the hypothesis is incorrect. If we are examining a set of data with a true or fixed value (as μ), the confidence interval approach can be used to determine whether or not the difference between the central value of the data $(\bar{X})$ and the fixed value (μ) is greater than expected from indeterminate error and, if so, whether the two values can be concluded to be different. In other words, we hypothesize that the two are the same (**null hypothesis**) and calculate the probability that this is not true. Earlier we noted that, for a high N value or when the true standard deviation is known,

$$\mu - \bar{X} = \frac{\pm z\sigma}{\sqrt{N}}$$

For a small set in which the estimate of standard deviation is calculated from the data,

$$\mu - \bar{X} = \frac{\pm ts}{\sqrt{N}}$$

If μ is the true value for a set of observations resulting in $\bar{X}$, then both come from the same distribution (null hypothesis) and their difference will be $< ts/\sqrt{N}$. However, if their difference is $> ts/\sqrt{N}$, then to a specified confidence level we may conclude that they are different. Often we will adopt a level of confidence such as 95% as our test. Another way to compare μ and $\bar{X}$ is to calculate t or z:

$$t = \frac{|\mu - \bar{X}|\sqrt{N}}{s} \quad \text{or} \quad z = \frac{|\mu - \bar{X}|\sqrt{N}}{\sigma} \qquad \text{(2-27)}$$

Then, using Table 2-5 or 2-7, we can determine at what level of confidence we may conclude that μ and $\bar{X}$ are from different distributions. From Equations 2-27, we see that t (or z) increases as $(\mu - \bar{X})$ and N increase. On the other hand, t (or z) increases as the standard deviation *decreases*. As t (or z) increases, our confidence increases that the group of observations resulting in $\bar{X}$ *do not* belong to a set from μ.

Note that as t increases, we can become more and more confident that the values are *different*, but for any value of t we cannot claim that $\bar{X}$ and μ are the same. As t decreases, we merely lose confidence that the two are different. As t approaches 0, we would conclude that because this particular analysis gives us no evidence that the two are different, they *may* be the same. The same argument applies to the use of z in place of t when σ is known. If several other analyses reach the same conclusion, we then become more confident that the sample resulting in $\bar{X}$ belongs to μ. In the scientific method, we postulate an hypothesis and then attempt to *disprove* it. A single inconsistency can disprove the hypothesis, whereas any consistent result merely *supports* the hypothesis—it does not prove the hypothesis correct. For analytical problems in which two numbers are compared, the null hypothesis postulates that they are the same; then statistical analysis is used in an attempt to disprove this hypothesis—that is, to prove that the two numbers are truly different because the observed difference is too great to be the result of indeterminate error. Examples 2-6 and 2-7 show how this method may be used.

EXAMPLE 2-6

Recently, a new method for the determination of polychlorinated biphenyls (PCBs) in a paper mill effluent was developed [J.J. Delfino and D.B. Easty, "Interlaboratory Study of the Determination of Polychlorinated Biphenyls in a Paper Mill Effluent," *Analytical Chemistry 51*:2235 (1979)]. The following table represents results from one set of experiments. Based on the true value of 1.50 mg/L, do these results show evidence of a determinate error associated with the new method?

Analyst	PCB (mg/L)	$(x - \bar{X})^2$
1	1.52	0.0025
2	1.41	0.0036
3	1.46	0.0001
4	1.70	0.0529
5	1.59	0.0144
6	1.26	0.0441
7	1.76	0.0841
8	1.05	0.1764
		$0.3781; \; s^2 = \dfrac{0.3781}{8-1} = 0.054$
$\bar{X}$ (mean)	1.47	
s	0.23	

Solution

The difference between $\bar{X}$ and the true mean, μ, is $1.50 - 1.47 = 0.03$ mg/L. Is this a significant difference?

$$\mu - \bar{X} = 0.03 = \frac{ts}{\sqrt{N}} = \frac{t \times 0.23}{\sqrt{8}} \qquad t = 0.37$$

From Table 2-7, we know that this value of t is below the 50% confidence level for 7 degrees of freedom. Therefore, based on these results, there is essentially *no* evidence of any determinate error.

EXAMPLE 2-7

Company A has patented a new, high-strength alloy that contains 2.0 to 3.0% tungsten (W). Analysis of a sample of Company B's alloy produced values of 3.25, 3.28, 3.13, and 3.19% W. Does Company B's alloy infringe on Company A's patent?

Solution

$$\bar{X} = \frac{3.25 + 3.28 + 3.13 + 3.19}{4} = 3.21\%$$

The difference between $\bar{X}$ and the maximum value covered by the patent is $3.21 - 3.00 = 0.21\%$. Is this a significant difference?

$$s = \sqrt{\frac{(0.04)^2 + (0.07)^2 + (0.08)^2 + (0.02)^2}{4 - 1}} = 0.07$$

$$\bar{X} - 3.0 = 0.21 = \frac{ts}{\sqrt{N}} = \frac{t \times 0.07}{\sqrt{4}}$$

$$t = \frac{0.21 \times \sqrt{4}}{0.07} = 6.0$$

Table 2-7 shows that with three degrees of freedom $t = 6$ is $>99\%$ confidence level. Therefore, we are more than 99% sure that Company B's alloy does *not* infringe on Company A's patent. There is less than a 1% probability that the true value of Company B's alloy contains less than 3% W. In fact, there is approximately a 0.5% chance that it contains $<3.00\%$ and a 0.5% chance that it contains $>(3.21 + 0.21) = 3.42\%$ W.

Often the comparison is not made between an experimental mean and a true value, but between two experimental means, $\bar{X}_1$ and $\bar{X}_2$. The same approach is used:

$$\mu_1 = \bar{X}_1 \pm \frac{ts}{\sqrt{N_1}}$$

$$\mu_2 = \bar{X}_2 \pm \frac{ts}{\sqrt{N_2}}$$

The null hypothesis is that $\mu_1 = \mu_2$, so that

$$\bar{X}_1 \pm \frac{ts}{\sqrt{N_1}} = \bar{X}_2 \pm \frac{ts}{\sqrt{N_2}}$$

and

$$\bar{X}_1 - \bar{X}_2 = \pm ts\sqrt{\frac{N_1 + N_2}{N_1 N_2}}$$

Note that it is assumed that s was the same for both analyses. Whenever the values of s differ, it is best to use a *pooled standard deviation* that utilizes the precision from both sets of data. A pooled standard deviation is calculated by summing all $(x - \bar{X})^2$, dividing by the degrees of freedom, then taking the square root. Calculation of a pooled standard deviation is shown in Example 2-8.

EXAMPLE 2-8

Two students analyze toothpaste samples for fluoride content using a fluoride ion–selective electrode. From the following data, can you conclude that they both analyzed the same brand of toothpaste?

	SnF_2 (%)	$|x - \bar{X}|$	$(x - \bar{X})^2$
Student A:			
	0.391	0.001	1×10^{-6}
	0.385	0.005	25×10^{-6}
	0.395	0.005	25×10^{-6}
$\bar{X} = 0.390$			51×10^{-6}
$s = 0.005$			
Student B:			
	0.385	0.011	121×10^{-6}
	0.368	0.006	36×10^{-6}
	0.372	0.002	4×10^{-6}
	0.370	0.004	16×10^{-6}
$\bar{X} = 0.374$			177×10^{-6}
$s = 0.008$			

Solution

Pooled standard deviation:

$$s = \sqrt{\frac{(51 \times 10^{-6}) + (177 \times 10^{-6})}{3 + 4 - 2}} = 0.007$$

We may compare the two means at the 95% confidence level ($t = 2.57$) as follows:

$$\bar{X}_A - \bar{X}_B = 0.390 - 0.374 = 0.016\% \; SnF_2$$

$$ts\sqrt{\frac{N_A + N_B}{N_A N_B}} = 2.57 \times 0.007 \sqrt{\frac{3 + 4}{3 \times 4}} = 0.014\% \; SnF_2$$

Because $0.016 > 0.014$, we are more than 95% confident that the two students used different brands of toothpaste.

Again, t for a particular confidence level may be used for decision-making, as in Example 2-8, or t may be calculated to determine the level of confidence at which the null hypothesis was incorrect.

Sometimes two experimental means are compared, but the analysis has been performed so often that the true standard deviation is known. This situation occurs in crime labs, clinical labs, and in commercial testing labs. Calculations are carried out using σ and z, as in Example 2-9.

The hair analysis in Example 2-9 suggests many analytical tests besides the analysis for zinc. Whenever t or z is small, we can only say that this particular test supports the null hypothesis that the two hairs are from the same head. But it will take only one test in which t or z is large to conclude that they are from different persons. Although several chemical analyses might support the null hypothesis, a simple visual observation might reveal that one is blond and the other is brunette!

EXAMPLE 2-9

Hair found at the scene of a crime is analyzed and compared with hair belonging to a suspect. Using atomic absorption analysis, a technique known to have a true standard deviation of 12 parts per million (ppm) zinc, the following results were obtained:

Hair at scene of crime: 196, 204 ppm $\bar{X}_1 = 200$ ppm
Hair from suspect: 193, 188, 195, 199 ppm $\bar{X}_2 = 194$ ppm

Are $\bar{X}_1$ and $\bar{X}_2$ sufficiently different to disprove the null hypothesis, thereby suggesting that the hairs are not from the same person?

Solution

$$\bar{X}_2 - \bar{X}_1 = 194 - 200 = -6 = \pm z\sigma \sqrt{\frac{N_1 + N_2}{N_1 N_2}} = \pm z \times 12 \sqrt{\frac{2 + 4}{2 \times 4}}$$

$$z = \frac{6}{12 \times 0.87} = 0.58$$

From Table 2-5, we know that there is $<50\%$ chance that these hairs came from different persons. Therefore, on the basis of this one test, we can only conclude that the results support (but do not prove) the hypothesis that the hair at the scene of the crime came from the suspect.

SUPPLEMENTARY READING

Articles Describing Analytical Methods or Applications

Dean, R.B., and W.J. Dixon, "Simplified Statistics for Small Numbers of Observations," *Analytical Chemistry*, 23:636 (1951).

Reference Books

Dixon, W.J., and F.J. Massey, Jr., *Introduction to Statistical Analysis*, 3rd ed. New York: McGraw-Hill Book Company, 1969.

Mandel, J., "Accuracy and Precision: Evaluation and Interpretation of Analytical Results," in *Treatise on Analytical Chemistry*, 2nd ed., I.M. Kolthoff and P.J. Elving, eds., pt. 1, vol. 1, chap. 5. New York: John Wiley and Sons, Inc., 1978.

Young, H.D., *Statistical Treatment of Experimental Data*, New York: McGraw-Hill Book Company, 1962.

PROBLEMS

2-1. Reduce the following experimental data to a single best estimate of the central value by calculating the mean.

   *a. Concentration of mercury (parts per million) in hog kidneys:
        12.0, 12.1, 25.2, 8.4, 20.5, 21.0

   b. Oxidant level (parts per million) in Los Angeles air on a smoggy day:
        0.21, 0.18, 0.19, 0.24

   c. Dissolved oxygen level (parts per million) in a water sample:
        8.1, 8.3, 7.9, 7.8

   d. Percentage of chloride in a salt sample:
        55.13, 55.27, 55.09, 54.98

   e. Determination of lead in wall paint (milligrams per square centimeter):
        0.52, 0.52, 0.44, 0.69, 0.47

   f. Sulfur dioxide content in Chicago atmosphere (parts per million):
        0.129, 0.131, 0.127, 0.132, 0.133

   g. Determination of lead in ruby glass (percentage):
        9.57, 9.83, 10.02, 9.66

2-2. Rework Problem 1, using the median in place of the mean.

2-3. Express the precision of the experimental data given in Problem 2-1 by calculating the range.

2-4. Express the precision of the experimental data given in Problem 2-1 by calculating the average deviation.

2-5. Express the precision of the experimental data given in Problem 2-1 by calculating the standard deviation.

2-6. Express the precision of the experimental data given in Problem 2-1 by calculating the variance.

2-7. Graph the normal error curve predicted from the data given in Problem 2-1 assuming $\sigma \cong s$ and $\mu \cong \bar{X}$.

2-8. Determine the 95% confidence interval for the data given in Problem 2-1.

2-9. Determine the 80% confidence interval for the data given in Problem 2-1.

2-10. Determine the 99% confidence interval for the data given in Problem 2-1.

2-11. Your first assignment with Amalgamated Metals is to analyze steel for manganese content, using a spectrophotometric technique. Results from your analysis are: 0.357, 0.359, 0.375, and 0.367% Mn. Construct a lab report as shown in Table 2-6 (or Table 2-9).

---

*Answers to problems marked with an asterisk will be found at the back of the book.

2-12. Your second assignment with Amalgamated Metals is to analyze a monel alloy for copper content, using a polarographic technique. Results from your analysis are: 29.2, 29.5, 29.6, and 29.0% Cu. Construct a lab report as shown in Table 2-6 (or Table 2-9).

2-13. Your third assignment with Amalgamated Metals is to analyze steel for nickel content, using a gravimetric technique. Results from your analysis are: 10.13, 10.16, 10.15, and 10.61% Ni. Construct a lab report as shown in Table 2-6 (or Table 2-9).

2-14. Filter paper used for a gravimetric iron analysis (weighed as Fe_2O_3) contains 0.4 mg ash. If this determinate error is not taken into account, what type of error is introduced (constant or proportional) and what error will be observed for the following sample weights, assuming the sample contains 12% Fe:
*a. 200 mg,
 b. 500 mg,
 c. 800 mg.

2-15. A gravimetric analysis for aluminum is conducted on a mineral sample containing 0.4% iron as an interference. If the aluminum is weighed as Al_2O_3 and contains the iron as Fe_2O_3, what type of error is introduced (constant or proportional) and what error will be observed for the following sample weights, assuming the sample contains 45% Al.
 a. 200 mg,
 b. 500 mg,
 c. 800 mg.

2-16. How many significant figures are there in
*a. 47.351   d. 48003.   g. 20.00
*b. 0.0036   e. 4.80×10^3 h. 6.02252×10^{23}
 c. 48000.   f. $\log x = 6.132$ i. 0.00002

2-17. What is the assumed uncertainty (± 1 in the last significant figure) in relative terms for each of the numbers in Problem 2-16.

*2-18. After conducting several hundred fluoride in water analyses using a fluoride ion-selective electrode, your lab has determined that the true standard deviation is 0.05 ppm of F^-. If analysis of a water sample shows it to contain 1.22 ppm of F^-, how many replicate analyses are needed to be 95% certain that the analyzed value is within 0.03 ppm of the true value?

2-19. An old classmate of yours works for a competing water analysis laboratory that prides itself on speedy analysis. If their true standard deviation is 0.09 ppm of F^-, how many replicate analyses are needed to be 95% certain that their analyzed value is within 0.03 ppm of the true value. (Compare this answer with the answer to Problem 2-18 to see the effects of poor precision.)

2-20. In all the data given in Problem 2-1, only one outlying result could be dropped on the basis of the Q test. Which one is it?

2-21. The following sets of results for glucose in blood were obtained. Which outlying results may be dropped on the basis of the Q test?
*a. 105, 109, 104 mg/dL
 b. 105, 109, 104, 105 mg/dL
 c. 97, 101, 105, 103 mg/dL
 d. 103, 103, 105 mg/dL
 e. 105, 101, 108, 106, 108, 105 mg/dL

*2-22. As chief purchasing agent for a cosmetics firm, you must decide which of two lots of palm oil should be purchased from a supplier. Palm oil that contains $25.0 \pm 0.4\%$ unsaturates is acceptable. Your testing division has sampled both of the lots for unsaturates and given you the following information.

What is your decision?

Lot	$\bar{X}$(%)	$s \cong \sigma$
A	25.2	0.1
B	25.0	0.2

2-23. Your first assignment at El Smello Perfume Co. is to determine the proper setting for filling the bottles. The company wishes to keep profits high by using as little perfume as possible, but a government regulation states that no more than 2.5% of the bottles may contain less than the stated volume. If the bottles you are filling are stated to contain 100 mL, what is the minimum volume you should use for filling if the following uncertainties are present in the procedure?

Uncertainty in thickness of bottle wall:	± 1.2 mL
Uncertainty in fill level:	± 1.6 mL

2-24. As chief purchasing agent for War-Toys, Ltd., you must decide which of two lots of plastic bullets should be purchased from a subcontractor. All bullets that are 0.150 ± 0.004 inches in diameter are acceptable. Your testing division has sampled both of the lots and given you the following information. What is your decision?

Lot	$\bar{X}$ (in.)	$s \cong \sigma$
A	0.152	0.001
B	0.150	0.002

2-25. A water sample containing calcium and magnesium is volumetrically analyzed by titration with EDTA. First the sum of calcium plus magnesium is determined; then a second titration is conducted to determine the amount of calcium. The amount of magnesium is found by subtracting the two amounts. If the uncertainty in each titration volume is ± 0.09 mL, determine the relative uncertainty in the calcium value and the relative uncertainty in the magnesium value with the following titration volumes:
*a.   45.36 mL (Ca + Mg titration), 21.73 mL (Ca titration)
 b.   39.71 mL (Ca + Mg titration), 27.14 mL (Ca titration)
 c.   48.16 mL (Ca + Mg titration), 41.77 mL (Ca titration)
 d.   41.22 mL (Ca + Mg titration), 10.13 mL (Ca titration)

2-26. Suppose the FDA reports that iodide in salt may be dangerous if daily intake exceeds 0.3 mg per day. Assuming the average salt consumption is 1.5 g per day, determine which of the following salt brands is more dangerous:

Brand	$\bar{X}$ (% iodide)	σ (% iodide)
X	1.82×10^{-2}	0.06×10^{-2}
Y	1.68×10^{-2}	0.16×10^{-2}

2-27. A classmate of yours has entered the wine business. At a wine-tasting party, she tells you that she is sure that one store is relabelling her wine as its own brand and charging premium prices. You tell her that, given a few cases of wine, you could determine whether or not the two are the same (during a wine-tasting party, people say almost anything). From the following results for alcohol content, what is your conclusion? (*Note:* First calculate a pooled standard deviation.)

Classmate's wine (% alcohol): 12.50, 12.34, 12.38, 12.33, 12.38, 12.41

Competitor's wine (% alcohol): 12.49, 12.62, 12.69, 12.64

2-28. The density of an alloy is determined by measuring the diameter, length and mass of a cylinder. Given the following experimental results, calculate the density and its uncertainty. Round your answer to the proper number of significant figures.

		Diameter (cm)	Length (cm)	Mass (g)
*a.	alloy *A*	1.03 ± 0.02	4.136 ± 0.004	27.328 ± 0.009
b.	alloy *B*	0.836 ± 0.004	2.671 ± 0.008	3.969 ± 0.003

*2-29. The mercury level in several tuna was determined by use of an atomic absorption technique. Determine the standard deviation for the procedure by pooling the following precision data.

Tuna	Hg Content (ppm)
1	1.65, 1.79, 1.55
2	0.50, 0.56, 0.56, 0.59
3	1.13, 1.17, 1.04, 1.09
4	1.88, 1.64, 1.69, 1.77, 1.45

2-30. A murder victim was found clutching some hair in his hand. Analysis of the hair for zinc level was compared with hair from the prime suspect, the butler. The following results involve a method that has a standard deviation of 18 ppm zinc. Did the butler do it?

Butler's hair:	253, 281, 268, 261 ppm of Zn
Hair found in victim's hand:	253, 237, 218, 222 ppm of Zn

2-31. An alleged drunk driver has his blood analyzed for alcohol content. The legal limit is 0.10%. Based on the following results, should charges be filed against the driver?

a. Use standard deviation calculated from data.

b. Use "true" standard deviation known from previous experience of 0.006% ethanol.

Blood alcohol content: 0.103, 0.108, 0.116, 0.113%

2-32. Some students using a new electroanalytical method for determining arsenic in insecticide obtain the following results. Pool the data to determine the standard deviation for the method.

Student *A*:   13.15, 13.18, 13.16, 13.19% As_2O_3

Student *B*:   9.64, 9.81, 9.69% As_2O_3

Student *C*:   11.19, 11.27, 11.28, 11.22, 11.31% As_2O_3

Chemical
Calculations

3

Let us review some elementary concepts of chemical equations and the quantitative information which can be deduced from them. When chemists wish to communicate information regarding chemical reactions which they have used or studied they will invariably use a *balanced chemical equation*. Suppose an analytical chemist has analyzed a sample of air for its ozone content by carrying out a chemical reaction in which the ozone reacted with potassium iodide in an aqueous solution to produce iodine. The amount of iodine produced was determined by carrying out a second reaction in which the iodine reacted with sodium thiosulfate. These long statements can be summarized very efficiently using chemical equations to describe these two reactions:

$$O_3 \text{ (air sample)} + 2KI(aq) + H_2O \rightleftharpoons I_2 + 2KOH(aq) + O_2(g) \qquad \textbf{(3-1)}$$

$$I_2 + 2Na_2S_2O_3(aq) \rightleftharpoons Na_2S_4O_6(aq) + 2NaI(aq) \qquad \textbf{(3-2)}$$

In addition to giving qualitative information, a balanced chemical equation contains much the same kind of quantitative information as does a recipe. The basic unit of such equations is the *mole*, which is 6.02×10^{23} particles (molecules, ions, electrons, or whatever). In other words, Equation 3-1 states that 1 mole of $O_3 + 2$ moles of $KI + 1$ mole of H_2O yield one mole of $I_2 + 2$ moles of $KOH + 1$ mole of O_2, *or*, using individual particles, 6.02×10^{23} molecules of $O_3 + 12.04 \times 10^{23}$ molecules of $KI + 6.02 \times 10^{23}$ molecules of H_2O yield 6.02×10^{23} molecules of $I_2 + 12.04 \times 10^{23}$ molecules of $KOH + 6.02 \times 10^{23}$ molecules of O_2.

To conduct the reaction in the lab, it is more useful to convert mole quantities into mass units (grams). To convert from moles to grams, it is necessary to calculate the molecular weight, MW (in grams per mole), from the atomic weights of the elemental constituents. Using molecular weights, Equation 3-1 may also be stated as 48 g of O_3 + 332 g of KI + 18 g of H_2O yield 254 g of I_2 + 112 g of KOH + 32 g of O_2. Although the total moles of products need not equal the moles of reactants there is a *conservation of mass* (in any chemical reaction matter is neither created nor destroyed). For this example, 398 g of products are produced from 398 g of reactants when the reaction is carried out as stated in Equation 3-1.

Many substances do not consist of discrete molecules (for example, KI in Equation 3-1). Therefore, we are using the term *molecular weight* in a broad sense. A more appropriate term, *gram formula weight* (FW), is widely used.[1] Numerically, formula weight is the same as the molecular weight—that is, FW(KI) = 166 g/mol—but doesn't carry the implication that potassium iodide consists of KI "molecules."

A mole can be a sizable amount of material (the 254 g of I_2 required for Equation 3-1 is more than half a pound). Analytical chemists usually do not work on such a large scale. Therefore, most examples and problems in this text will use milliquantities such as milligrams (mg), milliliters (mL), and millimoles (mmol). Formula weights remain the same numerically but are stated in milligrams per millimole (mg/mmol). These smaller units are more convenient because they seldom involve fractional quantities for typical analytical reactions. For example, it is more convenient to express sample weight as 23.1 mg rather than 0.0231 g or to express volume as 42.16 mL rather than 0.04216 L.

3-2 SOLUTIONS

CONCENTRATION

Because so much of analytical chemistry involves the use of solutions, we describe a solution in terms of the concentration of solute in the solvent. There are many concentration units in use, but a concentration is typically described as the amount of solute per unit volume of solution—that is, the volume of solvent *plus* solute. The convenience of this concentration definition is that the amount of solute can easily be found as the product of concentration (C) and volume (V):

$$\text{amount of solute} = \text{concentration} \times \text{volume} \qquad \textbf{(3-3)}$$

If we want to know the amount of solute in grams, we can use concentration as g/L:

$$\text{grams of solute} = C\ (\text{g/\cancel{L}}) \times V\ (\cancel{L}) \qquad \textbf{(3-4)}$$

---

[1] There is no reason to use different terms and abbreviations for the mass of one mole of any substance (atoms, molecules, electrons, or ions). It is now recommended that all be called *molar mass* and given the abbreviation M. However, the symbol M is still widely used to signify molar concentration. To avoid this ambiguity, FW (formula weight) will be used to denote molar mass in this text; F will be used to denote a concentration of one molar mass of substance per liter of solution.

Parts per Million (ppm) In the analysis of very dilute solutions, a concentration scale of milligrams per liter (mg/L) is often employed. Most solutions employ water as the solvent, which has a density of 1 g/mL (1000 g/L). Therefore, a solution concentration of 1 mg of solute per liter of aqueous solution contains 0.001 g of solute per 1000 g of solution, or one part solute to 10^6 (million) parts solution. Concentrations expressed as *parts per million* are commonly used in environmental analysis and are approximately equal to milligrams of solute per liter of solution. Concentrations of even more dilute solutions can be expressed as *parts per billion* (ppb), which equals micrograms of solute per liter (μg/L) for dilute aqueous solutions.

Concentrations of parts per million (or parts per billion) are also used in gas analysis (measurement of air pollution) but in this case "parts per million" refers to one part of substance to 10^6 parts of gas. There can be a confusion because the concentration will be numerically different depending on if it is one part by weight to 10^6 parts of gas, or one part by volume to 10^6 parts of gas.

Molarity and Formality More frequently, a chemical reaction is involved and the number of moles is desired. The concentration unit then is moles per liter of solution and is called the *molarity* (M) or *formality* (F):

$$\text{moles of solute} = M \text{ (moles/\cancel{L})} \times V \text{ (\cancel{L})} = FV \qquad \text{(3-5)}$$

In terms of milliquantities (note that the symbol and the numerical value for molarity and formality remain unchanged):

$$\text{mmol} = M \text{ (mmol/\cancel{mL})} \times V(\cancel{mL}) = FV \qquad \text{(3-6)}$$

Problems involving concentration units are given in Examples 3-1 and 3-2.

Percentage Concentration Another widely used concentration unit is *percentage concentration*. Unfortunately, unless the chemist specifies which scale is being used, there can be some confusion over the several different percentage concentration scales. Three common percentage concentration scales are

$$\text{weight percent (w/w\%)} = \frac{\text{weight of solute}}{\text{weight of solution}^2} \times 100 \, (\%)$$

$$\text{volume percent (v/v\%)} = \frac{\text{volume of solute}}{\text{volume of solution}} \times 100 \, (\%)$$

$$\text{weight--volume percent (w/v\%)} = \frac{\text{weight of solute}}{\text{volume of solution}} \times 100 \, (\%)$$

---

[2]This is one example of a concentration unit that is not on a per unit volume of solution basis. It is useful when weights of solution are measured—that is,

$$\text{weight of solute} = (\text{w/w\%}) \times \frac{\text{weight of solution}}{100(\%)}$$

EXAMPLE 3-1

(a) Calculate the formal concentration of potassium dichromate (used as an oxidizing agent in analytical reactions) in a solution that contains 16.5 g/L $K_2Cr_2O_7$.

(b) How many formula weights (moles) of $K_2Cr_2O_7$ will be present in 150 mL of this solution?

Solution

(a) The formal concentration is the number of formula weights per liter. Therefore, it is necessary to determine how many formula weights (moles) of $K_2Cr_2O_7$ are contained in 16.5 g. (Formula weights are given on inside back cover.)

$$FW(K_2Cr_2O_7) = 294 \text{ g/mol}$$

$$mol/L = \frac{16.5 \text{ g/L}}{294 \text{ g/mol}} = 0.0561 \ F$$

(b) Using Equation 3-5, we calculate moles as

$$mol = F \times V$$

$$mol = 0.0561 \times 0.150 = 8.42 \times 10^{-3}$$

(*Note:* In this example, milliquantities are more convenient.) Using Equation 3-6 for milliquantities (with V given in mL), we can write:

$$mmol = F \times V$$

$$= 0.0561 \times 150 = 8.42$$

p-FUNCTIONS

Many laymen have heard of pH and may even be aware that it is a measure of acidity. In general chemistry students are taught that pH is a logarithmic concentration scale:

$$pH = -\log [H_3O^+] \quad \text{(or, simply, } -\log [H^+]) \tag{3-7}$$

The symbol p has been used extensively to denote the operation of $-\log$ because it converts very small numbers with negative exponents to relatively small positive numbers (as an example, the p-function value for 6.3×10^{-8} is 7.20). Example 3-3 illustrates the calculation of a p-function, pCu. The p-function is also convenient for comparing or listing values of widely varying order of magnitude. For example, equilibrium constants, K values, for reactions may be tabulated using pK's, as in Appendix 2.

EXAMPLE 3-2

An analyst wishes to add 120 mg NaOH base to dissolve an organic acid sample using a 0.248 F NaOH solution. How many milliliters of the NaOH solution should be added to the organic acid?

Solution

First, the number of millimoles of NaOH required is calculated, based on the formula weight of NaOH. (Formula weights are given on inside back cover.)

$$FW (NaOH) = 40 \text{ mg/mmol}$$

$$\frac{120 \text{ mg}}{40 \text{ mg/mmol}} = 3.00 \text{ mmol}$$

Then, using Equation 3-6,

$$mmol = F \times V$$

$$3.00 = 0.248 \times V$$

$$V = \frac{3.00 \text{ mmol}}{0.248 \text{ mmol/mL}} = 12.1 \text{ mL}$$

EXAMPLE 3-3

A sewer stream is analyzed for copper. If the copper content is 0.15 ppm (mg/L), what is the pCu of the sewer stream?

Solution

To determine p-function, we must convert concentration from milligrams per liter to millimoles per milliliter. This is accomplished by dividing the concentration by the formula weight (63.5 for Cu)

$$\frac{0.15 \text{ mg/L}}{63.5 \text{ mg/mmol}} = 2.4 \times 10^{-3} \text{ mmol/L} = 2.4 \times 10^{-6} F$$

$$pCu = -\log [Cu] = -\log (2.4 \times 10^{-6}) = 5.62$$

In Section 2-2, we noted that, when logarithm quantities are given, only the mantissa (numbers to the right of the decimal point) is significant. The characteristic (numbers to the left of the decimal point) denotes only the exponent. In this example, the copper concentration contains two significant figures; therefore, the pCu should be expressed as two significant figures—that is, 5.62.

DENSITY

Analytical reactions are frequently carried out in dilute aqueous solutions in which the density can be assumed to be approximately 1 g/mL. However, when nonaqueous solvents or concentrated aqueous solutions are used, it may be necessary to know the density in order to determine the amount of solute present in a given amount of solution. If the concentration is given in weight percent (w/w%), it is necessary to multiply the weight percent by the density to determine the weight of solute per unit volume:

$$\text{concentration of solute} = \text{weight percent} \times \text{density}$$

$$\text{g/mL} = \frac{\text{w/w}\%}{100\%} \times \frac{\text{g}}{\text{mL}} \tag{3-8}$$

If formality (F) is desired, it is only necessary to divide the concentration in grams per milliliter by the formula weight (FW) of the solute and multiply by 1000 mL/L:

$$\text{formality} = \frac{\text{weight percent}}{100} \times \text{density} \times \frac{1}{\text{FW}} \times 1000 \text{ mL/L}$$

$$F = \frac{\text{w/w}\%}{100\%} \times \text{g/mL} \times \frac{1}{\text{g/mol}} \times 1000 \text{ mL/L}$$

A typical calculation is shown in Example 3-4.

DILUTION

Example 3-4 shows a two-step calculation for determining the amount of concentrated HNO_3 needed to prepare a dilute HNO_3 solution. The two steps can be combined into one basic equation, because in any dilution step the amount of solute must be conserved (assuming no sloppiness on the part of the chemist). In terms of moles:

$$\text{moles solute (concentrated solution)} = \text{moles solute (dilute solution)}$$

From Equation 3-5, we know that the number of moles $= FV$. Therefore,

$$FV \text{ (concentrated solution)} = FV \text{ (dilute solution)} \tag{3-9}$$

The same equation can be used to calculate millimoles of solute if V is given in milliliters.

The second part of Example 3-4 is then calculated, using Equation 3-9:

$$FV = FV \tag{3-10}$$

$$15.6 \, V = 2 \times 250$$

$$V = 32.1 \text{ mL}$$

EXAMPLE 3-4

(a) What is the formality of 69% HNO_3 (w/w%), which has a density of 1.42 g/mL?
(b) How would 250 mL of $2F$ HNO_3 be prepared from concentrated HNO_3 (69%)?

Solution

(a) The HNO_3 contains $\frac{69}{100}$ (g HNO_3/g soln). Therefore, the concentration of $HNO_3 = 0.69 \times 1.42 = 0.98$ g HNO_3/mL. Formality is the moles of solute per liter of solution. Therefore,

$$F = \frac{0.98 \text{ g } HNO_3/\text{mL} \times 1000 \text{ mL/L}}{FW\ (HNO_3)}$$

$$FW\ (HNO_3) = 63 \text{ g/mol}$$

$$F = \frac{980 \text{ g/L}}{63 \text{ g/mol}} = 15.6 \text{ mol/L}$$

(b) The amount of HNO_3 required can be found by using Equation 3-6:

$$\text{mmol } HNO_3 = F \times V = 2 \times 250 = 500$$

The volume of concentrated acid required is then calculated, again using Equation 3-6:

$$500 = F \times V = 15.6 \times V$$

$$V = 32.1 \text{ mL}$$

Therefore, the solution is prepared by adding 32.1 mL of concentrated HNO_3 solution to sufficient H_2O to make a total volume of 250 mL.

The simplified expression in Equation 3-10 can also be used for a chemical reaction in which the reactant is entered on one side and the product (or another reactant) on the other, *provided* an equal number of moles of reactant and product are involved (see Example 3-5). In a later section, another concentration unit will be introduced so that *all* chemical reactions can be handled by an equation as simple as Equation 3-10.

3-3   EQUIVALENTS

The balanced chemical equation can be used to calculate the quantitative relationship between a reactant and any other reactant or product of the reaction. For Equation 3-1, it would require 332 g of KI to reduce 48 g of ozone. The relationship is numerically simpler if moles are used (it requires 2 moles of KI to

EXAMPLE 3-5

One method for analyzing solutions containing calcium is to form a complex (1 mol of Ca to 1 mol of EDTA) by adding a known amount of an EDTA solution.

(a) If 50.0 mL of a hard-water sample requires 15.63 mL of a 0.0164 F EDTA solution, calculate the calcium formality in the hard-water sample.

(b) How many milligrams of calcium does the 50.0-mL sample of hard water contain?

(c) What is the concentration of calcium in the hard-water sample in terms of parts per million (milligrams per liter)?

Solution

(a) Because this reaction involves an equal number of moles of reactant and product, Equation 3-10 can be used:

$$F_{Ca} V_{Ca} = F_{EDTA} V_{EDTA}$$

$$F_{Ca} = \frac{0.0164 \times 15.63}{50.0} = 0.00513 \ F$$

(b)

$$0.00513 \ \frac{mmol}{mL} \times 50 \ mL \times 40 \ \frac{mg}{mmol} = 10.3 \ mg \ Ca$$

(c) Because 50 mL = 0.050 L,

$$\frac{10.3 \ mg \ Ca}{0.050 \ L} = 206 \ mg/L \cong 206 \ ppm$$

reduce 1 mole of ozone). But a balanced equation is still necessary to determine this mole relationship so that the calculation can be carried out. In Example 3-5, we could equate moles of calcium to moles of EDTA because a 1:1 mole ratio complex was formed. Problems involving stoichiometry can be considerably simplified if all reactions are put on a 1:1 basis for all reactants and products. The use of equivalents is a way to accomplish this goal.

REACTIVE SPECIES

The basic concept of equivalents is a definition of *reactive species* such that, for any reaction,

$$\text{equivalents } X = \text{equivalents } Y \qquad \qquad \textbf{(3-11)}$$

Here X and Y represent any two substances (reactant or product) involved in the chemical reaction. In the case of an analytical reaction, X and Y would be the analyte, and the reagent or product used as the measurement signal:

$$\text{equivalents analyte} = \text{equivalents reagent} = \text{equivalents product} \qquad \textbf{(3-12)}$$

If stated in milliquantities, Equation 3-12 would be

$$\text{meq analyte} = \text{meq reagent} = \text{meq product} \qquad \textbf{(3-13)}$$

Equations 3-12 and 3-13 are valid only when the analytical reagent has been added to the analyte in exact stoichiometric quantities. Methods for determining this stoichiometric point (the *equivalence point*) will be discussed later in the text.

Equations 3-12 and 3-13 demonstrate 1:1 relationships that will simplify calculations and even eliminate the need to write a balanced chemical equation. But first, to determine an equivalent, we will construct parallel definitions with moles and equivalents:

$$\text{moles} = \text{g/FW} \qquad \textbf{(3-14)}$$

$$\text{equivalents} = \text{g/EW} \qquad \textbf{(3-15)}$$

What is the equivalent weight (EW)? As with molecular weight, it is that weight which contains 6.02×10^{23} particles; in this case, the particle is the *reactive species*. Because molecular weights and formula weights are easy to calculate, it is convenient to use them for determining the equivalent weight:

$$\text{FW} = \text{g/mole} = \text{mg/mmol}$$

$$\text{EW} = \text{g/eq} = \text{mg/meq}$$

Therefore,

$$\text{EW} = \frac{\text{FW (g/mole)}}{n \text{ (eq/mole)}} \quad \text{or} \quad \frac{\text{(mg/mmol)}}{\text{(meq/mmol)}} \qquad \textbf{(3-16)}$$

The equivalent weight is simply the formula weight divided by some number (n), which represents the moles of reactive species in a mole of substance. When a reactive species is defined we can examine any substance to determine how many reactive species it contains—that number is n. It is fairly obvious that n will *normally* be a number ≥ 1 and usually an integer. Example 3-6 shows how a problem in converting one compound to another can be solved using reactive species and Equation 3-13. However, this method is most useful when solutions are employed, as will be demonstrated later in this chapter.

NORMALITY

A solution containing 1 mol of solute per liter of solution has a concentration of 1 F. In the same manner a solution containing one mole of reactive species per liter of solution will be defined as having a concentration of 1 *normal* (N). The terms

EXAMPLE 3-6

The phosphorus content of fertilizers is generally given as the percentage of phosphorus pentoxide (P_2O_5), even though the phosphorus is present in some other form. A 165-mg sample of fertilizer is decomposed with nitric acid and the phosphate ion precipitated as the quinoline salt of phosphomolybdic acid, $(C_9H_7N)_2 \cdot H_3PMo_{12}O_{40}$. The precipitate is then dried and weighed. If the final weight is 546 mg, calculate the percentage of P_2O_5 in the fertilizer sample.

Solution

Because we are concerned with phosphorus, we should define equivalents per mole as the number of moles of P per mole of substance. Analyte is P_2O_5 ($n = 2$) and product is $(C_9H_7N)_2 \cdot H_3PMo_{12}O_{40}$ ($n = 1$). According to Equation 3-13,

$$\text{meq analyte} = \text{meq product}$$

$$\text{meq analyte} = \frac{\text{mg analyte}}{\text{EW analyte}} = \frac{\text{mg analyte}}{\text{FW analyte}/n_{\text{analyte}}} = \frac{\text{mg analyte}}{141.9/2}$$

$$\text{meq product} = \frac{\text{mg product}}{\text{EW product}} = \frac{\text{mg product}}{\text{FW product}/n_{\text{product}}} = \frac{546}{2084/1}$$

Therefore,

$$\frac{\text{mg analyte}}{141.9/2} = \frac{546}{2084/1}$$

$$\text{mg analyte} = \frac{546}{2084/1} \times \frac{141.9}{2} = 18.6 \text{ mg } P_2O_5 \text{ in sample}$$

$$\% \ P_2O_5 \text{ in fertilizer} = \frac{18.6 \text{ mg } P_2O_5}{165 \text{ mg sample}} \times 100 \ (\%) = 11.3\%$$

equivalent, equivalent weight, and *normality* are analogous to *mole, molecular weight,* and *molarity.* These terms are summarized in Table 3-1. The former terms were devised to avoid ambiguity in situations where, for example, a 0.5 F solution could mean 0.5 moles of substance per liter or 0.5 moles of reactive species per liter. The number of equivalents a solution contains is given by Equation 3-17:

$$\text{equivalents} = N \ (\text{eq}/L) \times V \ (L) \tag{3-17}$$

Using milliquantities, this would be

$$\text{meq} = N \ (\text{meq}/mL) \times V \ (mL) \tag{3-18}$$

By using Equation 3-19, the normality of any solution can be calculated, provided

TABLE 3-1 Summary of Terms and Their Abbreviations

	Moles	Formula weight	Equivalents
weight of 6 × 10²³ units (6 × 10²⁰ for milliquantity)	molecular weight	formula weight	equivalent weight
abbreviation	MW	FW*	EW†
units	g/mol, mg/mmol	g/mol, mg/mmol	g/eq, mg/meq
amount of material	moles	moles	equivalents
abbreviation	mol	mol	eq
milliquantity	millimoles	millimoles	milliequivalents
abbreviation	mmol	mmol	meq
concentration	molar	formal	normal
abbreviation	$M^{\ddagger}$	F	N
units	mol/L, mmol/mL	mol/L, mmol/mL	eq/L, meq/mL

*FW denotes molar mass.
†EW is replaced by FW/n, as shown in Equation 3-16.
‡F is generally used instead of M for substances dissolved in solution. M is used for describing concentration of a specific species in solution.

that the formality and n (eq/mol) are known:

$$N \text{ (eq/L)} = n \text{ (eq/mol)} \times F \text{ (mol/L)} \tag{3-19}$$

Example 3-7 shows a typical calculation of this type. Equations 3-15 and 3-17 now allow us two ways to find the number of equivalents. There is no reason that both sides of an equivalent relationship (Equation 3-12 or 3-13) must be of the same form. For example, with the use of convenient milliquantities, an analyte and reagent can

---

EXAMPLE 3-7

Sulfuric acid (H_2SO_4), when used as an acid, contains 2 eq/mol. Calculate the normality of a 0.23 F H_2SO_4 solution.

Solution

The normality of a solution is its concentration in equivalents per liter or milliequivalents per milliliter. From Equation 3-19, we know that $N = n F$. Therefore,

$$N = 2 \text{ eq/mol} \times 0.23 \text{ mol/L} = 0.46 \text{ eq/L}$$

be related as follows:

$$\text{meq analyte} = \text{meq reagent}$$

$$\frac{\text{mg analyte}}{\text{EW(analyte)}} = \frac{\text{mg analyte}}{\text{FW(analyte)}/n} = N_{\text{reagent}} V_{\text{reagent}} \qquad \textbf{(3-20)}$$

If the analyte were a solution, we could as easily write:

$$\text{meq analyte} = \text{meq reagent}$$

$$N_{\text{analyte}} V_{\text{analyte}} = N_{\text{reagent}} V_{\text{reagent}} \qquad \textbf{(3-21)}$$

Equations 3-20 and 3-21 will be quite useful, even though they are merely extensions of Equation 3-13. Several other ways of expressing the equivalent relationship will be developed later. Before problems can be solved, however, it is necessary to determine n, the number of equivalents (moles of reactive species) per mole of substance. For this, it is useful to define four reaction types.

REACTION TYPES

There are many different reactions of interest to analytical chemists but they can be conveniently classified into four types:

1. Precipitation

2. Acid–base

3. Complexation

4. Redox

A precipitation reaction involves the combining of cation charge with anion charge to form an insoluble compound (the solubility is not the critical point, but merely reflects a specific analytical reaction). For example, Ag^+ combines with Cl^- to form AgCl, and Ba^{2+} combines with SO_4^{2-} to form $BaSO_4$. In the first case, any compound donating one Ag^+ or one Cl^- would have an n value of 1. In the second case, any compound donating one Ba^{2+} or one SO_4^{2-} would have an n value of 2 because *charge* is the reactive species. Table 3-2 presents several other examples.

Note that by using charge as the reactive species the n value will be the same whether the cation is used ($BaCl_2$ donates one barium cation of charge $+2$) or the anion is used ($BaCl_2$ donates *two* chloride anions of charge -1). This will hold true provided all the cation or anion charge is used.

The second reaction type is acid–base. For aqueous solutions typically used in analytical chemistry, an acid is any substance donating hydronium ions (H_3O^+, often written as H^+ and called "hydrogen ion" for simplicity) or consuming base. A base is any substance donating hydroxide ions (OH^-) or consuming acid. The reactive species are H_3O^+ and OH^-. This simple acid–base definition may be extended to substances not containing H^+ (Lewis acids) when reactions in other solvent systems are employed. As shown in Table 3-2, the n value for any substance

TABLE 3-2 Examples of Reaction Types

Substance	Analytical use	n(eq/mol)
precipitation reaction:		
reactive species: cation charge and anion charge		
$AgNO_3$	$Ag^+ \rightarrow AgBr(s)$	1
NaBr	$Br^- \rightarrow AgBr(s)$	1
$BaCl_2$	$Ba^{2+} \rightarrow BaSO_4(s)$	2
	$Cl^- \rightarrow AgCl(s)$	2
$Fe_2(SO_4)_3$	$SO_4^{2-} \rightarrow PbSO_4(s)$	6
	$Fe^{3+} \rightarrow Fe(OH)_3 \cdot xH_2O(s)$	6
$HClO_4$	$ClO_4^- \rightarrow KClO_4(s)$	1
acid–base reaction:		
reactive species: H_3O^+ and OH^-		
$HClO_4$	$HClO_4 \rightarrow H_3O^+$	1
H_2SO_4	$H_2SO_4 \rightarrow 2H_3O^+$	2
$Ca(OH)_2$	$Ca(OH)_2 \rightarrow 2OH^-$	2
HgO	$HgO + 4I^- + 2H_3O^+ \rightarrow HgI_4^{2-} + 3H_2O$	2
Na_2CO_3	$Na_2CO_3 + H_3O^+ \rightarrow 2Na^+ + HCO_3^- + H_2O$	1
	$Na_2CO_3 + 2H_3O^+ \rightarrow 2Na^+ + CO_2 + 3H_2O$	2
H_3PO_4	$H_3PO_4 + OH^- \rightarrow H_2PO_4^- + H_2O$	1
	$H_3PO_4 + 2OH^- \rightarrow HPO_4^{2-} + 2H_2O$	2
	$H_3PO_4 + 3OH^- \rightarrow PO_4^{3-} + 3H_2O$	3
complexation reaction:		
reactive species: electron pair donor and electron pair acceptor		
KI	$Hg^{2+} + 4I^- \rightarrow HgI_4^{2-}$	1
BaI_2	$Hg^{2+} + 4I^- \rightarrow HgI_4^{2-}$	2
$HgCl_2$	$Hg^{2+} + 4I^- \rightarrow HgI_4^{2-}$	4
$AgNO_3$	$Ag^+ + 2NH_3 \rightarrow Ag(NH_3)_2^+$	2
Ag_2SO_4	$Ag^+ + 2NH_3 \rightarrow Ag(NH_3)_2^+$	4
NH_3	$Cu^{2+} + 4NH_3 \rightarrow Cu(NH_3)_4^{2+}$	1
EDTA	$Ca^{2+} + EDTA^{4-} \rightarrow Ca(EDTA)^{2-}$	6
$CaCl_2$	$Ca^{2+} + EDTA^{4-} \rightarrow Ca(EDTA)^{2-}$	6
redox reaction		
reactive species: electron donor and electron acceptor		
$FeCl_2$	$Fe(II) \rightarrow Fe(III)$	1
$KMnO_4$	$Mn(VII) \rightarrow Mn(II)$	5
	$Mn(VII) \rightarrow Mn(IV)$	3
	$Mn(VII) \rightarrow Mn(VI)$	1
$K_2Cr_2O_7$	$Cr(VI) \rightarrow Cr(III)$	6
$Na_2S_2O_3$	$S_2O_3^{2-} \rightarrow S_4O_6^{2-}$	1
$HClO_4$	$Cl(VII) \rightarrow Cl^-$	8
I_2	$I_2 \rightarrow 2I^-$	2
As_2O_3	$As(III) \rightarrow As(V)$	4

acting as an acid or base is the number of acid (or base) reactive species the substance donates (or consumes, as in the case of HgO).

Table 3-2 shows that some substances may have more than one analytical use with different n values. It is important, therefore, to know how the substance will be used before any calculations can be carried out; however, as will be illustrated later, it is not necessary to write detailed balanced chemical reactions.

The third reaction type is complexation, in which the reactive species is an *electron pair*. One substance, (the *ligand*) donates an electron pair—that is, the electron density associated with this electron pair, to another substance, the acceptor. For this to occur, the acceptor substance must have vacant orbitals available for the electron pair. The acceptor is frequently a metal ion. Transition metal ions with low-lying vacant d, s, and p orbitals are especially good acceptors and will form strong complexes with ligands.

Table 3-2 presents some examples of complexation reactions and the n values for several substances. Iodide ion (I^-), ammonia (NH_3), and EDTA are ligands which donate one, one, and six electron pairs respectively. Mercury accepts four electron pairs when complexed with I^-, silver accepts two when complexed with NH_3, and calcium accepts six when complexed with EDTA. Many metal ions undergo more than one complexation reaction in which the n values may be different.

The fourth reaction type is oxidation–reduction, or *redox* for short. In a redox reaction, one substance is an electron donor (the reducing agent which is oxidized in the reaction); another is an electron acceptor (the oxidizing agent that is reduced in the reaction). The reactive species are the electrons that are transferred from the reducing agent to the oxidizing agent. The number of electrons transferred can be determined by noting oxidation state changes (see Table 3-2). The n value for any substance will be the oxidation state *change* for any element multiplied by the number of atoms of this element in the substance. For example, arsenic trioxide (As_2O_3) is used in a reaction in which arsenic is oxidized from the $+3$ oxidation state in As_2O_3 to the $+5$ oxidation state. The n value for As_2O_3 is 4 (oxidation state change of 2 times 2 arsenic atoms in As_2O_3).

From Table 3-2, note that, in order to determine the n value, only the oxidation state need be known, not the exact species. For example, Mn(VII) is MnO_4^- and not Mn^{7+}. Frequently, the exact species in solution is not known or several species may be present. The convention of giving oxidation as a roman numeral tells the reader the oxidation state of the element without making any claims with regard to exact species.

Table 3-2 indicates that, if used in different redox reactions, a substance may exhibit different n values. As in the case of acid–base reactions, it is important to know how a substance will be used, especially when stating the normality of a solution. A 0.1 F $KMnO_4$ solution will be 0.5 N if the $KMnO_4$ is reduced to Mn(II), 0.3 N if it is reduced to Mn(IV) and 0.1 N if it is reduced to Mn(VI). In each case, the normality was calculated according to Equation 3-19. For this reason, possible confusion can be avoided by including the formal concentration in addition to normality if there is any possibility that the solution will be used for reactions with different n values or that the solution may be used by someone else.

Note that perchloric acid ($HClO_4$) was used in three of the four reaction types described so far. If a solution of $HClO_4$ is prepared for use as an acid, its normality will be equal to its formality because $n = 1$. If this solution is now used to precipitate potassium perchlorate, the n value remains 1. However, if the $HClO_4$ is used as an oxidizing agent in which the chlorine atom is reduced from $+7$ to -1, the n value is 8 and the normality will equal 8 times the formality.

When redox reactions are involved, the number of equivalents can be determined by measuring the electrons transferred. Redox reactions may be carried out electrochemically and the amount of electrical charge (measured in coulombs, C) that is transferred may be measured. The charge on one electron, the reactive species, is 1.602×10^{-19} coulombs. Therefore, for one mole of electrons the charge transferred is:

$$\text{charge for 1 mole of electrons} = 1.6022 \times 10^{-19} \frac{\text{coulombs}}{\cancel{\text{electron}}} \times 6.022 \times 10^{23} \frac{\cancel{\text{electrons}}}{\text{mole}}$$

$$= 96{,}485 \text{ coulombs/mole of reactive species (C/eq)}$$

The quantity, 96,485 coulombs ($\sim 9.65 \times 10^4$), is called the Faraday constant (F). Therefore, the number of equivalents can be calculated from the number of coulombs of charge required to carry out the reaction:

$$\text{eq} = \frac{\text{charge }(\cancel{C})}{9.65 \times 10^4 \ (\cancel{C}/\text{eq})} \tag{3-22}$$

or in milliquantities

$$\text{meq} = \frac{\text{charge }(m\cancel{C})}{9.65 \times 10^4 \ (m\cancel{C}/\text{meq})} \tag{3-23}$$

Another method for determining equivalents is to measure volume when a gaseous reactant or product is involved. If the gas behaves nearly ideally, the volume of one mole of gas is 22.4L at STP (0°C, 1 atm pressure). The *equivalent volume* is then:

$$\text{equivalent volume (L/eq)} = \frac{22.4 \ (L/\cancel{\text{mol}})}{n \ (\text{eq}/\cancel{\text{mol}})} \tag{3-24}$$

Therefore, the number of equivalents can be calculated from the volume of gas consumed or produced:

$$\text{eq} = \frac{V \ (\cancel{L})}{22.4/n \ (\cancel{L}/\text{eq})} \tag{3-25}$$

Restated in milliquantities, this equation would be

$$\text{meq} = \frac{V \ (m\cancel{L})}{22.4/n \ (m\cancel{L}/\text{meq})} \tag{3-26}$$

SOLVING PROBLEMS WITH EQUIVALENTS

The fundamental equation needed for solving stoichiometric problems using equivalents (or milliequivalents) is Equation 3-12 (or 3-13). From that basic relationship, one may use any of the methods which have been presented for determining equivalents of analyte or reagent.

$$\text{eq analyte} = \frac{\text{mg analyte}}{\text{FW (analyte)}/n_{\text{analyte}}}; \text{eq reagent} = \frac{\text{mg reagent}}{\text{FW (reagent)}/n_{\text{reagent}}}$$

$$= N_{\text{analyte}} V_{\text{analyte}} \quad ; \qquad = N_{\text{reagent}} V_{\text{reagent}}$$

$$= n_{\text{analyte}} F_{\text{analyte}} V_{\text{analyte}}; \qquad = n_{\text{reagent}} F_{\text{reagent}} V_{\text{reagent}}$$

$$= \frac{\text{electrical charge (C)}}{9.65 \times 10^4}; \qquad = \frac{\text{electrical charge (C)}}{9.65 \times 10^4}$$

$$= \frac{V_{\text{analyte}}(g)}{22.4/n_{\text{analyte}}} \quad ; \qquad = \frac{V_{\text{reagent}}(g)}{22.4/n_{\text{reagent}}}$$

EXAMPLE 3-8

Standard solutions (those with accurately known concentrations) are often prepared to an approximate value and then *standardized* by carrying out an analytical reaction with another standard solution. Of a sodium hydroxide solution (NaOH) 25.00 mL requires 47.13 mL of 0.1613 N H_2SO_4. Calculate the normality of the NaOH solution.

Solution

Because both the analyte, NaOH, and the reagent, H_2SO_4, are solutions, the most useful form for expressing milliequivalents is with normality and volume:

$$\text{meq analyte} = \text{meq reagent}$$

$$N_{\text{analyte}} V_{\text{analyte}} = N_{\text{reagent}} V_{\text{reagent}}$$

$$N_{\text{analyte}} \times 25.00 = 0.1613 \times 47.13$$

$$N_{\text{analyte}} = 0.3041 \ N$$

Note: This is an acid–base reaction and $n = 1$ for NaOH and $n = 2$ for H_2SO_4. However, because the problem was worked entirely in equivalents, this information was not necessary. If the formal concentration is needed, Equation 3-9 would be used:

$$F_{\text{NaOH}} = \frac{N}{n} = 0.3041/1 = 0.3041 \ F$$

$$F_{H_2SO_4} = \frac{N}{n} = 0.1613/2 = 0.0806 \ F$$

The method used for determining equivalents of analyte need not be the same as that used for determining equivalents of reagent. Frequently, an amount of analyte is sought in mass units, but the reagent is a solution with a concentration involving volume units. In milliequivalents, this would be:

$$\text{meq analyte} = \text{meq reagent}$$

$$\frac{\text{mg analyte}}{\text{FW (analyte)}/n_{analyte}} = N_{reagent} V_{reagent}$$

Therefore,

$$\text{mg analyte} = N_{reagent} V_{reagent} \text{FW (analyte)}/n_{analyte}$$

The normality and volume of reagent required to carry out the reaction are known and the formula weight of the analyte is easily calculated. The value of n is determined by identifying the reaction type involved and noting how many reactive species are in the analyte. Some typical cases are described in Examples 3-8 to 3-10.

EXAMPLE 3-9

The iron content of an ore sample can be determined by dissolving the sample in HCl to form Fe(III) in solution, reducing to Fe(II) with $SnCl_2$, and then oxidizing to Fe(III) with a standard solution of $KMnO_4$. Given that a 453-mg sample of ore requires 45.2 mL of 0.1030 N $KMnO_4$ to reach the equivalence point, calculate the percentage of Fe_2O_3 in the ore.

Solution

The weight of the analyte as Fe_2O_3 is desired, but the reagent is in the form of a standard solution. Therefore, the most direct way to solve for Fe_2O_3 content is:

$$\text{meq analyte} = \text{meq reagent}$$

$$\frac{\text{mg } Fe_2O_3}{\text{FW } (Fe_2O_3)/n} = NV = 0.1030 \times 45.2 = 4.66$$

The only factor which is still needed is the n value for Fe_2O_3. The chemical reaction type is redox and the reactive species are the electrons transferred. Iron is oxidized from Fe(II) to Fe(III) by $KMnO_4$, therefore each iron has one equivalent per mole. A mole of Fe_2O_3 contains 2 moles of Fe and therefore contains 2 moles of reactive species. Therefore, the n value for Fe_2O_3 is 2. FW(Fe_2O_3) = 159.7, therefore,

$$\text{mg } Fe_2O_3 = 0.1030 \times 45.2 \times \frac{159.7}{2} = 372 \text{ mg}$$

$$\% Fe_2O_3 = \frac{372 \text{ mg } Fe_2O_3}{453 \text{ mg sample}} \times 100 \, (\%) = 82.2\%$$

EXAMPLE 3-10

The nickel content in a battery cathode is determined by forming a complex with cyanide ion according to the equation:

$$Ni^{2+} + 4CN^- \rightleftharpoons Ni(CN)_4^{2-}$$

A cathode sample weighing 160 mg was dissolved and required 38.3 mL of 0.137 N KCN to reach the equivalence point. Calculate the percentage of Ni_2O_3 in the cathode.

Solution

The weight of analyte as Ni_2O_3 is desired, but the reagent is in the form of a standard solution. Therefore, the most direct way to solve for Ni_2O_3 content is through Equation 3-13:

$$\text{meq analyte} = \text{meq reagent}$$

$$\frac{\text{mg } Ni_2O_3}{\text{FW } (Ni_2O_3)/n} = NV = 0.137 \times 38.3 = 5.25$$

The only factor still needed is the n value for Ni_2O_3. The chemical reaction type is complexation; the reactive species is the electron pair. Each CN^- donates one pair and, therefore, each Ni^{2+} accepts 4 pairs. A mole of Ni_2O_3 contains 2 mol of Ni and therefore 8 mol of reactive species—that is, $n (Ni_2O_3) = 8$ and FW $(Ni_2O_3) = 165$. Therefore,

$$\text{mg } Ni_2O_3 = 0.137 \times 38.3 \times \frac{165}{8} = 108 \text{ mg}$$

$$\% Ni_2O_3 = \frac{108 \text{ mg } Ni_2O_3}{160 \text{ mg sample}} \times 100 \, (\%) = 67.6\%$$

Note that in Example 3-9 there is no need to use the analyte species involved in the reaction (Fe^{2+}) for the calculation. The analyte species to be calculated (Fe_2O_3) can be solved for directly, which sometimes saves several steps. Also, there is no need to write a balanced chemical equation for each reaction as long as the n values can be assigned by noting the reaction type and the number of reactive species.

CALCULATIONS WITH MORE THAN ONE REAGENT

To this point, we have assumed that the analytical reaction involved one analyte and one reagent. However, at times, two or more reagents are needed, as seen in Example 3-11. These reagents may not all contribute the same reactive species; for example, one might be an acid, the other a base. For such situations, Equation 3-13

EXAMPLE 3-11

In order to analyze an insoluble organic acid, an analyst dissolved it in excess base and then added a standard acid solution (HCl) until the equivalence point was reached (a back-titration). Calculate the equivalent weight of the organic acid if 550 mg dissolved in 50.0 mL of 0.105 N NaOH required 17.3 mL of 0.102 N HCl to reach the equivalence point.

Solution

In this case, there are two acids (unknown organic and HCl) and one base, so the equation for reaction type 2 should be used when the system is at the equivalence point.

$$\text{total meq acid} = \text{total meq base}$$

$$\frac{\text{mg organic acid}}{\text{EW (organic acid)}} + N_{HCl} V_{HCl} = N_{NaOH} V_{NaOH}$$

$$\frac{550}{EW} + 0.102 \times 17.3 = 0.105 \times 50.0$$

$$EW = \frac{550}{(0.105 \times 50.0) - (0.102 \times 17.3)} = 158 \text{ mg/meq (or g/eq)}$$

may be generalized *when the system is at the equivalence point* as follows:

Reaction type 1: total meq cation charge = total meq anion charge

Reaction type 2: total meq acid = total meq base

Reaction type 3: total meq ligands = total meq electron pair acceptor

Reaction type 4: total meq e donor = total meq e acceptor

In Example 3-11, the analyte was an acid, one reagent was a base, and a second reagent was an acid. Any of the forms given previously could be used for each species involved.

  The same approach can be used if there are two analytes, but because the analyte is unknown there would be two unknowns with only one equation. Additional information must be obtained so that there will be as many independent equations as there are unknowns. Problems of this type require solving a set of simultaneous equations, as discussed in the following section.

PROBLEMS REQUIRING SIMULTANEOUS EQUATIONS

Some samples contain more than one analyte component. When a sample is to be analyzed for c components, it is necessary to make a minimum of c measurement signals. This will lead to a set of c simultaneous equations.

EXAMPLE 3-12

An alloy sample weighing 453.2 mg and containing only lead and silver is dissolved in nitric acid; the insoluble compounds $PbBr_2$ and $AgBr$ are precipitated with the addition of KBr. If the sample requires 41.44 mL of 0.1041 N KBr to reach the equivalence point, calculate the percentages of Pb and Ag in the alloy.

Solution

In this case, there are two analytes, Pb and Ag. One measurement signal is the weight of the sample:

$$\text{mg Pb} + \text{mg Ag} = 453.2$$

The second measurement signal is the meq of KBr required to precipitate $PbBr_2$ and AgBr.

$$\text{meq Pb} + \text{meq Ag} = \text{meq KBr}$$

$$\frac{\text{mg Pb}}{\text{FW (Pb)}/n_{Pb}} + \frac{\text{mg Ag}}{\text{FW (Ag)}/n_{Ag}} = NV = 0.1041 \times 41.44 = 4.314$$

Because this is a precipitation reaction, n for $Pb^{2+} = 2$ and n for $Ag^+ = 1$. In this case, the formula weights are simply atomic weights:

$$\frac{\text{mg Pb}}{207.2/2} + \frac{\text{mg Ag}}{107.9/1} = 4.314$$

We now have two equations and two unknowns, mg Pb and mg Ag. This set of equations may be solved by the method in Appendix 1-B or by substitution, as follows:

$$\text{mg Pb} = 453.2 - \text{mg Ag}$$

Therefore,

$$\frac{453.2 - \text{mg Ag}}{207.2/2} + \frac{\text{mg Ag}}{107.9/1} = 4.314$$

$$4.375 - 9.653 \times 10^{-3} \text{ mg Ag} + 9.268 \times 10^{-3} \text{ mg Ag} = 4.314$$

$$\text{mg Ag} = \frac{4.375 - 4.314}{(9.653 - 9.268) \times 10^{-3}} = 158 \text{ mg} = 35\% \text{ Ag}$$

$$\text{mg Pb} = 453.2 - 158 = 295 \text{ mg} = 65\% \text{ Pb}$$

Note: This is not a good method for high accuracy because subtraction of two numbers of nearly equal magnitude leads to a loss in significant figures. This occurred because Ag and Pb have nearly equal equivalent weights (107.9 and 103.6), even though the atomic weight of Pb is nearly twice that of Ag. Procedures involving such subtractions should be avoided if possible.

Suppose the sample contains only two components, analyte A_1 and analyte A_2. In this case, the sample weight itself (S) becomes one measurement signal because Equation 3-27 can be applied:

$$\text{mg } A_1 + \text{mg } A_2 = S \text{ (mg)} \qquad \textbf{(3-27)}$$

If the sample undergoes some analytical reaction in which the total milliequivalents of A_1 and A_2 are measured with reagent R, we may write:

$$\text{meq } A_1 + \text{meq } A_2 = R \text{ (meq)} \qquad \textbf{(3-28)}$$

In terms of analyte mass, we now have

$$\frac{\text{mg } A_1}{\text{EW } (A_1)} + \frac{\text{mg } A_2}{\text{EW } (A_2)} = R \qquad \textbf{(3-29)}$$

Both Equations 3-27 and 3-29 are simultaneous equations containing the two unknowns, mg A_1 and mg A_2. In principle, the problem is solved as shown in Example 3-12, although the algebra may become tedious, especially when three or more equations are involved. Appendix 1-B shows how sets of simultaneous equations can be solved quickly with some rudiments of matrix algebra and a desktop computer.

SUPPLEMENTARY READING

International Union of Pure and Applied Chemistry, *Compendium of Analytical Nomenclature*. Elmsford, N.Y.: Pergamon Press, Inc., 1978.

PROBLEMS

3-1. The following pollutants are commonly encountered in air and water environmental analyses. Calculate the formula weight for each.
  a. ozone, O_3
  *b. tetraethyllead (TEL), $Pb(C_2H_5)_4$
  c. sodium tripolyphosphate (STPP), $Na_5P_3O_{10}$
  d. phenol, C_6H_5OH
  e. sulfur dioxide, SO_2
  f. DDT, $C_{14}H_9Cl_5$

3-2. How many millimoles would be present if you had 150 mg of each of the pollutants in Problem 3-1?

3-3. How many micromoles would be present if you had 320 μg of each of the pollutants in Problem 3-1?

---

* Answers to problems marked with an asterisk will be found at the back of the book.

3-4. The following reactions are used in analytical procedures. Balance the chemical equations describing these reactions and calculate the amount (in milligrams) of reagent needed (second reactant in the equation) if 200 mg of analyte (first reactant in the equation) is present.

    a. $Pb^{2+} + I^- \rightleftharpoons PbI_2(s)$

   *b. $HgO + HCl \rightleftharpoons Hg^{2+} + H_2O + Cl^-$

    c. $NiO + CN^- + H_2O \rightleftharpoons Ni(CN)_4^{2-} + OH^-$

    d. $Fe^{2+} + MnO_4^- + H_3O^+ \rightleftharpoons Fe^{3+} + Mn^{2+} + H_2O$

    e. $Na_2S_2O_3 + I_2 \rightleftharpoons Na_2S_4O_6 + NaI$

    f. $O_3 + KI + H_3O^+ \rightleftharpoons I_3^- + H_2O + O_2 + K^+$

3-5. What type is each of the reactions in Problem 3-4?

3-6. Phosphate content in detergents can be expressed in several different forms.

    a. What is the percentage of phosphorus in each of the following: In P_2O_5? In $Na_5P_3O_{10}$?

    b. If a detergent contains 7.85% P, calculate its phosphate content as P_2O_5 and as $Na_5P_3O_{10}$.

3-7. A hemoglobin sample is found to contain 0.335% Fe. What is the minimum molecular weight of hemoglobin?

*3-8. What weight of AgCl would be obtained by decomposing 280 mg of an organic compound having the formula $C_6H_2Cl_4$ and precipitating all the chloride as AgCl?

3-9. Sulfate can be determined by precipitation as $BaSO_4$. How many mg of $BaCl_2 \cdot 2H_2O$ would be required to precipitate completely the sulfate in 350 mg K_2SO_4 and how many mg of $BaSO_4$ would be produced? The reaction takes the following form:

$$K_2SO_4 + BaCl_2 \cdot 2H_2O \rightleftharpoons BaSO_4 + 2KCl + 2H_2O$$

3-10. How would you prepare these solutions?

   *a. 500 mL of 1% (w/w) tincture of iodine (I_2 dissolved in ethanol).

    b. 500 mL of 1% (w/v) tincture of iodine

    c. 500 mL of 1% (v/v) tincture of iodine

3-11. How would you prepare these solutions?

   *a. 150 mL of 5% (w/w) saline (NaCl in H_2O)

    b. 150 mL of 5% (w/v) saline

3-12. What is the formal concentration for each of the following concentrated acid solutions (percentage by weight given)?

   *a. 38% HCl with a density of 1.19 g/mL

    b. 75% H_3PO_4 with a density of 1.58 g/mL

    c. 98% H_2SO_4 with a density of 1.82 g/mL

    d. 70% $HClO_4$ with a density of 1.67 g/mL

3-13. How would you prepare 500 mL of 0.1 N reagent solutions from the following concentrated solutions?

   *a. 36 N H_2SO_4

    b. 12.1 N HCl

    c. 15.6 N NH_3

3-14. A concentrated solution of NaOH (50% by weight) is frequently used to prepare dilute NaOH solutions.

    a. What weight of concentrated NaOH solution would be required to prepare 500 mL of 0.200 F NaOH?

b. What volume of concentrated NaOH would be required if the concentrated solution had a density of 1.52 g/mL?

3-15. How would you prepare the following solutions from pure solid reagents?
*a. 100 mL of 0.050 F $K_2Cr_2O_7$
b. 250 mL of 0.16 N NaOH
c. 500 mL of 0.025 F $AgNO_3$
d. 10 L of 2.4 F NaCl
e. 100 mL of 100 ppm Na^+ solution using NaCl
f. 200 mL of 2% (w/v) LiI in ethanol

3-16. Calculate the weight of reagent present in each of the following solutions:
*a. 250 mL of 0.107 F NaCl
b. 42.67 mL of 0.1163 F $AgNO_3$
c. 52.0 mL of 48.6% (w/w) NaOH with a density of 1.51 g/mL
d. 52.0 mL of 48.6% (w/v) NaOH

3-17. Calculate the number of milligrams of $K_2Cr_2O_7$ needed to prepare the following solutions:
*a. 80 mL of 0.0630 F $K_2Cr_2O_7$
b. 46.17 mL of 0.00167 F $K_2Cr_2O_7$
c. 0.75L of 1.12 F $K_2Cr_2O_7$
d. 250 mL of 4.6 × 10^{-4} F $K_2Cr_2O_7$

3-18. Calculate the millimoles of reagent present for each of the following solutions:
*a. 350 mL of 0.125 F $AgNO_3$
b. 23 mL of 0.278 F $KMnO_4$
c. 1.23 L of 0.00231 F EDTA
d. 125 mL of 5.3% (w/v) NaCl

3-19. A water sample is found to contain 0.00287 F $CaCO_3$.
a. What is the $CaCO_3$ content in parts per million (ppm)?
b. What is the calcium content in parts per million?

*3-20. A sample of air contains 0.130 $\mu g/L$ SO_2. Given that the density of the air sample is 1.19 g/L, calculate the SO_2 content in parts per million by weight.

3-21. A fluoridated water supply contains 1.0 mg/L F. What is its pF value?

3-22. Given that 250 mL of an HCl solution is pH 2.70, how many milligrams of HCl does it contain?

3-23. How many mL of 0.500 F H_2SO_4 are needed to prepare the following solutions?
*a. 250 mL of 0.100 F H_2SO_4
b. 50 mL of 0.078 F H_2SO_4
c. 4.7 L of 0.300 F H_2SO_4
d. 4.7 L of 0.0025 F H_2SO_4
e. 600 mL of 0.850 F H_2SO_4

3-24. Determine the n value (equivalents per mole or milliequivalents per millimole) for each of the following substances:
a. $RbNO_3$ precipitated as $RbClO_4$
*b. $Pb(NO_3)_2$ precipitated as $PbSO_4$
c. Cs_2CO_3 neutralized to CO_2 with HCl
d. SO_2 converted to H_2SO_4 and neutralized with NaOH
e. $Fe_2(SO_4)_3$ reduced to Fe^{2+}
f. KIO_4 reduced to KIO_3
g. $ZnCl_2$ complexed as $Zn(CN)_4^{2-}$
h. KCN complexed as $Zn(CN)_4^{2-}$

3-25. Calculate the equivalent weight (EW) for each of the substances in Problem 3-24.

3-26. What would be the normality for a 0.1 F solution of each of the substances in Problem 3-24?

3-27. How many milligrams of each of the substances in Problem 3-24 would be needed to prepare 250 mL of a 0.05 N solution?

3-28. A 940-mg sample of detergent is hydrolyzed in solution and phosphate is precipitated as $MgNH_4PO_4$. The precipitate is converted to $Mg_2P_2O_7$ by heating to high temperature. Given that the weight of $Mg_2P_2O_7$ is 213 mg, calculate the percentage of $Na_5P_3O_{10}$ in the detergent.

*3-29. A 50.0-mL solution of HIO_3 was standardized by titration with 0.145 N NaOH and required 45.8 mL. The HIO_3 solution was then used to analyze for iron, using the following redox reaction

$$HIO_3 + 4FeCl_2 + 5H_3O^+ + 6Cl^- \rightarrow 4FeCl_3 + ICl_2^- + 8H_2O$$

What is the normality of the HIO_3 when used in the iron titration?

3-30. a. Potassium dichromate $(K_2Cr_2O_7)$ is frequently used in redox reactions. If its reaction product is Cr(III), what is the n value for $K_2Cr_2O_7$ and what would be the normality of a 0.05 F solution?

b. It is also possible to reduce chromium to Cr(II). In this reduction reaction, what is the n value for $K_2Cr_2O_7$ and the normality of a 0.05 F solution?

c. In basic solution, $K_2Cr_2O_7$ hydrolyzes to form chromate ion, CrO_4^{2-}

$$Cr_2O_7^{2-} + 2OH^- \rightleftharpoons 2CrO_4^{2-} + H_2O$$

If the resulting CrO_4^{2-} is used to precipitate $PbCrO_4$, what is the n value for $K_2Cr_2O_7$ and the normality of a 0.05 F solution?

3-31. The concentration of a $KMnO_4$ solution can be determined by a redox reaction with sodium oxalate $(Na_2C_2O_4)$. Given that $KMnO_4$ oxidizes $Na_2C_2O_4$ to CO_2, calculate the normality of a $KMnO_4$ solution if 45.7 mL is needed to completely oxidize 347 mg of $Na_2C_2O_4$.

*3-32. In the Dumas method for determining nitrogen content in organic compounds, the sample is treated with CuO at high temperature. The nitrogen is converted to $N_2(g)$ and collected. Given that a 350-mg sample yields 42.2 mL N_2 at STP, calculate the percentage of nitrogen in the sample.

3-33. Calculate the equivalent weight of a metal if a 115-mg sample liberates 39.4 mL of H_2 at STP when treated with HCl.

$$2HCl + 2e \rightleftharpoons H_2 + 2Cl^-$$

*3-34. How many millicoulombs of electrical charge would be required to reduce 125 mg of Fe(III) to Fe(II)?

3-35. The amount of KI in "iodized salt" can be determined by oxidizing the iodide to iodine (I_2) electrochemically. If a 2.50-g sample of salt required 151mC of electrical charge to oxidize all of the iodide, calculate the percentage of KI in the salt sample.

*3-36. Seawater contains 400 ppm Ca^{2+}. How many milliliters of 0.025 F EDTA would be equivalent to the calcium in 20 mL of seawater? (EDTA forms 1:1 complexes with metal ions).

3-37. The concentration of a KI solution can be determined by adding excess $AgNO_3$ and back-titrating with KSCN. What is the normality of a KI solution if 100.0 mL of 0.0863 N $AgNO_3$ is added to a 50.0 mL sample and 15.2 mL of

0.0655 N KSCN was needed for the back-titration? (The details of the reaction are not needed because the normalities are given and normality of the sample is sought.)

*3-38. A 650-mg sample containing acetic acid (EW = 60.0) is dissolved in 50.0 mL of 0.106 N NaOH and back-titrated with 0.125 N HCl. If the back-titration required 19.3 mL, calculate the percentage of acetic acid in the sample.

*3-39. Pennies weigh 3.1 g and nickels weigh 5.0 g. If a mixture of pennies and nickels worth $3.25 weighs 587.5 g, how many of each coin does the mixture contain?

3-40. A mixture of halide salts weighing 400 mg is dissolved; the chloride and iodide present are precipitated as the silver salts. The silver halide precipitate weighs 297 mg. The precipitate is then ignited in a stream of Cl_2 gas, which converts the AgI in the mixture to AgCl. The weight of the precipitate is now 230 mg. Calculate the percentages of Cl and I in the sample.

3-41. An alloy sample containing only lead and tin weighs 560 mg. The sample is ignited in air to yield PbO and SnO_2. The oxide mixture weighs 626 mg. Calculate the percentages of Pb and Sn in the alloy.

Equilibrium Constant Calculations

4

In Chapter 3, chemical equations were used to describe chemical reactions and stoichiometric calculations were made. These calculations assumed that the reactions proceed to completion. For all practical purposes, this is true for most reactions of analytical use. However, all reactions will reach a state of dynamic equilibrium when the rate of the reaction in its forward direction equals the rate of the reaction in its backward direction. Frequently, a chemist must know where this equilibrium point lies in order to develop a successful analytical process—even when the reaction proceeds toward >99.9% completion (or conversely remains >99.9% as reactants). Details of the equilibrium state for particular reactions will be developed in Chapters 5 to 17, which deal with quantitation. However, in this chapter, we can make some generalized comments concerning equilibrium state calculations.

4-1 EQUILIBRIUM CONSTANT

The equilibrium state is described mathematically with the *equilibrium constant* expression which can be written for any balanced chemical equation. A reaction of the general type

$$a\text{A} + b\text{B} \rightleftharpoons c\text{C} + d\text{D} \tag{4-1}$$

will have the equilibrium constant expression

$$K = \frac{[\text{C}]^c[\text{D}]^d}{[\text{A}]^a[\text{B}]^b} \tag{4-2}$$

The equilibrium constant equals the products raised to their coefficient powers divided by the reactants raised to their coefficient powers. Before numerical calculations may proceed, the only problem remaining is to determine what values are to be inserted within the square brackets.

A very simple reaction occurring in the gaseous state could be stated as

$$2NO(g) + O_2(g) \rightleftharpoons 2NO_2(g)$$

This reaction is important to analytical chemists concerned with the analysis of air pollutants because it is one of the first steps in the formation of photochemical smog. The equilibrium constant expression for this reaction can be written

$$K = \frac{[NO_2]^2}{[NO]^2[O_2]} \tag{4-3}$$

The equilibrium state occurs when the rates of forward and backward reactions become equal. Because the rates of these reactions depend on the partial pressures of the gaseous species, the equilibrium expression becomes

$$K = \frac{p_{NO_2}^2}{p_{NO}^2 p_{O_2}} \tag{4-4}$$

Example 4-1 shows how the equilibrium constant can be calculated from equilibrium state information.

In Example 4-1, the equilibrium constant is quite large (1.6×10^{12}), indicating that the equilibrium point is far toward the product side—that is, the reactants (or at least one limiting reactant) will react essentially completely to form products.[1] The limiting reactant was NO, which has a partial pressure of only 10^{-6} atm at equilibrium. The main point is that when K is very large, the reaction goes essentially to completion consuming essentially all of the limiting reactant.

The concept of using partial pressures in atmospheres works well for reactions occurring in the gaseous state, but what about reactions occurring in solution or reactions involving solid materials?

---

[1]The reverse reaction may also be written as a chemical equation and in the form of an equilibrium constant expression:

$$2NO_2(g) \rightleftharpoons 2NO(g) + O_2(g)$$

$$K' = \frac{[NO]^2[O_2]}{[NO_2]^2}$$

It should be clear that $K' = 1/K$ and that, for the reaction in Example 4-1, the equilibrium constant for the reverse reaction is $1/1.6 \times 10^{12} = 6.2 \times 10^{-13}$ atm. This constant is quite small, indicating that, as now written, the equilibrium point lies near the reactant side—that is, small amounts of products will be formed from reactants.

EXAMPLE 4-1

An analysis of a gaseous mixture of NO, NO_2, and O_2 that has reached equilibrium at room temperature gives the following results:

$$NO_2 = 0.14 \text{ atm } (106 \text{ torr})$$

$$NO = 1.0 \times 10^{-6} \text{ atm } (7.6 \times 10^{-4} \text{ torr})$$

$$O_2 = 0.012 \text{ atm } (9.12 \text{ torr})$$

Calculate the equilibrium constant for the reaction:

$$2NO(g) + O_2(g) \rightleftharpoons 2NO_2(g)$$

Solution

Equation 4-4 is used to calculate the equilibrium constant:

$$K = \frac{[NO_2]^2}{[NO]^2[O_2]} = \frac{p_{NO_2}^2}{p_{NO}^2 p_{O_2}}$$

$$K_{atm} = \frac{(0.14)^2 \text{ atm}^2}{(1.0 \times 10^{-6})^2 \text{ atm}^2 \times (0.012) \text{ atm}} = 1.6 \times 10^{12} \text{ atm}^{-1}$$

or

$$K_{torr} = \frac{(106)^2 \text{ torr}^2}{(7.6 \times 10^{-4})^2 \text{ torr}^2 \times (9.2) \text{ torr}} = 2.1 \times 10^9 \text{ torr}^{-1}$$

This conversion of units can be accomplished simply by utilizing the conversion factor 1 atm = 760 torr:

$$K_{torr} = 1.6 \times 10^{12} \text{ atm}^{-1} \times \frac{1 \text{ atm}}{760 \text{ torr}} = 2.1 \times 10^9 \text{ torr}^{-1}$$

Other pressure units could be used. In S.I. units, the pressure would be expressed in pascals (Pa) with the conversion factor 1 atm = 1.01×10^5 Pa

$$K_{Pa} = 1.6 \times 10^{12} \text{ atm}^{-1} \times \frac{1 \text{ atm}}{1.01 \times 10^5 \text{ Pa}} = 1.6 \times 10^7 \text{ Pa}^{-1}$$

Therefore, because it appears to be a dimensional quantity, we must specify the units used to determine K. In Section 4-2, we shall see that, for gaseous components, it is fundamentally more correct to specify pressure units in atmospheres. However, for convenience, other units are occasionally used.

4-2 STANDARD STATES

In the equilibrium constant expression, atmospheres of pressure appeared reasonable as a measure of a gaseous material's ability to influence the rate of reaction. Now consider Equation 4-5:

$$\Delta G^0 = -RT \ln K \tag{4-5}$$

Here, ΔG^0 is the standard free energy change for the reaction, K is the equilibrium constant, R is the gas constant, and T is the absolute temperature. With this equation, the equilibrium constant may be calculated from the free energy change. The equilibrium constant occurs as a logarithmic term, but logarithms can only be taken of pure numbers—that is, the number must be dimensionless. In the equilibrium constant expression, the use of atmospheres was acceptable numerically, but in a more fundamental way a different measure is needed. All of the terms in the equilibrium constant expression are put on a *relative basis*; each species is compared to a reference state. A more exact method for expressing Equation 4-4 is:

$$K = \frac{(p_{NO_2}/p^0_{NO_2})^2}{(p_{NO}/p^0_{NO})^2(p_{O_2}/p^0_{O_2})} \tag{4-6}$$

Here p^0 denotes the pressure of the gas in the reference state called the *standard state*. In this form, K will be dimensionless because for every pressure term in the measured system there is one in the standard state. The standard state for gases is 1 atm pressure; therefore, all p^0 terms $= 1$ and Equation 4-6 reverts to Equation 4-4. Remember that all terms in the equilibrium constant expression are dimensionless and that we must define a standard state for each substance involved. However, it is not necessary to use the *same* standard state for each substance in the equilibrium constant expression. But the equilibrium expression must reflect each component's ability to influence the reaction relative to its ability when present in the standard state as was done in Equation 4-6. The ratio shown in Equation 4-6 for each substance is called the *activity*.

In Chapter 3, four reaction types were described. Now we may briefly examine them with regard to equilibrium constant expressions and the appropriate choice of standard states.

4-3 PRECIPITATION

Many substances have limited solubility in various solvents. Precipitation occurs if the substance is present at a concentration higher than its solubility. When the substance is ionic, this precipitation takes a form such as

$$A^{2+} + 2B^- \rightleftharpoons AB_2(s)$$

Here the solid AB_2 precipitates from solution. Similar expressions could be written for the formation of $AB(s)$, $A_2B(s)$, and so on.

The reverse reaction, representing the solubility of the substance, is commonly used to describe the equilibrium state:

$$AB_2(s) \rightleftharpoons A^{2+} + 2B^-$$

$$K = \frac{[A^{2+}][B^-]^2}{[AB_2(s)]}$$

The equilibrium constant equals the products raised to their coefficient powers divided by the reactant raised to its coefficient power. For example, lead iodide is a relatively insoluble compound of the type AB_2. Therefore, we may write

$$PbI_2 \rightleftharpoons Pb^{2+} + 2I^-$$

$$K = \frac{[Pb^{2+}][I^-]^2}{[PbI_2(s)]}$$

What standard states should be chosen for this system? First, consider the solid PbI_2. The most convenient standard state to choose for a solid substance such as PbI_2 is the *pure* solid substance itself. In other words, the activity of pure PbI_2 is equal to one. It doesn't matter how much pure PbI_2 is present as long as some is present when the system reaches equilibrium. Nor does it matter if other solid substances are present in the system, provided solid particles (crystallites) of *pure* PbI_2 exist. However, if the crystallites contain more than one substance in true solid solution, the activity of each will be less than in its standard state—that is, less than one. Fortunately, for most precipitations of analytical interest, the precipitating particles are essentially pure and their activity will be nearly, if not exactly, equal to one:

$$[PbI_2(s)] \cong 1$$

Because this assumption is traditionally made for solubility reactions in general, the equilibrium constant expression for AB_2 becomes

$$K_{sp} = [A^+][B^-]^2$$

or, in the case of PbI_2:

$$K_{sp} = [Pb^{2+}][I^-]^2$$

The equilibrium constant is given the symbol K_{sp} to denote a *solubility product constant* in which the solid reactant has an activity of one.

Now, what standard state should be adopted for ions in solution? In general, the ability of solutes to influence a reaction depends on their concentration, much the same as the activity of a gas depends on partial pressure. Therefore, a standard state based on some concentration scale would be appropriate. For example, if we use molarity, a 1 M solution could be adopted as the standard state, but this simple approach has some inconveniences.

The problem with using molarity (or formality) as a concentration scale is that the concentration is a function of temperature. As the temperature increases, the volume of solution increases and the molar concentration (moles of solute per liter of solution) will decrease. A more fundamental concentration unit is *molality* (m), which is defined as moles of solute per kilogram of solvent. A 1 m solution remains a 1 m solution no matter what the temperature, whereas a 1 M solution becomes "more dilute" as the temperature is raised. Therefore, when defining standard states for an activity scale, molality is a preferred concentration scale. But even defining the standard state of a solute as a 1 m solution poses difficulties.

For gases, 1 atm of pressure has been adopted as the standard state. Therefore, the activity of a gas would be numerically equal to its measured pressure. In other words, if the pressure were 0.02 atm, the activity of the gas would be 0.02 times that of the standard state.

Very dilute solutions, like gases, also behave in a nearly ideal fashion. The activity of an ion in a 10^{-4} m solution is very nearly 10 times its activity in a 10^{-5} m solution. If we define the standard state as a 1 m solution, then activity should equal concentration (that is, the activity of a 10^{-4} m solution would be 10^{-4}).

However, this equality is only true for ideal gases or ideal solutions in which all particles act independently. For ions in solution, it is no longer true that the activity of a solute at a concentration of 0.02 m will be 0.02 times the activity of a 1 m solution. In the more concentrated 1 m solution, ions are closer together and interact with each other, thereby decreasing their ability to influence the reaction. Consequently, for all dilute solutions, the activity would be considerably greater than that given by the concentration.

To circumvent this problem, chemists have defined the standard state for ions in solution as the *ideal* 1 m solution. The properties of very dilute solutions, which do behave nearly ideally, are used to predict the properties of a hypothetical ideal 1 m solution. By adopting this ideal 1 m solution as our standard state, the activity of any solute sufficiently dilute that it behaves ideally will be numerically equal to its concentration. For example, in a solution containing 0.02 m lead ion,

$$a_{Pb^{2+}} \simeq \frac{0.02}{1^*} = 0.02$$

Here, 1* refers to the activity of an ideal 1 m solution.

The standard state of a solute also depends on the solvent used; the solubility product constant is, therefore, a function of solvent. The solubility of ionic substances is greater in water than in less polar solvents. Therefore, solubility products have smaller values in such solvents than in water.

Most analytical work involving the precipitation of ionic substances is conducted in an aqueous medium, for which tables of solubility products are available (see Appendix 2-A). When dealing with dilute aqueous solutions, another approximation may be made. The density of the solution will be approximately 1 g/mL or

1 kg/L. Because, 1 kg of solvent will equal approximately 1 L, the molar concentration will essentially equal the molal concentration. Therefore, for a 0.02 M solution of lead ion,

$$0.02 \ M \simeq 0.02 \ m \cong 0.02 \ \text{(activity)}$$

Example 4-2 shows how the solubility product constant for PbI_2 can be determined from solubility information.

EXAMPLE 4-2

It is found that 1.30 mmol/L of PbI_2 will dissolve in H_2O at 25°C. What is the solubility product constant for PbI_2 in H_2O at 25°C?

Solution

$$PbI_2(s) \rightleftharpoons Pb^{2+} + 2I^-$$

$$K_{sp} = [Pb^{2+}][I^-]^2$$

The activities of Pb^{2+} and I^- will be approximately equal to their molar concentration (dilute aqueous solution); therefore,

Dissolution of 1.30 mmol/L PbI_2 yields 1.30 mmol/L Pb^{2+} + 2 × 1.30 mmol/L I^-

$$K_{sp} = (1.30 \times 10^{-3})(2.60 \times 10^{-3})^2 = 8.8 \times 10^{-9}$$

Because the terms within the square brackets are dimensionless activities (numerically equal to concentrations), the equilibrium constant also is dimensionless.

Finally, chemists need some way to describe more concentrated solutions that do not behave ideally. In a real solution, interactions between the solute particles are exceptionally strong when the solute particles are ionic. Even a real gas does not behave exactly as an ideal gas. Consequently, deviations from ideal behavior must sometimes be taken into account. The activity of a real gas is called its *fugacity*. The fugacity is nearly equal to the partial pressure of the gas, but may show marked deviation at high pressure or low temperature. As early as 1923 Peter Debye and Erich Hückel developed a theoretical equation to show the relationship between the activity of an ion and the *ionic strength*, μ, of the solution. The term ionic strength refers to the concentration of *all ions* in the solution and also takes into account their charge. Ionic strength and the activity of ions will be discussed in greater detail in Section 5-2.

A typical plot of an ion's activity as a function of concentration is shown in Figure 4-1. The line A has a slope of 1, representing ideal behavior in which the activity equals the molal concentration at all concentrations. Real solutions deviate from ideal behavior even at concentrations as low as $10^{-3} \ m$, and the activity of a 1 m solution is often <0.5. The deviation from ideal behavior is usually stated in the

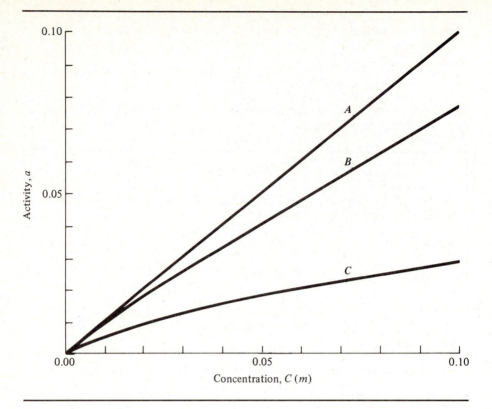

FIGURE 4-1   Activity as a function of concentration: (A) ideal solution for which $a = C$; (B) Na^+ activity in NaCl; (C) Ca^{2+} activity in $CaCl_2$.

form of an *activity coefficient, f*

$$f = \frac{a}{C} \qquad\qquad\qquad \textbf{(4-7)}$$

Here a is the activity of the species at the concentration C. For convenience, the concentration will generally be on the molar (formal) scale. Figure 4-2 is a plot of activity coefficients as a function of concentration. The activity coefficient approaches one (ideal behavior) as the solution becomes more and more dilute. Therefore, for reasonably accurate equilibrium state calculations, the activity for each species must be determined by using Equation 4-7 in the form

$$a = fC \qquad\qquad\qquad \textbf{(4-8)}$$

Example 4-3 shows how a more accurate value for the solubility product of PbI_2 can be calculated using activity coefficients.

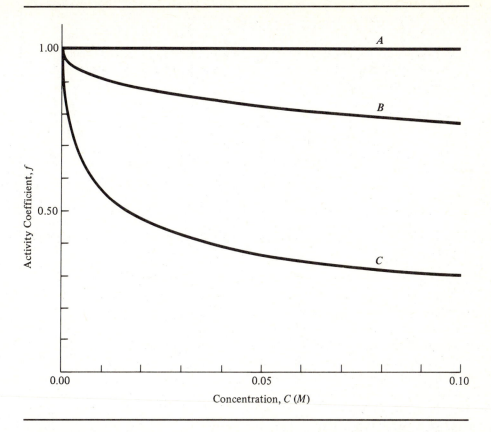

FIGURE 4-2 Activity coefficient as a function of concentration in the solution: (*A*) ideal solution for which $f = 1.00$ at all concentrations: (*B*) activity coefficient for Na^+ in NaCl solutions; (*C*) activity coefficient for Ca^{2+} in $CaCl_2$ solution.

The activity coefficient depends not only on the concentration of the ion in question, but on the total concentration of ions in the solution—that is, the ionic strength. Therefore, the solubility of PbI_2 depends on the concentration of other ions in the system, but the solubility product remains constant when activities are used. When the activity is only approximated by the concentration, as in Example 4-2, the solubility product calculated from concentrations is a function of the ionic strength. Therefore, it is important to know whether a solubility product (or any equilibrium constant) was measured at some finite concentration with no regard for activity coefficients or whether it is the equilibrium constant for the reaction at an infinite dilution in which all ions have activity coefficients of one.

The 20% difference between the answers in Examples 4-2 and 4-3 demonstrate the fact that equilibrium constant calculations only yield approximate answers (one or two significant figures) unless accurate information concerning ionic strength is

EXAMPLE 4-3

In Example 4-2, the solubility of PbI_2 was 1.30 mmol/L in H_2O at 25°C. Assuming that the activity coefficients for Pb^{2+} and I^- are 0.87 and 0.96 respectively, calculate the true solubility product constant.

Solution

$$[Pb^{2+}] = fC = 0.87 \times 1.30 \times 10^{-3} = 1.13 \times 10^{-3}$$

$$[I^-] = fC = 0.96 \times 2.60 \times 10^{-3} = 2.50 \times 10^{-3}$$

$$K_{sp} = [Pb^{2+}][I^-]^2$$

$$K_{sp} = (1.13 \times 10^{-3})(2.50 \times 10^{-3})^2 = 7.04 \times 10^{-9}$$

Note that this value is about 20% lower than the approximate value calculated in Example 4-2.

available. For analytical procedures, this information is usually unknown; therefore chemists often use the approximate calculation based on concentrations as a measure of activity (in effect making the approximation that $f = 1$). Except where specifically stated otherwise, all equilibrium state problems in this text should be worked using concentrations for activities.

4-4 ACID–BASE REACTIONS

The reaction between the hydronium ion (H_3O^+) and hydroxide ion (OH^-) in aqueous solution represents the elementary acid–base reaction in the solvent H_2O:

$$H_3O^+ + OH^- \rightleftharpoons 2H_2O$$

It is customary to write the equilibrium constant for the reverse reaction, the dissociation of water, as follows:

$$2H_2O \rightleftharpoons H_3O^+ + OH^-$$

$$K = \frac{[H_3O^+][OH^-]}{[H_2O]^2}$$

We have already discussed the choice of standard states for ions in solution. The *ideal* 1 *m* solution of H_3O^+ and 1 *m* solution of OH^- will be the standard states for these two species. For practical purposes, we will normally assume the molar concentration of these species to be numerically equal to their activity.

What about the standard state of the water itself? The most useful standard state will be *pure* water. If the system contains pure water, water will be in its standard state and its activity will equal one. In essentially all cases of analytical interest, the water is not pure but is actually the solvent for a solution. However, if the solution is dilute, the activity of the water will be nearly that of pure water and the difference can be neglected. As a demonstration of the magnitude of error involved with this approximation, the activity of H_2O may be equated with its mole fraction in the solution. For a 0.1 m solution of HCl, for example, the mole fraction H_2O is calculated as follows:

$$0.1 \ m \ \text{HCl} = \frac{0.1 \ \text{mol HCl}}{1000 \ \text{g} \ H_2O}$$

$$1000 \ \text{g} \ H_2O = \frac{1000 \ \text{g}}{18 \ \text{g/mol}} = 55.6 \ \text{moles}$$

The mole fraction of H_2O is calculated as

$$\frac{\text{mol} \ H_2O}{\text{mol} \ H_2O + \text{mol} \ H_3O^+ + \text{mol} \ Cl^-} = \frac{55.6}{55.6 + 0.1 + 0.1}$$

Therefore, the mole fraction of H_2O is

$$0.996 \cong \text{activity of} \ H_2O$$

Note that this calculation neglected the H_2O associated with the hydration of the ions, which is unimportant for this dilute solution, but can be critically important for concentrated solutions ($> 1 \ m$).

The equilibrium constant expression for water can then be simplified by making the assumption that the activity of H_2O is always equal to one:

$$K_w = [H_3O^+][OH^-] \tag{4-9}$$

Because the water dissociation constant, K_w, will apply to all aqueous solutions, if the activity (concentration) of H_3O^+ is known, the activity (concentration) of OH^- can be calculated. However, the numerical value of K_w (and equilibrium constants in general) will depend on the temperature. In Examples 4-4 and 4-5, the values given in Table 4-1 are used to make some typical calculations.

When a strong acid (completely ionized) is added to H_2O, the hydronium ion formed is in addition to that formed from the dissociation of water. However, when a weak acid, HA, is added to H_2O another, chemical equilibrium is involved:

$$HA + H_2O \rightleftharpoons H_3O^+ + A^-$$

$$K = \frac{[H_3O^+][A^-]}{[HA][H_2O]}$$

EXAMPLE 4-4

Calculate the activity of H_3O^+ in pure water at 25° and at 0°C.

Solution

In pure water, the only source of H_3O^+ is from the dissociation reaction of water:

$$2H_2O \rightleftharpoons H_3O^+ + OH^-$$

$$K_w = [H_3O^+][OH^-]$$

This reaction is also the only source of OH^- so that $[OH^-] = [H_3O^+]$ (assuming that the activity coefficients for H_3O^+ and OH^- are equal, which is a valid assumption for a solution of nearly pure water). Then

$$K_w = [H_3O^+][OH^-] = [H_3O^+]^2$$

From Table 4-1, at 25°C:

$$K_w = 1.01 \times 10^{-14} = [H_3O^+]^2$$

$$[H_3O^+] = 1.0 \times 10^{-7}$$

From Table 4-1, at 0°C:

$$K_w = 0.114 \times 10^{-14} = [H_3O^+]^2$$

$$[H_3O^+] = 3.4 \times 10^{-8}$$

Again, because the activity of H_2O will be assumed to be close to one, the dissociation of a weak acid will be in the form:

$$K_a = \frac{[H_3O^+][A^-]}{[HA]} \tag{4-10}$$

All of the species in the equilibrium expression are now solutes. Therefore, the ideal 1 m solution will be chosen as the standard state and the activity will be approximately equal to the concentration (molarity, formality, or molality for dilute solutions). It should be noted that the solute species HA (the undissociated acid) is not an ion and, therefore, will not exhibit strong interactions with itself or other solute species. This means that the activity coefficient is nearly one even in moderately concentrated solutions. Methods for solving problems involving weak acids will be mentioned briefly later in this chapter and in more detail in Chapter 9.

Some substances are basic in nature—that is, they generate OH^- when added to

EXAMPLE 4-5

Calculate the activity of OH^- in a solution of HCl at 50°C in which $[H_3O^+] = 2.5 \times 10^{-4}$.

Solution

In an aqueous solution it can always be assumed that the dissociation of H_2O is at equilibrium. Therefore,

$$K_w = [H_3O^+][OH^-]$$

$$[OH^-] = \frac{K_w}{[H_3O^+]}$$

From Table 4-1, at 50°C:

$$K_w = 5.47 \times 10^{-14}$$

Therefore,

$$[OH^-] = (5.47 \times 10^{-14})/(2.5 \times 10^{-4}) = 2.2 \times 10^{-10}$$

If it can be assumed that the ionic strength is not very high, the concentration of OH^- will be 2.2×10^{-10} M.

TABLE 4-1   Water Dissociation Constant

Temperature (°C)	K_w
0	0.114×10^{-14}
10	0.272×10^{-14}
25	1.01×10^{-14}
40	2.92×10^{-14}
50	5.47×10^{-14}
60	9.61×10^{-14}
100	49×10^{-14}

water. A strong base, such as NaOH, ionizes completely:

$$NaOH \rightarrow Na^+ + OH^-$$

A weak base only reacts partially with H_2O to produce OH^-. A base dissociation constant may be written in analogous fashion to the acid dissociation constant:

$$NH_3 + H_2O \rightleftharpoons NH_4^+ + OH^-$$
$$\text{ammonia} \qquad\qquad \text{ammonium} \atop \text{ion}$$

$$K_b = \frac{[NH_4^+][OH^-]}{[NH_3]}$$

Problem-solving techniques involving bases are similar to those involving weak acids.

4-5 COMPLEXATION REACTIONS

The reaction of a ligand (electron pair donor) with an electron pair acceptor can be written in the form of a chemical equation and its equilibrium constant expression. For example, the formation of the dark blue copper ammonia complex can be written as:

$$Cu^{2+} + 4NH_3 \rightleftharpoons Cu(NH_3)_4^{2+}$$

$$K_f = \frac{[Cu(NH_3)_4^{2+}]}{[Cu^{2+}][NH_3]^4}$$

The equilibrium constant (K_f) is called the *formation constant*. For the reverse direction, the equilibrium constant is called the *dissociation constant* (K_d):

$$Cu(NH_3)_4^{2+} \rightleftharpoons Cu^{2+} + 4NH_3$$

$$K_d = \frac{[Cu^{2+}][NH_3]^4}{[Cu(NH_3)_4^{2+}]}$$

Complexation reactions can be complicated by the fact that more than one complex may form with the same reactants. For example, Cu^{2+} and NH_3 form a series of complexes:

$$Cu^{2+} + NH_3 \rightleftharpoons Cu(NH_3)^{2+}$$

$$K_1 = \frac{[Cu(NH_3)^{2+}]}{[Cu^{2+}][NH_3]}$$

$$Cu(NH_3)^{2+} + NH_3 \rightleftharpoons Cu(NH_3)_2^{2+}$$

$$K_2 = \frac{[Cu(NH_3)_2^{2+}]}{[Cu(NH_3)^{2+}][NH_3]}$$

$$Cu(NH_3)_2^{2+} + NH_3 \rightleftharpoons Cu(NH_3)_3^{2+}$$

$$K_3 = \frac{[Cu(NH_3)_3^{2+}]}{[Cu(NH_3)_2^{2+}][NH_3]}$$

$$Cu(NH_3)_3^{2+} + NH_3 \rightleftharpoons Cu(NH_3)_4^{2+}$$

$$K_4 = \frac{[Cu(NH_3)_4^{2+}]}{[Cu(NH_3)_3^{2+}][NH_3]}$$

The overall formation constant for $Cu(NH_3)_4^{2+}$ can be calculated from the individual constants, K_1, K_2, K_3, and K_4:

$$K_f = K_1 K_2 K_3 K_4 = \frac{[Cu(NH_3)^{2+}]}{[Cu^{2+}][NH_3]} \frac{[Cu(NH_3)_2^{2+}]}{[Cu(NH_3)^{2+}][NH_3]}$$

$$\times \frac{[Cu(NH_3)_3^{2+}]}{[Cu(NH_3)_2^{2+}][NH_3]} \frac{[Cu(NH_3)_4^{2+}]}{[Cu(NH_3)_3^{2+}][NH_3]}$$

$$K_f = \frac{[Cu(NH_3)_4^{2+}]}{[Cu^{2+}](NH_3]^4}$$

An overall formation constant of successive complexation constants is given the symbol β_n where n refers to the number of ligands involved. For the preceding example, $\beta_4 = K_f = K_1 K_2 K_3 K_4$.

4-6   REDOX REACTIONS

Redox reactions involve the reaction between an electron donor (the reducing agent) and an electron acceptor (the oxidizing agent). The following is an example of a redox reaction and its corresponding equilibrium constant expression (assuming the activity of $H_2O \cong 1$):

$$5Fe^{2+} + MnO_4^- + 8H_3O^+ \rightleftharpoons 5Fe^{3+} + Mn^{2+} + 12H_2O$$

$$K = \frac{[Fe^{3+}]^5[Mn^{2+}][H_2O]^{12}}{[Fe^{2+}]^5[MnO_4^-][H_3O^+]^8} \cong \frac{[Fe^{3+}]^5[Mn^{2+}]}{[Fe^{2+}]^5[MnO_4^-][H_3O^+]^8}$$

The iron(II) is oxidized to iron(III) by the permanganate ion (MnO_4^-), which in turn is reduced to Mn^{2+}. Equilibrium constants for such reactions can be quite large ($K = 3 \times 10^{62}$ for this example) and can be calculated from the electrochemical potential for the reaction. Because electrochemical potentials are more useful for problem solving, they are usually tabulated (Appendix 3-A) instead of equilibrium constants. Methods for solving problems using electrochemical potentials will be studied in Chapter 11.

4-7 SEPARATION PROCESSES

Step 3 of the total analysis process involves the separation of the analyte from interfering substances. This separation step normally involves transfer of the analyte and/or the interferences from one physical phase (solid, liquid, or gas) to another. An equilibrium expression for a separation process can be written in the form:

$$A_1 \rightleftharpoons A_2 \tag{4-11}$$

species A in phase 1   species A in phase 2

$$K_D = \frac{[A_2]}{[A_1]}$$

Here K_D is a distribution constant (generally called a distribution coefficient), which indicates the extent to which A will be transferred from the first to the second phase. Note that the equilibrium constant is still "products over reactants." The activity of A in phase 1 will be relative to its standard state in phase 1, whereas the activity of A in phase 2 will be relative to its standard state in phase 2. Because A is some solute species, the standard state will be the ideal 1 m solution of A in the phase in question. A simple calculation involving a distribution coefficient is given in Example 4-6.

The distribution coefficient may be written for other types of transfer from one phase to another, and several will be discussed in Chapters 18 and 19.

EXAMPLE 4-6

An aqueous solution containing iodine is contacted with carbon tetrachloride (CCl_4). After equilibrium is established, the aqueous and CCl_4 phases are analyzed for iodine with the following results:

$$H_2O \text{ phase:} \qquad I_2 = 1.4 \times 10^{-5} \ F$$

$$CCl_4 \text{ phase:} \qquad I_2 = 1.2 \times 10^{-3} \ F$$

Determine the distribution coefficient for the transfer of I_2 from H_2O to CCl_4.

Solution

The distribution coefficient for $(I_2)_{H_2O} \rightleftharpoons (I_2)_{CCl_4}$ is:

$$K_D = \frac{[I_2]_{CCl_4}}{[I_2]_{H_2O}} = \frac{1.2 \times 10^{-3}}{1.4 \times 10^{-5}} = 86$$

4-8 PROBLEM-SOLVING TECHNIQUES

In later chapters, specific analytical problems will be described and mathematical methods useful for solving these problems will be developed. At this point, however, we can make some general comments concerning several different types of analytical problems involving equilibrium constant expressions.

A general reaction of the simple type

$$A + B \rightleftharpoons C + D \qquad\qquad (4\text{-}12)$$

has an equilibrium constant expression:

$$K = \frac{[C][D]}{[A][B]} \qquad\qquad (4\text{-}13)$$

If there are initially a mol/L of A and b mol/L of B, there will be x mol/L of C and x mol/L of D formed (because they both have the same coefficient) and the concentrations of A and B remaining will be $a - x$ and $b - x$ respectively. Therefore, the equilibrium constant expression will be:

initial:
$$
\begin{array}{ccccc}
a & b & 0 & 0 \\
A & + & B & \rightleftharpoons C + D
\end{array}
$$

equilibrium:
$$
\begin{array}{cccc}
a - x & b - x & x & x
\end{array}
$$

$$K = \frac{(x)(x)}{(a - x)(b - x)} \qquad\qquad (4\text{-}14)$$

This equation has one unknown, x, and is solvable using the quadratic equation. However, this method can be time-consuming and may require the use of several significant figures if x nearly equals a or b.

In solving problems of this type, it is useful to consider the magnitude of K to determine whether the equilibrium point favors products (large K) or reactants (small K). For a large K value, it will be mathematically easier to assume that the reaction goes to completion and that the reverse reaction then occurs to establish equilibrium. For Equation 4-12, the amount of C and D formed if the reaction goes to completion will be a or b, depending on which is smaller (limiting reactant). If we assume for the moment that a is smaller, then

completion:
$$
\begin{array}{cccc}
0 & b - a & a & a \\
A + & B & \rightleftharpoons C & + & D
\end{array}
$$

equilibrium:
$$
\begin{array}{cccc}
x & b - a + x & a - x & a - x
\end{array}
$$

$$K = \frac{(a - x)(a - x)}{(x)(b - a + x)} \qquad\qquad (4\text{-}15)$$

In general, Equation 4-14 should be used when K is small ($\ll 1$) and Equation 4-15 should be used when K is large ($\gg 1$). This will make x, the unknown quantity, small compared to the other variables, thereby minimizing loss of significant figures by the subtraction operation. Example 4-7 illustrates a typical equilibrium state problem solved by this approach.

In Example 4-7, note that x was so small compared to a or b (and even $b - a$), it could be neglected in the terms $(a - x)$, $(b - x)$, and $(b - a + x)$ without changing the value as much as 1%. With that in mind, the problems in Example 4-7 can be simplified considerably, as is shown in Example 4-8.

EXAMPLE 4-7

(a) For a chemical reaction of the type given in Equation 4-12, solve for the concentrations (or partial pressures if gases were involved) of A, B, C, and D if $K = 10^{-5}$.

(b) Solve for A, B, C, and D if $K = 10^2$. Assume that the system initially contained $A = 0.1$ and $B = 0.2$.

Solution

(a) For a small value of K, Equation 4-14 should be used:

$$K = 10^{-5} = \frac{(x)(x)}{(a - x)(b - x)} = \frac{x^2}{(0.1 - x)(0.2 - x)}$$

$$(2 \times 10^{-7}) - (3 \times 10^{-6})x + (10^{-5})x^2 = x^2$$

$$0.9999x^2 + (3 \times 10^{-6})x - (2 \times 10^{-7}) = 0$$

$$x = \frac{-3 \times 10^{-6} \pm \sqrt{(3 \times 10^{-6})^2 + 4 \times 0.9999 \times 2 \times 10^{-7}}}{2 \times 0.9999}$$

$$= \frac{-3 \times 10^{-6} \pm 8.94 \times 10^{-4}}{2.0}$$

$$x = 4.46 \times 10^{-4} \text{ (negative root without physical meaning)}$$

$$A \cong 0.1, \quad B \cong 0.2, \quad C = 4.46 \times 10^{-4}, \quad D = 4.46 \times 10^{-4}$$

(b) For a large value of K, Equation 4-15 should be used:

$$K = 10^2 = \frac{(0.1 - x)(0.1 - x)}{x(0.2 - 0.1 + x)}$$

$$10x + 100x^2 = 10^{-2} - 0.2x + x^2$$

$$99x^2 + 10.2x - 0.01 = 0$$

$$x = 9.7 \times 10^{-4}$$

$$A = 9.7 \times 10^{-4}, \quad B = 0.101, \quad C = 0.099, \quad D = 0.099$$

EXAMPLE 4-8

Solve the problems presented in Example 4-7, assuming that x is small compared to a, b, and $a - b$.

Solution

(a) $K = 10^{-5} = \dfrac{x^2}{(0.1 - x)(0.2 - x)} \cong \dfrac{x^2}{(0.1)(0.2)}$

$x^2 = 2 \times 10^{-7}$

$x = 4.47 \times 10^{-4}$ (compared with 4.46×10^{-4} from quadratic equation)

(b) $K = 10^2 = \dfrac{(0.1 - x)(0.1 - x)}{x(0.2 - 0.1 + x)} \cong \dfrac{(0.1)(0.1)}{x(0.1)}$

$x = \dfrac{10^{-2}}{10^1} = 10^{-3}$ (compared with 0.97×10^{-3} from quadratic equation)

4-9   ITERATIVE APPROACH

For problems in which x is not so small that it can be neglected entirely, we need not return to the quadratic equation, but can use the approach demonstrated in Example 4-8 as a starting point. Successive approximate calculations of the value of x can lead us closer and closer to the true value of x. This *iterative approach* is frequently used in computer programs for solving such equations. Because equilibrium state calculations have uncertainties of 1 to 10% (K values given to two or three significant figures and uncertainties involving activity coefficients), the value of x need only be calculated to two or three significant figures. In most problems, only one or two iterations are needed to reach a value of x within 1% of the true value, as can be seen in Example 4-9. The basic approach is to put known quantities and those containing x as a small part of the term—such as $(a - x)$—on the left side of the equation, and terms involving x on the right side. The left side is evaluated by letting $x = 0$ (or some other appropriate initial value) and then a first approximation to x is calculated. This value of x (x_1) is inserted in the left side of the equation and a second approximation to x (x_2) is calculated. This procedure is re-peated until the difference between x_n and x_{n-1} is less than the uncertainty inherent to the problem (data usually given to two or three significant figures). Appendix 1-C shows how a simple computer program can be written for a desktop computer to solve problems of this type in a matter of seconds.

EXAMPLE 4-9

The acid dissociation constant for chloroacetic acid ($ClCH_2COOH$) is 1.36×10^{-3}. Calculate $[H_3O^+]$ for a solution containing $0.020\ F\ ClCH_2COOH$, neglecting the small amount of H_3O^+ from the dissociation of H_2O and using an iterative approach.

Solution

The acid dissociation equation and its equilibrium constant are written, respectively, as

$$ClCH_2COOH + H_2O \rightleftharpoons H_3O^+ + ClCH_2COO^-$$

$$K_a = \frac{[H_3O^+][ClCH_2COO^-]}{[ClCH_2COOH]}$$

Before $ClCH_2COOH$ dissociates,

$$[H_3O^+] = [ClCH_2COO^-] = 0$$

$$[ClCH_2COOH] = 0.020$$

After $ClCH_2COOH$ dissociates,

$$[H_3O^+] = [ClCH_2COO^-] = x$$

$$[ClCH_2COOH] = 0.020 - x$$

Substitution into the equilibrium constant expression gives

$$1.36 \times 10^{-3} = \frac{(x)(x)}{(0.020 - x)} \qquad \text{or} \qquad (1.36 \times 10^{-3})(0.020 - x) = x^2$$

The iterative approach begins by assuming $x = 0$ in the term $(0.020 - x)$, then solves for x on the right side of the equation. This value of x is then substituted in $(0.020 - x)$ for the next approximation, and a second value of x on the right side is obtained.

x (left side)	x (right side)
0	5.22×10^{-3}
5.22×10^{-3}	4.48×10^{-3}
4.48×10^{-3}	4.59×10^{-3}
4.59×10^{-3}	4.58×10^{-3}

Therefore, $[H_3O^+] = 4.58 \times 10^{-3}$ (to three significant figures).

The solution to a quadratic equation can certainly be programmed as easily as an iterative approach. However, Equations 4-14 and 4-15 are quite simple. An equation of the type $2A + B \rightleftharpoons C$ is more difficult. Its equilibrium constant expression

EXAMPLE 4-9 (cont.)

These results, shown in Figure 4-3, demonstrate that the successive values of x alternate above and below the true answer, so that the true value must lie between 4.58×10^{-3} and 4.59×10^{-3}.

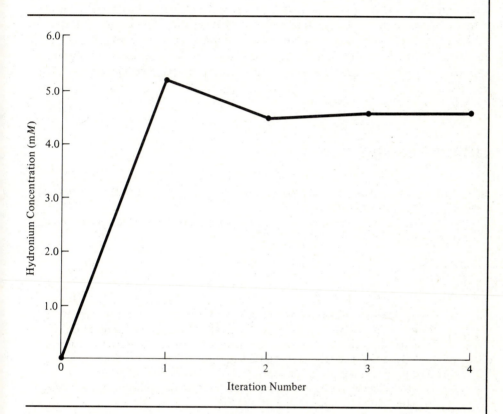

FIGURE 4-3 Iterative approach for Example 4-9. Hydronium ion concentration converges on the true value within a few iterations.

(assigning x on the basis of a small K value) is

$$K = \frac{[C]}{[A]^2[B]} = \frac{(x)}{(a - 2x)^2(b - x)}$$

This leads to a complicated third-order equation. Nonetheless, an equation of this type can be solved iteratively just as quickly as Equation 4-14 or 4-15.

The limits on the iterative approach involve the magnitude of x compared to terms such as $(a - x)$. To work well, x should be smaller than these other terms. In fact, if $(a - x)$ or another such term is smaller than x, the successive approximations may diverge rather than converge at the true value. When this happens, the variables should be reassigned, as illustrated in a comparison of Equation 4-15 with Equation 4-14. Sometimes it is necessary to choose an initial starting point value of x that is closer to the true value than the arbitrary value of 0. This does not mean that the answer must be known before the problem can be solved; it may be reasonably clear from the value of K what value of x can be expected. In later chapters, specific analytical procedures using some of these techniques will be discussed.

SUPPLEMENTARY READING

Butler, J.N., *Ionic Equilibrium: A Mathematical Approach*. Reading, MA: Addison-Wesley Publishing Company, Inc., 1964.
Moeller, T., and R. O'Connor, *Ions in Aqueous Systems*. New York: McGraw-Hill Book Company, 1972.

PROBLEMS

4-1. In the atmosphere, the air pollutant sulfur dioxide (SO_2) is slowly oxidized by O_2 to give sulfur trioxide (SO_3). SO_3 reacts readily with H_2O to produce H_2SO_4, resulting in "acid rain." Write a balanced chemical equation for the oxidation of SO_2 to SO_3 and give its equilibrium expression.

4-2. For the reaction given in Problem 4-1:
a. Calculate K, assuming that an equilibrium mixture contains 0.52 atm SO_3, 10^{-5} atm O_2, and 1.1×10^{-10} atm SO_2.
b. Calculate K for the reverse reaction.

*4-3. For the equilibrium conditions stated in Problem 4-2, calculate K assuming that the partial pressures have been expressed as torr instead of atmospheres.

4-4. For the equilibrium conditions given in Problem 4-2, calculate K using concentration units (mol/L). Assume the ideal gas equation holds and that the reaction mixture is at 25°C.

4-5. The following reactions are involved in various analytical procedures. Balance the equations, describe the reaction type, and write the equilibrium constant expression appropriate for each.
*a. $HC_2H_3O_2 + H_2O \rightleftharpoons H_3O^+ + C_2H_3O_2^-$
b. $Ca_3(PO_4)_2(s) \rightleftharpoons Ca^{2+} + PO_4^{3-}$
c. $Fe^{3+} + Sn^{2+} \rightleftharpoons Fe^{2+} + Sn^{4+}$

*Answers to problems marked with an asterisk will be found at the back of the book.

    d. $H_3O^+ + OH^- \rightleftharpoons H_2O$

    e. $Fe_2O_3(s) + H_3O^+ + Cl^- \rightleftharpoons FeCl_4^- + H_2O$

    f. $IO_3^- + SbCl_4^- + H_3O^+ + Cl^- \rightleftharpoons I_2 + SbCl_6^- + H_2O$

4-6. Equilibrium constants may be written in a variety of forms for a given reaction. Write the equilibrium constant expressions for the following two equations and derive the relationship between K_1 and K_2:

$$SO_2(g) + \tfrac{1}{2}O_2(g) \rightleftharpoons SO_3(g) \qquad K_1$$

$$2SO_3(g) \rightleftharpoons 2SO_2(g) + O_2(g) \qquad K_2$$

4-7. When solid $CaCO_3$ and other metal carbonates are heated to high temperature, they decompose and give off CO_2 gas, with the solid metal oxide remaining.

    a. Write the decomposition reaction and the equilibrium expression for $CaCO_3$.

    b. What standard states should be chosen for each of the reactants and products?

    c. What units will K apparently have?

4-8. The first step in the formation of nitrogen oxides (NO_x) as air pollutants is

$$N_2(g) + O_2(g) \rightleftharpoons 2NO(g)$$

    a. Assuming that the standard free energy change (ΔG^0) for this reaction at 298 K is $+41.42$ kcal/mol, calculate the equilibrium constant.

    b. Calculate the equilibrium pressure of NO at 298 K in air (that is, $P_{N_2} \cong 0.8$ atm, $P_{O_2} \cong 0.2$ atm).

4-9. The following compounds are relatively insoluble in water. Write equations for their solubility reactions and give the solubility product expression.

    *a. Barium sulfate, $BaSO_4$

    b. Calcium oxalate, CaC_2O_4

    c. Copper hydroxide, $Cu(OH)_2$

    d. Lanthanum iodate, $La(IO_3)_3$

    e. Magnesium ammonium phosphate, $MgNH_4PO_4$

    f. Silver sulfide, Ag_2S

4-10. Assuming that concentration is a reasonable estimate of activity, calculate the solubility products from the solubility of the following compounds in water:

    *a. $BaSO_4$, 1.14×10^{-5} F

    b. CaF_2, 2.31×10^{-4} F

    c. $CuCl$, 1.1 mF

    d. $PbCl_2$, 4.41 g/L

    e. AgI, 2.14 μg/L

    f. SrC_2O_4, 4.3 mg/100 mL

*4-11. Assuming that a solution containing 3.0×10^{-6} M OH^- and 6.5×10^{-4} M Cd^{2+} is in equilibrium with pure $Cd(OH)_2$ solid, calculate the solubility product for $Cd(OH)_2$.

4-12. The solubility product of mercuric sulfide (HgS) is 4×10^{-53}. Calculate the molar concentration of mercury (Hg^{2+}) and the total number of Hg^{2+} ions in a lake that is in contact with cinnabar (a mineral consisting of HgS) and that has a capacity of 2×10^{11} L.

*4-13. What iodide ion concentration is required to decrease the Pb^{2+} concentration of a solution to 2×10^{-6} M?

4-14. What sulfate ion concentration is required to decrease the Pb^{2+} concentration of a solution to 1.0 mg/L?

4-15. Calculate K_{sp} for $RaSO_4$, assuming that the solubility is 9.9×10^{-6} F in a solution of ionic strength 0.01, in which the activity coefficients are 0.67 for Ra^{2+} and 0.66 for SO_4^{2-}.

4-16. Which is more soluble in H_2O, PbI_2 or $PbSO_4$?

4-17. In some parts of the United States, tap water has a pH of about 7.5. Calculate the maximum concentration of Fe^{3+} in tap water if the solubility product of $Fe(OH)_3$ is 4×10^{-38}.

*4-18. The density of a 4.0% (w/w) HNO_3 solution is 1.02 g/mL. Calculate its molar and molal concentration.

4-19. The density of a 51% (w/w) H_2SO_4 solution is 1.41 g/mL. Calculate its molar and molal concentration.

4-20. Calculate the mole fraction H_2O in the following H_2SO_4 solutions:
   - *a. 0.5 m
   - b. 2 m
   - c. 2 F (1.12 g/L)
   - d. 50% (w/w)
   - e. 98% (w/w)
   - f. 18 F (1.84 g/L)

4-21. Calculate the pH of pure H_2O at the following temperatures:
   - a. 0°C
   - *b. 25°C
   - c. 50°C
   - d. 100°C

4-22. Calculate $[OH^-]$ in a solution of HCl in which $[H_3O^+] = 3.0 \times 10^{-5}$.
   - a. 0°C
   - *b. 25°C
   - c. 100°C

*4-23. Calculate the *concentration* of OH^- in 0.10 F HCl at 25°C. The activity coefficients at ionic strength 0.1 are 0.83 for H_3O^+ and 0.76 for OH^-.

4-24.
   - a. What is the $[H_3O^+]$ *activity* in a 0.10 F HCl solution at 25°C in which the activity coefficient for H_3O^+ is 0.83?
   - b. If pH is defined as $-\log [H_3O^+]$ *activity*, calculate the pH of 0.10 F HCl.

4-25. The following compounds are weak acids and dissociate slightly in aqueous solution. Write equations for their dissociation reactions and give the acid dissociation constant expression.
   - a. Acetic acid, $HC_2H_3O_2$
   - b. Benzoic acid, $HC_7H_5O_2$
   - c. Phthalic acid, $H_2C_8H_4O_4$ (two ionizable H^+, show both reactions)
   - d. Phosphoric acid, H_3PO_4 (three ionizable H^+, show all three reactions)

4-26. The salt of a weak acid can act like a base in water by reacting with H_2O (hydrolysis) to liberate OH^-. Write the hydrolysis equation for cyanide ion, CN^-, and give the equilibrium expression.

*4-27. Calculate the acid dissociation constant (using Equation 4-10) for lactic acid if at equilibrium at pH 4, the ratio of lactate ion to lactic acid is 1.38.

4-28. Calculate the acid dissociation constant (using Equation 4-10) for formic acid if a solution containing 0.02 F formic acid and 0.01 F sodium formate is pH 3.45.

4-29. Calculate the acid dissociation constant for benzoic acid if a solution containing 0.0030 F benzoic acid and 0.0023 F sodium benzoate is pH 4.00 with an ionic strength of 0.1. The activity coefficient for benzoate ion at this ionic strength is 0.8, but the activity coefficient for the neutral benzoic acid species can be assumed to be nearly 1. Remember that pH is $-\log [H_3O^+]$ *activity*.

4-30. Write balanced equations and equilibrium constant expressions for the overall formation (β_n) for the following complexes:
   a. AlF_6^{3-}
   b. $Ni(NH_3)_4^{2+}$
   c. $Zn(CN)_4^{2-}$
   d. CeF_2^+
   e. $Cu(EDTA)^{2-}$
   f. $Ag(NH_3)_2^+$

*4-31. Pb^{2+} forms a series of iodide complexes with the following formation constants: $K_1 = 18$, $K_2 = 35$, $K_3 = 4$, $K_4 = 3$. Write the formation reaction for each of these complexes and calculate the overall formation constant, β_4, for PbI_4^{2-}.

4-32. The stepwise formation constants for the bromide complexes of Bi^{3+} are: $K_1 = 2 \times 10^4$, $K_2 = 18$, $K_3 = 2$. Write the reactions for the stepwise *dissociation* reactions of $BiBr_3$ and calculate the stepwise *dissociation* constants.

4-33. The formation constant for the EDTA complex of magnesium (all EDTA complexes are 1:1) is 5.0×10^8. Calculate the equilibrium concentration of Mg^{2+} in a solution that initially contains:
   *a. $[Mg^{2+}] = 0.002 \ M$, $[EDTA] = 0.01 \ M$
   b. $[Mg^{2+}] = 0.005 \ M$, $[EDTA] = 0.005 \ M$
   c. $[EDTA] = 0.01 \ M$, $[Mg(EDTA)] = 0.002 \ M$

4-34. Silver(I) forms essentially one cyanide complex, $Ag(CN)_2^-$, with $\beta_2 = 7 \times 10^{19}$. Write the dissociation reaction for this complex and calculate the dissociation constant.

4-35. In Problem 4-34, the formation constant, β_2, for $Ag(CN)_2^-$ was given as 7×10^{19}. Calculate the equilibrium concentration of Ag^+ in a solution that initially contains:
   *a. $[Ag^+] = 2 \times 10^{-2} \ M$, $[CN^-] = 9 \times 10^{-2} \ M$
   b. $[Ag^+] = 2 \times 10^{-2} \ M$, $[CN^-] = 4 \times 10^{-2} \ M$
   c. $[Ag(CN)_2^-] = 5 \times 10^{-2} \ M$

4-36. In reasonably strong NH_3 solutions, Ag^+ forms predominantly $Ag(NH_3)_2^+$.
   a. Assuming that the stepwise constants are $K_1 = 2.3 \times 10^3$ and $K_2 = 6.9 \times 10^3$, calculate the overall formation constant, β_2, for $Ag(NH_3)_2^+$.
   b. Calculate the percentage of silver in the form of the simple Ag^+ (hydrated ion) in a 0.05 F NH_3 solution.

4-37. Redox reactions frequently have complicated stoichiometry, which leads to complicated equilibrium expressions. Balance the following redox equations and write the equilibrium constant expression for each.
   *a. $Fe^{2+} + Cr_2O_7^{2-} + H_3O^+ \rightleftharpoons Fe^{3+} + Cr^{3+} + H_2O$
   b. $Fe^{2+} + ClO_3^- + H_3O^+ \rightleftharpoons Fe^{3+} + Cl^- + H_2O$
   c. $C_2O_4^{2-} + MnO_4^- + H_3O^+ \rightleftharpoons CO_2 + Mn^{2+} + H_2O$
   d. $H_4IO_6^- + I^- \rightleftharpoons IO_3^- + I_2 + OH^- + H_2O$
   e. $C_3H_8O_3 + MnO_4^- + OH^- \rightleftharpoons CO_3^{2-} + MnO_4^{2-} + H_2O$

4-38. Iodine (I_2) can be removed from aqueous solutions by contacting the aqueous phase with a CCl_4 liquid phase.
   a. Write the reaction for this liquid–liquid separation process.
   b. Calculate the equilibrium constant (distribution coefficient, K_D) for the reaction assuming that at equilibrium $[I_2]_{H_2O} = 4.1 \times 10^{-4} \ F$ and $[I_2]_{CCl_4} = 0.035 \ F$.

*4-39. Formaldehyde can be transferred from an ether phase to an aqueous phase (liquid–liquid separation process) with a distribution coefficient, K_D, of 9.
   a. Write the reaction for this process and the expression for K_D.
   b. If at equilibrium the aqueous phase contains 0.0045 F formaldehyde, what will the formaldehyde concentration be in the ether phase?

*4-40. The synthesis of ammonia, NH_3, from hydrogen and nitrogen is an important industrial process that has a very small equilibrium constant at the temperature of operation as shown below:

$$3H_2(g) + N_2(g) \rightleftharpoons 2NH_3(g) \qquad K = 1.69 \times 10^{-4} \text{ atm}^{-2} \text{ at } 400°C$$

   a. Calculate the equilibrium partial pressure of NH_3 in a reaction mixture initially containing 5.00 atm H_2 and 2.00 atm N_2. Assume, based on the small K value, that the amount of H_2 and N_2 consumed is small.
   b. Use the iterative approach to obtain an answer accurate to three significant figures.
   c. What was the percent error introduced by using the approximation in part (a)?

4-41. Using the equilibrium constant information in Problem 4-40, how would you proceed to calculate the partial pressures of H_2 and N_2 produced in a reaction vessel that initially contained only NH_3? Calculate the equilibrium pressure for each of the species, assuming that the reaction vessel initially contained:
   a. 2.00 atm NH_3
   b. 0.04 atm NH_3

4-42. Based on the equilibrium constant given in Problem 4-40, calculate the free energy change for the formation of *one* mole of NH_3 at 400°C (673 K).

4-43. Water, although quite stable at room temperature, will thermally decompose at high temperatures to H_2 and O_2. In fact, this forms the basis for some solar energy conversion schemes:

$$2H_2O(g) \rightleftharpoons 2H_2(g) + O_2(g)$$

Using the iterative approach, calculate the equilibrium partial pressure for each of the species for the following conditions:
   *a. $K = 0.01$ atm, $[H_2O] = 0.1$ atm
   b. $K = 10$ atm, $[H_2O] = 0.1$ atm

4-44. The following reaction has an equilibrium constant of 1.59 at 986°C.

$$H_2(g) + CO_2(g) \rightleftharpoons H_2O(g) + CO(g)$$

Calculate the equilibrium partial pressure of each species if the original starting mixture contained:
   *a. $P_{H_2} = 0.10$ atm, $P_{CO_2} = 0.10$ atm
   b. $P_{H_2} = 0.10$ atm, $P_{CO_2} = 0.40$ atm
   c. $P_{H_2O} = 0.20$ atm, $P_{CO} = 0.20$ atm
   d. $P_{H_2} = 0.20$ atm, $P_{CO_2} = 0.40$ atm, $P_{H_2O} = 0.20$ atm

GRAVIMETRIC METHODS

Principles of Gravimetric Methods

5

5-1   MASS AS A MEASUREMENT SIGNAL

In Chapter 1, we outlined the five steps of the total analysis process: sampling, method, separation, quantitation, and evaluation. We will now focus on the quantitation step—that is, taking a quantitative measurement that has some functional relationship to the analyte in question. Measuring mass is relatively simple. Under proper conditions, this signal can be accurately related to the amount of analyte in the sample.

The relationship involved may be quite direct. Two examples from the environmental field illustrate this point. First, the concentration of dissolved solids in drinking water is one of the criteria for acceptability of the water supply. Water supplies containing over 500 ppm (mg/L) total dissolved solids should be rejected if more suitable supplies are available at reasonable cost. The amount of total dissolved solids is simply measured by adding a known volume of the water sample to a weighed container, evaporating the water, and then reweighing the container. The difference in weight represents total dissolved solids.

A second environmental analysis makes use of a mass measurement to determine the particulate level in air (measurement of dust particles). The most accurate and reliable method involves a gravimetric technique. A measured volume of the dust-laden air is drawn through a weighed filter; then the filter is reweighed. The weight gain represents the particulate level in the air sample. The dust level in a normal residential area is about 0.1 to 0.2 mg/m^3, but in industrial work areas, it may exceed 10 mg/m^3. The accepted standard for measuring particulates in air is the Hi-vol sampler, shown in Figure 5-1, which uses a glass-fiber filter for trapping dust particles.

113

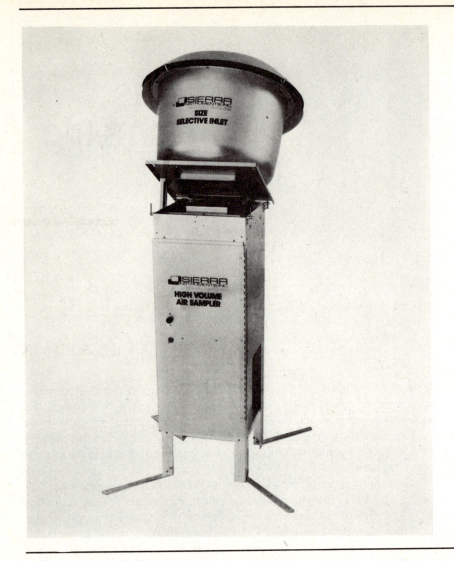

FIGURE 5-1   Hi-vol air sampler. Glass-fiber filters are used to obtain particulate samples with this mechanized unit.

A more common gravimetric procedure involves converting the analyte from a soluble to an insoluble form, then weighing it. Although we will be concerned primarily with analytes dissolved in aqueous solutions, this general procedure would also apply to an analyte in a gaseous "solution." For example, in the microdetermination of carbon and hydrogen in organic compounds, the sample is combusted in an atmosphere of pure oxygen (possibly in the presence of a catalyst such as CuO, $PbCrO_4$, or Co_3O_4) at 900°C. The carbon is converted to CO_2 while

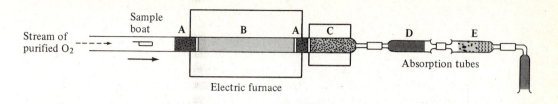

FIGURE 5-2   Carbon–hydrogen analysis with Pregl-type combustion apparatus. The sample is burned in a stream of pure O_2 and the combustion gases are carried through a series of combustion tube packings: (*A*) Ag to remove halogens and sulfur oxidation products, (*B*) $CuO/PbCrO_4$ (copper oxide/lead chromate) for complete oxidation of the sample, (*C*) PbO_2 to remove nitrogen oxides, (*D*) H_2O absorption tube containing $Mg(ClO_4)_2$ (Anhydrone), and (*E*) CO_2 absorption tube containing NaOH/asbestos.

hydrogen is converted to H_2O. As shown in Figure 5-2, the gaseous "solution" is passed through weighed tubes containing magnesium perchlorate $[Mg(ClO_4)_2]$, which absorbs H_2O, and tubes containing Ascarite (NaOH impregnated on asbestos), which absorbs the CO_2. The amount of carbon and hydrogen in the compound can be calculated from the gain in weight of the tubes, as in Example 5-1. The method is simple in principle, but it requires a skilled operator to obtain results of high accuracy. Most organic chemists do not attempt this analysis themselves; instead, they send samples to a commercial testing laboratory that specializes in this technique.

5-2   GRAVIMETRIC ANALYSIS REQUIREMENTS

In the more typical gravimetric procedure the analyte is dissolved in an aqueous solution (in some cases, a nonaqueous solution) and a reagent that forms an insoluble compound with the analyte is added. The weight of the insoluble compound constitutes the measurement signal. For purposes of discussion, the following four steps are required for a successful gravimetric procedure:

1. Identify an insoluble form.

2. Separate the analyte from any constituents that may interfere.

3. Wash the precipitate free of impurities and coprecipitants.

4. Convert the precipitate to a reliable weighing form.

IDENTIFY AN INSOLUBLE FORM

Because no substance is totally insoluble, we must find a form that, under the conditions specified for the analysis, has such a low solubility that the error introduced due to solubility is negligible. A measure of solubility for ionic substances

EXAMPLE 5-1

An organic compound is purified and then sent for carbon–hydrogen analysis. Assuming that a 15.24-mg sample is combusted and yields 22.36 mg CO_2 and 9.13 mg H_2O, determine the percentages of carbon and hydrogen in the compound. If the remaining percentage in the compound can be assumed to be oxygen, determine the empirical formula for the compound.

Solution

$$\text{meq C} = \text{meq CO}_2$$

$$\frac{\text{mgC}}{\text{FW (C)}/n_C} = \frac{\text{mgCO}_2}{\text{FW (CO}_2)/n_{CO_2}}$$

Define C as reactive species; that is, n is the number of carbon atoms in the substance: $n_C = n_{CO_2} = 1$

$$\frac{\text{mg C}}{12.01/1} = \frac{22.36}{44.01/1}$$

$$\text{mg C} = 6.102 \text{ mg} \qquad \% \text{ C} = \frac{6.102}{15.24} \times 100 \,(\%) = 40.04\%$$

$$\text{meq H} = \text{meq H}_2\text{O}$$

$$\frac{\text{mg H}}{\text{FW (H)}/n_H} = \frac{\text{mg H}_2\text{O}}{\text{FW (H}_2\text{O)}/n_{H_2O}}$$

Define H as the reactive species; that is, n is the number of hydrogen atoms in

---

is the solubility product, K_{sp}. Although no compound has $K_{sp} = 0$, it can be seen in Appendix 2-A that substances such as $Fe(OH)_3$ (for which $K_{sp} = 4 \times 10^{-38}$) and Ag_2S (for which $K_{sp} = 6 \times 10^{-50}$) exhibit sufficiently low solubility to meet this requirement. Even for a compound with a very low K_{sp}, such as $Fe(OH)_3$, experimental conditions must be specified. As shown in Example 5-2, a *quantitative precipitation* of iron as $Fe(OH)_3$ requires a pH greater than ~ 3.5.

Appendix 2-A indicates that many substances have solubility products of $\sim 10^{-10}$. Is this range sufficiently low to allow quantitative precipitation? Example 5-3 shows that the solubility of AgCl in pure water may be too high for highly accurate analytical results. (Even students in their first course of analytical chemistry are expected to obtain results within 0.2% of the true value.)

When activity coefficients are considered, the situation may be even worse. As mentioned in Chapter 4, Section 4-3, activities can be calculated from concentrations and activity coefficients with Equation 4-8. However, activity coefficients are a

EXAMPLE 5-1 (cont.)

substance: $n_H = 1, \quad n_{H_2O} = 2$

$$\frac{mg\ H}{1.008/1} = \frac{9.13}{18.02/2}$$

$$mg\ H = 1.021\ mg \qquad \%\ H = \frac{1.021}{15.24} \times 100\ (\%) = 6.70\%$$

If the remainder is oxygen,

$$\%\ 0 = 100.00 - 40.04 - 6.70 = 53.26\%$$

Empirical formula:

Element	$\left(\dfrac{g\ element}{100\ g\ compound} \div \dfrac{g\ element}{mol\ element}\right)$		$=$	$\left(\dfrac{mol\ element}{100\ g\ compound} \div \begin{array}{c}\textbf{Smallest Factor}\\ \textbf{(to normalize}\\ \textbf{based on one}\\ \textbf{atom/molecule}\\ \textbf{for lowest value)}\end{array}\right)$			$=$	$\dfrac{\textbf{atoms}}{\textbf{molecule}}$
C	40.04	$\div$ 12.01	$=$	3.33	$\div$	3.33	$=$	1
H	6.70	$\div$ 1.008	$=$	6.65	$\div$	3.33	$=$	2
O	53.26	$\div$ 16.00	$=$	3.33	$\div$	3.33	$=$	1

Therefore, the empirical formula is CH_2O (a carbohydrate whose molecular formula might be $C_6H_{12}O_6$)

---

function of ionic strength (μ) in the solution. Ionic strength can be calculated using Equation 5-1:

$$\mu = \tfrac{1}{2}\sum_i C_i Z_i^2 \tag{5-1}$$

Here C_i is the concentration (mol/L) and Z_i is the charge for each species i present in the solution. For a solute consisting of one $+1$ cation and one -1 anion, $\mu = C$.

The relationship between activity coefficient and ionic strength was determined by Debye and Hückel for dilute solutions. In addition to ionic strength, the activity coefficient for a particular ion depends on its charge (Z) and its size. For very dilute solutions at 25°C in which $\mu < 10^{-3}$, the activity coefficient may be calculated using Equation 5-2:

$$\log f \simeq -0.5Z^2\mu^{1/2} \tag{5-2}$$

EXAMPLE 5-2

Determine the minimum pH at which quantitative precipitation of iron as $Fe(OH)_3$ will take place.

Solution

In Appendix 2-A, the solubility product for $Fe(OH)_3$ is given as 4×10^{-38}. For any system containing pure $Fe(OH)_3$ at equilibrium, we can write

$$K_{sp} = 4 \times 10^{-38} = [Fe^{3+}][OH^-]^3$$

It would be natural to attempt to remove all the iron from solution, but obviously if we set $[Fe^{3+}] = 0$, then $[OH^-]$ approaches the limit of ∞. Therefore, the term *quantitative precipitation* must be defined to ensure that some level of Fe^{3+} will be acceptable. Suppose the original solution contained $0.01\ F\ Fe(III)$. How should quantitative precipitation be defined? Gravimetric procedures are inherently capable of high accuracy; errors of only 0.1% are typical. Therefore, a loss due to solubility of less than 0.1% would be unobservable because it is less than the indeterminate error of the procedure. A solubility of 0.1% would be

$$[Fe^{3+}] = 10^{-3}[Fe^{3+}]_0 = 10^{-3} \times 0.01 = 10^{-5}$$

where $[Fe^{3+}]_0$ is the initial concentration of iron in solution. Because of the uncertainties in the solubility product, activity effects, and so on, it is prudent to add an additional safety factor of 10, leading to

$$[Fe^{3+}] = 10^{-6}$$

For more concentrated solutions, it is convenient to use tabulated values for activity coefficients, such as those in Table 5-1.

Suppose the 100 mg of AgCl used in Example 5-3 is in contact with 200 mL of a solution of ionic strength 0.1. Table 5-1 indicates that the activity coefficients for Ag^+ and Cl^- at ionic strength 0.1 are 0.75 and 0.76 respectively. Therefore, using Equation 4-8,

$$K_{sp} = 1.8 \times 10^{-10} = [Ag^+][Cl^-]$$

$$[Ag^+] = C_{Ag}f_{Ag}$$

$$[Cl^-] = C_{Cl}f_{Cl}$$

$$1.8 \times 10^{-10} = C_{Ag}f_{Ag} \times C_{Cl}f_{Cl}$$

$$\frac{1.8 \times 10^{-10}}{0.75 \times 0.76} = 3.2 \times 10^{-10} = C_{Ag}C_{Cl}$$

The product of the concentrations of Ag^+ and Cl^- is now 3.2×10^{-10}. Following the procedure suggested in Example 5-3, the percentage error under these

EXAMPLE 5-2 (cont.)

With this value of Fe^{3+} as a measure of quantitative precipitation, the hydroxide ion concentration can now be calculated:

$$[OH^-]^3 = \frac{K_{sp}}{[Fe^{3+}]} = \frac{4 \times 10^{-38}}{10^{-6}} = 4 \times 10^{-32}$$

$$[OH^-] = 3.4 \times 10^{-11}$$

In any aqueous solution, the hydronium ion and hydroxide ion concentrations are related by the water dissociation constant:

$$K_w = [H_3O^+][OH^-]$$

At 25°C,

$$1 \times 10^{-14} = [H_3O^+][OH^-]$$

$$[H_3O^+] = \frac{1 \times 10^{-14}}{3.4 \times 10^{-11}} = 2.9 \times 10^{-4}$$

$$pH = -\log[H_3O^+] = 3.53$$

Remember that if the original solution contains a lower concentration of Fe(III), then $[Fe^{3+}]$ for quantitative precipitation will be correspondingly lower. This in turn will increase $[OH^-]$, decrease $[H_3O^+]$, and increase the minimum pH. For all common analytical situations, a pH of 3.5 to 4 will be high enough to effect quantitative precipitation.

---

conditions will be 0.51%. An accurate determination of chloride or silver using precipitation of AgCl is clearly impossible under these conditions. However, we can make use of the *common ion effect*. In Example 5-3, the solution contained stoichiometric amounts of Ag^+ and Cl^-. If we are conducting a chloride analysis we could add an *excess* of silver ion. The excess silver ion would increase $[Ag^+]$ and thereby decrease $[Cl^-]$ because the product of the two terms must equal K_{sp}. Conversely, for a silver analysis, an excess of chloride ion could be added to decrease $[Ag^+]$. This effect can drastically change the calculations in Example 5-3. Suppose that an excess silver ion concentration of 0.1 F is used in a chloride analysis. The chloride ion concentration will then be

$$[Cl^-] = \frac{K_{sp}}{[Ag^+]} = \frac{1.8 \times 10^{-10}}{0.1} = 1.8 \times 10^{-9}$$

This is four orders or magnitude smaller than that determined in Example 5-3. The determinate error due to solubility loss is now much smaller than the indeterminate error of the procedure.

EXAMPLE 5-3

Calculate the percentage error introduced in a chloride analysis, assuming an AgCl precipitate weighing 100 mg is in contact with 200 mL of pure H_2O.

Solution

The solubility product of AgCl is 1.8×10^{-10} (Appendix 2-A). Therefore,

$$K_{sp} = 1.8 \times 10^{-10} = [Ag^+][Cl^-]$$

The dissolution of AgCl is

$$AgCl(s) \rightleftharpoons Ag^+ + Cl^-$$

Therefore, in pure water or any solution in which there is no other source of Ag^+ and Cl^-,

$$[Ag^+] = [Cl^-] = \text{moles of AgCl dissolved per liter}$$
$$1.8 \times 10^{-10} = [Ag^+][Cl^-] = [Cl^-]^2$$
$$[Cl^-] = 1.34 \times 10^{-5} \text{ mmol/mL (assuming activity = concentration)}$$

The amount of AgCl dissolved is

$$1.34 \times 10^{-5} \frac{\text{mmol}}{\text{mL}} \times 200 \text{ mL} \times 143 \frac{\text{mg AgCl}}{\text{mmol}} = 0.38 \text{ mg}$$

The percentage error is

$$\frac{0.38 \text{ mg AgCl dissolved}}{100 \text{ mg AgCl total}} \times 100 \, (\%) = 0.38\%$$

Unfortunately, two problems are introduced when excess common ion is added to decrease solubility losses of the analyte. First, after filtration, the precipitate is wet with a solution containing the excess ion and must be washed before the analyte can be dried and weighed. Problems involved with washing precipitates will be discussed under Step 3 for a successful gravimetric procedure. At this point, however, note that if the precipitate is washed with 200 mL of pure water, we will be right back where we started in Example 5-3. The only difference is that the loss will be in the washing procedure instead of the precipitation step.

The second problem introduced with the addition of excess common ion involves multiple equilibria and arises when an analyst, having used the excess common ion effect for a chloride analysis, attempts to use this same approach for a silver analysis

TABLE 5-1 Activity Coefficients

Ion	Effective* diameter (Å)	Activity coefficient at indicated ionic strengths				
		0.001	0.005	0.01	0.05	0.1
H_3O^+	9	0.967	0.933	0.914	0.86	0.83
$Li^+, C_6H_5COO^-$	6	0.965	0.929	0.907	0.84	0.80
$Na^+, IO_3^-, HSO_3^-,$ $HCO_3^-, H_2PO_4^-,$ $H_2AsO_4^-, OAc^-$	4–4.5	0.964	0.928	0.902	0.82	0.78
$OH^-, F^-, SCN^-, HS^-,$ $ClO_3^-, ClO_4^-, BrO_3^-,$ IO_4^-, MnO_4^-	3.5	0.964	0.926	0.900	0.81	0.76
$K^+, Cl^-, Br^-, I^-, CN^-,$ $NO_2^-, NO_3^-, HCOO^-$	3	0.964	0.925	0.899	0.80	0.76
$Rb^+, Cs^+, Tl^+, Ag^+,$ NH_4^+	2.5	0.964	0.924	0.898	0.80	0.75
Mg^{2+}, Be^{2+}	8	0.872	0.755	0.69	0.52	0.45
$Ca^{2+}, Cu^{2+}, Zn^{2+}, Sn^{2+},$ $Mn^{2+}, Fe^{2+}, Ni^{2+},$ $Co^{2+}, phthalate^{2-}$	6	0.870	0.749	0.675	0.48	0.40
$Sr^{2+}, Ba^{2+}, Cd^{2+}, Hg^{2+},$ S^{2-}	5	0.868	0.744	0.67	0.46	0.38
$Pb^{2+}, CO_3^{2-}, SO_3^{2-},$ $C_2O_4^{2-}$	4.5	0.868	0.742	0.665	0.46	0.37
$Hg_2^{2+}, SO_4^{2-}, S_2O_3^{2-},$ CrO_4^{2-}, HPO_4^{2-}	4.0	0.867	0.740	0.660	0.44	0.36
$Al^{3+}, Fe^{3+}, Cr^{3+}, La^{3+},$ Ce^{3+}	9	0.738	0.54	0.44	0.24	0.18
$PO_4^{3-}, Fe(CN)_6^{3-}$	4	0.725	0.50	0.40	0.16	0.095
$Th^{4+}, Zr^{4+}, Ce^{4+}, Sn^{4+}$	11	0.588	0.35	0.255	0.10	0.065
$Fe(CN)_6^{4-}$	5	0.57	0.31	0.20	0.048	0.021

SOURCE: From J. Kielland, "Individual Activity Coefficients of Ions in Aqueous Solutions," *Journal of the American Chemical Society*, 59: 1675 (1937). Reproduced with permission.

*Although Equation 5-2 can be used to calculate activity coefficients in very dilute solutions when only ion charge and ionic strength are needed, the size of the ion also influences the activity coefficient in more concentrated solutions. For a given charge and ionic strength, the activity coefficient increases as the effective diameter increases.

by adding excess chloride ion. To decrease solubility losses of silver, the analyst adds a 1 *F* excess of chloride ion. Disappointingly, the solubility of silver is greater now than it is in pure water. How could this be? The solubility product is not the only chemical equilibrium involving silver and chloride ions that is occurring. Chloride ions are also ligands and can form complexes with silver according to Equations 5-3 and 5-4:

$$AgCl(s) + Cl^- \rightleftharpoons AgCl_2^-$$

$$K_1 = 2.0 \times 10^{-5} = \frac{[AgCl_2^-]}{[Cl^-]} \tag{5-3}$$

$$AgCl_2^- + Cl^- \rightleftharpoons AgCl_3^{2-}$$

$$K_2 = 1 = \frac{[AgCl_3^{2-}]}{[AgCl_2^-][Cl^-]} \tag{5-4}$$

The total solubility is the sum of *all* forms of silver remaining in solution:

$$\text{solubility} = [Ag^+] + [AgCl_2^-] + [AgCl_3^{2-}] \tag{5-5}$$

Equation 5-5 can be expressed in terms of the equilibrium constants, K_{sp}, K_1 and K_2 and $[Cl^-]$:

$$\text{solubility} = \frac{K_{sp}}{[Cl^-]} + K_1[Cl^-] + K_1 K_2 [Cl^-]^2 \tag{5-6}$$

Equation 5-6 shows that as $[Cl^-]$ is increased to decrease solubility from the first term on the right side of the equation, terms two and three will begin to increase.

To obtain a numerical answer, $[Cl^-]$ must be known. The dissolution of AgCl adds chloride, whereas the formation of $AgCl_2^-$ and $AgCl_3^{2-}$ consumes chloride. If the excess chloride concentration (the amount above stoichiometry) is f_{Cl}, then

$$[Cl^-] = f_{Cl} + [Ag^+] - [AgCl_2^-] - 2[AgCl_3^{2-}]$$

Or, in terms of the constants,

$$[Cl^-] = f_{Cl} + \frac{K_{sp}}{[Cl^-]} - K_1[Cl^-] - 2K_1 K_2 [Cl^-]^2 \tag{5-7}$$

The two unknowns ($[Cl^-]$ and solubility) in Equations 5-6 and 5-7 could be solved precisely through a laborious process. However, if a reasonable excess of chloride is added, f_{Cl} will be $\gg [Ag^+]$, $[AgCl_2^-]$, and $[AgCl_3^{2-}]$. Therefore, an approximation may be made:

$$[Cl^-] \cong f_{Cl}$$

This leads to

$$\text{solubility} = \frac{K_{sp}}{f_{Cl}} + K_1 f_{Cl} + K_1 K_2 f_{Cl}^2 \tag{5-8}$$

For $f_{Cl} = 1$, the solubility of silver chloride is

$$\text{solubility} = \frac{1.8 \times 10^{-10}}{1} + (2.0 \times 10^{-5} \times 1) + (2.0 \times 10^{-5} \times 1 \times 1^2) = 4 \times 10^{-5}$$

This is three times higher than the solubility in pure water calculated in Example 5-3. The assumption that $f_{Cl} \gg [Ag^+]$ $[AgCl_2^-]$, and $[AgCl_3^{2-}]$ was justified because $1 \gg 1.8 \times 10^{-10}$ and 2×10^{-5}.

When multiple equilibria are involved, as in the preceding situation, frequently an optimum concentration of reagent will give the minimum solubility of the analyte. Figure 5-3 shows the solubility of AgCl as a function of excess chloride concentration; the optimum value is near 0.003 F. Figure 5-3 shows that the optimum concentration of chloride ion would be considerably greater for the precipitation of $PbCl_2$. When developing an analytical procedure, the analytical chemist normally recommends this optimum concentration of precipitating reagent to minimize any determinate error arising from solubility losses. This optimum

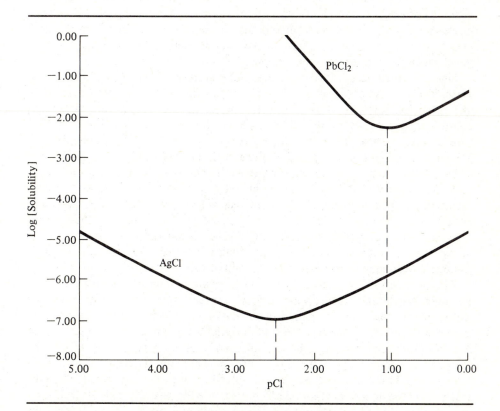

FIGURE 5-3   Solubility of AgCl and $PbCl_2$ as a function of $[Cl^-]$.

value can be calculated, given a knowledge of the equilibrium constants involved, using calculus as described in Appendix 1-D. For the values given in Equations 5-3 and 5-4, the optimum value is 0.003 F and the minimum solubility is $1.2 \times 10^{-7}\ F$ for AgCl.

Another equilibrium that contributes to total solubility is the solubility of the compound *without* ionization:

$$AgCl(s) \rightleftharpoons AgCl(aq) \tag{5-9}$$

If we can assume that AgCl(s) is in its standard state then:

$$K = [AgCl(aq)]$$

Therefore, the concentration of undissociated AgCl will always be the same, and K represents an intrinsic solubility. Fortunately, the intrinsic solubility is quite small for compounds of analytical interest. But for AgCl, K has been estimated to be somewhat greater than $10^{-7}\ F$. Addition of this solubility term would not change the shape of the curve in Figure 5-3, but would simply raise it by the value of K.

Figure 5-3 also shows that, even at the optimum chloride concentration, the solubility of $PbCl_2$ is considerably greater than that of AgCl and probably will not meet accuracy requirements for an analysis. Consequently, the analyst must find a method to decrease the solubility product of a moderately soluble compound so that quantitative precipitation can be attained.

One such method is to decrease the temperature at which precipitation takes place. The dissolution of all ionic compounds of limited solubility occurs with absorption of heat ($\Delta H > 0$). Therefore, increasing the temperature favors the dissolution direction for the solubility equilibrium (the Le Châtelier principle). Conversely, decreasing the temperature favors the reverse direction—precipitation. The effect of temperature on the solubility and K_{sp} for AgCl and $BaSO_4$ is shown in Table 5-2. For some compounds, such as AgCl, a decrease in temperature of only 25°C (from 25 to 0°C) decreases the solubility by a factor of more than two. Even so, the effect is only useful when the solubility is just slightly too high for the required accuracy. Other compounds, such as $BaSO_4$, exhibit little solubility change with temperature and because water freezes at 0°C, there is not a large temperature range at the analyst's disposal. Washing a precipitate with ice-cold water, however, is sometimes recommended to decrease solubility losses.

Another method for decreasing the solubility for an ionic compound is to use a nonaqueous solvent. Because water solvates ions so strongly, an ionic compound will be more soluble in water than in solvents that solvate ions less strongly. As shown in Table 5-3, this effect can be quite significant in decreasing the solubility and K_{sp}. However, remember that the solubility of all substances in the system may be markedly decreased and that the reagent itself may precipitate from the nonaqueous solvent. A mixture of water and organic solvent is often recommended. Then the solubility of the desired compound can be decreased to an acceptable level without approaching the solubility limits of the reagents and other constituents in the system. Table 5-3 shows how the solubility of $PbSO_4$, another slightly soluble compound, can be decreased by using ethanol–water mixtures. For $PbSO_4$ in pure water, K_{sp} is 100 times greater than that of AgCl; but with 30% ethanol, K_{sp} of $PbSO_4$ is less than that of AgCl in water.

Note that it is not valid to assume that all systems are always at equilibrium. The rate of precipitation may require minutes or even hours to establish equilibrium. Sometimes, the solubility product may be exceeded without forming a precipitate because the solution is *supersaturated*. A supersaturated solution may appear stable

---

TABLE 5-2   Solubility of AgCl and $BaSO_4$ As a Function of Temperature

Temperature	AgCl		$BaSO_4$	
(°C)	Solubility, F	K_{sp}	Solubility, F	K_{sp}
0	0.49×10^{-5}	0.24×10^{-10}	0.81×10^{-5}	0.66×10^{-10}
10	0.73×10^{-5}	0.54×10^{-10}	0.98×10^{-5}	0.97×10^{-10}
25	1.36×10^{-5}	1.85×10^{-10}	1.15×10^{-5}	1.34×10^{-10}
50	3.8×10^{-5}	14.2×10^{-10}	1.4×10^{-5}	2.1×10^{-10}
100	$15. \times 10^{-5}$	$215. \times 10^{-10}$	1.7×10^{-5}	2.8×10^{-10}

SOURCE: Calculated from data in H. Stephen and T. Stephen, eds., *Solubilities of Inorganic and Organic Compounds*, vol. 1 (New York: Macmillan, Inc., 1963).

TABLE 5-3   Effect of Solvent on Solubility

Solvent	$KClO_4$		$PbSO_4$	
	Solubility, F	K_{sp}	Solubility, F	K_{sp}
water	0.146	2×10^{-2}	1.5×10^{-4}	2×10^{-8}
methanol	7.6×10^{-4}	5.7×10^{-7}		
ethanol	8.7×10^{-5}	7.5×10^{-9}		
1-butanol	3.2×10^{-5}	1.1×10^{-9}		
ethanol–water mixtures (v/v% ethanol):				
10			5.4×10^{-5}	2.9×10^{-9}
20			2.1×10^{-5}	4.4×10^{-10}
30			7.6×10^{-6}	5.8×10^{-11}
50			1.6×10^{-6}	2.6×10^{-12}

SOURCE: Calculated from data in I.M. Kolthoff and E.B. Sandell, *Textbook of Quantitative Inorganic Analysis*, 3rd ed. (New York: Macmillan, Inc., 1952).

until a crystal of the compound whose solubility product has been exceeded is added, at which point precipitation begins.

Analytical chemists use any phenomenon that helps them solve a problem. A slow rate of precipitation is no exception. For example, both calcium and magnesium form insoluble oxalates, but the precipitation rate of magnesium oxalate is quite slow. Consequently, calcium can be separated from magnesium by adding oxalate and filtering off the calcium oxalate before magnesium oxalate starts to form. This procedure also illustrates another point: Although most analytical techniques demand slow, careful work, some demand rapid, skillful work to achieve high accuracy.

SEPARATE THE ANALYTE FROM ANY CONSTITUENTS THAT MAY INTERFERE

As outlined in Chapter 1, the third step in the total analysis process involves separation of the analyte from interferences. In a gravimetric procedure, the analyte is separated from the solution when it is precipitated as an insoluble compound. (In fact, as we will see in Chapter 11, precipitation can be used as a separation step with quantitation carried out by some other means.) Unfortunately, precipitating reagents are not highly specific and may form many insoluble compounds. If the solubility product of any substance other than the analyte is also exceeded, that substance will also precipitate, causing a positive determinate error. For example, consider the use of AgCl as an insoluble compound. For a silver analysis, the addition of chloride ion is reasonably selective, because there are not many insoluble chlorides. However, for a chloride analysis, the addition of silver ion is not very selective. Because all silver salts (except $AgNO_3$ and AgF) show limited solubility, the addition of silver ion will precipitate essentially all the anions in the solution. A successful use of AgCl for the determination of chloride requires that all anions other than nitrate, fluoride, and low concentrations of sulfate be absent or removed prior to precipitation.

Therefore, gravimetric procedures usually require that the analyst have prior knowledge of all components in the sample to ensure that no solubility products (other than the one for the compound formed from the analyte and the precipitating reagent) will be exceeded. Any interfering substances must be removed before the precipitating reagent is added. How these substances are removed depends on the analyst's skill and the nature of the substance. Sometimes the oxidation state of the interfering substance can be changed by adding an oxidizing or reducing agent, thereby significantly altering the solubility, as suggested in Table 5-4.

Even when an ion is a potential interference because its compound with the precipitating reagent has a limited solubility, it may still be possible to precipitate the analyte in the presence of this ion if the analyte compound is sufficiently less soluble. The concentration of the precipitating reagent must be controlled so that it is high enough to quantitatively precipitate the analyte, but not high enough to exceed the solubility product of the interfering substance. Knowledge of the solubility products involved allows the analytical chemist to determine the appropriate conditions for a successful analytical procedure, as shown in Example 5-4.

TABLE 5-4   Solubility Changes Resulting from Oxidation State Changes

Compound	Oxidation state	Solubility product
copper chloride	+1	1.2×10^{-6}
	+2	soluble
iron hydroxide	+2	$8 \ \times 10^{-16}$
	+3	$4 \ \times 10^{-38}$
mercury chloride	+1	1.3×10^{-18}
	+2	soluble

EXAMPLE 5-4

Determine the appropriate conditions for the precipitation of Ag^+ as the chloride in the presence of Pb^{2+} assuming the initial concentration of each is approximately 0.01 F.

Solution

The solubility products for $AgCl$ and $PbCl_2$ are given in Appendix 2-A:

$$K_{sp(Pb)} = 1.6 \times 10^{-5} = [Pb^{2+}][Cl^-]^2$$

$$K_{sp(Ag)} = 1.8 \times 10^{-10} = [Ag^+][Cl^-]$$

For quantitative precipitation of $AgCl$, the silver ion concentration should be decreased by a factor of 10^4 (see Example 5-2) so that $[Ag^+] = 0.01 \times 10^{-4} = 10^{-6}$. Therefore, the minimum concentration of *excess* chloride ion is

$$[Cl^-] = \frac{1.8 \times 10^{-10}}{10^{-6}} = 1.8 \times 10^{-4}$$

$PbCl_2$ will not precipitate until its solubility product is exceeded—that is, when the chloride concentration reaches

$$[Cl^-]^2 = \frac{1.6 \times 10^{-5}}{0.01} = 1.6 \times 10^{-3}$$

$$[Cl^-] = 4.0 \times 10^{-2}$$

Therefore, if the excess chloride ion concentration is maintained at a level between 1.8×10^{-4} and 4.0×10^{-2} F, quantitative precipitation of $AgCl$ will take place without any precipitation of $PbCl_2$.

In practice, the procedure suggested in Example 5-4 is difficult to conduct because higher concentrations of chloride ion must usually be added to the solution to precipitate the silver ion present—that is, a stoichiometric amount of chloride must be added over and above the required excess. In some regions of the solution, chloride concentrations may exceed 0.04 F and lead to some precipitation of $PbCl_2$. Even though the equilibrium calculations may indicate that any $PbCl_2$ formed should redissolve, this process may be slow. Separations of this type are more practical when the concentration of the precipitating reagent can always be controlled at the desired value throughout the precipitation reaction. This situation occurs when the precipitating ion is also involved in another equilibrium, such as the ionization of a weak acid. The simplest example of this is the precipitation of hydroxides at a controlled pH level, because the hydroxide ion concentration in an aqueous solution is determined by the hydronium ion concentration and the water dissociation constant:

$$[OH^-] = \frac{K_w}{[H_3O^+]}$$

Example 5-5 shows how $Fe(OH)_3$ can be precipitated in the presence of Mn^{2+} by controlling the pH with a buffer.

The hydroxide case is relatively simple because water has only one dissociation constant and its activity is assumed to be one throughout the process. When salts of other weak acids are used as precipitation reagents, the same calculation procedure applies but the calculations become more formidable, as shown in Examples 5-6 and 5-7.

WASH THE PRECIPITATE FREE OF IMPURITIES AND COPRECIPITANTS

After precipitation has taken place, the precipitate must be separated from the supernatant solution (which normally contains excess common ion) and the other constituents of the sample. A filtration process retains the precipitate on some type of filter, but the precipitate will still be wet with the supernatant solution. Drying and weighing the precipitate at this point would lead to significant positive determinate errors. Washing the precipitate with water will decrease this error, as shown in Figure 5-4; however, too much washing will introduce a negative error as some precipitate dissolves to maintain solubility equilibrium (K_{sp}). At the optimum wash volume (V_0), any positive error caused by impurities will be exactly offset by negative error caused by solubility. The correct amount of washing depends on the solubility product, the concentration, and the type of impurities present. The size of the precipitate particles, which depends on the temperature and rate of precipitation, is also important. Clearly, this profusion of uncontrolled variables makes it too difficult to calculate curves of the type shown in Figure 5-4. Appropriate volumes of wash solution are recommended in the lab manual; since these are based on the accumulated experiences of many analysts, the curve near V_0 should be sufficiently flat that little error will be introduced if somewhat smaller or larger volumes than V_0 are used.

EXAMPLE 5-5

Determine the appropriate conditions for the precipitation of Fe^{3+} as the hydroxide $(Fe(OH)_3)$ in the presence of Mn^{2+}, assuming the initial concentration of each is approximately 0.05 F.

Solution

The solubility products for the two hydroxides are given in Appendix 2-A:

$$K_{sp(Fe)} = 4 \times 10^{-38} = [Fe^{3+}][OH^-]^3$$
$$K_{sp(Mn)} = 1.9 \times 10^{-13} = [Mn^{2+}][OH^-]^2$$

For quantitative precipitation of $Fe(OH)_3$, the Fe^{3+} concentration should be decreased by a factor of 10^4 (see Example 5-2) so that $[Fe^{3+}] = 0.05 \times 10^{-4} = 5 \times 10^{-6}$. Therefore, the minimum concentration of hydroxide ion is

$$[OH^-]^3 = \frac{4 \times 10^{-38}}{5 \times 10^{-6}} = 8 \times 10^{-33}$$

$$[OH^-] = 2.0 \times 10^{-11}, [H_3O^+] = 5 \times 10^{-4}, pH = 3.30$$

$Mn(OH)_2$ will not precipitate until its solubility product is exceeded — that is, when the hydroxide ion concentration reaches

$$[OH^-]^2 = \frac{1.9 \times 10^{-13}}{0.05} = 3.8 \times 10^{-12}$$

$$[OH^-] = 1.9 \times 10^{-6}, [H_3O^+] = 5.1 \times 10^{-9}, pH = 8.29$$

Therefore, if a buffer solution of 3.30 to 8.29 pH is used, quantitative precipitation of Fe^{3+} will take place without precipitation of Mn^{2+}.

Often the precipitate remains contaminated with some impurity even after extensive washing. The impurity appears to have precipitated along with the analyte, even though its solubility product has not been exceeded and it would not have precipitated if the analyte had not been present. This phenomenon is called *coprecipitation*.

Coprecipitation may occur via several mechanisms. These may be classified into three types:

1. Surface adsorption

2. Inclusions

3. Occlusions

EXAMPLE 5-6

Determine the appropriate conditions for the precipitation of Pb^{2+} as the sulfide in the presence of Mn^{2+} assuming the initial concentration of each is approximately 0.1 F.

Solution

The solubility products for the two sulfides are given in Appendix 2-A:

$$K_{sp(Pb)} = 1 \times 10^{-28} = [Pb^{2+}][S^{2-}]$$

$$K_{sp(Mn)} = 3 \times 10^{-13} = [Mn^{2+}][S^{2-}]$$

For quantitative precipitation of PbS, the Pb^{2+} concentration must be decreased by a factor of 10^4 (see Example 5-2) so that $[Pb^{2+}] = 0.1 \times 10^{-4} = 10^{-5}$. Therefore, the minimum concentration of sulfide ion is

$$[S^{2-}] = \frac{1 \times 10^{-28}}{10^{-5}} = 1 \times 10^{-23}$$

The maximum concentration of sulfide which may be present before MnS starts to precipitate is

$$[S^{2-}] = \frac{3 \times 10^{-13}}{0.1} = 3 \times 10^{-12}$$

Therefore, the sulfide concentration must be maintained between 1×10^{-23} and 3×10^{-12} F. The sulfide concentration is controlled by pH through the following equilibria:

$$H_2S + H_2O \rightleftharpoons H_3O^+ + HS^- \qquad K_1 = 5.7 \times 10^{-8} = \frac{[H_3O^+][HS^-]}{[H_2S]}$$

$$HS^- + H_2O \rightleftharpoons H_3O^+ + S^{2-} \qquad K_2 = 1.2 \times 10^{-15} = \frac{[H_3O^+][S^{2-}]}{[HS^-]}$$

Surface Adsorption   When a precipitate of an ionic solid forms, the constituent ions fill lattice sites until the limiting reactant, the analyte, is depleted. Excess common ion will continue to occupy sites on the surface of the precipitating particle forming an adsorbed layer. The charge of the particles will be the same as that of the excess common ion. For example, in Figure 5-5, AgCl is being precipitated in the presence of excess Ag^+ (as in the determination of chloride in a sample). The size of the precipitating particle will depend on both the type of crystal forming and the speed at which the analyte concentration is depleted. In many cases, the particles may be colloidal—so small that they would pass through the pores of ordinary filters. Because the particles carry the same charge, they repel each other and will not easily coagulate into larger particles. However, the adsorbed ions will also attract an

EXAMPLE 5-6 (cont.)

An overall dissociation of H_2S can be written as the sum of these two equilibria:

$$H_2S + 2H_2O \rightleftharpoons 2H_3O^+ + S^{2-} \qquad K = K_1K_2 = \frac{[H_3O^+]^2[S^{2-}]}{[H_2S]}$$

$$K_1K_2 = 6.8 \times 10^{-23}$$

Solutions saturated with H_2S have concentrations of approximately 0.1 F. Therefore, to control the sulfide concentration at 1×10^{-23}, the pH must be

$$[H_3O^+]^2 = \frac{6.8 \times 10^{-23} \times 0.1}{1 \times 10^{-23}} = 0.68$$

$$[H_3O^+] = 0.82 \qquad pH = 0.08$$

The highest pH that can be tolerated before precipitation of MnS occurs is:

$$[H_3O^+]^2 = \frac{6.8 \times 10^{-23} \times 0.1}{3 \times 10^{-12}} = 2.27 \times 10^{-12}$$

$$[H_3O^+] = 1.5 \times 10^{-6} \qquad pH = 5.82$$

Therefore, lead can be quantitatively precipitated in the presence of Mn^{2+} by using acidic conditions (a pH of 1 is commonly recommended).

The solubility products of metal sulfides vary over many orders of magnitude (as shown in Appendix 2-A). Consequently, metal ions can often be separated through the use of sulfide precipitations.

ion of opposite charge, such as the nitrate shown in Figure 5-5. When the solution contains a high concentration of these solute ions (high ionic strength) there will be a sufficient number very near the precipitating particle to neutralize the charge. In that case, precipitating particles are able to coagulate and, thereby, grow to such size that filtration is possible. Weighing the AgCl precipitate will show a positive error and an analysis of it will reveal the presence of $AgNO_3$. Silver nitrate is very soluble, yet it appears to have coprecipitated with the AgCl. This phenomenon is coprecipitation caused by surface adsorption. Because it is a surface effect, the amount of coprecipitation will depend on the total surface area. In terms of percentage error, the error due to surface adsorption will depend on the ratio of surface area to volume of precipitate. The smaller the particle size, the larger this ratio will be; therefore, if

EXAMPLE 5-7

Determine the appropriate conditions for the precipitation of Ca^{2+} as the oxalate in the presence of Mg^{2+}, assuming the initial concentration of each is approximately 0.01 F.

Solution

The solubility products for the two oxalates are given in Appendix 2-A:

$$K_{sp(Ca)} = 2.3 \times 10^{-9} = [Ca^{2+}][C_2O_4^{2-}]$$

$$K_{sp(Mg)} = 8.6 \times 10^{-5} = [Mg^{2+}][C_2O_4^{2-}]$$

For quantitative precipitation of CaC_2O_4, the Ca^{2+} concentration must be decreased by a factor of 10^4 (see Example 5-2) so that $[Ca^{2+}] = 0.01 \times 10^{-4} = 10^{-6}$. Therefore, the minimum concentration of oxalate ion is

$$[C_2O_4^{2-}] = \frac{2.3 \times 10^{-9}}{10^{-6}} = 2.3 \times 10^{-3}$$

The maximum concentration of oxalate which may be present before MgC_2O_4 starts to precipitate is

$$[C_2O_4^{2-}] = \frac{8.6 \times 10^{-5}}{0.01} = 8.6 \times 10^{-3}$$

Therefore, the oxalate concentration must be maintained between 2.3×10^{-3} and 8.6×10^{-3} F. The oxalate concentration is controlled by pH through the following equilibria:

$$H_2C_2O_4 + H_2O \rightleftharpoons H_3O^+ + HC_2O_4^- \qquad K_1 = 5.36 \times 10^{-2} = \frac{[H_3O^+][HC_2O_4^-]}{[H_2C_2O_4]}$$

$$HC_2O_4^- + H_2O \rightleftharpoons H_3O^+ + C_2O_4^{2-} \qquad K_2 = 5.42 \times 10^{-5} = \frac{[H_3O^+][C_2O_4^{2-}]}{[HC_2O_4^-]}$$

If the formal concentration of oxalate (in all its forms) is 0.1 F, then

$$f_{H_2C_2O_4} = 0.1 = [H_2C_2O_4] + [HC_2O_4^-] + [C_2O_4^{2-}]$$

Therefore, when $[C_2O_4^{2-}] = 2.3 \times 10^{-3}$,

$$[H_2C_2O_4] + [HC_2O_4^-] = 0.1 - 2.3 \times 10^{-3} = 9.77 \times 10^{-2} \qquad \textbf{(5-10)}$$

EXAMPLE 5-7 (cont.)

$$K_2 = 5.42 \times 10^{-5} = \frac{[H_3O^+][2.3 \times 10^{-3}]}{[HC_2O_4^-]} \qquad \text{(5-11)}$$

$$K_1 = 5.36 \times 10^{-2} = \frac{[H_3O^+][HC_2O_4^-]}{[H_2C_2O_4]} \qquad \text{(5-12)}$$

Thus we have three equations (5-10, 5-11, 5-12) and three unknowns. The concentration of $[H_3O^+]$ can be eliminated by dividing K_2 by K_1.

$$\frac{5.42 \times 10^{-5}}{5.36 \times 10^{-2}} = \frac{2.3 \times 10^{-3}[H_2C_2O_4]}{[HC_2O_4^-]^2}$$

From Equation 5-10, we know that $[HC_2O_4^-] = 0.0977 - [H_2C_2O_4]$. Therefore,

$$[H_2C_2O_4] = \frac{5.42 \times 10^{-5}}{5.36 \times 10^{-2} \times 2.3 \times 10^{-3}} (0.0977 - [H_2C_2O_4])^2$$

The concentration of $[H_2C_2O_4]$ can be solved by using an iterative approach (shown in Appendix 1-C) or the quadratic formula. The results are:

$$[H_2C_2O_4] = 3.9 \times 10^{-3}$$

$$[HC_2O_4^-] = 0.094$$

$$[H_3O^+] = 2.2 \times 10^{-3} \qquad pH = 2.66$$

The same calculation carried out using $[C_2O_4^{2-}] = 8.6 \times 10^{-3}$ leads to

$$[H_2C_2O_4] = 9.6 \times 10^{-4}$$

$$[HC_2O_4] = 0.090$$

$$[H_3O^+] = 5.67 \times 10^{-4} \qquad pH = 3.25$$

Therefore, if the pH is kept between 2.66 and 3.25, it should be possible to precipitate CaC_2O_4 quantitatively in the presence of magnesium. However, this range is very narrow for practical control and is affected by the concentrations of Ca, Mg, and excess formal concentration of oxalate. Moreover, it does not account for some positive and negative aspects of kinetics. The negative aspect is that calcium does not quantitatively precipitate in such acidic solutions, especially in the presence of magnesium. The positive aspect is that, by raising the pH to 5 to 7, calcium can be quantitatively precipitated; magnesium precipitates so slowly that filtration can be accomplished before the CaC_2O_4 is appreciably contaminated with MgC_2O_4.

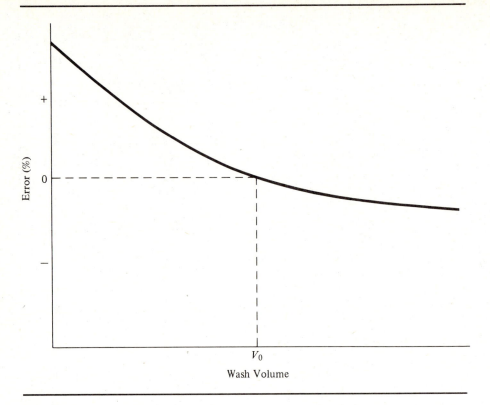

FIGURE 5-4   Washing Errors: Too little washing ($<V_0$) of precipitate introduces positive determinate error, whereas too much washing ($>V_0$) introduces negative determinate error.

the gravimetric procedure is to be successful, it is crucial to produce particles large enough to be filtered and to decrease surface adsorption effects.

Digestion is one of the best methods to increase particle size. The precipitate and the supernatant solution are heated (typically 15 minutes) and/or kept at room temperature for several hours. The dynamic equilibrium is maintained during this time, with the precipitate dissolving at the same rate it reprecipitates. This precipitation process is much slower than the initial precipitation, which allows the particles to grow in size, thereby decreasing the surface-to-volume ratio.

Washing the precipitate can also help decrease surface adsorption because the adsorption process is another chemical equilibrium. For example, using AgCl, we have

$$Ag^+(aq) + AgCl \rightleftharpoons Ag^+ \text{ (adsorbed on AgCl)} \qquad (5\text{-}13)$$

As the concentration of Ag^+ in the wash solution decreases, the amount of adsorption decreases (the Le Châtelier principle). However, this method poses two

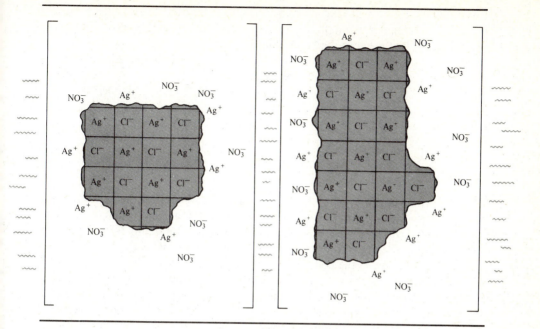

FIGURE 5-5 Surface adsorption type of coprecipitation. The excess common ion (Ag$^+$) is adsorbed on AgCl particle. The positive charge is neutralized by counterions (NO$_3^-$) in the solution surrounding the particles (double layer). The charged layer surrounding the particles repels and, therefore, inhibits coagulation of the precipitating particles.

problems. First, when the wash solution ceases to have any appreciable excess common ion, the solubility of the precipitate will increase and a significant fraction of the precipitate may dissolve in the wash water. Second, coagulation, too, is a reversible process. The particles coagulated because a counterion (NO$_3^-$ in Figure 5-5) was present to neutralize the charge. If pure water is used for washing, coagulated particles may break up and re-form as colloid-sized particles that will pass through the pores of the filter. This process, called *peptization*, can be prevented if an electrolyte is added to the wash water. Obviously, it must be an electrolyte that will not cause a determinate error by increasing the weight of the precipitate. In the AgCl example, HNO$_3$ may be added to the wash water because it will be volatilized in the drying step and therefore cause no error.

The adsorption equilibrium shown in Equation 5-13 suggests other methods to decrease surface adsorption. The adsorption process releases energy ($\Delta H < 0$); in other words, it requires energy to remove an adsorbed ion from the surface ($\Delta H > 0$). Therefore, a higher temperature will favor the removal of adsorbed ions (the Le Châtelier principle) and precipitation at higher temperature will decrease surface adsorption. This method clearly will not be useful if the compound's solubility product is too high at the elevated temperature. Moreover, Equation 5-13 shows that the amount of adsorption is directly proportional to the concentration of excess common ion in solution. Consequently, precipitation from dilute solution will help decrease surface adsorption. Again, the solubility product may require a minimum concentration to prevent unacceptably low results caused by the compound's solubility.

Sometimes all of these remedies still do not satisfactorily decrease coprecipitation due to surface adsorption. Precipitates of $Fe(OH)_3$ are notoriously able to coprecipitate other metal hydroxides on the surface. One method for decreasing this potential error turns coprecipitation to the analyst's advantage. A wash solution containing NH_4OH can be used to establish the following equilibrium:

$$NH_4^+(aq) + OH_{ads}^- - M_{counter}^+ \rightleftharpoons M^+(aq) + OH_{ads}^- - NH_{4\ counter}^+ \qquad (5\text{-}14)$$

The ammonium ions compete with the coprecipitated M^+ ions as the counterion for the adsorbed layer of OH^-. Although this competition normally favors the metal ions, continual washing of the precipitate with solutions containing NH_4OH slowly decreases the concentration of M^+ both on the precipitate and in the wash solution. The competition favors the metal ions because the counterion that forms the least soluble compound with the adsorbed ion is the most likely incorporated. Remember, to qualify as coprecipitation, the solubility product cannot be exceeded. In other words, the compound with the solubility product that is most nearly met will be the preferred coprecipitate. When the M^+ counterions are replaced with NH_4^+, coprecipitation still occurs—but now the coprecipitate is NH_4OH, which will be volatilized during the drying procedure.

Finally, a brute force method can be used to decrease coprecipitation caused by surface adsorption. This method is called *reprecipitation* or *double precipitation*. According to Equation 5-13, the amount of surface adsorption (and amount of counterion) depends on the concentration of this ion in solution. Normally, only a small fraction of the interfering ion in the original solution will coprecipitate; if the precipitate is filtered and then dissolved, the concentration of the interfering ion will be much lower in the new solution. A second precipitation can now be carried out with a corresponding lower amount of the interfering ion coprecipitating. However, because precipitation, digestion, and filtration are such slow processes, this method should be avoided if at all possible.

Inclusions   The second mechanism leading to coprecipitation is *inclusions*. As shown in Figure 5-6, these interferences are incorporated into the precipitating crystal either as substitute elements on a lattice site or as "extra" elements between sites called *interstitials*. This process can be expressed in the form of a chemical equilibrium using the coprecipitation of $PbSO_4$ in $BaSO_4$ as an example:

$$Pb^{2+}(aq) + Ba_{Ba}^{2+} \rightleftharpoons Pb_{Ba}^{2+} + Ba^{2+}(aq) \qquad (5\text{-}15)$$

Here Ba_{Ba}^{2+} indicates a Ba^{2+} ion on a barium ion site in the $BaSO_4$ crystal; Pb_{Ba}^{2+} indicates a Pb^{2+} ion on a barium ion site in the $BaSO_4$ crystal. Because this is an equilibrium process, digestion or slow precipitation will not decrease the concentration of inclusions. A fast non-equilibrium precipitation, on the other hand, may

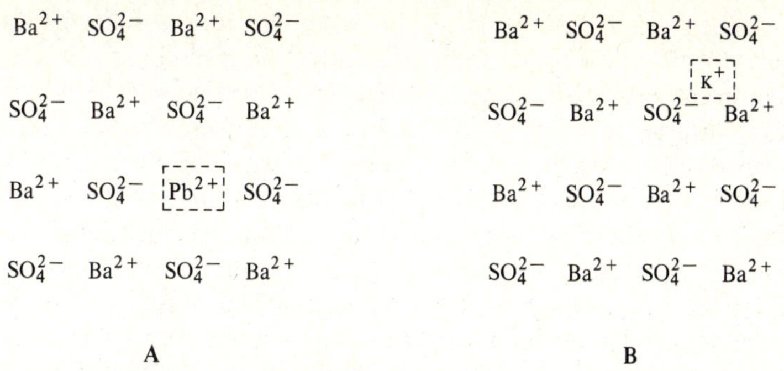

Ba^{2+} SO_4^{2-} Ba^{2+} SO_4^{2-} Ba^{2+} SO_4^{2-} Ba^{2+} SO_4^{2-}

SO_4^{2-} Ba^{2+} SO_4^{2-} Ba^{2+} SO_4^{2-} Ba^{2+} SO_4^{2-} $\boxed{K^+}$ Ba^{2+}

Ba^{2+} SO_4^{2-} $\boxed{Pb^{2+}}$ SO_4^{2-} Ba^{2+} SO_4^{2-} Ba^{2+} SO_4^{2-}

SO_4^{2-} Ba^{2+} SO_4^{2-} Ba^{2+} SO_4^{2-} Ba^{2+} SO_4^{2-} Ba^{2+}

A   B

FIGURE 5-6 Inclusion type of coprecipitation:
(A) Solid solution formed by interference (Pb^{2+}) substituting for lattice ion (Ba^{2+}).
(B) Solid solution formed by interference (K^+) filling an interstitial site in lattice.

result in fewer inclusions. Inclusions occur throughout the crystal, not only on the surface; therefore, changes in particle size will not affect the extent of inclusions. Fortunately, there are few cases of analytical importance in which inclusions play a role; but when they do, the errors can be serious. In the case of surface adsorption, the number of adsorption sites is relatively small compared to the number of ions in the crystallite (unless the particle size is extremely small); in the case of inclusions, all lattice or interstitial sites are available for the interfering ion.

Washing a precipitate containing inclusions will not help because—unlike the interferences involved in surface adsorption—they are essentially all within the bulk crystallite and not exposed to the surface. In fact, with the exception of reprecipitation, the analyst can do little to decrease coprecipitation occurring via inclusions. Equation 5-15 shows that the amount of inclusion (in this case, lead ion on a barium site) is directly proportional to the lead ion concentration in solution. Normally, only a small percentage of the interfering ion will coprecipitate. Therefore, dissolving the precipitate in a solution containing no interfering ion will result in a much lower concentration of this ion than was present in the original solution. A second precipitation from the second solution will lead to an even purer precipitate.

Analytical chemists, always on the lookout for useful phenomena, have employed coprecipitation to handle trace quantities of materials. Compounds in concentrations of 10^{-10} to 10^{-15} F are usually too soluble to be quantitatively precipitated. However, if a chemically similar compound is precipitated, the trace material may be coprecipitated via inclusions. This not only gives the analytical chemist a "handle" for trace materials; it provides information regarding the chemical properties of the trace material. For example, Madame Curie was able to coprecipitate trace amounts of $RaSO_4$ with $BaSO_4$, thereby learning that the new element was in the same chemical family as barium. In more recent years, the fact that low concentrations of radioactive tracers can be handled by coprecipitation techniques has been of great use to chemists studying the actinide elements.

Occlusions The third coprecipitation mechanism, *occlusions*, are "trapped" materials. When precipitates form quickly, equilibrium conditions cannot be maintained; droplets of solution may lose contact with the bulk solution because they are completely surrounded by the precipitate (see Figure 5-7). It may even be difficult to volatilize the trapped solvent in the occlusion, which would lead to a serious positive error. Occlusions form under non-equilibrium conditions. Therefore, they can be minimized by conducting the precipitation under near-equilibrium conditions. Two excellent methods to prevent or eliminate occlusions are: (1) slow precipitation by slow addition of dilute reagent to a dilute solution of the analyte and (2) digestion, during which time trapped material may eventually be exposed to and return to the supernatant solution. Washing a precipitate containing occlusions will not eliminate them, because the occlusion is not in contact with the surface.

Table 5-5 summarizes the three mechanisms of coprecipitation and the methods for decreasing their effect on a gravimetric determination. The extent of coprecipitation and, indeed, even the ability to separate the precipitate from the supernatant solution (filterability) depends on the nature of precipitate. If the precipitate is colloidal or near colloidal, it will pass through the filter and be lost just as surely as it

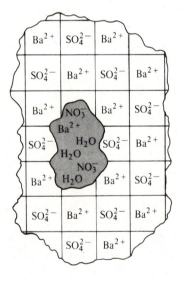

FIGURE 5-7 Occlusion type of coprecipitation:
Material (Ba^{2+}, NO_3^-, H_2O) is trapped inside a precipitating particle ($BaSO_4$) when precipitate growth too rapid to maintain equilibrium conditions.

TABLE 5-5 Coprecipitation and Methods for Prevention and Elimination

Coprecipitation mechanism	Helpful procedures	Procedures having little effect

surface adsorption	digestion
	use of dilute solutions
	precipitation at high
	    temperatures
	washing
	conversion to volatile
	    form
	reprecipitation	

inclusion	reprecipitation	digestion washing use of dilute solutions use of high temperatures slow addition of reagents
occlusion	digestion use of dilute solutions slow addition of reagents	washing

---

would if it were soluble. Large particles are preferred because (1) they are easier to filter, (2) they have a lower surface/volume ratio to decrease surface adsorption, and (3) they form when precipitation is slow—near equilibrium conditions that prevent the formation of occlusions.

What determines the rate of precipitation? What factors determine whether small or large particles will be formed? Both questions can be answered using a concept of *relative supersaturation*. The relative supersaturation is the ratio of excess concentration of solute ($Q - S$ where Q is solute concentration) at any moment during precipitation to the equilibrium solubility (S):

$$\text{relative supersaturation} = \frac{(Q - S)}{S} \qquad (5\text{-}16)$$

When the solution becomes supersaturated, a precipitating particle may start to form; this initial formation is called *nucleation*. Then, precipitation may continue as new nuclei are formed or the existing one grows. The rate of nucleation depends on relative supersaturation to some power, but growth of a particle is proportional (first power) to relative supersaturation. Therefore, when relative supersaturation is high (either Q greatly exceeds S or S is very small), nucleation is favored and many small particles are formed; when relative supersaturation is low, crystallite growth is favored and fewer, but larger particles result. Although the analyst may have some control over Q, little control can be exerted over S. Consequently, some precipitates with very low K_{sp} values—such as $Fe(OH)_3$—will always form extremely small particles.

One method to carry out precipitation under conditions of low relative supersaturation is to use *precipitation from homogeneous solution*. In this technique, the precipitating reagent is generated by a slow chemical reaction; therefore, Q is never significantly larger than S. This method leads to larger crystallites and purer precipitates. For example, OH^- can be generated using the hydrolysis of urea:

$$(H_2N)_2CO + 3H_2O \rightleftharpoons CO_2 + 2NH_4^+ + 2OH^- \qquad \textbf{(5-17)}$$

The reaction occurs so slowly that, even near boiling, a few hours may be required to generate sufficient OH^- for a quantitative precipitation. Other examples will be mentioned in Chapter 6.

CONVERT THE PRECIPITATE TO A RELIABLE WEIGHING FORM

So far we have (1) found a compound sufficiently insoluble to allow quantitative precipitation; (2) eliminated (if necessary) any other substance whose solubility products would be exceeded; and (3) found methods to prevent or eliminate coprecipitation so that our precipitate is pure. However, it will still be wet with solvent (wash solution) and must be dried before a measurement of mass will be meaningful. Drying the precipitate may seem straightforward, but problems can arise. For example, the precipitate may be nonstoichiometric or may start to decompose when dried at an elevated temperature. Even a very simple compound such as AgCl faces the problem of photodecomposition—that is, if drying is carried out in the presence of sunlight, the following reaction occurs.

$$AgCl \xrightarrow{h\nu} Ag^0 + \tfrac{1}{2}Cl_2 \qquad \textbf{(5-18)}$$

Some of the chlorine is lost and a negative error will result. Too little drying will result in a positive error because not all solvent has been removed; too prolonged drying will result in a negative error. However, if the AgCl precipitate is handled without exposure to any ultraviolet radiation, it can be dried to "constant weight," indicating the complete removal of solvent.

The analyte need not be weighed in the form in which it was precipitated. Compounds that tend to decompose during drying can sometimes be converted to some more stable compound for weighing. For example, $Fe(OH)_3 \cdot xH_2O$ is a useful form for the determination of iron with an extremely low solubility product. However, the water of hydration cannot be removed without some decomposition of $Fe(OH)_3$ occurring to the oxide form, Fe_2O_3. The precipitate is, therefore, converted to the stable form Fe_2O_3 by ignition at a high temperature (1000°C). The analyst must carefully control the ignition temperature, because at temperatures below 1000°C some water may be retained but at temperatures above 1200°C the Fe_2O_3 starts to be reduced to Fe_3O_4.

Another example of conversion for weighing purposes is encountered with the precipitation of $MgNH_4PO_4$ used in the determination of phosphorus (or magnesium). Drying leads to some loss of NH_3, and ignition at 1100°C is generally used to convert $MgNH_4PO_4$ to the reliable weighing form, $Mg_2P_2O_7$:

$$2(MgNH_4PO_4 \cdot 6H_2O) \xrightleftharpoons{1100°C} Mg_2P_2O_7 + 2NH_3 + 13H_2O \qquad (5\text{-}19)$$

After a reliable weighing form is obtained, the weight constitutes the measurement signal, S; from it, the amount of analyte, A, can be calculated:

$$S = kA$$

$$A = \frac{S}{k} = k^{-1}S$$

The value of k^{-1} is the ratio of formula weight FW (A) to FW (S), assuming each contains the same number of atoms of the analyte element common to A and S. When this assumption is not valid, we can use the concept of equivalents by defining the common element being converted from A to S as our reactive species. For any

---

EXAMPLE 5-8

Calculate the percentage of bromide in a sample weighing 354 mg that yields a dried precipitate of AgBr weighing 187 mg.

Solution

The gravimetric factor can be calculated using Equation 5-20, and from this equation the amount of analyte, Br, can then be calculated.

$$mg\ Br = \frac{FW\ (Br)/n_A}{FW\ (AgBr)/n_S} \times mg\ AgBr$$

$$n_A = \text{number of Br atoms in analyte, Br} = 1$$

$$n_S = \text{number of Br atoms in measurement signal form, AgBr} = 1$$

$$FW\ (Br) = 79.9$$

$$FW\ (AgBr) = 187.8$$

$$mg\ Br = \frac{79.9/1}{187.8/1} \times 187 = 0.4255 \times 187 = 79.6\ mg$$

$$\%\ Br = \frac{79.6}{354} \times 100\ (\%) = 22.5\%$$

Alternatively, Table 5-6 lists the gravimetric factor as 0.4255. Therefore,

$$mg\ Br = G_f \times mg\ AgBr = 0.4255 \times 187 = 79.6\ mg$$

species, the number of atoms of the common element it contains will be n (eq/mol):

$$\text{eq } A = \text{eq } S \quad \text{or in milliquantities:} \quad \text{meq } A = \text{meq } S$$

$$\text{FW} \frac{\text{mg } A}{(A)/n_A} = \frac{\text{mg } S}{\text{FW}(S)/n_S}$$

$$\text{mg } A = \frac{\text{FW}(A)/n_A}{\text{FW}(S)/n_S} \times \text{mg } S \tag{5-20}$$

Accordingly, the term $[\text{FW}(A)/n_A]/[\text{FW}(S)/n_S]$ is k^{-1} and is called the *gravimetric factor*. Examples 5-8 and 5-9 show some typical calculations using gravimetric factors. However, if a table of gravimetric factors (such as Table 5-6) is not available, Equation 5-20 can be used just as easily to calculate the amount of analyte A present from measurement signal S. Finally, the required percentage of A is readily calculated from mg A:

$$\% \, A = \frac{\text{mg } A}{\text{mg sample}} \times 100 \, (\%) \tag{5-21}$$

EXAMPLE 5-9

After a sample weighing 865 mg is dissolved, $Fe(OH)_3 \cdot xH_2O$ is precipitated and ignited to yield 262 mg Fe_2O_3. Calculate the percentage of Fe_3O_4 in the ore sample.

Solution

The gravimetric factor can be calculated using Equation 5-20. Then the amount of analyte Fe_3O_4 can be calculated.

$$\text{mg } Fe_3O_4 = \frac{\text{FW}(Fe_3O_4)/n_A}{\text{FW}(Fe_2O_3)/n_S} \times \text{mg } Fe_2O_3$$

$$n_A = \text{number of Fe atoms in analyte, } Fe_3O_4 = 3$$

$$n_S = \text{number of Fe atoms in measurement signal form, } Fe_2O_3 = 2$$

$$\text{FW}(Fe_3O_4) = 231.5$$

$$\text{FW}(Fe_2O_3) = 159.7$$

$$\text{mg } Fe_3O_4 = \frac{231.5/3}{159.7/2} \times 262 = 0.966 \times 262 = 253 \text{ mg}$$

$$\% \, Fe_3O_4 = \frac{253}{865} \times 100 \, (\%) = 29.2\%$$

Alternatively, Table 5-6 lists the gravimetric factor as 0.966. Therefore,

$$\text{mg } Fe_3O_4 = G_f \times \text{mg } Fe_2O_3 = 0.966 \times 262 = 253 \text{ mg}$$

Occasionally (more often on examinations than in the laboratory), two analytes are weighed together. Because this results in one equation with two unknowns, additional information is necessary. If it is known that the two analytes constitute the entire sample, the sample weight itself constitutes a second measurement signal and the two simultaneous equations can be used to solve for the amount of each analyte, as shown in Example 3-12. If this assumption is not justified, another measurement must be made, as shown in Example 5-10.

TABLE 5-6 Gravimetric Factors, G_f

Sought, A	Weighed, S	Gravimetric factor $= G_f$
Cl	AgCl	$\dfrac{\text{FW (Cl)}/1}{\text{FW (AgCl)}/1} = 0.2474$
Br	AgBr	$\dfrac{\text{FW (Br)}/1}{\text{FW (AgBr)}/1} = 0.4255$
I	AgI	$\dfrac{\text{FW (I)}/1}{\text{FW (AgI)}/1} = 0.5405$
Fe	Fe_2O_3	$\dfrac{\text{FW (Fe)}/1}{\text{FW (Fe}_2\text{O}_3)/2} = 0.6994$
Fe_3O_4	Fe_2O_3	$\dfrac{\text{FW (Fe}_3\text{O}_4)/3}{\text{FW (Fe}_2\text{O}_3)/2} = 0.966$
P	$Mg_2P_2O_7$	$\dfrac{\text{FW (P)}/1}{\text{FW (Mg}_2\text{P}_2\text{O}_7)/2} = 0.2783$
P_2O_5	$Mg_2P_2O_7$	$\dfrac{\text{FW (P}_2\text{O}_5)/2}{\text{FW (Mg}_2\text{P}_2\text{O}_7)/2} = 0.6378$
$Na_5P_3O_{10}$	$Mg_2P_2O_7$	$\dfrac{\text{FW (Na}_5\text{P}_3\text{O}_{10})/3}{\text{FW (Mg}_2\text{P}_2\text{O}_7)/2} = 1.102$
S	$BaSO_4$	$\dfrac{\text{FW (S)}/1}{\text{FW (BaSO}_4)/1} = 0.1374$
SO_2	$BaSO_4$	$\dfrac{\text{FW (SO}_2)/1}{\text{FW (BaSO}_4)/1} = 0.2745$
Ba	$BaSO_4$	$\dfrac{\text{FW (Ba)}/1}{\text{FW (BaSO}_4)/1} = 0.5884$
Ni	$Ni(C_4H_7N_2O_2)_2$	$\dfrac{\text{FW (Ni)}/1}{\text{FW (Ni(C}_4\text{H}_7\text{N}_2\text{O}_2)_2)/1} = 0.2032$

EXAMPLE 5-10

An alloy sample weighing 570 mg contains aluminum, magnesium, and other metals. After the sample has been dissolved and treated to prevent interference by the other metals, the aluminum and magnesium are precipitated with 8-hydroxyquinoline. After filtration and drying, the mixture of $Al(C_9H_6NO)_3$ and $Mg(C_9H_6NO)_2$ weighs 961 mg. The mixture of dried precipitates is then ignited to convert them to Al_2O_3 and MgO, which weighs 117 mg. Calculate the percentages of Al and Mg in the alloy.

Solution

Because there are two analytes, Al and Mg, two measurement signals are needed:

$$\text{mg } Al(C_9H_6NO)_3 + \text{mg } Mg(C_9H_6NO)_2 = 961 \text{ mg} \tag{5-22}$$

and

$$\text{mg } Al_2O_3 + \text{mg MgO} = 117 \text{ mg} \tag{5-23}$$

There are still four unknowns, but the equivalents of aluminum and magnesium are conserved in the ignition step. Therefore,

$$\text{meq Al} = \frac{\text{mg } Al(C_9H_6NO)_3}{\text{FW }(Al(C_9H_6NO)_3)/1} = \frac{\text{mg } Al_2O_3}{\text{FW }(Al_2O_3)/2} \tag{5-24}$$

$$\text{meq Mg} = \frac{\text{mg } Mg(C_9H_6NO)_2}{\text{FW }(Mg(C_9H_6NO)_2)/1} = \frac{\text{mg MgO}}{\text{FW (MgO)}/1} \tag{5-25}$$

With these four equations, the weight of each product can be obtained. Algebraic substitution would be possible, but tedious. The simple method suggested in Appendix 1-B operates on all four simultaneous equations. We can then construct a table with each of the four unknowns in a column on the left side and the constant terms on the right side:

Unknown	mg $Al(C_9H_6NO)_3$	mg Al_2O_3	mg $Mg(C_9H_6NO)_2$	mg MgO		Constant
Eq. 5-22	1	0	1	0	=	961
Eq. 5-23	0	1	0	1	=	117
Eq. 5-24	0.00218	−0.0196	0	0	=	0
Eq. 5-25	0	0	0.00321	−0.0248	=	0

EXAMPLE 5-10 (cont.)

A desk-top calculator gives the answers to this 4 × 4 matrix:

$$Al(C_9H_6NO)_3 = 406 \text{ mg}$$
$$Al_2O_3 = 45.1 \text{ mg}$$
$$Mg(C_9H_6NO)_2 = 555 \text{ mg}$$
$$MgO = 71.9 \text{ mg}$$

From the weight of $Al(C_9H_6NO)_3$ or Al_2O_3, we obtain the amount of Al present:

$$\frac{\text{mg Al}}{\text{FW (Al)}/1} = \frac{45.1}{102/2}$$

$$\text{mg Al} = 0.884 \times 27.0 = 23.9 \text{ mg}$$

$$\% \text{ Al} = \frac{23.9}{570} \times 100 \ (\%) = 4.19\%$$

In the same way, the percentage of Mg is obtained:

$$\frac{\text{mg Mg}}{\text{FW (Mg)}/1} = \frac{71.9}{40.3/1}$$

$$\text{mg Mg} = 1.784 \times 24.3 = 43.4 \text{ mg}$$

$$\% \text{ Mg} = \frac{43.4}{570} \times 100 \ (\%) = 7.61\%$$

5-3 SUMMARY OF THE GRAVIMETRIC METHOD

Converting an analyte to a known composition and weighing it can be one of the most accurate analytical methods available to the analytical chemist. The measurement signal (S) is directly proportional to the analyte (A) and the proportionality constant (k) is known accurately from stoichiometric considerations (the gravimetric factor $= k^{-1}$). The only exceptions to this straightforward scheme are non-stoichiometric compounds, which are used only occasionally. Only the weight of sample and the weight of the final weighing form are measured. The analytical balance is one of the most precise and accurate tools in the laboratory and even taking into consideration the operation of weighing by difference, weighing errors will be less than 0.1%. The typical analytical balance is sensitive to ± 0.1 mg; therefore, a minimum weight of 100 mg for samples and precipitates is required. However, for smaller weights, microbalances are available.

Although the accuracy of gravimetric procedures is inherently high (typically 0.1 to 0.2%), the actual accuracy achieved for a particular gravimetric procedure will depend on factors such as the solubility product involved, the presence of interferences, the extent of coprecipitation, and any uncertainty in the composition of the final weighing form. Some sources of determinate error in a gravimetric procedure are listed in Table 5-7. Note that most of the sources lead to negative errors. An analyst must take the utmost care in each step to avoid accumulating a series of fairly small errors that will produce a significantly low result. Situations in which interferences may be present or severe coprecipitation may occur will lead to unrealistically high results.

TABLE 5-7   Sources of Determinate Error in Typical Gravimetric Procedures

Procedure or phenomenon	Direction of error
1. incomplete drying of sample	−
2. incomplete transfer of sample	−
3. solubility due to K_{sp} and complexes formed	−
4. insufficient amount of precipitant added	−*
5. interfering substance present that also precipitates	+
6. incomplete transfer of precipitate	−
7. coprecipitation	+†
8. washing (too little → too much)	+ → −
9. peptization	−
10. incomplete drying of precipitate	+
11. photodecomposition (silver halides)	−

*An insufficient quantity of precipitant can produce a very large negative error and *must* be avoided to obtain acceptable results.
† Coprecipitation normally involves an extraneous ion that will cause a positive error, but it could involve occlusions of analyte that will cause a negative error (if the equivalent weight of the occluded form is less than the equivalent weight of the weighing form).

Gravimetric procedures are not highly selective. If a metal ion analyte is precipitated with an anion such as OH^-, S^{2-}, or $C_2O_4^{2-}$, many other metal ions are also precipitated with this anion. However, some selectivity can be gained through the control of the pH and/or the concentration of precipitating reagent. Due to the large amounts of other constituents that will become possible interferences because they have solubility products that will be exceeded or because they will be coprecipitated, gravimetric procedures are not usually used for analytes present in trace quantities ($<0.1\%$).

The slowness of gravimetric procedures is usually considered a drawback of the technique. Many hours are spent digesting, filtering, and drying precipitates. Instrumental methods, some of which will be discussed later in this text, can produce a measurement signal within a few minutes. However, before ruling out a gravimetric procedure as the method of choice (step 2 in the total analysis process), two factors should be considered: First, much of the time involved is "clock" time, not "operator" time. The analyst can carry out other analyses while digestion, drying, and other processes are taking place. Second, the fact that the functional relationship between measurement signal and analyte amount is known exactly from theory means that the proportionality constant, k, need not be measured. Many instrumental methods require extensive calibration before the functional relationship (k) between measurement signal and analyte amount can be determined. If this calibration time can be prorated among numerous samples, calibration time *per sample* will be small. However, the gravimetric procedure may require less time per sample when only one or a small number of analyses are required.

SUPPLEMENTARY READING

Erdey, L., *Gravimetric Analysis*. Elmsford, N.Y.: Pergamon Press, Inc., 1965.

Ingram, G., and M. Lonsdale, "Organic Analysis," in *Treatise on Analytical Chemistry*, I.M. Kolthoff aand P.J. Elving, eds, pt. 2, vol. 2. New York: Interscience Publishers, Inc., 1965.

Salutsky, M.L., "Precipitates: Their Formation, Properties, and Purity," in *Treatise on Analytical Chemistry*, I.M. Kolthoff and P.J. Elving, eds., pt. 1, vol. 1. New York: Interscience Publishers, Inc., 1959.

PROBLEMS

5-1. Several water samples are analyzed for dissolved solids by weighing the residue after evaporation. Based on the following data, calculate the *total dissolved solids* value for each and determine which of the water samples do not meet the 500-ppm maximum for suitable water supplies.
  *a. 21.2 mg of residue from 40-mL sample
   b.   8.7 mg of residue from 80-mL sample
   c. 38.7 mg of residue from 80-mL sample
   d. 10.7 mg of residue from 20-mL sample

---

*Answers to problems marked with an asterisk will be found at the back of the book.

*5-2. An organic compound is purified and then sent for carbon–hydrogen analysis. If a 12.67-mg sample is combusted and yields 27.87 mg of CO_2 and 15.10 mg of H_2O, what are the percentages of C and H in the compound? Assuming the remaining percentage in the compound is oxygen, determine the empirical formula for the compound.

5-3. An organic compound is purified and then sent for carbon–hydrogen analysis. If an 18.19-mg sample is combusted and yields 56.49 mg of CO_2 and 25.01 mg of H_2O, what are the percentages of C and H in the compound? What conclusions can you draw concerning the structure of the compound?

5-4. Using the data in Table 5-2, calculate the solubility of AgCl (milligrams per liter) as a function of temperature. Plot the data graphically.

5-5. Using the data in Table 5-2, calculate the solubility of $BaSO_4$ (milligrams per liter) as a function of temperature. Plot the data graphically.

5-6. Using the data in Table 5-3, calculate the solubility of $PbSO_4$ (milligrams per liter) as a function of ethanol content in the solvent. Plot the data graphically, beginning with pure water solvent.

5-7. Calculate the solubility (milligrams per liter) of the following compounds in H_2O, based on their solubility products listed in Appendix 2-A:
- *a. $BaSO_4$
- b. PbI_2
- c. Hg_2Cl_2 (forms Hg_2^{2+} + $2Cl^-$)
- d. Ag_2CrO_4
- e.   TlCl

5-8. Calculate the solubility products for the following compounds in H_2O based on the solubility data given:
- *a.   AgBr (7.2×10^{-7} F)
- b.   AgSCN (0.17 mg/L)
- c. PbI_2 (1.2×10^{-3} F)
- d. $Fe(OH)_2$ (0.53 mg/L)
- e. $La(IO_3)_3$ (6.9×10^{-4} F)
- f. $MgNH_4PO_4$ (9.2 mg/L)

5-9. Calculate the solubility (in milligrams per liter) of $Cd(OH)_2$ in solutions with the following pH values:
- *a.   pH 11
- b.   pH 9
- c.   pH 7

*5-10. Lead(II) in a solution was precipitated as $PbSO_4$. The volume of solution was 150 mL and contained 0.01 F excess sulfate ion. The precipitate was filtered, washed with 20 mL of H_2O, and when dried weighed 450 mg. What was the percentage loss from solubility?

5-11. a.   The relatively low solubility of PbI_2 can be used to determine lead content gravimetrically. Determine the percentage error that would occur from solubility losses if precipitation were carried out in 300 mL of solution containing 0.01 F *excess* iodide and 50 mL of water was used to wash the precipitate. The final weight of PbI_2 was 750 mg.

   b.   What suggestions would you make to decrease this error?

5-12. Zinc(II) in a solution was precipitated as $Zn(OH)_2$. The volume of the solution was 500 mL and precipitation was carried out at pH 9.0. The precipitate was filtered, washed with 150 mL H_2O, ignited to form ZnO, and weighed 300 mg. What was the percentage loss?

5-13. What would be the percentage error in a sulfate analysis if the 300 mL of supernatant liquid contained 5×10^{-5} F Ba^{2+} after precipitation was complete and 120 mL of water was subsequently used to wash the precipitate, which weighed 140 mg after drying?

*5-14. Calculate the minimum excess iodide concentration that must be present to ensure quantitative precipitation of PbI_2, assuming an initial Pb^{2+} concentration of 0.02 F.

5-15. Calculate the minimum pH for quantitative precipitation of $Pb(OH)_2$, assuming an initial Pb^{2+} concentration of 0.001 F.

5-16. Because ionic solutions must contain cations as well as anions, individual activity coefficients cannot be measured directly. However, a mean activity coefficient, $f_\pm (f_\pm = \sqrt{f_+ f_-})$, can be calculated from activity data. Calculate $f_\pm$ for Ag^+ and Cl^- in 0.1 F $NaNO_3$ from the following solubility data for AgCl.

Solution	Solubility, F
water	1.35×10^{-5}
0.1 F $NaNO_3$	1.79×10^{-5}

*5-17. Compare the solubility of $Mn(OH)_2$ in water with its solubility in a solution of ionic strength 0.1.

5-18. What is the solubility loss if $BaSO_4$ is precipitated in 600 mL of a solution at ionic strength 0.1 and containing an excess sulfate ion *concentration* of 0.001 F?

*5-19. What is the solubility loss if $SrSO_4$ is precipitated in 200 mL of a solution at ionic strength 0.1 and containing an excess sulfate ion *concentration* of 0.01 F?

5-20. In Problem 5-19, what would be the calculated solubility loss if activity coefficients were ignored?

5-21. $CaSO_4$ is considerably more soluble than $BaSO_4$ and has a $K_{sp} \sim 10^4$ times greater. If $[SO_4^{2-}]$ could be controlled accurately, would it be possible to specify a $[SO_4^{2-}]$ at which $BaSO_4$ could be precipitated quantitatively in the presence of Ca^{2+}? Assume initial concentrations of 0.1 F for both $[Ca^{2+}]$ and $[Ba^{2+}]$.

*5-22. Determine the pH range over which Cu^{2+} can be quantitatively separated from Mn^{2+} by precipitation of $Cu(OH)_2$. Assume initial concentration of 0.01 F Cu^{2+} and 0.01 F Mn^{2+}.

5-23. Determine the pH range over which Zn^{2+} can be quantitatively separated from Mn^{2+} by precipitation of ZnS. Assume initial concentrations of 0.01 F Zn^{2+} and 0.01 F Mn^{2+} and that the solution is saturated with respect to H_2S (0.1 F).

5-24. Determine the pH range in which 0.01 F Cd^{2+} may be separated from 0.01 F Fe^{2+} by precipitation of the sulfide in a solution saturated with H_2S (0.1 F).

5-25. Assuming that magnesium ammonium phosphate ($MgNH_4PO_4$) has a solubility of 6×10^{-5} F in water, calculate the solubility in 1 F NH_4Cl. What assumptions or approximations need to be made?

*5-26. AgSCN is an insoluble compound frequently used in precipitation titrations (see Chapter 8). However, SCN^- also forms a complex with Ag^+, as follows:

$$Ag^+ + 2\,SCN^- \rightleftharpoons Ag(SCN)_2^- \qquad \beta_2 = 2.5 \times 10^8$$

Calculate the solubility (as milligrams per liter) of AgSCN in 0.1 F KSCN.

5-27. Calculate the solubility of MCl in 0.1 F HCl.

$$K_{sp} = 1 \times 10^{-9}$$
$$M^+ + 2Cl^- \rightleftharpoons MCl_2^- \qquad \beta_2 = 1 \times 10^7$$
$$M^+ + 3Cl^- \rightleftharpoons MCl_3^{2-} \qquad \beta_3 = 1 \times 10^5$$

5-28. The solubility of compounds is increased when complexation occurs. This solubility is greatly increased when the compound dissolves in solution containing large amounts of the complexing ion. However, even in water the solubility will be increased. For solutions containing very low concentrations of complexing ion, only the first complex (K_1) will be important. Based on K_{sp} (Appendix 2A) and K_1 as follows, calculate the solubility of $PbBr_2$ in H_2O and compare it with the solubility calculated if K_1 is neglected.

$$Pb^{2+} + Br^- \rightleftharpoons PbBr^+ \qquad K_1 = 45$$

5-29. The solubility of metal hydroxides may be limited by the OH^- formed by the dissociation of water. That is, for very slightly soluble hydroxides, the $[OH^-]$ primarily comes from water dissociation. For accounting purposes we can say:

$$[OH^-]_{total} = [OH^-]_{solubility} + [OH^-]_{H_2O}$$

For very insoluble hydroxides, $[OH^-]_{total} \cong [OH^-]_{H_2O} \cong 10^{-7} F$. For more soluble hydroxides, $[OH^-]_{total} \cong [OH^-]_{solubility}$. Calculate which term predominates for the following hydroxides and the solubility of each compound in H_2O.
  a. $Fe(OH)_3$
  *b. $Fe(OH)_2$
  c. $Zn(OH)_2$

*5-30. Lead halides are insoluble, similar to silver halides. Because of their greater solubility and the existence of complex ions, it is essential to use the correct amount of excess halide ion when analyzing for lead. Using K_{sp} and the following equilibrium constant, calculate the optimum bromide ion concentration for a lead analysis—that is, the concentration that gives the minimum solubility of $PbBr_2$.

$$PbBr_2 + Br^- \rightleftharpoons PbBr_3^- \qquad K = 3.6 \times 10^{-6}$$

5-31. a. Although lead chloride is reasonably insoluble, using the common ion effect is difficult because $PbCl_3^-$ forms in excess chloride. Using K_{sp} and the following constant, determine the optimum chloride concentration to minimize lead left in solution.

$$PbCl_2 + Cl^- \rightleftharpoons PbCl_3^- \qquad K = 4.2 \times 10^{-2}$$

  b. Using this chloride concentration, determine the amount of lead (in milligrams) left in 50 mL of solution.

5-32. Calculate gravimetric factors for the following analyses:

Sought	Weighed
*a. Al_2O_3	$Al(C_9H_6NO)_3$
b. $C_6H_6Cl_6$	$AgCl$
c. Al_2O_3	Al_2O_3
d. Pb_3O_4	$PbCrO_4$

5-33. Several weighing forms are available when calcium is precipitated as CaC_2O_4. Calculate the gravimetric factors for each if the percentage of calcium is sought. Show which one has the highest sensitivity—that is, the greatest precipitate weight for a given amount of calcium. Weighing forms:

$$CaC_2O_4, \quad CaO, \quad CaSO_4, \quad CaF_2$$

5-34. Comment on the suggestion that a precipitate should almost always be washed with a large volume of cold distilled water.

*5-35. A 563-mg sample containing only $CaCO_3$ and $SrCO_3$ is dissolved in acid; $Na_2C_2O_4$ is added to precipitate CaC_2O_4 and SrC_2O_4. The oxalate precipitate is filtered and dried to yield 703 mg of the anhydrous mixed oxalates. Calculate the percentages of $CaCO_3$ and $SrCO_3$ in the mixture.

5-36. An alloy sample weighing 154.2 mg contains iron, manganese, and other metals. After dissolving the sample and treating it to prevent interference by the other metals, the iron and manganese are precipitated with 8-hydroxyquinoline. After filtration and drying, the mixture of $Fe(C_9H_6NO)_3$ and $Mn(C_9H_6NO)_2$ weighs 983 mg. The precipitate is redissolved and the amount of 8-hydroxyquinoline is coulometrically determined to be 5.972 mmol. Calculate the percentages of Fe and Mn in the alloy sample.

Gravimetric Applications

6

The summary of the gravimetric method presented in Section 5-3 can be condensed as follows: "The gravimetric method is slow but may be highly accurate when used to analyze for major constituents in a sample." The development of fast, reasonably accurate instrumental methods has reduced the number of gravimetric applications. However, even when an instrumental method is available, the analyst may not have access to the instrument or may not wish to spend the time necessary for calibration when only one or two samples are to be analyzed. Therefore, older gravimetric methods are still used, even when a newer method is available.

For certain kinds of analysis, the inherent lack of specificity of gravimetric methods makes them particularly advantageous. In Section 5-1, the gravimetric methods for determination of dissolved solids in water and particulates in air were described. In that case, the gravimetric method was adopted because it does not differentiate on the basis of chemical properties, but merely yields the total weight of analyte in the sample. Similarly, in a clinical lab method for determining the total amount of protein, all of the protein is precipitated when acetone is added. Other analytical methods may measure only a particular component of the protein in the sample.

6-1 INORGANIC PRECIPITATING REAGENTS

Many classical gravimetric methods were developed for the analysis of metal ions by the precipitation of some insoluble metal halide (MX). These methods can be surveyed by considering the various X anions that have been commonly used. Following this survey, some other examples of gravimetric procedures are described.

HALIDES

Because most metal halides are soluble, the addition of halide ion to a sample in solution will selectively precipitate those few metal ions with halides of low solubility. Appendix 2-A lists most of them. With chloride as a precipitating reagent, this procedure is essentially selective for silver ion. Copper(I), lead(II), thallium(I), and mercury(I) have limited solubility as chlorides, but Tl(I), and Pb(II) may be kept in solution, particularly if precipitation is carried out at an elevated temperature. Interference from Cu(I) and Hg(I) can be prevented by oxidation to Cu(II) and Hg(II), which have soluble chlorides.

Adding HCl to a solution containing Ag^+ is a highly accurate method of precipitating silver ion, but the excess amount of HCl must be carefully controlled to avoid the formation of significant amounts of soluble chlorocomplexes:

$$AgCl + Cl^- \rightleftharpoons AgCl_2^-$$
$$AgCl_2^- + Cl^- \rightleftharpoons AgCl_3^{2-}$$

This problem was described in detail in Section 5-2 (see Figure 5-3). (The reverse procedure—the determination of chloride in a sample by the addition of $AgNO_3$ as a precipitating reagent—is also possible. A detailed procedure for this is given in the laboratory manual.)

AgCl starts to precipitate as soon as the precipitating reagent is added, but the particles are colloidal size. Because they may pass through the pores of the filter, the precipitate must be digested (heated to near the boiling point for a few minutes) and aged (left standing a few hours) to form a curdy coagulate. Because coprecipitation is usually negligible, normal washing of the precipitate results in a very pure material. However, pure H_2O should not be used for washing because, as was described in Section 5-2, the particles can decoagulate and be washed through the filter. This peptization process is avoided in the gravimetric chloride analysis by adding a little HNO_3 (a volatile electrolyte) to the wash water.

The AgCl precipitate, after being dried to remove water, serves as the reliable weighing form. However, AgCl, especially while it is still moist, must be protected from sunlight to prevent photodecomposition (Chapter 5). Even with these precautions, the final AgCl product will normally be tinged with purple (indicating the presence of neutral silver atoms), but the error introduced with a trace of surface photodecomposition is negligible.

The determination of bromide, iodide, or thiocyanate (SCN^-) can be accomplished gravimetrically in a process similar to that used for chloride by the addition of silver nitrate as a precipitating reagent.

Although solubility products for TlI, PbI_2, and CuI are small enough for quantitative precipitation, errors due to the formation of iodocomplexes can be severe unless the iodide concentration is carefully controlled. Several other methods for determining the presence of these metal ions are more convenient. The precipitation of mercury(I) as Hg_2Cl_2 also involves the problem of solubility losses due to the formation of chlorocomplexes; moreover, the precipitate is difficult to dry because Hg_2Cl_2 is quite volatile.

HYDROXIDES

Many metal ions are insoluble in basic solution because the metal hydroxide solubility product is low (see Appendix 2-A). The wide range of solubility product values makes it possible to separate metal ions by controlling the pH in a region where one metal hydroxide is insoluble while others are soluble (see Example 5-5).

The formation of a hydroxide precipitate is generally carried out by the addition of ammonia to an acidic solution until the desired pH level is reached. The determination of iron(III) is typical; in fact, as shown in Table 6-1, several other elements can be determined in this way. Because the solubility product for iron(III) hydroxide is 4×10^{-38}, normal precipitation conditions lead to high relative supersaturation. As discussed in Chapter 5, this results in very small particle size. The finely divided precipitate will readily clog the pores of filters, making filtering time inordinately long. Digestion increases filtering speed by increasing the particle size but does not make the procedure so simple and fast as the precipitation of AgCl. The high surface area strongly coprecipitates other constituents from the solution and holds water tenaciously. In fact, the only way to eliminate water is to heat the precipitate until the hydroxide decomposes to the oxide. Temperatures near $1000°C$ produce the simple oxides as the reliable weighing forms. It is probably more realistic to think of the precipitate as a hydrous oxide, $Fe_2O_3 \cdot xH_2O$, rather than a hydrated hydroxide, $Fe(OH)_3 \cdot xH_2O$, because there is no difference in the water molecules involved as implied by the chemical formula for hydrated hydroxide.

In addition to the physical problems associated with hydroxide precipitates, these gravimetric procedures are complicated by other considerations. In the

TABLE 6-1   Survey of Inorganic Precipitants for Inorganic Analytes

Type	Analyte	Precipitating Reagent
halides	Ag^+, Hg_2^{2+}	HCl
	Cl^-, Br^-, I^-, SCN^-	$AgNO_3$
	F^-	$NaCl + Pb(NO_3)_2$
hydroxides (hydrous oxides)	Fe^{3+}, Al^{3+}, Cr^{3+}, Sc^{3+}, Be^{2+}, Ga^{3+}, Zr^{4+}	NH_3
	Rh^{4+}, Ru^{4+}, Os^{4+}, Ir^{4+}	$NaHCO_3$
	Sn^{4+}	HNO_3
sulfides	Cu^{2+}, Zn^{2+}, Bi^{3+}	H_2S
	Hg^{2+}, Co^{2+}	$(NH_4)_2S$
sulfates	Ba^{2+}, Pb^{2+}, Sr^{2+}	H_2SO_4
	SO_4^{2-}	$BaCl_2$
oxalates	Ca^{2+}, Sr^{2+}, Th^{4+}	$H_2C_2O_4$
phosphates	Mg^{2+}, Mn^{2+}, Zn^{2+}, Zr^{4+}	$(NH_4)_2 HPO_4$
	PO_4^{3-}	$MgCl_2 + NH_4Cl$

determination of iron, any iron(II) must be oxidized to iron(III) before precipitation. This is accomplished through the addition of nitric acid before the addition of ammonia. Oxidation proceeds according to Equation 6-1:

$$3Fe^{2+} + NO_3^- + 4H_3O^+ \rightleftharpoons 3Fe^{3+} + NO + 6H_2O \qquad \text{(6-1)}$$

The hydrous oxide must be ignited at a temperature high enough to eliminate all water ($\sim 1000°C$), but not so high that reduction to Fe_3O_4 occurs:

$$6Fe_2O_3 \overset{\Delta}{\rightleftharpoons} 4Fe_3O_4 + O_2 \qquad \text{(6-2)}$$

As can be seen from Equation 6-2, reduction will introduce a negative error. Other compounds, of course, may require different ignition temperatures; Al_2O_3, for example, is so hygroscopic that a minimum ignition temperature of 1200°C is required to prevent a positive determinate error.

Because so many metal ions form insoluble hydroxides, it is difficult to ensure the absence of interferences. In addition to aluminum, no chromium(III), titanium, or zirconium may be present when iron is the analyte. Also, anions that form insoluble iron salts such as arsenate, phosphate, silicate, and vanadate must be absent. Silica may contaminate the iron oxide precipitate and may originate in the sample or from the glassware used in the procedure. All these interferences will introduce a positive determinate error. Any substances that form complexes with iron and keep it in solution will act as interferences leading to negative determinate errors. Examples include citrate, tartrate, and fluoride ions.

Even divalent metal ions with solubility products that are large enough to allow quantitative precipitation of iron(III), chromium(III), and aluminum(III)—that is, which are not exceeded at pH 3.5 to 4.0—can introduce significant positive determinate errors because they will coprecipitate when present in large concentrations. Under such conditions, the most effective method for quantitative separation of these ions is double precipitation.

Another way to increase particle size and decrease coprecipitation is to employ precipitation from a homogeneous solution, as mentioned in Chapter 5. Hydroxide ions can be generated slowly in solution by the hydrolysis of urea in boiling solution, according to Equation 5-17. This technique is employed in the laboratory manual. Precipitation from homogeneous solution (frequently called *homogeneous precipitation*) maintains conditions of low supersaturation and may enable the analyst to achieve satisfactory results with a single precipitation, whereas a double precipitation may be necessary if the conventional procedure is followed. Precipitation from homogeneous solution can be used with several other gravimetric methods, many of which are summarized in Table 6-2.

SULFIDES

Metal sulfides form another class of insoluble compounds that are occasionally used for gravimetric procedures. They are precipitated through the addition of hydrogen

TABLE 6-2 Precipitations from Homogeneous Solution

Precipitant generated	Reagent	Generation reaction*	Analytes
OH^-	urea	$(NH_2)_2CO + 3H_2O \rightleftharpoons CO_2 + 2NH_4^+ + 2OH^-$	Al, Fe(III), Ga, Sn, Th, Zn
S^{2-}	thioacetamide	$CH_3CSNH_2 + H_2O \rightleftharpoons CH_3CONH_2 + H_2S$	As, Bi, Cd, Cu, Hg, Mn, Mo, Pb, Sb, Sn
SO_4^{2-}	dimethyl sulfate	$(CH_3O)_2SO_2 + 4H_2O \rightleftharpoons 2CH_3OH + SO_4^{2-} + 2H_3O^+$	Ba, Ca, Pb, Sr
$C_2O_4^{2-}$	dimethyl oxalate	$(CH_3)_2C_2O_4 + 2H_2O \rightleftharpoons 2CH_3OH + H_2C_2O_4$	Ac, Am, Ca, rare earths, Th
PO_4^{3-}	triethyl phosphate	$(C_2H_5O)_3PO + 3H_2O \rightleftharpoons 3C_2H_5OH + H_3PO_4$	Hf, Zr
CO_3^{2-}	trichloroacetic acid	$Cl_3CCOOH + 2OH^- \rightleftharpoons CHCl_3 + CO_3^{2-} + H_2O$	Ba, Ra, rare earths
8-hydroxyquinoline	8-acetoxyquinoline	$C_{11}H_9NO_2 + H_2O \rightleftharpoons C_2H_4O_2 + C_9H_7NO$	Al, Mg, U, Zn

*The precipitant is indicated by boldface type.

sulfide (H_2S) to the solution or from homogeneous solution by the hydrolysis of thioacetamide (C_2H_5NS):

$$CH_3-\overset{\overset{\textstyle S}{\|}}{C}-NH_2 + H_2O \rightleftharpoons CH_3-\overset{\overset{\textstyle O}{\|}}{C}-NH_2 + H_2S$$

Separations of metal ions are possible by control of pH because sulfide is the anion of the weak acid, hydrogen sulfide. A typical calculation was given in Example 5-6. Sulfides have been used more for separation purposes (qualitative analysis) than as a weighing form in quantitative analysis. It is difficult to obtain pure, stoichiometric sulfides and difficult to dry them without introducing decomposition. In some instances, ignition to the oxide can provide a reliable weighing form. The determination of zinc is one example:

$$\text{precipitation:} \quad Zn^{2+} + S^{2-} \rightleftharpoons ZnS(s)$$

$$\text{ignition:} \quad ZnS + \frac{3}{2}O_2 \overset{\Delta}{\rightleftharpoons} ZnO + SO_2$$

Other problems involving sulfide precipitates can frustrate the analytical chemist or can be used to advantage. Equilibrium calculations show that some metal ions such as zinc should precipitate readily from acidic solution (pH 1). However, the sulfide ion concentration will be only $\sim 10^{-19}$ M, and the kinetics for the precipitation are so slow that zinc ion will remain in a supersaturated condition for a considerable time. This slow precipitation rate prevents precipitation of ZnS from strongly acidic solution but allows other metal ions—for example, Hg(II) or Cu(II)—to be precipitated as the sulfide in the presence of zinc. However, zinc sulfide exhibits a phenomenon known as *postprecipitation*. If another sulfide precipitate is in contact with a supersaturated solution of zinc sulfide, the ZnS will start to precipitate out on the other sulfide. This is not coprecipitation because it does not occur at the time of precipitation; if the precipitate is filtered rapidly, the other sulfide will be relatively free of ZnS.

SULFATES

Appendix 2-A shows that several metal sulfates have solubility products low enough to be useful for gravimetric procedures. The determination of barium or, conversely, the determination of sulfate by the addition of $BaCl_2$ is typical.

The most serious problem with precipitates of barium sulfate is that they may coprecipitate many ions which may be present. In fact, coprecipitation of the $BaCl_2$ precipitating reagent (sulfate analysis) itself may lead to a positive determinate error of several tenths of one percent. The coprecipitation type can be primarily inclusions which, as seen from Table 5-5, can only be combated with double precipitation techniques. However, because $BaSO_4$ is only soluble in hot concentrated H_2SO_4, a reprecipitation procedure is not very convenient. Moreover, unless precautions are taken to avoid high supersaturation, occlusions will cause large positive errors.

BaSO$_4$ is normally precipitated from a weakly acidic solution (0.05 F HCl) because the coarse-grained precipitates formed under acidic conditions are easier to filter than the finer precipitates formed from neutral solution. Note that the second dissociation of H$_2$SO$_4$ is not strong (that is, it is not completely dissociated) and has a dissociation constant of 1.2×10^{-2}. Consequently, the addition of strong acid will increase the solubility of BaSO$_4$ according to Equation 6-3

$$BaSO_4 + H_3O^+ \rightleftharpoons HSO_4^- + Ba^{2+} + H_2O \qquad \text{(6-3)}$$

Example 6-1 illustrates this point.

One way to improve filterability and, at the same time, decrease coprecipitation (at least occlusions and surface adsorption) is to precipitate BaSO$_4$ from homogeneous solution. As Table 6-2 indicates, SO$_4^{2-}$ can be generated by the hydrolysis of dimethyl sulfate. This method is useful for the determination of barium, calcium, strontium, and lead. However, the analysis of sulfate involves the problems inherent to the traditional method—that is, the addition of dilute BaCl$_2$ to a slightly acidic dilute solution containing sulfate analyte near the boiling point. Results are usually in error by several tenths of one percent, although occasionally compensation of positive and negative determinate errors yields highly accurate results.

Other members of the alkaline earth family form insoluble sulfates and may be determined gravimetrically. Because the solubility product for SrSO$_4$, 3.2×10^{-7}, is quite high for quantitative precipitation, a mixed solvent of alcohol and water must be used to decrease the solubility to an acceptable value. Because there are several other methods for determining calcium, precipitation of CaSO$_4$ ($K_{sp} = 1.2 \times 10^{-6}$) is rarely used.

OXALATES

In the previous section, we noted that calcium was rarely determined using the precipitation of CaSO$_4$ because the solubility product involved is so large. An alternative gravimetric method for the determination of calcium is to use the precipitation of calcium oxalate (CaC$_2$O$_4$), which has a solubility product of 2.3×10^{-9}, with oxalic acid (H$_2$C$_2$O$_4$). Oxalic acid is usually considered to be an organic compound, but the oxalate salts are ionic and similar to carbonate salts, which are generally classified as inorganic compounds. Therefore, oxalates will be described here.

Many samples containing calcium also contain magnesium, and the solubility product of magnesium oxalate is small enough to be exceeded under typical conditions. As mentioned in Chapter 5, equilibrium calculations do not tell the whole story; the kinetics are such that calcium oxalate can be quantitatively precipitated around pH 5 in the presence of magnesium ion without significant interference, provided filtration is carried out rapidly. In other words, magnesium oxalate (MgC$_2$O$_4$) will remain in a supersaturated state long enough to remove the calcium oxalate. However, when high concentrations of magnesium are present,

EXAMPLE 6-1

Calculate the solubility of $BaSO_4$ in 100 mL of 0.2 F HCl, and compare this with the solubility in 100 mL of a neutral solution.

Solution

In neutral solution, the solubility is given simply by

$$BaSO_4(s) \rightleftharpoons Ba^{2+} + SO_4^{2-}$$

$$K_{sp} = 1.3 \times 10^{-10} = [Ba^{2+}][SO_4^{2-}]$$

$$[Ba^{2+}] = [SO_4^{2-}] = solubility$$

Therefore,

$$solubility = \sqrt{1.3 \times 10^{-10}} = 1.1 \times 10^{-5} \ F \ (mmol/mL)$$

$$1.1 \times 10^{-5} \frac{mmol}{mL} \times 233 \frac{mg}{mmol} \times 100 \ mL = 0.26 \ mg \ in \ 100 \ mL$$

In acidic solution, the additional equilibrium involving dissociation of HSO_4^- is important:

$$HSO_4^- + H_2O \rightleftharpoons H_3O^+ + SO_4^{2-}$$

$$K_2 = 1.2 \times 10^{-2} = \frac{[H_3O^+][SO_4^{2-}]}{[HSO_4^-]}$$

MgC_2O_4 may start to precipitate or coprecipitate on the CaC_2O_4, thereby necessitating a double or even triple precipitation procedure.

The use of CaC_2O_4 also complicates the process of producing a reliable weighing form. The compound precipitated contains one *water of hydration*, $CaC_2O_4 \cdot H_2O$, but it is difficult, if not impossible, to remove adsorbed and occluded water at 100 to 120°C without losing some water of hydration to produce the anhydrous salt, CaC_2O_4. Therefore, it is usually recommended that the oxalate be converted to another form for weighing. Several options are available. Heating to 400 to 500°C converts calcium oxalate to calcium carbonate ($CaCO_3$) according to Equation 6-4. The $CaCO_3$ may then be used as the weighing form.

$$CaC_2O_4 \cdot H_2O \rightleftharpoons CaCO_3 + H_2O + CO \qquad (6-4)$$

At higher temperatures, the carbonate begins to decompose. If ignition is carried out

Example 6-1 (cont.)

In 0.2 F HCl, $[H_3O^+] = 0.2$. Therefore,

$$\frac{[SO_4^{2-}]}{[HSO_4^-]} = \frac{1.2 \times 10^{-2}}{0.2} = 0.06$$

When $BaSO_4$ dissolves the sulfate must exist as SO_4^{2-} or as HSO_4^-. Therefore,

$$\text{solubility} = [Ba^{2+}] = [SO_4^{2-}] + [HSO_4^-] = [SO_4^{2-}] + \frac{[SO_4^{2-}]}{0.06}$$

Returning to the solubility product, we conclude:

$$1.3 \times 10^{-10} = [Ba^{2+}][SO_4^{2-}]$$

$$1.3 \times 10^{-10} = \left([SO_4^{2-}] + \frac{[SO_4^{2-}]}{0.06}\right)[SO_4^{2-}]$$

$$1.3 \times 10^{-10} = (1 + 17)[SO_4^{2-}]^2$$

$$[SO_4^{2-}] = 2.7 \times 10^{-6}$$

$$[HSO_4^-] = \frac{2.7 \times 10^{-6}}{0.06} = 4.5 \times 10^{-5}$$

$$[Ba^{2+}] = \text{solubility} = 2.7 \times 10^{-6} + 4.5 \times 10^{-5} = 4.8 \times 10^{-5} \, F \,(\text{mmol/mL})$$

$$4.8 \times 10^{-5} \frac{\text{mmol}}{\text{mL}} \times 233 \frac{\text{mg}}{\text{mmol}} \times 100 \,\text{mL} = 1.1 \text{ mg in 100 mL}$$

The solubility, therefore, is four times greater in 0.2 F HCl than it is in neutral solution.

at 1000 to 1200°C, complete conversion takes place (Equation 6-5) and the resulting calcium oxide (CaO) may be used as the weighing form:

$$CaCO_3 \rightleftharpoons CaO + CO_2 \qquad (6\text{-}5)$$

Other options include treatment of the CaC_2O_4 with dilute sulfuric acid (H_2SO_4) to convert it to $CaSO_4$. It has also been proposed that calcium fluoride (CaF_2) can be used as a reliable weighing form by treatment of the oxalate with dilute solutions of hydrogen fluoride (HF). However, most analysts prefer to avoid the use of HF, which is highly irritating and corrosive.

Precipitation of calcium oxalate may also be used as a means of separating calcium from the solution and establishing a 1:1 mole ratio between calcium ion and oxalate ion. The precipitate can then be dissolved in hydrochloric acid, after which the quantitation step can be carried out by determining the amount of oxalate present with a redox volumetric procedure (see Section 11-7). This procedure saves the time-consuming steps of bringing the precipitate to a reliable weighing form. Thus, precipitation procedures may be used for step 3 in the total analysis process (separation from interferences), whereas step 4 (quantitation) is accomplished by some means other than weighing.

6-2   ORGANIC PRECIPITATING REAGENTS

The inorganic precipitating reagents described in Section 6-1 are not very selective. However, the wide range of organic compounds available and the large number of existing insoluble *organometallic compounds* suggest that perhaps high selectivity can be engineered into the organic structure so that analytical chemists would have a precipitating reagent for iron, one for calcium, one for nickel, and so on. Although such high selectivity is not usually attained, through control of the organic structure and the precipitating conditions (for example, pH, concentration of reagents, and the use of masking agents to eliminate interferences), several useful gravimetric procedures have been developed. Investigations in this area are continuing, so several more procedures may be available in the future.

The organic reagent may form a simple saltlike precipitate, as previously discussed regarding oxalates, which can also be classified as organic reagents. Another example is sodium tetraphenylboron, $NaB(C_6H_5)_4$. The large tetra-phenylboron anion, $B(C_6H_5)_4^-$, forms insoluble salts with large cations and is highly selective for NH_4^+ and K^+. Precipitation is carried out in cold, acid solution, after which drying to constant weight is easily accomplished at 105 to 120°C, the only interferences being Rb^+, Cs^+, and Hg^{2+}. A mixture containing both NH_4^+ and K^+ can be analyzed by determining their sum; then, after treatment with base to expel NH_3, the remaining K^+ can be precipitated alone as $KB(C_6H_5)_4$.

More often, the organic reagent forms a nonionic complex with the metal ion. The complexes normally involve more than one electron pair from the organic ligand and are called *chelates*. The formation of chelate complexes in solution is the basis for many volumetric analytical techniques and will be discussed in Chapter 10. An example that involves an almost completely specific organic precipitating reagent is the determination of nickel using as a reagent dimethylglyoxime, which has the following structure:

$$CH_3-\underset{\underset{\underset{OH}{N}}{\|}}{C}-\underset{\underset{\underset{OH}{N}}{\|}}{C}-CH_3$$

Nonbonding electron pairs on the nitrogen atoms form a five-membered ring chelate to nickel with a 2:1 stoichiometry, as follows:

The chemical reaction can be written as

$$Ni^{2+} + 2C_4H_8N_2O_2 + 2H_2O \rightleftharpoons Ni(C_4H_7N_2O_2)_2(s) + 2H_3O^+$$

The low gravimetric factor of 0.2032 (see Table 5-6) suggests that small amounts of nickel may be determined accurately. Actually, only small amounts of nickel (~ 30 mg) can be determined easily because the bright-red precipitate is so bulky. In addition to the bulky nature of the precipitate, which makes it hard to filter, other subtleties must be considered to carry out a successful procedure. The precipitation must be carried out in a weak ammonia solution and tartaric acid is used to mask possible interference from iron(III) and chromium(III), which would precipitate as the hydrous oxides. The only other metal ions that are potential interferences are Pd(II) and Bi(III), which also form insoluble compounds with dimethylglyoxime. However, the palladium compound is so insoluble that it can be precipitated at pH 1, whereas the bismuth compound does not precipitate until pH 11. Consequently, it is quite simple to check for the presence of palladium while the solution is still acidic, then to increase the pH to approximately 8 by adding ammonia to precipitate nickel without interference from bismuth. Moreover, organic complexing agents such as dimethylglyoxime are usually not very water soluble; as reagents, they are often added as an alcoholic solution. However, the solubility of nickel dimethylglyoxime increases in alcohol–water solvents; therefore, large additions of the alcoholic reagent solution must be avoided to prevent negative errors due to solubility. At the same time, the solubility of dimethylglyoxime is decreased when added to the aqueous solution containing the analyte; at this point, it may start to precipitate, causing a positive error. Because copper(II) and cobalt(II) form soluble complexes with dimethylglyoxime, they are not direct interferences; however, when they are present, additional dimethylglyoxime may be necessary to ensure a sufficient concentration for quantitative precipitation of the nickel. Unfortunately, the addition of more dimethylglyoxime dissolved in alcohol may increase the solubility of the nickel dimethylglyoxime, thereby introducing a negative error. Thus even though dimethylglyoxime is a nearly specific reagent for nickel, the analyst must still have some knowledge of the sample composition to ascertain the optimum experimental conditions for an accurate analysis.

Finally, nonionic organic complexes do not adsorb water readily; in fact, they are hydrophobic. This makes drying easy (110°C is sufficient), but it also gives the precipitate of nickel dimethylglyoxime the physical property of *creeping*—it appears to crawl out of the filter and can be lost if not carefully monitored. After all of these factors have been considered, the gravimetric procedure for nickel with dimethylglyoxime is one of the best methods available for an accurate determination.

Its high selectivity for nickel has made dimethylglyoxime important as an organic precipitating reagent. Another precipitating reagent, 8-hydroxyquinoline (also called oxine) has gained widespread use because it is a general precipitating reagent. With the exception of the alkalies, almost every metal ion in the periodic table can be precipitated with 8-hydroxyquinoline. A typical reaction is with aluminum (III):

$$Al^{3+} + 3H_2O + 3 \quad \text{(structure)} \quad \rightleftharpoons \quad Al\left(\text{(structure)}\right)_3 + 3H_3O^+ \qquad \textbf{(6-6)}$$

Trivalent metal ions (Al^{3+}, Bi^{3+}, Sc^{3+}, Ce^{3+}, Fe^{3+}) form 3:1 complexes with 8-hydroxyquinoline, whereas divalent metal ions (Cd^{2+}, Cu^{2+}, Mg^{2+}, Ni^{2+}, Pb^{2+}, Sr^{2+}, Zn^{2+}) form 2:1 complexes. Again, Equation 6-6 shows that the equilibrium constant will be pH-dependent, and control of pH allows some selectivity to be gained. In addition, it is possible to add masking agents—such as EDTA, cyanide, and tartrate—that will form soluble complexes with some interferences, thereby allowing more selectivity for precipitation with 8-hydroxyquinoline.

Another way of using 8-hydroxyquinoline precipitates is in an indirect measurement, as is frequently done with calcium oxalate precipitates. If the stoichiometry between 8-hydroxyquinoline and metal is exact (for example, 3:1 for aluminum), then a determination of the 8-hydroxyquinoline will, in effect, determine the amount of metal present. Volumetric and electroanalytical methods for determining 8-hydroxyquinoline with bromine are relatively fast and avoid any weighing form problems.

The organic compounds precipitated frequently are not sufficiently stable or their stoichiometry is not exact enough to act as the weighing form. In such cases, ignition to the metal oxide is often recommended. However, this loses one of the advantages of organic precipitating reagents—high formula weight. Remember (see Chapter 5) that a high formula weight for the weighing form will result in a low gravimetric factor and that the gravimetric factor is k^{-1} where k is the sensitivity for the method. Example 6-2 shows this effect. Other weighing forms are sometimes recommended. For example, α-nitroso-β-naphthol is a highly specific reagent for cobalt. However, its precipitate is of uncertain composition: Ignition in oxygen gives the oxide Co_3O_4, but ignition in a hydrogen atmosphere yields elemental cobalt as the reliable weighing form.

The hundreds of organic precipitating agents proposed in chemical literature have properties ranging between the high selectivity of dimethylglyoxime and the all-purpose nature of 8-hydroxyquinoline. Table 6-3 lists some examples of organic precipitating reagents and their use in gravimetric analysis.

EXAMPLE 6-2

Compare the gravimetric factors (G_f) for aluminum and magnesium, assuming (1) that they are precipitated with 8-hydroxyquinoline and weighed directly, and (2) that the precipitates are ignited to Al_2O_3 and MgO and weighed as the oxides. Assume the substance sought is the elemental form.

Solution

The gravimetric factor can be calculated using Equation 5-20:

$$G_f = \frac{FW\ (A)/n_A}{FW\ (S)/n_S}$$

where A is the analyte and S is the measurement signal (the weighing form). The formula weights are given in Appendix 4.

Sought, A	Weighed, S	Gravimetric Factor, G_f	Sensitivity, $1/G_f$
Al	$Al(C_9H_6NO)_3$	$\dfrac{26.98154/1}{459.44/1} = 0.05873$	17.03
Al	Al_2O_3	$\dfrac{26.98154/1}{101.96/2} = 0.5293$	1.889
Mg	$Mg(C_9H_6NO)_2$	$\dfrac{24.305/1}{312.61/1} = 0.07775$	12.86
Mg	MgO	$\dfrac{24.305/1}{40.30/1} = 0.6031$	1.658

The gravimetric factor is increased by nearly an order of magnitude (conversely, the sensitivity is decreased by nearly an order of magnitude) when the weighing form is the oxide instead of the 8-hydroxyquinoline compound itself. Therefore, for high sensitivity, the preferred weighing form would be the oxine.

6-3 PRECIPITATING REAGENTS FOR ORGANIC ANALYSIS

Section 6-2 discussed the use of organic reagents for determining inorganic metal ions. Reagents can also be used to precipitate organic compounds as a basis for a gravimetric procedure. In some cases, a specific organic compound is involved. Aspirin (acetylsalicylic acid), for example, can be hydrolyzed to salicylic acid, which

TABLE 6-3　Examples of Organic Precipitating Reagents

Reagents	Structure	Analytes	Conditions
Selective reagents:			
dimethylglyoxime	$\text{CH}_3\text{C}\overset{\text{NOH}}{\underset{}{}}$——$\overset{\text{NOH}}{\underset{}{}}\text{CCH}_3$	Ni(II) Pd(II)	NH_3 HCl
α-benzoinoxime (cupron)	$C_6H_5\overset{\text{OH}}{\underset{}{}}\text{CH}$——$\overset{\text{NOH}}{\underset{}{}}\text{C}C_6H_5$	Cu(II) Mo(VI), W(VI)	NH_3, tartrate acid
α-nitroso-β-naphthol	(N=O, OH on naphthol)	Co(II), Fe(III), Pd(II)	weak acid
sodium tetraphenylboron	$NaB(C_6H_5)_4$	K^+, Rb^+, Cs^+, NH_4^+ (large +1 cations)	acid

General reagents:

Reagent	Structure	Elements	Conditions
8-hydroxyquinoline (oxine)		Al, Bi, Cd, Co, Cu, Ga, Hf, Fe, In, Hg, Mo, Ni, Nb, Pa, Pd, Ag, Ta, Ti, Th, W, U, Zn, Zr	pH 4–5
		Al, Be, Bi, Cd, Cu, Ga(I), Hf, Fe, In, Mg, Mn, Hg, Nb, Pd, Sc, Ta, Ti, Th, U, Zn, Zr, rare earths	NH_3
ammonium nitrosophenyl hydroxylamine (cupferron)		Fe, Hg, Nb, Ta, W, Zr	strong acid
		Sb, Bi, Ga, Fe, Mo, Pd, Sn, Ta, Ti, V, W, Zr, rare earths	weak acid
anthranilic acid		Cu, Cd, Ni	acetic acid
		Co, Pb, Zn	neutral

in turn is treated with an iodine reagent in basic solution according to Equation 6-7:

$$2 \text{(COO}^- \text{, OH)} + 6I_2 + 8OH^- \rightleftharpoons O = \text{...} = O + 8I^- + 2HCO_3^- + 6H_2O \quad (6\text{-}7)$$

The yellow precipitate tetraiodophenylenequinone, can be employed as the reliable weighing form for the analysis of aspirin. In other cases, the method is selective for a particular functional group. For example, nicotine and other alkaloids as analytes may be precipitated with heteropoly acids (such as silicotungstic acid, $H_4SiW_{12}O_{40}$) and ignited to yield the mixed oxides, $SiO_2 + 12WO_3$. This is an interesting example because the weighing form contains none of the analyte. As with the indirect method for calcium (by determining the amount of oxalate present in the precipitate), the reliability of the 1:1 stoichiometric ratio between nicotine and silicotungstic acid is the critical factor.

6-4 REDUCING AGENTS AS PRECIPITATING REAGENTS

Except for the few elements that react with water (alkali metals, alkaline earths, F_2, Cl_2), elemental substances are relatively insoluble in water. This is especially true of metallic elements. Consequently, if a metal ion is reduced to its elemental state, a metallic analyte can be precipitated from solution. However, because metallic ions are generally very difficult to reduce to the metallic state with chemical reducing agents, this technique is only used for a few easily reduced metal ions.

Selenium and tellurium, which can be easily reduced, are frequently determined by this method. Selenous acid (H_2SeO_3) and tellurous acid (H_2TeO_3) are reduced to the elemental state with SO_2 according to Equation 6-8:

$$H_2SeO_3(\text{or } H_2TeO_3) + 2SO_2 + 5H_2O \rightleftharpoons 2SO_4^{2-} + 4H_3O^+ + Se \text{ (or Te)} \quad (6\text{-}8)$$

If the elements are in their higher oxidation state ($+6$), either they must first be reduced through boiling with HCl or the elemental form must be precipitated using a stronger reducing agent such as hydrazine hydrochloride ($N_2H_4\cdot 2HCl$). For example, the reduction of telluric acid using hydrazine hydrochloride proceeds as shown in Equation 6-9:

$$2H_6TeO_6 + 3(N_2H_4\cdot 2HCl) \rightleftharpoons 12H_2O + 6HCl + 3N_2 + 2Te \quad (6\text{-}9)$$

Because tellurium will not precipitate from concentrated HCl solutions but selenium is quantitatively precipitated under these conditions, the use of SO_2 permits the separation of selenium from tellurium. After the selenium has been filtered out, the remaining solution can be diluted with H_2O, thereby allowing the tellurium to precipitate.

TABLE 6-4   Reducing Agents as Precipitating Reagents

Elemental form of analyte	Reducing agents
Se	SO_2, $N_2H_4 \cdot HCl$, NH_2OH
Te	SO_2, $N_2H_4 \cdot HCl$, NH_2OH
Au	SO_2, $H_2C_2O_4$, $NaNO_2$
Re	H_2
Ir	H_2
Pt	HCOOH
Rh	$TiCl_2$
Hg	$SnCl_2$
Os	H_2
Ru	H_2

Other analytes that can be reduced to the elemental state with chemical reducing agents are shown in Table 6-4. In this analytical method, the measurement signal *equals* the analyte value and, obviously, the gravimetric factor is 1.

6-5   ELECTROGRAVIMETRIC ANALYSIS

In the previous section, we noted that most elements are insoluble in water in their elemental state, but that in only a few cases can a chemical reducing agent produce the needed reduction. Redox reactions can be carried out at electrodes; at the *cathode*, the electrons themselves act as the reducing agent. Therefore, an analyst can "precipitate" copper from solution by reducing Cu^{2+} to Cu^0 at the cathode, as follows:

$$Cu^{2+} + 2e \rightleftharpoons Cu^0(s) \qquad \textbf{(6-10)}$$

The "precipitation" is an electroplating process. Under the correct conditions, the copper will adhere so tightly to the cathode that the electrode may be washed, dried, and weighed to yield the amount of copper that was present.

Figure 6-1 shows a schematic electrolysis cell that could be used for the determination of copper(II) in solution. In this process, the large platinum gauze cathode is cleaned, dried, and weighed. Then the electrodes are immersed in a 1 *F* H_2SO_4 solution containing the copper(II) analyte. The solution must be stirred vigorously to transport the copper ions from all points in the solution to the electrode surface where reduction takes place. The stirring may be accomplished with a magnetic or a motor-driven stirrer, or the anode itself may be rotated mechanically.

The voltage applied to the cell from the power supply can be adjusted to the appropriate value (usually about 2 V is needed to reach desired current value) with

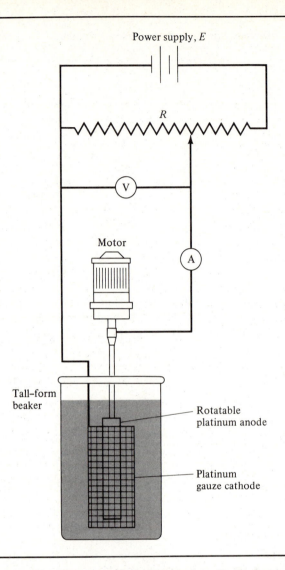

Power supply, E

R

V

Motor

A

Tall–form beaker

Rotatable platinum anode

Platinum gauze cathode

FIGURE 6-1 Schematic arrangement for electrolysis cell: (*A*) Ammeter to measure electrical current. (*V*) Voltmeter to measure voltage applied. (*R*) Adjustable resistance (rheostat) for varying the applied voltage. (*Note:* A regulated power supply will perform the role of E and R in most analytical labs.)

an adjustable resistance, R. The copper plate must be tightly adherent or some of it may be lost when the cathode is removed from solution. For this reason, the current density (current per unit electrode area) must be controlled; values of 5 to 50 mA/cm^2 are typical. Higher current densities encourage the growth of dendrites ("whiskers" of copper metal), which can be easily dislodged from the cathode, thereby causing negative determinate errors.

The evolution of H_2 gas at the cathode can also lead to an unsatisfactory copper deposit. As the applied voltage is raised to increase the current density (and bring about a faster analysis), the potential of the cathode will become sufficiently negative to allow the reduction of H_3O^+ according to Equation 6-11:

$$2H_3O^+ + 2e \rightleftharpoons H_2 + 2H_2O \qquad (6-11)$$

If a gas is evolved at the electrode while a metal is plating out, gas bubbles will be trapped in the plated material, causing a spongy deposit that is unsuitable for an electrogravimetric determination. Gas evolution may be prevented by adding a *cathodic depolarizer* to the solution. The depolarizer is a material that is more readily reduced than H_3O^+ but that does not lead to a gas or weighable product on the cathode. Nitric acid, which is reduced according to Equation 6-12, is frequently used:

$$NO_3^- + 10H_3O^+ + 8e \rightleftharpoons NH_4^+ + 13H_2O \qquad (6-12)$$

The solution composition also helps determine the physical nature of the electroplate. Reduction of complex ions usually results in smooth, adherent coatings that are superior in quality to those produced by the reduction of the simple metal ion (surrounded by water of hydration). Although acceptable electroplates of copper can be obtained from Cu^{2+} reduction, the use of a nitric acid solution for the determination of silver would cause dendritic growth. However, an excellent smooth adherent electroplate of silver can be obtained by reducing silver in the form of the cyanide complex, $Ag(CN)_2^-$, or ammonia complex, $Ag(NH_3)_2^+$.

The electrochemical cell (see Figure 6-1) may suggest this is an electroanalytical method of analysis. However, because the quantitation step involves a measurement of mass, the method is gravimetric in the strict sense. Alternatively, a chemist could determine the number of electrons needed to carry out Equation 6-10 by measuring the coulombs of electrical charge that passed through the cell. This would be a *coulometric* method of analysis (see Chapter 16). It must be demonstrated (or assumed) that no other electron-consuming reaction, including H_2 evolution, took place at the cathode. Any other electrochemical reaction would be an interference in the coulometric method, but only reactions that result in weighable products on the cathode would interfere with the electrogravimetric method. This last point was more important when instruments for controlling the potential at the electrode were less readily available than they are now. Then the potential was often sufficient to reduce water to hydrogen gas (plus OH^-), but resulted in no error in the electrogravimetric method.

Controlled potential is also useful for electrogravimetric procedures because separation can be achieved at the same time. Suppose a mixture of Cu^{2+}, Cd^{2+}, and Zn^{2+} is to be analyzed. Appendix 3-A shows the following potentials for these reductions (on the hydrogen reference scale):

$$Cu^{2+} + 2e \rightleftharpoons Cu \qquad E^0 = +0.337 \text{ V}$$

$$2H_3O^+ + 2e \rightleftharpoons H_2 + 2H_2O \qquad E^0 = 0.000 \text{ V}$$

$$Cd^{2+} + 2e \rightleftharpoons Cd \qquad E^0 = -0.403 \text{ V}$$

$$Zn^{2+} + 2e^- \rightleftharpoons Zn \qquad E^0 = -0.763 \text{ V}$$

A cathode controlled below $+0.337$ V will be sufficiently negative to reduce Cu^{2+}, but not negative enough to reduce Cd^{2+} or Zn^{2+}. To ensure complete reduction, the cathode would be controlled at $\sim +0.1$ V. After all the copper was reduced to copper metal and determined by weighing the electrode, the potential could be decreased to -0.6 V which would be sufficiently negative to reduce Cd^{2+} but not Zn^{2+}. The four potentials previously listed show that when the cathode potential is dropped below 0 V, the reduction of the hydronium ion may take place. This reduction could be prevented if basic solutions were used; however, because most metal ions form insoluble hydroxides, complexing agents would then be needed. A more effective method for preventing reduction of hydronium ions or water itself at the thermodynamic value is to carry out the reduction at a mercury pool cathode. Hydrogen evolution at mercury is kinetically slow. (An electrochemical reaction has a reaction rate constant like any other chemical reaction, but has the added property of being potential-dependent.) Therefore, additional applied voltage is necessary to bring about this reduction. The additional voltage, called *overvoltage*, is about 1 V for reduction of H_3O^+ or H_2O at a mercury cathode. This makes it possible to reduce metal ions at potentials of less than zero at the cathode without interference from hydrogen evolution. However, a mercury pool is more difficult to weigh than a platinum gauze cathode (as shown in Figure 6-1); coulometric methods are usually used for reductions carried out at mercury cathodes.

After completion of the second reduction, the potential could be decreased to -1 V to reduce Zn^{2+}. With potential control, the electron can be an analytical reagent with fine-tuning capability. High selectivity is possible *provided* that the electrochemical potentials of the substances present differ by more than ~ 0.2 V. However, even when the potentials (see Appendix 3-A) do not differ by a sufficient amount, it is often possible to shift the potentials and increase the difference by complexing the ions with a ligand that forms stronger complexes with one component than with the other. These techniques will be discussed in more detail when electroanalytical methods are presented in Chapters 15 to 17.

Table 6-5 summarizes some of the elements that can be analyzed through electrogravimetric methods. Note that not all electrogravimetric determinations

---

TABLE 6-5 Electrogravimetric Analysis

reduction to elemental state at the cathode:

Ag, Bi, Cd, Co, Cu, In, Ni, Re, Sb, Sn, Zn

oxidation at the anode:

Pb as PbO_2
Co as Co_2O_3
Cl as AgCl (at Ag anode)
Mn as MnO_2
Tl as Tl_2O_3

need be reductions at the cathode. Oxidation of Pb^{2+}, for example, leads to the production of PbO_2 at the anode; this product may be weighed as a measurement signal for the determination of lead, as illustrated in Example 6-3. Electrolysis is carried out in strong ($\sim$2 F) HNO_3 solution containing a small amount of H_2SO_4 to make the deposit more adherent. Voltage is applied and electrolysis is carried out for one to two hours to ensure complete oxidation. This is considerably longer than the time required for most metal ion reductions. When oxidation is complete, the

---

EXAMPLE 6-3

The amount of lead in a brass sample was determined by dissolving a 1.653-g sample in 150 mL HNO_3. After the hydrous tin oxide was filtered off, the HNO_3 concentration was adjusted to 2 F (total volume = 180 mL) and electrolysis was carried out at an applied voltage of 2.1 V. The average current was 1.75 A. The platinum anode weighed 10.1943 g before electrolysis and 10.2857 g after electrolysis. Calculate the percentage of lead in the brass sample.

Solution

The anode deposit is PbO_2. Its weight is calculated as follows:

$$\text{anode} + PbO_2: \quad 10.2857\,g$$

$$\text{anode}: \quad \underline{10.1943\,g}$$

$$PbO_2 \text{ deposit}: \quad 0.0914\,g = 91.4\,mg$$

The amount of lead present is calculated as follows:

$$\text{mg Pb} = \frac{\text{FW (Pb)}}{\text{FW (PbO}_2)} \times \text{mg PbO}_2$$

$$= \frac{207.2}{239.2} \times 91.4 = 0.866 \times 91.4 = 79.2\,mg$$

$$\% \text{ Pb in brass} = \frac{79.2\,mg\,Pb}{1{,}653\,mg\,brass} \times 100\,(\%) = 4.79\%$$

Note: From experience, chemists know that the PbO_2 is not exactly stoichiometric and that the gravimetric factor of 0.864 should be used instead of 0.866. If the empirical gravimetric factor is used, the percentage of lead will be

$$\% \text{ Pb in brass} = \frac{0.864 \times 91.4 \times 100\,(\%)}{1{,}653} = 4.78\%$$

Note also that, aside from the sample weight, the calculation required only one measurement signal—the weight of PbO_2 deposited on the anode. The volume, applied voltage, and current values were not needed for calculation.

electrode is removed. To prevent PbO_2 from falling off, the electrode is washed with water while still connected to the power supply. The anode is rinsed with acetone to remove water and dried at 120°C.

6-6 THERMOGRAVIMETRIC ANALYSIS

As mentioned in Chapter 5, organic compounds can be decomposed by heating in air; this produces CO_2 and H_2O, which can be weighed as the basis for a gravimetric analysis. The concept of heating a sample to observe weight changes is the principle of *thermogravimetric analysis* (TGA). Temperature is scanned at a carefully controlled rate of a few degrees per minute while the same weight is monitored continuously, as shown in Figure 6-2. The temperature at which the weight changes yields qualitative information concerning the reaction; the weight change itself can be used as a measurement signal to determine the amount of analyte present.

Figure 6-3 depicts a typical thermogram. The weight of the $CaC_2O_4 \cdot H_2O$ decreases in several stages. First, the water of hydration is lost, corresponding to 2.46 mg (12.3%) in the temperature range 100–250°C. Then, between 400° and 500°C, carbon monoxide is evolved and the sample loses another 3.84 mg (19.2%). Finally, above 700°C, another weight loss of 6.02 mg (30.1%) occurs as CO_2 is evolved, resulting in the final product, CaO. The chemical reactions are summarized

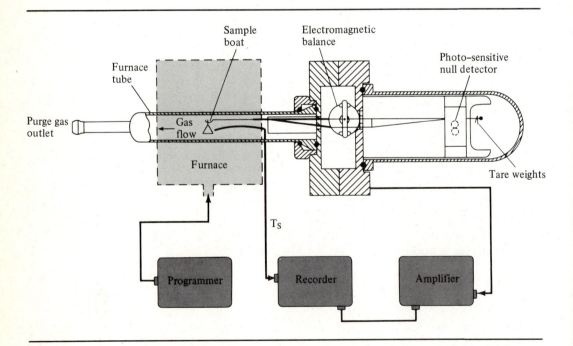

FIGURE 6-2 Schematic arrangement for thermogravimetric analysis.

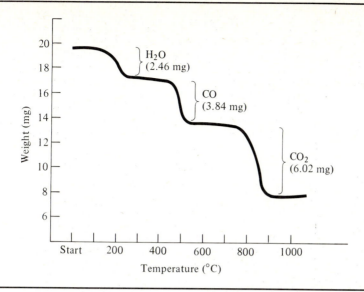

FIGURE 6-3 Thermogravimetric analysis of $CaC_2O_4 \cdot H_2O$ heated at 6°C per minute.

as follows:

$$CaC_2O_4 \cdot H_2O \xrightleftharpoons[]{100-250°C} CaC_2O_4 \xrightleftharpoons[]{400-500°C} CaCO_3 \xrightleftharpoons[]{700-850°C} CaO$$

$$+H_2O\uparrow \qquad\qquad +CO\uparrow \qquad\qquad +CO_2\uparrow$$

$$\text{(6-13)}$$

One of the most important uses for thermogravimetric analysis is the determination of which chemical reactions occur at various temperatures. A qualitative picture is thereby obtained, although the weight changes must be measured quantitatively to deduce which chemical changes have occurred. For example, the 19.2% weight loss that occurred between 400° and 500°C must correspond to 19.2% of the original formula weight for $CaC_2O_4 \cdot H_2O$ (FW = 146). Therefore, the product being evolved has a formula weight of $0.192 \times 146 = 28.0$, which corresponds to CO.

Clearly, if a compound with a known thermogram occurs in combination with inert materials, the composition of the mixture can be obtained from the percentage weight change that occurs relative to the percentage weight change observed with the pure compound:

$$\text{wt \% component A} = \frac{\text{\% wt change for mixture}}{\text{\% wt change for pure A}} \times 100 \, (\%) \qquad \text{(6-14)}$$

If more than one component shows weight changes and if these changes occur in different temperature regions, then the composition can be determined as shown in Example 6-4.

EXAMPLE 6-4

A thermogram of pure $CaCO_3$ shows a single weight loss around 800°C, correspond-ing to 44%. A thermogram of pure $MgCO_3$, on the other hand, shows a two-step weight loss reaching a plateau value above 450°, corresponding to 52%. The thermograms are shown in Figure 6-4.

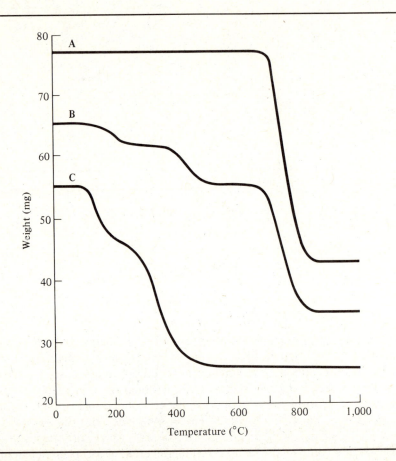

FIGURE 6-4   Thermograms for $CaCO_3$ and $MgCO_3$: (A) Thermogram for pure $CaCO_3$. (B) Thermogram for limestone sample containing both $CaCO_3$ and $MgCO_3$. (C) Thermogram of pure $MgCO_3$.

EXAMPLE 6-4 (cont.)

(a) Determine the reactions that account for these weight losses.
(b) Calculate the composition of a 65-mg limestone sample that exhibits the thermogram shown in Figure 6-4. The sample weighs 56 mg at 500°C and 36 mg at 900°C.

Solution

(a) The formula weight of $CaCO_3$ is 100. The weight loss at 800° corresponds to 0.44 × 100 = 44 (FW of evolved product). Therefore, because FW (CO_2) = 44,

$$CaCO_3 \rightleftharpoons CaO + CO_2\uparrow$$

The formula weight of $MgCO_3$ is 84.3. The weight loss at 450° corresponds to 0.52 × 84.3 = 44 (FW of evolved product). Therefore, because FW (CO_2) = 44,

$$MgCO_3 \rightleftharpoons MgO + CO_2\uparrow$$

(b) The weight loss due to $MgCO_3$ = 65 − 56 = 9 mg:

$$\% \text{ loss} = \frac{9}{65} \times 100 \, (\%) = 14\%$$

The weight loss due to $CaCO_3$ = 56 − 36 = 20 mg:

$$\% \text{ loss} = \frac{20}{65} \times 100 \, (\%) = 31\%$$

From Equation 6-14,

$$\% \, MgCO_3 \text{ in limestone} = \frac{14}{52} \times 100 \, (\%) = 27\%$$

$$\% \, CaCO_3 \text{ in limestone} = \frac{31}{44} \times 100 \, (\%) = 70\%$$

$$\% \text{ inert impurities} = 100 - (27 + 70) = 3\%$$

The fact that weight *gains* may also occur if the sample reacts with air in the furnace compartment can be useful for analytical purposes. However, if it is an interference to the desired analysis, the sample can be heated in an inert atmosphere or in a sealed sample tube.

6-7   VOLATILIZATION TECHNIQUES

The thermogravimetric technique discussed in the previous section used the weight loss of the sample as a measurement signal. It is also possible to collect the evolved product and weigh it as a measurement signal. This technique was described in Chapter 5, using the example of microdetermination of carbon and hydrogen in an organic compound. Carbon is oxidized to CO_2 and hydrogen to H_2O. These products are collected in absorption tubes (see Figure 5-2), page 115, and the weight gains for the tubes are measured. These measurement signals are used to determine the percentages of carbon and hydrogen in the compound.

This technique has other applications. For example, the properties of alloys are strongly dependent on the amount of carbon present in the metal. If a metal sample is burned in oxygen, the carbon is converted to CO_2 and can be collected in a weighed bulb packed with ascarite. From the weight of CO_2 collected, the percentage of carbon in the sample is calculated as shown in Example 6-5.

Halogens in organic compounds can be liberated in the elemental form (Cl_2, Br_2, I_2) by careful treatment with highly corrosive, fuming HNO_3. The liberated halogen is treated with Ag^+, which forms the insoluble silver halide. (Because AgF is soluble, fluorine content cannot be determined through this method.) The halogen

---

EXAMPLE 6-5

A National Bureau of Standards nichrome sample (20% Cr–77% Ni) weighing 1.365 g was burned in oxygen and the CO_2 collected. The weight of the collecting bulb increased by 2.2 mg. Calculate the percentage of carbon in the nichrome sample.

Solution

The fraction of carbon in CO_2 is

$$\frac{12.01}{44.01} = 0.2729 \text{ (gravimetric factor)}$$

Therefore, the percentage of carbon in nichrome is

$$\frac{0.2729 \text{ mg C/mg CO}_2 \times 2.2 \text{ mg CO}_2}{1{,}365 \text{ mg nichrome}} \times 100 \, (\%) = 0.044\%$$

content in the organic compound can be calculated from the weight of silver halide, as shown in Example 6-6.

EXAMPLE 6-6

An organic compound containing chlorine is treated with fuming HNO_3 and the chlorine evolved reacts with silver ion to form AgCl. The sample of organic compound weighed 120 mg, and the AgCl weighed 153 mg. Calculate the percentage of chlorine in the organic compound and the minimum molecular weight for the compound.

Solution

The gravimetric factor for the determination of Cl from the weight of AgCl is 0.2474. Therefore,

$$mg\ Cl = \frac{0.2474\ mg\ Cl}{mg\ AgCl} \times 153\ mg\ AgCl = 37.85\ mg\ Cl$$

$$\%\ Cl = \frac{37.85\ mg\ Cl}{120\ mg\ sample} \times 100\ (\%) = 31.5\%$$

In terms of molecular weight (MW), the percentage of chlorine is given by

$$\%\ Cl = \frac{35.453 n_{Cl}}{MW} \times 100\ (\%)$$

where n_{Cl} is the number of chlorine atoms in the compound. Therefore,

$$MW = \frac{35.453 n_{Cl}}{\%\ Cl} \times 100\ (\%) = \frac{35.453 n_{Cl}}{31.5} \times 100\ (\%) = 113 n_{Cl}$$

The minimum molecular weight is reached when $n_{Cl} = 1$ and the molecular weight $= 113$. For example, chlorobenzene with the formula C_6H_5Cl fits the data. If the compound contains two chlorine atoms, the molecular weight would be 226, and so on.

Sometimes everything in the sample can be volatilized except the analyte, which is left in a reliable weighing form. For example, in the determination of metal content in an organic compound, a small amount of sulfuric acid is added to the sample, which is then burned in air. Everything is volatilized except the metal, which remains as the metal sulfate. If the compound contains more than one metal, the determination is incomplete because the weight of metal sulfates then constitutes only one measurement signal with at least two unknown analyte concentrations. Alternatively, if no H_2SO_4 is used, it is possible to convert some metals to the oxide form; and noble metals, such as silver, may remain in the elemental form after the organic material has been burned in air.

6-8 OTHER ANALYTICAL USES FOR PRECIPITATION REACTIONS

Other analytical uses for precipitation reactions will be explored in later chapters. The precipitation step may constitute only step 3 in the total analysis process—that is, it may be only a means to separate the analyte from interferences. Qualitative analysis schemes have been developed around sulfide precipitations to separate metals ions into various groups. Final quantitative analysis can then be carried out for each component by some convenient technique.

The precipitation step may establish a known stoichiometric ratio between cation and anion, thereby allowing an indirect determination of the analyte via the amount of the counter-ion present. For example, calcium may be determined by precipitating CaC_2O_4, which establishes a $1:1$ mole ratio between calcium and oxalate. The oxalate may then be determined by a volumetric technique. From the amount of oxalate present, the amount of calcium can be calculated. A precipitation reaction can also be used in a volumetric method, as will be discussed in more detail in Chapter 8. Moreover, some volumetric methods utilize the formation or disappearance of a precipitate to signal the endpoint for the titration.

SUPPLEMENTARY READING

Daniels, T., *Thermal Analysis*. New York: John Wiley & Sons, Inc., 1973.
Gordon, L., M.L. Salutsky, and H.H. Willard, *Precipitation from Homogeneous Solution*. New York: John Wiley & Sons, Inc., 1959.

PROBLEMS

6-1. How are the four requirements for a successful gravimetric analysis met in the analysis of chloride by precipitation of AgCl?

6-2. How are the four requirements for a successful gravimetric analysis met in the analysis of iron by precipitation of $Fe(OH)_3 \cdot xH_2O$?

6-3. How were the three types of coprecipitation processes minimized in the gravimetric chloride analytical procedure?

6-4. Explain why the following procedures were used in the gravimetric chloride analysis. If these procedures were not followed, would a positive or negative error be introduced?

a.   Silver nitrate added slowly and with good stirring.

b.   Precipitate heated to boiling and stored for several hours before filtering.

c.   Nitric acid added to wash water.

d.   Silver chloride precipitate not exposed to sunlight.

6-5. Using the gravimetric analysis of chloride as an example, discuss the *total analysis* concept, pointing out where errors could arise at each step.

6-6. How will the following procedures affect precision and accuracy?
  a. Forget to zero the balance before weighing chloride samples.
  b. Filter AgCl immediately without digestion.
  c. Switch trial 1 and trial 3 filter crucibles when weighing AgCl (use AgCl weight for trial 1 with sample weight for trial 3 and vice versa).

*6-7. Recalculate Example 6-1 to include activity coefficients in ionic strength 0.2: $f_{H_3O^+} = 0.81$, $f_{Ba^{2+}} = 0.34$, $f_{SO_4^{2-}} = 0.32$, $f_{HSO_4^-} = 0.75$.

*6-8. A H_2SO_4 solution is standardized by determining its SO_4^{2-} content gravimetrically as $BaSO_4$. What is the normality of the H_2SO_4 as an acid if a 50.00-mL sample yielded 422.3 mg $BaSO_4$?

6-9. a. In what pH range could aluminum hydroxide be quantitatively precipitated in the presence of zinc? Assume initial concentrations of 0.1 F and that removal of 99.9% of the aluminum is sufficient.
  b. What types of coprecipitation might be encountered and how might this affect the analysis?

6-10. A typical soluble chloride sample may contain 40 to 60% Cl. How many milliliters of 0.30 F $AgNO_3$ are needed to ensure complete precipitation if the sample weighs 400 mg?

*6-11. What weight of urea must be added to generate sufficient OH^- for precipitation of aluminum as $Al(OH)_3 \cdot xH_2O$, if samples may contain 20 to 30% $Al_2(SO_4)_3$ and weigh 1.2 g?

6-12. What weight of thioacetamide must be added to ensure complete precipitation of bismuth as Bi_2S_3, if samples may contain 10 to 20% Bi_2O_3 and weigh 750 mg?

6-13. A certain steel alloy may contain 5 to 10% Ni. What volume of 1% (by weight) ethanolic dimethylglyoxime ($C_4H_8N_2O_2$) is needed to ensure complete precipitation, assuming the sample weighs 300 mg?

*6-14. Seawater contains about 885 ppm (mg/L) S. Assuming that this is in the form of SO_4^{2-}, what weight of $BaSO_4$ precipitate would be expected from a 100-mL seawater sample?

6-15. The arsenic content in an insecticide can be determined by digesting the sample in acid and precipitating it as $MgNH_4AsO_4$ (analogous to determination of P as $MgNH_4PO_4$). The precipitate is then ignited and weighed as $Mg_2As_2O_7$. Calculate the gravimetric factor, and report arsenic content as percentage As_2O_3 in a 1.356-g sample that yields 127.3 mg of $Mg_2As_2O_7$.

*6-16. The nicotine in a tobacco sample is extracted and then precipitated with silicotungstic acid ($H_4SiW_{12}O_{40}$). The 1:1 adduct is ignited and weighed as $SiO_2 + 12 WO_3$. If a 728-mg tobacco sample yields 214 mg of $SiO_2 + 12 WO_3$, calculate the percentage of nicotine ($C_{10}H_{14}N_2$) in the tobacco.

6-17. A 400-mg tablet containing milk of magnesia was dissolved in acid; then $(NH_4)_2HPO_4$ was added to precipitate Mg^{2+} as $MgNH_4PO_4$. After washing, the precipitate was ignited, converting it to the reliable weighing form, $Mg_2P_2O_7$. The final weight was 305 mg. Calculate the percentage of $Mg(OH)_2$ in the tablet.

*6-18. A 740-mg sample of insecticide containing DDT ($C_{14}H_9Cl_5$) is digested with fuming HNO_3, and the chlorine evolved is reacted with Ag^+ to produce 253 mg AgCl. Calculate the percentage of DDT in the insecticide.

6-19. Twenty aspirin tablets weigh a total of 7.03 g. The tablets are ground to a powder, mixed, and a 275-mg sample taken for analysis. The aspirin ($C_9H_8O_4$) is converted to salicylic acid and precipitated as tetraiodophenylenequinone

---

*Answers to problems marked with an asterisk will be found at the back of the book.

$(C_6H_2I_2O)_2$, shown in Equation 6-7 (page 168). The dried precipitate weighs 490 mg. Calculate the average grains (64.8 mg/grain) of aspirin per tablet.

*6-20. An antidandruff shampoo was analyzed by treating the sample with $HClO_4$–HNO_3 mixture to decompose organic matter. This treatment was followed by precipitation of Zn(II) as $ZnNH_4PO_4$ and ignition to $Zn_2P_2O_7$ as the weighing form. A 5.613-g sample yielded 62.8 mg $Zn_2P_2O_7$. Calculate the percentage of zinc pyrithione ($C_{10}H_8N_2O_2S_2Zn$) in the shampoo.

6-21. A deodorant is analyzed for its zinc phenolsulfonate ($C_{12}H_{10}O_8S_2Zn$) content by decomposing the organic matter and precipitating Zn(II) with 8-hydroxyquinoline. The precipitate is dried at a temperature low enough to prevent decomposition, then weighed as $Zn(C_9H_6NO)_2$. A 5.63-g sample yielded 127.3 mg of precipitate. Calculate the percentage of zinc phenolsulfonate in the deodorant.

*6-22. The phosphorus content in a fertilizer can be determined gravimetrically by dissolving the sample and precipitating phosphorus as $MgNH_4PO_4$. The precipitate is then ignited to $Mg_2P_2O_7$ as the reliable weighing form. If a 1.53-g sample yields 316 mg $Mg_2P_2O_7$, what is the P_2O_5 content of the fertilizer? If the fertilizer actually contains superphosphate, $Ca(H_2PO_4)_2 \cdot H_2O$, calculate its content in the fertilizer.

6-23. The saccharine ($C_7H_5NO_3S$) content in a diet sweetener can be determined by oxidizing the sulfur to sulfate and precipitating it as $BaSO_4$. A 102-mg sample yielded 128 mg $BaSO_4$. Calculate the saccharine content.

*6-24. A foot powder containing zinc and weighing 2.536 g was decomposed by digestion in $HClO_4$–HNO_3 mixture and precipitated as $ZnNH_4PO_4$. The precipitate was ignited to $Zn_2P_2O_7$, the reliable weighing form. The final weight was 182.3 mg. Calculate the percentage of zinc oxide in the sample.

6-25. Fifteen diet supplement pills weigh a total of 18.7 g. The pills are ground to a powder and thoroughly mixed. A sample of the powder weighing 11.4 g is dissolved in hot HNO_3, which converts the iron to Fe(III). Hydrated iron oxide is precipitated and ignited to yield 198 mg Fe_2O_3. Calculate the average iron content in the pills as milligrams of Fe per pill.

6-26. The iron content of spinach can be determined gravimetrically by precipitating it with 8-hydroxyquinoline (oxine). With careful drying, it may be weighed as $Fe(C_9H_6NO)_3$. Assuming a 40.5-g sample yields 12.7 mg of Fe(III) oxinate, calculate the iron content of spinach. How many ounces (28.3 g/ounce) of spinach will provide the minimum daily requirement of 12 mg Fe?

*6-27. Before the advent of fast instrumental methods such as atomic absorption spectroscopy, analysis of sodium was quite difficult. One gravimetric method employed the slightly soluble compound $NaZn(UO_2)_3(C_2H_3O_2)_9 \cdot 9H_2O$. Calculate the gravimetric factor for this compound for the analysis of Na; then calculate the weight of the precipitate obtained for a sample containing only 10 mg of Na.

6-28. In Problem 6-27, the determination of sodium as sodium zinc uranyl acetate was described. Calculate the Na content in blood serum as milligrams per 100 mL, assuming a 2.00-mL serum sample yields 462 mg of $NaZn(UO_2)_3$ $(C_2H_3O_2)_9 \cdot 9H_2O$.

*6-29. There are very few insoluble compounds of K^+. Some large anions, such as tetraphenyl borate, $B(C_6H_5)_4^-$, form reasonably insoluble salts. Calculate the K content in blood serum as milligrams per 100 mL, assuming a 25.00-mL serum sample yields 44.6 mg of $KB(C_6H_5)_4$.

6-30. A mixture of $BaCl_2 \cdot 2H_2O$ and NaCl is analyzed by measuring the amount of H_2O evolved when the sample is heated to 120°C. A sample weighs 455.3 mg initially and 414.7 mg after drying. Calculate the percentage of $BaCl_2 \cdot 2H_2O$ in the sample.

6-31. The tin content in an organotin compound is determined by carefully treating it with HNO_3 and heating it to volatilize the organic material and leave a residue of SnO_2. A 215-mg sample yielded 127 mg of residue. Calculate the molecular weight of the compound, assuming it contains one tin atom.

6-32. A steel sample weighing 1.083 g was burned in oxygen and the CO_2 evolved was collected and weighed. Assuming the collecting bulb gained 4.63 mg, calculate the percentage of C in the steel.

*6-33. A brass sample weighing 1.379 g was analyzed for its Cu and Pb content by electrogravimetry. The cathode gained 997 mg and the anode gained 74.3 mg. Calculate the percentages of Cu and Pb in the sample.

6.34. A mineral sample containing malachite $[CuCO_3 \cdot Cu(OH)_2]$ and weighing 875 mg was dissolved in acid and electrolyzed to deposit copper on the cathode. Assuming the cathode gained 158 mg, calculate the malachite content in the sample.

*6-35. A thermogram of a $FeSO_4$ sample weighing 48.7 mg loses 23.0 mg between 500 and 560°C. Identify the residue product.

6-36. Gypsum, $CaSO_4 \cdot 2H_2O$ can be analyzed by thermogravimetry. Assuming a 14.6-mg sample loses 2.3 mg at 90°C, determine the reaction product. If a mineral sample weighing 29.6 mg loses 2.2 mg at 90°C, what is its gypsum content?

6-37. A sample of silicate mineral weighing 540.1 mg yields 162.3 mg of NaCl + KCl. When these chlorides are dissolved and treated with $AgNO_3$, the resulting precipitate of AgCl weighs 351.4 mg. Calculate the percentages of Na_2O and K_2O in the mineral.

6-38. A 900-mg sample containing only $Ca(ClO_4)_2$ and $CaCl_2$ was treated with oxalic acid to precipitate the calcium. The precipitate was ignited and the resulting calcium oxide weighed 306 mg. Calculate the percentage of each salt in the sample.

*6-39. A sample of carbonate rock weighing 1.250 g yields a precipitate of the hydrated oxides of Fe and Al. These are filtered off and ignited. The resulting Fe_2O_3 and Al_2O_3 are found to weigh 117.5 mg. On a separate sample of the rock, a volumetric method shows 3.22% Fe. Calculate the percentage of Al in the rock.

6.40. Polychlorobiphenyls (PCB's) have gained widespread attention in environmental analysis. Their manufacture involves the following reaction:

$$C_{12}H_{10} + nCl_2 \xrightarrow{\text{Fe catalyst}} C_{12}H_{10-n}Cl_n + nHCl$$

In this reaction, a mixture of products is obtained with various amounts of chlorine substitution. If the product mixture is analyzed by decomposition followed by precipitation of Cl^- as AgCl, develop an expression for the average value of n from the sample weight and AgCl weight. Based on this expression, calculate n for a sample that weighs 115 mg and yields 282 mg of AgCl.

VOLUMETRIC METHODS

PART THREE

Principles of Volumetric Analysis

7

7-1 VOLUME AS A MEASUREMENT SIGNAL

The analytical possibilities that arise when mass is used as the measurement signal were presented in Part Two. Let us now examine the use of volume as a measurement signal. Volumetric methods include any procedure in which a measurement of volume is related to the analyte. For example, a volume change measured in a gaseous or liquid system (measuring the volume of a solid has little analytical value) might represent the consumption of an analyte or reagent or the formation of product. This type of volumetric analysis most often involves consumption or generation of a gas. If the gas is considered ideal, the milliequivalents of analyte can be related to the measured gas as follows:

$$\text{meq analyte} = \text{meq gas} = \frac{V \text{ (mL at STP)}}{22.4/n} \qquad \textbf{(7-1)}$$

When the gas is measured at conditions other than STP, the volume must be corrected or the molar volume must be used for the gas at the measurement conditions. The gas might be a reagent that reacts with the analyte, a product of the reaction, or the analyte itself. Some examples are discussed later in this chapter.

A more common volumetric method of analysis involves adding a known volume of reagent solution to the analyte to carry out a chemical reaction. The carefully controlled addition of reagent to the analyte in such a procedure is called *titration*, and this special class of analysis is commonly called *titrimetry*. Volumetric analyses employing titrations are widely used; in fact, most of the discussion in Part Three is devoted to titrimetry.

7-2   VOLUMETRIC ANALYSIS REQUIREMENTS

A titration may involve the same chemical reaction as a gravimetric procedure. The determination of chloride ion as AgCl provides an example of similarities and differences between volumetric and gravimetric procedures. In the gravimetric procedure, a solution of $AgNO_3$ is added to a solution containing the chloride analyte. With each addition of $AgNO_3$, more precipitate forms until the reaction shown in Equation 7-2 is complete:

$$Cl^- \text{ (analyte)} + Ag^+ \text{ (precipitating reagent)} \rightleftharpoons AgCl(s). \qquad (7\text{-}2)$$

In the gravimetric procedure, an excess of $AgNO_3$ is added to provide excess common ion, thereby reducing solubility losses to an acceptable level. If the point at which precipitation ceased were noted, it would be unnecessary to add excess $AgNO_3$ or to digest, filter, wash, dry, and weigh the AgCl precipitate to determine the amount of chloride analyte present. The milliequivalents of $AgNO_3$ added to this point would be specified as follows:

$$\text{meq } AgNO_3 = N_{AgNO_3} V_{AgNO_3} \qquad (7\text{-}3)$$

and

$$\text{meq Cl} = \text{meq } AgNO_3 = N_{AgNO_3} V_{AgNO_3} \qquad (7\text{-}4)$$

Whereas the gravimetric procedure requires several additional steps (and therefore requires more time), the volumetric procedure requires a more carefully measured addition of $AgNO_3$ reagent. In addition, the $AgNO_3$ reagent—which can be just an approximate solution (such as 5%) for the gravimetric procedure—must have an accurately known concentration in the volumetric procedure. Finally, knowing the exact point at which the reaction is complete is not essential in the gravimetric procedure because excess $AgNO_3$ is to be added anyway; in the volumetric procedure, however, some convenient method for detecting the equivalence point is needed. Therefore, the requirements for a successful volumetric procedure (titration) can be summarized as follows:

1. Reagent solution of known concentration (standard solution)

2. Accurate measurement of volume

3. Endpoint signal available

4. Known relationship between endpoint and equivalence point

STANDARD SOLUTION

Volumetric methods are nearly as accurate as gravimetric methods; errors of a few tenths of 1% are typical. Therefore, the concentration of the reagent solution must be known to that level of accuracy. Although the concentration value could be stated in any of the units discussed in Chapter 3, normality (equivalents per liter or milliequivalents per milliliter) is commonly used. However, the use of formality with EDTA solutions is convenient because this complexing agent forms 1:1 complexes with metal ions. A reagent solution of accurately known concentration is called a *standard solution*.

We may attempt to prepare a standard solution by dissolving a weighed amount of the solid reagent in a known volume of solution. (It is the final volume that must be known, not the volume of solvent added to the reagent.) To ascertain the final concentration, the analyst must weigh the reagent to within 0.1% and measure the volume of solution to within 0.1%. These requirements are not difficult to meet with an analytical balance and a *volumetric flask*. However, it is much more difficult to locate a reagent that maintains 100.0 ± 0.1% purity for a reasonable storage time. Materials that meet these requirements for high purity and stability (stable in regard to decomposition as well as weight changes associated with the gain or loss of water) are called *primary standards*. Standard solutions prepared by dissolving a known amount of primary standard material in a solvent and diluted to an accurately known volume are said to be prepared *determinately*.

Because so few materials meet the requirements for a primary standard, many volumetric methods employ solutions containing other reagents. These solutions must be *standardized* with a primary standard material to determine the concentration accurately. Thus primary standard materials have two uses in the preparation of standard solutions:

1. Primary standard materials may be used to prepare standard solutions determinately—that is, dissolving an accurately weighed amount in an accurately known volume of final solution.

2. Primary standard materials may be used to standardize standard solutions— that is, determining the concentration of a reagent solution by titrating an accurately weighed amount of primary standard with the reagent solution.

Examples 7-1 and 7-2 show how a primary standard material may be used in these two ways. To be useful reagents for titrations, standard solutions must meet as closely as possible the following requirements:

1. A standard solution must contain a primary standard reagent or be able to be standardized with a primary standard material.

2. A standard solution must be stable for a reasonable length of time—that is, its concentration may change only slowly during storage.

3. The reagent should react quickly with analytes.

4. The reagent must react with analytes with a definite stoichiometry.

5. The reagent must react nearly completely with analytes; in other words, the equilibrium constant for the reaction must be high.

6. A signal must be available to detect the equivalence point in the reaction.

7. The reagent must be sufficiently soluble to allow preparation of solutions of desired concentration. Concentrations are typically 0.1 N.

Standard solutions are frequently prepared with ordinary reagents and then standardized with primary standard materials because primary standards, to be used for preparing standard solutions determinately, must meet the previously listed

EXAMPLE 7-1

Silver nitrate is available as a primary standard material. Describe how 250 mL of 0.1200 N $AgNO_3$ should be prepared determinately. Assume that the reagent will be used for precipitation reactions such as the precipitation of chloride ion.

Solution

$$\text{meq} = NV$$

$$\text{meq } AgNO_3 = 0.1200 \times 250 = 30.00$$

$$\text{meq } AgNO_3 = \frac{\text{mg } AgNO_3}{\text{FW } (AgNO_3)/n} \qquad n = 1 \text{ because } AgNO_3 \text{ contains a charge of one}$$
$$\text{(see Chapter 3)}$$

$$\text{mg } AgNO_3 = 30.00 \times \frac{169.9}{1} = 5097 \text{ (5.097 g)}$$

Therefore, 5.097 g of primary standard $AgNO_3$ should be dissolved in H_2O and, after dissolution, the solution should be diluted to exactly 250 mL in a volumetric flask.

Suppose the analyst weighs out 5.103 g of $AgNO_3$; the 250-mL volumetric flask has been calibrated to contain only 249.3 mL. Calculate the normality of the $AgNO_3$ solution.

$$\text{meq } AgNO_3 = \frac{\text{mg } AgNO_3}{\text{FW } (AgNO_3)/n} = \frac{5103}{169.9/1} = 30.04$$

$$N_{AgNO_3} = \frac{\text{meq}}{\text{mL}} = \frac{30.04}{249.3} = 0.1205 \ N$$

seven requirements *and* meet the following requirements for a primary standard material:

1. A primary standard material must exhibit high purity, typically $100.00 \pm 0.05\%$. (It may appear "impossible" to be 100.05% pure, but purity must be determined by some analytical procedure involving uncertainty—that is, indeterminate error. In addition, some impurities may give the compound a lower equivalent weight, thereby causing the assayed primary standard to appear more than 100% pure.) Some method of assaying to verify the purity must be available.

2. A primary standard must exhibit high stability, preferably for years under proper storage conditions. It must be stable with respect to both spontaneous decomposition and reaction with moisture or air. Primary standards preferably should not contain water of hydration; if they do, it could be impossible to dry and weigh the material without incurring a gain or loss of water.

3. High equivalent weight is preferable when used for standardizing solutions, because the amount of primary standard used per trial in standardization is

---

EXAMPLE 7-2

An analyst wishes to prepare a standard solution of NaCl to use in the titration of solutions containing silver ion. Approximately 3 g of NaCl are dissolved in about 500 mL of H_2O. The NaCl solution is standardized by titrating samples of primary standard $AgNO_3$. Calculate the normality of the NaCl solution, assuming that a 685.2-mg sample of $AgNO_3$ requires 38.23 mL of the NaCl solution in the titration.

Solution

First, we may calculate the approximate concentration of the NaCl solution as a useful check.

$$\sim N_{NaCl} = \frac{meq}{mL} = \frac{mg\ NaCl/[FW\ (NaCl)/n]}{mL} = \frac{3000/(58.44/1)}{500}$$

$$= 0.103\ N$$

The accurate value must be determined using the primary standard:

$$meq\ NaCl = meq\ AgNO_3$$

$$NV = \frac{mg\ AgNO_3}{FW\ (AgNO_3)/n} = \frac{685.2}{169.9/1} = 4.033$$

$$exact\ N_{NaCl} = \frac{4.033}{V} = \frac{4.033}{38.23} = 0.1055\ N$$

---

typically 3 to 5 meq. If the equivalent weight were < 50, the weight per trial would be < 150 to 250 mg and weighing uncertainties with normal analytical balances would become significant. However, this requirement is not important if the primary standard is to be used for preparing a standard solution determinately; in fact, a low equivalent weight would help meet the following requirement because less would be needed to prepare the standard solution.

4. The primary standard must be reasonably priced and generally available. This is a practical consideration; in time of inflation, nothing has a "reasonable" cost. However, given a choice, a lab director would prefer to purchase the less expensive reagent and would also prefer one that could be obtained on short notice.

Clearly, some materials might meet the four requirements for a primary standard without meeting the seven requirements for a standard solution. The primary standard must be stable for years in a closed bottle in the dry state, but may react with air when dissolved in water. The primary standard may react slowly or incompletely (low equilibrium constant) with many analytes but may react reasonably well in one particular standardization reaction. For example, Na_2CO_3 is a primary standard base often used to standardize HCl solutions. Because it is the salt of a weak acid, it usually does not have a large equilibrium constant for its reaction with a weak acid but reacts satisfactorily in titration with a strong acid such as HCl. Many primary standard materials exhibit no convenient signals to indicate completion of the reaction. Finally, the solubility of the primary standard may be limited. Although this problem can be overcome when the primary standard is used for standardizing solutions (for example, a large volume of sample solution, could be used, a mixed solvent might be acceptable for the sample solution, or a back-titration technique could be used, as will be discussed in Section 7-3), its limited solubility could prohibit the use of a primary standard as a standard solution.

Some materials do not meet the first requirement for a primary standard material, but they can be *assayed* (purity level determined) and meet all other requirements. For example, some sodium chloride may have an assay value of 99.86%. This value can then be used in calculations to adjust for the fact that it was not 100.00% pure, as shown in Example 7-3. Materials of lesser (but accurately

---

EXAMPLE 7-3

Sodium chloride with an assay of 99.86% is used to standardize a $AgNO_3$ solution. Assuming that 241.5 mg of the NaCl requires 47.56 mL of the $AgNO_3$ solution to reach the equivalence point, calculate the normality of the $AgNO_3$ solution.

Solution

$$meq\ NaCl = meq\ AgNO_3$$

$$\frac{mg\ NaCl}{FW\ (NaCl)/n} = N_{AgNO_3}\ V_{AgNO_3}$$

$$mg\ NaCl = 241.5\ \text{mg Sample} \times 0.9986\ \frac{mg\ NaCl}{mg\ sample} = 241.2\ mg$$

$$\frac{241.2}{58.44/1} = N_{AgNO_3} \times 47.56$$

$$N_{AgNO_3} = 0.08677\ N$$

For convenience, the equivalent weight is sometimes adjusted to compensate for the assay value; then the assay value is not needed again for calculations. In this example, the NaCl equivalent weight would be

$$EW = \frac{mg\ NaCl/meq}{mg\ NaCl/mg\ sample} = \frac{58.44}{0.9986} = 58.52\ \frac{mg\ sample}{meq}$$

In the preceding calculation,

$$\frac{mg\ sample}{EW} = N_{AgNO_3} V_{AgNO_3}$$

$$N_{AgNO_3} = \frac{241.5}{58.52 \times 47.56} = 0.08677\ N$$

known) purity are called *secondary standards*. They could be included as primary standards if the first requirement were expanded to include materials of "high purity, typically $100.00 \pm 0.05\%$ *or* reliably known purity to within $\pm 0.1\%$." Table 7-1 lists several materials that are available as primary standards and the reaction type for which they are useful.

The *titer* is another concentration scale occasionally used for standard solutions, especially for the determination of one particular analyte. The titer of a solution is the number of milligrams of analyte equivalent to 1 mL of the solution, as shown in Example 7-4. When the titer is given, the amount of analyte is calculated easily from the volume of standard reagent solution used in the titration:

$$\text{mg analyte} = \text{titer} \left(\frac{\text{mg analyte}}{\text{mL reagent solution}} \right) V \ (\text{mL reagent solution}) \qquad \textbf{(7-5)}$$

EXAMPLE 7-4

A $KMnO_4$ solution is to be used for determining Fe_2O_3 content of iron ores.
(a) If the concentration of the $KMnO_4$ is $0.1314 \ N$, determine its titer value for Fe_2O_3.
(b) Calculate the percentage of Fe_2O_3 in an iron ore sample weighing 484.3 mg, assuming that it required 37.45 mL of the $KMnO_4$ solution to reach the equivalence point.

Solution

(a)
$$\text{meq } Fe_2O_3 = \text{meq } KMnO_4$$

$$\frac{\text{mg } Fe_2O_3}{\text{FW } (Fe_2O_3)/n} = NV$$

(*Note:* In this titration, iron is oxidized from $+2$ to $+3$. Therefore, $n = 1$ for each iron atom. Fe_2O_3 contains two iron atoms; consequently, $n = 2$.)
Because titer is defined as milligrams of analyte per mL of reagent solution, the volume value is 1.

$$\frac{\text{mg } Fe_2O_3}{159.7/2} = 0.1314 \times 1$$

$$\frac{\text{mg } Fe_2O_3}{\text{mL reagent}} = \text{titer} = \frac{159.7}{2} \times 0.1314 \times 1 = 10.49 \left(\frac{\text{mg } Fe_2O_3}{\text{mL } KMnO_4} \right)$$

(b)
$$\text{mg } Fe_2O_3 = 10.49 \ \frac{\text{mg } Fe_2O_3}{\text{mL } KMnO_4} \times 37.45 \ \text{mL} = 392.9 \ \text{mg}$$

$$\% \ Fe_2O_3 = \frac{392.9 \ \text{mg } Fe_2O_3}{484.3 \ \text{mg sample}} \times 100(\%) = 81.13\%$$

TABLE 7-1   Primary Standards

Reaction type	Primary standards available	Use
precipitation	$AgNO_3$	prepare solution determinately, standardize KSCN, NaCl
acid–base	Na_2CO_3	standardize acid
	$Na_2B_4O_7 \cdot 10H_2O$	standardize acid
	constant-boiling HCl	prepare solution determinately
	$KH(IO_3)_2$	standardize base
	benzoic acid	standardize base
	$H_2C_2O_4 \cdot 2H_2O$	standardize base
	potassium hydrogen phthalate	standardize base, standardize $HClO_4$ (acetic acid solvent)
redox	$H_2C_2O_4 \cdot 2H_2O$	standardize oxidant $(KMnO_4)$
	$Na_2C_2O_4$	standardize oxidant $(KMnO_4)$
	$K_2Cr_2O_7$	prepare solution determinately, standardize reductant $(Na_2S_2O_3)$
	KIO_3	prepare solution determinately, standardize reductant $(Na_2S_2O_3)$
	$Ce(NO_3)_4 \cdot 2NH_4NO_3$	prepare solution determinately
	As_2O_3	prepare solution determinately, standardize oxidant (Ce^{4+})
	KI	standardize oxidant (Ce^{4+})
	$Na_2S_2O_3$	prepare solution determinately, standardize I_3^- solution
	Fe wire	standardize oxidant $(KMnO_4)$
	$FeSO_4 \cdot (NH_4)_2SO_4 \cdot 6H_2O^*$	prepare solution determinately, standardize oxidant $(KMnO_4)$
complexation	disodium salt of EDTA[†]	prepare solution determinately, standardize $MgCl_2$ solution
	$CaCO_3$	standardize EDTA solution

* Ferrous ammonium sulfate (Mohr's salt) is a useful reagent, but it does not strictly meet primary standard requirements for purity. It can be used when accuracy of a few tenths of 1% is sufficient.
† The acid form of EDTA is available as a primary standard at high cost. More often, the disodium salt reagent is used although it does not strictly meet the primary standard requirement for purity. It can be used when accuracy of a few tenths of 1% is sufficient.

MEASUREMENT OF VOLUME

The second requirement for a successful volumetric analysis is the accurate measurement of volume. For the gravimetric method for determining chloride using $AgNO_3$ as the precipitating reagent, a dropper could be used with little attention paid to the number of drops required. In the volumetric method for this same analysis, the volume of $AgNO_3$ required must be measured carefully. The standard laboratory glassware for measuring reagent volume in titrations is called a *buret*. In volumetric methods in which gas volumes are measured, a similar piece of glassware—a gas buret—is used.

In a typical titration, volumes of reagent are 30 to 50 mL. To measure a volume of 30 mL to within 0.1%, the uncertainty for this measurement must be $< \pm 0.03$ mL (about one drop). Because there is just as much uncertainty in reading the initial level as the final level, a buret must be read to ± 0.02 mL to reach the desired uncertainty level. Remember that, as stated in Equation 2-4, the overall variance of a process $\sigma^2_{process}$, is equal to the sum of the individual variances. In this case, there is an initial volume uncertainty (σ^2_i) and a final volume uncertainty (σ^2_f). The square root of the total variance (σ^2_t) is then a measure of the overall uncertainty:

$$\sigma^2_t = \sigma^2_i + \sigma^2_f = (0.02)^2 + (0.02)^2 = 0.0008 \text{ mL}^2$$

$$\sigma_t = \sqrt{\sigma^2_t} = \sqrt{0.0008} = 0.028 \approx 0.03 \text{ mL} \tag{7-6}$$

For some titrations, the use of 30 to 50 mL of reagent solution is impossible or inconvenient; in such case, other volume-measuring glassware is needed. Burets are available in 2 to 100-mL sizes. Because reading smaller burets entails somewhat larger *relative* uncertainties, analyses using these burets will have larger percentage errors (several tenths of 1%).

There are volume-measuring devices other than burets. Automatic titrators may use a motor-driven syringe that delivers perhaps only 1 mL with an uncertainty of ± 0.001 mL (0.1%). This approach uses considerably less reagent solution than in a conventional titration, essentially no volume change occurs in the sample solution, and the motor-driven syringe can be coupled to the signal used to stop the titration. However, an automatic titrator costs several thousand dollars; a 50-mL buret costs only a few dollars.

ENDPOINT SIGNAL

When the gravimetric procedure for chloride analysis was carried out, the point when precipitation ceased could be noted with a fair amount of confidence; then an excess amount was added in any case. In fact, after the precipitate is digested, the supernatant solution must be checked once more through the addition of a drop of $AgNO_3$ solution to verify that sufficient $AgNO_3$ has been added. Detecting the formation of AgCl precipitate with successive drops of $AgNO_3$ reagent solution would involve a lengthy process. Therefore, to use this procedure as a volumetric method of analysis, some signal that tells the analyst to stop adding reagent and

TABLE 7-2   Endpoint Signals

Phenomena involved	Signal observed	Example
acid–base equilibria	color change	color indicator, such as phenolphthalein or methyl red (acid–base titration)
redox equilibria	color change	color indicator, such as Na-diphenylamine sulfonate (titration of Fe^{2+} with $Cr_2O_7^{2-}$)
excess reagent	color change	$KMnO_4$ titrant
complexation equilibria	color change	color indicator, such as eriochrome black T (titration of Mg^{2+} with EDTA)
precipitation/complexation equilibria	color change	red $Fe(SCN)^{2+}$ "Volhard" (titration of Ag^+ with KSCN)
adsorption equilibria	color change	dichlorofluorescein "Fajans" (titration of Cl^- with $AgNO_3$)
precipitation equilibria	color change (second precipitation)	brown Ag_2CrO_4 "Mohr" (titration of Cl^- with $AgNO_3$)
specific complex formation	color change	blue starch–I_3^- (titration of I_3^- with $Na_2S_2O_3$)
precipitation/complexation equilibria	precipitation starts	$Ag[Ag(CN)_2](s)$ (titration of CN^- with $AgNO_3$)
precipitation	precipitation ends	"clear point" (titration of Cl^- or I^- with $AgNO_3$)
precipitation/complexation equilibria	precipitation disappears	dissolution of $AgI(s)$ (titration of Ni^{2+} with KCN)
membrane potential	potential change	glass electrode (acid–base titration)
membrane potential	potential change	ion-selective electrode (titration of Ca^{2+} with EDTA)
solution potential	potential change	Pt electrode (redox titration, such as Fe^{2+} with $K_2Cr_2O_7$)
oxidation or reduction of excess titrant	current change	amperometric endpoint, such as titration of Pb^{2+} with $K_2Cr_2O_7$
conductivity of ions in system	conductivity change	conductometric titration, such as weak acid with NaOH
heat of reaction	temperature change	thermometric titration, such as boric acid with NaOH

make the final reading on the buret must be available. This is the *endpoint signal*. As a simple example, an inexperienced analyst might add reagent from a buret until an experienced analyst said, "Stop, that's it." These words would be the endpoint signal. Unfortunately or fortunately (as the case may be), chemists seldom have someone looking over their shoulder to tell when to stop adding reagent (*titrant*). The system itself must somehow warn us when to stop. Analytical chemists have been ingenious at discovering phenomena that signal the analyst to stop adding titrant because the reaction is complete. Table 7-2 lists several phenomena and signals but is by no means complete.

ENDPOINT VERSUS EQUIVALENCE POINT

The endpoint signal tells the analyst to stop adding titrant and record the volume. The total number of milliequivalents of reagent added will equal NV. However, this value will equal the total milliequivalents of analyte present *only* if the endpoint occurs precisely at the equivalence point. The endpoint signal frequently occurs at some point other than the equivalence point, in which case the analyst must know how to correct for this difference. In general, the difference between endpoint and equivalence point will vary somewhat from trial to trial, thereby introducing another indeterminate error to the procedure. However, it is important to determine the expected difference between endpoint and equivalence point; failure to correct for this difference could introduce determinate error in the procedure.

A simple case occurs when a finite amount of titrant is required to produce a color change. For example, when $KMnO_4$ is the titrant, no color change occurs at the equivalence point; a certain amount of $KMnO_4$ must be added to produce the pink color of excess MnO_4^- ion. To correct for this potential error, the analyst determines how much $KMnO_4$ must be added to a *blank* solution to produce the endpoint signal. The volume of the blank solution should approximately equal that of the sample solution (at the endpoint) and should contain no substance that might react with MnO_4^-. Water can be used if the sample solution is colorless, although a blank solution of the same background color as the sample solution is preferable. A typical correction is shown in Example 7-5.

When color indicators are used, as in an acid–base titration, the endpoint signal will occur at a specific pH level. If this pH level differs from that at the equivalence point, an endpoint error results. Calculation of the equivalence point pH is critical to the choice of a suitable indicator. However, no indicators may be available for the particular pH required, in which case a correction may be necessary. This type of calculation is discussed in Section 9-2.

In addition, because the perception of color changes may vary from analyst to analyst, an endpoint error could be present even if the published color change pH matches the equivalence point pH. In fact, indicators for acid–base titrations change color not at a particular pH but over a range of pH (1 to 2 pH units). If the equivalence point region is sharply defined—a large change in pH with a small volume of titrant—the indicator will make a sharp transition from its base-form color to its acid-form color (or vice versa). On the other hand, a poorly defined equivalence point region will require a significant volume of titrant to convert the indicator from one color to the other.

EXAMPLE 7-5

The titanium content of an alloy is determined by dissolving the sample, reducing the titanium to the $+3$ state, and titrating with $KMnO_4$ (which oxidizes titanium to the $+4$ state). Calculate the percentage of titanium in an alloy sample weighing 403.2 mg if 27.45 mL of 0.02631 N $KMnO_4$ were required to reach the endpoint. A blank solution requires 0.08 mL of the $KMnO_4$ solution to produce the endpoint signal. What percentage error would result if the blank correction were not made?

Solution

The equivalence point must be passed with excess $KMnO_4$ to produce the endpoint signal. The amount required is determined to be 0.08 mL from the blank solution.

$$V_{eq\ pt} + V_{blank} = V_{endpt}$$

$$V_{eq\ pt} = 27.45 - 0.08 = 27.37\ \text{mL}$$

$$\text{meq Ti} = \text{meq } KMnO_4$$

$$\frac{\text{mg Ti}}{\text{FW (Ti)}/n} = N_{KMnO_4} V_{KMnO_4} \qquad Ti^{3+} \rightarrow Ti^{4+};\ \text{therefore, } n = 1$$

$$\frac{\text{mg Ti}}{47.90/1} = 0.02631 \times 27.37$$

$$\text{mg Ti} = 34.49\ \text{mg} \qquad \%\ \text{Ti} = \frac{34.49\ \text{mg Ti}}{403.2\ \text{mg sample}} \times 100\ (\%) = 8.55\%$$

If no blank correction is made,

$$\frac{\text{mg Ti}}{47.90} = 0.02631 \times 27.45$$

$$\text{mg Ti} = 34.59\ \text{mg} \qquad \%\ \text{Ti} = \frac{34.59}{403.2} \times 100\ (\%) = 8.58\%$$

$$\%\ \text{error} = \frac{34.59 - 34.49}{34.49} \times 100\ (\%) = +0.29\%$$

Note: The percentage error can be calculated from the volumes alone:

$$\frac{27.45 - 27.37}{27.37} \times 100\ (\%) = +0.29\%$$

Color indicators for redox titrations are similar to acid-base indicators except that they change color when the solution potential reaches a certain value. If the solution potential differs significantly from the equivalence point potential, an endpoint error requiring correction will occur.

One subtle way of correcting for an endpoint error is to standardize the titrant with a known amount of analyte (if the analyte is available in the pure form as a primary standard or as an assayed sample). For example, calcium content in minerals can be determined by titration with EDTA, using hydroxynaphthol blue indicator. To avoid endpoint error, the EDTA can be standardized with primary standard $CaCO_3$. A similar example is the determination of soda ash (Na_2CO_3) by titration with HCl that has been standardized with primary standard Na_2CO_3, using the same endpoint signal.

7-3   BACK-TITRATION

In several instances, the direct titration of an analyte with a reagent is not feasible. For example, the titration of an insoluble organic acid with NaOH is not practical because the analyst would have to wait for complete reaction to take place after each addition of NaOH. To circumvent this problem, the analyst must add an excess amount of NaOH, allow the reaction to reach completion, and titrate the excess NaOH with a standard solution of HCl: The system has gone from being acid, past the equivalence point to the basic side (excess base), and then *back* to the equivalence point. The final titration to the equivalence point is called a *back-titration*, and calculations for this type were given in Example 3-11.

The preceding example is essentially a case of slow kinetics due to the heterogeneity (*solid* acid with base *solution*) of the reaction. Bromination of an organic compound is another example of slow kinetics, as shown in Example 7-6.

Slow kinetics is not the only reason for using a back-titration method: Oxidizing agents, such as $Cr_2O_7^{2-}$, Ce^{4+}, and even MnO_4^-, are more stable in solution than are most reducing agents because reducing agents are susceptible to oxidation by atmospheric oxygen. Therefore, if the analyte is an oxidizing agent, it may be more convenient and more accurate to add an excess amount of reducing agent in the form of a solid primary standard and then back-titrate the excess reducing agent with a standard solution of an oxidizing agent (Example 7-7).

Another reason for employing a back-titration is that the endpoint signal may be readily available, easy to detect, or close to the equivalence point in the back-titration direction. The permanganate endpoint is one example; for in addition to the stability problem mentioned above, reducing agents as a general rule are not intensely colored, whereas $KMnO_4$ is a deep purple and quite easy to detect shortly after the equivalence point.

The determination of heavy metals with EDTA can be carried out by adding excess EDTA and back-titrating with a $MgCl_2$ standard solution, with eriochrome black T as the indicator. This indicator forms red complexes with many metal ions, but if the metal-indicator-complex formation constant is not in the correct range, the endpoint will not correspond to the equivalence point.

EXAMPLE 7-6

The phenol content in an industrial discharge is determined by addition of excess KBr and a known amount of $KBrO_3$ to produce Br_2:

$$BrO_3^- + 5Br^- + 6H_3O^+ \rightleftharpoons 3Br_2 + 9H_2O$$

The bromine reacts with phenol according to the following equation

The excess bromine is back-titrated with As_2O_3; α-naphthoflavone is used as an indicator. Calculate the phenol content (ppm) in a 500-mL waste-stream sample if 35.0 mg of $KBrO_3$ is added and 21.63 mL of 0.0496 N As_2O_3 solution is required to reach the equivalence point in the back-titration.

Solution

$$\text{meq oxidants} = \text{meq reductants}$$

$$\text{meq } KBrO_3 = \text{meq } As_2O_3 + \text{meq phenol}$$

$$\frac{\text{mg } KBrO_3}{\text{FW } (KBrO_3)/n} = NV + \frac{\text{mg phenol}}{\text{FW (phenol)}/n}$$

$$n_{KBrO_3} = 6 \, (BrO_3^- \rightarrow 3Br_2)$$

$$n_{phenol} = 6 \, (\text{consumes } 3Br_2)$$

Therefore,

$$\frac{35.0}{167.0/6} = 0.0496 \times 21.63 + \frac{\text{mg phenol}}{94.11/6}$$

$$\text{mg phenol} = (1.257 - 1.073)15.68 = 2.89 \text{ mg}$$

$$\text{phenol content in waste stream} = \frac{2.89 \text{ mg}}{0.5 \text{ L}} = 5.78 \text{ mg/L (ppm)}$$

EXAMPLE 7-7

A chromite sample weighing 1.782 g was analyzed for its chromium content by first fusing it with sodium peroxide and then treating with H_2O to solubilize it and oxidize the chromium to Cr(VI). The sample was then acidified, and there was added 2.746 g of ferrous ammonium sulfate (FW = 392.1; $n = 1$), which reduced chromium back to Cr(III). The excess Fe(II) was back-titrated with 10.14 mL of 0.04033 N $K_2Cr_2O_7$. Calculate the percentage of chromium, as Cr_2O_3, in the sample.

Solution

$$\text{meq oxidants} = \text{meq reductants}$$

$$\text{meq } Cr_2O_3 + \text{meq } K_2Cr_2O_7 = \text{meq Fe}$$

$$\frac{\text{mg } Cr_2O_3}{\text{FW } (Cr_2O_3)/n} + N_{K_2Cr_2O_7} V_{K_2Cr_2O_7} = \frac{\text{mg ferrous ammonium sulfate}}{\text{FW (ferrous ammonium sulfate)}/n}$$

n for $Cr_2O_3 = 6[Cr(VI) \rightarrow Cr(III) = 3, \text{ and 3 for each } Cr \times 2Cr = 6]$

n for ferrous ammonium sulfate $= 1$ [Fe(II) $\rightarrow$ Fe(III) $= 1$]

$$\frac{\text{mg } Cr_2O_3}{152.0/6} + 0.04033 \times 10.14 = \frac{2746}{392.1}$$

$$\text{mg } Cr_2O_3 = (7.003 - 0.409)25.33 = 167.1 \text{ mg}$$

$$\% \ Cr_2O_3 = \frac{167.1 \text{ mg } Cr_2O_3}{1782 \text{ mg sample}} \times 100 \ (\%) = 9.37\%$$

7-4   GAS ANALYSIS

The volume of a gas can be a very sensitive measurement signal. If it is assumed that the ideal-gas law is obeyed, the volume of gas is given by Equation 7-7:

$$PV = nRT$$

$$V = \left(\frac{RT}{P}\right)n \tag{7-7}$$

That is, with a sensitivity factor of RT/P, the volume of gas is directly proportional to the number of moles present. To achieve high sensitivity, measurement of the gas can be carried out at low pressure (high vacuum). Example 7-8 shows how the sensitivity factor can be calculated.

In modern high-vacuum gas analysis the measurement signal is actually a pressure in a known volume so that Equation 7-7 should be written as Equation 7-8

$$PV = (RT)n \tag{7-8}$$

EXAMPLE 7-8

Calculate the sensitivity factor for the determination of hydrogen (as H_2O) in metals if the volume is measured at 10^{-3} torr (mm Hg) and 25°C.

Solution

$$H_2 + \tfrac{1}{2}O_2 \rightarrow H_2O$$

Therefore,

$$n_{H_2} = n_{H_2O}$$

$$\text{sensitivity factor} = \frac{RT}{P} \text{ (from Equation 7-7)}$$

$$= \frac{(0.082 \text{ L-atm/mol-deg}) \times 298 \text{ K}}{10^{-3} \text{ torr}/760 \text{ torr/atm}}$$

$$= 1.86 \times 10^7 \text{ L/mol, or } 1.86 \times 10^{10} \text{ mL/mol}$$

The PV product is measured and is conveniently stated in units of torr–milliliters (that is, 1 mL of gas at 1 torr pressure). A typical calculation is given in Example 7-9.

One important area of application for gas analysis is the determination of gases in metals. The presence of oxygen, hydrogen, or nitrogen in metals may give the metals poor physical properties; thus a knowledge of the concentrations of included gases is critical. Most metals contain carbon, oxygen, hydrogen, and nitrogen in either a free or a combined state. When such metals are heated to high temperature under vacuum, gas is evolved that contains carbon as CO or CO_2, hydrogen as H_2 or H_2O, and nitrogen as N_2. The determination of hydrogen, nitrogen, and oxygen can be made by adding some graphite powder to a chunk of metal sample so that, when the sample is fused under vacuum, all oxides will be reduced to CO while nitrogen is liberated as N_2 and hydrogen as H_2. The PV product of evolved gases gives the total CO + H_2 + N_2. Hydrogen is then oxidized to H_2O and absorbed in $MgClO_4$, while CO is oxidized to CO_2 and condensed in a liquid-nitrogen trap. The PV product of the remaining N_2 is measured, CO_2 is released from the liquid-nitrogen trap, and the PV product of CO_2 + N_2 is measured. The amount of hydrogen is determined by difference.

In recent years, analysis of evolved gases has been simplified through the use of chromatographic techniques (Chapter 19). Gas chromatography is well suited for this type of analysis because it separates the various gases evolved from the sample and then quantitatively measures each gaseous component.

The determination of impurities in metals is just one example of gas analysis, and this topic will not be pursued further here. It serves to show that volumetric methods include more than volume of a titrant and that liquid and gas techniques actually

EXAMPLE 7-9

Calculate the hydrogen content in a titanium alloy if fusion of a 2.00-g sample evolves sufficient H_2O to fill a 200-mL volume at a pressure of 1.52×10^{-3} torr at 25°C

Solution

$$PV = (RT)n$$

$$\frac{1.52 \times 10^{-3} \text{ torr}}{760 \text{ torr/atm}} \times 0.200 \text{ L} = (0.082 \frac{\text{L-atm}}{\text{mol-deg}} \times 298 \text{ K})n$$

$$n = \frac{1.52 \times 10^{-3} \times 0.200}{760 \times 0.082 \times 298} = 1.64 \times 10^{-8} \text{ mol } H_2O$$

and from Example 7-8,

$$n_{H_2} = n_{H_2O}$$

Therefore,

$$\text{wt } H_2 = 1.64 \times 10^{-8} \text{ mol} \times 2.016 \text{ g/mol} = 3.30 \times 10^{-8} \text{ g}$$

$$H_2 \text{ content} = \frac{3.30 \times 10^{-8} \text{ g } H_2}{2.00 \text{ g sample}} \times 10^6 = 0.017 \text{ ppm}$$

require measurement of a product: In titrimetry, the NV product is measured, whereas in gas analysis, the PV product is measured. In both cases, the product is directly proportional to the amount of analyte in the sample.

7-5   GAS-PHASE TITRATIONS

A volumetric titration normally involves measurement of concentration and volume of liquid solutions. It is also possible to measure these same properties for gaseous solutions. Instead of volume *per se*, the typical *gas-phase titration* involves measurement of flow rates (liters per minute). The equivalence relation becomes

$$\frac{\text{eq analyte}}{\text{min}} = \frac{\text{eq reagent}}{\text{min}} \qquad (7\text{-}9)$$

$$N_a\left(\frac{\text{eq}}{L}\right)F_a\left(\frac{L}{\text{min}}\right) = N_r\left(\frac{\text{eq}}{L}\right)F_r\left(\frac{L}{\text{min}}\right) \qquad (7\text{-}10)$$

where F_a and F_r respectively represent the flow rates of the analyte and reagent. A detector senses the presence of excess analyte or excess reagent, and the flow rate of reagent (or analyte) is adjusted until the endpoint is reached and Equation 7-10 can be used for calculation. Finding the endpoint in this way is not always accurate because of slow kinetics near the equivalence point or because the equilibrium constant is not very large, so that measurable amounts of analyte and reagent still remain at the equivalence point. A more accurate method involves a "titration curve," in which detector response as a function of reagent flow rate is measured. The concentration of analyte can be determined from the slope of the line or from the endpoint obtained by extrapolation to zero detector response.

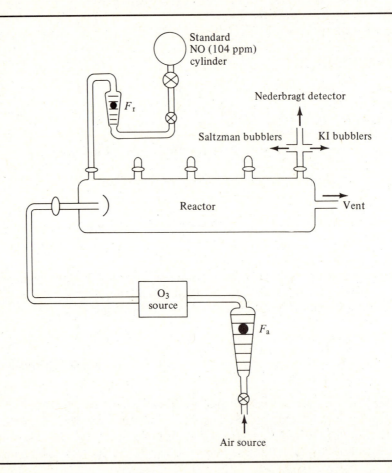

FIGURE 7-1 Schematic diagram for the gas-phase titration of ozone with nitric oxide. Flow rates of an O_3 analyte (F_a) and the NO reagent (F_r) are measured, and the presence of excess ozone is measured with the chemiluminescent detector (Nederbragt detector).

SOURCE: [From Hodgeson J.A., R.E. Baumgardner, B.E. Martin, and K.A. Rehme, "Stoichiometry in the Neutral Iodometric Procedure for Ozone by Gas-Phase Titration with Nitric Oxide," *Analytical Chemistry*, *43*: 1123 (1971).]

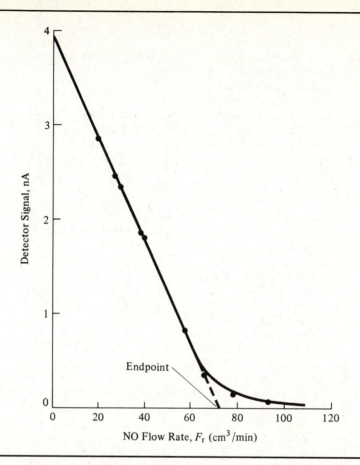

FIGURE 7-2   Titration curve for the titration of O_3 with NO. Detector response is measured as a function of NO flow rate. The endpoint is marked.

SOURCE: [From Hodgeson J.A., R.E. Baumgardner, B.E. Martin, and K.A. Rehme, "Stoichiometry in the Neutral Iodometric Procedure for Ozone by Gas-Phase Titration with Nitric Oxide," *Analytical Chemistry, 43*: 1123 (1971).]

An example of a gas-phase titration important in the environmental field is the determination of ozone (O_3) with nitric oxide (NO). The reaction is

$$O_3 + NO \rightleftharpoons O_2 + NO_2 \qquad (7\text{-}11)$$

A diagram of the apparatus is shown in Figure 7-1, and a titration curve is shown in Figure 7-2.

SUPPLEMENTARY READING

Kolthoff, I.M., V.A. Stenger, and R. Belcher, *Volumetric Analysis* (3 vols.). New York: Interscience Publishers, Inc., 1942-1957.

PROBLEMS

7-1. Determine the n value for each of the following primary standards, and calculate their equivalent weights:
*a. $AgNO_3$ (precipitation)
b. $AgNO_3$ [$Ag(CN)_2^-$ complexation]
c. Na_2CO_3 (acid–base to HCO_3^- equivalence point)
d. Na_2CO_3 (acid–base to CO_2 equivalence point)
e. $KHC_8H_4O_4$ (acid–base)
*f. $Na_2C_2O_4$ (redox to CO_2)
g. $K_2Cr_2O_7$ (redox to Cr^{3+})
h. $Ce(NO_3)_4 \cdot 2NH_4 NO_3$ (redox to Ce^{3+})
i. KI (redox to I_2)
j. Fe wire (redox Fe^{2+} to Fe^{3+})

7-2. Determine the n value for each of the following titrants, and calculate their equivalent weights:
*a. $KSCN$ (precipitation)
b. $KSCN$ [$Hg(SCN)_2$ complexation]
c. H_2SO_4 (acid–base)
d. $NaOH$ (acid–base)
*e. $KMnO_4$ (redox to Mn^{2+})
f. $KMnO_4$ (redox to MnO_2)
g. $KMnO_4$ (redox to MnO_4^{2-})
h. $Na_2S_2O_3$ (redox to $S_4O_6^{2-}$)
i. $KBrO_3$ (redox to Br^-)

7-3. Calculate the normality of the following solutions if they each contain 2.50 g per 500 mL:
*a. $K_2Cr_2O_7$ (redox to Cr^{3+})
b. H_2SO_4 (acid–base)
c. $KMnO_4$ (redox to Mn^{2+})
d. $AgNO_3$ (precipitation)
e. $Na_2S_2O_3$ (redox to $S_4O_6^{2-}$)
f. As_2O_3 (redox to AsO_4^{3-})

7-4. What is the difference between "primary standard" and "standard solution"?

7-5. Describe the two uses for primary standard materials in preparing standard solutions, and give an example for each.

*7-6. A NaOH solution is standardized against benzoic acid ($HC_7H_5O_2$). If a 452.3-mg sample of benzoic acid requires 39.67 mL of NaOH to reach the equivalence point, calculate the normality of the NaOH.

7-7. Calculate the normality of a $K_2Cr_2O_7$ solution prepared determinately if 2.639 g of primary standard material is dissolved in H_2O and diluted to 500 mL in a volumetric flask. The solution will be used as a titrant in redox reactions in which chromium is reduced to Cr^{3+}.

7-8. A $KMnO_4$ solution is standardized against oxalic acid ($H_2C_2O_4 \cdot 2H_2O$), in which the oxalic acid is oxidized to CO_2 while $KMnO_4$ is reduced to Mn^{2+}. If a 251.8-mg sample of oxalic acid requires 37.18 mL of $KMnO_4$ to reach the endpoint, calculate the normality of the $KMnO_4$.

---

*Answers to problems marked with an asterisk will be found at back of the book.

*7-9.   In Problem 7-8, a blank correction should have been made for the amount of $KMnO_4$ needed to produce a visible color after the equivalence point. If it is found that 0.04 mL is required, recalculate the normality of the $KMnO_4$ and calculate the determinate error that is introduced if the blank correction were neglected.

7-10.   Oxalic acid ($H_2C_2O_4 \cdot 2H_2O$) is frequently used as a primary standard for standardizing $KMnO_4$ for which $n = 2$ (oxalic acid is oxidized to CO_2). What would be its n value if the oxalic acid were used to standardize NaOH solutions or $CaCl_2$, in which CaC_2O_4 is precipitated?

7-11.   Analysis for Cl^- and other halides may often be carried out either gravimetrically or volumetrically by precipitation of the silver halide. Compare the inherent precision of these two methods using the following estimates: sample weight, 250 mg for both methods (± 0.2 mg for each weighing operation); AgCl weight, 600 mg (± 0.3 mg for each weighing operation); $AgNO_3$ concentration, $\pm 0.1\%$; $AgNO_3$ volume, 40 mL (± 0.03 mL for each volume reading); endpoint uncertainty, ± 0.04 mL. What additional factors should be considered when choosing between these two analytical methods?

7-12.   A $KMnO_4$ solution is standardized against Fe wire, which is first dissolved in acid to give Fe^{2+}. In this titration, $KMnO_4$ is reduced to Mn^{2+} while Fe^{2+} is oxidized to Fe^{3+}. Calculate the normality of the $KMnO_4$ if 36.17 mL is required to titrate a 210.6-mg sample of Fe wire. What would the normality of the $KMnO_4$ solution be if it is then used to titrate Mn^{2+}, in which $KMnO_4$ is reduced to MnO_2?

*7-13. Potassium hydrogen phthalate ($KHC_8H_4O_4$, abbreviated KHP) with an assay value of 99.76% is used to standardize a NaOH solution. If a 963.2-mg sample of KHP requires 37.16 mL of NaOH to reach the equivalence point, calculate the normality of the NaOH.

7-14. $CaCO_3$ with an assay value of 99.65% is used to standardize a solution of EDTA. If a 121.6-mg sample of $CaCO_3$ requires 48.63 mL of EDTA to reach the equivalence point for this 1:1 complexation reaction, calculate the formality of the EDTA and its CaO titer.

*7-15. A $K_2Cr_2O_7$ standard solution is prepared by dissolving 1.287 g of primary standard material in H_2O and diluting to 250 mL in a volumetric flask. What is its titer value in the determination of Fe_2O_3? In the redox titration, $K_2Cr_2O_7$ is reduced to Cr(III) while Fe(II) is oxidized to Fe(III).

7-16.   The $K_2Cr_2O_7$ standard solution described in Problem 7-15 is used to determine the iron content in an ore sample. Calculate the percentage of Fe_2O_3 if a 563.2-mg sample requires 43.64 mL of $K_2Cr_2O_7$ to reach the equivalence point.

7-17.   A $AgNO_3$ solution has a KCl titer value of 7.83 mg/mL. Calculate its normality.

7-18.   The $AgNO_3$ standard solution described in Problem 7-17 is used to determine the KCl content in impure KCl. If a 300-mg sample requires 38.13 mL of $AgNO_3$ to reach the equivalence point, determine the purity of the sample.

7-19.   When small amounts of analyte are present, the analyst may use a smaller buret with less uncertainty in reading volume. Alternatively, he may use a less concentrated standard solution. However, in the latter case, the endpoint uncertainty will increase; that is, it takes more volume to observe the endpoint signal. Compare these two approaches given the following information:

Small buret 10-mL volume with ± 0.01 mL each volume reading, endpoint uncertainty ± 0.02 mL

Dilute titrant 50-mL volume with ± 0.03 mL each volume reading, endpoint uncertainty ± 0.10 mL

7-20. Describe the difference between equivalence point and endpoint in a volumetric analysis.

7-21. In Example 7-5, the $KMnO_4$ blank titration carried out with the appropriate volume of H_2O is 0.08 mL. If the titanium alloy contains some Cr that is present as light-green Cr^{3+} in the titrated solution, what determinate error will be introduced? How can this endpoint error be corrected?

*7-22. Suppose that a base analyte in 100 mL total volume is to be titrated with 0.15 N HCl and that the equivalence point is theoretically at pH 5. If a color indicator is used whose endpoint signal is at pH 4, what is the endpoint correction?

7-23. For the situation described in Problem 7-22, suppose the base analyte to be a strong base. What would be the endpoint correction?

7-24. The blue starch–iodine endpoint signal can be observed when the excess iodine concentration is 10^{-6} F (this value varies greatly, depending on conditions). What would the endpoint correction be if the sample volume is 300 mL and the titrant is 0.02 N I_2? Calculate the relative determinate error introduced if this endpoint correction is neglected and the titration volume is 25 mL.

7-25. In some titrations with an amperometric endpoint, the signal is a linear increase in current after the equivalence point. The endpoint is found by extrapolation back to zero current. However, there may be a residual current, which constitutes an endpoint error if left uncorrected. Calculate the endpoint error if the residual current is 0.2 μA in the following titration:

8-Hydroxyquinoline (mL)	Current (μA)
22.20	5.0
22.45	7.0
22.70	9.0

*7-26. An organic mixture weighing 2.416 g and containing benzoic acid ($HC_7H_5O_2$) is extracted with 100.0 mL of 0.1513 N NaOH. After extraction, the excess base is titrated with 0.1063 N H_2SO_4, and 41.67 mL is required to reach the equivalence point. What is the benzoic acid content in the mixture?

7-27. The concentration of hypochlorite ion (OCl^-) in liquid bleach can be determined by a back-titration method in which the bleach first oxidizes iodide ion to iodine while OCl^- is reduced to Cl^-. Then excess $Na_2S_2O_3$ is added to reduce the iodine, followed by titration of the excess $Na_2S_2O_3$ with electrogenerated I_2. If a 1.00-mL sample of bleach is added to 50.0 mL of 0.0301 N $Na_2S_2O_3$ and 2.65×10^4 mC of electrical charge is required to reach the equivalence point, calculate the percentage of NaOCl (weight per volume) in the bleach.

*7-28. The amount of zinc in a zinc alloy is determined by measuring the volume of H_2 evolved when the sample is treated with HCl. If a 486-mg sample liberates 58.7 mL at 738 torr pressure and 23°C, calculate the percentage of Zn in the sample.

7-29. Calculate the hydrogen content (ppm) in a sample of zirconium if fusion of a 1.63-g sample evolves sufficient H_2O to fill a 200-mL volume at a pressure of 2.69×10^{-3} torr at 30°C.

7-30. The ozone content in a polluted atmosphere is determined by a gas-phase titration with NO. If the flow rates at equivalence are $NO = 1.63 \times 10^{-3}$ L/min and sample = 0.375 L/min and if the NO concentration is 4.11×10^{-6} mol/L, calculate the O_3 concentration as parts per million by volume (assume STP conditions).

Precipitation Titrations

8

The volumetric determination of chloride by the addition of a standard solution of $AgNO_3$ was used as an example for comparing a volumetric procedure with a gravimetric procedure. We now examine the titration in detail as a way of presenting the theoretical basis for a precipitation titration. One of our primary considerations will be to show how various endpoint signals may be used (see Table 7-2) for this titration. Following this discussion, a brief survey of other precipitation titrations will show how the theoretical principles may be applied to other situations of analytical interest.

8-1 THEORETICAL PRINCIPLES: THE TITRATION CURVE

When titrant is added to the solution containing analyte, the concentration of analyte will decrease with each addition. At the same time, the concentration of titrant in the solution will increase. A knowledge of the concentrations as a function of titrant volume added to the solution will help us to devise possible endpoint signals, calculate the expected accuracy, and be aware of potential interferences. A plot of concentration (or some other property related to the concentration, such as electrochemical potential) versus a titrant volume is called a *titration curve*. Titration curves will be deduced from theoretical principles for each of the four reaction types in Chapters 8 to 11, beginning with a precipitation titration of Cl^- with Ag^+ in this chapter.

Initially, before any $AgNO_3$ titrant is added, the concentration of chloride will be some initial value being sought in the analysis. For purposes of deriving a theoretical titration curve, let us assume a convenient value such as 0.1 F. At this point, the Ag^+ concentration will be zero, and no AgCl will be present in the system.

The addition of even a small amount of $AgNO_3$ will cause AgCl to precipitate, provided that K_{sp} has been exceeded; from this point on, the concentrations of Ag^+ and Cl^- are related by the solubility-product equilibrium constant:

$$Ag^+ \text{ (titrant)} + Cl^- \text{ (analyte)} \rightleftharpoons AgCl(s) \tag{8-1}$$

$$K_{sp} = 1.8 \times 10^{-10} = [Ag^+][Cl^-] \tag{8-2}$$

If we assume that the chemical reaction given in Equation 8-1 goes essentially to completion, the amount of chloride remaining will be

$$\text{mmol } Cl^- = (\text{mmol } Cl^-)_0 - \text{mmol } Ag^+ \tag{8-3}$$

where $(\text{mmol } Cl^-)_0$ is the number of millimoles of chloride present initially. The concentration of chloride will be

$$[Cl^-] = \frac{\text{mmol } Cl^-}{mL} = \frac{(\text{mmol } Cl^-)_0 - \text{mmol } Ag^+}{V_0 + V_{Ag}} \tag{8-4}$$

where V_0 is the initial volume and V_{Ag} is the volume of $AgNO_3$ added. In other words, $V_0 + V_{Ag}$ equals the total volume. For convenience the amount of analyte present initially and the amount of titrant added will be given in terms of formal concentration, which changes Equation 8-4 to Equation 8-5

$$[Cl^-] = \frac{F_{Cl}V_{Cl} - F_{Ag}V_{Ag}}{V_{Cl} + V_{Ag}} \tag{8-5}$$

The concentration of Ag^+ can be calculated at any point using the solubility product given in Equation 8-2

$$[Ag^+] = \frac{K_{sp}}{[Cl]} = \frac{1.8 \times 10^{-10}}{(F_{Cl}V_{Cl} - F_{Ag}V_{Ag})/(V_{Cl} + V_{Ag})} \tag{8-6}$$

When the equivalence point is reached (note that, with the 1:1 stoichiometry involved in this example, the equivalence point occurs when the moles of Ag^+ equal the moles of Cl^-), Equation 8-5 predicts that $[Cl^-] = 0$ because $F_{Cl}V_{Cl} = F_{Ag}V_{Ag}$. We know that that cannot be true because the solubility product must be satisfied. Our assumption that the reaction goes to completion cannot be justified at the equivalence point, and Equation 8-2 must be used. At the equivalence point, we can visualize that all Ag^+ and Cl^- form AgCl and that then some AgCl dissolves to establish the equilibrium. Therefore, at the equivalence point,

$$[Ag^+] = [Cl^-]$$

$$1.8 \times 10^{-10} = [Ag^+][Cl^-] = [Cl^-]^2$$

$$[Cl^-] = \sqrt{K_{sp}}$$

$$= [Ag^+] = 1.3 \times 10^{-5}$$

TABLE 8-1 Methods for Calculating Precipitation Titration Curve*

Region	Method	Analyte	Titrant
Initial ($V_T = 0$)	—	F_A	0
$0 < V_T < V_{eq\ pt}$	calculate excess analyte; use K_{sp} to determine titrant	$[A] = \dfrac{F_A V_A - F_T V_T}{V_A + V_T}$	$[T] = \dfrac{K_{sp}}{[A]}$
$V_{eq\ pt}$	solubility of product	$[A] = \sqrt{K_{sp}}$	$[T] = \sqrt{K_{sp}}$
$V_T > V_{eq\ pt}$	calculate excess titrant; use K_{sp} to determine analyte	$[A] = \dfrac{K_{sp}}{[T]}$	$[T] = \dfrac{F_T V_T - F_A V_A}{V_A + V_T}$

*Assuming analyte (A) and titrant (T) react with a 1:1 mole ratio. V_T = volume of titrant added.

After the equivalence point, Equation 8-5 would predict a *negative* concentration for $[Cl^-]$, which is clearly unrealistic. In actual fact, the chloride concentration drops to a very low value, controlled by the amount of excess $AgNO_3$ titrant added. If we now assume that all Ag^+ up to the equivalence point is precipitated as $AgCl$, then the concentration of Ag^+ will be given by Equations 8-7 and 8-8, which are analogous to Equations 8-4 and 8-5.

$$[Ag^+] = \frac{mmol\ Ag^+}{mL} = \frac{mmol\ Ag^+ - (mmol\ Cl^-)_0}{V_0 + V_{Ag}} \tag{8-7}$$

$$= \frac{F_{Ag} V_{Ag} - F_{Cl} V_{Cl}}{V_{Cl} + V_{Ag}} \tag{8-8}$$

The concentration of Cl^- can now be calculated from Equations 8-8 and 8-2:

$$[Cl^-] = \frac{K_{sp}}{[Ag^+]} = \frac{1.8 \times 10^{-10}}{(F_{Ag} V_{Ag} - F_{Cl} V_{Cl})/(V_{Cl} + V_{Ag})} \tag{8-9}$$

These three methods needed to calculate the titration curve are summarized in Table 8-1. Example 8-1 shows how a typical titration curve for the titration of chloride with $AgNO_3$ can be calculated. The results are displayed in Figure 8-1 on a linear scale and in Figure 8-2 on a logarithmic scale using p-functions.

8-2 ENDPOINT SIGNALS

Table 8-2 indicates that, at the equivalence point, the concentrations of chloride and silver ion are both $1.3 \times 10^{-5}\ M$. So what is needed is some type of endpoint signal that will trigger at this concentration or be triggered by the two concentrations'

EXAMPLE 8-1

Calculate a titration curve for the titration of 100 mL of 0.05 F chloride with 0.10 F $AgNO_3$. Determine $[Cl^-]$, $[Ag^+]$, pCl, and pAg at the following titrant volumes: 0, 10, 25, 40, 48, 49, 50, 51, 52, 60, and 100 mL.

Solution

At $V_{Ag} = 0$, $[Cl^-] = 0.05$ and $[Ag^+] = 0$ (because no Ag^+ has been added, pAg is not meaningful)

$$pCl = -\log 0.05 = 1.30$$

The equivalence point volume is determined by

$$mmol\ Ag^+ = mmol\ Cl^- \quad (1:1\ stoichiometry)$$

$$F_{Ag}V_{eq\ pt} = F_{Cl}V_{Cl}$$

$$V_{eq\ pt} = \frac{F_{Cl}V_{Cl}}{F_{Ag}} = \frac{0.05 \times 100}{0.1} = 50\ mL$$

For titrant volumes < 50 mL, the concentration of chloride is given by Equation 8-5. For example, with 10 mL added,

$$[Cl^-] = \frac{F_{Cl}V_{Cl} - F_{Ag}V_{Ag}}{V_{Cl} + V_{Ag}} = \frac{0.05 \times 100 - 0.1 \times 10}{100 + 10} = 3.6 \times 10^{-2}$$

The concentration of Ag^+ is calculated from Equation 8-6:

$$[Ag^+] = \frac{1.8 \times 10^{-10}}{[Cl^-]} = \frac{1.8 \times 10^{-10}}{3.6 \times 10^{-2}} = 5.0 \times 10^{-9}$$

This same calculation for $[Cl^-]$ and $[Ag^+]$ is carried out for the addition of 20, 40, 48,

being equal; we also know that, for all practical purposes, precipitation ceases after the equivalence point.

That last possibility can be made the basis for an endpoint signal; it is called the "clear point." Unfortunately, such a signal is quite slow because the analyst must wait after each addition of $AgNO_3$ to see whether additional precipitation occurs with the next drop. It can be practical when the amount of chloride is known approximately and the titration is merely to verify that the appropriate concentration is present (as in a process-control analysis). The method is aided by the fact that adsorption effects change sign at the equivalence point; that is, before the equivalence point, excess Cl^- will be predominantly adsorbed, while after the equivalence point, excess Ag^+ will be predominantly adsorbed. The adsorbed ions tend to keep the precipitate suspended in solution. At the equivalence point the switch in adsorption allows particles of AgCl to coagulate and settle to the bottom.

EXAMPLE 8-1 (cont.)

and 49 mL of 0.10 F AgNO$_3$. The results are given in Table 8-2. pCl and pAg are calculated by the definition of a p-function given in Chapter 3:

$$pCl = -\log[Cl^-] \qquad pAg = -\log[Ag^+]$$

The concentration of chloride and silver ion at the equivalence point (50 mL) was already given as

$$\text{equivalence pt: } [Cl^-] = [Ag^+] = \sqrt{K_{sp}} = 1.3 \times 10^{-5}$$

After the equivalence point, the concentration of silver ion is given by Equation 8-8. For example, with 51 mL added,

$$[Ag^+] = \frac{F_{Ag}V_{Ag} - F_{Cl}V_{Cl}}{V_{Cl} + V_{Ag}} = \frac{0.1 \times 51 - 0.05 \times 100}{100 + 51} = 6.6 \times 10^{-4}$$

The concentration of chloride is calculated from Equation 8-9:

$$[Cl^-] = \frac{K_{sp}}{[Ag^+]} = \frac{1.8 \times 10^{-10}}{6.6 \times 10^{-4}} = 2.7 \times 10^{-7}$$

Note that the concentration of chloride at 51 mL (1 mL after the equivalence point) is essentially the same as the concentration of silver ion at 49 mL (1 mL before the equivalence point). Likewise, the concentration of silver ion at 51 mL is essentially the same as the concentration of chloride at 49 mL. The slight difference results from the fact that there is a small volume change from 149 mL to 151 mL when these two points are compared in the titration curve. This symmetry around the equivalence point is clearly seen when p-functions are plotted (Figure 8-2) and offers a convenient check that the calculations had been carried out correctly.

Calculation of $[Cl^-]$ and $[Ag^+]$ at 52, 60, and 100 mL is carried out in the same fashion as was done for 51 mL of AgNO$_3$ added and completes the titration curve.

The supernatant solution has a tendency to become clear, and this clear point is readily discerned by an experienced analyst. The method works especially well in the titration of iodide with AgNO$_3$.

Methods are now available for measuring pCl and pAg with ion-selective electrodes (see Chapter 15) so that the titration curves shown in Figure 8-2 can be obtained experimentally. Because of uncertainty in K_{sp}, which will depend on the ionic strength of the solution at the equivalence point, the endpoint signal is usually taken as the point of greatest slope (in terms of the calculus, the first derivative is a maximum, and the second derivative is zero). There are some practical problems connected with the use of ion-selective electrodes to furnish an endpoint signal: Ion-selective electrodes cost around $200 each—in addition to some type of meter for recording the potential—and are not available in every laboratory. The activity of silver ion or chloride ion (when AgCl is present in the system) may also be measured

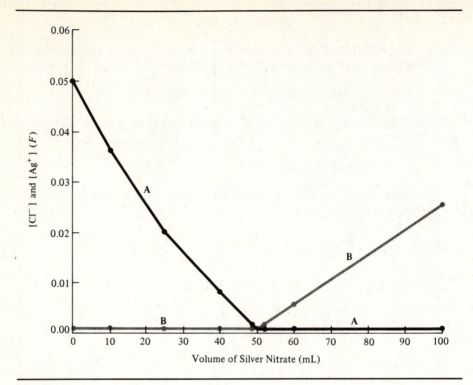

FIGURE 8-1   Titration of 100 mL of 0.05 F chloride with 0.10 F AgNO$_3$. Linear concentration scale: x = [Cl$^-$]; o = [Ag$^+$].

TABLE 8-2   Values of [Cl$^-$], [Ag$^+$], pCl, and pAg for Titration of 100 mL of 0.05 F chloride with 0.10 F AgNO$_3$

Volume of 0.10 F AgNO$_3$	[Cl$^-$]	pCl	[Ag$^+$]	pAg
0	0.050	1.30	0	—
10	0.036	1.44	5.0×10^{-9}	8.30
25	0.020	1.70	9.0×10^{-9}	8.05
40	7.1×10^{-3}	2.15	2.5×10^{-8}	7.60
48	1.35×10^{-3}	2.87	1.3×10^{-7}	6.88
49	6.7×10^{-4}	3.17	2.7×10^{-7}	6.57
50 (eq pt)	1.3×10^{-5}	4.87	1.3×10^{-5}	4.87
51	2.7×10^{-7}	6.57	6.6×10^{-4}	3.18
52	1.4×10^{-7}	6.86	1.3×10^{-3}	2.88
60	2.9×10^{-8}	7.54	6.3×10^{-3}	2.20
100	7.2×10^{-9}	8.14	0.025	1.60

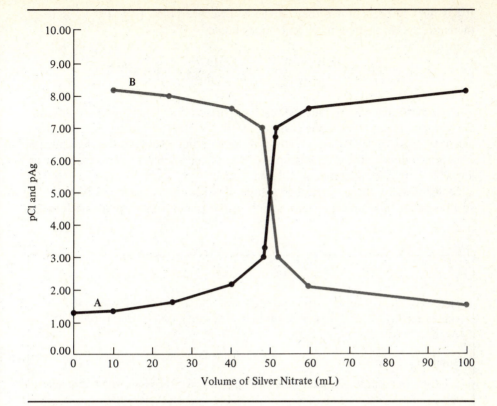

FIGURE 8-2   Titration of 100 mL of 0.05 F chloride with 0.20 F AgNO$_3$. Logarithmic concentration scale (p-function): A = pCl; B = pAg.

with a silver electrode. As with an ion-selective electrode, electrochemical potential is measured at several points throughout the titration, which takes considerable time. Therefore, endpoint methods that are faster and less costly are widely used.

One such method is to employ another compound that will *start* to precipitate at the equivalence point and can be detected by eye. Silver chromate meets these requirements, and its use is called the "Mohr titration" (K.F. Mohr published a paper describing this method in 1855). The concept is simple: Some chromate ion, CrO_4^{2-}, is added to the analyte solution, and after an amount of AgNO$_3$ equivalent to the chloride present has been added, the next addition of AgNO$_3$ forms the reddish-brown precipitate of Ag_2CrO_4.

The concentration of chromate ion needed so that precipitation of Ag_2CrO_4 begins at the equivalence point can be calculated with the solubility products for AgCl and Ag_2CrO_4:

$$AgCl: \quad K_{sp} = 1.8 \times 10^{-10} = [Ag^+][Cl^-]$$

$$Ag_2CrO_4: \quad K_{sp} = 1.2 \times 10^{-12} = [Ag^+]^2[CrO_4^{2-}]$$

It has already been shown that, at the equivalence point, $[Ag^+] = 1.3 \times 10^{-5}$; therefore, the chromate ion concentration should be

$$[CrO_4^{2-}] = \frac{1.2 \times 10^{-12}}{[Ag^+]^2} = \frac{1.2 \times 10^{-12}}{(1.3 \times 10^{-5})^2} = 6.7 \times 10^{-3}$$

This concentration of chromate, however, imparts such an intense yellow color that the endpoint is obscured. Concentrations of $< 5 \times 10^{-3}$ M are usually used. This means that $[Ag^+]$ will be $> 1.3 \times 10^{-5}$ at the endpoint, and a positive determinate error is introduced. However, Example 8-2 shows that, even with a concentration as low as 2×10^{-3} M, the error introduced is small.

Unfortunately, there is another source of error that must be considered: A small—but not insignificant—amount of Ag_2CrO_4 must be precipitated before it is

---

EXAMPLE 8-2

Determine the percentage error introduced by the use of a chromate indicator concentration of 2×10^{-3} M in the titration of 50 mL of 0.05 F chloride with 0.05 F $AgNO_3$.

Solution

Ag_2CrO_4 will start to precipitate when its K_{sp} is just met. Therefore, with a chromate ion concentration of 2×10^{-3} M,

$$1.2 \times 10^{-12} = [Ag^+]^2[CrO_4^{2-}]$$

$$[Ag^+]^2 = \frac{1.2 \times 10^{-12}}{[CrO_4^{2-}]} = \frac{1.2 \times 10^{-12}}{2 \times 10^{-3}} = 6 \times 10^{-10}$$

$$[Ag^+] = 2.4 \times 10^{-5}$$

Thus additional $AgNO_3$ must be added to reach the precipitation point for Ag_2CrO_4. This amount is

$$\text{Excess } [Ag^+] = [Ag^+] \text{ endpoint} - [Ag^+] \text{ equivalence point}$$

$$= 2.4 \times 10^{-5} - 1.3 \times 10^{-5} = 1.1 \times 10^{-5} \ M$$

The amount of excess $Ag^+ = MV = 1.1 \times 10^{-5}$ mmol/mL $\times (50 + 50)$ mL $= 1.1 \times 10^{-3}$ mmol

$$\text{percentage error} = \frac{\text{mmol excess } Ag^+ \times 100 \ (\%)}{\text{mmol } Ag^+ \text{ equivalence point}}$$

$$= \frac{+1.1 \times 10^{-3} \times 10^2}{0.05 \times 50} = +0.044\%$$

EXAMPLE 8-3

(a) Determine the percentage error introduced in the titration of 50 mL of 0.05 F chloride with 0.05 F AgNO$_3$ if it is necessary that a concentration of excess silver ion equal to 4×10^{-5} M be present before precipitation of Ag$_2$CrO$_4$ is observed.

(b) What volume of AgNO$_3$ will be expected for the blank titration?

Solution

(a) At the equivalence point, volume $= 50 + 50 = 100$ mL.

$$\text{mmol silver needed} = MV = 4 \times 10^{-5}\ \text{mmol/mL} \times 100\ \text{mL} = 4 \times 10^{-3}\ \text{mmol}$$

$$\text{percentage error} = \frac{\text{mmol excess Ag}^+ \times 100\ (\%)}{\text{mmol Ag}^+ \text{ equivalence point}}$$

$$= \frac{+4 \times 10^{-3} \times 10^2}{0.05 \times 50} = +0.16\%$$

(b)
$$\text{mmol Ag}^+ = FV$$

$$4 \times 10^{-3} = 0.05\ V$$

$$V = 0.08\ \text{mL}$$

---

visible to the eye. This amount can be determined by a blank titration: Silver nitrate is added to a volume of H$_2$O equal to the volume expected at the equivalence point, with some CaCO$_3$ present and playing the role of the white AgCl that will be present in the actual titration. The volume of AgNO$_3$ required to observe the reddish-brown Ag$_2$CrO$_4$ is the blank value, which is subtracted from the observed endpoint. The determinate error introduced if this correction is not taken into account can be significant, as shown in Example 8-3.

The Mohr titration is limited to neutral or slightly basic solutions. In acidic solution, the concentration of chromate is decreased by the following equilibria:

$$CrO_4^{2-} + H_3O^+ \rightleftharpoons HCrO_4^- + H_2O$$
$$2CrO_4^{2-} + 2H_3O^+ \rightleftharpoons Cr_2O_7^{2-} + 3H_2O$$

If the solution becomes too basic ($>$ pH 10), Ag$_2$O will start to precipitate:

$$2Ag^+ + 2OH^- \rightleftharpoons 2AgOH(s) \rightleftharpoons Ag_2O(s) + H_2O$$

Another endpoint technique, based on the adsorption properties of AgCl, may be used in weakly acidic solution. In Chapter 5, we discussed that AgCl precipitates will adsorb excess common ion and that this adsorbed charge will be neutralized by a counterion present in solution. In the titration of chloride ion with AgNO$_3$, there will be adsorbed Cl^- throughout the titration up to the equivalence point. After the equivalence point, the excess common ion is Ag^+, and it will be the adsorbed ion, as shown in Figure 8-3. If an anionic dye, D^-, is added as an indicator, it will act as the counterion so that a coprecipitation of AgD will occur. If AgD is colored, the white precipitate will turn color just past the equivalence point. Examples of such a dye are fluorescein and dichlorofluorescein. The color effect is twofold: Before the equivalence point, the fluoresceinate ion in solution is pale green, a color that decreases at the equivalence point when the fluoresceinate ion is adsorbed on the AgCl particles; at the equivalence point, the AgCl particles turn from white to pink because of the coprecipitation of silver fluoresceinate. It is truly an adsorption coprecipitation phenomenon because the solubility product of silver fluoresceinate is not exceeded. This method of determining the endpoint is called the "Fajans method" in recognition of the extensive research K. Fajans did in the 1920s on the subject of adsorption effects.

There are some practical considerations in the use of an adsorption indicator that must be taken into account to ensure high accuracy and a sharp, reliable endpoint. First, the dye is a weak acid and will exist predominantly in the un-ionized protonated form in strongly acidic solution ($pH < pK_a$ of the indicator). For this reason, dichlorofluorescein is usually recommended over fluorescein because it is a stronger acid and can be used in solutions as acidic as pH 3. A second consideration

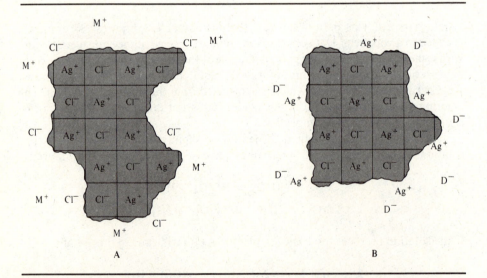

FIGURE 8-3 Mechanism for adsorption indicator endpoint signal. (**A**) Before equivalence point, excess Cl^- present with metal cation (M^+) as countercharge; precipitate surface is white. (**B**) After equivalence point, excess Ag^+ present with dye anion (D^-) as countercharge; precipitate surface is pink.

is that the endpoint depends on a large surface area, and throughout the titration the AgCl precipitate should be kept in colloidal form if possible. (Note how this requirement is just the opposite of the goals for a successful gravimetric determination of chloride or silver ion by precipitation of AgCl.) Procedures recommend addition of dextrin to keep the AgCl particles from coagulating. Noncoagulation is accompanied, however, by the deleterious effect that colloidal AgCl—especially with adsorbed dye—is rapidly photodecomposed, as shown in Equation 5-18. Photodecomposition will turn the AgCl gray and obscure the endpoint. Therefore, sunlight must be avoided to prevent photodecomposition.

When these factors have been taken into account, the adsorption indicator method is rapid and capable of high accuracy for the determination of chloride with $AgNO_3$. We can also titrate bromide, iodide, and thiocyanate (SCN^-) ions with $AgNO_3$, using eosin (tetrabromofluorescein) as an adsorption indicator. Eosin is a stronger acid than even dichlorofluorescein so that it can be used in solutions as low as pH 2. The structures of these adsorption indicators are shown in Figure 8-4. Unfortunately, eosin cannot be used for the determination of chloride because the eosinate ion is more strongly adsorbed than is chloride ion itself. This means that the red color of silver eosinate develops well before the equivalence point and results in a significant negative determinate error.

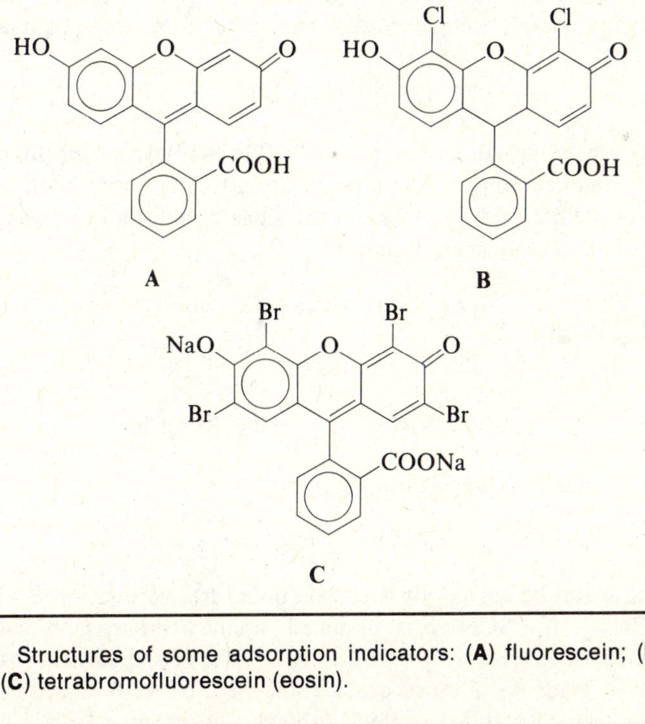

FIGURE 8-4 Structures of some adsorption indicators: **(A)** fluorescein; **(B)** dichloro-fluorescein; **(C)** tetrabromofluorescein (eosin).

None of the color endpoint methods described so far is useful in strongly acidic solution. In acidic solution, the preferred color endpoint method is that of back-titration, in which excess $AgNO_3$ is added to the chloride sample, followed by titration of the excess with a standard solution of KSCN. The titration of silver ion with KSCN is based on the precipitation of insoluble AgSCN and is called the "Volhard titration" after J. Volhard who published the method in 1874:

$$Ag^+ + SCN^- \rightleftharpoons AgSCN(s)$$

$$K_{sp} = 1.1 \times 10^{-12} = [Ag^+][SCN^-] \tag{8-10}$$

The endpoint signal commonly used is the formation of the red iron thiocyanate complex, $Fe(SCN)^{2+}$, which has as the equilibrium constant for its formation reaction

$$Fe^{3+} + SCN^- \rightleftharpoons Fe(SCN)^{2+}$$

$$K_f = 1.4 \times 10^2 = \frac{[Fe(SCN)^{2+}]}{[Fe^{3+}][SCN^-]} \tag{8-11}$$

To obtain an endpoint at the equivalence point (no endpoint error), it is necessary to calculate the correct concentration of Fe^{3+} so that the red $Fe(SCN)^{2+}$ will be visible to the analyst when a stoichiometric amount of KSCN has been added to the Ag^+ sample. This would normally be when $[Ag^+] = [SCN^-]$, but in this case some of the thiocyanate ion is complexed with iron. Therefore, at the equivalence point,

$$[Ag^+] = [SCN^-] + [Fe(SCN)^{2+}] \tag{8-12}$$

We now have three equations (Equations 8-10 to 8-12) with four unknowns, Ag^+, SCN^-, Fe^{3+}, and $Fe(SCN)^{2+}$. All that is needed is to know how much $Fe(SCN)^{2+}$ is required to be visible, and this concentration has been found to be approximately 6.4×10^{-6} M. Therefore, from Equation 8-12,

$$[Ag^+] = [SCN^-] + 6.4 \times 10^{-6}$$

and substituting for $[Ag^+]$ in Equation 8-10,

$$1.1 \times 10^{-12} = ([SCN^-] + 6.4 \times 10^{-6})[SCN^-] \tag{8-13}$$

leads to the quadratic expression

$$[SCN^-]^2 + 6.4 \times 10^{-6}[SCN^-] - 1.1 \times 10^{-12} = 0 \tag{8-14}$$

This quadratic can be solved, or it can be noted from Equation 8-13 that $6.4 \times 10^{-6} > [SCN^-]$. If $[SCN^-]$ is assumed small compared to 6.4×10^{-6} in Equation 8-13, then $[SCN^-] = 1.72 \times 10^{-7}$ M. The quadratic expression in Equation 8-14 leads to a more exact value of 1.67×10^{-7} M, but only two significant figures are justified so that both calculations give 1.7×10^{-7} M as the answer. Now the concentration of Fe^{3+} can be calculated from Equation 8-11:

$$1.4 \times 10^2 = \frac{6.4 \times 10^{-6}}{[Fe^{3+}](1.7 \times 10^{-7})} ; \qquad [Fe^{3+}] = 0.27 \ M$$

Just as in the Mohr titration, in which the theoretical concentration of chromate ion was too high for a well-defined endpoint, this concentration of Fe^{3+} imparts too much color for the Volhard endpoint. A concentration of 0.01 M is typically recommended, and as Example 8-4 shows, even a concentration of 0.005 M introduces an insignificant endpoint error.

When the titration of chloride is attempted with the Volhard technique, a problem arises because AgCl is more soluble than AgSCN. As KSCN is added in the back-titration, some chloride ion is displaced from AgCl, as shown in Equation 8-15:

$$AgCl(s) + SCN^- \rightleftharpoons AgSCN(s) + Cl^- \tag{8-15}$$

EXAMPLE 8-4

Determine the percentage error in the titration of 50 mL of 0.10 F Ag^+ with 0.10 F KSCN, using the Volhard endpoint method, if the Fe^{3+} concentration is only 0.005 M.

Solution

If the $[Fe^{3+}]$ is less than the theoretical value required (0.27 M), then more KSCN will need to be added to produce a concentration of 6.4×10^{-6} M for $Fe(SCN)^{2+}$. The concentration of thiocyanate required can be calculated from Equation 8-11:

$$1.4 \times 10^2 = \frac{[Fe(SCN)^{2+}]}{[Fe^{3+}][SCN^-]}$$

$$[SCN^-] = \frac{6.4 \times 10^{-6}}{(0.005)(1.4 \times 10^2)} = 9.1 \times 10^{-6}$$

The concentration of thiocyanate at the equivalence point was found by solving Equation 8-13 and was 1.7×10^{-7}. Therefore, the excess $[SCN^-]$ is

$$\text{excess } [SCN^-] = 9.1 \times 10^{-6} - 1.7 \times 10^{-7} = 8.9 \times 10^{-6}$$

$$\text{mmol of excess } SCN^- = MV = 8.9 \times 10^{-6} \text{ mmol/mL} \times 100 \text{ mL} = 8.9 \times 10^{-4}$$

$$\text{percentage error} = +\frac{\text{mmol excess } SCN^-}{\text{mmol } SCN^- \text{ at equivalence point}} \times 100 \ (\%)$$

$$= +\frac{8.9 \times 10^{-4}}{0.10 \times 50} \times 100 \cong +0.02\%$$

Thus even a concentration of $Fe(SCN)^{2+} < 1/50$ the theoretical value introduces no significant error when 0.10 F solutions are titrated.

This displacement will begin to occur near the equivalence point as the concentration of thiocyanate increases significantly. The analyst will observe a fading endpoint because the thiocyanate ion required to form the red $Fe(SCN)^{2+}$ complex will be consumed according to Equation 8-15. The endpoint would be observed late and would not be sharp. The determinate error would be *negative* because this is a back-titration.

$$meq \text{ cation charge} = meq \text{ anion charge}$$

$$meq \ Ag^+ = meq \ Cl^- + meq \ SCN^-$$

$$meq \ Cl^- = meq \ Ag^+ - meq \ SCN^- \qquad \textbf{(8-16)}$$

Equation 8-16 shows that, if the milliequivalents of thiocyanate have a positive error (that is, the endpoint is after the equivalence point), then the milliequivalents of chloride will have a negative error.

One technique commonly used to avoid this problem is the addition of some nitrobenzene (an organic chemical immiscible with water) to the solution before the back-titration is carried out. The nitrobenzene coats the AgCl particles and prevents the reaction of Equation 8-15 from taking place. A sharp, permanent endpoint is observed with little endpoint error.

The Volhard method can easily determine bromide because AgBr is less soluble than AgSCN; the reaction analogous to that of Equation 8-15 does not occur to any appreciable extent. The determination of iodide is also possible with the Volhard method, but Fe^{3+} must not be added until excess $AgNO_3$ is present; that is, essentially all the iodide is precipitated as AgI because the reaction given in Equation 8-17 can occur:

$$2Fe^{3+} + 2I^- \rightleftharpoons 2Fe^{2+} + I_2 \qquad \textbf{(8-17)}$$

This reaction consumes iodide, and Fe^{2+} does not form a colored complex with SCN^-; so the analysis cannot be carried out if Equation 8-17 occurs. In the presence of excess silver ion and with the extremely low solubility product of AgI ($K_{sp} = 8.3 \times 10^{-17}$), the concentration of free iodide is essentially zero. For example, if the excess Ag^+ concentration is 0.01,

$$[I^-] = \frac{K_{sp}}{[Ag^+]} = \frac{8.3 \times 10^{-17}}{10^{-2}} = 8.3 \times 10^{-15} \ M$$

8-3   OTHER PRECIPITATION TITRATIONS

The foregoing discussion focused on the titration of halide ions (chloride in particular) with $AgNO_3$, and it was also shown that silver ion could be titrated with KSCN by the Volhard method. The Volhard method, moreover, opens up the possibility of titrating any anion that forms an insoluble silver salt. This possibility is quite useful because almost all anions form insoluble silver salts. If the silver compound is less soluble than AgSCN (as is the case with AgBr), there is no problem in the back-titration with KSCN. However, if the compound is more soluble than AgSCN (as is AgCl), the silver compound must be removed (by filtration or by coating with nitrobenzene as described for the chloride determination) before the back-titration. Several examples are given in Table 8-3. Clearly the method is not very specific, and the analyst must be certain that there are present no anions other than the analyte to form insoluble silver salts.

The methods described above are useful for chloride, bromide, and iodide. What about fluoride? Silver fluoride is quite soluble so that a precipitation titration with $AgNO_3$ is not possible. However, there is one indirect technique based on the Volhard method: The compound PbClF is reasonably insoluble so that a quantitative precipitation of fluoride ion can be accomplished by adding an excess of Pb^{2+} and Cl^-. The PbClF is filtered and dissolved in acid, excess Ag^+ is added to precipitate AgCl, and the excess Ag^+ is titrated with KSCN. The titration actually determines the amount of chloride present, but a 1:1 mole ratio between F^- and Cl^- was established in the formation of PbClF. Example 8-5 shows a typical calculation.

Another method for the determination of fluoride by a precipitation titration is titration with $Th(NO_3)_4$. ThF_4 has a solubility (<0.2 g/L) sufficiently low to allow a successful titration. Excess Th^{4+} is detected by the formation of a complex with sodium alizarin monosulfonate. The color change is not readily detected by eye;

EXAMPLE 8-5

The fluoride content of 250 mL of well-water sample is determined by the precipitation of PbClF followed by the Volhard method for chloride. If 10.00 mL of 0.0215 N $AgNO_3$ is added to the dissolved PbClF and 6.13 mL of 0.0178 N KSCN is required to reach the $Fe(SCN)^{2+}$ endpoint, calculate the fluoride content in parts per million.

Solution

$$\text{meq anion charge} = \text{meq cation charge}$$

$$\text{meq } Cl^- + \text{meq } SCN^- = \text{meq } Ag^+$$

$$\text{meq } Cl^- = \text{meq } Ag^+ - \text{meq } SCN^-$$

$$= N_{Ag}V_{Ag} - N_{SCN}V_{SCN}$$

$$= 0.0215 \times 10.00 - 0.0178 \times 6.13$$

$$= 0.106 = \text{meq } F \quad (1\!:\!1 \text{ stoichiometry in PbClF})$$

$$0.106 = \frac{\text{mg F}}{\text{FW (F)}/n} = \frac{\text{mg F}}{19.0/1}$$

$$\text{mg F} = 2.01 \text{ in 250-mL sample}$$

$$\frac{\text{mg F}}{\text{L}} = \frac{2.01 \text{ mg}}{0.25 \text{ L}} = 8.04 \text{ mg/L (ppm)}$$

TABLE 8-3 Examples of Precipitation Titration Methods

Analyte	Reagent(s)*	Precipitate(s)*	Endpoint method
acetylenes	$AgNO_3/KSCN$	Ag acetylide/AgSCN	Volhard†
AsO_4^{3-}	$AgNO_3/KSCN$	$Ag_3AsO_4/AgSCN$	Volhard
Br^-	$AgNO_3$	$AgBr$	Mohr or adsorption indicator
	$AgNO_3/KSCN$	$AgBr/AgSCN$	Volhard
Cl^-	$AgNO_3$	$AgCl$	Mohr or adsorption indicator
	$AgNO_3/KSCN$	$AgCl/AgSCN$	Volhard†
CNO^-	$AgNO_3/KSCN$	$AgCNO/AgSCN$	Volhard
CO_3^{2-}	$AgNO_3/KSCN$	$Ag_2CO_3/AgSCN$	Volhard†
$C_2O_4^{2-}$	$AgNO_3/KSCN$	$Ag_2C_2O_4/AgSCN$	Volhard†
	Pb acetate	PbC_2O_4	Adsorption indicator–fluorescein
CrO_4^{2-}	$AgNO_3/KSCN$	$Ag_2CrO_4/AgSCN$	Volhard†
epoxides	$(HCl)AgNO_3/KSCN$	$AgCl/AgSCN$	Volhard†
F^-	$(Pb^{2+}, Cl^-)AgNO_3/KSCN$	$(PbClF)AgCl/AgSCN$	Volhard†
	$Th(NO_3)_4$	ThF_4	sodium alizarin monosulfonate (spectrophotometric endpoint) or ion-selective electrode
Hg_2^{2+}	NaCl	Hg_2Cl_2	bromophenol blue
I^-	$AgNO_3$	AgI	adsorption indicator–eosin
	$AgNO_3/KSCN$	$AgI/AgSCN$	Volhard
K^+	$NaB(C_6H_5)_4/AgNO_3/KSCN$	$KB(C_6H_5)_4/AgB(C_6H_5)_4/AgSCN$	Volhard
mercaptans (RSH)	$AgNO_3$	RSAg	potentiometric
MoO_4^{2-}	$Pb(NO_3)_2$	$PbMoO_4$	adsorption indicator–eosin
PO_4^{3-}	$AgNO_3/KSCN$	$Ag_3PO_4/AgSCN$	Volhard†
	Pb acetate	$Pb_3(PO_4)_2$	adsorption indicator–dibromofluorescein
S^{2-}	$AgNO_3/KSCN$	$Ag_2S/AgSCN$	Volhard†
SCN^-	$AgNO_3$	AgSCN	Volhard
SO_4^{2-}	$Pb(NO_3)_2$	$PbSO_4$	erythrosin B indicator
	$Ba(ClO_4)_2$	$BaSO_4$	dimethylsulfonazo III indicator
Zn^{2+}	$K_4Fe(CN)_6$	$K_2Zn_3[Fe(CN)_6]_2$	diphenylamine indicator

*All entries showing / indicate a back-titration. Excess of the first reagent is added, and back-titration with the second reagent is carried out by the endpoint method listed.
† Remove silver salt before back-titrating with KSCN

so—especially for microgram quantities of fluoride—the color change is measured with a spectrophotometer (Section 13-3) to determine the endpoint. The method has been used to determine fluorides in plant-tissue, air, and water samples. Larger quantities of fluoride can be titrated with $Th(NO_3)_4$, using methylthymol blue (turns dark blue in the presence of Th^{4+}) as an indicator. A fluoride ion-selective electrode (Section 15-3) can also be used for an endpoint signal in the titration of fluoride with Th^{4+}, La^{3+}, or Eu^{3+}.

Some organic compounds can be analyzed by the Volhard method. For example, acetylenes react with $AgNO_3$ in ammonia solution according to Equation 8-18:

$$2AgNO_3 + HC\equiv CR \rightleftharpoons AgC\equiv CR \cdot AgNO_3 + HNO_3 \qquad \text{(8-18)}$$

The silver acetylide product is insoluble and may be removed from the solution containing the excess silver ion (the acetylide should not be allowed to dry because it is explosive), which is then acidified and titrated with KSCN. Example 8-6 illustrates the method of calculation.

EXAMPLE 8-6

The acetylene (C_2H_2) content of a gas sample is determined by bubbling 5.00 L of the gas through a solution containing 50.0 mL of 0.1052 N $AgNO_3$. After removal of the silver acetylide, the excess $AgNO_3$ is titrated with 0.1215 N KSCN, and 29.18 mL is required to reach the $Fe(SCN)^{2+}$ endpoint. If the gas has a density of 0.850 g/L, calculate the acetylene content in percentage by weight

Solution

$$\text{meq anion charge} = \text{meq cation charge}$$

$$\text{meq } C_2H_2 + \text{meq KSCN} = \text{meq } AgNO_3$$

$$\frac{\text{mg } C_2H_2}{FW\ (C_2H_2)/n} + N_{SCN}V_{SCN} = N_{Ag}V_{Ag}$$

From Equation 8-18, it is seen that each C_2H_2 combines with $2Ag^+$; therefore, $n = 2$.

$$\frac{\text{mg } C_2H_2}{26.02/2} + 0.1215 \times 29.18 = 0.1052 \times 50.0$$

$$\text{mg } C_2H_2 = 22.31 \text{ in } 5.00 \text{ L of gas}$$

$$\frac{\text{mg } C_2H_2}{L} = \frac{22.31}{5.00}$$

$$= 4.46 \text{ mg/L}$$

(3 significant figures because 5.00 L has only 3)

$$\% C_2H_2 = \frac{4.46 \text{ mg/L } C_2H_2}{850 \text{ mg/L gas}} \times 100\ (\%) = 0.52\%$$

Another example is the determination of epoxide groups by the hydrochlorination reaction

$$R_1C \underset{O}{\overset{\diagup\diagdown}{-}} CR_2 + HCl \rightleftharpoons R_1 \underset{OH}{\overset{|}{C}} - \underset{Cl}{\overset{|}{C}}R_2 \qquad (8\text{-}19)$$

According to Equation 8-19, 1 mol of HCl is consumed for each mole of epoxide present. The excess chloride can be determined by the Volhard technique, or—alternatively—the excess acid can be determined by acid-base titration (Chapter 9).

One more example from the field of organic chemistry is the determination of mercaptans using the reaction with Ag^+:

$$RSH + AgNO_3 \rightleftharpoons RSAg + HNO_3 \qquad (8\text{-}20)$$

The reaction is sufficiently fast to be carried out directly by a potentiometric endpoint method (Chapter 15). Alternatively, excess $AgNO_3$ can be added and back-titrated with KSCN, but this method gives less satisfactory results because the silver mercaptide coprecipitates some of the excess silver ion.

By precipitation titrations, it is possible to titrate many other metal cations as well as anions. Frequently, there are no indicators that yield satisfactory color-change endpoint signals, and the analyst must resort to somewhat slower electrochemical endpoint signals (potentiometric or amperometric). On occasion, an indicator for a different reaction type may be employed. For example, zinc ion can be determined by titration with potassium hexacyanoferrate(II) (also known as potassium ferrocyanide). The reaction is

$$2K_4Fe(CN)_6 + 3Zn^{2+} \rightleftharpoons K_2Zn_3[Fe(CN)_6]_2 + 6K^+ \qquad (8\text{-}21)$$

Note that this equation provides us with an example in which not all the anion charge of the titrant is combining with the cation charge of the analyte. For this reaction, Zn^{2+} has an n value of 2, while $K_4Fe(CN)_6$ has an n value of 3; each $Fe(CN)_6^{4-}$ furnishes 3 negative charges in the formation of the precipitate. The fourth negative charge on $Fe(CN)_6^{4-}$ remains neutralized by K^+.

The indicator recommended for this titration is diphenylamine, which is normally an indicator for a redox titration (Section 11-4). It changes color according to the electrochemical potential of the solution. When the solution contains some hexacyanoferrate(III) (ferricyanide) the indicator will be in its highly colored, violet, oxidized form. After the equivalence point, excess hexacyanoferrate(II) is present, and the ratio $[Fe(CN)_6^{3-}]/[Fe(CN)_6^{4-}]$ drops rapidly along with the electrochemical potential. The indicator is converted to its colorless, reduced form and results in a sharp endpoint signal.

Precipitation titrations are less often recommended than are the other titrimetric methods. Many of the metal ions that can be titrated in a precipitation method can also be titrated more conveniently with EDTA in a complexation method, and they will be discussed in Chapter 10.

SUPPLEMENTARY READING

Ayers, C., "Argentometric Methods," in *Comprehensive Analytical Chemistry*, C.L. Wilson and D.W. Wilson, eds., vol. 1B. New York: Elsevier North-Holland, Inc., 1960.

Coetzee, J.F., "Equilibria in Precipitation Reactions, and Precipitation Lines," in *Treatise on Analytical Chemistry*, I.M. Kolthoff and P.J. Elving, eds., pt. 1, vol. 1. New York: Interscience Publishers, Inc., 1959.

PROBLEMS

8-1. A solution of $AgNO_3$ is prepared by dissolving 8.916 g in H_2O and diluting to 500 mL in a volumetric flask. What is its titer for the following analytes?

*a. Br^-

b. $CaCl_2$

c. $C_2H_2Cl_2$

d. As_2O_3 (the precipitate is Ag_3AsO_4)

8-2. Using the data in Table 5-2, calculate pAg at the equivalence point for the titration of Cl^- with $AgNO_3$ at each of the temperatures listed. Does this change present any problems in carrying out precipitation titrations at elevated temperatures?

8-3. Using the data in Table 5-2, calculate pBa at the equivalence point for the titration of SO_4^{2-} with $BaCl_2$ at each of the temperatures listed.

8-4. A calcium ion-selective electrode is to be used for the following precipitation titrations. Calculate the volume and pCa at the equivalence point.

*a. Titration of 50 mL of 0.10 F $C_2O_4^{2-}$ with 0.10 F $CaCl_2$

b. Titration of 100 mL of 0.20 F F^- with 0.40 F $CaCl_2$

c. Titration of 25 mL of 0.50 F SO_4^{2-} with 0.20 F $CaCl_2$

8-5. Douglasite is a mineral with the formula $2KCl \cdot FeCl_2 \cdot 2H_2O$. If it can be assumed that this is the only source of Cl^-, calculate the percentage of douglasite in a 386-mg sample if a titration with 0.1276 N $AgNO_3$ requires 31.96 mL to reach the equivalence point.

*8-6. The amount of acetylene (C_2H_2) in a gas sample can be determined by scrubbing it with ammoniacal $AgNO_3$, in which the following reaction takes place:

$$2Ag^+ + C_2H_2 + 2H_2O \rightleftharpoons Ag_2C_2(s) + 2H_3O^+$$

If a 4.85-L sample was bubbled through 150.0 mL of 0.0641 N $AgNO_3$ and the excess Ag^+ was titrated with 31.6 mL of 0.0752 N KSCN, determine the acetylene content of the gas as milligrams per liter.

8-7. The phosphorus content in a 2.00-mL liquid-fertilizer sample was determined by converting the phosphorus to phosphate and precipitating Ag_3PO_4 with the addition of 50.0 mL of 0.1163 N $AgNO_3$. The excess $AgNO_3$ was titrated with 0.1245 N KSCN by the Volhard technique. If 10.63 mL was required to reach the equivalence point, what is the percentage of P_2O_5 (weight per volume) in the liquid fertilizer?

8-8. The arsenic content in an insecticide sample weighing 783 mg was determined by treatment with HNO_3 to convert arsenic to As(V), followed by precipitation of Ag_3AsO_4 with the addition of 20.0 mL of 0.0863 N $AgNO_3$. The excess $AgNO_3$

---

*Answers to problems marked with an asterisk will be found at the back of the book.

was titrated with 0.0741 N KSCN by the Volhard technique. If 15.17 mL was required to reach the equivalence point, what is the percentage of As_2O_3 in the insecticide?

*8-9. An antidandruff shampoo containing zinc pyrithione was analyzed for its zinc content by digestion of a 2.654-g sample with acid to form Zn^{2+} and titration with 0.0261 F $K_4Fe(CN)_6$. If 4.72 mL was required to reach the diphenylamine endpoint signal, calculate the percentage of Zn in the shampoo.

8-10. Many chlorinated organic compounds will precipitate AgCl when titrated with $AgNO_3$. Excess $AgNO_3$ can then be titrated with KSCN, using the Volhard method. One example is chloroacetic acid ($HC_2H_2O_2Cl$), which is used as a preservative in fruit juice. The chloroacetic acid in a 200-mL sample is extracted into ether and then transferred back to an aqueous NaOH solution. The solution is then acidified, and 30 mL of a $AgNO_3$ solution is added. After filtration and washing, the filtrate is titrated with 0.0429 N KSCN, and 7.26 mL is required to reach the endpoint. A blank run using H_2O in place of fruit juice requires 39.71 mL of KSCN. Calculate chloroacetic acid content as milligrams per 100 mL.

8-11. Other halogenated organic compounds may be analyzed as described in Problem 8-10. For example, the disinfectant iodoform (CHI_3) can be decomposed and AgI precipitated. If a 925-mg sample is treated with 50.0 mL of 0.0613 N $AgNO_3$ and the excess Ag^+ required 17.32 mL of 0.0467 N KSCN calculate the percentage of iodoform in the sample.

*8-12. Mercaptans (RSH) are added to natural gas to impart a distinctive odor. If the gas sample is scrubbed with a solution containing $AgNO_3$, a precipitate of RSAg is formed. The excess $AgNO_3$ can then be titrated with KSCN. If a 120.0-L gas sample (density, 0.73 g/L) is bubbled through a solution containing 10.00 mL of 0.00467 N $AgNO_3$ and 5.67 mL of 0.00311 N KSCN is required for the back-titration, calculate parts per million of CH_3SH in the gas sample.

8-13. The trace fluoride content in vegetation can be determined by ashing a relatively large sample plus CaO to form CaF_2, followed by treatment with acid and SiO_2 to produce volatile SiF_4. The SiF_4 is then distilled and collected in a receiving vessel, and the F^- is titrated with $Th(NO_3)_4$, using alizarin red as the indicator. If a 72.5-g sample requires 4.17 mL of 0.0113 N $Th(NO_3)_4$, calculate parts per million of F^- in the sample.

8-14. The chloride content in a 353-mg sample is determined by the Volhard method. If 50.00 mL of 0.1063 N $AgNO_3$ is added to the sample and 35.86 mL of 0.1216 N KSCN is required for the back-titration, determine the uncertainty in the percentage of Cl if the standard solutions have an uncertainty of $\pm 0.1\%$ and all volume readings (initial and final) have an uncertainty of ± 0.03 mL.

*8-15. The endpoint uncertainty for a titration with a potentiometric endpoint depends on the slope at the endpoint (Δ mV/Δ mL) and the uncertainty in reading potential. For the chloride titration described in Example 8-1, the slope at the endpoint is ~ 100 mV/mL. If the uncertainty in reading potential is ± 2 mV, what is the endpoint uncertainty in percent?

8-16. Calculate the equilibrium constant for Equation 8-15.

8-17. The titration of Br^- with $AgNO_3$ can be carried out by the Volhard method (back-titration with KSCN) without removing AgBr, as is necessary for AgCl. That is, the displacement reaction analogous to Equation 8-15 does not proceed to any measurable extent. Calculate the equilibrium constant for the displacement reaction.

*8-18. Ag_3AsO_4 is a dark-red compound that can easily be detected when it precipitates. Calculate the proper $[AsO_4^{3-}]$ so that Ag_3AsO_4 would begin to precipitate at the equivalence point in the AgCl titration, a Mohr-type endpoint signal. What problem does this present?

8-19. An analyst wishes to design a precipitation titration of SO_4^{2-} with $Pb(NO_3)_2$ and notes that the yellow PbI_2 is somewhat more soluble. Since the $PbSO_4$ titration product is white, a Mohr-type endpoint signal might be feasible. Calculate the I^- concentration so that PbI_2 will begin to form at the equivalence point. What problems might be encountered?

8-20. Iodate ion, IO_3^-, forms an insoluble silver salt, which could be the basis for a precipitation titration. If a Mohr endpoint is contemplated, calculate $[CrO_4^{2-}]$ so that Ag_2CrO_4 will begin to precipitate at the equivalence point. Does this pose any problems?

8-21. For the titration of IO_3^- with $AgNO_3$ described in Problem 8-20, suppose the actual concentration of CrO_4^{2-} is 1×10^{-3} M. Calculate the endpoint error introduced in the titration of 50 mL of 0.10 F IO_3^- with 0.10 F $AgNO_3$.

*8-22. A mixture of UF_3 and UF_4 weighing 530 mg is treated with acid, and HF is distilled from the mixture. The HF is collected in a solution containing Pb^{2+} and Cl^- to precipitate PbClF. The PbClF precipitate is filtered, washed, and then dissolved in HNO_3. The resulting Cl^- is titrated with Ag^+, using dichlorofluorescein indicator, after partial neutralization. If 37.16 mL of 0.1613 N $AgNO_3$ is required to reach the endpoint, calculate the fluorine content as the value of x in the formula UF_x.

8-23. A mixture of potassium salts, including KBr and KI, weighs 875 mg and is dissolved in 250 mL total volume. A 50.0-mL aliquot is titrated with 0.0517 N $AgNO_3$ and requires 21.63 mL to reach the adsorption indicator endpoint. A second 50.0-mL aliquot is treated with an oxidant to form I_2, which is extracted with CCl_4. The remaining aqueous phase containing Br^- is titrated with 0.0517 N $AgNO_3$ and now requires 7.17 mL to reach the endpoint. Calculate the percentages of KBr and KI in the mixture.

8-24. Calculate titration curves for the following titrations for the same volumes as given in Example 8-1. Determine $[Cl^-]$, $[Ag^+]$, pCl, and pAg, and plot pCl as a function of volume. Compare with Figure 8-2.
  *a. Titration of 100 mL of 0.2 F chloride with 0.4 F $AgNO_3$
  b. Titration of 100 mL of 0.005 F chloride with 0.01 F $AgNO_3$
  c. Titration of 100 mL of 0.0005 F chloride with 0.001 F $AgNO_3$

8-25. Calculate a titration curve for the titration of 50 mL of 0.10 F KBr with 0.10 F $AgNO_3$. Determine $[Br^-]$, $[Ag^+]$, pBr, and pAg at the following titrant volumes: 0, 10, 25, 40, 49, 50, 51, 60, and 100 mL. Plot the data as in Figure 8-2.

8-26. The concentration (activity) of Pb^{2+} can be measured with a lead ion-selective electrode. Calculate pPb as a function of volume for the following titrations, and show the titration curve graphically. Use titrant volumes of 0, 10, 20, 24, 25, 26, 30, and 50 mL.
  a. Titration of 50 mL of 0.05 F CrO_4^{2-} with 0.10 F $Pb(NO_3)_2$
  b. Titration of 25 mL of 0.10 F SO_4^{2-} with 0.10 F $Pb(NO_3)_2$
  c. Titration of 25 mL of 0.20 F I^- with 0.10 F $Pb(NO_3)_2$

Acid–Base Titrations

9

The second reaction type described in Chapter 3 was the reaction of an acid with a base. In aqueous solutions, this reaction can be formulated as

$$H_3O^+ + OH^- \rightleftharpoons 2H_2O \qquad \text{(9-1)}$$

Any substance that furnishes H_3O^+ or consumes OH^- is an acid, while any substance that furnishes OH^- or consumes H_3O^+ is a base. These definitions can be broadened when dealing with nonaqueous solutions, but the distinction between acid–base and complexation reactions becomes blurred if we adopt a completely generalized Lewis acid–Lewis base definition.

As in Chapter 8, let us first determine a titration curve for the addition of a base titrant to an acid analyte. It will be convenient to calculate $[H_3O^+]$ after each addition of titrant and plot pH as a function of titrant volume. Then we can use this information to devise satisfactory endpoint signals for acid–base titrations. Finally, we shall consider polyprotic acids, nonaqueous titrations, and some practical applications.

9-1 THEORETICAL PRINCIPLES: THE TITRATION CURVE

INITIAL POINT

First, let us calculate $[H_3O^+]$ in a solution containing the acid analyte. If an acid, HA, can dissociate to furnish H_3O^+ according to Equation 9-2,

$$HA + H_2O \rightleftharpoons H_3O^+ + A^- \qquad \text{(9-2)}$$

there will be an equilibrium constant called the *acid dissociation constant*

$$K_a = \frac{[H_3O^+][A^-]}{[HA]} \tag{9-3}$$

in which $[H_2O]$ has been assumed to be ~ 1, that is, nearly pure water.

To solve for all species in the system, it is necessary to have as many independent equations as there are species. In the case of HA in water, the species are $[H_3O^+]$, $[OH^-]$, $[HA]$, and $[A^-]$. Therefore, four equations are needed, and Equation 9-3 is one of them.

In aqueous solutions, the dissociation of water is always in equilibrium so that Equation 9-4 can be used as a second equation:

$$K_w = [H_3O^+][OH^-] \tag{9-4}$$

There are no other equilibria involved; therefore, no other equilibrium constants can be written.

After all chemical equilibria have been accounted for, an additional equation can be written: a *charge balance*. The solution must be electrically neutral so that the sum of anion charge must equal the cation charge. This is given in Equation 9-5 for the solution of HA in water (charge balance: cation charge = anion charge):

$$[H_3O^+] = [A^-] + [OH^-] \tag{9-5}$$

Each term in Equation 9-5 has a coefficient equal to the charge (all are 1 in Equation 9-5) because it is the molar concentration of charge that is balanced. This aspect can be seen more clearly if the charge-balance equation is written for an acid such as H_3PO_4 (Equation 9-6):

$$[H_3O^+] = [H_2PO_4^-] + 2[HPO_4^{2-}] + 3[PO_4^{3-}] + [OH^-] \tag{9-6}$$

Equation 9-5 represents the third equation of the four mentioned above, and only one more is needed. It becomes evident to see what is needed if we ask the question "What is the pH of an acetic acid solution?" The response should be "How much acetic acid is present?" In other words, we have to know the concentration of acid before we can calculate $[H_3O^+]$ or any of the four species present. The concentration of acid present is called the "formal" concentration, and it is *not* equal to $[HA]$ because some of the acid will dissociate according to Equation 9-3. However, the acid must be present in either its dissociated or undissociated form, and this can be expressed as a *mass balance*, shown in Equation 9-7:

$$\text{mass balance (acid): } f_{HA} = [HA] + [A^-] \tag{9-7}$$

There will be a mass-balance equation for each substance added to the system, and this will complete the set of equations needed to solve for the concentrations of all species. Note that charge balance and mass balance really involve concentrations, while equilibrium expressions involve activities that we shall only approximate as concentrations.

We now have four equations (Equations 9-3 to 9-5 and 9-7) and four unknowns ($[H_3O^+]$, $[OH^-]$, $[HA]$, and $[A^-]$). In principle, the problem is solved; but in practice, we end up with the third-order equation for $[H_3O^+]$ shown in Equation 9-8, which is difficult to solve.

$$[H_3O^+]^3 + K_a[H_3O^+]^2 - (K_a f_{HA} + K_w)[H_3O^+] - K_a K_w = 0 \qquad \text{(9-8)}$$

The problem can be simplified, however, if we consider the chemistry involved.

Suppose that HA is a strong acid; that is, it is completely dissociated. There will be no acid dissociation equilibrium; so Equation 9-3 is no longer applicable ($[HA] \to 0$). However, with the knowledge that $[HA] = 0$, there are only three species. The mass balance equation becomes

$$f_{HA} = [A^-]$$

and the charge-balance equation, Equation 9-5, can be combined with Equation 9-4 to give

$$[H_3O^+] = f_{HA} + [OH^-] = f_{HA} + \frac{K_w}{[H_3O^+]}$$

For almost any situation of analytical interest, $f_{HA} \gg K_w/[H_3O^+]$ so that

$$[H_3O^+] \cong f_{HA}$$

$$[OH^-] \cong \frac{K_w}{f_{HA}}$$

Thus the concentrations of all three species have been calculated. In fact, the entire titration curve for the titration of a strong acid with a strong base can be simply derived by assuming that $[H_3O^+] = f_{HA} - f_b$, where f_b is the concentration of base added. A formal concentration in the final solution may be calculated from the amount added by realizing that $f = $ mmol added/total volume and that mmol added $= FV$(added). This approach is used in Example 9-1, with the resulting titration curve shown in Figure 9-1.

Now let us turn our attention to the case in which HA is a weak acid. The charge-balance equation, Equation 9-5, can also be viewed as the sum of two dissociations. The dissociation of HA gives one H_3O^+ and one A^-, while the dissociation of H_2O gives one H_3O^+ and one OH^-. If HA is about to be titrated with base, it must be a stronger acid than H_2O is; therefore, $[A^-] \gg [OH^-]$, and

$$[H_3O^+] \cong [A^-] \qquad \text{(9-9)}$$

With this approximation, Equation 9-9 can be used in place of Equation 9-5, and the four equations can be solved in terms of $[H_3O^+]$:

$$K_a = \frac{[H_3O^+][A^-]}{[HA]} = \frac{[H_3O^+][H_3O^+]}{f_{HA} - [A^-]} = \frac{[H_3O^+]^2}{f_{HA} - [H_3O^+]} \qquad \text{(9-10)}$$

$$[H_3O^+]^2 + K_a[H_3O^+] - f_{HA}K_a = 0 \qquad \text{(9-11)}$$

EXAMPLE 9-1

Derive the titration curve for the titration of 50 mL of 0.10 F HCl with 0.20 F NaOH. Calculate the pH after the addition of 0, 10, 20, 24, 25, 26, 30, and 50 mL of base.

Solution

0 mL

$$[H_3O^+] = f_{HA} = 0.10 \qquad pH = 1.00$$

10 mL

$$[H_3O^+] = f_{HA} - f_b = \frac{F_{HA}V_{HA}}{V_{HA} + V_b} - \frac{F_b V_b}{V_{HA} + V_b} = \frac{F_{HA}V_{HA} - F_b V_b}{V_{HA} + V_b}$$

$$= \frac{0.10 \times 50 - 0.20 \times 10}{50 + 10} = 0.050 \qquad pH = 1.30$$

20 mL

$$[H_3O^+] = \frac{0.10 \times 50 - 0.20 \times 20}{50 + 20} = 0.014 \qquad pH = 1.85$$

24 mL

$$[H_3O^+] = \frac{0.10 \times 50 - 0.20 \times 24}{50 + 24} = 0.0027 \qquad pH = 2.57$$

25 mL

$f_{HA} = f_b$. Therefore, this is the equivalence point. Since all the HCl has been neutralized with NaOH, the solution is the same as a NaCl solution in which the only source of H_3O^+ is from dissociation of water:

$$[H_3O^+] = \sqrt{K_w} = 10^{-7} \qquad pH = 7.00$$

Equation 9-11 can also be obtained from Equation 9-8 if we recognize that the last two terms will be insignificant whenever $K_a f_{HA} \gg K_w$ and f_{HA} is some reasonable concentration. Equation 9-11 can be solved by use of the quadratic formula, as shown in Equation 9-12 (only the positive root has physical meaning):

$$[H_3O^+] = \frac{-K_a + \sqrt{K_a^2 + 4f_{HA}K_a}}{2} \qquad (9\text{-}12)$$

However, even solving the quadratic equation can be tedious and offers numerous opportunities for calculation error. Another approximation, which

EXAMPLE 9-1 (cont.)

26 mL

There is now excess base; so the concentration of OH^- is determined in the same way that excess acid was calculated:

$$[OH^-] = f_b - f_{HA} = \frac{F_b V_b - F_{HA} V_{HA}}{V_{HA} + V_b}$$

$$[OH^-] = \frac{0.20 \times 26 - 0.10 \times 50}{50 + 26} = 0.0026$$

$$[H_3O^+] = \frac{K_w}{[OH^-]} = \frac{1 \times 10^{-14}}{0.0026} = 3.8 \times 10^{-12} \qquad pH = 11.41$$

30 mL

$$[OH^-] = \frac{0.20 \times 30 - 0.10 \times 50}{50 + 30} = 0.0125$$

$$[H_3O^+] = \frac{1 \times 10^{-14}}{0.0125} = 8.0 \times 10^{-13} \qquad pH = 12.10$$

50 mL

$$[OH^-] = \frac{0.20 \times 50 - 0.10 \times 50}{50 + 50} = 0.05$$

$$[H_3O^+] = \frac{1 \times 10^{-14}}{0.05} = 2.0 \times 10^{-13} \qquad pH = 12.70$$

The titration curve is shown in Figure 9-1.

allows a rapid solution for $[H_3O^+]$, can be made. In Equation 9-10, if a *weak* acid is to be titrated, then only a small fraction will be dissociated; so $f_A \gg [H_3O^+]$, and

$$K_a \cong \frac{[H_3O^+]^2}{f_{HA}}$$

$$[H_3O^+] \cong \sqrt{K_a f_{HA}} \qquad \qquad \text{(9-13)}$$

What if $[H_3O^+]$ turns out to be a significant fraction of f_{HA} so that the approximation $f_{HA} \gg [H_3O^+]$ is not really valid? In such cases, an iterative

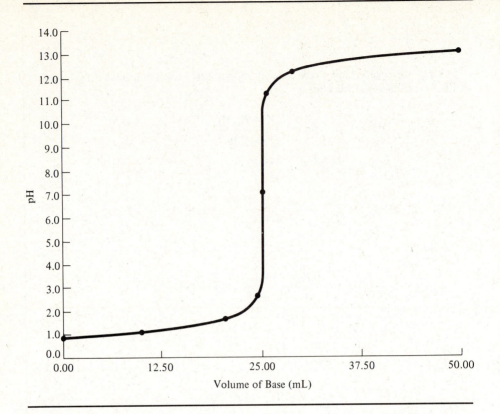

FIGURE 9-1 Titration curve for the titration of 0.10 F HCl with 0.20 F NaOH. Points calculated in Example 9-1 are marked as ●.

approach can be used by rewriting Equation 9-10 as shown in Equation 9-14:

$$[H_3O^+] = \sqrt{K_a(f_{HA} - [H_3O^+])} \qquad (9\text{-}14)$$

We solve for $[H_3O^+]$ on the left side by letting $[H_3O^+] = 0$ on the right side. This first approximation to $[H_3O^+]$ is used on the right side, and a second approximation to $[H_3O^+]$ is solved for. Usually within a few iterations, the values of $[H_3O^+]$ on the left side and on the right side will be equal to three significant figures, which is ample for equilibrium calculations. This iterative technique is shown in Appendix 1-C. Equation 9-14 represents the calculation of $[H_3O^+]$ in a solution of weak acid whether we assume $[H_3O^+] < f_{HA}$ or use the iterative approach or the quadratic formula given in Equation 9-12.

What about the other three species? Equation 9-4 can be used to calculate $[OH^-]$. We have already assumed that $[H_3O^+] \cong [A^-]$, and our approximation that $[A^-] \gg [OH^-]$ can now be checked. If not justified (a very rare situation), the full third-order Equation 9-8 must be solved. Finally, $[HA] = f_{HA} - [A^-]$ from Equation 9-7 completes the calculations.

BUFFER REGION

The titration begins with the addition of some NaOH titrant. There is now one more species, $[Na^+]$, because NaOH is a strong base completely dissociating to give Na^+ and OH^-. Therefore, one more equation is needed, and this is fulfilled by another mass-balance equation. The amount of base added must be specified to solve for the concentrations of all species. This amount of base will be the formal concentration of base, f_b. It alone determines the sodium ion concentration:

$$f_b = [Na^+] \tag{9-15}$$

The other four equations used before (Equations 9-3 to 9-5 and 9-7) are still valid *except* that the charge-balance Equation 9-5 must include the new ion, Na^+:

$$[H_3O^+] + [Na^+] = [A^-] + [OH^-] \tag{9-16}$$

The five equations (Equations 9-3, 9-4, 9-7, 9-15, and 9-16) can now be employed to solve for the five species in solution ($[H_3O^+]$, $[OH^-]$, $[HA]$, $[A^-]$, and $[Na^+]$). This situation again leads to a third-order equation, which is difficult to solve in the exact form given in Equation 9-17:

$$[H_3O^+]^3 + (K_a + f_b)[H_3O^+]^2 + (K_a f_b - K_a f_{HA} - K_w)[H_3O^+] - K_a K_w = 0 \tag{9-17}$$

The situation is greatly simplified by the recognition that, for any reasonable amount of weak acid in the system, $[A^-] \gg [OH^-]$ and that, as soon as any significant amount of base is added, $[Na^+] \gg [H_3O^+]$. Therefore, the charge balance in Equation 9-16 becomes (with the incorporation of Equation 9-15)

$$f_b = [Na^+] \cong [A^-] \tag{9-18}$$

Combining Equations 9-18 and 9-7, we obtain

$$f_{HA} = [HA] + [A^-] = [HA] + f_b$$

$$[HA] = f_{HA} - f_b \tag{9-19}$$

Equation 9-19 mathematically states what is chemically expected for a reaction that nearly goes to completion. That is, the amount of free undissociated acid is equal to the total acid in the system minus the amount neutralized by base, Equations 9-18 and 9-19 can now be incorporated into Equation 9-3, that for the acid dissociation constant

$$K_a = \frac{[H_3O^+][A^-]}{[HA]} = \frac{[H_3O^+]f_b}{f_{HA} - f_b}$$

or

$$[H_3O^+] = \frac{K_a(f_{HA} - f_b)}{f_b} \tag{9-20}$$

EXAMPLE 9-2

(a) Calculate the pH of a solution containing 0.15 F acetic acid and 0.05 F sodium hydroxide.
(b) Calculate the pH change that occurs when 30 mL of 0.1 F NaOH is added to 200 mL of the buffer solution just described. Compare this change in pH with that which would occur when 30 mL of 0.1 F NaOH is added to 200 mL of 3.5×10^{-5} F HCl.

Solution

(a) $$[H_3O^+] = \frac{K_a(f_{HA} - f_b)}{f_b} \qquad K_a = 1.75 \times 10^{-5} \quad \text{(Appendix 2-B)}$$

$$[H_3O^+] = \frac{1.75 \times 10^{-5}(0.15 - 0.05)}{(0.05)} = 3.5 \times 10^{-5}$$

$$pH = 4.46$$

Note that this is the same situation as that of a solution containing 0.10 F acetic acid and 0.05 F sodium acetate. Buffer solutions contain macro quantities of both the weak acid (HA) and its conjugate base (A⁻).

(b). The addition of NaOH will neutralize some of the HA and thereby increase the concentration of base-form acetate ion:

$$\text{mmol NaOH} = FV = 0.1 \times 30 = 3$$

$$\text{Therefore, mmol HA remaining} = 0.10 \times 200 - 3 = 17$$

$$\text{mmol A}^- = 0.05 \times 200 + 3 = 13$$

If the assumptions made to simplify Equation 9-16 turn out to be valid, all five species can now be calculated.

$$[H_3O^+] = \frac{K_a(f_{HA} - f_b)}{f_b}$$

$$[OH^-] = \frac{K_w}{[H_3O^+]} = \frac{K_w f_b}{K_a(f_{HA} - f_b)}$$

$$[HA] = f_{HA} - f_b$$

$$[A^-] = f_b$$

$$[Na^+] = f_b$$

We can obtain Equation 9-20 from the exact expression given in Equation 9-17 by noting that usually terms 1, 2, 6, and 7 ($[H_3O^+]^3$, $K_a[H_3O^+]^2$, $K_w[H_3O^+]$, and $K_a K_w$) are small and therefore can be neglected. However, the smallness of the terms

EXAMPLE 9-2 (cont.)

These quantities can now be used in the expression for the acid dissociation constant

$$[H_3O^+] = \frac{K_a[HA]}{[A^-]}$$

$$[H_3O^+] = \frac{1.75 \times 10^{-5}(17/230)}{13/230}$$

(Note that total volume cancels out and that the same answer would be obtained if some extra H_2O were added.)

$$[H_3O^+] = 2.29 \times 10^{-5} \qquad pH = 4.64; \Delta pH = +0.18$$

When 30 mL of 0.1 F NaOH is added to 200 mL of 3.5×10^{-5} F HCl, some NaOH neutralizes the HCl while the rest simply exists as free $[OH^-]$:

neutralization: $F_{HCl}V_{HCl} = F_{NaOH}V_{NaOH}$

$$V_{NaOH} = \frac{3.5 \times 10^{-5} \times 200}{0.1} = 0.07 \text{ mL}$$

excess base: $30 - 0.07 = 29.93 \text{ mL}$

$$[OH^-] = \frac{mmol}{mL} = \frac{0.1 \times 29.93}{230} = 0.013$$

$$[H_3O^+] = \frac{K_w}{[OH^-]} = \frac{10^{-14}}{0.013} = 7.7 \times 10^{-13} \qquad pH = 12.11$$

$$\Delta pH = 12.11 - 4.46 = +7.65$$

is not obvious, and frequently, students ask, "How do I know which assumptions will be valid?" or "What terms in an equation will be small?" Thinking about the chemistry involved will help but does not guarantee success. The best advice is to solve the problem with whatever approximations are expedient and then check to verify that they were justified.

Equation 9-20 will be valid after the first addition of NaOH (assuming it is more than a few drops) up to the equivalence point, when $f_{HA} = f_b$. This part of the titration curve is called the *buffer region* because, when $f_{HA} - f_b$ and f_b are both reasonably large (for example, both greater than 0.001 F), it takes the addition of a considerable amount of acid or base to change the pH appreciably. This is shown in Example 9-2.

A measure of how resistant a buffer solution is toward change in pH is its *buffer capacity*. The buffer capacity is defined as the number of equivalents of strong acid or strong base needed to change the pH of 1 L of the buffer solution by one pH unit. A calculation of buffer capacity is given in Example 9-3.

EXAMPLE 9-3

Calculate the pH and buffer capacity of a buffer containing 0.4 F formic acid and 0.6 F sodium formate.

Solution

$$[H_3O^+] = \frac{K_a[HA]}{[A]} = 1.77 \times 10^{-4}\frac{0.4}{0.6}$$

$$[H_3O^+] = 1.18 \times 10^{-4} \qquad pH\ 3.93$$

To determine the buffer capacity, we must calculate how much strong base must be added to 1 L of the buffer to increase the pH to 4.93 or how much strong acid to decrease the pH to 2.93

$$\text{strong-base addition:} \quad pH = 4.93 \qquad [H_3O^+] = 1.18 \times 10^{-5}$$

$$[H_3O^+] = 1.18 \times 10^{-5} = 1.77 \times 10^{-4}\frac{[HA]}{[A^-]}$$

Since we have 1 L, the moles present equal the molar concentrations. The addition of x equivalents (moles of OH^-) needed to change the pH one unit can therefore be calculated:

$$1.18 \times 10^{-5} = 1.77 \times 10^{-4}\frac{(0.4 - x)}{(0.6 + x)}$$

$$0.067(0.6 + x) = 0.4 - x$$

$$1.067x = 0.40 - 0.04 = 0.36$$

$$x = 0.337 = \text{buffer capacity}$$

EQUIVALENCE POINT

The buffer region calculation in Equation 9-20 is valid from the first addition of base to the weak acid until the equivalence point (or very nearly so). At the equivalence point, $f_{HA} - f_b = 0$, and therefore the concentration of $[H_3O^+]$ must be made by another approach. At the equivalence point, the solution is equivalent to a solution of the sodium salt of the weak acid with a formal concentration of f_{NaA}. The important chemical equilibrium involves the dissociation of the conjugate base of the weak acid (A^-) to yield OH^-, as shown in Equation 9-21:

$$A^- + H_2O \rightleftharpoons HA + OH^- \tag{9-21}$$

It is sometimes called the *hydrolysis* of the salt form of the weak acid, but this

EXAMPLE 9-3 (cont.)

strong-acid addition: pH = 2.93 $[H_3O^+] = 1.18 \times 10^{-3}$

Proceeding as above,

$$[H_3O^+] = \frac{K_a(0.4 + x)}{(0.6 - x)}$$

$$6.67(0.6 - x) = 0.4 + x$$

$$7.67x = 4 - 0.4 = 3.6$$

$$x = 0.469 = \text{buffer capacity}$$

It can be seen that the buffer has more capacity toward addition of acid than it does toward addition of base. This will be the case whenever there is a greater concentration of the conjugate base form than of the acid form. Buffer capacities are generally quoted in terms of base addition; and without information stating otherwise, it should be assumed that a buffer capacity value refers to base addition.

Buffer capacity defined as above means a finite change in pH (ΔpH = 1). If calculus is used, the slope of the curve of pH versus moles of acid (or base) can be determined. The slope, $d(\text{pH})/d(\text{mol})$, measures the buffering capacity of the system, lower slopes indicating a better buffer, that is, showing less change of pH with acid or base addition. The slope of the curve will be the same for base addition as it is for acid addition so that defining buffer capacity in this way (*buffer index*) is not ambiguous.

reaction is not different from the dissociation of any weak base, such as NH_3.

$$NH_3 + H_2O \rightleftharpoons NH_4^+ + OH^- \qquad \textbf{(9-22)}$$

where NH_4^+ is the conjugate acid of NH_3. The equilibrium constant for Equation 9-21 can be calculated from K_a and K_w as shown in Equation 9-23:

$$K_b = \frac{[HA][OH^-]}{[A^-][H_2O]} \cong \frac{[HA][OH^-]}{[A^-]}$$

$$\frac{K_w}{K_a} = \frac{[H_3O^+][OH^-]}{[H_3O^+][A^-]/[HA]} = \frac{[HA][OH^-]}{[A^-]} = K_b \qquad \textbf{(9-23)}$$

For convenience both K_a and K_b values are tabulated in Appendix 2-B.

If we wish to be rigorous, there are the same five species as in the buffer region and five equations (Equations 9-4, 9-16, and 9-23 to 9-25).

$$f_b = f_{NaA} = [Na^+] \tag{9-24}$$

$$f_{HA} = f_{NaA} = [A^-] + [HA] \tag{9-25}$$

These can be combined to give the third-order equation shown in Equation 9-26:

$$K_b[H_3O^+]^3 + (K_b f_{NaA} + K_w)[H_3O^+]^2 - K_b K_w[H_3O^+] - K_w^2 = 0 \tag{9-26}$$

Equation 9-26 is equivalent to Equation 9-17 when it is recognized that $f_{HA} = f_{NaA}$ at the equivalence point and that $K_b = K_w/K_a$. To simplify Equation 9-26, we generally assume that dissociation of NaA is the primary source of HA and OH^-. Therefore, they are essentially equal,

$$[HA] \cong [OH^-]$$

and the system can be solved for $[OH^-]$

$$K_b = \frac{[HA][OH^-]}{[A^-]} = \frac{[OH^-][OH^-]}{f_{NaA} - [HA]} = \frac{[OH^-]^2}{f_{NaA} - [OH^-]} \tag{9-27}$$

Equation 9-27 leads to an expression analogous to the dissociation of the weak acid at the initial point, that is Equation 9-28,

$$[OH^-] = \sqrt{K_b(f_{NaA} - [OH^-])} = \sqrt{\frac{K_w}{K_a}(f_{NaA} - [OH^-])} \tag{9-28}$$

and we can solve it for $[OH^-]$ by assuming it to be small compared with f_{NaA} (generally true) and using an iterative routine if necessary.

The factor K_b^{-1} is the equilibrium constant for the neutralization reaction involved in the titration

$$HA + OH^- \rightleftharpoons A^- + H_2O \qquad K = K_b^{-1} = \frac{K_a}{K_w}$$

The feasibility of any titration depends on the magnitude of K_b^{-1} (or K_a/K_w) and should be $> 10^6$ for a successful titration. This means that weak acids with $K_a < 10^{-8}$ will not exhibit titration curves satisfactory for analytical purposes. This criterion for titration feasibility also depends on f_{HA}, which has been assumed to be about 0.1 F.

EXCESS BASE

After the equivalence point, the addition of strong base will increase $[OH^-]$ and repress the dissociation in Equation 9-21. Essentially all OH^- in the system is from the excess base and can be calculated, as shown in Equation 9-29, as it was for the strong acid–strong base titration curve derived in Example 9-1.

$$[OH^-] = f_b - f_{HA} \qquad (9\text{-}29)$$

Normally, problems are presented in terms of a formal concentration of the base titrant solution (F_b) and the volume of titrant added (V_b) so that

$$f_b = \frac{F_b V_b}{V_{HA} + V_b}$$

In a similar fashion, the formal concentration of HA is

$$f_{HA} = \frac{F_{HA} V_{HA}}{V_{HA} + V_b}$$

and Equation 9-29 becomes

$$[OH^-] = \frac{F_b V_b - F_{HA} V_{HA}}{V_{HA} + V_b} \qquad (9\text{-}30)$$

The numerator of Equation 9-30 is the millimoles of excess base, while the denominator is the total volume of the system.

SUMMARY OF TITRATION CURVE

The four regions of the titration curve are summarized in Table 9-1, and the equations derived are used to compute a titration curve for a typical case shown in Example 9-4.

It can be seen from Table 9-2 that the change in pH from 49 to 50 mL is almost exactly the same as the change from 50 to 51 mL. This symmetry around the equivalence point is clear in Figure 9-2 and offers a means of checking the calculations since each of the three points was calculated by a different method. Symmetry around the equivalence point was also noted for the precipitation titration curve developed in Section 8-1.

Computer programs for calculating acid–base titration curves can be written, and Appendix 1-E illustrates one way of accomplishing the task with the approach of Example 9-4. Subroutines are called up for the four regions given in Table 9-1. A plot of the points taken every increment of titrant volume $(V_{eq\,pt}/50)$ is shown in

EXAMPLE 9-4

Determine a titration curve (pH versus V_b) for the titration of 100 mL of 0.05 F acetic acid with 0.10 F NaOH. Calculate the pH for the addition of 0, 10, 25, 40, 48, 49, 50, 51, 52, 60, and 100 mL of the base titrant.

Solution

0 mL

The initial point is calculated using

$$[H_3O^+] = \sqrt{K_a(f_{HA} - [H_3O^+])} = \sqrt{1.75 \times 10^{-5}(0.05 - [H_3O^+])}$$

The first approximation is obtained by letting $[H_3O^+] = 0$ on the right side.

$$[H_3O^+] = \sqrt{1.75 \times 10^{-5} \times 0.05} = \sqrt{8.75 \times 10^{-7}} = 9.4 \times 10^{-4}$$

which is $<2\%$ of 0.05. Therefore, the approximation was justified. The iterative approach can be used to refine the value:

$$[H_3O^+] = \sqrt{1.75 \times 10^{-5}(0.05 - 9.4 \times 10^{-4})} = 9.3 \times 10^{-4}$$

In either case, pH = 3.03.

10 mL

The equivalence point volume, $V_{eq\,pt}$ can be calculated with milliequivalents (or millimoles since $n = 1$ for both NaOH and acetic acid)

$$\text{mmol HA} = \text{mmol NaOH}$$

$$F_{HA}V_{HA} = F_b V_{eq\,pt}$$

$$V_{eq\,pt} = \frac{F_{HA}V_{HA}}{F_b} = \frac{0.05 \times 100}{0.1} = 50 \text{ mL}$$

The buffer region points are then $V_b = 10, 25, 40, 48$, and 49 mL. The pH after the addition of 10 mL of 0.10 F NaOH is

$$[H_3O^+] = \frac{K_a(f_{HA} - f_b)}{f_b}$$

$$[H_3O^+] = 1.75 \times 10^{-5} \frac{\left(\dfrac{F_{HA}V_{HA}}{V_{HA} + V_b} - \dfrac{F_b V_b}{V_{HA} + V_b}\right)}{\dfrac{F_b V_b}{V_{HA} + V_b}}$$

$$[H_3O^+] = 1.75 \times 10^{-5} \frac{(0.05 \times 100 - 0.10 \times 10)}{0.10 \times 10}$$

$$[H_3O^+] = 1.75 \times 10^{-5} \frac{4}{1} = 7.0 \times 10^{-5} \qquad \text{pH} = 4.15$$

EXAMPLE 9-4 (cont.)

The pH at each of the buffer region points is determined in the same way and listed in Table 9-2. Note that, at the midpoint of the titration (25 mL), the pH = pK_a.

50 mL

The equivalence point pH is calculated by using

$$[OH^-] = \sqrt{\frac{K_w}{K_a}(f_{NaA} - [OH^-])}$$

$$f_{NaA} = \frac{F_{HA}V_{HA}}{V_{HA} + V_b} = \frac{0.05 \times 100}{100 + 50} = 0.033$$

Alternatively,

$$f_{NaA} = \frac{F_b V_b}{V_{HA} + V_b} = \frac{0.10 \times 50}{100 + 50} = 0.033$$

$$[OH^-] = \sqrt{\frac{10^{-14}}{1.75 \times 10^{-5}}(0.033 - [OH^-])}$$

The first approximation is obtained by letting $[OH^-]$ on the right side = 0

$$[OH^-] = \sqrt{\frac{10^{-14}}{1.75 \times 10^{-5}} \times 0.033} = 4.34 \times 10^{-6}$$

This is clearly ≪0.033 and no iteration is needed.

$$[H_3O^+] = \frac{K_w}{[OH^-]} = \frac{10^{-14}}{4.34 \times 10^{-6}} = 2.3 \times 10^{-9}$$

$$pH = 8.64$$

51 mL

pH after the equivalence point is calculated with

$$[OH^-] = \frac{F_b V_b - F_{HA}V_{HA}}{V_{HA} + V_b} = \frac{0.10 \times 51 - 0.05 \times 100}{100 + 51} = 6.62 \times 10^{-4}$$

$$[H_3O^+] = \frac{K_w}{[OH^-]} = \frac{10^{-14}}{6.62 \times 10^{-4}} = 1.5 \times 10^{-11}$$

$$pH = 10.82$$

The pH at each of the points after the equivalence point is determined in the same way and listed in Table 9-2. The titration curve is shown in Figure 9-2.

TABLE 9-1 Summary of Acid-Base Titration Curve Calculation (Weak Acid with strong base)

Region	Volume of titrant, V_b	Equation
initial point	0	$[H_3O^+] = \sqrt{K_a(f_{HA} - [H_3O^+])}$
buffer region	$0 < V_b < V_{eq\ pt}$	$[H_3O^+] = \dfrac{K_a(f_{HA} - f_b)}{f_b}$
equivalence point	$V_{eq\ pt}$	$[OH^-] = \sqrt{\dfrac{K_w}{K_a}(f_{NaA} - [OH^-])}$
		$[H_3O^+] = \dfrac{K_w}{[OH^-]}$
excess base	$> V_{eq\ pt}$	$[OH^-] = f_b - f_{HA} = \dfrac{F_b V_b - F_{HA} V_{HA}}{V_{HA} + V_b}$
		$[H_3O^+] = \dfrac{K_w}{[OH]}$

TABLE 9-2 Summary of Titration Curve Calculation in Example 9-4

V_b(mL)	$[H_3O^+]$	pH
0	9.3×10^{-4}	3.03
10	7.0×10^{-5}	4.15
25	1.75×10^{-5}	4.76
40	4.4×10^{-6}	5.36
48	7.3×10^{-7}	6.14
49	3.6×10^{-7}	6.45
50	2.3×10^{-9}	8.64
51	1.5×10^{-11}	10.82
52	7.6×10^{-12}	11.12
60	1.6×10^{-12}	11.80
100	4.0×10^{-13}	12.40

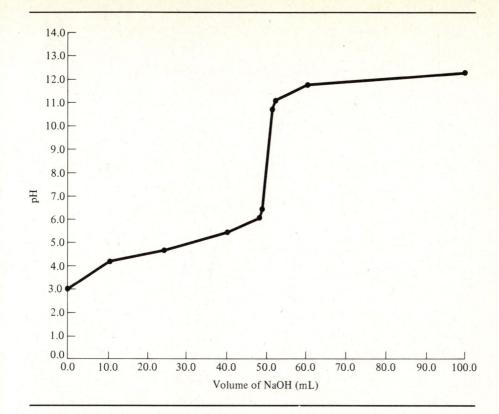

FIGURE 9-2   Titration curve for the titration of 100 mL of 0.05 *F* acetic acid with 0.10 *F* NaOH. Points calculated in Example 9-4 are marked as ●.

Figure 9-3 for the titration of 0.1 *F* lactic acid with 0.2 *F* NaOH. This method of calculation gives very few data points near the equivalence point because the pH changes so rapidly in this region.

   Another way of calculating the acid–base titration curve is to take points every increment of pH (reversing independent and dependent variables) and find the volume of titrant that corresponds to that pH. This method has certain advantages. The exact Equation 9-17 was difficult to solve because it is third order in $[H_3O^+]$. However, if we know $[H_3O^+]$, the expression is only first order in titrant volume. Thus the exact equation can be solved for titrant volume with known pH values (negative titrant volumes are disregarded), and a program of this type is given in Appendix 1-F. A plot of the same titration of lactic acid with this program is given in Figure 9-4. Now there are many data points near the equivalence point. Also, with the first addition of titrant, there appears to be a dip in pH on Figure 9-3 but not on Figure 9-4. The approximation that this small addition of titrant put the system into the buffer region was not justified. The exact solution makes no approximations and shows a smooth curve right down to zero titrant volume. The difference in pH for

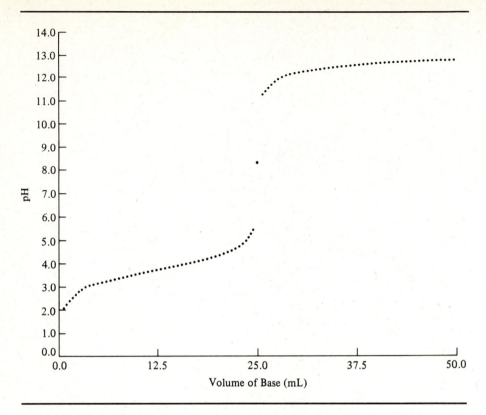

FIGURE 9-3 Titration curve for titration of 0.1 F lactic acid ($K_a = 1.37 \times 10^{-4}$) with 0.2 F NaOH, calculated with the computer program shown in Appendix 1-E. Subroutines perform the calculation in the same way as in Example 9-4.

this first point is ~0.3 pH units. This difference is insignificant in terms of a titration curve, especially when one realizes that the pH change near the equivalence point is the critical feature for analytical application. One last comment regarding these two programs: The exact-solution program uses the same equation for every point and therefore can be shorter than the program using the four subroutines can be.

9-2 ENDPOINT SIGNALS

What is needed for our titration of acetic acid in Example 9-4 is an endpoint signal that triggers when the pH reaches 8.64. The glass electrode responds to $[H_3O^+]$, and measurement of pH can be carried out using a pH meter (Section 15-1). The pH is recorded with every increment of titrant volume, and the volume corresponding to pH 8.64 may be taken as the endpoint.

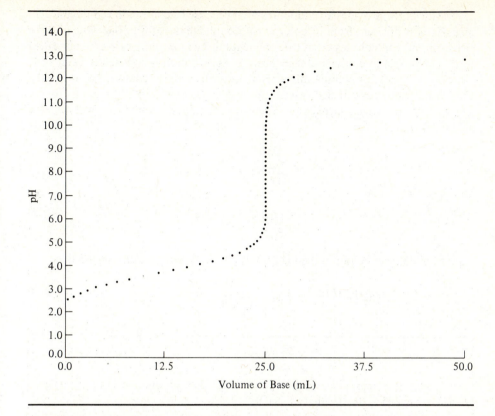

FIGURE 9-4 Titration curve for titration of 0.1 F lactic acid ($K_a = 1.37 \times 10^{-4}$) with 0.2 F NaOH, calculated with the computer program shown in Appendix 1-F. Exact Equation 9-17 used for calculation with points taken at equal increments of pH.

The calculated equivalence point pH neglected activity coefficients and assumed that a certain concentration of acid was present. In the actual experiment, the pH values may be somewhat different. Therefore, it is usually recommended that the endpoint be taken as the point of greatest slope (first derivative reaches a maximum) or the inflection point (second derivative equals zero). For typical weak acids ($K_a > 10^{-8}$) at reasonable concentrations ($> 10^{-2}$ F), all methods of reading the endpoint from the plot of pH versus volume will give reasonably consistent results with little endpoint error.

Recording pH with every volume increment is slow, and color indicators are frequently substituted. The rationale for acid–base endpoint indicators is that they are also weak acids or bases. For example, if the indicator (HIn) is a weak acid,

$$HIn + H_2O \rightleftharpoons H_3O^+ + In^- \tag{9-31}$$

$$K_{HIn} = \frac{[H_3O^+][In^-]}{[HIn]} \tag{9-32}$$

The indicator provides an endpoint signal because its color in the conjugate acid form, HIn, is different from its color in the conjugate base form, In$^-$. One of these two forms may be colorless, as is true of phenolphthalein (colorless in conjugate acid form, red-purple in conjugate base form). If the solution is sufficiently acidic, the indicator will exist predominantly as HIn and will be in its acid-form color. When the solution becomes sufficiently basic, the indicator converts predominantly to In$^-$ and will turn to its base-form color. As a rule of thumb, the human eye will perceive acid form when

$$\frac{[\text{HIn}]}{[\text{In}^-]} \geq 10$$

and base form when

$$\frac{[\text{HIn}]}{[\text{In}^-]} \leq 0.1$$

Equation 9-32 can be put in the logarithmic form shown in Equation 9-33:

$$\log[\text{H}_3\text{O}^+] = \log K_{\text{HIn}} + \log\frac{[\text{HIn}]}{[\text{In}^-]} \qquad \text{(9-33)}$$

EXAMPLE 9-5

Recommend an appropriate indicator from Table 9-3 for the titration of 0.1 F glycolic acid ($K_a = 1.48 \times 10^{-4}$) with 0.1 F NaOH.

Solution

The pH at the equivalence point is calculated from Equation 9-28.

$$[\text{OH}^-] = \sqrt{\frac{K_w}{K_a}\left(f_{\text{NaA}} - [\text{OH}^-]\right)} = \sqrt{\frac{10^{-14}}{1.48 \times 10^{-4}}(0.05 - [\text{OH}^-])}$$

(Note the dilution factor; that is, $f_{\text{NaA}} = 0.1/2 = 0.05$.)

For a first approximation assume that $[\text{OH}^-] = 0$ on the right side (small compared to 0.05).

$$[\text{OH}^-] = \sqrt{\frac{10^{-14}}{1.48 \times 10^{-4}} \times 0.05} = 1.8 \times 10^{-6} \text{ (approximation justified:}$$
$$1.8 \times 10^{-6} \ll 0.05)$$

$$[\text{H}_3\text{O}^+] = \frac{K_w}{[\text{OH}^-]} = \frac{10^{-14}}{1.8 \times 10^{-6}} = 5.4 \times 10^{-9}$$

$$\text{pH} = 8.26$$

From Table 9-3, we choose an indicator whose $pK_{\text{HIn}} = 8.26 \pm 1$. The indicator that matches most closely is cresol purple, but the more common phenolphthalein is within the acceptable range too.

or

$$pH = pK_{HIn} - \log\frac{[HIn]}{[In^-]}$$

The acid form will be observed in solutions in which

$$pH = pK_{HIn} - \log 10 = pK_{HIn} - 1 \text{ (or lower)}$$

and the base form will be observed in solutions in which

$$pH = pK_{HIn} - \log 0.1 = pK_{HIn} + 1 \text{ (or above)}$$

A transition from acid color to base color will occur in the pH range

$$pH = pK_{HIn} \pm 1 \qquad\qquad\qquad\qquad\qquad \textbf{(9-34)}$$

Therefore, the analyst should choose an indicator whose pK_{HIn} is within one pH unit of the pH at the equivalence point. This is worked out in Example 9-5.

Suppose the analyst chooses the wrong indicator. For instance, a bottle of methyl red is handy and works well for the titration of HCl with NaOH; so why not use it for the titration of glycolic acid given in Example 9-5? As Table 9-3 shows, the

TABLE 9-3   Acid–Base Indicators

Indicator name	Type*	pK	pH transition interval	Color Acid form	Color Base form
thymol blue[†]	acid	1.65	1.2– 2.8	red	yellow
dimethyl yellow	base	3.25	2.9– 4.0	red	yellow
methyl orange	base	3.46	3.1– 4.4	red	orange-yellow
bromophenol blue	acid	4.10	3.0– 4.6	yellow	purple
bromocresol green	acid	4.90	3.8– 5.4	yellow	blue
methyl red	base	5.00	4.4– 6.2	red	yellow
chlorophenol red	acid	6.25	4.8– 6.4	yellow	red
p-nitrophenol	acid	7.15	5.6– 7.6	colorless	yellow
bromothymol blue	acid	7.30	6.0– 7.6	yellow	blue
phenol red	acid	8.00	6.4– 8.2	yellow	red
m-cresol purple	acid	8.32	7.6– 9.2	yellow	purple-red
thymol blue[†]	acid	9.20	8.0– 9.6	yellow	blue
phenolphthalein	acid	9.5	8.0– 9.8	colorless	purple
thymolphthalein	acid	9.7	9.3–10.5	colorless	blue

SOURCE: Data from E. Bishop, ed., *Indicators* (Elmsford, NY: Pergamon Press, Inc., 1972).

*Acid indicator $HIn + H_2O \rightleftharpoons H_3O^+ + In^-$; when $[HIn] = [In^-]$, $pH = pK_{HIn} = pK$ in table. Base indicator: $In + H_2O \rightleftharpoons HIn^+ + OH^-$; when $[HIn^+] = [In]$, $pOH = pK_{In}$, $pH = 14 - pK_{In} = pK$ in table.
†Note that thymol blue has two color transitions: one in acidic solution, the other in basic solution.

EXAMPLE 9-6

Calculate the endpoint error for the titration described in Example 9-5 if methyl red indicator is used as the endpoint signal. (Assume that the color change is observed when pH = 5.3.)

Solution

The endpoint signal will occur too soon because pH 5.3 is lower than the equivalence point pH of 8.26 calculated in Example 9-5. The system will still be essentially in its buffer region.

$$K_a = \frac{[H_3O^+][A^-]}{[HA]}$$

$$1.48 \times 10^{-4} = 5 \times 10^{-6} \frac{[A^-]}{[HA]}$$

$$\frac{[HA]}{[A^-]} = \frac{5 \times 10^{-6}}{1.48 \times 10^{-4}} = 0.034 \qquad [HA] = 0.034[A^-]$$

The percentage of total acid remaining in the undissociated (and untitrated) form is

$$\% \,[HA] = \frac{[HA]}{[HA] + [A^-]} \times 100\,(\%) = \frac{0.034[A^-]}{0.034[A^-] + [A^-]} \times 100\,(\%) = 3.3\%$$

The error will be approximately −3.3%. This calculation will not be exact very near the equivalence point because it assumes that [HA] = 0 at the equivalence point. *Very* near the equivalence point, the error can be calculated by determining the difference between [HA] at the endpoint and [HA] at the equivalence point. However, that is not strictly accurate because the extent of dissociation will shift when the system is not at the equivalence point. The exact expression given in Equation 9-17 should be used *if* the problem deserves that much attention. Indicators change over a range of pH, and endpoint error calculations are usually accurate to a factor of only 2 or so. Solution of Equation 9-17 for $[H_3O^+] = 5 \times 10^{-6}$ leads to the same answer of 3.3%, to two significant figures.

endpoint will be reached in the pH range 4.4 to 6.2—considerably more acidic than the equivalence point pH of 8.16 calculated in Example 9-5. The resultant determinate error is calculated in Example 9-6.

9-3 WEAK BASES

There are many substances that produce OH^- in an aqueous solution. Strong bases, such as NaOH, dissociate completely to give a concentration of hydroxide ion equal to nf_b, where n is the number of hydroxides per mole (equivalents per mole) and f_b is the formal concentration of base. Other substances are only weakly dissociated and are called weak bases. In general, they produce hydroxide ion by a dissociation reaction, shown in Equation 9-35:

$$B + H_2O \rightleftharpoons BH^+ + OH^- \tag{9-35}$$

$$K_b = \frac{[BH^+][OH^-]}{[B]} \tag{9-36}$$

Inspection of Equation 9-35 shows that it is fundamentally the same as the reaction cited at the equivalence point of the weak-acid titration given in Equation 9-21. In Equation 9-35, BH^+ is the conjugate acid of B in the same way that HA is the conjugate acid of A^-. The conjugate acid BH^+ will have a dissociation reaction and equilibrium constant as follows:

$$BH^+ + H_2O \rightleftharpoons H_3O^+ + B \tag{9-37}$$

$$K_a = \frac{[H_3O^+][B]}{[BH^+]} \tag{9-38}$$

The two equilibrium constants are related by Equation 9-23. This can be expressed in the form

$$K_a K_b = K_w$$

or

$$pK_a + pK_b = pK_w = 14 \tag{9-39}$$

Thus, if we are given pK_a for the system, it is easy to calculate pK_b. For convenience, both are given in Appendix 2-B. In general, those substances which we normally call weak acids have $K_a > K_b$ ($pK_a < pK_b$), while those substances which we call weak bases have $K_b > K_a$ ($pK_b < pK_a$); but this is not always true. Consider, for example, hydrocyanic acid, HCN.

$$HCN + H_2O \rightleftharpoons H_3O^+ + CN^- \qquad pK_a = 9.21 \qquad pK_b = 4.79$$

Compare it with ammonia, NH_3.

$$NH_4^+ + H_2O \rightleftharpoons H_3O^+ + NH_3 \qquad pK_a = 9.26 \qquad pK_b = 4.74$$

We call HCN a weak acid and Na^+CN^- a salt, whereas NH_3 is thought of as a base and $NH_4^+Cl^-$ a salt. Actually NH_4^+ is just as strong an acid as HCN, and CN^- is just as strong a base as NH_3. Example 9-7 shows this fact by calculating the pH of 0.1 F

EXAMPLE 9-7

Calculate the pH of 0.1 F solutions of HCN, NH_4Cl, NH_4OH, and NaCN.

Solution

HCN

$$HCN + H_2O \rightleftharpoons H_3O^+ + CN^-$$

$$K_a = 6.2 \times 10^{-10} = \frac{[H_3O^+][CN]}{[HCN]}$$

$$[H_3O^+] = \sqrt{K_a(f_{HA} - [H_3O^+])} \cong \sqrt{K_a f_{HA}} = 7.9 \times 10^{-6} = [CN^-]$$

$$pH = 5.10$$

The approximation that $[H_3O^+] \ll f_{HA}$ is justified because $7.9 \times 10^{-6} \ll 0.1$. The assumption that $[CN^-] \gg [OH^-]$ is still valid because $7.9 \times 10^{-6} \gg 1.3 \times 10^{-9}$.

NH₄Cl

$$NH_4^+ + H_2O \rightleftharpoons H_3O^+ + NH_3$$

$$K_a = 5.56 \times 10^{-10} = \frac{[H_3O^+][NH_3]}{[NH_4^+]}$$

$$[H_3O^+] = \sqrt{K_a(f_{HA} - [H_3O^+])} \cong \sqrt{K_a f_{HA}} = 7.5 \times 10^{-6} = [NH_3]$$

$$pH = 5.13$$

The approximations that $[H_3O^+] \ll f_{HA}$ and $[NH_3] \gg [OH^-]$ are justified as in the HCN case.

---

solutions of HCN, NH_4Cl, NH_4OH, and NaCN, the point being that NH_4OH and NaCN are such strong bases that they can be titrated with strong acids equally well and will show almost identical titration curves. On the other hand, NH_4Cl and HCN are such weak acids that they cannot be titrated satisfactorily in aqueous solutions.

Another reason for demonstrating the parallelism between weak acids and weak bases is that it is clearly not necessary to go through the calculation of a titration curve for a weak base with a strong acid. The initial point will be calculated in the same way as is the equivalence point in the weak-acid titration. Then all points up to the equivalence point will be in a buffer region. The equivalence point will be calculated in the same way as is the initial point in the weak-acid titration; and finally excess-strong-acid points will be calculated analogously to the excess-strong-base points in the weak-acid titration. This is summarized in Table 9-4.

EXAMPLE 9-7 (cont.)

NH_4OH (or NH_3)

$$NH_3 + H_2O \rightleftharpoons NH_4^+ + OH^-$$

$$K_b = 1.8 \times 10^{-5} = \frac{[NH_4^+][OH^-]}{[NH_3]}$$

$$[OH^-] = \sqrt{K_b(f_b - [OH^-])} \cong \sqrt{K_b f_b} = 1.35 \times 10^{-3}$$

$$[H_3O^+] = \frac{10^{-14}}{1.35 \times 10^{-3}} = 7.4 \times 10^{-12}$$

$$pH = 11.13$$

The approximation that $[OH^-] \ll f_b$ is valid to two significant figures because $1.35 \times 10^{-3} \ll 0.1$.

NaCN

$$CN^- + H_2O \rightleftharpoons HCN + OH^-$$

$$K_b = 1.6 \times 10^{-5} = \frac{[HCN][OH^-]}{[CN^-]}$$

$$[OH^-] = \sqrt{K_b(f_b - [OH^-])} \cong \sqrt{K_b f_b} = 1.27 \times 10^{-3}$$

$$[H_3O^+] = \frac{10^{-14}}{1.27 \times 10^{-3}} = 7.9 \times 10^{-12}$$

$$pH = 11.10$$

The approximation that $[OH^-] \ll f_b$ holds as with the NH_3 case.

9-4   POLYPROTIC ACIDS

COMPOSITION AS A FUNCTION OF PH

There are several substances that contain more than one ionizable hydrogen atom. A series of acid dissociation equilibrium equations may be written: In general,

$$H_nA + H_2O \rightleftharpoons H_3O^+ + H_{n-1}A^- \qquad K_1 = \frac{[H_3O^+][H_{n-1}A^-]}{[H_nA]}$$

$$H_{n-1}A^- + H_2O \rightleftharpoons H_3O^+ + H_{n-2}A^{2-} \qquad K_2 = \frac{[H_3O^+][H_{n-2}A^{2-}]}{[H_{n-1}A^-]}$$

$$HA^{1-n} + H_2O \rightleftharpoons H_3O^+ + A^{n-} \qquad K_n = \frac{[H_3O^+][A^{n-}]}{[HA^{1-n}]}$$

TABLE 9-4   Summary of Acid-Base Titration Curve Calculation (Weak Base with Strong Acid)

Region	Volume of titrant, V_a	Equation
initial point	0	$[OH^-] = \sqrt{K_b(f_b - [OH^-])}$
		$[H_3O^+] = \dfrac{K_w}{[OH^-]}$
buffer region	$0 < V_a < V_{eq\,pt}$	$[OH^-] = \dfrac{K_b(f_b - f_a)}{f_a}$
		$[H_3O^+] = \dfrac{K_w}{[OH^-]}$
equivalence point	$V_{eq\,pt}$	$[H_3O^+] = \sqrt{K_a(f_{BH^+} - [H_3O^+])}$
		$\left(note: K_a = \dfrac{K_w}{K_b}\right)$
excess acid	$> V_{eq\,pt}$	$[H_3O^+] = f_a - f_b = \dfrac{F_a V_a - F_b V_b}{V_a + V_b}$

A specific example:

$$H_3PO_4 + H_2O \rightleftharpoons H_3O^+ + H_2PO_4^- \qquad K_1 = 7.1 \times 10^{-3} = \frac{[H_3O^+][H_2PO_4^-]}{[H_3PO_4]} \qquad (9\text{-}40)$$

$$H_2PO_4^- + H_2O \rightleftharpoons H_3O^+ + HPO_4^{2-} \qquad K_2 = 6.3 \times 10^{-8} = \frac{[H_3O^+][HPO_4^{2-}]_4}{[H_2PO_4^-]} \qquad (9\text{-}41)$$

$$HPO_4^{2-} + H_2O \rightleftharpoons H_3O^+ + PO_4^{3-} \qquad K_3 = 4.2 \times 10^{-13} = \frac{[H_3O^+][PO_4^{3-}]}{[HPO_4^{2-}]} \qquad (9\text{-}42)$$

A rigorous approach to solving a titration curve for the titration of H_3PO_4 with NaOH would involve seven species: $[H_3O^+]$, $[OH^-]$, $[H_3PO_4]$, $[H_2PO_4^-]$, $[HPO_4^{2-}]$, $[PO_4^{3-}]$, and $[Na^+]$. The seven simultaneous equations required are Equations 9-4 (dissociation of water), 9-40 to 9-42 (three acid dissociation constants for H_3PO_4), and 9-15 (mass balance for NaOH addition), plus the charge balance and mass balance for total phosphoric acid present shown in Equations 9-43 and 9-44.

$$\text{charge balance:} \quad [\text{H}_3\text{O}^+] + [\text{Na}^+] = [\text{H}_2\text{PO}_4^-] + 2[\text{HPO}_4^{2-}]$$
$$+ 3[\text{PO}_4^{3-}] + [\text{OH}^-] \tag{9-43}$$

$$\text{acid mass balance:} \quad f_{\text{H}_3\text{A}} = [\text{H}_3\text{PO}_4] + [\text{H}_2\text{PO}_4^-] + [\text{HPO}_4^{2-}] + [\text{PO}_4^{3-}] \tag{9-44}$$

This situation is hopelessly complicated, and therefore to obtain reasonable answers in a reasonable length of time requires making intelligent approximations. How do we go about making the right approximations? The key to answering this question is shown in Figure 9-5, in which the fraction, α, of phosphoric acid existing in each phosphate species is plotted against pH. When any substance X may exist in more than one form $(x_1, x_2, x_3, \ldots, x_n)$, the fraction, α, for any form—for example, x_1—is defined as

$$\alpha_1 \equiv \frac{[x_1]}{[x]} = \frac{[x_1]}{\sum\limits_{i=1}^{i=n} [x_i]}$$

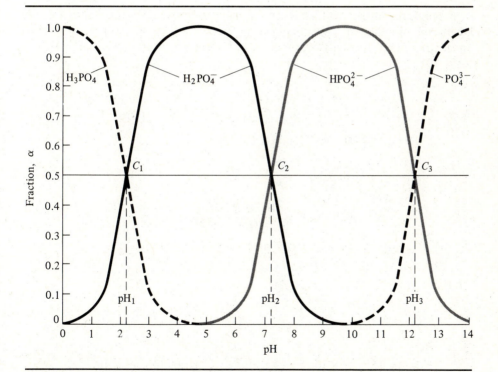

FIGURE 9-5   Fraction (α) of phosphoric acid existing as various species as a function of pH. Crossover points (C_1, C_2, C_3) define regions for predominant species.

Curves of this type will be observed for any acid; only the number of species possible (four for H_3PO_4) and values for α at each pH will differ. What is needed is to define these curves more quantitatively, and one simple way of doing this is to calculate the crossover points, C_1, C_2, and C_3. At C_1, $\alpha_{H_3PO_4} = \alpha_{H_2PO_4^-}$, and therefore $[H_3PO_4] = [H_2PO_4^-]$. If that is true, then

$$K_1 = 7.1 \times 10^{-3} = \frac{[H_3O^+][H_2PO_4^-]}{[H_3PO_4]} = [H_3O^+] \qquad pH = 2.15 \qquad \text{(9-45)}$$

At C_2, $\alpha_{H_2PO_4^-} = \alpha_{HPO_4^{2-}}$, and therefore $[H_2PO_4^-] = [HPO_4^{2-}]$, and

$$K_2 = 6.3 \times 10^{-8} = \frac{[H_3O^+][HPO_4^{2-}]}{[H_2PO_4^-]} = [H_3O^+] \qquad pH = 7.20 \qquad \text{(9-46)}$$

At C_3, $\alpha_{HPO_4^{2-}} = \alpha_{PO_4^{3-}}$, and therefore $[HPO_4^{2-}] = [PO_4^{3-}]$, and

$$K_3 = 4.2 \times 10^{-13} = \frac{[H_3O^+][PO_4^{3-}]}{[HPO_4^{2-}]} = [H_3O^+] \qquad pH = 12.38 \qquad \text{(9-47)}$$

It can be readily shown that, when the acid dissociation constants differ by any significant amount, all other weak-acid species will be present at very low concentration at the crossover point. For example, at C_1, with $[H_3O^+] = 7.1 \times 10^{-3}$,

$$K_2 = 6.3 \times 10^{-8} = \frac{[H_3O^+][HPO_4^{2-}]}{[H_2PO_4^-]} = \frac{7.1 \times 10^{-3}[HPO_4^{2-}]}{[H_2PO_4^-]}$$

$$[HPO_4^{2-}] = 8.9 \times 10^{-6}[H_2PO_4^-]$$

and

$$K_3 = 4.2 \times 10^{-13} = \frac{[H_3O^+][PO_4^{3-}]}{[HPO_4^{2-}]} = \frac{7.1 \times 10^{-3}[PO_4^{3-}]}{[HPO_4^{2-}]}$$

$$[PO_4^{3-}] = 5.9 \times 10^{-11}[HPO_4^{2-}] = 5.3 \times 10^{-16}[H_2PO_4^-]$$

Thus at C_1,

$$\alpha_{H_3PO_4} = 0.5 \qquad\qquad\qquad\qquad [H_3PO_4] = 0.5f_{H_3A}$$

$$\alpha_{H_2PO_4^-} = 0.5 \qquad\qquad\qquad\qquad [H_2PO_4^-] = 0.5f_{H_3A}$$

$$\alpha_{HPO_4^{2-}} = 0.5 \times 8.9 \times 10^{-6} = 4.4 \times 10^{-6} \qquad [HPO_4^{2-}] = 4.4 \times 10^{-6}f_{H_3A}$$

$$\alpha_{PO_4^{3-}} = 0.5 \times 5.3 \times 10^{-16} = 2.6 \times 10^{-16} \qquad [PO_4^{3-}] = 2.6 \times 10^{-16}f_{H_3A}$$

Without our solving any complicated equations, the concentrations of all phos-

EXAMPLE 9-8

Calculate the concentrations of all four phosphate species for a 0.05 F H_3PO_4 solution whose pH has been adjusted to 1.5.

Solution

$$\text{at pH 1.5: } [H_3O^+] = 3.16 \times 10^{-2}$$

Since this pH is to the left of C_1 on Figure 9-5, we use K_1 to determine $[H_2PO_4^-]$, knowing that $[H_3PO_4]$ will be the major species.

$$K_1 = 7.1 \times 10^{-3} = \frac{[H_3O^+][H_2PO_4^-]}{[H_3PO_4]} = \frac{3.16 \times 10^{-2}[H_2PO_4^-]}{[H_3PO_4]}$$

$$[H_2PO_4^-] = 0.225[H_3PO_4]$$

These two species must account for essentially all the formal concentration because it was shown that, at the crossover point, $[HPO_4^{2-}]$ and $[PO_4^{3-}]$ were small.

$$f_{H_3A} = 0.05 \cong [H_3PO_4] + [H_2PO_4] = [H_3PO_4] + 0.225[H_3PO_4]$$

$$[H_3PO_4] = \frac{0.05}{1.225} = 0.041 \ M$$

$$[H_2PO_4^-] = 0.225[H_3PO_4] = 0.009 \ M$$

The small concentrations of $[HPO_4^{2-}]$ and $[PO_4^{3-}]$ can be calculated by using K_2 and K_3:

$$K_2 = 6.3 \times 10^{-8} = \frac{[H_3O^+][HPO_4^{2-}]}{[H_2PO_4^-]}$$

$$[HPO_4^{2-}] = \frac{6.3 \times 10^{-8}[H_2PO_4^-]}{[H_3O^+]} = \frac{6.3 \times 10^{-8} \times 0.009}{3.16 \times 10^{-2}} = 1.8 \times 10^{-8} \ M$$

$$K_3 = 4.2 \times 10^{-13} = \frac{[H_3O^+][PO_4^{3-}]}{[HPO_4^{2-}]}$$

$$[PO_4^{3-}] = \frac{4.2 \times 10^{-13} \times 1.8 \times 10^{-8}}{3.16 \times 10^{-2}} = 2.4 \times 10^{-19} \ M$$

phate species have been calculated at pH 2.15. For all pH values less than 2.15, H_3PO_4 will be the predominant species, and the amount of $H_2PO_4^-$ can be calculated with K_1. This is shown in Example 9-8.

When pH is given, the concentrations of all species can be calculated by using Figure 9-5 to determine in which region the solution is. The values of α can be determined by starting with the major species, as was done in Example 9-8, or by using mathematical formulas derived as follows:

$$\alpha_{H_3PO_4} = \frac{[H_3PO_4]}{[H_3PO_4] + [H_2PO_4^-] + [HPO_4^{2-}] + [PO_4^{3-}]}$$

$$\alpha_{H_2PO_4^-} = \frac{[H_2PO_4^-]}{[H_3PO_4] + [H_2PO_4^-] + [HPO_4^{2-}] + [PO_4^{3-}]}$$

$$\alpha_{HPO_4^{2-}} = \frac{[HPO_4^{2-}]}{[H_3PO_4] + [H_2PO_4^-] + [HPO_4^{2-}] + [PO_4^{3-}]}$$

$$\alpha_{PO_4^{3-}} = \frac{[PO_4^{3-}]}{[H_3PO_4] + [H_2PO_4^-] + [HPO_4^{2-}] + [PO_4^{3-}]}$$

$$\alpha_{H_3PO_4} + \alpha_{H_2PO_4^-} + \alpha_{HPO_4^{2-}} + \alpha_{PO_4^{3-}} = 1$$

To convert these expressions for α values to equations containing only K_a's and $[H_3O^+]$, solve for each species in terms of one of them, such as $[H_3PO_4]$.

$$[H_2PO_4^-] = \frac{K_1[H_3PO_4]}{[H_3O^+]}$$

$$[HPO_4^{2-}] = \frac{K_1 K_2[H_3PO_4]}{[H_3O^+]^2}$$

$$[PO_4^{3-}] = \frac{K_1 K_2 K_3[H_3PO_4]}{[H_3O^+]^3}$$

Therefore,

$$\alpha_{H_3PO_4} = \frac{[H_3PO_4]}{\begin{array}{c}[H_3PO_4] + K_1[H_3PO_4]/[H_3O^+] \\ + K_1 K_2[H_3PO_4]/[H_3O^+]^2 + K_1 K_2 K_3[H_3PO_4]/[H_3O^+]^3\end{array}}$$

which simplifies to

$$\alpha_{H_3PO_4} = \frac{[H_3O^+]^3}{[H_3O^+]^3 + K_1[H_3O^+]^2 + K_1 K_2[H_3O^+] + K_1 K_2 K_3} \qquad (9\text{-}48)$$

In the same way α values for the other species are obtained:

$$\alpha_{H_2PO_4^-} = \frac{K_1[H_3O^+]^2}{[H_3O^+]^3 + K_1[H_3O^+]^2 + K_1 K_2[H_3O^+] + K_1 K_2 K_3} \qquad (9\text{-}49)$$

$$\alpha_{HPO_4^{2-}} = \frac{K_1 K_2[H_3O^+]}{[H_3O^+]^3 + K_1[H_3O^+]^2 + K_1 K_2[H_3O^+] + K_1 K_2 K_3} \qquad (9\text{-}50)$$

$$\alpha_{PO_4^{3-}} = \frac{K_1 K_2 K_3}{[H_3O^+]^3 + K_1[H_3O^+]^2 + K_1 K_2[H_3O^+] + K_1 K_2 K_3} \qquad (9\text{-}51)$$

Equations 9-48 to 9-51 can be used to calculate the concentrations of all species in a solution of known pH, as shown in Example 9-9.

EXAMPLE 9-9

Calculate the concentrations of all phosphate species in a 0.03 F H_3PO_4 solution at pH 8.

Solution

To solve this problem we may use Equations 9-48 to 9-51.

$$\alpha_{H_3PO_4} = \frac{(10^{-8})^3}{(10^{-8})^3 + 7.1 \times 10^{-3}(10^{-8})^2 + 7.1 \times 10^{-3} \times 6.3 \times 10^{-8} \times 10^{-8}}$$
$$+ 7.1 \times 10^{-3} \times 6.3 \times 10^{-8} \times 4.2 \times 10^{-13}$$

$$= \frac{10^{-24}}{5.18 \times 10^{-18}} = 1.93 \times 10^{-7}$$

$$[H_3PO_4] = 1.93 \times 10^{-7} \times 0.03 = 5.8 \times 10^{-9} \ M$$

$$\alpha_{H_2PO_4^-} = \frac{7.1 \times 10^{-3}(10^{-8})^2}{5.18 \times 10^{-18}} = 0.137$$

$$[H_2PO_4^-] = 0.137 \times 0.03 = 4.1 \times 10^{-3} \ M$$

$$\alpha_{HPO_4^{2-}} = \frac{7.1 \times 10^{-3} \times 6.3 \times 10^{-8} \times 10^{-8}}{5.18 \times 10^{-18}} = 0.863$$

$$[HPO_4^{2-}] = 0.863 \times 0.03 = 2.59 \times 10^{-2} \ M$$

$$\alpha_{PO_4^{3-}} = \frac{7.1 \times 10^{-3} \times 6.3 \times 10^{-8} \times 4.2 \times 10^{-13}}{5.18 \times 10^{-18}} = 3.62 \times 10^{-5}$$

$$[PO_4^{3-}] = 3.62 \times 10^{-5} \times 0.03 = 1.1 \times 10^{-6} \ M$$

Note from Figure 9-5 that we would predict that at pH 8 the major species would be HPO_4^{2-} and that the only other species with a significant concentration would be $H_2PO_4^-$.

TITRATION CURVE

The previous discussion showed how the concentration of each phosphate species could be determined at a particular pH. This approach could be used in constructing a titration curve, as was shown in Appendix 1-F, by calculating the volume of titrant needed to give a particular pH.

If we examine the various regions in the titration of a multiprotic acid, we see that only one new problem arises. Let us qualitatively go through the titration of H_3PO_4 with NaOH to emphasize the correspondence with our previous titration curve.

Initially, only the first dissociation of H_3PO_4 is important, and the pH is determined exactly as was done in Equation 9-14. With the first addition of NaOH, a buffer region is established, the $[H_2PO_4^-]/[H_3PO_4]$ ratio determining the pH just as was done in Equation 9-20 with $K_1 = K_a$. The problem occurs at the equivalence point. $H_2PO_4^-$ may dissociate (hydrolyze) in the same way as A^- in Equation 9-21 *but* $H_2PO_4^-$ may also dissociate to form HPO_4^{2-} and H_3O^+. In other words, $H_2PO_4^-$ may act both as a base and as an acid. Its relative strength for these two reactions will determine the exact value of $[H_3O^+]$ at the equivalence point. If we assume these two reactions to proceed to about the same extent, then $[H_3PO_4] \cong [HPO_4^{2-}]$, and $[H_3O^+]$ can be determined with K_1 and K_2:

$$K_1K_2 = \frac{[H_3O^+][H_2PO_4^-]}{[H_3PO_4]} \times \frac{[H_3O^+][HPO_4^{2-}]}{[H_2PO_4^-]} \qquad \text{(9-52)}$$

$$[H_3O^+] = \sqrt{K_1K_2} \qquad \text{(9-53)}$$

$$pH = \frac{pK_1 + pK_2}{2} \qquad \text{(9-54)}$$

In other words, the equivalence point pH would be just the average of pK_1 and pK_2 and would be exactly halfway between the first two crossover points shown in Figure 9-5. This approach is appealing in its simplicity and leads to a simple equation (Equation 9-53) for calculating $[H_3O^+]$ at the first equivalence point.

We can make a more detailed analysis of the situation at the first equivalence point by noting all the reactions that generate or consume H_3O^+, starting with the predominant species, $H_2PO_4^-$, and neglecting any subsequent dissociation of HPO_4^{2-} to give PO_4^{3-}. The equations are

$$H_2PO_4^- + H_2O \rightleftharpoons (H_3O^+)_1 + HPO_4^{2-} \qquad \text{(9-55)}$$

$$(H_3O^+)_1 = [HPO_4^{2-}]$$

$$H_2PO_4^- + (H_3O^+)_2 \rightleftharpoons H_2O + H_3PO_4 \qquad \text{(9-56)}$$

$$(H_3O^+)_2 = -[H_3PO_4]$$

$$2H_2O \rightleftharpoons (H_3O^+)_3 + OH^- \qquad \text{(9-57)}$$

$$(H_3O^+)_3 = [OH^-]$$

Thus, starting with $H_2PO_4^-$ and undissociated H_2O, the total concentration of H_3O^+ is given by Equation 9-58 (*proton balance*):

$$[H_3O^+] = (H_3O^+)_1 + (H_3O^+)_2 + (H_3O^+)_3 = [HPO_4^{2-}] - [H_3PO_4] + [OH^-] \quad \text{(9-58)}$$

Each of the phosphate species can be put in terms of the predominating species, $H_2PO_4^-$, using K_1 and K_2:

$$[H_3O^+] = \frac{K_2[H_2PO_4^-]}{[H_3O^+]} - \frac{[H_3O^+][H_2PO_4^-]}{K_1} + \frac{K_w}{[H_3O^+]} \qquad \text{(9-59)}$$

Since $H_2PO_4^-$ is the predominant species, we can make the assumption that $[H_2PO_4^-] \cong f_1$ (formal concentration of phosphoric acid at the first equivalence point). After some algebraic manipulation, Equation 9-60 is obtained.

$$[H_3O^+] = \sqrt{\frac{K_1 K_2 f_1 + K_1 K_w}{K_1 + f_1}} \qquad (9\text{-}60)$$

Equation 9-60 is a little more accurate than Equation 9-53, but it still involves the approximations stated above. Inspection of Equation 9-60 shows that in many situations $f_1 > K_1$ and $K_2 f_1 > K_w$. If so, Equation 9-60 transforms to Equation 9-53.

$$[H_3O] = \sqrt{\frac{K_1 K_2 f_1 + K_1 K_w}{K_1 + f_1}} \cong \sqrt{\frac{K_1 K_2 f_1}{f_1}} = \sqrt{K_1 K_2}$$

In the case of H_3PO_4, if $f_1 \geq 0.1$, Equation 9-53 may be used with little error, as shown in Example 9-10.

EXAMPLE 9-10

Calculate the pH at the first equivalence point in the titration of (a) 0.2 F H_3PO_4 with 0.2 F NaOH and (b) 0.05 F H_3PO_4 with 0.1 F NaOH.

Solution

(a) *Approximate method*

$$[H_3O^+] = \sqrt{K_1 K_2}$$
$$= \sqrt{7.1 \times 10^{-3} \times 6.3 \times 10^{-8}} = 2.1 \times 10^{-5} \qquad pH = 4.67$$

More exact method

$$[H_3O^+] = \sqrt{\frac{K_1 K_2 f_1 + K_1 K_w}{K_1 + f_1}} \qquad f_1 = 0.1 \text{ (dilution factor)}$$

$$= \sqrt{\frac{7.1 \times 10^{-3} \times 6.3 \times 10^{-8} \times 0.1 + 7.1 \times 10^{-3} \times 10^{-14}}{7.1 \times 10^{-3} + 0.1}}$$

$$= 2.0 \times 10^{-5} \qquad pH = 4.69$$

(b) *Approximate method*

$$[H_3O^+] = \sqrt{K_1 K_2} = 2.1 \times 10^{-5} \qquad pH = 4.67$$

More exact method

$$f_1 = 0.05 \times \tfrac{2}{3} = 0.033$$

$$[H_3O^+] = \sqrt{\frac{7.1 \times 10^{-3} \times 6.3 \times 10^{-8} \times 0.033 + 7.1 \times 10^{-3} \times 10^{-14}}{7.1 \times 10^{-3} + 0.033}}$$

$$= 1.9 \times 10^{-5} \qquad pH = 4.72$$

After the first equivalence point, continued addition of NaOH leads to a new buffer region determined by the $[HPO_4^{2-}]/[H_2PO_4^-]$ ratio. When the second equivalence point is reached,

$$[H_3O^+] = \sqrt{\frac{K_2K_3f_2 + K_2K_w}{K_2 + f_2}} \cong \sqrt{K_2K_3} \qquad (9\text{-}61)$$

In this case, $f_2 \gg K_2$ for any reasonable concentration, but $K_3f_2 > K_w$ only if $f_2 > 0.024$.

After the second equivalence point, a third buffer region will begin. At the third equivalence point, there is no further dissociation; that is, PO_4^{3-} acts only as a base, and $[H_3O^+]$ can be determined in principle as in Equation 9-28. However, in practice, phosphate ion is such a strong base that the titration curve will not show any appreciable "break." In fact, the calculated value for $[OH^-]$, if $f_3 = 0.1$, is 0.038, which is probably not much lower than the concentration of OH^- in the strong-base titrant.

After the last equivalence point, continued addition of NaOH will result in excess base, and $[OH^-]$ is calculated as in Equation 9-30. However, it must be remembered that in the case of a multiprotic acid there is not a 1:1 mole ratio between acid and base; so equivalents must be used or, more simply, the volume of *excess* base in Equation 9-62:

$$[OH^-] = \frac{F_b V_{excess}}{V_{HA} + V_b} \qquad (9\text{-}62)$$

9-5 ACID–BASE TITRATIONS IN NONAQUEOUS SOLVENTS

It is generally assumed throughout this text that the solvent for any solution is water. Most—but certainly not all—analytical procedures are carried out in an aqueous medium. Titrations in nonaqueous solvents have been reported for all reaction types, but the greatest use is in acid–base reactions. The strength of an acid or a base—and hence its ability to be titrated successfully—depends on the solvent. Equation 9-2 expressed the competition for protons in an aqueous medium, and this situation is generalized to any solvent, HS, in Equation 9-63.

$$HA + HOH \rightleftharpoons H_3O^+ + A^-$$

$$HA + HS \rightleftharpoons H_2S^+ + A^- \qquad (9\text{-}63)$$

The equilibrium constant for the reaction shown in Equation 9-63 will depend on both HA and HS. Some solvents, such as liquid NH_3 or ethylenediamine, are more basic than water is. That is, they more readily accept protons from acidic solutes so that the equilibrium constant in Equation 9-63 is higher. These solvents are useful for the titration of very weak acids.

On the other hand, some solvents, such as glacial acetic acid, are more acidic than water is; that is, they more readily tend to donate protons than to accept them, as shown in Equation 9-64.

$$B + HS \rightleftharpoons HB^+ + S^- \tag{9-64}$$

Substances dissolved in glacial acetic acid will be weaker acids but stronger bases. Glacial acetic acid, therefore, is a useful solvent for titrating very weak bases.

To decide whether or not a titration will succeed in a particular solvent, we must know K_a and K_b (or one of these equilibrium constants and K_s, solvent dissociation constant, which is K_w for H_2O).

$$\text{solvent dissociation (autoprotolysis):} \quad 2HS \rightleftharpoons H_2S^+ + S^- \tag{9-65}$$

A number of K_s values are known, and a few are tabulated in Table 9-5. The three constants K_a, K_b, and K_s are related by the equation

$$K_a K_b = K_s = K_w \text{ (aqueous solutions)}$$

Unfortunately, K_a and K_b values have not been determined for many substances in nonaqueous solvents so that feasibility of titration is often found empirically.

In general, solvents with low K_s values will have superior usefulness for acid–base titrations because the solvent dissociation establishes the range of K_a and K_b values that are theoretically titratable. This can be seen by writing the neutralization reaction and determining its equilibrium constant:

$$\text{titration of weak acid:} \quad HA + S^- \rightleftharpoons A^- + HS$$

$$K = \frac{[A^-]}{[HA][S^-]} = \frac{1}{K_b} = \frac{K_a}{K_s}$$

$$\text{titration of weak base:} \quad B + H_2S^+ \rightleftharpoons HB^+ + HS$$

$$K = \frac{[HB^+]}{[B][H_2S^+]} = \frac{1}{K_a} = \frac{K_b}{K_s}$$

As K_s decreases, the equilibrium constant for the titration reaction will increase; consequently, titration feasibility increases.

Not all solvents are capable of dissociating as shown in Equation 9-65. Some are completely inert (aprotic), such as benzene. Others are proton acceptors (basic in nature) but show no acidic properties; examples include ketones, ethers, and esters. These solvents do not exhibit the competition reactions described in Equations 9-63 and 9-64—which can be an advantage (higher equilibrium constant for the neutralization reaction)—but unfortunately, most inorganic solutes are sparingly soluble in them. Their usefulness in titrations often appears in the form of mixed solvents, that is, aprotic solvent mixed with polar solvent.

In addition to satisfactory K_a and K_b values, some available endpoint signal is needed for a successful nonaqueous titration. Acid–base indicators may be used, but

TABLE 9-5   Solvent Dissociation Constants (Autoprotolysis)

Solvent	K_s at 25°C	pK_s
water	1×10^{-14}	14.0
methanol	2×10^{-17}	16.7
ethanol	8×10^{-20}	19.1
solvents more acidic than H_2O:		
acetic acid	3.6×10^{-15}	14.4
formic acid	6×10^{-7}	6.2
sulfuric acid	1.4×10^{-4}	3.9
solvents more basic than H_2O:		
ammonia	1×10^{-33} (-50°C)	33.0
ethylenediamine	5×10^{-16}	15.3

their K_{HIn} values will in general be considerably different from those in aqueous solutions. Trial and error must be used to find proper indicators. Crystal violet(methyl violet) is an indicator useful for the titration of bases in glacial acetic acid with $HClO_4$ as the titrant. As an alternative, a potentiometric endpoint method may be used. A glass electrode will be responsive to the acidity of the medium in which it is immersed, but it will not measure the actual hydrogen ion activity, as it does in water. Glass electrodes will be discussed in more detail in Chapter 15. Titration curves of the shape seen in Figure 9-2 will be observed, and the inflection point as an endpoint signal is generally close to the equivalence point.

Finally, there must be available a primary standard for the determinate preparation of standard solutions or for the standardization of solutions. One widely used material is potassium hydrogen phthalate ($KHC_8H_4O_4$). It is used to standardize $HClO_4$ in glacial acetic acid solvent. It is interesting to note from Table 7-1 that this material is sufficiently acidic in water to be used for standardizing bases, whereas, in glacial acetic acid, it is sufficiently basic for standardizing acids ($HClO_4$). The reactions are

$$H_2O \text{ solvent:}   KHC_8H_4O_4 + OH^- \rightleftharpoons K^+ + C_8H_4O_4^{2-} + H_2O$$

$$CH_3COOH \text{ solvent:}   KHC_8H_4O_4 + CH_3COOH_2^+ \rightleftharpoons K^+ + H_2C_8H_4O_4 + CH_3COOH$$

The range of possible acid–base titrations in nonaqueous solvents is so broad that entire books have been devoted to this subject (see the supplementary reading list at the end of this chapter). For example, titrations of bases in glacial acetic acid with $HClO_4$ include alkaloids, amines, amides, amino acids, antibiotics, antihistamines, and vitamin B's. Titrations of acids in ethylenediamine or dimethylformamide include organic acids, phenols, sulfonamides and barbiturates. The titration of nicotine in tobacco is described in the laboratory manual.

9-6   APPLICATION OF ACID–BASE TITRATIONS

Acid–base titrations are used in so many areas that it is impossible to summarize them. Instead, a few general comments will be made concerning those standard solutions which are typically used, followed by some procedures used in the analysis of air pollutants. Finally, a few other analytical applications are described to show a little of the scope of acid–base titrations.

STANDARD ACIDS

Many analytes are weak acids or bases and therefore must be titrated with a strong base or acid. The titration of a weak acid with a weak base will have a small equilibrium constant and, as a result, will show a very shallow break in the titration curve at the equivalence point. For example, the titration of acetic acid (HOAc) with ammonia has an equilibrium constant

$$HOAc + NH_3 \rightleftharpoons NH_4^+ + OAc^-$$

$$K = \frac{[NH_4^+][OAc^-]}{[HOAc][NH_3]}$$

$$K = \frac{K_a K_b}{K_w} = \frac{[H_3O^+][OAc^-]}{[HOAc]} \cdot \frac{[NH_4^+][OH^-]}{[NH_3]} \cdot \frac{1}{[H_3O^+][OH^-]}$$

$$pK = pK_a + pK_b - pK_w = 4.76 + 4.74 - 14 = -4.5$$

$K = 3.2 \times 10^4$ (a value of $> 10^6$ is needed for a satisfactory titration curve)

Standard solutions of acids are normally those of HCl, H_2SO_4, or $HClO_4$, with HCl the most common. They are prepared to an approximate value by dilution of the concentrated acid and then standardized against a primary standard. It is possible to prepare standard solutions determinately by measuring the density of the concentrated acid and using published tables to find an accurate concentration value. A weighed amount of the acid is then diluted to a known volume. Hydrochloric acid solutions can also be prepared determinately from a constant-boiling HCl solution. When a solution of HCl is distilled, the distillate will eventually reach a fixed composition (around $6\ F$), depending on atmospheric pressure. A liter of concentrated HCl is boiled with the first three-quarters of distillate being discarded. Then the next 200 mL of distillate is collected as constant-boiling HCl, which has a concentration of about $6\ F$. If the distillation is carried out at 760 torr, 180.2 g of distillate will contain exactly 1 mol of HCl.

More often the diluted acid solution is standardized with a base primary standard, such as Na_2CO_3. Because carbonate is the base form of CO_2 (commonly denoted as carbonic acid, H_2CO_3) there will be two equivalence points:

$$CO_3^{2-} + H_3O^+ \rightleftharpoons HCO_3^- + H_2O \qquad pH_{eq\ pt} = 8.4 \qquad \textbf{(9-66)}$$

$$HCO_3^- + H_3O^+ \rightleftharpoons CO_2 + 2H_2O \qquad pH_{eq\ pt} = 4.0 \qquad \textbf{(9-67)}$$

The second break in the titration curve (Figure 9-6) is bigger than the first and therefore is the preferred equivalence point. Indicators that show a color transition near pH 4, such as bromocresol green or methyl orange, give endpoints close to the equivalence point. The endpoint can be sharpened by stopping the titration just as the indicator starts to change and boiling the solution to eliminate CO_2. The pH will increase because the concentration of acid form (CO_2) has decreased. Addition of acid titrant will now show a sharp transition because of the large pH change (as shown in Figure 9-6).

Another way of standardizing HCl solutions is that of gravimetric determination by precipitation of AgCl. A precipitation titration against primary standard $AgNO_3$ could also be used (see Section 8-1).

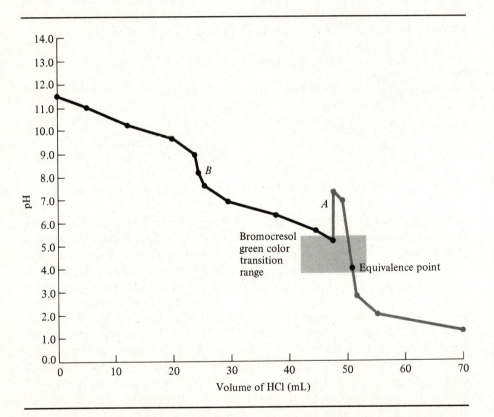

FIGURE 9-6 Titration of 50 mL of 0.1 F Na_2CO_3 with 0.2 F HCl. (**A**) Solution boiled after addition of 48 mL (when the bromocresol green indicator is just beginning to turn green) to eliminate dissolved CO_2 (H_2CO_3). The actual equivalence point is near the limit of the color transition region for bromocresol green. (**B**) Note shallow titration break at the first equivalence point for titration of CO_3^{2-} to HCO_3^-.

STANDARD BASES

The most commonly used base standard solution is NaOH. Occasionally, KOH or $Ba(OH)_2$ is also employed. None of these materials can be purchased as a primary standard; so solutions must be standardized. The main problem with obtaining pure NaOH (and bases in general) is that the atmosphere contains CO_2, which readily reacts with the base to form the carbonate. Even fresh bottles of solid NaOH will contain 1 to 2% of Na_2CO_3. For the preparation of NaOH solutions containing little or no Na_2CO_3, concentrated solutions of NaOH (50%) are used as the starting material because Na_2CO_3 has a very low solubility in concentrated NaOH and will precipitate, leaving the supernatant solution carbonate free. Water for dilution must be boiled first to eliminate dissolved CO_2, and if the NaOH solution is to be kept for any length of time, it should be stored in a bottle that prevents CO_2 from entering. Glass is attacked by NaOH so that polyethylene bottles are normally used.

After the NaOH solution has been prepared, absorption of CO_2 would not change the normality, provided that the pH 4 equivalence point is used as in Equation 9-67. In other words, each CO_2 consumes two hydroxide ions to form CO_3^{2-}, while each CO_3^{2-} consumes two hydronium ions to form CO_2. Unfortunately, many analytes are weak acids, and therefore the equivalence point will be nearer pH 9, with phenolphthalein as a typical indicator for endpoint signal. In this case, carbonate has $n = 1$, and the NaOH will be weaker than it was before the CO_2 entered. This is called a *carbonate error* and can be determined by titrating a strong acid, such as HCl, and noting the difference between the phenolphthalein endpoint (pH 9) and the bromocresol green endpoint (pH 4).

$Ba(OH)_2$ has the advantage that $BaCO_3$ is quite insoluble, and solutions will be carbonate free. However, if any CO_2 enters the solution, the concentration will decrease as $BaCO_3$ precipitates out. Also, several other barium salts are insoluble, especially $BaSO_4$, which might be encountered in analytical situations. The formation of a precipitate can obscure the endpoint signal.

Several primary standard acids are available (see Table 7-1), but probably the most common is potassium hydrogen phthalate ($KHC_8H_4O_4$). A review of the requirements for a primary standard material listed in Section 7-2 shows that it is ideal; for it is readily available in high-purity form and is nonhygroscopic, and its 204.2 equivalent weight is very high, minimizing uncertainty in weighing. Other primary standards suitable for standardizing NaOH solutions are listed in Table 7-1.

ANALYSIS OF AIR POLLUTANTS

To show the wide range of applications for acid–base titrations, we now survey some of the procedures developed for the analysis of air pollutants. Of course, highly sensitive instrumental methods are utilized for pollutants present in trace quantities, but titration methods are useful for long-term averaging or for highly polluted gases, such as stack gases at industrial or power-generating plants.

Sulfur dioxide is one pollutant that can be analyzed by an acid–base titration.

The gas sample is bubbled through an absorption solution containing hydrogen peroxide, H_2O_2. The SO_2 reacts according to Equation 9-68,

$$SO_2 + H_2O_2 \rightleftharpoons H_2SO_4 \tag{9-68}$$

The sulfuric acid is titrated with a standard solution of 0.01 N Na_2CO_3 to pH 4.5. Example 9-11 shows how SO_2 level is calculated from the results.

The analysis of nitrogen-containing air pollutants is another area well suited to acid–base titrimetry. Ammonia as NH_3 or NH_4^+ salts is often present in air, especially in the vicinity of nitrogen-containing organic substances, because it is the normal end product of decomposition. Ammonia is also widely used in the chemical industry and may therefore be encountered in the atmosphere in and around industrial plants. In general, a measured volume of air containing ammonia is bubbled through a dilute sulfuric acid solution (for example, 0.0005 N). The excess acid can be back-titrated; or as illustrated in Example 9-12, the volume of air necessary to neutralize the H_2SO_4 solution is measured by including a color indicator in the absorbing solution.

Waste gases from nitric acid plants can be analyzed by an acid–base titration method for their content of nitrogen oxides (NO_x). The sample is collected in a large evacuated flask, H_2O or dilute H_2O_2 is added, and the mixture is allowed to stand for several hours. As shown in Equation 9-69, HNO_3 is formed, which is titrated

---

EXAMPLE 9-11

A quantity, 1.3 m^3, of air (density = 1.19 g/L) containing SO_2 is bubbled through 50 mL of an absorption solution containing 0.3% H_2O_2. Titration of the solution with 0.0232 N Na_2CO_3 requires 15.7 mL to reach the endpoint. Calculate the SO_2 content in parts per million by weight.

Solution

$$\text{meq } SO_2 = \text{meq } Na_2CO_3$$

$$\frac{\text{mg } SO_2}{\text{FW }(SO_2)/n} = NV$$

From Equation 9-68:

$$SO_2 \rightarrow H_2SO_4 \rightarrow 2H_3O^+; \text{ therefore, } n = 2$$

$$\frac{\text{mg } SO_2}{64.1/2} = 0.232 \times 15.7$$

$$\text{mg } SO_2 = 11.7 \text{ mg in } 1.3 \text{ m}^3 \quad (1 \text{ m}^3 = 10^3 \text{ L})$$

$$\text{ppm} = \frac{\text{mg } SO_2}{\text{kg air}} = \frac{11.7 \text{ mg}}{1.3 \text{ m}^3 \times 10^3 \text{ L/m}^3 \times 1.19 \text{ g/L} \times 10^{-3} \text{ kg/g}} = 7.56 \text{ ppm}$$

EXAMPLE 9-12

A sample of air from a fertilizer factory is bubbled at about 1 L/min through a solution containing 20.0 mL of 0.00052 N H_2SO_4 until the bromophenol blue indicator turns blue. If the volume of air necessary to neutralize the H_2SO_4 is 10.2 L, calculate the ammonia concentration in milligrams of NH_3 per cubic meter.

Solution

$$\text{meq } NH_3 = \text{meq } H_2SO_4$$

$$\frac{\text{mg } NH_3}{\text{FW } (NH_3)/n} = NV$$

$$NH_3 + H_3O^+ \rightleftharpoons NH_4^+ + H_2O; \text{ therefore, } n = 1$$

$$\frac{\text{mg } NH_3}{17.0/1} = 0.00052 \times 20.0$$

$$\text{mg } NH_3 = 0.177 \text{ in } 10.2 \text{ L}$$

$$\frac{0.177 \text{ mg}}{10.2 \text{ L}} \times 1000 \text{ L/m}^3 = 17.3 \text{ mg/m}^3$$

---

with 0.01 N NaOH using methyl red indicator.

$$2NO_2 + H_2O_2 \rightleftharpoons 2HNO_3 \tag{9-69}$$

So much carbon dioxide occurs in the atmosphere that its concentration is relatively easy to determine by a titration method. The air sample is shaken with a known amount of $Ba(OH)_2$, which forms $BaCO_3$. The excess $Ba(OH)_2$ and $BaCO_3$ are titrated with HCl to the phenolphthalein endpoint, according to Equation 9-66.

$$Ba(OH)_2 + 2H_3O^+ \rightleftharpoons Ba^{2+} + 4H_2O$$

$$BaCO_3 + H_3O^+ \rightleftharpoons Ba^{2+} + HCO_3^- + H_2O$$

Thus the $Ba(OH)_2$ has two equivalents per mole, while the $BaCO_3$ only has one. Actually, CO_2 acts both as an acid—when reacting with $Ba(OH)_2$ to form $BaCO_3$ ($n = 2$)—and as a base—in the form of $BaCO_3$ ($n = 1$) when reacting with HCl. The equivalence relationship can be written as

$$\text{meq base} = \text{meq acid}$$

$$\text{meq } Ba(OH)_2 + \text{meq}_b \ CO_2 = \text{meq HCl} + \text{meq}_a \ CO_2$$

$$\text{meq } Ba(OH)_2 + \frac{\text{mg } CO_2}{\text{FW } (CO_2)/1} = \text{meq HCl} + \frac{\text{mg } CO_2}{\text{FW } (CO_2)/2}$$

$$\text{meq } Ba(OH)_2 - \text{meq HCl} = \frac{\text{mg } CO_2}{\text{FW } (CO_2)/1}$$

KJELDAHL METHOD FOR NITROGEN

Acid–base titrations are widely used in the determination of nonmetal composition of materials. Elements such as nitrogen, sulfur, carbon, phosphorus, and the halogens are readily converted to acidic or basic species and then titrated. The discussion of air pollutants described a few such applications. The determination of nitrogen in organic compounds by the *Kjeldahl method* offers another common example. It is a standard method for the determination of nitrogen in proteins, including blood, meat, and grain.

In the Kjeldahl method, the sample is oxidized with sulfuric acid in a long-necked flask, as shown in Figure 9-7. During the oxidation step, carbon is converted to CO_2, hydrogen to H_2O, and nitrogen to NH_4^+. The successful conversion of nitrogen to NH_4^+ depends on the original state of the nitrogen. Amines, amides, and other reduced-state nitrogen compounds are readily converted, while organic compounds containing oxidized nitrogen, such as nitrates, nitrites, and nitro compounds, will form N_2 or nitrogen oxides. This can be prevented by reducing the nitrogen before treatment with H_2SO_4. One useful reducing agent is salicylic acid with sodium thiosulfate. Inorganic nitrates and nitrites can be reduced to NH_3 with Devarda's

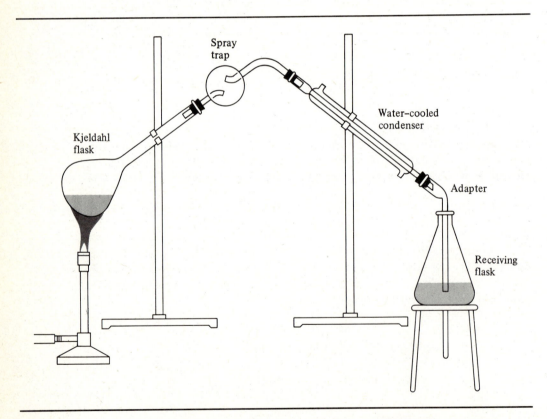

FIGURE 9-7 Apparatus for carrying out the Kjehldahl method for the determination of nitrogen.

alloy (50% Cu, 45% Al, 5% Zn); no sulfuric acid treatment is then needed. The sulfuric acid oxidation is slow and may require more than 1 hr. Raising the boiling point of H_2SO_4 by addition of a salt is sometimes recommended to increase the rate of oxidation. Catalysts, such as mercury, copper, or selenium, can also be used to hasten the process.

After oxidation is complete, a concentrated NaOH solution is carefully (!) poured down the side of the flask to form a layer at the bottom of the diluted sulfuric acid mixture. The distillation apparatus shown in figure 9-7 is connected, and the two layers are carefully mixed. For distillation of the NH_3, the resulting basic mixture is heated with a spray trap to prevent carryover of NaOH droplets into the receiving flask. The NH_3 is collected in excess standard acid and back-titrated with base. The endpoint is observed around pH 4 so that titration of NH_4^+, the conjugate acid of NH_3, does not begin. Alternatively, the NH_3 is collected in boric acid to form

$$HBO_2 + NH_3 \rightleftharpoons NH_4^+ + BO_2^-$$

The borate ion, the conjugate base of HBO_2, is strong enough ($pK_b = 4.77$, about the same as NH_3) to be titrated with HCl. Again, an endpoint signal in the acidic range must be used. Example 9-13 shows a typical calculation.

EXAMPLE 9-13

The nitrogen content of a dried blood sample is to be determined by the Kjeldahl method. A 1.246-g sample is treated with 25 mL of concentrated H_2SO_4, 10 g K_2SO_4, and 0.1 g Se for 3 hr at the boiling point, diluted with 250 mL of H_2O, and made basic with 50 mL of 50% NaOH solution. The mixture is distilled, with the NH_3 collected in 50 mL of 4% boric acid. The boric acid solution is titrated with 0.1065 N HCl and requires 37.65 mL to reach the bromocresol green endpoint. Calculate the percentage of nitrogen in the blood sample.

Solution

$$\text{meq N} = \text{meq HCl}$$

$$\frac{\text{mg N}}{\text{FW (N)}/n} = NV$$

$$N \rightarrow NH_3 \rightarrow BO_2^- \text{ (consumes one } H_3O^+\text{); therefore, } n = 1$$

$$\frac{\text{mg N}}{14.0/1} = 0.1065 \times 37.65$$

$$\text{mg N} = 56.1$$

$$\% \text{ N} = \frac{56.1 \text{ mg N}}{1246 \text{ mg sample}} \times 100 \, (\%) = 4.50\%$$

Note that the amounts of H_2SO_4, NaOH, and boric acid did not enter the calculation.

OTHER EXAMPLES OF ACID–BASE TITRATIONS

Sulfur in organic and biological materials can be determined by the method described for SO_2 in air: The sample is burned in a stream of oxygen, and the SO_2 is collected in a dilute solution of H_2O_2. In a similar fashion, carbon can be converted to CO_2 and collected in $Ba(OH)_2$, forming the insoluble $BaCO_3$. The excess $Ba(OH)_2$ is then titrated with HCl.

The HCO_3^- content of the blood can be an important diagnostic parameter during a patient's illness. The HCO_3^- is determined by the analyst's adding excess HCl to convert HCO_3^- to CO_2, swirling to allow the CO_2 to escape, and then back-titrating with NaOH.

Various organic functional groups are sufficiently acidic or basic to be determined by acid–base titration. Sulfonic acids are normally quite strong, and most carboxylic acids have pK_a's of 3 to 7, which is satisfactory for titration with base. Frequently their low solubility requires a back-titration method, in which the acid is dissolved in excess base and then back-titrated with standard acid. As an alternative, the acid may be titrated in a nonaqueous or mixed solvent, such as ethanol, in which it has sufficient solubility for titration.

Aliphatic amines are similar to ammonia, with pK_b's of 3 to 6, which is sufficiently basic for titration in aqueous solution. Aromatic amines, such as aniline and other very weak amines, can be titrated in a nonaqueous solvent such as glacial acetic acid, as was described previously.

Esters cannot be titrated directly but may be saponified with excess base.

$$RCOOR' + OH^- \rightleftharpoons RCOO^- + R'OH$$

After saponification, the excess base is titrated with a standard acid solution.

Compounds containing carbonyl groups also cannot be titrated directly but may be determined by their reaction with hydroxylamine hydrochloride to form an oxime.

$$\overset{\displaystyle O}{\overset{\|}{R-C-R'}} + NH_2OH \cdot HCl \rightleftharpoons \overset{\displaystyle NOH}{\overset{\|}{R-C-R'}} + HCl + H_2O$$

where R' is hydrogen for an aldehyde. The HCl produced is titrated with NaOH, but a potentiometric endpoint technique must be used. $NH_2OH \cdot HCl$ is a strong enough acid ($pK_a = 6$) so that the endpoint region is not sufficiently sharp for a color indicator.

There are, of course, many inorganic applications of acid–base titrations besides determination of pollutants in air, water, or solid wastes. Many industrial processes are carried out in an acidic or basic medium, and process control will require a continuous or frequent analysis of the medium. One such example is the control of electroplating-bath composition for satisfactory plating, which is often critically dependent on acid content.

SUPPLEMENTARY READING

ARTICLES DESCRIBING ANALYTICAL METHODS OR APPLICATIONS

Willis, C.J., "Another Approach to Titration Curves," *Journal of Chemical Education*, 58: 659, 1981.

REFERENCE BOOKS

Bishop, E., ed., *Indicators*. Elmsford, N.Y.: Pergamon Press, Inc., 1972.

Kucharsky, J., and L. Safarik, *Titrations in Non-aqueous Solvents*. New York: Elsevier North-Holland, Inc., 1965.

Rosenthal, D., and P. Zuman, "Acid-Base Equilibria, Buffers, and Titrations in Water," in 2nd ed., I.M. Kolthoff and P.J. Elving, eds., *Treatise on Analytical Chemistry*, pt. 1, vol. 2, chap. 18. New York: John Wiley & Sons, Inc., 1979.

PROBLEMS

9-1. What are $[H_3O^+]$ and pH for the following solutions?
  *a. 0.003 F HCl
   b. 0.0001 F H_2SO_4
   c. 0.25 F NaOH
   d. 10^{-9} F $HClO_4$
   e. 10^{-9} F KOH
   f. 5×10^{-5} F NaOH

9-2. What are the $[H_3O^+]$ activity and pH in 0.050 F HCl at ionic strength 0.05 (see Table 5-1 for activity coefficients)?

9-3. What are $[H_3O^+]$ and pH for the following solutions?
  *a. 0.2 F acetic acid
   b. 0.03 F formic acid
   c. 0.5 F lactic acid
   d. 2 F acetic acid

9-4. What are $[H_3O^+]$ and pH for the following solutions?
  *a. 0.05 F NH_3
   b. 0.6 F ethanolamine
   c. 0.01 F methylamine
   d. 0.1 F pyridine

*9-5. Calculate the pH of a 0.001 F benzoic acid solution.

9-6. Calculate the pH of a solution 0.08 F in Na_2CO_3.

9-7. What is $[H_3O^+]$ and pH for the following solutions?
  *a. 0.03 F sodium acetate
   b. 1 F sodium benzoate
   c. 0.01 F sodium formate
   d. 2×10^{-3} F sodium lactate

---

*Answers to problems marked with an asterisk will be found at the back of the book.

9-8. What is the $[H_3O^+]$ and pH of 0.001 F HCN?

9-9. HgO is available as a primary standard and can be used to standardize acids when dissolved in a solution containing KI:

$$HgO(s) + 4I^- + H_2O \rightleftharpoons HgI_4^{2-} + 2OH^-$$

If 37.84 mL of an HCl solution is required to titrate 457 mg of HgO, calculate the normality of the HCl.

9-10. A NaOH solution is standardized by titrating a 223.6-mg sample of primary standard Na_2CO_3. If 39.17 mL was required to reach the equivalence point (pH 4), calculate its titer for the following analytes:
*a. acetic acid
  b. SO_2 (oxidized to H_2SO_4)
  c. vitamin C (L-ascorbic acid)
  d. H_3PO_4 (1st equivalence point)
*e. H_3PO_4 (2nd equivalence point)
  f. P_2O_5 (converted to H_3PO_4, 2nd equivalence point)

9-11. A solution contains 0.2 F formic acid and 0.05 F sodium formate. Calculate the pH change that occurs when the following are added to 100 mL of this solution:
  a. 100 mL of H_2O
*b. 100 mL of 0.2 F NaOH
  c. 100 mL of 0.1 F NaOH
  d. 50 mL of 0.15 F NaOH
  e. 50 mL of 0.1 F HCl
  f. 200 mL of 0.05 F HCl
  g. 200 mL of 0.2 F NaOH

9-12. Calculate $[H_3O^+]$ and pH for the following buffers:
*a. 0.25 F acetic acid + 0.75 F sodium acetate
  b. 50 mL of 0.1 F acetic acid + 50 mL of 0.02 F NaOH
  c. 400 mg of sodium acetate dissolved in 100 mL of 0.1 F acetic acid
  d. 0.1 F boric acid + 0.2 F sodium borate
  e. 0.2 F Na_2CO_3 + 0.3 F $NaHCO_3$

9-13. The pH of human blood is maintained at 7.40 with a HCO_3^-/CO_2 buffer. What must the ratio $[HCO_3^-]/[CO_2]$ be to provide the proper pH?

9-14. If the HCO_3^- concentration in a blood sample is 162 mg per 100 mL, calculate the CO_2 concentration in blood as milligrams per 100 mL, using the data given in Problem 9-13.

*9-15. What is the buffer capacity of a solution containing 0.2 F sodium acetate and 0.5 F acetic acid?

9-16. What is the buffer capacity of a solution containing 0.3 F sodium formate and 0.2 F formic acid?

9-17. Derive titration curves for the following strong-acid–strong-base titrations. Calculate the pH after the addition of 0, 10, 20, 24, 25, 26, 30, and 50 mL of base, and plot the data.
  a. 25 mL of 0.01 F HCl with 0.01 F NaOH
  b. 50 mL of 0.05 F HCl with 0.05 F $Ba(OH)_2$
  c. 10 mL of 0.25 F $HClO_4$ with 0.1 F KOH

9-18. Derive titration curves for the following weak-acid–strong-base titrations. Calculate the pH after the addition of 0, 10, 20, 24, 25, 26, 30, and 50 mL of base, and plot the data.
*a. 50 mL of 0.1 F formic acid with 0.2 F NaOH

    b.   25 mL of 0.01 F chloroacetic acid with 0.01 F NaOH

    c.   25 mL of 0.5 F propanoic acid with 0.5 F KOH

9-19. Derive titration curves for the following weak-base–strong-acid titrations. Calculate the pH after the addition of 0, 10, 20, 24, 25, 26, 30, and 50 mL of acid, and plot the data.

    *a.   50 mL of 0.05 F NH_3 with 0.1 F HCl

    b.   25 mL of 0.1 F trimethylamine with 0.1 F HCl

9-20. Derive a titration curve for the following titration: 50 mL of 0.1 F Na_2CO_3 with 0.2 F HCl. Calculate pH after the addition of 0, 10, 20, 24, 25, 26, 30, 40, 49, 50, 51, 60, and 100 mL of titrant. Plot the data, and compare the two titration breaks.

9-21. What factors are important in the selection of an indicator for an acid–base titration?

9-22. Recommend color indicators for each of the titrations posed in Problems 9-18 to 9-20.

9-23. A quick measure of the titration break (and thus the feasibility of an acid–base titration) is the pH change that occurs between 90% and 100% of the equivalence point volume added. If this pH change must be >2 (see the third example in Appendix 1-E), determine which of the following titrations are feasible:

    *a.   0.02 F acetic acid with 0.02 F NaOH

    b. 2×10^{-4} F HCl with 2×10^{-4} F NaOH

    c.   0.1 F HCN with 0.1 F NaOH

9-24. Your first job as a biochemist is to determine the formic acid content in red ants. If you decide to titrate the acid using a color indicator, choose the best indicator from Table 9-3. Assume the acid concentration to be about 0.01 F.

*9-25. A 250-mg sample of an organic acid was titrated with 0.106 N NaOH and required 45.0 mL to reach the equivalence point. To this solution was added 25.0 mL of 0.120 N HCl, and the pH was 6.30. Calculate the pK_a and the equivalent weight of the organic acid.

9-26. Titration of 650 mg of an unknown weak acid required 45.0 mL of 0.124 N NaOH to reach the equivalence point. The analyst then added 20.0 mL of 0.110 N HCl and found the pH to be 6.30. Calculate the equivalent weight and the pK_a for the unknown acid.

9-27. What are the principal carbonate species at the following pH's?

    *a.   pH 2

    b.   pH 6.35

    c.   pH 8

    d.   pH 10.33

    e.   pH 12

9-28. Calculate the concentration of all species present in 0.2 F H_2CO_3 brought to pH 7 with the addition of NaOH, assuming no volume change.

9-29. What are the α values for all oxalic acid species at the following pH's?

    a.   pH 0

    *b.   pH 1

    c.   pH 3

    d.   pH 4.27

    e.   pH 5

9-30. Calculate the pH at the first equivalence point for the titration of 0.05 F phosphorous acid with 0.1 F NaOH. From Table 9-3, what indicator would you recommend for this titration?

9-31. What is the pH at the first and second equivalence points for the titration of 50 mL of 0.1 F maleic acid with 0.2 F NaOH?

9-32. Adipic acid is a diprotic acid with $pK_1 = 4.41$ and $pK_2 = 5.28$. How would you design an analytical procedure for the analysis of adipic acid by titration with NaOH? That is, would you attempt to use the first equivalence point or the second equivalence point and what indicator would you recommend?

*9-33. Citric acid is a triprotic acid in water. Calculate α for each citrate species as a function of pH and plot the values.

9-34. Write autoprotolysis reactions for the following solvents:
   a. H_2O
   b. methanol (CH_3OH)
   c. acetic acid $(HC_2H_3O_2)$
   d. H_2SO_4
   e. NH_3

9-35. If the titration reaction must have an equilibrium constant $> 10^7$ for a successful titration, calculate the minimum value of K_a (or K_b) in each of the solvents listed in Table 9-5.

*9-36. The amount of nicotine is determined in a 3.00-g sample of cigarette tobacco by extraction with a benzene–chloroform solvent, followed by a nonaqueous titration with 0.0463 N $HClO_4$ in glacial acetic acid. If 15.63 mL is required to reach the crystal violet endpoint signal, calculate the percentage of nicotine $(C_{10}H_{14}N_2$, with both nitrogens titratable) in the tobacco.

9-37. How are the four requirements for a successful volumetric procedure met in the titration of Na_2CO_3 with HCl?

9-38. You have accepted a job as an analyst for the vinegar division of Vin Aigre & Co. For the analysis of vinegar for acetic acid content, using an acid–base titration, describe how you would accomplish each step of the total analysis process (see Chapter 1).

9-39. A 356-mg sample of a sulfur-containing organic compound was burned in a stream of O_2, and the resulting SO_2 was absorbed in a solution of H_2O_2 $(H_2O_2 + SO_2 \rightleftharpoons H_2SO_4)$. The acid produced was titrated with 42.17 mL of 0.1127 N KOH. Calculate the percentage of S in the compound.

9-40. Vitamin C (L-ascorbic acid, FW = 176) is a weak acid, which may be titrated with base. Suppose a 500-mg tablet is dissolved in 100 mL of H_2O and titrated with 0.100 N NaOH. What is the equivalence point volume and the pH at the equivalence point?

*9-41. Suppose that thymolphthalein is used as the indicator for the vitamin C titration in Problem 9-40 and that the endpoint signal occurs at pH 9.6. Calculate the endpoint error for this titration.

9-42. Suppose that bromothymol blue is used as the indicator for the vitamin C titration in Problem 9-40 and that the endpoint signal occurs at pH 6.8. Calculate the endpoint error for this titration.

9-43. The vitamin C content in vitamin tablets is determined by dissolving three tablets in hot water and titrating with 0.0563 N NaOH. If the titration requires 29.77 mL, with a pH meter to observe the endpoint, what is the average vitamin C content per tablet?

9-44. A mineral sample containing dolomite $(CaCO_3 \cdot MgCO_3)$ and weighing 865 mg is treated with 10.00 mL of 1.542 N HCl to dissolve the carbonate material. The resulting mixture is boiled to remove CO_2, and the excess HCl is titrated with 0.2163 N NaOH. If 26.03 mL is required to reach the methyl red endpoint, calculate the percentage of dolomite in the sample.

*9-45. A waste-gas sample (3.0 L) from a HNO_3 plant is collected in dilute H_2O_2 to convert NO_2 to HNO_3 (see Equation 9-69). The resulting HNO_3 is then titrated with 0.00863 N NaOH, and 6.72 mL is required to reach the methyl red endpoint signal. Assuming the waste gas has a density of 1.20 g/L, calculate the nitrogen oxide content as parts per million of NO_2 (milligrams of NO_2 per kilogram of air).

9-46. The CO content in blood can be determined by placing a diluted and acidified sample in a Conway chamber, in which CO diffuses into a $PdCl_2$ solution. Any CO present reacts as follows:

$$Pd^{2+} + 3H_2O + CO \rightleftharpoons Pd + 2H_3O^+ + CO_2$$

The resulting H_3O^+ is titrated with 0.0200 N NaOH. A blank is carried through the procedure, using H_2O in place of blood. Calculate volume percent (milliliters of CO per 100 mL of blood) if a 1.3-mL blood sample requires a titration of 0.71 mL, while the blank requires 0.54 mL. Assume CO to be an ideal gas at STP.

9-47. A sample of impure NaOH, which has been exposed to air, is analyzed by titrating a 188.5-mg sample with 0.1065 N HCl. The volume required to reach the phenolphthalein endpoint is 39.19 mL, while the volume required to reach the bromocresol green endpoint is 40.67 mL. Calculate the percentages of NaOH and Na_2CO_3 in the sample.

9-48. The nitrogen content in an organic fertilizer is determined by the Kjeldahl method, in which the distilled NH_3 is collected in 50 mL of 4% boric acid. The solution is then titrated with 0.1171 N HCl. With the data below, calculate the mean and median value for N in the fertilizer:

Trial	Sample (mg)	Titration volume (mL)
1	650	30.63
2	672	31.35
3	635	30.18

9-49. The same fertilizer as in Problem 9-48 is analyzed in another laboratory by the Kjeldahl method but with the variation that the distilled NH_3 is collected in 50.00 mL of 0.1096 N HCl and back-titrated with 0.1206 N NaOH. With the data below, calculate the mean and median value for N in the fertilizer.

Trial	Sample (mg)	Back-titration volume (mL)
1	582	19.17
2	510	22.13
3	596	18.63

9-50. Do the data presented in Problems 9-48 and 9-49 provide evidence that the two methods give different results?

9-51. A 75-mL sample containing HCl is titrated with 0.01 N NaOH, and 25 mL is required to reach the methyl red endpoint signal. If the endpoint uncertainty is pH 5 ± 1 and the volume reading uncertainty is ± 0.02 mL (each reading), calculate the overall uncertainty in the milliequivalents of HCl present in the sample.

Complexation Titrations

10

It has been suggested at the opening of Chapter 9 that complexation reactions can be viewed as acid–base reactions if one uses a Lewis acid–base definition. The metal ion forming a complex by accepting electron pairs from the ligand is a Lewis acid, while the electron-pair donor, the ligand, would be the Lewis base. However, in the description of these reactions in terms of titration curves and analytical applications, it is clearer to separate the two reaction types.

10-1 LIGANDS

There are many possible ligands capable of forming complexes with metal ions. It should be kept in mind that water itself will act as a ligand by donating electron-pair density on the oxygen atom to a metal ion in solution. Thus all complexation reactions in aqueous solution are really a competition between H_2O and the ligand, L, for the metal ion, as shown in Equation 10-1.

$$M(H_2O)_6^{n+} + 6L^{l-} \rightleftharpoons ML_6^{n-6l} + 6H_2O \tag{10-1}$$

For example, the formation of hexacyanoferrate(II) ion can be written as

$$Fe(H_2O)_6^{2+} + 6CN^- \rightleftharpoons Fe(CN)_6^{4-} + 6H_2O \tag{10-2}$$

Ligands, such as H_2O, CN^-, NH_3, and halide ions, that donate one pair of electrons are called *monodentate* ligands. There are a few important analytical applications using these inorganic monodentate ligands. However, the great rise in popularity of *complexation titrations* (or *complexometric methods* or *titrations*) has been due to the development of organic complexing agents that donate more than one electron pair: Ethylenediamine, for example, donates two electron pairs—one

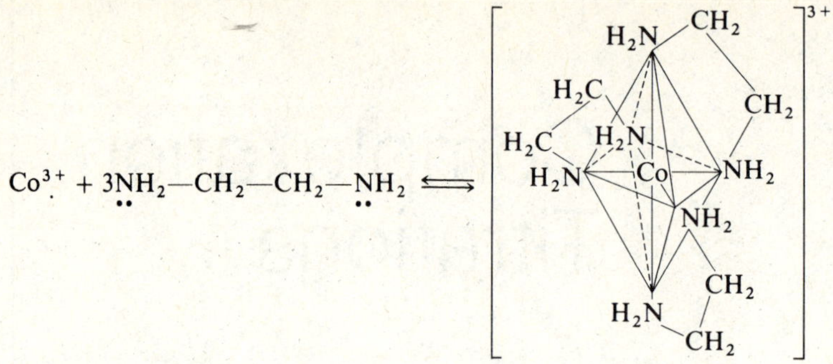

$$Co^{3+} + 3NH_2-CH_2-CH_2-NH_2 \rightleftharpoons$$

FIGURE 10-1 Formation of tris(ethylenediamine)cobalt(III) complex.

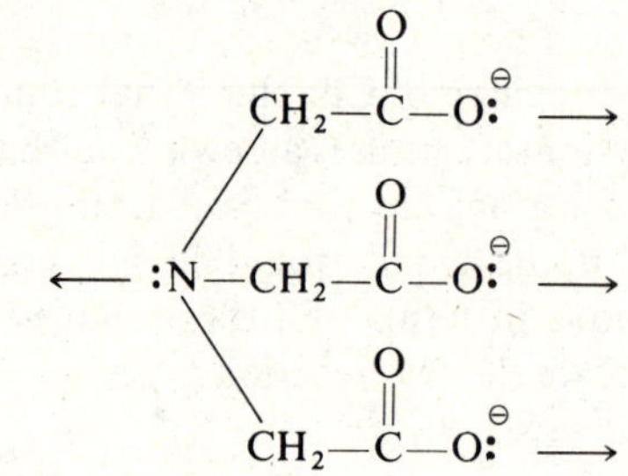

FIGURE 10-2 Nitrilotriacetic acid (NTA) in its anion form can donate four electron pairs to a metal ion, forming a tetradentate chelate.

on each nitrogen atom, to form a *chelate*[1], as shown in Figure 10-1. Ethylenediamine is a *bidentate* ligand. Ligands donating three, four, five, or six electron pairs are termed tri–, tetra–, penta–, and hexadentate. An important tetradentate ligand is nitrilotriacetic acid (abbreviated as NTA), which can donate three electron pairs from oxygen (the hydrogen atoms bonded to oxygen atoms are acidic, and the metal ion competes with H^+ for these sites) and one electron pair from the nitrogen, as illustrated in Figure 10-2. This material was developed as a substitute for phosphates in detergents because it complexes Ca^{2+}, Mg^{2+}, and other ions present in hard water. But concern has arisen because it is so effective a complexing agent that it can solubilize highly toxic heavy metal ions, increasing their concentration in water supplies; moreover, there is the possible toxicity of NTA and its degradation products.

---

[1]A chelate complex is one in which a ligand attaches with two or more electron pair donors. The word "chelate" is from the Greek word *chele*, meaning "claw".

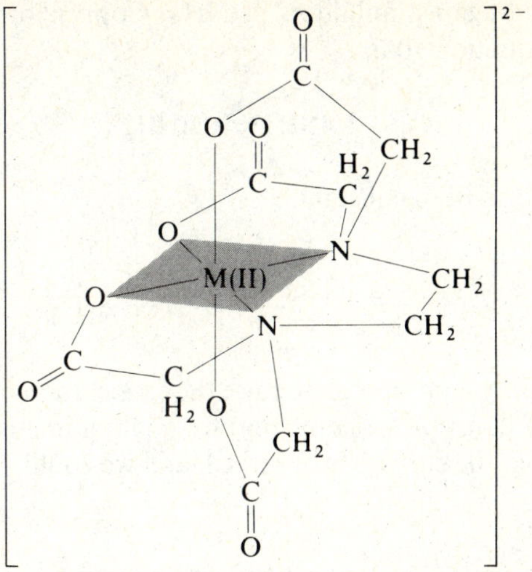

FIGURE 10-3   Structure of a metal–EDTA complex. EDTA forms hexadentate chelates.

For analytical purposes, by far the most widely used complexing agent for complexation titrations is ethylenediaminetetraacetic acid, also known as ethylene-dinitrilotetraacetate and given the abbreviation EDTA. EDTA has the structure

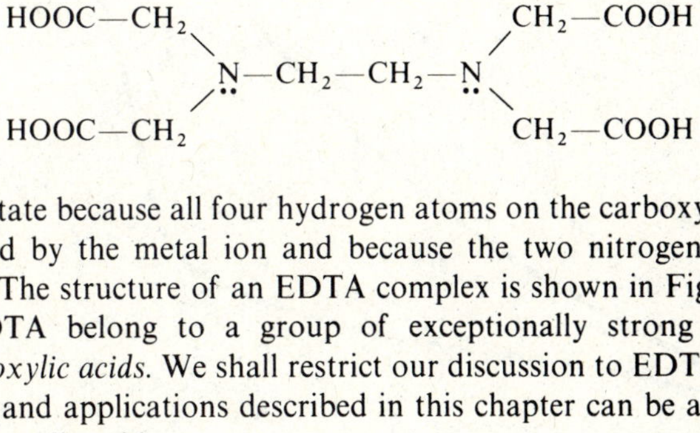

and is hexadentate because all four hydrogen atoms on the carboxylic acid groups can be replaced by the metal ion and because the two nitrogen atoms donate electron pairs. The structure of an EDTA complex is shown in Figure 10-3. Both NTA and EDTA belong to a group of exceptionally strong ligands called *aminopolycarboxylic acids.* We shall restrict our discussion to EDTA, but many of the techniques and applications described in this chapter can be applied to other aminopolycarboxylic acids.

10-2   THEORETICAL PRINCIPLES: THE TITRATION CURVE

As with the previous reaction types, let us derive a titration curve for the addition of ligand to a metal ion. The concentration of uncomplexed metal ion (really the complex formed with H_2O, but this aspect will be neglected for simplicity) will be calculated as a function of titrant volume.

For example, a titrant solution of NH_3 could be added to a Cu^{2+} analyte, and the pCu could be plotted against milliliters of NH_3. Copper forms a deep-blue complex according to Equation 10-3,

$$Cu^{2+} + 4NH_3 \rightleftharpoons Cu(NH_3)_4^{2+} \tag{10-3}$$

and has an overall formation constant

$$K_f = \beta_4 = 2 \times 10^{13} = \frac{[Cu(NH_3)_4^{2+}]}{[Cu^{2+}][NH_3]^4} \tag{10-4}$$

This K value is certainly high enough to raise the expectation of a good break in the titration curve at the equivalence point, and we might anticipate curve A shown in Figure 10-4. However, the curve B is observed, and we could then ask, "What went wrong?"

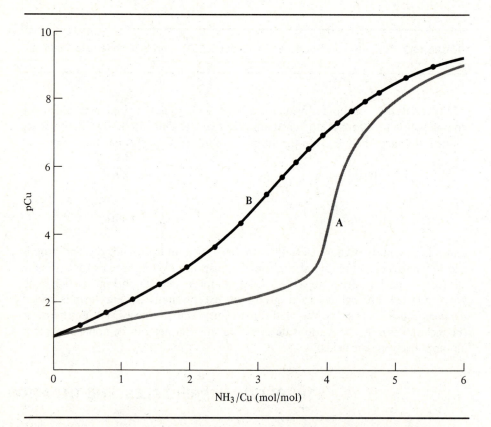

FIGURE 10-4   Titration of Cu^{2+} with NH_3. **(A)** Titration curve expected if copper formed only $Cu(NH_3)_4^{2+}$ complex. **(B)** Titration curve actually observed, showing a series of copper–ammonia complexes.

The difficulty is that copper ion forms not just one complex with NH_3 but a series of complexes, which were shown in Section 4-5. In other words, there is a titration-curve break at 1, 2, 3, and 4 mol of NH_3 per mole of copper. Unfortunately, there is insufficient differentiation among the formation constants to cause one complex to form before the next begins. The overlap is frequently observed with monodentate ligands and explains why their use in analytical applications is limited. EDTA, on the other hand, forms only 1:1 complexes with metal ions and will show one break in the titration curve. In addition, EDTA forms strong complexes (high formation constants) with essentially all metal ions so that the break will be sharp. Only alkali metal ions do not form sufficiently strong EDTA complexes to be useful for analytical purposes.

Before tackling a curve of an EDTA titration, we shall find it helpful to introduce a few concepts to facilitate calculations. First, it is necessary to realize that EDTA is a polyprotic acid with four ionizable hydrogens. These can ionize sequentially as shown in Equations 10-5 to 10-8, which use the symbol Y to stand for the completely ionized anion form of EDTA, $C_{10}H_{12}N_2O_8^{4-}$.

$$H_4Y + H_2O \rightleftharpoons H_3O^+ + H_3Y^- \qquad K_1 = 1.00 \times 10^{-2} \qquad \textbf{(10-5)}$$

$$H_3Y^- + H_2O \rightleftharpoons H_3O^+ + H_2Y^{2-} \qquad K_2 = 2.16 \times 10^{-3} \qquad \textbf{(10-6)}$$

$$H_2Y^{2-} + H_2O \rightleftharpoons H_3O^+ + HY^{3-} \qquad K_3 = 6.92 \times 10^{-7} \qquad \textbf{(10-7)}$$

$$HY^{3-} + H_2O \rightleftharpoons H_3O^+ + Y^{4-} \qquad K_4 = 5.50 \times 10^{-11} \qquad \textbf{(10-8)}$$

The composition of EDTA solutions as a function of pH, of the type described for phosphoric acid in Section 9-4, is shown in Figure 10-5. These equilibria show that five EDTA species must be considered. In addition, in strong-acid solution, the nitrogen atoms can be protonated, leading to H_5Y^+ and H_6Y^{2+} species. Most EDTA titrations are carried out in weakly acidic to weakly basic solution, which means that the predominant species will be H_2Y^{2-} and HY^{3-}. The formation of a metal–EDTA complex can then be better considered as

in weak acid: $M^{n+} + H_2Y^{2-} + 2H_2O \rightleftharpoons MY^{n-4} + 2H_3O^+$

in weak base: $M^{n+} + H_3Y^- + 3OH^- \rightleftharpoons MY^{n-4} + 3H_2O$

However, for tabulation purposes, the formation constant is normally given as in Equation 10-9

$$M^{n+} + Y^{4-} \rightleftharpoons MY^{n-4} \qquad K_{MY} = \frac{[MY^{n-4}]}{[M^{n+}][Y^{4-}]} \qquad \textbf{(10-9)}$$

Formation constants for several EDTA complexes are given in Appendix 2-C. It can be seen that they are frequently $> 10^{10}$, but since this constant is based on Y^{4-}, the fraction of metal ion complexed will depend on pH. The pH dictates what fraction of EDTA will be in the Y^{4-} form ($\alpha_{Y^{4-}}$), as shown in Equation 10-10,

$$\alpha_{Y^{4-}} = \frac{[Y^{4-}]}{[EDTA]} \qquad \textbf{(10-10)}$$

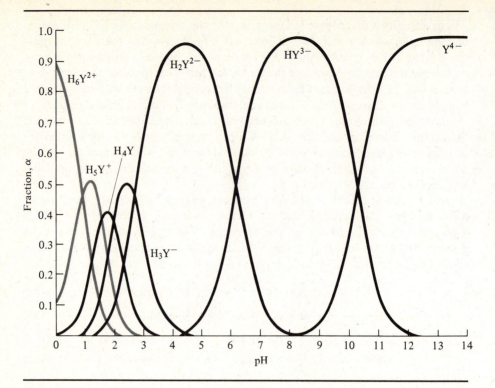

FIGURE 10-5 Fraction (α) of EDTA existing as various species as a function of pH. In strong acid (pH < 2), protonation of nitrogen begins to occur with formation of H_5Y^+ and H_6Y^{2+} species.

where [EDTA] is the formal concentration of all forms of EDTA not complexed with metal ion. From Equations 10-9 and 10-10 the following relationship is obtained:

$$\frac{[MY^{n-4}]}{[M^{n+}]} = K_{MY}[Y^{4-}] = K_{MY}\alpha_{Y^{4-}}[EDTA] \tag{10-11}$$

The term $\alpha_{Y^{4-}}$ can be calculated as was shown in Section 9-4 for a polyprotic acid (Equations 9-48 to 9-51). Thus we can write

$$\alpha_{Y^{4-}} = \frac{K_1 K_2 K_3 K_4}{[H_3O^+]^4 + K_1[H_3O^+]^3 + K_1K_2[H_3O^+]^2 + K_1K_2K_3[H_3O^+] + K_1K_2K_3K_4}$$

The values of $\alpha_{Y^{4-}}$ as a function of pH are given in Table 10-1. Only above pH 11 is Y^{4-} the predominant species.

pH	$\alpha_{Y^{4-}}$

TABLE 10-1 $\alpha_{Y^{4-}}$ **Values As a function of pH for EDTA**

pH	$\alpha_{Y^{4-}}$
3	2.5×10^{-11}
4	3.6×10^{-9}
5	3.5×10^{-7}
6	2.2×10^{-5}
7	4.8×10^{-4}
8	5.4×10^{-3}
9	5.2×10^{-2}
10	3.5×10^{-1}
11	8.5×10^{-1}
12	9.8×10^{-1}

Instead of calculating $[Y^{4-}]$ for each point in the titration, we shall find it simpler to realize that $[MY^{n-4}]/[M^{n+}]$ can be calculated from K_{MY} and $[EDTA]$ if we use Equation 10-11 and $\alpha_{Y^{4-}}$ from Table 10-1. A formation constant dependent on pH conditions can be defined and given the term *conditional formation constant*. It is essentially the effective formation constant for the metal complex at the pH during the titration, and it is shown in Equation 10-12.

$$\text{conditional formation constant:} \quad K' = \alpha_{Y^{4-}} K_{MY} = \frac{[\cancel{Y^{4-}}][MY^{n-4}]}{[EDTA][M^{n+}][\cancel{Y^{4-}}]}$$

$$= \frac{[MY^{n-4}]}{[EDTA][M^{n+}]} \tag{10-12}$$

Now let us look at the basic calculations for an EDTA titration curve, using Cu^{2+} at pH 5 as an example. The titration curve will be derived as pCu versus volume of EDTA titrant.

INITIAL POINT

The Cu^{2+} concentration before any EDTA is added will be just f_{Cu}.

$0 < V_{EDTA} < V_{eqpt}$

Up to the equivalence point, it is generally assumed that the complexation reaction is essentially complete; that is, no Cu^{2+} is formed from dissociation of the Cu^{2-}

complex. Therefore,

$$[Cu^{2+}] = f_{Cu} - f_{EDTA} = \frac{F_{Cu}V_{Cu} - F_{EDTA}V_{EDTA}}{V_{Cu} + V_{EDTA}} \tag{10-13}$$

using the same nomenclature as with acid–base titrations. The formal concentrations in the solution at that point are f_{Cu} and f_{EDTA}, while F_{Cu} and F_{EDTA} are the formal concentrations of the original analyte solution and the EDTA standard solution, respectively. Equation 10-13 states that $[Cu^{2+}]$ is simply the copper ion still in excess.

EQUIVALENCE POINT

At the equivalence point, Equation 10-13 goes to zero because there is no excess copper ion. Therefore, the dissociation of the copper complex is the *only* source of Cu^{2+} and EDTA (in all its *uncomplexed* forms).

$$[Cu^{2+}] = [EDTA]$$

With the conditional formation constant (pH 5), it is seen that

$$K' = \alpha_{Y^{4-}} K_{MY} = 3.5 \times 10^{-7} \times 6.3 \times 10^{18} = \frac{[CuY^{2-}]}{[EDTA][Cu^{2+}]} = \frac{[CuY^{2-}]}{[Cu^{2+}]^2} \tag{10-14}$$

$$[Cu^{2+}]^2 = \frac{[CuY^{2-}]}{2.2 \times 10^{12}} \tag{10-15}$$

The concentration of CuY^{2-} will essentially be f_{Cu} (or f_{EDTA}) at the equivalence point (with the assumption that little dissociation will occur, which must be true for any successful titration).

$$[CuY^{2-}] = \frac{F_{Cu}V_{Cu}}{V_{Cu} + V_{eq\ pt}} = \frac{F_{EDTA}V_{eq\ pt}}{V_{Cu} + V_{eq\ pt}} \tag{10-16}$$

$V_{EDTA} > V_{eq\ pt}$

After the equivalence point, we again assume negligible dissociation of the complex and that the only source of uncomplexed EDTA is the excess EDTA.

$$[EDTA] = f_{EDTA} - f_{Cu} = \frac{F_{EDTA}V_{EDTA} - F_{Cu}V_{Cu}}{V_{Cu} + V_{EDTA}} \tag{10-17}$$

The concentration of CuY^{2-} changes only from dilution:

$$[CuY^{2-}) = \frac{F_{Cu}V_{Cu}}{V_{Cu} + V_{EDTA}} \tag{10-18}$$

From the values of [EDTA] and $[CuY^{2-}]$, the concentration of Cu^{2+} can be calculated, using the conditional formation constant.

$$K' = \alpha_{Y^{4-}} K_{MY} = \frac{[CuY^{2-}]}{[EDTA][Cu^{2+}]}$$

$$[Cu^{2+}] = \frac{[CuY^{2-}]}{\alpha_{Y^{4-}} K_{MY}[EDTA]} = \frac{[CuY^{2-}]}{2.2 \times 10^{12}[EDTA]} \qquad (10\text{-}19)$$

SUMMARY OF TITRATION-CURVE CALCULATIONS

All points on the titration curve can now be calculated, as shown in Example 10-1.

The sharp break seen in Figure 10-6 is a consequence of the high conditional formation constant of 2.2×10^{12} for the titration reaction. Equation 10-12 shows

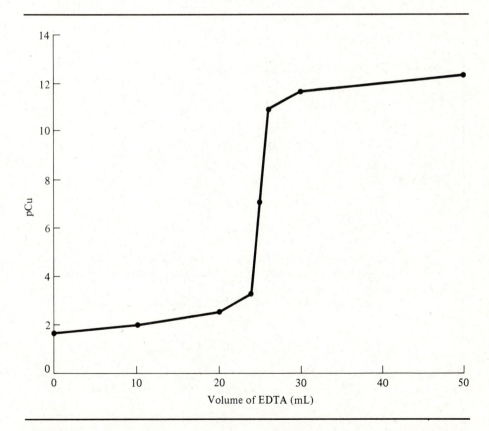

FIGURE 10-6 Titration curve for the titration of 50 mL of 0.02 F Cu^{2+} with 0.04 F EDTA. Points are calculated in Example 10-1.

EXAMPLE 10-1

Derive the titration curve for the titration of 50 mL of 0.020 F Cu^{2+} at pH 5 by calculating pCu for the addition of 0, 10, 20, 24, 25, 26, 30, and 50 mL of 0.040 F EDTA.

Solution

0 mL

At the initial point

$$[Cu^{2+}] = f_{Cu} = 0.020$$
$$pCu = 1.70$$

10 mL

$$[Cu^{2+}] = \frac{F_{Cu}V_{Cu} - F_{EDTA}V_{EDTA}}{V_{Cu} + V_{EDTA}} = \frac{0.020 \times 50 - 0.040 \times 10}{50 + 10} = 0.01$$
$$pCu = 2.00$$

All points $< V_{eq\,pt}$ are calculated in the same way.

20 mL

$$[Cu^{2+}] = 2.9 \times 10^{-3}$$
$$pCu = 2.54$$

24 mL

$$[Cu^{2+}] = 5.4 \times 10^{-4}$$
$$pCu = 3.27$$

At this point, the concentration of excess Cu^{2+} is quite small, and it is prudent to check the assumption that the amount of Cu^{2+} formed by dissociation of the CuY^{2-} is still small enough to be neglected.

The true concentration of Cu^{2+} will be the sum of excess Cu^{2+} and Cu^{2+} from dissociation:

$$[Cu^{2+}] = [Cu^{2+}]_{excess} + [Cu^{2+}]_{dissoc}$$

For each CuY^{2-} that dissociates, a free EDTA (in one of its forms) is produced along with the Cu^{2+}, and this is the only source of free EDTA at this point in the titration:

$$[Cu^{2+}]_{dissoc} = [EDTA]$$

EXAMPLE 10-1 (cont.)

Therefore,

$$[Cu^{2+}] = [Cu^{2+}]_{excess} + [EDTA] = 5.4 \times 10^{-4} + [EDTA]$$

The concentration of CuY^{2-} at this point is essentially equal to the amount of EDTA added:

$$[CuY^{2-}] = \frac{F_{EDTA} V_{EDTA}}{V_{Cu} + V_{EDTA}} = \frac{0.040 \times 24}{50 + 24} = 1.3 \times 10^{-2}$$

From the expression for the conditional formation constant, we obtain

$$K' = 2.2 \times 10^{12} = \frac{[CuY^{2-}]}{[EDTA][Cu^{2+}]} = \frac{1.3 \times 10^{-2}}{[EDTA](5.4 \times 10^{-4} + [EDTA])}$$

This is a quadratic, which may be solved directly or iteratively by a first approximation that EDTA $\ll 5.4 \times 10^{-4}$. In this case,

$$[EDTA] = \frac{1.3 \times 10^{-2}}{2.2 \times 10^{12} \times 5.4 \times 10^{-4}} = 1.1 \times 10^{-11}$$

which is very small compared to 5.4×10^{-4}. This means

$$[Cu^{2+}] = 5.4 \times 10^{-4} + 1.1 \times 10^{-11} \cong 5.4 \times 10^{-4}$$

and we have now been justified in our assumption that the only significant source of Cu^{2+} is the excess Cu^{2+}.

25 mL

$$F_{Cu} V_{Cu} = F_{EDTA} V_{EDTA}$$

Therefore, this is the equivalence point.

$$[Cu^{2+}]^2 = \frac{[CuY^{2-}]}{K'} = \frac{[CuY^{2-}]}{2.2 \times 10^{12}}$$

$$[CuY^{2-}] = \frac{F_{Cu} V_{Cu}}{V_{Cu} + V_{eq\,pt}} = \frac{0.020 \times 50}{50 + 25} = 1.3 \times 10^{-2}$$

$$[Cu^{2+}]^2 = \frac{1.3 \times 10^{-2}}{2.2 \times 10^{12}} = 6.1 \times 10^{-15}$$

$$[Cu^{2+}] = 7.8 \times 10^{-8}$$

$$pCu = 7.11$$

EXAMPLE 10-1 (cont.)

26 mL

After the equivalence point, excess EDTA is first calculated from Equation 10-17:

$$[EDTA] = \frac{F_{EDTA}V_{EDTA} - F_{Cu}V_{Cu}}{V_{Cu} + V_{EDTA}} = \frac{0.040 \times 26 - 0.020 \times 50}{50 + 26} = 5.3 \times 10^{-4}$$

$$[CuY^{2-}] = \frac{F_{Cu}V_{Cu}}{V_{Cu} + V_{EDTA}} = \frac{0.020 \times 50}{50 + 26} = 1.3 \times 10^{-2}$$

$$[Cu^{2+}] = \frac{CuY^{2-}}{K'[EDTA]} = \frac{1.3 \times 10^{-2}}{2.2 \times 10^{12} \times 5.3 \times 10^{-4}} = 1.1 \times 10^{-11}$$

$$pCu = 10.95$$

All points $> V_{eq\ pt}$ are calculated in the same way.

30 mL

$$[EDTA] = 2.5 \times 10^{-3}$$
$$[CuY^{2-}] = 1.25 \times 10^{-2}$$
$$[Cu^{2+}] = 2.3 \times 10^{-12}$$
$$pCu = 11.64$$

50 mL

$$[EDTA] = 1.0 \times 10^{-2}$$
$$[CuY^{2-}] = 1.0 \times 10^{-2}$$
$$[Cu^{2+}] = 4.5 \times 10^{-13}$$
$$pCu = 12.34$$

The results are plotted in Figure 10-6 and show a very sharp break in the equivalence point region. Note that the curve is symmetrical through the equivalence point, with ΔpCu from 24 to 25 mL essentially equal to ΔpCu from 25 to 26 mL.

that the conditional formation constant depends on K_{MY}, the formation constant for the metal complex, and $\alpha_{Y^{4-}}$, the fraction of uncomplexed EDTA existing as Y^{4-}. The latter factor depends on pH (Table 10-1) so that the break in the titration curve will become sharper as K_{MY} increases and as the pH increases. These two effects can be seen in Figure 10-7, in which titration curves are shown for various conditional formation constants. The same conditional formation constant (and therefore the same titration curve) can be attained with a metal ion of relatively low K_{MY} at high pH as with one of relatively high K_{MY} at low pH.

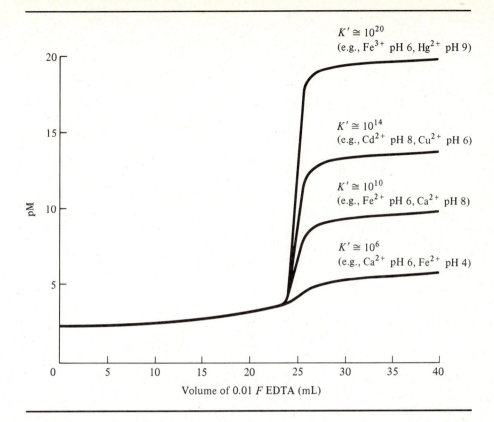

FIGURE 10-7 Titration curves, having various conditional formation constants, for metal ions (50 mL of 0.005 F) with 0.01 F EDTA.

A certain amount of break is necessary to obtain accurate analytical results. Thus there is a minimum pH for the successful EDTA titration of any metal ion, with higher pH's required for metal ions having low K_{MY} values. This is seen in Figure 10-8.

SPECIFICITY OF EDTA TITRATIONS

It may appear that EDTA titrations would not be very useful because essentially all metal ions form complexes. In other words, the method lacks specificity. However, the use of pH, as shown in Figure 10-8, offers a means of titrating one metal ion (of high K_{MY}) in the presence of another (of low K_{MY}).

Another method of increasing specificity is to add another complexing agent, which will compete with EDTA for the metal ion. In essence, this technique removes free metal ion from the system (by forming a complex with the new ligand), just as decreasing the pH removed free Y^{4-} from the system. The titration of the metal ion

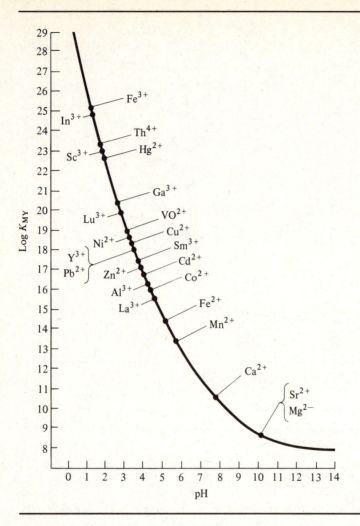

FIGURE 10-8   Minimum pH needed for satisfactory titration of various cations with EDTA.

will now depend on a conditional formation constant involving three factors K_{MY}, $\alpha_{Y^{4-}}$, and α_M, the fraction of metal ion that is uncomplexed. This relationship is given in Equation 10-20.

$$\text{conditional formation constant:} \quad K' = \alpha_{Y^{4-}} \alpha_M K_{MY} \tag{10-20}$$

Just as in the case of $\alpha_{Y^{4-}}$, in which EDTA complexed with metal ion was not considered, α_M does not include metal ion complexed with EDTA. That is, α_M is the fraction of all the metal ion *not* complexed with the EDTA that is in the free form (not complexed with the new ligand added to the system). For example, if the titration of Cu^{2+} described in Example 10-1 was buffered with an acetic acid–sodium acetate buffer to maintain pH 5, the conditional formation constant would have to include the effects of the copper acetate complex. α_M is calculated in Example 10-2 to be 7.9×10^{-2}.

With the new conditional formation constant, the titration curve can be calculated as in Example 10-1. For all points up to the equivalence point, it must be remembered that $[Cu^{2+}] = \alpha_M(f_{Cu} - f_{EDTA})$. That is, the concentration of copper not complexed with EDTA must be multiplied by α_M to obtain the concentration of

---

EXAMPLE 10-2

Calculate the conditional formation constant for copper–EDTA in a solution containing 0.18 F sodium acetate and 0.10 F acetic acid. The formation constant for $CuOAc^+$ is 62, and that for $Cu(OAc)_2$ is 14.

Solution

$$\alpha_M = \frac{[Cu^{2+}]}{[Cu^{2+}] + [CuOAc^+] + [Cu(OAc)_2]} \quad \begin{array}{l}\text{(fraction of copper not}\\\text{complexed with EDTA as } Cu^{2+})\end{array}$$

$$K_1 = \frac{[CuOAc^+]}{[Cu^{2+}][OAc^-]}$$

$$K_2 = \frac{[Cu(OAc)_2]}{[CuOAc^+][OAc^-]} = \frac{[Cu(OAc)_2]}{K_1[Cu^{2+}][OAc^-]^2}$$

Therefore,

$$\alpha_M = \frac{[Cu^{2+}]}{[Cu^{2+}] + K_1[Cu^{2+}][OAc^-] + K_1K_2[Cu^{2+}][OAc^-]^2}$$

$$= \frac{1}{1 + K_1[OAc^-] + K_1K_2[OAc^-]^2} = \frac{1}{1 + (62 \times 0.18) + 14 \times (0.18)^2}$$

$$= 7.9 \times 10^{-2}$$

$$K_a = 1.75 \times 10^{-5} = \frac{[H_3O^+][OAc^-]}{[HOAc]}$$

$$[H_3O^+] = 1.75 \times 10^{-5}\frac{[HOAC]}{[OAc^-]} = 1.75 \times 10^{-5} \times \frac{0.1}{0.18} \cong 1 \times 10^{-5} \qquad pH = 5$$

$$\text{at pH 5 } \alpha_{Y^{4-}} = 3.5 \times 10^{-7} \text{ (Table 10-1)}$$

$$K_{MY} = 6.3 \times 10^{18} \text{ (Appendix 2-C)}$$

Therefore,

$$K' = \alpha_Y^{4-} \alpha_M K_{MY} = 3.5 \times 10^{-7} \times 7.9 \times 10^{-2} \times 6.3 \times 10^{18} = 1.75 \times 10^{11}$$

free copper ion, Cu^{2+}. This will have the effect of raising the titration curve for all points up to and including the equivalence point. However, points after the equivalence point will not change from those determined in Example 10-1 because α_M cancels out as shown below:

$$K' = \alpha_{Y^4-} \cancel{\alpha_M} K_{MY} = \frac{[CuY^{2-}]}{[EDTA][Cu]} = \frac{[CuY^{2-}]}{[EDTA][Cu^{2+}]/\cancel{\alpha_M}}$$

where [Cu] represents all copper not complexed with EDTA. Then

$$[Cu^{2+}] = \frac{[CuY^{2-}]}{\alpha_{Y^4-} K_{MY} [EDTA]}$$

which is the same as Equation 10-19. Because the titration curve is raised before the equivalence point and not affected for points after the equivalence point, the presence of a competing complex will decrease the titration break. Metal ions that form strong complexes with the new ligand may no longer be titratable with EDTA. Such effects can therefore be used to increase specificity for EDTA titrations.

Addition of another ligand also prevents precipitation of the analyte as the hydroxide when the EDTA titration must be carried out in neutral or basic solution (recall that Fe^{3+} is quantitatively precipitated as the hydrous oxide at pH 4 and above).

10-3 ENDPOINT SIGNALS

To observe the titration curve shown in Figure 10-6, we must have a probe that responds to $[Cu^{2+}]$ or, better yet, pCu. The glass electrode in acid–base titrations had been noted to respond to H_3O^+ so that pH could be measured. In recent years many other ion-selective electrodes have been developed, and they will be discussed in Section 15-3. There is a copper electrode available, and so it is possible to measure pCu during the titration with EDTA and plot the titration curve. The inflection point on this curve (point of greatest slope) would be used as the endpoint signal and would occur very near the equivalence point.

Measurement of titration curves is slow, and probes of the type described are expensive compared to color indicators. The work of G. Schwarzenbach in the 1940s showed not only the ability of EDTA and other ligands to form strong complexes with metal ions but also the availability of other reagents that could perform the role of color indicators. It was from this pioneering work that was developed the very active field of complexometric titrations with organic chelating agents.

We have seen that indicators for acid–base titrations are weak acids (or bases) themselves, and the same is true for complexation indicators. In addition to color changes arising from conversion of a conjugate acid form to its conjugate base, the metal ion of interest is involved in chemical equilibria with associated color changes, and thus the indicators are called *metallochromic*. For example, one color indicator widely used in EDTA titrations is *eriochrome black T*, which has the acid–base equilibria

$(10-21)$

This can be written in an abbreviated fashion:

$$\underset{\text{red}}{H_2In^-} + H_2O \rightleftharpoons \underset{\text{blue}}{HIn^{2-}} + H_3O^+$$

At pH ~ 12 or above, HIn^{2-} will also dissociate:

$$\underset{\text{blue}}{HIn^{2-}} + H_2O \rightleftharpoons \underset{\text{orange}}{In^{3-}} + H_3O^+$$

The dissociation constant given in Equation 10-21 shows that, at pH 7 or above (see Example 10-3), the indicator will appear blue. When magnesium ion is added, a new equilibrium is involved.

$$Mg^{2+} + \underset{\text{blue}}{HIn^{2-}} + H_2O \rightleftharpoons \underset{\text{red}}{MgIn^-} + H_3O^+ \qquad \textbf{(10-22)}$$

This equilibrium is also pH dependent and will favor the red $MgIn^-$ form at high pH. The normal range for use is pH 10 to 11. If a magnesium ion solution contains Mg^{2+} and $MgIn^-$, the addition of EDTA will first form the EDTA complex with uncomplexed Mg^{2+}:

$$Mg^{2+} + HY^{3-} + H_2O \rightleftharpoons MgY^{2-} + H_3O^+ \qquad \textbf{(10-23)}$$

Then, when Mg^{2+} is essentially all complexed, EDTA will begin to compete for magnesium in the eriochrome black T complex:

$$\underset{\text{red}}{MgIn^-} + HY^{3-} \rightleftharpoons MgY^{2-} + \underset{\text{blue}}{HIn^{2-}} \qquad \textbf{(10-24)}$$

EXAMPLE 10-3

Calculate the minimum pH at which eriochrome black T will appear to be pure blue.

Solution

$$K_a = 5 \times 10^{-7} = \frac{[\text{HIn}^{2-}][\text{H}_3\text{O}^+]}{[\text{H}_2\text{In}^-]}$$

Pure blue will be perceived when the indicator is predominantly ($>90\%$) in the HIn^{2-} form. Therefore,

$$\frac{[\text{HIn}^{2-}]}{[\text{H}_2\text{In}^-]} \cong 10$$

$$[\text{H}_3\text{O}^+] = \frac{5 \times 10^{-7}}{10} = 5 \times 10^{-8} \qquad \text{pH} = 7.30$$

The solution will begin to change from red to blue through shades of purple. However, not until *all* the MgIn^- has been converted to MgY^{2-} will the equivalence point be reached; that is, meq Mg = meq EDTA, and so the correct endpoint signal is the final pure blue color.

Many other metal ions form red complexes with eriochrome black T, but Equation 10-24 shows that, if a metal ion forms too strong a complex with the indicator, it may take additional EDTA to displace the metal, and the endpoint will be late. On the other hand, if the metal indicator complex is weak compared with the EDTA complex, the color change may be premature. Because of this critical balance between the metal–EDTA formation constant and the metal–indicator formation constant, only a few metal ions can be titrated directly with eriochrome black T as the indicator. Magnesium can be titrated successfully at pH 10, but premature endpoints are encountered at pH 8 and late endpoints at pH 11. Thus the minimum pH given in Figure 10-8 is only one factor for a successful EDTA titration. When metallochromic indicators are used, their equilibria must also be considered and will further limit the range of acceptable conditions for titration.

In cases where a suitable indicator is not available but the metal ion forms a very strong EDTA complex, it is possible to use a back-titration technique. Excess EDTA is added to the solution containing metal ion, and then a titration is carried out with a standard solution of MgCl_2 with eriochrome black T as the indicator. In this case, the endpoint that most closely matches the equivalence point is the first tinge of purple that indicates the formation of MgIn^-. An example is given in Example 10-4.

Calcium does not give a completely satisfactory endpoint with eriochrome black T and does not form a sufficiently strong EDTA complex for the back-titration technique. There are some other metallochromic indicators, such as hydroxy naphthol blue and murexide, that can be used, as noted in Table 10-2. The pH of the

---

EXAMPLE 10-4

The nickel content of a plating solution is determined by diluting a 1.00-mL sample with H_2O buffered to pH 10 with an ammonia–ammonium chloride buffer and treated with 50.0 mL of 0.0496 F EDTA. The excess EDTA is back-titrated with 0.102 F $MgCl_2$ and requires 4.73 mL to reach the eriochrome black T endpoint. Calculate the nickel content as grams per liter of $NiSO_4 \cdot 6H_2O$.

Solution

$$\text{mmol } NiSO_4 \cdot 6H_2O + \text{mmol } MgCl_2 = \text{mmol EDTA}$$

$$\frac{\text{mg } NiSO_4 \cdot 6H_2O}{\text{FW } (NiSO_4 \cdot 6H_2O)} + F_{Mg}V_{Mg} = F_{EDTA}V_{EDTA}$$

$$\frac{\text{mg } NiSO_4 \cdot 6H_2O}{262.9} + 0.102 \times 4.73 = 0.0496 \times 50.0$$

$$\text{mg } NiSO_4 \cdot 6H_2O = 525 \text{ mg in 1 mL}$$

$$\text{nickel content} = \frac{0.525 \text{ g}}{0.001 \text{ L}} = 525 \text{ g/L}$$

---

titration is around 12, which eliminates magnesium interference at the same time because, at this high pH, $Mg(OH)_2$ precipitates.

There is another titration technique, employing eriochrome black T as an indicator, that can be used for determining calcium at pH 9 to 10. This technique is called a *substitution titration*. A small, unmeasured amount of the magnesium–EDTA complex is added to the calcium sample along with the eriochrome black T indicator. Calcium forms a somewhat stronger EDTA complex than magnesium does; so the equilibrium in Equation 10-25 lies somewhat to the right.

$$Ca^{2+} + MgY^{2-} \rightleftharpoons CaY^{2-} + Mg^{2+} \tag{10-25}$$

The free Mg^{2+} is now able to form the red complex with eriochrome black T, as given in Equation 10-22. EDTA is added and will complex all Ca^{2+} first, followed by any remaining Mg^{2+}. Finally, EDTA will displace magnesium from the indicator complex as shown in Equation 10-24. The endpoint is the pure blue color, indicating that essentially all the magnesium has been converted to the EDTA complex.

It may appear that this endpoint will not correspond to the equivalence point because both magnesium and calcium are titrated. However, a look at the stoichiometric relations shows

$$\text{mmol EDTA} = \text{mmol Ca} + \text{mmol Mg}$$

(For EDTA titrations, all reactions are 1:1 so that millimoles can be used in place of

TABLE 10-2 Metallochromic Indicators

Indicator	Structure	Dissociation constants and colors of free indicator species	Colors of metal–indicator complexes	Applications
eriochrome black T	(H_2In^-)	H_2In^- (red); $pK_{a2} = 6.4$ HIn^{2-} (blue); $pK_{a3} = 11.5$ In^{3-} (orange)	wine red	direct titration of Ba, Ca, Cd, In, Pb, Mg, Mn, Sc, Sr, Tl, Zn, and lanthanides back titration of Al, Ba, Bi, Ca, Co, Cr, Fe, Ga, Pb, Mn, Hg, Ni, Pd, Sc, Tl, V displacement titration of Ba, Ca, Cu, Au, Pb, Hg, Pd, Sr
calmagite	(H_2In^-)	H_2In^- (red); $pK_{a2} = 8.1$ HIn^{2-} (blue); $pK_{a3} = 12.4$ In^{3-} (orange)	wine-red	titrations performed with eriochrome black T as indicator may be carried out equally well with calmagite
PAN	(HIn)	HIn (orange-red); $pK_a = 12.3$ In^- (pink)	red	direct titration of Cd, Cu, In, Sc, Tl, Zn back titration of Cu, Ga, Fe, Pb, Ni, Sc, Sn, Zn displacement titration of Al, Ca, Co, Ga, In, Fe, Pb, Mg, Mn, Hg, Ni, V, Zn

Indicator	Structure	Acid–base forms and colors	Color with metal	Applications
murexide		H_4In^- (red-violet); $pK_{a2} = 9.2$ H_3In^{2-} (violet); $pK_{a3} = 10.9$ H_2In^{3-} (blue)	red with Ca^{2+} yellow with Co^{2+}, Ni^{2+}, and Cu^{2+}	direct titration of Ca, Co, Cu, Ni back titration of Ca, Cr, Ga displacement titration of Au, Pd, Ag
pyrocatechol violet		H_4In (red); $pK_{a1} = 0.2$ H_3In^- (yellow); $pK_{a2} = 7.8$ H_2In^{2-} (violet); $pK_{a3} = 9.8$ HIn^{3-} (red-purple); $pK_{a4} = 11.7$	blue, except red with Th(IV)	direct titration of Al, Bi, Cd, Co, Ga, Fe, Pb, Mg, Mn, Ni, Th, Zn back titration of Al, Bi, Ga, In, Fe, Ni, Pd, Sn, Th, Ti
hydroxy naphthol blue		H_2In^{3-} (red-violet); $pK_{a4} = 6.4$ HIn^{4-} (blue); $pK_{a5} = 12.9$ In^{5-} (pink)	red-violet	direct titration of Ca
xylenol orange		H_3I^{3-} (yellow); $pK_{a4} = 6.4$ H_2I^{4-} (red); $pK_{a5} = 10.4$ HI^{5-} (red); $pK_{a6} = 12.3$	red (varies from pink to red to purple, depending on metal)	direct titration of Bi, Cd, Co, Cu, Hg, Pb, Y, Zn, rare earths

milliequivalents.) The total mmol EDTA = mmol MgY added + mmol EDTA as standard solution. Also, mmol Mg = mmol MgY added:

$$\text{mmol MgY} + \text{mmol EDTA soln} = \text{mmol Ca} + \text{mmol MgY}$$

Thus

$$\text{mmol EDTA soln} = \text{mmol Ca}$$

There is one other titration technique that is sometimes used for EDTA reactions. The formation of the metal complex liberates H_3O^+, as shown in Equation 10-26.

$$M^{n+} + H_2Y^{2-} + 2H_2O \rightleftharpoons MY^{n-4} + 2H_3O^+ \qquad \textbf{(10-26)}$$

The H_3O^+ is then titrated with NaOH as an acid–base titration. The initial pH of the EDTA and the endpoint pH must be the same so that no complications arise from association or dissociation of H_2Y^{2-}. Because all metal ions will liberate $2H_3O^+$, they will all exhibit an n value of 2.

10-4 APPLICATIONS

It was pointed out earlier in this chapter that complexation titrations involving monodentate ligands are not generally applied because a series of complexes is usually encountered, with little or no breaks in the titration curve. However, there are a few such reactions that are useful; they represent some of the oldest titration methods known. One example is the determination of iodide with Hg^{2+} by the formation of the HgI_4^{2-} complex. This titration was published in 1832[2], and although not often used for the determination of iodide today, the complex formation reaction is used as the basis for an acid–base titration. This is shown in Equation 10-27.

$$HgO + 4I^- + H_2O \rightleftharpoons HgI_4^{2-} + 2OH^- \qquad \textbf{(10-27)}$$

From Equation 10-27 it is seen that HgO generates two hydroxide ions when it forms HgI_4^{2-}. Because this reaction proceeds essentially to completion and HgO can be obtained in a pure form, mercury(II) oxide is a useful primary standard for standardizing acids.

A more useful complexation reaction involving a monodentate ligand is the Liebig method for the determination of cyanide ion, reported in 1851.[3] The titration reaction is

$$Ag^+ + 2CN^- \rightleftharpoons Ag(CN)_2^- \qquad \textbf{(10-28)}$$

The system can act as its own endpoint signal because, when excess $AgNO_3$ titrant is added, a white precipitate of $Ag[Ag(CN)_2]$ forms:

$$Ag(CN)_2^- + Ag^+ \rightleftharpoons Ag[Ag(CN)_2](s) \qquad \textbf{(10-29)}$$

---

[2] Marozeau, J. *Journal de Pharmacie et de Chimie, 18*: 302 (1832).
[3] von Liebig, J., *Annalen der Chemie, 77*: 102 (1851).

However, increased sensitivity and a sharper endpoint are obtained if some iodide is added to the system in the presence of ammonia. With this modification, silver iodide is precipitated in place of $Ag[Ag(CN)_2]$. Silver iodide is less soluble than the cyanide compound is, and its yellowish precipitate is more readily perceived because it remains colloidally dispersed. This determination of cyanide is useful for the analysis of electroplating baths containing cyanide ion.

The Liebig method can be applied to the determination of silver in photographic films by dissolving the sample in excess standard cyanide solution and back-titrating with standard $AgNO_3$.

Other cyanide complexes have found use in complexation titrations because usually one stoichiometric ratio is highly preferred. For example, addition of cyanide to copper(II) forms $Cu(CN)_4^{2-}$, nickel(II) forms $Ni(CN)_4^{2-}$, and mercury(II) forms $Hg(CN)_2$. The last-named complex is an example of a neutral complex and is frequently encountered with Hg(II). A standard solution of $Hg(NO_3)_2$ can also titrate Br^-, Cl^-, and SCN^-, with the formation of neutral HgX_2 complexes. The reverse titration, that is, the determination of Hg(II), can be accomplished by titration with KSCN, using the $FeSCN^{2+}$ endpoint as described in Section 8-2 for the Volhard method.

Applications involving EDTA and other chelating agents are prevalent, and their numbers are increasing every year. In addition to its simple 1:1 stoichiometry with all metal ions and the availability of metallochromic indicators, EDTA can also be obtained as a primary standard material. Thus standard solutions can be prepared determinately and require no standardization. EDTA is an acid and forms a series of salts,

$$H_4Y \rightarrow H_3Y^- \rightarrow H_2Y^{2-} \rightarrow HY^{3-} \rightarrow Y^{4-}$$

The acid form, H_4Y, is available as a primary standard but is not very soluble in water. At the other end of the spectrum, Na_4Y is a strong base and, like NaOH, cannot be obtained in a high-purity form. The most useful form is the disodium salt of EDTA, $Na_2C_{10}H_{14}N_2O_8 \cdot 2H_2O$, which is available as an analytical reagent of high purity of nearly primary standard quality. The anhydrous compound is also available. The dihydrate may be dried very gently at 80°C for a short period of time to remove excess moisture *without* loss of the two waters of hydration.

If high-purity material is not available or if an old standard solution needs to be restandardized, EDTA solutions can be standardized with primary standard $CaCO_3$. A weighed amount of $CaCO_3$ is dissolved in acid, buffered to the alkaline pH required for the endpoint indicator, and then titrated with the EDTA solution. Two methods are feasible: the use of either a metallochromic indicator appropriate for calcium alone (such as hydroxy naphthol blue) or a substitution titration involving addition of the magnesium–EDTA complex with eriochrome black T indicator.

For back-titrations in which an excess of EDTA is added to the metal ion analyte, a standard solution of $MgSO_4$ or $MgCl_2$ is generally employed, provided that a basic solution (pH 10) is acceptable. For back-titrations in acid solution, a zinc salt or a copper salt solution is recommended.

One of the primary uses for EDTA titrations is in the analysis of alkaline-earth metal ions, particularly Ca^{2+} and Mg^{2+}. These ions are found in many systems, including inorganic, organic, and biochemical. Choice of analytical methods is more limited than for transition metal ions, which typically exhibit several oxidation states in solution (hence redox reactions frequently form the basis of analytical procedures for transition metals).

Determination of water hardness commonly employs an EDTA titration. Water hardness is caused by the presence of metal ions that form insoluble carbonates ("scale" on pipe walls) and precipitate soap ("bathtub ring"). Calcium and magnesium are the main sources of these metal ions although some water supplies contain small amounts of other metal ions, such as Zn^{2+} and Fe^{3+}. Because EDTA complexes with all these metal ions, the milliequivalents (or millimoles) required in the titration represent the total amount of water hardness. For all practical purposes, the titration also measures the total amount of Ca^{2+} and Mg^{2+} in the water. For convenience, the result is frequently stated as parts per million of $CaCO_3$, as shown in Example 10-5.

Expressing water hardness as parts per million of $CaCO_3$ assumes that Ca^{2+} is the only metal ion present (aside from alkali metal ions). This assumption is not really valid for most water supplies, and a true chemical analysis must differentiate calcium from magnesium. This can be accomplished by performing a second EDTA titration at pH 12 to 13, with an indicator appropriate for calcium, such as hydroxy naphthol blue. At this high pH, magnesium forms $Mg(OH)_2$ and will not complex

---

EXAMPLE 10-5

A 50.00-mL sample of well water is acidified with HCl, boiled to remove CO_2, and neutralized with NaOH. The solution is buffered to pH 10 and titrated with 0.01204 F EDTA, using eriochrome black T as the indicator. If the titration requires 31.63 mL to reach the pure blue endpoint, calculate the water hardness as parts per million of $CaCO_3$.

Solution

$$\text{mmol } CaCO_3 = \text{mmol EDTA}$$

$$\frac{\text{mg } CaCO_3}{\text{FW }(CaCO_3)} = FV = 0.01204 \times 31.63$$

$$\frac{\text{mg } CaCO_3}{100.1} = 0.3808 \text{ mmol}$$

$$\text{mg } CaCO_3 = 38.12 \text{ mg in 50-mL sample}$$

$$\text{ppm} = \frac{\text{mg}}{\text{L}} = \frac{38.12}{0.05000} = 762.4 \text{ mg/L} = 762.4 \text{ ppm}$$

(*Note:* The water hardness may be a result of Ca^{2+}, Mg^{2+}, and other metal ions, and therefore expressing the result as parts per million of $CaCO_3$ is only for convenience. More exactly, the analysis shows that the water contains 0.3808 mmol of metal ions titratable by EDTA per 50 mL (7.616×10^{-3} F).

with EDTA provided that the titration is carried out relatively quickly to prevent either coprecipitation of $Ca(OH)_2$ or formation of the magnesium–EDTA complex by the reaction given in Equation 10-30.

$$Mg(OH)_2 + Y^{4-} \rightleftharpoons MgY^{2-} + 2OH^-$$ (10-30)

Note that Y^{4-} was given as the form of EDTA because above pH 12 over 98% will be the completely deprotonated form (see Table 10-1). Thus the second titration will give the calcium value alone, and from this value and the total Ca + Mg found in the first titration, the concentration of magnesium may be determined by difference. This is shown in Example 10-6.

This method of determining calcium and magnesium may be applied to mineral samples such as limestone although the presence of other metal ions (for example,

---

EXAMPLE 10-6

A 100-mL hard-water sample was buffered to pH 10 and titrated with 0.0150 F EDTA, using eriochrome black T indicator. A volume of 36.3 mL was required to turn the indicator blue. A second 100-mL sample was buffered to pH 12 to precipitate $Mg(OH)_2$ and then titrated with 0.0150 F EDTA, using hydroxy naphthol blue indicator. A volume of 28.6 mL was required to reach the endpoint. Calculate parts per million of $CaCO_3$ and $MgCO_3$ in the water.

Solution

In the second titration, only calcium was complexed with EDTA.

$$\text{mmol } CaCO_3 = \text{mmol EDTA}$$

$$\frac{\text{mg } CaCO_3}{FW \, (CaCO_3)} = FV = 0.0150 \times 28.6 = 0.429 \text{ mmol}$$

$$\text{mg } CaCO_3 = 100.1 \times 0.429 = 42.9 \text{ mg in 100 mL}$$

$$\text{ppm} = \frac{\text{mg}}{L} = \frac{42.9}{0.100} = 429 \text{ ppm } CaCO_3$$

In the first titration, both calcium and magnesium were complexed with EDTA.

$$\text{mmol } CaCO_3 + \text{mmol } MgCO_3 = \text{mmol EDTA}$$

From the second titration, mmol $CaCO_3/100$ mL sample = 0.429; therefore,

$$\frac{\text{mg } MgCO_3}{FW \, (MgCO_3)} = FV - \text{mmol } CaCO_3 = 0.0150 \times 36.3 - 0.429 \text{ mmol}$$

$$\text{mg } MgCO_3 = 84.3 \times 0.116 = 9.7 \text{ mg in 100 mL}$$

$$\text{ppm} = \frac{\text{mg}}{L} = \frac{9.7}{0.100} = 97 \text{ ppm } MgCO_3$$

TABLE 10-3   Examples of EDTA Titrations

Species determined	Procedure	Typical materials analyzed
Al(III)	add slight excess of standard EDTA; adjust pH to 6.5; dissolve salicylic acid indicator in the solution; back-titrate unreacted EDTA with standard iron(III) solution	alloys, cryolite ores, cracking catalysts, clays
Bi(III)	adjust pH of sample solution to 2.5; add pyrocatechol violet indicator; titrate with EDTA	alloys, pharmaceuticals
Ca(II)	add NaOH to neutral sample solution to obtain pH > 12; add murexide; titrate with EDTA	pharmaceuticals, phosphate rocks, water, biological fluids
Cd(II) or Zn(II)	to neutral sample solution, add NH_3–NH_4^+ buffer; add eriochrome black T indicator, and titrate with EDTA	plating baths, alloys
Co(II)	adjust pH of sample solution to 6; add murexide indicator; add ammonia to obtain orange color of cobalt–murexide complex; titrate with EDTA	paint driers, magnet alloys, cemented carbides
Cu(II)	adjust pH of sample solution to 8 with NH_3 and NH_4Cl; add murexide indicator, and titrate with EDTA	alloys, ores, electro-plating baths
Fe(III)	adjust pH of sample solution to 4; dissolve salicylic acid indicator in the solution; titrate with EDTA	hemoglobin, limestone, cement, tungsten alloys, paper pulp, boiler scale
Ga(III)	to neutral sample solution, add NH_3 and NH_4^+ to reach pH 9; add eriochrome black T indicator and a known excess of standard EDTA; back-titrate the unreacted EDTA with a standard manganese(II) solution	
Hf(IV) or Zr(IV)	[same procedure as for Al(III)]	alloys, ores, paint driers, sand
Hg(II)	to the sample solution, add a known excess of standard magnesium(II)–EDTA solution; neutralize the solution; add NH_3–NH_4^+ buffer and eriochrome black T indicator; perform substitution titration of free magnesium(II) with EDTA	mercury-containing pharmaceuticals, organomercury compounds, ores

TABLE 10-3 (cont.)

Species determined	Procedure	Typical materials analyzed
In(III)	add tartaric acid to an acidic sample solution; neutralize the solution; add NH_3–NH_4^+ buffer and eriochrome black T indicator; heat solution to boiling, and titrate with EDTA	
Mg(II)	[same procedure as for Cd(II)]	pharmaceuticals, aluminum alloys, gunpowder, soil, plant materials, water, biological fluids
Mn(II)	add triethanolamine and ascorbic acid to acidic sample solution; neutralize the solution; add NH_3–NH_4^+ buffer; add eriochrome black T, and titrate with EDTA	metallurgical slags, silicate rocks, alloys, ferromanganese
Ni(II)	add murexide indicator to neutral sample solution; add NH_3 to obtain orange color of nickel–murexide complex; titrate with EDTA	electroplating baths, manganese catalysts, Alnico, ferrites
Pb(II)	[same procedure as for In(III), except that a room-temperature titration is performed]	gasoline, ores, paints, alloys, pharmaceuticals
Sn(IV)	to strongly acidic sample solution, add known excess of standard EDTA; buffer the solution at pH 5; heat solution to 75°C, add pyrocatechol violet indicator, and titrate unreacted EDTA with standard Zn(II) solution	alloys, electroplating baths
Th(IV)	adjust pH of sample solution to 2; add pyrocatechol violet indicator; warm solution to 40°C, and titrate with EDTA	reactor fuels, ores, minerals, glasses, alloys
Tl(III)	[same procedure as for Hg(II)]	alloys

SOURCE: Information excerpted from G. Schwarzenbach and H. Flaschka, *Complexometric Titrations* (H.M.N.H. Irving, trans.), 2nd ed. (London: Methuen & Co. Ltd., 1969).
Peters, D.G., J.M. Hayes and G.M. Hieftje, *Chemical Separations and Measurements,* Philadelphia: W.B. Saunders Co., 1974.

Fe^{3+}) can be an interference. Masking agents such as CN^- are frequently used to prevent such interferences.

In the biochemical field, EDTA titrations for calcium and magnesium (or just their sum if that is all that is required) are used in the analysis of urine, blood serum, eggshells, milk, and many other substances.

Biological fluids may be analyzed for sugar content if the sample is treated with an excess of alkaline copper(II). The copper(II) is reduced to Cu_2O by the sugar, and the excess copper(II) is back-titrated with EDTA. This concept of carrying out a reaction with excess metal ion and back-titration with EDTA can be applied to the analysis of several organic compounds that form precipitates with mercury(II). A few examples are nicotinamide, phenobarbital, and ethyl gallate. Several other compounds, such as caffeine, quinine, codeine, and thiamine, are precipitated as a salt of the anion BiI_4^-, and the excess bismuth(III) is titrated with EDTA. Some appreciation of the wide range of applications for EDTA titrations can be gained from Table 10-3.

SUPPLEMENTARY READING

Přibil, R., *Analytical Applications of EDTA and Related Compounds.* Elmsford, N.Y.: Pergamon Press, Inc., 1972.

Ringbom, A., *Complexation in Analytical Chemistry.* New York: John Wiley & Sons, Inc., 1963.

Ringbom, A., and E. Wanninen, "Complexation Reactions," in *Treatise on Analytical Chemistry,* 2nd ed., I.M. Kolthoff and P.J. Elving, eds., pt. 1, vol. 2, chap. 20. New York: John Wiley & Sons, Inc., 1979.

Welcher, F.J., *The Analytical Uses of Ethylenediaminetetraacetic acid.* New York: D. Van Nostrand Co., 1958.

PROBLEMS

10-1. Based on the definition that n is the moles of reacting species per mole of compound, calculate n and equivalent weight for the following (complex formed in brackets):

*a. HgO $[HgI_4^{2-}]$

b. $AgNO_3$ $[Ag(CN)_2^-]$

c. KCN $[Ag(CN)_2^-]$

d. Fe_2O_3 $[Fe-EDTA^-]$

*e. Ni_2O_3 $[Ni(CN)_4^{2-}]$

f. $CaBr_2$ $[PbBr_4^{2-}]$

10-2. Write reactions for the formation of the following complexes and expressions for the stepwise and overall formation equilibrium constants:

*a. $Hg(SCN)_2$

b. Co(ethylenediamine)$_3^{3+}$

c. $Ag(CN)_2^-$

---

*Answers to problems marked with an asterisk will be found at the back of the book.

       d. $Zn(EDTA)^{2-}$

       e. HgI_4^{2-}

10-3. a.   A 1.2000-g sample of soluble chloride was analyzed by titration with mercuric nitrate (the product is un-ionized mercuric chloride complex, $HgCl_2$). If the titration required 28.0 mL of 0.100 F mercuric nitrate, calculate the percentage chloride in the sample.

      b.   Qualitatively, what might have been a possible endpoint signal for this titration?

10-4. How were the four requirements for a successful volumetric procedure met in the determination of calcium by titration with EDTA?

10-5. How will the following procedures affect precision and accuracy?

      a.   Use incorrect formula weight for EDTA in calcium analysis.

      b.   Use first purple tinge in Ca titration as endpoint signal.

10-6. A standard solution of EDTA (FW = 372) is prepared determinately by dissolving 1.093 g in a volume of 250 mL. Calculate its titer value for the following analytes:

      *a. $CaCO_3$

      b. $MgCO_3$

      c. Fe_2O_3

      d. Pb_3O_4

*10-7. If 250-mg samples are used in the EDTA method, calculate the formality of EDTA so that milliliters of EDTA equals percentage of Ca in the sample.

10-8. If 250-mg samples are used in the EDTA method, calculate the formality of EDTA so that milliliters of EDTA equals percentage of CaO in the sample.

10-9. A solution of $MgCl_2$ is standardized by titrating 25.00-mL aliquots with 0.01162 F EDTA, using eriochrome black T for an endpoint signal. Titration volumes are 39.83, 40.03, and 39.97 mL. Calculate the formality of the $MgCl_2$ solution and its standard deviation.

10-10. Using the dissociation constants for EDTA (Equations 10-5 to 10-8), calculate α_{Y4-} values at the following pH values:

      a.   pH 2

      b.   pH 6

      c.   pH 10

      d.   pH 12

10-11. Nitrilotriacetic acid (NTA), shown in Figure 10-2, is an effective chelating agent; it is a triprotic acid with dissociation constants given in Appendix 2-B. Construct a plot of α versus pH for NTA analogous to Figure 10-5.

10-12. Calculate the conditional formation constant for the following simple metal ions with EDTA:

      *a. Ca^{2+} at pH 12

      b. Ca^{2+} at pH 10

      c. Fe^{3+} at pH 10

      d. Cu^{2+} at pH 10

      e. Mg^{2+} at pH 6

10-13. Using the data in Appendix 2-C, calculate the conditional formation constant for lead–EDTA in a solution containing 1 F NaI at pH 7.

10-14. Referring to Section 9-4 for a discussion of α values, calculate the fraction of EDTA as the 2– anion (H_2Y^{2-}) at the following pH values. Compare with Figure 10-5.

      *a.   pH 4

      b.   pH 7

      c.   pH 10

10-15. Using the data in Appendixes 2-B and 2-C, calculate the conditional formation constant for zinc–EDTA in a solution containing 0.25 F NH_3 and 0.45 F NH_4NO_3.

10-16. Calculate pM at the equivalence point for the following titrations with 0.02 F EDTA:

*a. 0.02 F Ca^{2+} at pH 12

b. 0.002 F Ca^{2+} at pH 12

c. 0.02 F Zn^{2+} at pH 8

d. 0.02 F Fe^{3+} at pH 5

*10-17. Derive the titration curve for the titration of 50 mL of 0.02 F Ca^{2+} with 0.02 F EDTA at pH 10. Calculate pCa for the addition of 0, 10, 25, 40, 48, 50, 52, 60, and 100 mL.

10-18. Derive the titration curve for the titration of 50 mL of 0.02 F Fe^{3+} with 0.02 F EDTA at pH 5. Calculate pFe for the addition of 0, 10, 25, 40, 48, 50, 52, 60, and 100 mL.

10-19. A quick measure for the titration break is to calculate pM at 98% and at 102% of the equivalence point volume. For a potentially successful titration, the ΔpM must be > 2. Which of the following titrations meet this criterion?

*a. 0.01 F Ca^{2+} with 0.01 F EDTA at pH 8

b. 0.02 F Zn^{2+} with 0.02 F EDTA at pH 6

c. 0.02 F Mg^{2+} with 0.02 F EDTA at pH 8

d. 0.001 F Zn^{2+} with 0.001 F EDTA at pH 6

e. 0.001 F Zn^{2+} with 0.001 F EDTA at pH 4

10-20. Of a hard-water sample 50.0 mL is added to 50.0 mL of 0.032 F EDTA at pH 4.83. The resulting solution is titrated with 0.0216 N NaOH, and 18.2 mL is required to reach the original pH value of 4.83. Calculate the water hardness as parts per million of $CaCO_3$.

*10-21. An eggshell is treated with NaOH to remove the membrane, dried, and found to weigh 5.131 g. The shell is then dissolved in 25 mL of 6 F HCl and diluted to 250 mL in a volumetric flask. A 10.00-mL aliquot is buffered to pH 10 and titrated with 0.04916 F EDTA. If 40.78 mL is required to reach the calmagite endpoint signal, calculate percentage of $CaCO_3$ in the eggshell.

10-22. A powdered-milk sample weighing 968 mg is added to 10 mL of H_2O. Then 20.00 mL of 0.0421 F EDTA is added. The solution is adjusted to pH 12 and titrated with 0.0242 F Ca^{2+} ($CaCO_3$ dissolved in HCl). If 21.09 mL is required to reach the first purple tint indicative of the endpoint signal, calculate percentage of Ca in the milk sample.

10-23. A 1.062-g sample of dried calamine lotion is dissolved in acid and diluted to 250 mL. A 10.00-mL aliquot is taken for the analysis of zinc. KF is added to mask the Fe present, which would interfere; and after adjustment to pH 10, the Zn^{2+} is titrated with 0.0143 F EDTA. To reach the calmagite endpoint signal 35.6 mL is required. Then a 50.00-mL aliquot is taken, buffered, and titrated with 0.00286 F ZnY^{2-}. In this titration, Fe^{3+} replaces Zn^{2+},

$$Fe^{3+} + ZnY^{2-} \rightleftharpoons FeY^- + Zn^{2+}$$

and 2.19 mL is required to reach the endpoint signal. Calculate the percentages of ZnO and Fe_2O_3 in the calamine lotion.

10-24. The calcium level in small blood-serum samples can be determined with an EDTA titration, using a microburet. If a 100-μL serum sample is titrated with 0.00122 F EDTA and 0.237 mL is required to reach the endpoint signal, calculate milligrams of Ca per 100 mL of serum. Does this fall in the normal range of 9 to 11 mg/100 mL?

*10-25. The chloride level in blood serum can be determined with a complexation titration with $Hg(NO_3)_2$:

$$2Cl^- + Hg^{2+} \rightleftharpoons HgCl_2$$

A 0.300-mL serum sample is diluted, acidified, and titrated with 0.0113 F $Hg(NO_3)_2$. If 1.34 mL is required to reach the diphenylcarbazone endpoint, calculate milligrams of Cl^- per 100 mL serum.

10-26. Mercury(II) can be titrated with KSCN, using the Volhard endpoint signal. The titration reaction is a complexation,

$$Hg^{2+} + 2SCN^- \rightleftharpoons Hg(SCN)_2$$

while the endpoint signal involves formation of the red $FeSCN^{2+}$ complex. If a 432-mg sample of impure $HgCl_2$ required 37.16 mL of 0.0816 N KSCN to reach the endpoint, calculate the purity of the $HgCl_2$.

10-27. A 5.00-mL sample of a copper electroplating-bath solution containing CN^- is diluted and neutralized and has 5 mL of 10% KI added. The resulting solution is titrated with 0.1032 F $AgNO_3$, and 35.4 mL is required to reach the endpoint signal, the appearance of a faint turbidity. Calculate the cyanide content of the bath as ounces per gallon of NaCN (1 oz = 28.35 g; 1 gal = 3.785 L).

*10-28. A 168.3-mg sample of an alloy containing only magnesium and zinc is dissolved in acid, buffered to pH 10 with NH_3–NH_4Cl, and titrated with 0.0917 F EDTA. If 44.23 mL is required to reach the eriochrome black T endpoint signal, calculate percentages of Mg and Zn in the alloy.

10-29. Caffeine in analgesic tablets can be determined by using a complexation back-titration method by precipitation of the protonated salt with BiI_4^-:

$$(C_3H_{10}N_4O_2)H^+ + BiI_4^- \rightleftharpoons (C_3H_{10}N_4O_2)HBiI_4$$

An analgesic tablet weighing 325 mg is dissolved in H_2O, interferences precipitated out, and diluted to 50 mL in a volumetric flask. To a 20.00-mL aliquot, 5.00 mL of 0.263 F $KBiI_4$ is added, and the precipitate is filtered from the solution. The filtrate containing excess Bi(III) is titrated with 0.0472 F EDTA at pH 5, and 18.89 mL is required to reach the endpoint signal, the disappearance of the yellow BiI_4^- species. Calculate percentage of caffeine in the tablet.

*10-30. A 100-mL hard-water sample was buffered to pH 10 and titrated with 0.0200 F EDTA, using eriochrome black T indicator. A volume of 32.5 mL was required to turn the indicator pure blue. A second 100-mL sample was buffered to pH 12 to precipitate $Mg(OH)_2$ and then titrated with 0.0200 F EDTA, using hydroxy naphthol blue indicator. A volume of 25.3 mL was required to titrate the calcium. Calculate parts per million (milligrams per liter) of $CaCO_3$ and $MgCO_3$ in the water.

10-31. For the titrations in Problem 10-30, calculate the uncertainty in $CaCO_3$ and $MgCO_3$ values if the *overall* uncertainty was ±0.10 mL in each of the titrations.

10-32. A cadmium ion-selective electrode is used as an endpoint signal for the titration of Cd^{2+} with EDTA. The slope at the endpoint is 50 mV/mL, and the potential measurement has an uncertainty of ±3mV. If the titration volume is 28 mL and each volume reading has an uncertainty of ±0.03 mL, calculate the overall relative uncertainty (percentage). What is the major contribution to the overall uncertainty?

Redox Titrations

11

We now come to the fourth reaction type: oxidation–reduction, or redox for short. In Chapter 3, it was stated that the reactive species in this reaction type is the electron, which is transferred from the reductant to the oxidant. Most elements are capable of exhibiting more than one oxidation state, and therefore volumetric methods based on redox reactions are widely used.

As with the other three reaction types, the first objective will be to derive a titration curve from theoretical principles. However, it is customary to describe redox reactions in electrochemical terms because the transfer of electrons may also be carried out in an electrochemical cell, the reductant and oxidant never coming into contact with each other. Thus it is necessary to discuss terms and concepts of electrochemistry before tackling the titration curve, where, instead of the reactant concentration, the electrochemical potential will be plotted as a function of titrant volume. These terms and concepts will also provide the background needed for the electroanalytical methods discussed in Part Five.

11-1 BASIC TERMS

One way of introducing electrochemical terminology is to consider electrical energy as given in Equation 11-1.

$$\text{electrical energy} = \text{voltage} \times \text{charge} = VQ \tag{11-1}$$

This is analogous to mechanical energy as force times distance or gas-expansion energy as pressure times volume. The unit for voltage is the *volt* (V), while the unit for electrical charge is the *coulomb* (C). The unit of electrical energy then is equivalent to one V-C and is called the *joule* (J).

Normally electrical charge is not measured; instead, flow of charge, coulombs per second, is measured. This flow of charge is called the *current*, and a current of 1 C/sec is called the *ampere* (A), generally said "amps." If an electrochemical process is carried out at constant current, I (or if I represents the average current), then the electrical charge involved is given by Equation 11-2:

$$Q \text{ (C)} = I \text{ (C/sec)} t \text{ (sec)} \tag{11-2}$$

If Equations 11-1 and 11-2 are combined, we obtain

$$\text{electrical energy (J)} = VQ = VIt \tag{11-3}$$

The VI product is another important electrical term, called the *power*, and is given the unit of *watt* (W). A joule of electrical energy is then one W-sec. When we purchase electrical energy from the local utility company, it is computed in power times time, but the larger unit of kilowatt-hours (kWh) is used in place of watt-seconds. It takes 3.6×10^6 W-sec (J) to equal one kWh.

For theoretical discussions and calculations, the voltage of an electrochemical cell is given as the ideal thermodynamic value and is called the *cell potential*, E. In the ideal cell, no current is flowing (open-circuit conditions). As soon as current flows, the cell voltage, V, becomes less than E, and the electrical energy produced is less. The maximum electrical energy available is then EQ, in which charge must be transferred at an infinitely slow rate ($I \to 0$). Electrochemical processes can be measured under conditions where current is extremely small and nearly achieve thermodynamically reversible conditions.

If 1 mol of reaction is carried out, the total charge transferred is given by Equation 11-4:

$$Q \left(\frac{\text{C}}{\text{mol}} \right) = n \left(\frac{\text{equivalents}}{\text{mol}} \right) F \left(\frac{\text{C}}{\text{equivalent}} \right) \tag{11-4}$$

The Faraday constant ($F = 96{,}485 = 9.65 \times 10^4$) is the charge of 1 mol of electrons. The maximum electrical energy that can be obtained then is

$$\text{maximum electrical energy} = EnF \text{ (commonly written as } nFE) \tag{11-5}$$

In thermodynamics, the maximum useful energy (work) that can be obtained from a system is the free energy change, ΔG, and it should therefore equal nFE. One modification must be incorporated in the final equation because thermodynamics and electrochemistry were established independently with their own ground rules. In thermodynamics, a spontaneous process is conceived as a system going from a high energy state to a lower one. Therefore, $\Delta G < 0$. On the other hand, electrochemists conceive of a spontaneously operating cell as having a positive voltage, $V > 0$. Thus an inversion in sign is needed to relate the two fields, as shown in Equation 11-6.

$$\Delta G = -nFE \tag{11-6}$$

NERNST EQUATION

To relate electrochemical potentials to activities (concentrations) of species in the system, we can draw on the thermodynamics relationship involving free energy change and activities, namely

$$\Delta G = \Delta G^0 + RT \ln \mathbf{Q} \tag{11-7}$$

where $\mathbf{Q}$ = activities of products raised to their coefficient powers divided by activities of reactants raised to their coefficient powers (the same form as in the equilibrium-constant expression), and ΔG^0 = free energy change when all species are in their standard state ($a = 1$).

By combining Equations 11-6 and 11-7, we can see that

$$-nFE = -nFE^0 + RT \ln \mathbf{Q}$$

$$E = E^0 - \frac{RT}{nF} \ln \mathbf{Q} \tag{11-8}$$

At 25°C and using base 10 logarithms,

$$E = E^0 - \frac{0.059}{n} \log \mathbf{Q} \tag{11-9}$$

Equations 11-8 and 11-9 are forms of the *Nernst equation* and give us a relationship between electrochemical potentials and activities of species in the reaction. E^0, as with ΔG^0, refers to the electrochemical potential for the reaction when all species are in their standard state ($a = 1$).

Suppose we are interested in the titration of Fe^{2+} with Ce^{4+}, as shown in Equation 11-10.

$$Fe^{2+} \text{ (analyte)} + Ce^{4+} \text{ (titrant)} \rightleftharpoons Fe^{3+} + Ce^{3+} \tag{11-10}$$

The Nernst equation for this reaction at 25°C is given by Equation 11-11:

$$E = E^0 - \frac{0.059}{n} \log \frac{[Fe^{3+}][Ce^{3+}]}{[Fe^{2+}][Ce^{4+}]} \tag{11-11}$$

Since iron is oxidized from +2 to +3 it loses one electron; so $n = 1$. Alternatively, cerium is reduced from +4 to +3; so it gains one electron.

When solutions of Fe^{2+} and Ce^{4+} are mixed, the reaction quickly reaches equilibrium. At equilibrium, $\Delta G = 0$; so $E = 0$. Also, at equilibrium, $\mathbf{Q} = K$, the equilibrium constant. This is shown in Equation 11-12:

$$\text{at equilibrium:} \quad E = 0 = E^0 - \frac{0.059}{n} \log K$$

Therefore,

$$E^0 = \frac{0.059}{n} \log K = \frac{0.059}{1} \log K \tag{11-12}$$

With a knowledge of E^0, one can calculate the equilibrium constant for any redox reaction, as shown in Example 11-1.

One beautiful aspect of electrochemical reactions is that the reactants can be separated and allowed to react under control of the investigator by his placing

EXAMPLE 11-1

Determine the equilibrium constant at 25°C for the titration reaction given in Equation 11-10. The standard electrochemical potential, E^0, for this reaction is $+0.93$ V.

Solution

From Equation 11-12, we calculate

$$E^0 = \frac{0.059}{1} \log K$$

$$\log K = \frac{+0.93}{0.059} = 15.8$$

$$K = 5.8 \times 10^{15} = \frac{[Fe^{3+}][Ce^{3+}]}{[Fe^{2+}][Ce^{4+}]}$$

Recall that, for the other reaction types discussed previously, a K value of $> 10^6$ was usually sufficient for a satisfactory titration. Therefore, we conclude that the proposed redox titration should give a sharp break at the equivalence point, and this will be demonstrated later in the chapter.

them in separate electrode compartments, as shown in Figure 11-1. The electrochemical potential can be measured with a voltmeter, and by the closing of switch S, electrons can flow from the left electrode compartment to the right electrode compartment. The reactions are

$$\text{left side (oxidation occurs):} \quad Fe^{2+} \rightleftharpoons Fe^{3+} + e \qquad \textbf{(11-13)}$$

$$\text{right side (reduction occurs):} \quad Ce^{4+} + e \rightleftharpoons Ce^{3+} \qquad \textbf{(11-14)}$$

The salt bridge shown allows ionic charge to move from one compartment to the other to maintain electroneutrality. As electrons are removed from the left side, the solution increases in positive charge. This positive charge can be compensated for by cations moving into the salt bridge and/or anions moving from the salt bridge to the left compartment. In the right-hand compartment, cations may enter from the salt bridge, and/or anions may migrate into the salt bridge to preserve charge balance.

This experimental setup gives us a way to measure E^0. If a solution containing Fe^{2+} ($a = 1$) and Fe^{3+} ($a = 1$) is placed in the left compartment and a solution containing Ce^{3+} ($a = 1$) and Ce^{4+} ($a = 1$) is placed in the right compartment, the measured voltage (with S open so that no current flows) will be E^0. In fact, just preparing solutions in which $[Fe^{2+}] = [Fe^{3+}]$ and in which $[Ce^{3+}] = [Ce^{4+}]$ would also measure E^0 because in Equation 11-11 the logarithm term would become $\log 1 = 0$, and $E = E^0$. Note that a single solution containing unit-activity Fe^{2+}, Fe^{3+}, Ce^{3+}, and Ce^{4+} would react rapidly to reach equilibrium, but by keeping the iron couple separated from the cerium couple, these measurements can be made.

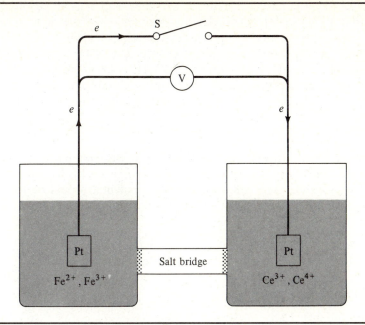

FIGURE 11-1 Electrochemical cell for carrying out redox reaction. When the switch (S) is closed, electrons will flow from the left-hand compartment to the right-hand compartment, resulting in oxidation of Fe^{2+} to Fe^{3+} in the left compartment. The salt bridge allows for flow of ionic charge to maintain electroneutrality.

HALF-CELL POTENTIALS

The possibility of separating the iron couple from the cerium couple leads to the concept that there is an iron-couple potential and a cerium-couple potential. What is measured is the *potential difference* between the two.

$$E_{cell} = E_{right} - E_{left} = E_{reduction} - E_{oxidation} = E_{Ce} - E_{Fe} \qquad \text{(11-15)}$$

E_{Ce} and E_{Fe} are the half-cell potentials, and together they form the cell potential, which is actually measured. By convention, the cell potential is given as the right side minus the left side, and the cell reaction involves reduction of the right-hand couple and oxidation of the left-hand couple. It would make life simple if E_{Ce} and E_{Fe} could be measured directly, but all electrochemical potential measurements involve the difference between two half-cells (electrochemical couples). Therefore, there is no absolute value for a half-cell potential.[1]

---

[1] In recent years, efforts have been made to relate half-cell potentials to the physicists' energy level of an electron in vacuum. The best value for the hydrogen couple is approximately −4.5 V on this scale. However, nearly all electrochemical measurements in the scientific literature still use the hydrogen couple as the reference point, arbitrarily adopting this half-cell as 0.000 V. All values in this book are on this scale.

A simple way around this dilemma is to compare all half-cell potentials to one that is arbitrarily adopted as the reference point and given the value of zero. This approach is seen in everyday life, in which altitude is given in feet (or meters) above sea level: Calling sea level zero is merely a convenient reference point, and there is nothing absolute about saying a certain location is $+600$ ft and another is $+400$ ft; but there is an absolute difference between them of 200 ft. In the electrochemical world, the hydrogen couple in its standard state is given the value of zero:

$$H_3O^+ (a = 1) + e \rightleftharpoons \tfrac{1}{2}H_2 (g; p = 1 \text{ atm}) + H_2O \qquad E^0_{H^+/H_2} \equiv 0.000 \text{ V} \qquad \textbf{(11-16)}$$

Therefore, to measure any half-cell potential, we need to put it in the right-hand compartment and a *normal hydrogen electrode* (*NHE*) in the left-hand compartment. For example, for our iron couple, its standard potential would be measured as shown in Figure 11-2. A shorthand method of describing this cell is

$$\text{Pt, } H_2 \text{ (1 atm)} | H_3O^+ (a = 1) \| Fe^{2+} (a = 1), Fe^{3+} (a = 1) | \text{Pt}$$

in which the symbol | represents the electrode/electrolyte phase boundary, while the symbol ‖ represents the salt bridge that connects the two compartments. From Equation 11-15, we see that

$$E_{cell} = E_{right} - E_{left} = E^0_{Fe} - E^0_{H^+/H_2} = E^0_{Fe} - 0 = E^0_{Fe}$$

The cell potential measured in the lab for this cell is $+0.771$ V. For convenience, we shall use E^0_{Fe} to represent the Fe^{3+}/Fe^{2+} couple; but to avoid ambiguity with other

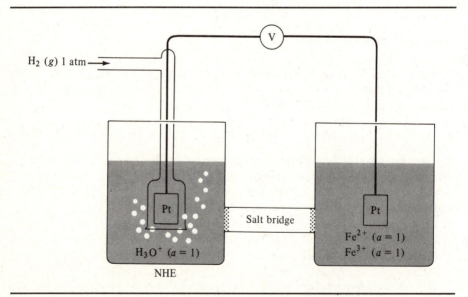

FIGURE 11-2 Measurement of the Fe^{3+}/Fe^{2+} half-cell potential, using the normal hydrogen electrode (NHE) as the reference.

possible iron couples, such as Fe^{2+}/Fe, it would be more accurate to write $E^0_{Fe^{3+}/Fe^{2+}}$.

There is considerable confusion regarding signs and conventions in electro-chemistry. We have, so far, inserted a minus sign in Equation 11-6 to make electrochemistry consistent with thermodynamics and stated that the cell potential was the right side minus the left side instead of the other way around. At this point, it should be remembered that there is *no* ambiguity regarding the $+0.771$ V measured for the cell described above. The iron couple is at a positive potential compared with the hydrogen couple. It also means that Equation 11-17 will proceed spontaneously in the direction written if switch S is closed:

$$\tfrac{1}{2}H_2 + H_2O + Fe^{3+} \xrightarrow{\text{spontaneous}} H_3O^+ + Fe^{2+} \qquad (11\text{-}17)$$

Thus the $+0.771$ V can be attributed to the half-reaction

$$Fe^{3+} + e \rightleftharpoons Fe^{2+} \qquad E^0_{Fe} = +0.771 \text{ V vs. NHE} \qquad (11\text{-}18)$$

When the same experiment is carried out with the Ce^{4+}/Ce^{3+} couple in $HClO_4$, it is found that

$$Ce^{4+} + e \rightleftharpoons Ce^{3+} \qquad E^0_{Ce} = +1.70 \text{ V vs. NHE} \qquad (11\text{-}19)$$

The electrochemical cell reaction for our proposed titration was given in Equation 11-15, and therefore, E^0 for the cell is $+1.70 - (+0.771) \cong +0.93$ V, the value used in Example 11-1.

The convention of quoting all half-cells as reduction potentials was adopted fairly recently (in 1953) at the International Union of Pure and Applied Chemistry (IUPAC) in Stockholm. The Stockholm convention is widely accepted now and is used in this text. Unfortunately, one of the most important reference books on electrochemical potentials still used today[2] is based on quoting oxidation potentials in which

$$Fe^{2+} \rightleftharpoons Fe^{3+} + e \qquad E^0_{Fe} = -0.771 \text{ V vs. NHE}$$

Students should be aware that this possible confusion still lurks in the literature and should check which convention is being used whenever they consult a source.

One might wonder why it is useful to tabulate half-cell potentials instead of just quoting the cell potential of interest, that is, $+0.93$ V for our titration reaction: One reaction requires two half-cell measurements, whereas one direct measurement of the cell in question would be sufficient. Yet one tremendous advantage of quoting half-cell potentials can be seen when we consider the large number of possible oxidants and reductants available. With n oxidants and m reductants, the total number of cell potentials would be $n \times m$, whereas the number of half-cell potentials would be $n + m$. For example, if $n = 80$ and $m = 100$, there would be 8000 possible cell potentials and only 180 half-cell potentials.

---

[2] W.M. Latimer, *The Oxidation States of the Elements and Their Potentials in Aqueous Solutions*, 2nd ed. (Englewood Cliffs, N.J.: Prentice-Hall, Inc., 1952).

At this point, we need to return to the question of concentration effects. All standard half-cell potentials are E^0 values—all species in their standard state ($a = 1$). A Nernst-type equation can be written for each half-cell though, and this will allow calculation of the half-cell potential for any set of conditions. Suppose that in Figure 11-1 the right-hand compartment had contained 0.1 F Ce^{3+} and 10^{-3} F Ce^{4+}. Then

$$Ce^{4+} + e \rightleftharpoons Ce^{3+} \qquad E^0_{Ce} = +1.70 \text{ V vs. NHE}$$

$$E_{Ce} = E^0_{Ce} - \frac{0.059}{1} \log \frac{[Ce^{3+}]}{[Ce^{4+}]}$$

$$= +1.70 - \frac{0.059}{1} \log \frac{0.1}{10^{-3}} = +1.70 - 0.12 = +1.58 \text{ V vs. NHE}$$

Now we are in a position to calculate the cell potential under any conditions, given the standard half-cell potentials. For example, if the left-hand compartment contains 0.02 F Fe^{2+} and 0.12 F Fe^{3+},

$$Fe^{3+} + e \rightleftharpoons Fe^{2+}$$

(note all half-cells written as reduction), and

$$E_{Fe} = E^0_{Fe} - \frac{0.059}{1} \log \frac{[Fe^{2+}]}{[Fe^{3+}]}$$

$$= +0.771 - \frac{0.059}{1} \log \frac{0.02}{0.12} = +0.771 - (-0.046) = +0.817 \text{ V vs. NHE}$$

From Equation 11-15,

$$E_{cell} = E_{right} - E_{left} = E_{Ce} - E_{Fe}$$

$$= +1.58 - (+0.817) = +0.76 \text{ V}$$

Because half-cells refer to one electrode in the electrochemical cell, they are normally called *electrode potentials*, as distinguished from the total cell potential.

Another practical aspect of electrochemical measurements is that the normal hydrogen electrode is inconvenient to prepare and use and that therefore most electrode potentials are measured compared to some other *reference electrode*. One common reference electrode is the *saturated calomel electrode (SCE)*; it has the potential of $+0.242$ V vs. NHE. Example 11-2 shows how the potential of a solution containing an Fe^{3+}/Fe^{2+} couple can be measured by a platinum indicator electrode and a saturated calomel electrode as reference. This reference electrode is constructed with a salt bridge in the tip so that in most cases it may be immersed directly into the solution to be measured.

Although there are still more details to be discussed in Chapter 15, it is now possible to construct the titration curve for the titration of Fe^{2+} with Ce^{4+}.

EXAMPLE 11-2

Determine the cell potential for the cell shown in Figure 11-3 if the solution contains $0.1 \ F \ Fe^{2+}$ and $10^{-4} \ F \ Fe^{3+}$.

Solution

$$E_{cell} = E_{right} - E_{left} = E_{SCE} - E_{Pt} \qquad E_{SCE} = +0.242 \ V \ vs. \ NHE$$

$$E_{Pt} = E_{Fe} \ (i.e., \ Pt \ indicator \ electrode \ measures \ Fe \ couple \ potential)$$

$$E_{Fe} = E_{Fe}^0 - \frac{0.059}{1} \log \frac{[Fe^{2+}]}{[Fe^{3+}]}$$

$$= +0.771 - \frac{0.059}{1} \log \frac{0.1}{10^{-4}} = +0.771 - 0.177 = +0.594 \ V \ vs. \ NHE$$

Therefore,

$$E_{cell} = +0.242 - (+0.594) = -0.352 \ V$$

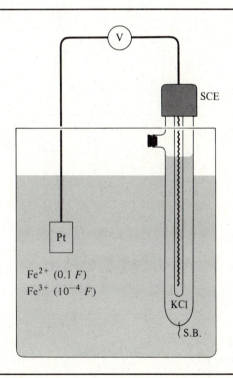

FIGURE 11-3 Measurement of Fe^{3+}/Fe^{2+} half cell with a saturated calomel reference electrode (SCE). Salt bridge (S.B.) now consists of a fiber thread wetted with saturated KCl and connecting the reference electrode to the measurement (indicator) electrode. The inner cell of SCE is filled with Hg, Hg_2Cl_2, and KCl to establish the Hg_2Cl_2/Hg couple as reference.
Shorthand notation for cell shown above:
Pt|Fe^{2+} (0.1 F), Fe^{3+} ($10^{-4} \ F$)||SCE

11-2   THEORETICAL PRINCIPLES: THE TITRATION CURVE

As mentioned earlier, the titration curve for a redox reaction is usually plotted as solution potential versus titrant volume. In the case of an Fe^{2+} analyte and Ce^{4+} titrant, we need to calculate the electrochemical potential of the analyte solution with each addition of Ce^{4+}. Figure 11-4 shows a typical experiment setup. The electrochemical cell consists of a platinum indicator electrode, which measures the solution potential compared with a saturated calomel reference potential. The

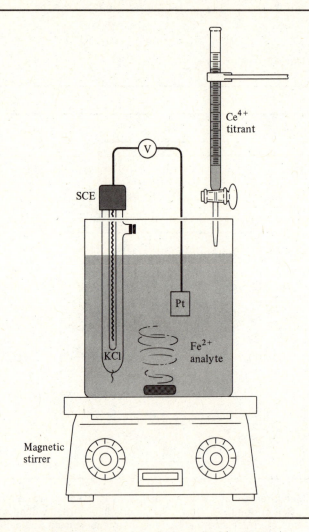

FIGURE 11-4   Experimental setup for the titration of Fe^{2+} with Ce^{4+}.
Shorthand notation for cell shown above:
SCE ‖ Fe^{2+} (analyte) | Pt

measured cell potential will be

$$E_{cell} = E_{right} - E_{left} = E_{Pt} - E_{SCE}$$

It is important to realize that, *if* the solution has come to equilibrium, all half-cells *in the analyte solution* (the SCE is isolated from the analyte solution by the salt bridge in its tip) will be at the *same* potential. That is, the potential difference between any of them is zero (at equilibrium, $\Delta G = 0$ and $E = 0$). This means we can use whichever couple is useful for calculating the solution potential. More explicitly, according to the Nernst equation, we need to know the concentrations of both oxidized and reduced forms of one couple to calculate E.

We now proceed with the titration of 50 mL of 0.05 F Fe^{2+} with 0.10 F Ce^{4+}.

INITIAL POINT

Interestingly enough, the very first point on the titration curve presents a problem unlike that of any of the other reaction-type titration curves. The system contains no cerium yet; so it is impossible to use the cerium couple. The system contains Fe^{2+}, but as Equation 11-20 shows,

$$Fe^{3+} + e \rightleftharpoons Fe^{2+} \qquad E^0_{Fe} = +0.771 \text{ V vs. NHE}$$

(again note that we write the half-cell as a reduction even though the titration reaction is the oxidation of Fe^{2+})

$$E_{Pt} = E_{Fe} = E^0_{Fe} - \frac{0.059}{1} \log \frac{[Fe^{2+}]}{[Fe^{3+}]} \qquad \text{(11-20)}$$

we still cannot calculate E because Fe^{3+} should equal zero at this point:

$$E_{Fe} = +0.771 - 0.059 \log \frac{0.05}{0} \rightarrow -\infty$$

The potential will not go running offscale toward $-\infty$ but, instead, will be determined by some small amount of Fe^{3+} in the system. Iron(II) is readily oxidized by air to give iron(III), and thus the longer the solution is kept in contact with air, the more positive the potential will become. If the titration is going to be quantitatively accurate, then $[Fe^{3+}]$ must be $\leq [Fe^{2+}]/1000$; otherwise an error of $\geq 0.1\%$ will occur. If we use this value as an upper limit for $[Fe^{3+}]$, then

$$[Fe^{3+}] \leq \frac{[Fe^{2+}]}{1000} = \frac{0.05}{1000} = 5 \times 10^{-5}$$

$$E_{Fe} \leq +0.771 - 0.059 \log \frac{0.05}{5 \times 10^{-5}} = +0.771 - 0.177 = +0.594 \text{ V vs. NHE}$$

Thus the initial point in a redox titration is undefined, but it is possible to calculate a maximum value based on quantitative accuracy. This same idea was presented in gravimetric procedures, when it was shown that *complete* precipitation of an ion could not occur and that some acceptable value for quantitative removal had to be assumed.

$0 < V_{Ce} < V_{eq\ pt}$

For all points on the titration curve after the first addition of Ce^{4+} up to the equivalence point, the system contains both Fe^{2+} and Fe^{3+}. Therefore, Equation 11-20 can be used to calculate E_{Fe}, which in turn equals E_{Pt}.

For example, if 10 mL of Ce^{4+} is added, it can be assumed with an equilibrium constant of 5.8×10^{15} (Example 11-1) that the titration goes essentially to completion. Therefore,

$$Fe^{2+} + Ce^{4+} \rightleftharpoons Fe^{3+} + Ce^{3+}$$

Every millimole of Ce^{4+} added will decrease Fe^{2+} and increase Fe^{3+} the same amount:

$$[Fe^{2+}] = \frac{F_{Fe}V_{Fe} - F_{Ce}V_{Ce}}{V_{Fe} + V_{Ce}} = \frac{0.05 \times 50 - 0.10 \times 10}{50 + 10} = 2.5 \times 10^{-2}$$

$$[Fe^{3+}] = \frac{F_{Ce}V_{Ce}}{V_{Fe} + V_{Ce}} = \frac{0.10 \times 10}{50 + 10} = 1.7 \times 10^{-2}$$

From Equation 11-20,

$$E_{Pt} = E_{Fe} = +0.771 - 0.059 \log \frac{2.5 \times 10^{-2}}{1.7 \times 10^{-2}}$$

$$= +0.771 - 0.010 = +0.761 \text{ V vs. NHE}$$

It is useful to realize that for all points in this region the iron couple provides a "redox buffer." That is, the macroquantities (for example, $>0.001\ F$) of both oxidation states maintain a solution potential near E_{Fe}^0, and the solution potential does not change rapidly with titrant volume. This is analogous to the pH buffer region established when both conjugate-acid and conjugate-base forms of a weak acid or weak base are present in a system.

It is not necessary in this region to calculate the actual concentrations of Fe^{2+} and Fe^{3+} because the solution potential is determined by the $[Fe^{2+}]/[Fe^{3+}]$ ratio. Calculation of the ratio will be used in Example 11-4.

EQUIVALENCE POINT

At the equivalence point, we encounter another problem. If the reaction truly goes to completion, then $[Fe^{2+}]$ and $[Ce^{4+}]$ equal zero, and neither couple can be used to calculate the solution potential. It is clear that, even with the large equilibrium constant of 5.8×10^{15}, there must be some small concentration of Fe^{2+} and Ce^{4+} present, and this concentration can be calculated using K.

Such an exercise is laborious and turns out to be unnecessary. As was stated earlier, at equilibrium both half-cell potentials, E_{Fe} and E_{Ce}, must be equal. Therefore, we can state

$$E_{Fe} = E^0_{Fe} - 0.059 \log \frac{[Fe^{2+}]}{[Fe^{3+}]} \qquad \text{(11-21)}$$

$$E_{Ce} = E^0_{Ce} - 0.059 \log \frac{[Ce^{3+}]}{[Ce^{4+}]} \qquad \text{(11-22)}$$

$$E_{eq\,pt} = E_{Fe} = E_{Ce} \qquad \text{(11-23)}$$

The sum of Equations 11-21 and 11-22 gives

$$E_{Fe} + E_{Ce} = 2E_{eq\,pt} = E^0_{Fe} + E^0_{Ce} - 0.059 \log \frac{[Fe^{2+}][Ce^{3+}]}{[Fe^{3+}][Ce^{4+}]} \qquad \text{(11-24)}$$

At the equivalence point,

$$[Fe^{2+}] = [Ce^{4+}] \quad \text{and} \quad [Fe^{3+}] = [Ce^{3+}]$$

Therefore,

$$\log \frac{[Fe^{2+}][Ce^{3+}]}{[Fe^{3+}][Ce^{4+}]} = \log 1 = 0$$

and

$$2E_{eq\,pt} = E^0_{Fe} + E^0_{Ce}$$

$$E_{eq\,pt} = \frac{E^0_{Fe} + E^0_{Ce}}{2} = \frac{+0.771 + 1.70}{2} = +1.236 \text{ V}$$

This simple calculation, taking the average of the two E^0 values, gives the same $E_{eq\,pt}$ as would be calculated with the equilibrium constant. Whenever the n values for the two half-cells are not equal, it is necessary to take a weighted average, as shown in Example 11-3.

EXCESS CE⁴⁺

After the equivalence point, all the iron exists predominantly as Fe^{3+}. The small amount of Fe^{2+} would have to be calculated, using the equilibrium constant, before the solution potential could be determined, using the Fe^{3+}/Fe^{2+} half-cell. However,

---

EXAMPLE 11-3

Calculate the solution potential at the equivalence point for the titration of Sn^{2+} with Ce^{4+}. The reaction is

$$Sn^{2+} + 2Ce^{4+} \rightleftharpoons Sn^{4+} + 2Ce^{3+}$$

Solution

$$E_{Sn} = E_{Sn}^0 - \frac{0.059}{2} \log \frac{[Sn^{2+}]}{[Sn^{4+}]}$$

$$E_{Ce} = E_{Ce}^0 - \frac{0.059}{1} \log \frac{[Ce^{3+}]}{[Ce^{4+}]}$$

$$E_{eq\,pt} = E_{Sn} = E_{Ce}$$

To combine logarithmic terms when E_{Sn} and E_{Ce} are summed, it is necessary to multiply E_{Sn} by 2:

$$2E_{Sn} = 2E_{Sn}^0 - \frac{0.059 \times \cancel{2}}{\cancel{2}} \log \frac{[Sn^{2+}]}{[Sn^{4+}]}$$

$$= 2E_{Sn}^0 - \frac{0.059}{1} \log \frac{[Sn^{2+}]}{[Sn^{4+}]}$$

---

we now have information concerning $[Ce^{3+}]$ and $[Ce^{4+}]$ so that the solution potential can be calculated more simply by the Ce^{4+}/Ce^{3+} half-cell.

$$E_{Pt} = E_{Ce} = E_{Ce}^0 - \frac{0.059}{1} \log \frac{[Ce^{3+}]}{[Ce^{4+}]} \tag{11-25}$$

For example, with the addition of 30 mL of 0.10 F Ce^{4+}, we can assume (based on the large equilibrium constant for the titration reaction) that 25 mL ($V_{eq\,pt}$) has been reduced to Ce^{3+}, while the remaining 5 mL represents excess Ce^{4+}. This, then, is another redox buffer system, and little change in potential will be observed with further additions of Ce^{4+}.

$$[Ce^{3+}] = \frac{F_{Ce}V_{eq\,pt}}{V_{Fe} + V_{Ce}} = \frac{0.10 \times 25}{50 + 30} = 3.1 \times 10^{-2}$$

$$[Ce^{4+}] = \frac{F_{Ce}(V_{Ce} - V_{eq\,pt})}{V_{Fe} + V_{Ce}} = \frac{0.10(30 - 25)}{50 + 30} = 6.2 \times 10^{-3}$$

EXAMPLE 11–3 (cont.)

Addition gives

$$3E_{eq\,pt} = 2E^0_{Sn} + E^0_{Ce} - \frac{0.059}{1} \log \frac{[Sn^{2+}][Ce^{3+}]}{[Sn^{4+}][Ce^{4+}]}$$

At the equivalence point, there is twice as much Ce as Sn:

$$[Ce^{3+}] = 2[Sn^{4+}] \qquad [Ce^{4+}] = 2[Sn^{2+}]$$

$$\log \frac{[Sn^{2+}][Ce^{3+}]}{[Sn^{4+}][Ce^{4+}]} = \log \frac{[Sn^{2+}]2[Sn^{4+}]}{[Sn^{4+}]2[Sn^{2+}]} = \log 1 = 0$$

Therefore,

$$3E_{eq\,pt} = 2E^0_{Sn} + E^0_{Ce}$$

$$E_{eq\,pt} = \frac{2E^0_{Sn} + E^0_{Ce}}{3}$$

That is, the E^0 values have been weighted by the n value for the couple. The numerical values for this example are

$$E^0_{Ce} = +1.70 \text{ V vs. NHE}$$

$$E^0_{Sn} = +0.154 \text{ V vs. NHE}$$

$$E_{eq\,pt} = \frac{2(+0.154) + 1.70}{3} = +0.669 \text{ V vs. NHE}$$

$$E_{Pt} = E_{Ce} = +1.70 - 0.059 \log \frac{3.1 \times 10^{-2}}{6.2 \times 10^{-3}}$$

$$= +1.70 - 0.041 = +1.66 \text{ V vs. NHE}$$

As in the case of the iron-couple buffer region, only the ratio of concentrations is needed—which can be calculated from the volume of titrant used. That is, of the 30 mL of 0.10 E Ce^{4+} added, 25 mL was reduced to Ce^{3+}, while the 5-mL excess remains as Ce^{4+}. Thus

$$\frac{[Ce^{3+}]}{[Ce^{4+}]} = \frac{25}{(30 - 25)} = 5$$

$$E_{Pt} = E_{Ce} = +1.70 - 0.059 \log 5$$

$$= +1.70 - 0.041 = +1.66 \text{ V vs. NHE}$$

This approach saves time and is used in Example 11-4, in which all the points needed to draw the titration curve are calculated.

EXAMPLE 11-4

Construct the titration curve for the titration of 50 mL of 0.05 F Fe^{2+} with 0.1 F Ce^{4+} by calculating the potential of the cell shown in Figure 11-4 after the addition of 0, 5, 10, 12.5, 20, 23, 24, 25, 26, 30, and 50 mL of titrant.

Solution

0 mL

The initial point is undefined, but E_{Pt} should be $< +0.594$ V vs. NHE, as discussed earlier in Section 11-2:

$$E_{cell} = E_{Pt} - E_{SCE} < +0.594 - (+0.242) = +0.352 \text{ V}$$

$0 < V_{Ce} < V_{eq\,pt}$

As discussed in Section 11-2, only the ratio of iron concentrations is needed for points before the equivalence point, and this is simply given by the fraction of Fe^{2+} that has been oxidized. The equivalence relation states (we recognize the 1:1 stoichiometry involved) that at the equivalence point

$$F_{Fe}V_{Fe} = F_{Ce}F_{eq\,pt}$$

$$V_{eq\,pt} = \frac{0.05_i \times 50}{0.10} = 25 \text{ mL}$$

Therefore, when 5 mL of 0.10 F Ce^{4+} has been added, 5 mL of titrant has oxidized Fe^{2+} to Fe^{3+}, and there remains Fe^{2+} corresponding to 20 mL of titrant. Thus

$$\frac{[Fe^{2+}]}{[Fe^{3+}]} = \frac{(25-5)}{5} = 4$$

$$E_{Pt} = +0.771 - 0.059 \log 4 = +0.771 - 0.036 = +0.735 \text{ V vs. NHE}$$

$$E_{cell} = E_{Pt} - E_{SCE} = +0.735 - (+0.242) = +0.493 \text{ V}$$

The same approach is used for the addition of 10, 12.5, 20, 23, and 24 mL of titrant, and the numerical values are given in Table 11-1. Note that, at the midpoint of the titration ($V_{Ce} = V_{eq\,pt}/2$), the logarithmic term goes to zero and $E_{Pt} = E_{Fe}^0$. This is analogous to an acid–base titration in which $pH = pK_a$ at the midpoint.

$V_{eq\,pt}$ (25 mL)

$$E_{Pt} = E_{eq\,pt} = \frac{E_{Fe}^0 + E_{Ce}^0}{2} = \frac{+0.771 + 1.70}{2} = +1.236 \text{ V vs. NHE}$$

$$E_{cell} = 1.236 - (+0.242) = +0.994 \text{ V}$$

EXAMPLE 11-4 (cont.)

$V_{Ce} < V_{eq\,pt}$

With excess Ce^{4+}, the system is a Ce^{4+}/Ce^{3+} redox buffer, and the ratio of cerium concentrations determines the potential. At 26 mL,

$$\frac{[Ce^{3+}]}{[Ce^{4+}]} = \frac{25}{(26-25)} = 25$$

$$E_{Pt} = +1.70 - 0.059 \log 25 = +1.70 - 0.082 = +1.62 \text{ V vs. NHE}$$

$$E_{cell} = +1.62 - (+0.242) = +1.38 \text{ V}$$

The same approach is used for the addition of 30 and 50 mL of titrant, and the numerical values are also given in Table 11-1. Note that, when $V_{Ce} = 2V_{eq\,pt}$, the logarithmic term goes to zero and $E_{Pt} = E_{Ce}^0$, analogous to the midpoint calculation. Because E_{Ce}^0 is given only to ± 0.01 V, all potentials after the equivalence point have been rounded off to ± 0.01 V.

The titration curve points given in Table 11-1 are plotted in Figure 11-5. Note that the shape of the curve is the same whether we plot E_{Pt} or E_{cell}, but the y-axis values will be different.

Many commercial potentiometers have electrode connections designed so that the reference electrode is the right-hand electrode in the cell. In such cases,

$$E_{cell} = E_{SCE} - E_{Pt}$$

and cell values will be inverted with regard to sign. For example, at the equivalence point,

$$E_{cell} = E_{right} - E_{left} = E_{SCE} - E_{Pt}$$

$$E_{cell} = +0.242 - (+1.236) = -0.994 \text{ V}$$

One final point that should be observed from the titration curve: As with the other reaction types, it is symmetrical around the equivalence point when the analyte and titrant n values are equal. That is, the potential changes 0.384 V from 24 to 25 mL ($V_{eq\,pt}$) of titrant and 0.384 V from 25 to 26 mL. This can be seen in Figure 11-5. The symmetry provides a check because the three points have been calculated by three different methods.

A simple computer program for carrying out a redox titration curve calculation is given in Appendix 1-G.

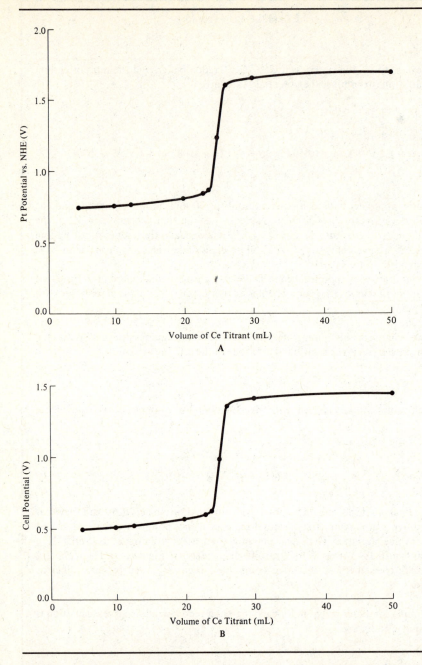

FIGURE 11-5 Titration curve for the titration of 50 mL of 0.05 F Fe^{2+} with 0.1 F Ce^{4+}. **(A)** E_{Pt} as a function of titrant volume. **(B)** E_{cell} as a function of titrant volume. The experimental points are calculated in Example 11-4. See Appendix 1-G for computer program to generate titration curves and examples.

TABLE 11-1 Titration Curve Values for Titration of 50 mL of 0.05 F Fe^{2+} with 0.10 F Ce^{4+}, SCE Reference Electrode

Volume of titrant	E_{Pt} vs. NHE	E_{cell} (V)
0	—	—
5	+0.735	+0.493
10	+0.761	+0.519
12.5	+0.771	+0.529
20	+0.807	+0.565
23	+0.834	+0.592
24	+0.852	+0.610
25	+1.236	+0.994
26	+1.62	+1.38
30	+1.66	+1.42
50	+1.70	+1.46

11-3 FORMAL POTENTIALS

To calculate the titration curve, we had to know the standard potential, E^0, for each of the couples involved. The standard potential is the potential that would be observed when all the species are in their standard state (unit activity) compared with the NHE. Frequently, analytes are in solutions in which concentration is not a good approximation to activity. For example, it may be necessary to carry out a titration in a highly concentrated acidic solution, in which activity coefficients are considerably different from 1. Another situation occurs when complex ions are formed so that α_M, the fraction of the metal ion in the free form (aquo species), is low. In these cases, it is convenient to use *formal potentials*. A formal potential is the potential observed (relative to NHE) when each of the species of the redox couple is present in 1 F concentration. It is then assumed that the activity of each species in this medium will be directly proportional to its formal concentration. In this way, the potential for the couple can be calculated for any concentrations, as shown in Example 11-5.

The formal potential for a metal ion couple is generally (but not always) *lower* than the standard potential. This is true because, if the medium forms complexes, the formation constant is usually greater for the higher oxidation state. Even when activity coefficients are the only consideration, the formal potential is generally less than the standard potential because solute interactions are more significant as the charge on the ion increases (see Table 5-1). For example, for a solution containing

EXAMPLE 11-5

Calculate the potential of a 1 F H_2SO_4 solution containing 10^{-3} F Ce^{4+} and 0.05 F Ce^{3+}

Solution

Because of the high ionic strength of this solution and the existence of Ce(IV) sulfate complexes, it is necessary to use the formal potential of the Ce(IV)/Ce(III) couple in 1 F H_2SO_4. Appendix 3-A gives the value as $+1.44$ V vs. NHE. Therefore,

$$E = E^0_{f,Ce} - \frac{0.059}{1} \log \frac{[Ce^{3+}]}{[Ce^{4+}]}$$

in which $[Ce^{3+}]$ and $[Ce^{4+}]$ refer to their formal concentrations in 1 F H_2SO_4.

$$E = +1.44 - 0.059 \log \frac{0.05}{10^{-3}} = +1.44 - 0.10 = +1.34 \text{ V vs. NHE}$$

Fe^{2+} and Fe^{3+} with activities given by Equation 4-8,

$$[Fe^{2+}] = f_{Fe^{2+}} C_{Fe^{2+}}$$

$$[Fe^{3+}] = f_{Fe^{3+}} C_{Fe^{3+}}$$

$$E^0_{f,Fe} = E^0_{Fe} - \frac{0.059}{1} \log \frac{[Fe^{2+}]}{[Fe^{3+}]} \qquad \text{(11-26)}$$

From Equation 11-26 and the definition that $C_{Fe^{2+}} = C_{Fe^{3+}} = 1$ for the formal potential conditions,

$$E^0_{f,Fe} = E^0_{Fe} - 0.059 \log \frac{f_{Fe^{2+}} \times 1}{f_{Fe^{3+}} \times 1}$$

Table 5-1 shows that, even in a solution of ionic strength of only 0.1, the activity coefficient for Fe^{2+} is 0.40, while that of Fe^{3+} is 0.18. Therefore,

$$E^0_{f,Fe} = E^0_{Fe} - 0.059 \log \frac{0.40 \times 1}{0.18 \times 1} = E^0_{Fe} - 0.020$$

$$= +0.771 - 0.020 = +0.751 \text{ V vs. NHE}$$

The effect in a more concentrated 1 F solution can be greater. For example, the Fe(III)/Fe(II) couple formal potential in 1 F $HClO_4$, in which complexation is negligible, is $+0.732$ V, 0.039 V less than E^0.

11-4 ENDPOINT SIGNALS

The titration curve calculated above was based on the premise that the solution potential could be measured in the laboratory, as shown in Figure 11-4. The calculated equivalence point potential could be used as the endpoint signal, or the point of steepest slope (maximum first derivative or zero second derivative) could be adopted. The latter is preferred because the standard potentials may not be strictly appropriate for the analyte solution and the formal potentials may not be known. However, measurement of an entire titration curve to determine the endpoint in this way is slow and inconvenient. As with the other reaction types, it is useful to find indicators that change color at or near the equivalence point to provide an endpoint signal.

The simplest color indicator is the titrant itself. If it is sufficiently intense in color, it will impart its color to the solution as soon as it is in excess. The most common example is $KMnO_4$, which is used as an oxidant and was discussed in Section 7-2. Because a certain excess is needed for the analyst to observe the $KMnO_4$ color in the solution, a blank correction is usually made (see Example 7-5).

A variation on this concept of excess analyte's providing an endpoint signal is the addition of a complexing agent that forms a colored complex with excess analyte. This idea is realized in the intense blue starch–iodine complex. If iodine is being used as an oxidizing agent, the system will not change color (preferably remaining colorless) until excess iodine is present. Iodine—actually I_3^- in solutions containing I^-—imparts a yellow color, but this color is not intense enough for a satisfactory endpoint signal (marginally acceptable in a colorless analyte solution). In the presence of starch, a small excess of iodine in the system will produce the sharp blue endpoint color. The reverse reaction, the disappearance of I_3^-, can also be carried out successfully. For example, the following reaction may be used analytically for the determination of copper:

$$2Cu^{2+} + 5I^- \rightleftharpoons 2CuI(s) + I_3^- \qquad \text{(11-27)}$$

The I_3^- produced is titrated with sodium thiosulfate, $Na_2S_2O_3$, and the disappearance of the yellow I_3^- is observed. When the color is nearly gone, starch is added to intensify the color, and the endpoint signal is taken as the final disappearance of the blue starch–iodine color. A typical calculation is given in Example 11-6. It is possible to detect I_3^- color at a concentration of $5 \times 10^{-6}\ F$ under ideal conditions, but the starch–iodine complex is detectable when present at a concentration of $2 \times 10^{-7}\ F$, which makes endpoint corrections unnecessary for typical titration conditions.

The number of analytes that can be detected directly is very limited, and what is needed is an indicator that changes color at a certain solution potential in the same way that an acid–base indicator changes color at a certain pH. Continuing this analogy, the ideal situation would be a series of redox indicators that change color at various potentials so that we could pick one that changes color at the equivalence

EXAMPLE 11-6

A 952-mg sample of copper ore is dissolved in a concentrated $HCl–HNO_3$ mixture and evaporated with H_2SO_4 to eliminate excess HCl and HNO_3. The sulfuric acid is neutralized, and 4 g of KI dissolved in 10 mL of water is added. The resulting I_3^- (see Equation 11-27) is titrated with 0.0761 N $Na_2S_2O_3$ to the starch–iodine endpoint and requires 32.7 mL. The titration reaction is

$$I_3^- + 2S_2O_3^{2-} \rightleftharpoons 3I^- + S_4O_6^{2-}$$

Calculate the percentage of copper in the ore.

Solution

$$meq \ Cu = meq \ S_2O_3^{2-}$$

$$\frac{mg \ Cu}{FW \ (Cu)/n} = NV$$

According to Equation 11-27,

$$Cu^{2+} \rightarrow Cu^{1+} \ (in \ CuI) \qquad n = 1$$

$$\frac{mg \ Cu}{63.5/1} = 0.0761 \times 32.7$$

$$mg \ Cu = 158 \ mg$$

$$\% \ Cu = \frac{158 \ mg \ Cu}{952 \ mg \ sample} \times 100 \ (\%) = 16.6\%$$

Note that the amount of KI added is not needed for calculation. The I_3^- produced is merely an intermediate step—required because $S_2O_3^{2-}$ does not reduce Cu^{2+} directly.

---

point potential. This ideal situation has been nearly realized through the development of several redox indicators, some of which are listed in Table 11-2.

The basic concept of a redox indicator is that it exhibits two oxidation states of different colors. At least one of these states is intensely colored so that a sharp endpoint signal may be detected. The color transition region can be calculated in a way similar to that of the acid–base indicator system. That is, the redox indicator will show its reduced-form color when

$$\frac{[red]}{[ox]} \geq 10$$

TABLE 11-2 Examples of Redox Indicators

Indicator name	Color		E_{ind}^0 vs. NHE (V)
	Reduced form	Oxidized form	
indigo 5-monosulfonic acid	colorless	blue	+0.26
phenosafranine	colorless	red	+0.28
indigo 5,5',7,7'-tetrasulfonic acid	colorless	blue	+0.36
new methylene blue	colorless	blue	+0.46
methylene blue	colorless	blue	+0.53
diphenylbenzidine	colorless	violet	+0.76
barium diphenylamine sulfonate	colorless	red-violet	+0.84
erioglaucine A	green	red	+1.00
tris (1,10-phenanthroline) iron(II) sulfate (known as ferroin)	red	pale blue	+1.06
tris(5-nitro-1,10-phenanthroline) iron(II) sulfate (known as nitroferroin)	red	pale blue	+1.25

and will show its oxidized-form color when

$$\frac{[\text{red}]}{[\text{ox}]} \leq 0.1$$

These rules of thumb can now be combined with the Nernst expression for the redox indicator to predict the color transition region:

$$\text{ox} + n e \rightleftharpoons \text{red}$$

$$E = E_{ind}^0 - \frac{0.059}{n} \log \frac{[\text{red}]}{[\text{ox}]}$$

Reduced form will be observed when

$$E_{red} = E_{ind}^0 - \frac{0.059}{n} \log 10 = E_{ind}^0 - \frac{0.059}{n}$$

Oxidized form will be observed when

$$E_{ox} = E_{ind}^0 - \frac{0.059}{n} \log 0.1 = E_{ind}^0 + \frac{0.059}{n}$$

The entire color transition region is then

$$E_{endpt} = E_{ind}^0 \pm \frac{0.059}{n}$$

Most redox indicator dyes have $n = 2$ so that the transition will occur from about 30 mV below E_{ind}^0 to 30 mV above E_{ind}^0.

For the selection of the most appropriate redox indicator, the equivalence point potential for the titration reaction is first calculated, and then the redox indicator whose E^0 most closely matches this potential is chosen. The procedure is shown in Example 11-7. Note that it is assumed that the system is at chemical equilibrium; that is, the indicator, the analyte, and the titrant couples are all at the same potential. This points out another requirement for the redox indicator: It must reach equilibrium quickly.

EXAMPLE 11-7

Choose an appropriate redox indicator for the titration of 0.05 N Fe^{2+} with 0.12 N Ce^{4+} in 1 F H_2SO_4.

Solution

First it is necessary to find the formal potentials for the Fe(III)/Fe(II) couple and the Ce(IV)/Ce(III) couple in 1 F H_2SO_4. These values are (Appendix 3-A)

$$E_{f,Fe}^0 = +0.68 \text{ V vs. NHE}$$

$$E_{f,Ce}^0 = +1.44 \text{ V vs. NHE}$$

$$E_{eq\,pt} = \frac{E_{f,Fe}^0 + E_{f,Ce}^0}{2} = \frac{+0.68 + 1.44}{2} = +1.06 \text{ V vs. NHE}$$

From Table 11-2, it is readily seen that tris(1,10-phenanthroline) iron(II) sulfate (ferroin) has an E^0 value of $+1.06$ V, which matches the equivalence point exactly.

Note that no concentrations (for example, Fe^{2+} and Ce^{4+}) were required for the calculation. This is generally true *provided that* no H_3O^+ is involved in the half-cell (for example, $MnO_4^- + 8H_3O^+ + 5e \rightleftharpoons Mn^{2+} + 12H_2O$) or changes in association (for example, $Cr_2O_7^{2-} + 14H_3O^+ + 6e \rightleftharpoons 2Cr^{3+} + 21H_2O$) take place. These latter cases are a little more involved and will be somewhat concentration dependent.

It is not always possible to find a redox indicator whose E^0 is within 30 mV of the equivalence point potential. However, many redox titration curves are so sharp, as shown in Figure 11-5, that even a mismatch in potentials by > 100 mV will not cause a serious endpoint error, as illustrated in Example 11-8.

EXAMPLE 11-8

Suppose, in the titration described in Example 11-7, that the analyst had decided to use barium diphenylamine sulfonate as the redox indicator. What endpoint error would have resulted?

Solution

From Example 11-7,

$$E_{eq\,pt} = +1.06 \text{ V vs. NHE}$$

From Table 11-2,

$$E_{endpt} = +0.84 \pm 0.030 \text{ V vs. NHE}$$

For purposes of calculation, let us assume the endpoint to have been observed at $+0.84$ V vs. NHE (220 mV before the equivalence point). Because the endpoint potential is too low, the system will still be in the iron couple region so that

$$E_{endpt} = E_{f,Fe}^0 - \frac{0.059}{1} \log \frac{[Fe^{2+}]}{[Fe^{3+}]}$$

$$+0.84 = +0.68 - 0.059 \log \frac{[Fe^{2+}]}{[Fe^{3+}]}$$

$$\log \frac{[Fe^{2+}]}{[Fe^{3+}]} = \frac{+0.68 - 0.84}{0.059} = -2.71$$

$$\frac{[Fe^{2+}]}{[Fe^{3+}]} = 1.94 \times 10^{-3}$$

Therefore, endpoint error $= -0.2\%$

The fraction of iron still remaining as untitrated Fe(II) is, therefore,

$$\frac{[Fe^{2+}]}{[Fe^{2+}] + [Fe^{3+}]} = \frac{[Fe^{2+}]}{[Fe^{2+}] + [Fe^{2+}]/1.94 \times 10^{-3}} = \frac{1}{1 + 515}$$

$$= 1.94 \times 10^{-3} \simeq 0.2\%$$

Note that this calculation assumes that the amount of Fe(II) at the true equivalence point would be zero. This assumption will be acceptable for all titration reactions with large equilibrium constants (large difference in E^0 values for analyte and titrant) and for reasonable concentrations ($> 10^{-3}$ F).

11-5 REDOX TITRATION FEASIBILITY

It can be seen in Figure 11-5 that the plateau region before the equivalence point is determined by E^0 for the analyte. The plateau region after the equivalence point is determined by E^0 for the titrant. Because of the redox buffering, most of the potential change occurs in the equivalence point region so that

$$\text{titration break} \cong E^0_{\text{titrant}} - E^0_{\text{analyte}}$$

The question is how small the titration break can be and still represent a satisfactory titration. Some examples of different titration breaks are shown in Figure 11-6, assuming $n = 1$ for the titration reaction. Clearly, the break is not well defined for <0.2 V difference in E^0 values. With a difference of 0.2 V, the break is readily perceived when the entire titration curve is plotted as in Figure 11-6. With redox indicators, however, the color transition region (~ 59 mV for typical eye response to

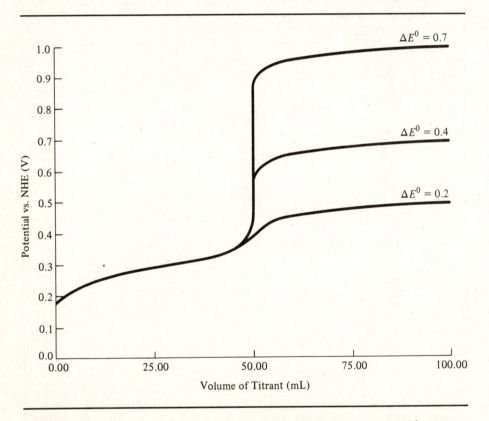

FIGURE 11-6   Titration curve breaks for various $E^0_{\text{titrant}} - E^0_{\text{analyte}}$ values (ΔE^0). Analyte is assumed to be 50 mL of 0.10 N, and $E^0_{\text{analyte}} = +0.3$ V vs. NHE; the n value is assumed to be 1 for both analyte and titrant.

color change) represents a range of $\sim 12\%$ (6% before the equivalence point to 6% after the equivalence point) so that an accurate titration is not possible. If $n > 1$, the break will be sharper, and the color transition region would be $\sim 0.5\%$ for a difference of 0.2 V in E^0 values and $n = 2$. In general, titration breaks of >0.4 V are needed when redox indicators are used but can be somewhat less if the analyst takes the time to measure the entire titration curve (potentiometric endpoint signal).

11-6 MULTICOMPONENT SYSTEMS

What happens when more than one analyte is present? For example, suppose an analyte solution of Fe^{2+} also contained Sn^{2+} and is titrated with Ce^{+4}. From the titration curve is it possible to determine the amount of each analyte?

The addition of Ce^{4+} to the analyte solution can oxidize Fe^{2+} or Sn^{2+}:

$$Fe^{2+} + Ce^{4+} \rightleftharpoons Fe^{3+} + Ce^{3+} \qquad K = 5.8 \times 10^{15} \text{ (see Example 11-1)}$$

$$Sn^{2+} + 2Ce^{4+} \rightleftharpoons Sn^{4+} + 2Ce^{3+} \qquad K = 3.5 \times 10^{52} \text{ (see Example 11-9)}$$

EXAMPLE 11-9

Calculate the equilibrium constant at 25°C for the titration reaction of Sn^{2+} with Ce^{4+}.

Solution

$$Sn^{2+} + 2Ce^{4+} \rightleftharpoons Sn^{4+} + 2Ce^{3+}$$

$$E^0_{cell} = E^0_{reduction} - E^0_{oxidation} = E^0_{Ce} - E^0_{Sn} = +1.70 - (+0.154)$$

$$= +1.55V \ (\pm 0.01 \text{ V})$$

$$= \frac{0.059}{n} \log K$$

$$\log K = \frac{+1.55}{0.059/2} = 52.5$$

$$K = 10^{52.5} = 3.5 \times 10^{52}$$

The second reaction has a much higher equilibrium constant (larger cell potential) and thus will be driven to completion in preference to the oxidation of Fe^{2+}. Therefore, all the titration curve points up to (but not including) the equivalence point will be the same as those for the simple titration of Sn^{2+} with Ce^{4+}. That is,

$$E_{Pt} = E_{Sn} = E^0_{Sn} - \frac{0.059}{2} \log \frac{[Sn^{2+}]}{[Sn^4]}$$

After all the Sn^{2+} has been oxidized, further addition of Ce^{4+} will begin to oxidize Fe^{2+}, and the titration curve points can be calculated as in Example 11-4. The equivalence point potential will again be about $+1.236$ V vs. NHE, and all points past the second equivalence point will be given by the Ce^{4+}/Ce^{3+} couple

$$E_{Pt} = E_{Ce} = E_{Ce}^0 - \frac{0.059}{1} \log \frac{[Ce^{3+}]}{[Ce^{4+}]}$$

That leaves only one point in question, the potential at the first equivalence point.

If we assume that the potential will be the same as for the simple titration of Sn^{2+} with Ce^{4+}, $+0.669$ V vs. NHE as calculated in Example 11-3, an examination of the Fe^{3+}/Fe^{2+} couple shows the problem:

$$E_{Fe} = E_{Fe}^0 - \frac{0.059}{1} \log \frac{[Fe^{2+}]}{[Fe^{3+}]}$$

$$+0.669 = +0.771 - 0.059 \log \frac{[Fe^{2+}]}{[Fe^{3+}]}$$

$$\frac{[Fe^{2+}]}{[Fe^{3+}]} = 53.6$$

$$[Fe^{3+}] = 0.0187[Fe^{2+}]$$

In other words, almost 2% of the iron would be oxidized to Fe^{3+} at $+0.669$ V vs. NHE. What we want is the potential of a solution in which a stoichiometric amount of Ce^{4+} has been added equivalent to the Sn^{2+}. This situation can be thought of as a solution prepared by adding Fe^{2+}, Sn^{4+}, and Ce^{3+}; that is, none of the Fe^{2+} has been oxidized, all the tin has been oxidized to Sn^{4+}, and all the cerium has been reduced to Ce^{3+}. The equilibrium between Fe^{2+} and Sn^{4+} is shown in Equation 11-28:

$$Sn^{4+} + 2Fe^{2+} \rightleftharpoons Sn^{2+} + 2Fe^{3+} \tag{11-28}$$

But this is exactly the same situation that would prevail at the equivalence point if we were titrating Sn^{2+} with Fe^{3+}. Therefore, the potential can be calculated in the same way that the titration equivalence point potential was calculated in Example 11-3:

$$E_{eq\,pt} = \frac{2E_{Sn}^0 + E_{Fe}^0}{3} = \frac{2(+0.154) + 0.771}{3} = +0.360 \text{ V vs. NHE}$$

Thus the equivalence point potential for the first analyte will be the same as it would have been if it had been titrated with the second analyte (in its oxidized form for oxidant titrants). A satisfactory titration break at this point will occur, provided that the titration of the first analyte with the second analyte is feasible.

The titration curve for the titration of Sn^{2+} and Fe^{2+} with Ce^{4+} is shown in Figure 11-7 and is calculated in Example 11-10.

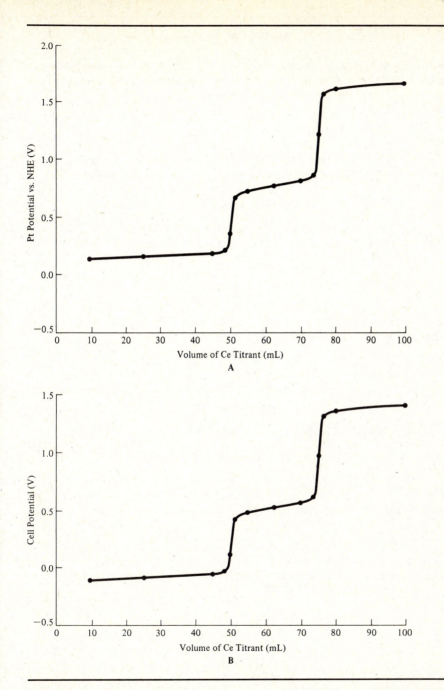

FIGURE 11-7 Titration curve for the titration with 0.10 F Ce^{4+} of a solution containing 25 mL of 0.10 F Fe^{2+} and 25 mL of 0.10 F Sn^{2+}. **(A)** E_{Pt} as a function of titrant volume. **(B)** E_{cell} as a function of titrant volume. Experimental points calculated in Example 11-10.

EXAMPLE 11-10

Construct the titration curve for the titration with $0.10\ F\ Ce^{4+}$ of a solution containing 25 mL of $0.10\ F\ Fe^{2+}$ and 25 mL of $0.10\ F\ Sn^{2+}$ by calculating the potential of the cell shown in Figure 11-4 after the addition of 0, 10, 25, 45, 49, 50, 51, 55, 62.5, 70, 74, 75, 76, 80, and 100 mL of titrant.

Solution

It is instructive to check where the equivalence point volumes will be for this titration before attempting to calculate solution potentials. First Sn^{2+} will be oxidized:

$$N_{Sn}V_{Sn} = N_{Ce}V_{eq\ pt,\ 1}$$

Note that in this example milliequivalents should be used because the n values are not all the same. For Sn (oxidized from $+2$ to $+4$), $n = 2$. Therefore,

$$0.20 \times 25 = 0.10 \times V_{eq\ pt,\ 1}$$

$$V_{eq\ pt,\ 1} = \frac{0.20 \times 25}{0.10} = 50\ mL$$

The amount of Ce^{4+} needed for the oxidation of Fe^{2+} is

$$N_{Fe}V_{Fe} = N_{Ce}V_{Ce}$$

$$0.10 \times 25 = 0.10 \times V_{Ce}$$

$$V_{Ce} = 25\ mL$$

The volume at the second equivalence point will then be

$$V_{eq\ pt,\ 2} = V_{eq\ pt,\ 1} + V_{Ce} = 50 + 25 = 75\ mL$$

0 mL

As in Example 11-5, the initial point is undefined.

$0 < V_{Ce} < V_{eq\ pt,\ 1}$

Points <50 mL will involve oxidation of Sn^{2+} and will be given by the Sn^{4+}/Sn^{2+} couple. Because 50 mL is required to oxidize all the Sn^{2+}, the amounts of Sn^{2+} and Sn^{4+} after 10 mL of $0.10\ F\ Ce^{4+}$ added are given by

$$\frac{[Sn^{2+}]}{[Sn^{4+}]} = \frac{50 - 10}{10} = 4$$

$$E_{Pt} = E_{Sn}^0 - \frac{0.059}{2} \log \frac{[Sn^{2+}]}{[Sn^{4+}]}$$

$$E_{Pt} = +0.154 - \frac{0.059}{2} \log 4 = +0.136\ V\ vs.\ NHE$$

$$E_{cell} = E_{Pt} - E_{SCE} = +0.136 - (+0.242) = -0.106\ V$$

EXAMPLE 11–10 (cont.)

The same approach is used for the addition of 25, 45, and 49 mL of titrant, and the numerical values are

mL	E_{Pt} vs. NHE (V)	E_{cell} (V)
25	+0.154	−0.088
45	+0.182	−0.060
49	+0.204	−0.038

50 mL (first equivalence point)

$$E_{Pt} = E_{eq\ pt,\ 1} = \frac{2E_{Sn}^0 + E_{Fe}^0}{3} = \frac{2(+0.154) + 0.771}{3} = +0.360 \text{ V vs. NHE}$$

$$E_{cell} = E_{Pt} - E_{SCE} = +0.360 - (+0.242) = +0.118 \text{ V}$$

$V_{eq\ pt,\ 1} < V_{Ce} < V_{eq\ pt,\ 2}$

Points that lie between 50 and 75 mL will be given by the Fe^{3+}/Fe^{2+} couple, as in Example 11-4. At 51 mL,

$$\frac{[Fe^{2+}]}{[Fe^{3+}]} = \frac{25 - 1}{1} = 24$$

$$E_{Pt} = +0.771 - \frac{0.059}{1} \log 24 = +0.690 \text{ V vs. NHE}$$

$$E_{cell} = E_{Pt} - E_{SCE} = +0.690 - (+0.242) = +0.448 \text{ V}$$

The same approach is used for the addition of 55, 62.5, 70, and 74 mL of titrant and corresponds to 5, 12.5, 20, and 24 mL in Example 11-4. The numerical values are

mL	E_{Pt} vs. NHE (V)	E_{cell} (V)
55	+0.736	+0.494
62.5	+0.771	+0.529
70	+0.807	+0.565
74	+0.852	+0.610

75 mL (second equivalence point)

This corresponds to 25 mL in Example 11-4 except that $[Ce^{3+}]$ formed during oxidation of Sn^{2+} increases the $[Ce^{3+}]/[Ce^{4+}]$ ratio by a factor of 3 and will decrease

Example 11-10 continued

EXAMPLE 11–10 (cont.)

equivalence point potential by $-(0.059 \log 3)/2 = -0.014$ V. Therefore,

$$E_{Pt} = +1.222 \text{ V vs. NHE} \qquad E_{cell} = +0.980 \text{ V}$$

$V_{Ce} > V_{eq\,pt,2}$

After the second equivalence point, the system is a Ce^{4+}/Ce^{3+} couple redox buffer and potentials are calculated as in Example 11-4. For example, 76 mL corresponds to 26 mL in Example 11-4 except that $[Ce^{3+}]$ formed during oxidation of Sn^{2+} increases the $[Ce^{3+}]/[Ce^{4+}]$ ratio by a factor of 3 and will decrease solution potential by $-0.059 \log 3 = -0.028$ V $\simeq -0.03$ V. Therefore,

$$E_{Pt} = +1.59 \text{ V vs. NHE} \qquad E_{cell} = 1.35 \text{ V}$$

The same approach is used for the addition of 80 and 100 mL of titrant and corresponds to 30 and 50 mL in Example 11-4. The numerical values are

mL	E_{Pt} vs. NHE (V)	E_{cell} (V)
80	$+1.63$	$+1.39$
100	$+1.67$	$+1.43$

The complete titration curve is plotted in Figure 11-7.

11-7 APPLICATIONS

With each reaction type so far, there has been a rather limited choice of titrants available, or perhaps it would be more accurate to say that only a few titrants find widespread use. Precipitation titrations most often use $AgNO_3$ as a titrant, or by addition of excess $AgNO_3$, back-titration with KSCN is used. Acid–base titrations generally employ a strong acid (for example, HCl, $HClO_4$, or H_2SO_4) or a strong base (for example, NaOH, $Ba(OH)_2$, or KOH) as a titrant. Rarely will it make any difference *which* strong acid or *which* strong base is used for a particular application; so in a sense there are only two widely used titrants: strong acid and strong base. And we saw in Chapter 10 that there are many complexing agents available but that by far the most common one in use today is EDTA. Even with this limited range of titrants, the number of analytes that can be determined is large and quite varied in scope.

When we enter the realm of redox titrations, a whole new ballgame opens up. There are several oxidants available as titrants as well as several reductants. Each of them has different characteristics. They all have their own E^0 value, which characterizes their thermodynamic ability to oxidize or reduce analytes. Beyond that, the rate of reaction may vary markedly from titrant to titrant for a particular analyte; and even the stoichiometry itself (n value for the analyte) may vary. The endpoint signal may depend on both the titrant and the analyte. Finally, the methods used to prepare, store, and standardize titrants are diverse.

With such diversity, it is easy to see why there are probably more redox titration methods available than for any other reaction type. Any analyte that can exhibit more than one oxidation state is a candidate for a redox titration. To limit our discussion, only a few of the many titrants in use will be explored after we present a few general comments.

In theory, an analyst could use a titrant to reduce an analyte as easily as a titrant to oxidize an analyte. In practice, reductants are less often used primarily because they are capable of reacting with oxygen in air. In addition, strong reducing agents, such as Ti(III), V(II), and Cr(II), are capable of reducing H_3O^+ in an aqueous solution so that, even with protection from air, they are not stable for long periods of time. Probably the most common reducing agent as a titrant is $Na_2S_2O_3$, which is almost exclusively utilized in the reduction of I_3^- (Equation 11-29).

$$2S_2O_3^{2-} + I_3^- \rightleftharpoons S_4O_6^{2-} + 3I^- \tag{11-29}$$

Another factor to be considered is the oxidation state of the analyte. Often during dissolution of the sample or separation from interferences, a mixture of oxidation states occurs. Sometimes the analyst is interested in knowing how much of each oxidation state is present; but more often he or she seeks the total amount of material present without regard to the oxidation state. For example, dissolution of an insecticide containing arsenic may give a mixture of As(III) and As(V); however, the analyst is required to know only the total arsenic content in the insecticide. In such cases, the analyst is to ensure that all the analyte is in one oxidation state. Reagents are available that can oxidize or reduce analytes to establish the existence of one oxidation state. Clearly, the reagents must be removable from the system, or else they may be titrated along with the analyte. Because most redox titrations employ oxidant titrants, it is generally necessary to reduce the analyte to one of its lower oxidation states before titration. The following subsection surveys some of the reagents used to adjust the analyte's oxidation state.

ADJUSTMENT OF ANALYTE'S OXIDATION STATE

It was stated above that most redox titrations employ oxidizing agents as titrants; therefore, adjustment of an analyte's oxidation state before titration is usually accomplished by reduction. It was also mentioned that the reagent used must be removable, or else it too could react with the titrant. One of the most effective removal ways is the use of a *metal reductor column*. The list of standard potentials in Appendix 3-A shows that many metals are strong reducing agents (low E^0 values), but the two most common are zinc (in the *Jones reductor*) and silver (in the *Walden reductor*).

The analyte solution is poured through a glass column filled with granules (or coarse powder) of the metal, as shown in Figure 11-8. The perforated plate and glass wool prevent particles of the metal from reaching the exit flask so that effective separation of metal from analyte is accomplished. For example, if a mixture of Fe^{2+} and Fe^{3+} is present, the zinc in a Jones reductor will reduce all the Fe^{3+} according to Equation 11-30.

$$2Fe^{3+} + Zn \rightleftharpoons 2Fe^{2+} + Zn^{2+} \qquad \text{(11-30)}$$

Zinc is a fairly strong reducing agent ($E^0 = -0.763$ V vs. NHE for Zn^{2+}/Zn couple), and because $E^0 < 0$, it is capable of reducing H_3O^+ (Equation 11-31):

$$2H_3O^+ + Zn \rightleftharpoons H_2 + 2H_2O + Zn^{2+} \qquad \text{(11-31)}$$

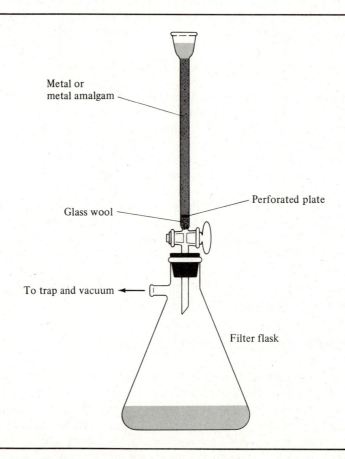

Metal or
metal amalgam

Perforated plate

Glass wool

To trap and vacuum

Filter flask

FIGURE 11-8   Metal reductor column.

Pouring an acidic analyte solution through a zinc metal column would result in copious amounts of Zn^{2+} and H_2 formed and possibly sufficient neutralization of acid to increase the pH to the point where metal hydroxides would precipitate. This disadvantage is prevented by amalgamating the zinc with mercury. Treatment of the zinc with $HgCl_2$ or elemental mercury in a mild acid forms a Zn(Hg) coating on the particles. The reaction with H_3O^+ to form H_2 becomes kinetically very slow and allows the use of zinc as a reducing agent without producing appreciable amounts of H_2.

As the simple Ag^+/Ag couple ($E^0 = +0.799$ V vs. NHE), silver is not an effective reducing agent, but its ability can be improved when the analyte solutions contain HCl. In this case the couple becomes

$$AgCl + e \rightleftharpoons Ag + Cl^- \qquad E^0 = +0.222 \text{ V vs. NHE}$$

The E^0 is now low enough to reduce several metal ions, as shown in Table 11-3. Because the potential of the Walden reductor is considerably higher than that of the Jones reductor, it is a weaker reducing agent and will not reduce so many analyte species. For example, Table 11-3 shows that neither Ti(IV) nor Cr(III) is reduced in a Walden reductor but that both are reduced to Ti(III) and Cr(II) in a Jones reductor. The final oxidation state may also be different. Vanadium (V) is reduced to V(IV) in the mild Walden reductor but is reduced to V(II) in the Jones reductor. These differences are sometimes useful in the analysis of mixtures. For example, a mixture of iron(III) and titanium(IV) can be treated with a Walden reductor to form Fe(II), which can then be titrated with Ce^{4+} to determine the iron content. Another portion of the sample can be treated with a Jones reductor to form Fe(II) and Ti(III), which

---

TABLE 11-3   Reductions with Metal Reductor Columns

Walden reductor $Ag(s) + Cl^- \rightleftharpoons AgCl(s) + e$	Jones reductor $Zn(s) \rightleftharpoons Zn^{2+} + 2e$
Cr^{3+} not reduced	$Cr^{3+} + e \rightleftharpoons Cr^{2+}$
$Cu^{2+} + e \rightleftharpoons Cu^+$ (in 2 F HCl)	Cu^{2+} reduced to metallic copper
$Fe^{3+} + e \rightleftharpoons Fe^{2+}$	$Fe^{3+} + e \rightleftharpoons Fe^{2+}$
$H_2MoO_4 + 2H_3O^+ + e \rightleftharpoons MoO_2^+ + 4H_2O$ (in 2 F HCl)	$H_2MoO_4 + 6H_3O^+ + 3e \rightleftharpoons Mo^{3+} + 10H_2O$
TiO^{2+} not reduced	$TiO^{2+} + 2H_3O^+ + e \rightleftharpoons Ti^{3+} + 3H_2O$
$UO_2^{2+} + 4H_3O^+ + 2e \rightleftharpoons U^{4+} + 6H_2O$	$UO_2^{2+} + 4H_3O^+ + 2e \rightleftharpoons U^{4+} + 6H_2O^*$
	$UO_2^{2+} + 4H_3O^+ + 3e \rightleftharpoons U^{3+} + 6H_2O^*$
$VO_2^+ + 2H_3O^+ + e \rightleftharpoons VO^{2+} + 3H_2O$	$VO_2^+ + 4H_3O^+ + 3e \rightleftharpoons V^{2+} + 6H_2O$

SOURCE: Adapted from I.M. Kolthoff and R. Belcher, *Volumetric Analysis*, vol. 3. p. 12 (New York: Interscience Publishers, Inc., 1957). With permission.

---

*A mixture of U(III) and U(IV) is obtained. The U(III) may be oxidized to U(IV) by bubbling in air for a few minutes.

can be titrated with Ce^{4+} to determine the total iron plus titanium. By difference, the amount of titanium can be obtained, as shown in Example 11-11.

Reducing agents other than metals are sometimes used to adjust the analyte's oxidation state. The gaseous reactants H_2S and SO_2 are mild reducing agents similar in potential to the Walden reductor. Their potentials in acidic solution are

$$S + 2H_3O^+ + 2e \rightleftharpoons H_2S + 2H_2O \qquad E^0 = +0.141 \text{ V vs. NHE}$$

$$SO_4^{2-} + 4H_3O^+ + 2e \rightleftharpoons SO_2 + 6H_2O \qquad E^0 = +0.17 \text{ V vs. NHE}$$

Excess reagent can be removed from the acidified analyte solution by boiling.

Adjustment of an analyte's oxidation state is occasionally carried out by oxidation. Oxidizing reagents are especially useful in oxidizing species with high E^0 values that cannot be easily titrated in their lower oxidation state. For example,

---

EXAMPLE 11-11

A sample containing both iron(III) and titanium(IV) was analyzed by dissolving a 1.75-g sample in acid and diluting to exactly 250 mL in a volumetric flask. A 50-mL aliquot of the solution was passed through a silver (Walden) reductor, and titration with 0.0750 N Ce^{4+} required 18.2 mL. Another 50-mL aliquot was then passed through a zinc (Jones) reductor, and titration with 0.0750 N Ce^{4+} required 46.2 mL. Calculate the percentages of Fe_2O_3 and TiO_2 in the sample.

Solution

According to Table 11-3, the Walden reductor reduces only Fe(III) to Fe(II) and does not reduce Ti(IV). Therefore, titration of the analyte solution after this treatment determines the iron content:

$$\text{meq } Fe_2O_3 = \text{meq Ce}$$

$$\frac{\text{mg } Fe_2O_3}{\text{FW } (Fe_2O_3)/n} = NV$$

Iron is oxidized from $+2$ to $+3$ in the titration so that $n = 1$ for each iron atom. Therefore, $n = 2$ for Fe_2O_3.

$$\frac{\text{mg } Fe_2O_3}{159.7/2} = 0.0750 \times 18.2 = 1.365 \text{ meq}$$

$$\text{mg } Fe_2O_3 = \frac{1.365 \times 159.7}{2} = 109 \text{ mg in 50-mL aliquot}$$

$$\text{mgFe}_2O_3 \text{ (sample)} = 109 \text{ mg (aliquot)} \times \frac{250 \text{ mL (sample)}}{50 \text{ mL (aliquot)}} = 545 \text{ mg (sample)}$$

$$\% \ Fe_2O_3 = \frac{545 \text{ mg } Fe_2O_3}{1750 \text{ mg sample}} \times 100 \ (\%) = 31.1\%$$

Mn^{2+} can be oxidized to MnO_4^- with sodium bismuthate (approximate formula is $NaBiO_3$). Sodium bismuthate is sufficiently insoluble so that any excess can be removed from the analyte solution by filtration.

Potassium peroxodisulfate $(K_2S_2O_8)$ and silver(II) oxide are two other powerful oxidizing agents, which can, along with sodium bismuthate, carry out the following oxidations:

$$Mn^{2+} \to MnO_4^- (+2 \to +7)$$

$$Ce^{3+} \to Ce^{4+} (+3 \to +4)$$

$$Cr^{3+} \to Cr_2O_7^{2-} (+3 \to +6)$$

$$VO^{2+} \to VO_2^+ (+4 \to +5)$$

EXAMPLE 11–11 (cont.)

According to Table 11-3, the Jones reductor reduces Fe(III) to Fe(II) and Ti(IV) to Ti(III). Therefore, titration of the analyte solution after this treatment determines the sum of iron plus titanium:

$$meq\ Fe_2O_3 + meq\ TiO_2 = meq\ Ce$$

$$meq\ Fe_2O_3 + \frac{mg\ TiO_2}{FW\ (TiO_2)/n} = NV$$

Titanium is oxidized from (III) to (IV) in the titration so that $n = 1$. The meq Fe_2O_3 is known to be 1.365 from the previous titration.

$$1.365 + \frac{mg\ TiO_2}{79.9/1} = 0.0750 \times 46.2 = 3.465$$

$$mg\ TiO_2 = (3.465 - 1.365)79.9 = 168\ mg\ in\ 50\text{-mL aliquot}$$

$$mg\ TiO_2\ (sample) = 168\ mg\ (aliquot) \times \frac{250\ mL\ (sample)}{50\ mL\ (aliquot)} = 840\ mg\ (sample)$$

$$\%\ TiO_2 = \frac{840\ mg\ TiO_2}{1750\ mg\ sample} \times 100\ (\%) = 48.0\%$$

Excess reagent is decomposed by water with heating, according to Equations 11-32 and 11-33:

$$2S_2O_8^{2-} + 6H_2O \rightleftharpoons 4SO_4^{2-} + O_2 + 4H_3O^+ \qquad \text{(11-32)}$$

$$4Ag^{2+} + 6H_2O \rightleftharpoons 4Ag^+ + O_2 + 4H_3O^+ \qquad \text{(11-33)}$$

Armed with these reagents to adjust the analyte's oxidation state prior to titration, the analyst must now consider which titrant to use.

POTASSIUM PERMANGANATE

Probably the most common redox titrant is the strong oxidant potassium permanganate, $KMnO_4$. One reason for its popularity is that, with its intense purple color, it provides its own endpoint signal. A drop of $0.1\ N\ KMnO_4$ will impart a distinctly visible pink color to 100 mL of H_2O. The only major obstacle to this endpoint signal is a colored analyte solution's obscuring the presence of excess MnO_4^-. In addition to its built-in endpoint signal, $KMnO_4$ has the advantage of ready availability, low cost, and ability to react with a large range of inorganic and organic analytes. However, against these advantages, $KMnO_4$ cannot be obtained sufficiently pure for determinate preparation of standard solutions, and therefore its solutions must be standardized with a primary standard. Also, its powerful oxidizing ability enables it to oxidize Cl^-, essentially all organic impurities that might be present, and even H_2O itself, according to Equation 11-34:

$$4MnO_4^- + 2H_2O \rightleftharpoons 4MnO_2 + 3O_2 + 4OH^- \qquad \text{(11-34)}$$

The cell potential for this reaction is $+0.187$ V in $1\ F\ OH^-$ (and even higher in neutral solutions), which indicates it should proceed spontaneously as written. Fortunately, the reaction is exceedingly slow, and $KMnO_4$ solutions can be stable for reasonable lengths of time. However, MnO_2 (and also many other substances) acts as a catalyst for this reaction; so it is important to ensure that the fresh solution of $KMnO_4$ is free from MnO_2. This is accomplished by heating the $KMnO_4$ solution when it is prepared to complete all reaction with traces of organic matter and then filtering to remove any MnO_2 before standardization.

Being in group VIIB of the periodic table, the highest oxidation state of manganese is $+7$, as in MnO_4^-, but it can exhibit several lower oxidation states. Three lower states of analytical importance are $+2$, $+4$, and $+6$. The most frequently used analytical reaction of MnO_4^- occurs in acid solution, given in Equation 11-35:

$$MnO_4^- + 8H_3O^+ + 5e \rightleftharpoons Mn^{2+} + 12H_2O \qquad E^0 = +1.51 \text{ V vs. NHE} \quad \text{(11-35)}$$

In this reaction, manganese is reduced from $+7$ to $+2$ so that $n = 5$.

In neutral solution, MnO_4^- is reduced to MnO_2 according to Equation 11-36:

$$MnO_4^- + 4H_3O^+ + 3e \rightleftharpoons MnO_2 + 6H_2O \qquad E^0 = +1.695 \text{ V vs. NHE} \quad \text{(11-36)}$$

In this reaction, manganese is reduced from $+7$ to $+4$ so that $n = 3$. The E^0 value is observed when all species are in their standard state and would include H_3O^+ in Equation 11-36. The formal potential in solutions of pH 7 would be only $+1.144$ V (calculation of a formal potential at pH 7 is illustrated later, in Example 11-14).

Finally, in strong-base solution, MnO_4^- is reduced to MnO_4^{2-} according to Equation 11-37:

$$MnO_4^- + e \rightleftharpoons MnO_4^{2-} \qquad E^0 = +0.564 \text{ V vs. NHE} \qquad \textbf{(11-37)}$$

In this reaction, manganese is reduced from $+7$ to $+6$ so that $n = 1$.

Equations 11-35 to 11-37 show that the strength of a given $KMnO_4$ solution in terms of normality (milliequivalents per milliliter) will depend on the reaction used. A 0.1 F $KMnO_4$ solution will be 0.5 N if reaction Equation 11-35 is employed, 0.3 N for reaction 11-36 and 0.1 N for reaction 11-37. Thus an analyst using a $KMnO_4$ solution prepared by someone else or standardized by a reaction different from the one used in the analysis must be careful in calculating the normality (as shown in Example 11-12).

Standardization of $KMnO_4$ is generally accomplished by using primary standard sodium oxalate or oxalic acid in an acidic solution. The reaction is

$$2MnO_4^- + 5H_2C_2O_4 + 6H_3O^+ \rightleftharpoons 2Mn^{2+} + 10CO_2(g) + 14H_2O \qquad \textbf{(11-38)}$$

This method has the advantage that all species are colorless, resulting in a sharp endpoint signal when excess MnO_4^- is present, but it suffers the disadvantage of being kinetically slow. The reaction is actually autocatalytic; that is, one of the products—Mn^{2+} in this case—acts as a catalyst. Thus the first addition of $KMnO_4$ reacts very slowly, but the rate increases as the concentration of Mn^{2+} increases. To improve the kinetics, the titration is carried out at 60 to 90°C. It has also been recommended (Fowler and Bright procedure) that 90 to 95% of the $KMnO_4$ should be added before heating the oxalate solution to prevent the possible air oxidation of $H_2C_2O_4$. Details are given in the laboratory manual.

Several other primary standards are available for the standardization of $KMnO_4$, including As_2O_3, iron wire, and potassium iodide. Ferrous ammonium sulfate does not meet primary standard requirements, but it is available in reasonably high purity for less exacting analytical work and is convenient to use.

Essentially all inorganic analytes are titrated in acidic medium, in which MnO_4^- is reduced to Mn^{2+} (Example 11-12b illustrates one exception to this rule), and several examples are given in Table 11-4.

One very useful application is the analysis of iron ores for iron content. Iron ores normally contain Fe_2O_3 or Fe_3O_4 and must be adjusted to the $+2$ state after dissolution in strong HCl. A metal reductor column can be used, but with the Jones reductor, there is the danger of reducing other components in the iron ore, such as titanium, vanadium, and chromium. With the milder Walden reductor, only vanadium will interfere. Another recommended reducing agent for iron ores is $SnCl_2$ and details are given in the laboratory manual.

A further useful and interesting application of $KMnO_4$ is the indirect determination of calcium (as in limestone). Alkali and alkaline-earth metals do not exhibit more than one oxidation state in solution; so redox methods are not normally

EXAMPLE 11-12

(a) A solution of $KMnO_4$ is standardized with primary standard sodium oxalate ($Na_2C_2O_4$, FW = 134.0). If a 282.0-mg sample of $Na_2C_2O_4$ requires 35.87 mL of $KMnO_4$ to reach the equivalence point, determine the normality of the $KMnO_4$ solution.

(b) If the $KMnO_4$ is now used in a Volhard determination for manganese,

$$2MnO_4^- + 3Mn^{2+} + 4OH^- \rightleftharpoons 5MnO_2(s) + 2H_2O$$

calculate the percentage of manganese in a mineral specimen if a 487.4-mg sample requires 45.73 mL of $KMnO_4$ to reach the equivalence point.

Solution

(a)
$$meq \ KMnO_4 = meq \ Na_2C_2O_4$$

$$NV = \frac{Na_2C_2O_4}{FW \ (Na_2C_2O_4)/n}$$

The reaction of $Na_2C_2O_4$ is

$$C_2O_4^{2-} \rightarrow 2CO_2 + 2e \qquad n = 2$$

$$N \times 35.87 = \frac{282.0}{134.0/2}$$

$$\text{normality of } KMnO_4 = 0.1173 \ N$$

(b)
$$meq \ Mn = meq \ KMnO_4$$

$$\frac{mg \ Mn}{FW \ (Mn)/n} = NV$$

In this reaction, MnO_4^- is reduced to MnO_2 so that $n = 3$. However, in the standardization, $n = 5$. Therefore, the $KMnO_4$ is only $\frac{3}{5}$ as strong, and $N = 0.1173 \times \frac{3}{5} = 0.07038 \ N$. The manganese analyte is oxidized from $+2$ to $+4$; therefore, $n = 2$.

$$\frac{mg \ Mn}{54.94/2} = 0.07038 \times 45.73$$

$$mg \ Mn = 88.41$$

$$\% \ Mn = \frac{88.41 \ mg \ Mn}{487.4 \ mg \ sample} \times 100 \ (\%) = 18.14\%$$

available. However, calcium can be precipitated quantitatively as calcium oxalate (CaC_2O_4); and after removal from the excess oxalate by filtration, the precipitate can be dissolved and titrated with $KMnO_4$ according to Equation 11-38. The 1:1 stoichiometry between oxalate and calcium allows calculation of the amount of calcium present from the determination of oxalate in the precipitate. Clearly, it is essential to devise experimental conditions that ensure this 1:1 stoichiometry. Formation of $Ca(OH)_2$ must be prevented by precipitation from an acidic solution (pH 4). Coprecipitation of sodium and magnesium oxalates can cause a positive error if their content is high in the sample. Under these conditions, a double precipitation will be required. Details are given in the laboratory manual.

Several organic compounds can be determined with $KMnO_4$ in basic solution (Equation 11-37). Unfortunately, organic compounds react slowly with $KMnO_4$, as we saw in the standardization reaction with oxalic acid, so that back-titration methods are generally recommended. For example, glycerol (glycerin) reacts according to Equation 11-39:

$$\begin{array}{ccc} \text{OH} & \text{OH} & \text{OH} \\ | & | & | \\ \end{array}$$
$$CH_2\!-\!CH\!-\!CH_2 + 14MnO_4^- + 20OH^- \rightleftharpoons 3CO_3^{2-} + 14MnO_4^{2-} + 14H_2O \qquad \text{(11-39)}$$

After this reaction is completed, the solution is acidified, and all higher oxidation states of manganese are determined. The difference between this value and the original amount of $KMnO_4$ represents glycerol content in the sample. A similar method can be employed for the determination of formic acid, which has the advantage that acetic acid, commonly present along with formic acid, does not interfere.

These techniques are useful for other organic acids, sugars, ethylene glycol (antifreeze), and methanol.

POTASSIUM DICHROMATE

Potassium dichromate ($K_2Cr_2O_7$) is a strong oxidizing agent, similar to $KMnO_4$, which exhibits only one important reduced oxidation state:

$$Cr_2O_7^{2-} + 14H_3O^+ + 6e \rightleftharpoons 2Cr^{3+} + 21H_2O \qquad E^0 = +1.33 \text{ V vs. NHE} \quad \text{(11-40)}$$

It is seen from Equation 11-40 that chromium is reduced from +6 to +3 so that $n = 3$ for each Cr. Because $K_2Cr_2O_7$ contains two chromium atoms, its n value is 6, and its equivalent weight equals the formula weight divided by 6.

The E^0 value of $+1.33$ V is a little less than that of $KMnO_4$ in acid solution, and this difference becomes even greater when formal potentials are considered. In 1 F HCl the E_f^0 for the $Cr_2O_7^{2-}/Cr^{3+}$ couple is only $+1.0$ V vs. NHE. This limits the usefulness of $Cr_2O_7^{2-}$ but, on the other hand, means that $Cr_2O_7^{2-}$ does not oxidize Cl^- and exhibits much greater stability in aqueous solution than $KMnO_4$ does.

Potassium dichromate can be obtained as a primary standard reagent; so standard solutions may be prepared determinately and stored for long periods of time.

TABLE 11-4   Applications of $KMnO_4$ in Acidic Media*

Analyte	Analyte half-reaction	Experimental conditions
As(III)	$HAsO_2 + 4H_2O \rightleftharpoons H_3AsO_4 + 2H_3O^+ + 2e$	$0.5\ F$ HCl with ICl catalyst
Br^-	$2Br^- \rightleftharpoons Br_2 + 2e$	boiling $2\ F\ H_2SO_4$ to expel Br_2
C_2H_5OH	$C_2H_5OH + 5H_2O \rightleftharpoons CH_3COOH + 4H_3O^+ + 4e$	add excess $KMnO_4$ in $0.5\ F\ H_2SO_4$; let stand 24 hr, and titrate excess $KMnO_4$
Fe(II)	$Fe^{2+} \rightleftharpoons Fe^{3+} + e$	reduce Fe(III) to Fe(II) in Jones reductor or with Sn(II); titrate in $1\ F\ H_2SO_4$ (preferred over HCl) with Zimmerman-Reinhardt reagent added.
$H_2C_2O_4$	$H_2C_2O_4 + 2H_2O \rightleftharpoons 2CO_2 + 2H_3O^+ + 2e$	$1\ F\ H_2SO_4$ at 70°C, or, if possible, add nearly equivalent amount of $KMnO_4$ at 25°C; then complete titration at 60°C
H_2O_2	$H_2O_2 + 2H_2O \rightleftharpoons O_2 + 2H_3O^+ + 2e$	$1\ F$ HCl or $1\ F\ H_2SO_4$; Mn(II) catalyst may be used
Metal oxalates [Bi(III), Ca(II), Cd(II), Co(II), Mg(II), Ni(II), Pb(II), rare earths(III), Sn(II), Zn(II)]	$H_2C_2O_4 + 2H_2O \rightleftharpoons 2CO_2 + 2H_3O^+ + 2e$	precipitate metal ion as oxalate, filter, wash, and dissolve in $1\ F\ H_2SO_4$, and titrate as for $H_2C_2O_4$
Mn(II)	$Mn^{2+} \rightleftharpoons Mn^{3+} + e$	$2\ F$ HCl plus 6 g NH_4F at 10°C; MnO_4^- reduced to Mn^{3+}
Mo(III)	$Mo^{3+} + 6H_2O \rightleftharpoons MoO_2^{2+} + 4H_3O^+ + 3e$	reduce Mo(VI) to Mo(III) in Jones reductor; add excess Fe^{3+}, and titrate Fe^{2+} formed in $1\ F\ H_2SO_4$ plus Zimmerman-Reinhardt reagent

TABLE 11-4 (cont.)

Analyte	Analyte half-reaction	Experimental conditions
NO_2^-	$NO_2^- + 3H_2O \rightleftharpoons NO_3^- + 2H_3O^+ + 2e$	add excess $KMnO_4$ in $0.25\ F\ H_2SO_4$ solution; titrate excess $KMnO_4$ iodometrically
Sb(III)	$SbCl_4^- + 2Cl^- \rightleftharpoons SbCl_6^- + 2e$	reduce Sb(V) to Sb(III) with SO_2; titrate in $2\ F$ HCl
Sn(II)	$SnCl_4^{2-} + 2Cl^- \rightleftharpoons SnCl_6^{2-} + 2e$	reduce Sn(IV) in $6\ F$ HCl with Bi reductor column at $45°C$; add excess Fe^{3+}, or titrate directly in absence of air
Ti(III)	$Ti^{3+} + 3H_2O \rightleftharpoons TiO^{2+} + 2H_3O^+ + e$	reduce Ti(IV) to Ti(III) in Jones reductor; add excess Fe^{3+}, or titrate directly in absence of air
Tl(I)	$Tl^+ \rightleftharpoons Tl^{3+} + 2e$	$1\ F$ HCl plus 3 g NaF; MnO_4^- reduced to Mn^{3+}
U(IV)	$U^{4+} + 6H_2O \rightleftharpoons UO_2^{2+} + 4H_3O^+ + 2e$	reduce U(VI) to mixture of U(III) and U(IV) in Jones reductor; aerate for 5 min to form U(IV) only; titrate in $1\ F\ H_2SO_4$
V(IV)	$VO^{2+} + 3H_2O \rightleftharpoons VO_2^+ + 2H_3O^+ + e$	reduce V(V) to V(IV) in Bi reductor column; titrate at $80°C$ in $1\ F\ H_2SO_4$; add Zimmerman-Reinhardt reagent if Cl^- present
W(V)	$WO_2^+ \rightleftharpoons WO_2^{2+} + e$	reduce W(VI) to W(V) in $1\ F\ H_3PO_4$ with Cd reductor column; titrate in absence of air

*Half-reaction for MnO_4^- in acidic solution is $MnO_4^- + 8H_3O^+ + 5e \rightleftharpoons Mn^{2+} + 12H_2O$ except where noted in experimental conditions.

Potassium dichromate is colored orange but is not sufficiently intense to provide its own endpoint signal, especially in the presence of the green Cr^{3+} ion, which would be present at the equivalence point. Redox indicators are normally used, especially barium diphenylamine sulfonate (Table 11-2).

One useful application of $K_2Cr_2O_7$ as a titrant is the analysis of iron. It was pointed out earlier that iron ores are usually dissolved in HCl, and the high concentration of Cl^- in the analyte solution represents a possible interference when $KMnO_4$ is the titrant. However, the formal potential difference between the $Cr_2O_7^{2-}/Cr^{3+}$ and Fe^{3+}/Fe^{2+} couples is not sufficiently great for a sharp endpoint signal and does not occur at the equivalence point when barium diphenylamine sulfonate redox indicator is used. The situation is improved when phosphoric acid (H_3PO_4) is added to the analyte solution—which lowers E_f^0 for the iron(III)/iron(II) couple. Several analytical procedures are available in which the analyte reacts with excess Fe^{2+} and the excess Fe^{2+} is back-titrated with $K_2Cr_2O_7$. In fact, this type of procedure can be used for the analysis of $Cr_2O_7^{2-}$ itself in electroplating baths. A question then arises: "Why not titrate the $Cr_2O_7^{2-}$ in the plating bath directly with a standard Fe^{2+} solution?" There are two main reasons why the direct titration is not normally used: Standard Fe^{2+} solutions are not air stable and must be standardized frequently; the redox indicator is not stable in a large excess of $Cr_2O_7^{2-}$ and is irreversibly oxidized (however, the endpoint is reversible with a small excess of $Cr_2O_7^{2-}$, as would be present just past the equivalence point when titrating with $K_2Cr_2O_7$).

Potassium dichromate has one other disadvantage that limits its usefulness: It reacts slowly with many analytes, especially organic compounds.

Potassium dichromate can also be a primary standard for standardizing solutions of reductants. In the case of sodium thiosulfate solutions, an excess of iodide is added to a known amount of $K_2Cr_2O_7$, which forms a known amount of I_3^-. The I_3^- is then titrated with sodium thiosulfate, a procedure that will be discussed later in this chapter.

CERIUM(IV)

The titration of Fe^{2+} with Ce^{4+} has earlier been discussed in detail from a theoretical standpoint. Cerium(IV) is a powerful oxidant of even higher potential in $HClO_4$ than is $K_2Cr_2O_7$ or $KMnO_4$. The formal potential for the $Ce(IV)/Ce(III)$ couple in HNO_3 or H_2SO_4 is considerably lower and points out the complexity of cerium solutions. Complexes form with both $Ce(III)$ and $Ce(IV)$, and activity coefficients for the highly charged Ce^{3+} and Ce^{4+} will be far removed from one in solutions of analytical interest. Thus the E^0 value in the previous discussion is really the E_f^0 for the $Ce(IV)/Ce(III)$ couple in 1 F $HClO_4$, which probably lies close to the true E^0 value when all species are at unit activity. Because most solutions will not contain the Ce^{4+} species, it is more accurate to use the term $Ce(IV)$, which indicates only the presence of the $+4$ oxidation state.

Cerium(IV) standard solutions can be prepared from several commercially available salts; and cerium(IV) ammonium nitrate [ammonium hexanitratocerate(IV)], with the formula $(NH_4)_2Ce(NO_3)_6$, is available as a primary standard.

Solutions must contain a high concentration of acid to prevent precipitation of hydrous ceric oxide. Sulfuric acid is usually used because the formal potential is significantly lower, leading to greater stability. Solutions of Ce(IV) in HNO_3 or $HClO_4$ will react with water photochemically; that is, Ce(IV) absorbs light and is more reactive in this excited state. Cerium(IV), like MnO_4^- oxidizes Cl^-, and solutions cannot be *stored* in the presence of HCl. Titration of analyte solutions containing some Cl^- can be carried out because the rate of reaction with Cl^- is slow.

Standardization of Ce(IV) solutions can be carried out with the same primary standards used for $KMnO_4$, namely, sodium oxalate, iron wire, and As_2O_3. The last-named standard is frequently recommended, but the reaction is so slow that a catalyst must be used; in HCl, ICl is an effective catalyst, while in H_2SO_4 solutions, OsO_4 is generally recommended. As with $KMnO_4$, ferrous ammonium sulfate (Mohr's salt) is a convenient reagent of near primary standard quality for standardization.

Cerium (IV) is yellow, while cerium(III) is colorless; but the color change at the equivalence point is not sharp enough to act as an endpoint signal. Redox indicators are generally used, with ferroin (see Table 11-2) especially popular. Other phenanthrolines are also recommended. These indicators have potentials of about +1.0 to +1.2 V vs. NHE.

In addition to inorganic analytes that may be oxidized, Ce(IV) solutions are quite useful for quantitative organic analysis. Excess Ce(IV) in 4 F $HClO_4$ may be added, and after reaction is completed, the excess Ce(IV) is back-titrated with a standard oxalic acid solution. A second method is to add excess Ce(IV) in 12 F H_2SO_4 and boil. The excess Ce(IV) is then back-titrated with oxalic acid or Fe(II). This second method causes many organic compounds to be completely oxidized to CO_2. For example, tartaric acid is oxidized in 12 F H_2SO_4 according to Equation 11-41,

$$\underset{\substack{OH \\ }}{\overset{O}{\diagdown}}C-\underset{\substack{| \\ OH}}{CH}-\underset{\substack{| \\ OH}}{CH}-C\underset{\substack{OH}}{\overset{O}{\diagup}} + 10Ce^{4+} + 12H_2O \rightleftharpoons 4CO_2 + 10Ce^{3+} + 10H_3O^+$$

$$(11\text{-}41)$$

while the milder treatment in 4 F $HClO_4$ of the first method reacts according to Equation 11-42:

$$\underset{\substack{HO \\ }}{\overset{O}{\diagdown}}C-\underset{\substack{| \\ OH}}{CH}-\underset{\substack{| \\ OH}}{CH}-C\underset{\substack{OH}}{\overset{O}{\diagup}} + 6Ce^{4+} + 8H_2O \rightleftharpoons$$

$$2CO_2 + 2HC\underset{\substack{OH}}{\overset{O}{\diagup}} + 6Ce^{3+} + 6H_3O^+$$

$$(11\text{-}42)$$

Thus the n value for tartaric acid is 10 if Equation 11-41 is used, while the n value is 6 if Equation 11-42 is used.

IODINE

Iodine is another common oxidizing agent for redox titrations, but it is considerably less oxidizing than $KMnO_4$, $K_2Cr_2O_7$, or Ce(IV). The half-cell potential for the I_2/I^- couple is

$$I_2(s) + 2e \rightleftharpoons 2I^- \qquad E^0 = +0.5345 \text{ V vs. NHE} \qquad \text{(11-43)}$$

Equation 11-43 uses solid iodine as the standard state because of the low solubility of iodine in H_2O (*around* 10^{-3} *F*). This low solubility would be an obstacle to its use in analysis, but it can be increased tremendously by the formation of triiodide ion, I_3^-,

$$I_2(aq) + I^- \rightleftharpoons I_3^- \qquad K_f = 7.1 \times 10^2 \qquad \text{(11-44)}$$

Thus, when we speak of iodine as a redox reagent, it usually means that the actual species involved is I_3^-. The potential for the I_3^-/I^- couple is just slightly higher than the potential given in Equation 11-43.

$$I_3^- + 2e \rightleftharpoons 3I^- \qquad E^0 = +0.536 \text{ V vs. NHE} \qquad \text{(11-45)}$$

Iodine(I_2) crystals can be obtained sufficiently pure for preparing standard solutions determinately. Solutions are prepared by adding a weighed quantity of I_2 to a fairly concentrated KI solution (to increase the rate of dissolution), followed by dilution to a known volume. A typical iodine titrant would contain 0.05 *F* iodine (0.1 *N* because $n = 2$, as given in Equation 11-43 or 11-45) and 0.25 *F* KI. Example 11-13 shows that the concentration of $I_2(aq)$ is quite small in such a solution.

However, countering this advantage is the fact that the vapor pressure of iodine is significant even when most of the iodine is present as I_3^- so that iodine solutions tend to become weaker with storage. Therefore iodine solutions must be restandardized frequently. Another problem encountered with storage of triiodide solutions is that the excess I^- is a reducing agent and capable of reacting with O_2 in air:

$$6I^- + O_2 + 4H_3O^+ \rightleftharpoons 2I_3^- + 6H_2O \qquad \text{(11-46)}$$

Equation 11-46 shows that the concentration of I_3^- will *increase* with storage in contact with air. This reaction is quite slow but is catalyzed by light so that iodine solutions should be kept in the dark. Because iodine solutions are not very stable, many of the possible titrations are now carried out electrochemically. I_3^- is generated at an anode as the titrant, and the electrical charge (coulombs) necessary to carry out this *coulometric titration* is measured instead of volume from a buret and titrant concentration. Coulometric titrations will be discussed in Chapter 16.

Iodine solutions can be standardized with arsenious oxide, As_2O_3, or with sodium thiosulfate. Arsenious oxide dissolves in base to form arsenite, as shown in Equation 11-47:

$$As_2O_3 + 4OH^- \rightleftharpoons 2HAsO_3^{2-} + H_2O \qquad \text{(11-47)}$$

EXAMPLE 11-13

Calculate the concentration of iodine, $I_2(aq)$, in an iodine titrant containing 0.05 F iodine and 0.25 F KI.

Solution

Equation 11-44 shows that, for each mole of I_2 added, 1 mol of I^- will be consumed in the formation of I_3^-. Therefore,

initial: 0.05   0.25   0

$$I_2 \; + \; I^- \; \rightleftharpoons \; I_3^-$$

final: x $0.20 + x$ $0.05 - x$

where x is the small amount of iodine remaining to maintain equilibrium.

$$K = 7.1 \times 10^2 = \frac{[I_3^-]}{[I_2][I^-]}$$

$[I_2]$ represents dissolved iodine as I_2, *provided* that its solubility is not exceeded (about 10^{-3} F in H_2O).

$$7.1 \times 10^2 = \frac{0.05 - x}{(x)(0.20 + x)}$$

With the reasonably large K value, we can assume as a first approximation that $x \ll 0.05$. Under these conditions,

$$7.1 \times 10^2 \cong \frac{0.05}{(x)(0.20)}$$

$$[I_2] = x = 3.5 \times 10^{-4}$$

This value is much smaller than 0.05 so that our assumption was justified. The value is also smaller than the solubility limit.

Arsenic in $HAsO_3^{2-}$ is in the $+3$ oxidation state and can be oxidized to $+5$ in neutral solution:

$$H_3AsO_3 + I_2 + 3H_2O \rightleftharpoons H_3AsO_4 + 2I^- + 2H_3O^+ \qquad \textbf{(11-48)}$$

The reason for using a neutral solution (pH 7 to 8, maintained with a CO_2/bicarbonate buffer) is that, in acid solution, H_3AsO_4 is a stronger oxidant than I_3^- is ($E^0 = +0.559$ V vs. NHE compared with $+0.536$ V for I_3^-). However, H_3O^+ is involved in the half-cell reaction

$$H_3AsO_4 + 2H_3O^+ + 2e \rightleftharpoons H_3AsO_3 + 3H_2O \qquad E^0 = +0.559 \text{ V vs. NHE} \quad \textbf{(11-49)}$$

so that a decrease in $[H_3O^+]$ will decrease the potential for this couple. At pH 7, the potential is below that of the I_3^-/I^- couple, as shown in Example 11-14. At pH's higher than 11 other problems are encountered with this reaction. First, I_3^- disproportionates into iodide and hypoiodous acid,

$$I_3^- + OH^- \rightleftharpoons 2I^- + HOI \qquad (11\text{-}50)$$

and, second, As(III) is easily oxidized by O_2 in air. Therefore, the standardization titration conditions are recommended to be pH 7 to 9.

EXAMPLE 11-14

Determine the formal potential of the H_3AsO_4/H_3AsO_3 couple at pH 7, and compare it with the I_3^-/I^- couple.

Solution

$$H_3AsO_4 + 2H_3O^+ + 2e \rightleftharpoons H_3AsO_3 + 3H_2O \qquad E^0 = +0.559 \text{ V vs. NHE}$$

The Nernst equation for this half-cell is

$$E = E^0 - \frac{0.059}{2} \log \frac{[H_3AsO_3]}{[H_3AsO_4][H_3O^+]^2}$$

The formal potential conditions at pH 7 are

$$[H_3AsO_3] = [H_3AsO_4] = 1$$
$$[H_3O^+] = 10^{-7}$$

$$E^0_{f,As} = +0.559 - \frac{0.059}{2} \log \frac{1}{1 \times (10^{-7})^2} = +0.146 \text{ V vs. NHE}$$

The potential of the I_3^-/I^- couple is not pH dependent. Therefore,

$$E^0_{f,I} = E^0_I = +0.536 \text{ V vs. NHE}$$

At pH 7, I_3^- is a stronger oxidant than H_3AsO_4 is, while at pH 0, H_3AsO_4 is the stronger oxidant.

The pH dependence can be seen more clearly by writing the Nernst equation in the form

$$E^0_{f,As} = E^0_{As} - \frac{0.059}{2} \log \frac{[H_3AsO_3]}{[H_3AsO_4]} - \frac{0.059}{2} \log \frac{1}{[H_3O^+]^2}$$

$$= +0.559 + \frac{0.059}{2} \log [H_3O^+]^2 = +0.559 - 0.059 \text{ pH}$$

Standardization with sodium thiosulfate introduces one of the few reductants that are commonly used for redox titrations. It is most frequently used for reduction of I_3^- according to Equation 11-51, which takes place in the pH range 0 to 7:

$$I_3^- + 2S_2O_3^{2-} \rightleftharpoons 3I^- + S_4O_6^{2-} \tag{11-51}$$

The product, tetrathionate ion, displays an unusual formal oxidation state for sulfur of $+2.5$. Therefore, each sulfur in $S_2O_3^{2-}$ increases in oxidation state from $+2$ to $+2.5$ so that $n = \frac{1}{2}$ for each sulfur, and $n = 1$ for $S_2O_3^{2-}$. Most oxidizing agents oxidize $S_2O_3^{2-}$ to sulfite ($+4$ oxidation state) or sulfate ($+6$ oxidation state). Even I_3^- oxidizes $S_2O_3^{2-}$ to SO_4^{2-} in slightly basic solution (pH 8 to 9). At higher pH's the disproportionation reaction (Equation 11-50) complicates the situation.

Iodine solutions may also provide their own endpoint signal because the I_3^- is sufficiently colored to be observed when a slight excess exists. However, this color is of marginal intensity and is not adequate for colored analyte solutions. More often, starch is added to the analyte solution so that excess I_3^- will form a starch–I_3^- complex, which has an intense blue color perceptible at concentrations as low as $2 \times 10^{-7} F$. For back-titrations in which I_3^- is being titrated with sodium thiosulfate, the starch should not be added until near the equivalence point (yellow color of I_3^- nearly gone) because starch is decomposed and may also start to coagulate in the presence of I_3^-. Details on the preparation and preservation of starch solutions are given in the laboratory manual. With a standard solution of I_3^-, several inorganic as well as organic analytes can be titrated; many examples can be found in the supplementary readings to this chapter.

It was mentioned above that sodium thiosulfate is a common reductant for use in reducing I_3^-. In addition to its use for standardization of I_3^- solutions, sodium thiosulfate finds widespread use for indirect iodine methods (*iodometric methods*). These involve an analyte that oxidizes I^- to I_3^-; then the I_3^- formed is titrated with $S_2O_3^{2-}$. The direct titration of the analyte with $S_2O_3^{2-}$ is usually not possible because the stoichiometry may not be exact, the reaction is slow, or no convenient endpoint signal is available.

Sodium thiosulfate is available sufficiently pure as $Na_2S_2O_3 \cdot 5H_2O$ for the determinate preparation of standard solutions. However, solutions are not stable for long periods of time, and standardization is usually carried out with $K_2Cr_2O_7$ or KIO_3 (both of which react with I^- to produce I_3^-) as primary standards.

The stability of thiosulfate solutions has been the object of many investigations. Factors that play a role include pH, bacteria, light, and traces of heavy-metal impurities. In acid solutions, thiosulfate disproportionates to elemental sulfur and bisulfite:

$$S_2O_3^{2-} + H_3O^+ \rightleftharpoons S + HSO_3^- + H_2O \tag{11-52}$$

The HSO_3^- formed is also oxidized by I_3^-, but the product is sulfate (bisulfate),

$$HSO_3^- + I_3^- + 3H_2O \rightleftharpoons HSO_4^- + 3I^- + 2H_3O^+ \tag{11-53}$$

which means $n = 2$ for any thiosulfate decomposed. Thus the normality would show an increase with time. But titrations of I_3^- with $S_2O_3^{2-}$ can be carried out in acidic solution because the reaction between I_3^- and $S_2O_3^{2-}$ is much faster than is the disproportionation reaction shown in Equation 11-52.

In more neutral solutions, the primary cause of thiosulfate decomposition is the presence of sulfur-metabolizing bacteria, which convert $S_2O_3^{2-}$ to sulfur, sulfite, and sulfate. To avoid this, sodium thiosulfate is dissolved in boiled water, and substances that inhibit bacterial growth, such as chloroform or HgI_2, are added. Sunlight and trace impurities, such as Cu(II), also catalyze the decomposition of $S_2O_3^{2-}$. However, sodium thiosulfate solutions prepared under sterile conditions and kept at pH 9 or 10 exhibit great stability.

Iodometric methods find widespread application. The analysis for copper was already mentioned (Equation 11-27), in which Cu(II) is reduced to Cu(I) by I^- with the formation of I_3^-. The I_3^- is titrated with $Na_2S_2O_3$ to determine the amount of copper analyte present in the sample (Example 11-6). Two other applications are of importance in the environmental field: the determination of *dissolved oxygen* (usually reported as "D.O.") in water and the determination of ozone (O_3) in air.

The iodometric method for determining dissolved oxygen is called the "Winkler method." In essence, O_2 is reduced to H_2O, while $S_2O_3^{2-}$ is oxidized to $S_4O_6^{2-}$, but the direct titration is not feasible because of slow kinetics. In the Winkler method, an excess of Mn(II) is added to the water sample plus base to carry out Equation 11-54:

$$4Mn(OH)_2(s) + O_2 + 2H_2O \rightleftharpoons 4Mn(OH)_3(s) \qquad \textbf{(11-54)}$$

An excess of iodide is also present so that, when the mixture is acidified, Equation 11-55 takes place:

$$2Mn(OH)_3(s) + 3I^- + 6H_3O^+ \rightleftharpoons I_3^- + 12H_2O + 2Mn^{2+} \qquad \textbf{(11-55)}$$

The resulting I_3^- is then titrated with $Na_2S_2O_3$. Because O_2 has been reduced from oxidation state zero to -2 in H_2O, the n value for each oxygen atom is 2; so $n = 4$ for O_2.

There are many methods for the determination of ozone in air, and certainly in field locations some type of instrumental method that gives a continuous record of the ozone level is preferred over a manual titration method. However, the titration method is a reference method against which other methods are checked and calibrated. Thus an air-monitoring instrument could be calibrated by using a set-up as shown in Figure 11-9. The reaction of ozone with iodide is given in Equation 11-56:

$$3I^- + O_3 + H_2O \rightleftharpoons I_3^- + O_2 + 2OH^- \qquad \textbf{(11-56)}$$

The resulting I_3^- is then titrated with $Na_2S_2O_3$. Each O_3 yields one I_3^- so that $n = 2$.

This method has been used for many years by the Los Angeles Air Pollution Control District and, in more recent years, by numerous other agencies that monitor air quality. The Environmental Protection Agency (EPA) decided that, since OH^- is formed in the reaction (Equation 11-56), it would be better to substitute a buffered iodide solution for the unbuffered one of the Los Angeles District, and they used an

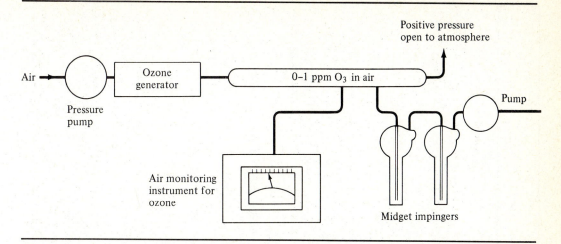

FIGURE 11-9 Calibration by the iodometric method of an air-monitoring instrument for ozone.

instrumental endpoint instead of the visual one. It gradually became apparent that the unbuffered method gave results 30% lower than those from the new buffered method. A committee reviewed the problem in 1974, checking samples by both methods and by two UV light analyzers as a cross check. Although the buffered method was accurate in dry air, it produced errors of $+25\%$ in normally humid air samples; the old unbuffered method produced errors of only -5%. So much for progress! Because of these uncertainties, the reference method has been changed to the UV technique, but the iodometric method is still in use because of its convenience.

It should be noted that the iodometric method really measures total oxidant in air because any oxidant that can oxidize I^- to I_3^- will be an interference to the ozone determination. For example, NO_2 normally present in polluted air is reduced to NO, but the reaction is slow. On the other hand, SO_2 acts as a reducing agent and will reduce I_3^- back to I^-, causing a negative interference. Such interferences must be considered if the analytical results are to be meaningful in terms of O_3 concentration in the air.

Several examples of indirect iodine methods are given in Table 11-5.

POTASSIUM IODATE

There are many other oxidants that find use as titrants in redox reactions. Potassium iodate (KIO_3) is available as a primary standard material, and its solutions are stable for long periods of time. It oxidizes many analytes while it is reduced to iodine. An interesting aspect of this reagent is that, after iodine is formed in strong HCl (3 to 9 F), continued addition of KIO_3 oxidizes the iodine to the $+1$ oxidation state as the species ICl_2^-. The endpoint signal is taken as the disappearance of all I_2, and

TABLE 11-5   Examples of Indirect Iodine (Iodometric) Methods

Analyte	Analyte reaction with I^-	n	Experimental conditions
As(V)	$H_3AsO_4 + 3I^- + 2H_3O^+ \rightleftharpoons H_3AsO_3 + I_3^- + 3H_2O$	2	6 F HCl; exclude air with $NaHCO_3$; stand 5 min
Au(III)	$AuCl_3 + 4I^- \rightleftharpoons AuI + I_3^- + 3Cl^-$	2	neutral solution
Br_2	$Br_2 + 3I^- \rightleftharpoons 2Br^- + I_3^-$	2	dilute acid
BrO_3^-	$BrO_3^- + 9I^- + 6H_3O^+ \rightleftharpoons Br^- + 3I_3^- + 9H_2O$	6	0.5 F HCl or H_2SO_4
Ce(IV)	$2Ce^{4+} + 3I^- \rightleftharpoons 2Ce^{3+} + I_3^-$	1	1 F H_2SO_4
Cl_2	$Cl_2 + 3I^- \rightleftharpoons 2Cl^- + I_3^-$	2	dilute acid
Co(II)	$2Co(OH)_3 + 3I^- + 6H_3O^+ \rightleftharpoons 2Co^{2+} + I_3^- + 12H_2O$	1	oxidize Co(II) with sodium perborate in basic solution; boil 10 min, excluding air; acidify with H_2SO_4
$Cr_2O_7^{2-}$	$Cr_2O_7^{2-} + 9I^- + 14H_3O^+ \rightleftharpoons 2Cr^{3+} + 3I_3^- + 21H_2O$	6	0.3 F HCl; keep 6 min in the dark before titrating
Cu(II)	$2Cu^{2+} + 5I^- \rightleftharpoons Cu_2I_2 + I_3^-$	1	dilute acetic acid; add NH_4SCN near endpoint
Fe(III)	$2Fe^{3+} + 3I^- \rightleftharpoons 2Fe^{2+} + I_3^-$	1	0.1 F HCl; stand 10 min
$Fe(CN)_6^{3-}$	$2Fe(CN)_6^{3-} + 3I^- \rightleftharpoons 2Fe(CN)_6^{4-} + I_3^-$	1	2 F HCl
H_2O_2	$H_2O_2 + 3I^- + 2H_3O^+ \rightleftharpoons I_3^- + 4H_2O$	2	0.5 F H_2SO_4 with $(NH_4)_2MoO_4$ catalyst
IO_3^-	$IO_3^- + 8I^- + 6H_3O^+ \rightleftharpoons 3I_3^- + 9H_2O$	6	0.3 F HCl
IO_4^-	$IO_4^- + 3I^- + H_2O \rightleftharpoons IO_3^- + I_3^- + 2OH^-$	2	borate buffer (pH 8–9)
Metal chromates [Ba(II), Pb(II), Sr(II), Tl(I)]	$Cr_2O_7^{2-} + 9I^- + 14H_3O^+ \rightleftharpoons 2Cr^{3+} + 3I_3^- + 21H_2O$	$\dfrac{3z*}{2}$	precipitate metal chromate in neutral solution; wash; dissolve in 1 F HCl
MnO_2	$MnO_2 + 3I^- + 4H_3O^+ \rightleftharpoons Mn^{2+} + I_3^- + 6H_2O$	2	1 F H_3PO_4; exclude air with $NaHCO_3$
MnO_4^-	$2MnO_4^- + 15I^- + 16H_3O^+ \rightleftharpoons 2Mn^{2+} + 5I_3^- + 24H_2O$	5	0.1 F HCl or H_2SO_4
O_2	$O_2 + 4Mn(OH)_2 + 2H_2O \rightleftharpoons 4Mn(OH)_3$; $4Mn(OH)_3 + 6I^- + 12H_3O^+ \rightleftharpoons 4Mn^{2+} + 2I_3^- + 24H_2O$	4	reduce O_2 with Mn^{2+} in basic KI solution; wait 1 min; acidify with H_2SO_4
O_3	$O_3 + 3I^- + H_2O \rightleftharpoons O_2 + I_3^- + 2OH^-$	2	absorb O_3 in neutral KI solution; acidify with 1 F H_2SO_4
OCl^-	$OCl^- + 3I^- + 2H_3O^+ \rightleftharpoons Cl^- + I_3^- + 3H_2O$	2	0.5 F H_2SO_4
Sb(V)	$SbCl_6^- + 3I^- \rightleftharpoons SbCl_4^- + I_3^- + 2Cl^-$	2	5 F HCl

*z is cation charge of the metal ion.

therefore the net reaction has been the reduction of KIO_3 (in which iodine is in the $+5$ oxidation state) to ICl_2^-. Thus the n value for KIO_3 is 4. Starch cannot be used as an endpoint signal in the strong acid solution, but an extraction method can be used. Iodine has high solubility and intense color in many organic solvents immiscible with water, such as CCl_4. If a small amount of CCl_4 is added to the system, the presence of iodine in the CCl_4 layer can be seen by its violet color. After each addition of titrant, the two layers are shaken vigorously in a stoppered flask, and the endpoint signal is the disappearance of the violet color in the CCl_4 layer.

PERIODIC ACID

We have seen iodine used analytically in the following oxidation states: -1 (I^-), 0 (I_2), $+1$ (ICl_2^-), and $+5$ (KIO_3). A fifth oxidation state, $+7$, is also used. Periodic acid, H_5IO_6, is an oxidant useful for several highly selective reactions with organic functional groups. Solutions of H_5IO_6 are standardized by reaction with excess I^- to form I_3^-:

$$H_5IO_6 + 3I^- \rightleftharpoons IO_3^- + I_3^- + OH^- + 2H_2O \qquad \text{(11-57)}$$

The I_3^- formed is then titrated with thiosulfate or arsenite. Periodic acid will attack organic compounds that have functional groups (alcohol, aldehyde, or ketone) on *adjacent* carbon atoms. The carbon–carbon bond between these groups is ruptured, and oxidation occurs. A hydroxyl group is oxidized to an aldehyde or ketone, and a carbonyl group is oxidized to a carboxylic acid. The reaction is fairly slow; so a direct titration is not possible. Excess H_5IO_6 is added, and after reaction is complete, the excess H_5IO_6 is determined as in the standardization reaction. One simple example is the analysis of antifreeze for ethylene glycol content:

$$\begin{array}{c} \text{OH} \quad \text{OH} \\ | \qquad | \\ CH_2\!-\!CH_2 + H_5IO_6 \end{array} \rightleftharpoons 2HC\!\!\begin{array}{c}\diagup O \\ \diagdown H\end{array} + IO_3^- + H_3O^+ + 2H_2O \qquad \text{(11-58)}$$

Because one H_5IO_6 is consumed by each ethylene glycol in Equation 11-58 and each H_5IO_6 produces two iodine atoms as I_3^- in Equation 11-57, the n value for ethylene glycol is 2. The same answer is obtained by assigning formal oxidation-state numbers. Carbon is -1 in ethylene glycol, while it is 0 in formaldehyde (HCHO). Therefore, $n = 1$ for each carbon, and $n = 2$ for ethylene glycol because it contains two carbon atoms.

POTASSIUM BROMATE

Potassium bromate is another oxidant frequent in carrying out organic reactions. Direct titrations are rarely used, and most often bromine is produced by adding excess Br^- to $KBrO_3$ in an acid solution:

$$BrO_3^- + 5Br^- + 6H_3O^+ \rightleftharpoons 3Br_2(aq) + 9H_2O \qquad \text{(11-59)}$$

Thus each BrO_3^- produces $3Br_2$ so that $n = 6$ for $KBrO_3$. Bromine reacts with many organic compounds by addition (ethylene for example),

$$(11-60)$$

or by substitution (phenol for example),

$$(11-61)$$

Equation 11-61 illustrates a useful reaction for environmental analysis. Phenol is frequently encountered in industrial waste streams and can be analyzed by bromination reaction. Bromine is added, and after reaction is complete (about $\frac{1}{2}$ hr), the excess bromine is back-titrated with arsenious acid (H_3AsO_3). Alternatively, I^- can be added, which forms I_3^-, and this may be titrated with $S_2O_3^{2-}$. Because three bromine molecules (Br_2) are consumed by phenol, $n = 6$.

REDUCTANTS AS TITRANTS

It was noted earlier in this chapter that most redox titrations involve oxidants because reductants tend to be air oxidized. The only exception to this rule is sodium thiosulfate, discussed along with iodometric methods. However, a few reductants are used on occasion, as shown in Table 11-6. They have been listed in order of *increasing* power as reductant, that is, decreasing E^0 values. The last three are capable of reducing H_2O to H_2.

TABLE 11-6   Reductants as Titrants

Titrant	Half-cell reaction	E^0 vs. NHE (V)
Fe(II)	$Fe^{3+} + e \rightleftharpoons Fe^{2+}$	$+0.771$
ascorbic acid	$C_6H_6O_6 + 2H_3O^+ + 2e \rightleftharpoons C_6H_8O_6 + 2H_2O$	$\sim +0.3$
Sn(II)	$Sn^{4+} + 2e \rightleftharpoons Sn^{2+}$	$+0.154$
Ti(III)	$TiO^{2+} + 2H_3O^+ + e \rightleftharpoons Ti^{3+} + 3H_2O$	$+0.10$
V(II)	$V^{3+} + e \rightleftharpoons V^{2+}$	-0.255
Cr(II)	$Cr^{3+} + e \rightleftharpoons Cr^{2+}$	-0.41

increasing reducing power →

The most stable, because it is the weakest reductant, is Fe^{2+}. Solutions can be prepared in 1 F H_2SO_4, using Mohr's salt, $Fe(NH_4)_2(SO_4)_2 \cdot 6H_2O$, which is available at a purity approaching primary standard. As a solid reagent, it is resistant to air oxidation. However, in neutral or in strong acid, the Fe(II) reacts quickly with O_2. In a dilute acid such as 1 F H_2SO_4, Fe(II) is reasonably stable although, for highest accuracy, the titrant solution should be restandardized against primary standard $K_2Cr_2O_7$ each time (or day) it is used.

One interesting analysis that can be accomplished with Fe^{2+} is the titration of gold in solution as Au^{3+}. Gold(III) is reduced to elemental gold in the titration, so the n value is 3.

The problems of instability can be circumvented by generating the reductant at an electrode that is the basis for coulometric titrations (Chapter 16). Many of the reactions carried out coulometrically may also be accomplished with a classical redox titration, using a standard solution of the reductant.

SUPPLEMENTARY READING

Berka, A., J. Vulterin, and J. Zýka, *Newer Redox Titrants*. Elmsford, N.Y.: Pergamon Press, Inc., 1965.

Davis, D.G., "Potentiometric Titrations," in *Standard Methods of Chemical Analysis*, 6th ed. F.J. Welcher, ed., New York: D. Van Nostrand Company, 1966.

Latimer, W.M., *Oxidation Potentials*, 2nd ed., Englewood Cliffs, N.J.: Prentice-Hall, Inc., 1952.

PROBLEMS

11-1. How many coulombs of charge are required to carry out the following reactions?
  *a. Deposit 63.55 g of Cu from a solution containing $CuSO_4$
  b. Deposit 163 mg of Pb from a solution containing $Pb(NO_3)_2$
  c. Completely reduce 50 mL of a 0.106 F Fe^{3+} solution to Fe^{2+}

11-2. Balance the following redox reactions occurring in acid solution; add H_3O^+ and/or H_2O if necessary:
  *a. $UO_2^{2+} + Ti^{3+} \rightleftharpoons U^{4+} + TiO^{2+}$
  b. $MnO_4^- + Fe^{2+} \rightleftharpoons Mn^{2+} + Fe^{3+}$
  c. $Cr_2O_7^{2-} + Sn^{2+} \rightleftharpoons Sn^{4+} + Cr^{3+}$
  d. $I^- + H_2O_2 \rightleftharpoons I_3^-$
  e. $I^- + H_5IO_6 \rightleftharpoons I_3^-$
  f. $I^- + Cr_2O_7^{2-} \rightleftharpoons I_3^- + Cr^{3+}$
  g. $UO_2^{2+} + Zn \rightleftharpoons U^{3+} + Zn^{2+}$

*11-3. For the reactions given in Problem 11-2, state which species is the oxidant and which is the reductant.

*11-4. For the reactions given in Problem 11-2, calculate the coulombs of charge necessary to carry out 1 mol of reaction.

---

*Answers to problems marked with an asterisk will be found at the back of the book.

*11-5. For the reactions given in Problem 11-2, and using E^0 values given in Appendix 3-A, calculate the electrical energy produced in carrying out 1 mol of reaction.

*11-6. Write the Nernst equation for the reactions given in Problem 11-2.

*11-7. Using E^0 values given in Appendix 3-A, calculate the equilibrium constants for the reactions given in Problem 11-2.

11-8. A hydrogen–oxygen fuel cell may be written as

$$H_2, M|KOH|M, O_2$$

where M represents the electrode material.
a. Write the half-cell reactions and the total cell reaction.
b. If the gases are at 1 atm pressure, calculate the theoretical potential of the fuel cell from the free energy of formation of $H_2O = -237$ kJ/mol.

*11-9. Suppose the fuel cell described in Problem 11-8 operated under the following conditions: $H_2(p = 4.2$ atm), $O_2(p = 5$ atm), and KOH (3 F); calculate the theoretical potential.

11-10. Analysis for oxygen pressure in exhaust gases (say, from an automobile) can be carried out using an oxygen monitor:

$$O_2(\text{exhaust}), Pt|ZrO_2\text{–}CaO|Pt, O_2(0.2 \text{ atm, air})$$

where ZrO_2–CaO is a calcia stabilized zirconia solid electrolyte. If the monitor gave a reading of $+0.350$ V at a temperature of 850°C ($2.303 \, RT/F = 0.223$, not 0.059!), calculate the O_2 pressure in the exhaust. (*Hint:* Don't forget to determine the n value for the reduction of O_2 to O^{2-}.)

11-11. A platinum indicator electrode is immersed in the following solutions. Calculate the solution (indicator electrode) potential on the normal hydrogen electrode (NHE) scale.
  *a. Fe^{2+} (0.25 F), Fe^{3+} (0.05 F)
  b. Ce^{3+} (0.01 F), Ce^{4+} (0.05 F) in 1 F H_2SO_4
  c. pH 4.3 and saturated with H_2 at 1 atm
  d. Cr^{3+} (0.2 F), $Cr_2O_7^{2-}$ (10^{-3} F), pH 1.00
  e. Mn^{2+} (10^{-5} F), MnO_4^- (0.1 F), pH 2.30
  f. U^{+4} (0.035 F), UO_2^{2+} (0.055 F), 2×10^{-3} F H_3O^+

*11-12. For the solutions given in Problem 11-11, calculate the indicator electrode potential on the saturated calomel electrode (SCE) scale. This is equivalent to: SCE || solution | Pt.

11-13. a. Which of the solutions in Problem 11-11 is the most oxidizing?
  b. Which of the solutions in Problem 11-11 is the most reducing?

11-14. Ozone is sometimes used for water purification because it is a strong oxidant and its reaction product is nontoxic O_2. Calculate the potential of a solution containing O_3 ($p = 10^{-3}$ atm), saturated with air ($O_2 = 0.2$ atm), and at pH 5.3.

11-15. Fe^{2+} is a mild reducing agent, reasonably stable in air, and used as a titrant. Which of the following analytes could be titrated with Fe^{2+}, that is, show a large enough titration break for accurate endpoint detection?
  *a. Au^{3+}
  b. I_3^-
  c. MnO_4^- (pH 0)
  d. Ag^+
  e. Pd^{2+}
  f. Br_2 (aq)

*11-16. Using E^0 for the Ag^+/Ag couple and E^0 for the $AgBr/Ag$ couple, show how the solubility product, K_{sp}, for AgBr can be calculated from the schematic cell: $Ag|Ag^+||Br^-, AgBr|Ag$.

11-17. Using E^0 for the Hg_2^{2+}/Hg couple, and E^0 for the Hg_2Cl_2/Hg couple, show how the solubility product, K_{sp}, for Hg_2Cl_2 can be calculated from the schematic cell: $Hg|Hg_2^{2+}||Cl^-, Hg_2Cl_2|Hg$.

11-18. Calculate the equivalence point potential for the following titrations:
   *a. V^{2+} with Ce^{4+} (V^{2+} oxidized to V^{3+})
   b. VO^{2+} with Ti^{3+} (pH 0)
   c. Sn^{2+} with MnO_4^- (pH 0)
   d. Ti^{3+} with Fe^{3+} (pH 0)
   e. Fe^{2+} with MnO_4^- (pH 1)

11-19. Recommend a redox indicator for each of the titrations given in Problem 11-18.

*11-20. Derive a titration curve for each of the titrations given in Problem 11-18 if the analyte solution contains 50 mL of 0.1 N analyte and the titrant concentration is 0.1 N. Calculate the solution potential after the addition of 10, 25, 40, 48, 49, 50, 51, 52, 60, and 100 mL of titrant.

11-21. Derive a titration curve for each of the titrations given in Problem 11-18, using a saturated calomel reference (SCE) and the same conditions given in Problem 11-20.

*11-22. Calculate the endpoint error in the titration of Ti^{3+} with Fe^{3+} in 1 F H_2SO_4 if 1-naphthol-2-sulfonic acid indophenol ($E_{ind}^0 = +0.54$ V) is used as the redox indicator.

11-23. A cerium(IV) solution was standardized by primary standard As_2O_3. A 212.4-mg sample of As_2O_3 is dissolved in NaOH and then acidified with H_2SO_4; and with OsO_4 as a catalyst, 45.79 mL was required to reach the endpoint. Calculate the normality of the cerium(IV) solution.

11-24. A commercial hydrogen peroxide solution is analyzed by titration with cerium(IV). A 2.00-mL sample is diluted to 50.0 mL, and a 10.00-mL aliquot is titrated in acid solution. If 43.15 mL of 0.1789 N Ce(IV) is required to reach the endpoint, calculate the strength of the peroxide as weight per volume percent.

*11-25. A 451.0-mg sample containing tartaric acid is oxidized with Ce (IV) in 12 F H_2SO_4, which converts tartaric acid to CO_2, as shown in Equation 11-41. If 50.0 mL of 0.1504 N Ce(IV) is used for the oxidation and the excess Ce(IV) requires 10.17 mL of 0.1006 N Fe(II) to reach the endpoint, calculate the percentage of tartaric acid in the sample.

11-26. An insecticide sample weighing 1.063 g is digested in acid, arsenic reduced to As(III), buffered to pH 7, and titrated with I_3^-. If 26.45 mL of 0.0564 N I_3^- is required to reach the starch–iodine endpoint, calculate percentage of As_2O_3 in the insecticide.

11-27. An iron-ore sample weighing 597 mg is dissolved in HCl, reduced with $SnCl_2$, and titrated with $KMnO_4$. If 47.35 mL of 0.1075N $KMnO_4$ is required to reach the endpoint, calculate percentage of Fe_2O_3 in the iron ore.

*11-28. A $KMnO_4$ solution is standardized with sodium oxalate ($Na_2C_2O_4$). If a 271.2-mg sample of $Na_2C_2O_4$ requires 40.21 mL of the $KMnO_4$ solution to reach the endpoint, calculate the normality of the $KMnO_4$ solution.

11-29. What would be the normality of the $KMnO_4$ solution given in Problem 11-28 if it were used for
   a. analysis of Mn(II) in neutral solution?
   b. analysis of glycerol in basic solution?

11-30. What is the Fe_2O_3 titer for the $KMnO_4$ solution given in Problem 11-28?

11-31. A limestone sample weighing 453 mg is dissolved in acid, oxalic acid added, pH adjusted to 4, and the resulting precipitate of CaC_2O_4 collected on a filter. After washing, the precipitate is dissolved in acid and titrated with $0.1757\ N$ $KMnO_4$. If 48.75 mL was required to reach the equivalence point, calculate percentage of $CaCO_3$ in the limestone.

*11-32. Even when metal ions do not exhibit redox reactions in solution, it is still possible to devise a redox titration method by forming a stoichiometric compound with a species that can be oxidized or reduced. This was seen in Problem 11-31 for the determination of calcium. Another variation on this theme is the determination of sodium by precipitating $NaZn(UO_2)_2(OAc)_9 \cdot 6H_2O$. This precipitate can then be dissolved in acid, passed through a Jones reductor to reduce UO_2^{2+} to U^{3+}, and air oxidized to U^{4+}. The U^{4+} is titrated with $KMnO_4$, which returns the uranium to UO_2^{2+}. If a 459-mg sample requires 39.57 mL of $0.1215\ N$ $KMnO_4$ to reach the endpoint, calculate the percentage of Na in the sample.

11-33. A 951-mg steel sample containing manganese is dissolved in acid and oxidized to MnO_4^- with $NaBiO_3$. After removal of excess $NaBiO_3$, 1.432 g of ferrous ammonium sulfate (FW = 392) is added, which reduces the MnO_4^- to Mn^{2+}. The excess Fe^{2+} is titrated with $0.1063\ N$ $K_2Cr_2O_7$ and requires 22.16 mL to reach the endpoint, using barium diphenylamine sulfonate indicator. Calculate the percentage of Mn in the steel.

11-34. Dissolved oxygen content of a water sample is determined by the Winkler method. A 100-mL sample is treated with NaI, NaOH, and Mn(II). The resulting precipitate containing $Mn(OH)_3$ is dissolved in H_2SO_4, releasing I_3^-. The I_3^- is titrated with $0.0157\ N$ $Na_2S_2O_3$ and requires 5.25 mL to reach the endpoint. Calculate parts per million (milligrams per liter) of dissolved oxygen.

*11-35. Ozone content in a polluted atmosphere is determined iodometrically. The air sample is bubbled through an iodide solution (buffered or unbuffered?) at a rate of 2.5 L/min for 2.0 hr. The resulting I_3^- is titrated with $0.0016\ N$ $Na_2S_2O_3$, and 5.43 mL is required to reach the endpoint. If the density of the air sample is 1.19 g/L, calculate ozone content as parts per million.

11-36. Carbon monoxide, CO, content in a busy tunnel can be analyzed by a redox method involving oxidation by I_2O_5 at 150°C:

$$I_2O_5 + 5CO \rightleftharpoons 5CO_2 + I_2$$

The iodine is removed from the tube containing I_2O_5 and then titrated with $Na_2S_2O_3$. A sample of tunnel air is passed through the I_2O_5 column at a rate of 1.63 L/min for 45 min, and the resulting I_2 requires 5.05 mL of $0.0113\ N$ $Na_2S_2O_3$ to reach the endpoint. If the density of the air sample is 1.18 g/L, calculate the CO content as parts per million.

11-37. Commercial liquid bleach contains sodium hypochlorite, which can be analyzed iodometrically by the following reaction:

$$OCl^- + 3I^- + 2H_3O^+ \rightleftharpoons Cl^- + I_3^- + 3H_2O$$

The resulting I_3^- is then titrated with $Na_2S_2O_3$. If a 5.00-mL sample of bleach produces sufficient I_3^- to require 38.56 mL of $0.1986\ N$ $Na_2S_2O_3$ to reach the starch endpoint, calculate the NaOCl content as weight per volume percent.

11-38. The use of chromium-plated "tin" cans in place of tin-coated cans has been under development for several years. The thickness of the chromium film can be determined by dissolving the sample in acid and oxidizing the resulting Cr^{3+} with ammonium peroxodisulfate:

$$3S_2O_8^{2-} + 2Cr^{3+} + 21H_2O \rightleftharpoons Cr_2O_7^{2-} + 14H_3O^+ + 6SO_4^{2-}$$

After removal of excess $S_2O_8^{2-}$, an excess of ferrous ammonium sulfate (FW = 392) is added, and the excess is back-titrated with $K_2Cr_2O_7$. If a 20-cm^2 sample is treated by this process, 250 mg of ferrous ammonium sulfate is added, and the back-titration requires 10.63 mL of 0.0417 N $K_2Cr_2O_7$ to reach the barium diphenylamine sulfonate endpoint, calculate the average thickness of chromium on the steel.

11-39. If all the conditions are kept constant, including the amount of ferrous ammonium sulfate, for the chromium analysis given in Problem 11-38 and the replicate values for the volume of 0.0417 N $K_2Cr_2O_7$ are 10.63, 10.58, and 10.79 mL, calculate the 95% confidence interval for the thickness of the chromium plate.

*11-40. Antimony content in a stibnite ore is determined by dissolution of the sample and treatment of a strongly acidic solution (3 F HCl) with KIO_3:

$$2IO_3^- + 5SbCl_4^- + 12H_3O^+ + 10Cl^- \rightleftharpoons I_2 + 5SbCl_6^- + 18H_2O$$

Continued addition of KIO_3 oxidizes the I_2 to ICl_2^-:

$$IO_3^- + 2I_2 + 10Cl^- + 6H_3O^+ \rightleftharpoons 5ICl_2^- + 9H_2O$$

If a 327-mg sample of the ore requires 35.63 mL of 0.0214 F (note F given, not N) KIO_3 to reach the endpoint, calculate the percentage of Sb_2S_3 in the ore.

11-41. Stainless steel is analyzed for iron and chromium content by using metal reductor columns and titration with $K_2Cr_2O_7$. A 1.775-g sample is dissolved in HCl and diluted to 250 mL in a volumetric flask. A 50.0-mL aliquot is passed through a Walden reductor and titrated with 35.17 mL of 0.1207 N $K_2Cr_2O_7$, using barium diphenylamine sulfonate as a redox indicator. A second 50-mL aliquot is passed through a Jones reductor and collected in 0.1 F Fe^{3+} (which is reduced to Fe^{2+} by Cr^{2+}). Titration with 0.1207 N $K_2Cr_2O_7$ requires 41.77 mL to reach the endpoint. Calculate percentages of iron and chromium in the stainless steel.

11-42. Aluminum can be determined in a redox titration by precipitation with 8-hydroxyquinoline (Q). The precipitate (AlQ_3) is filtered, washed, and redissolved. The Q is brominated, using $KBrO_3$ titrant containing excess Br^-:

$$BrO_3^- + 5Br^- + 6H_3O^+ \rightleftharpoons 3Br_2(aq) + 9H_2O$$

If a 65.3-mg sample requires 35.89 mL of 0.1508 N $KBrO_3$ to reach the endpoint (first appearance of excess Br_2), calculate percentage of Al_2O_3 in the sample.

*11-43. Glucose ($C_6H_{12}O_6$) content can be determined in a redox titration by addition of NaOH and an excess of I_3^-. After reaction

$$C_6H_{12}O_6 + 3OH^- + I_3^- \rightleftharpoons C_6H_{11}O_7^- + 2H_2O + 3I^-$$

is completed, the solution is acidified, and the excess I_3^- is back-titrated with $Na_2S_2O_3$. A 452-mg sugar sample was treated with 50.0 mL of 0.1052 N I_3^- and the back-titration required 23.7 mL of 0.0867 N $Na_2S_2O_3$. Calculate the glucose percentage in the sugar sample.

11-44. Nitrobenzene (FW = 123) can be reduced with Ti^{3+} in a neutral solution according to the following reaction:

$$\underset{NO_2}{\bigcirc} + 6Ti^{3+} + 10H_2O \rightleftharpoons \underset{NH_2}{\bigcirc} + 6TiO^{2+} + 6H_3O^+$$

After reduction is complete, the solution is acidified, and the excess Ti^{3+} is back-titrated with Fe^{3+}. A 253-mg sample containing nitrobenzene was treated with 25.0 mL of 0.187 N Ti^{3+} and the back-titration required 5.13 mL of 0.155 N Fe^{3+} titrant. Calculate the percentage of nitrobenzene content.

11-45. Thioglycolic acid (FW = 92.1) is sometimes used in hair-setting preparations and can be analyzed by its reaction with iodine

$$2HSCH_2-COOH + I_2 \rightleftharpoons HOOC-CH_2-S-S-CH_2-COOH + 2HI$$

If a 562-mg sample required 15.7 mL of 0.0317 N I_2 titrant, calculate percentage of thioglycolic acid in the sample.

11-46. A 2.00-mL sample of antifreeze is diluted to 50 mL, and a 5.00-mL aliquot is analyzed for ethylene glycol content by oxidation with 30.00 mL of a Na_5IO_6 solution, as shown in Equation 11-58. The excess periodate is treated with I^- (Equation 11-57) to produce I_3^-. Then the I_3^- is titrated with 0.1215 N As_2O_3 solution and requires 12.73 mL to reach the endpoint. If another 30.00 mL of the Na_5IO_6 solution is titrated directly and requires 45.17 mL of the As_2O_3 titrant to reach the endpoint, calculate the weight per volume percent of ethylene glycol in the antifreeze sample.

11-47. Tetraethyllead (TEL) content in gasoline can be determined by diluting the sample with methanol and titrating directly with I_3^-. The titration reaction is

$$Pb(C_2H_5)_4 + I_3^- \rightleftharpoons PbI(C_2H_5)_3 + C_2H_5I + I^-$$

If a 25.0-mL (6.6×10^{-3}-gal) sample is diluted to 200 mL with methanol and requires 2.17 mL of 0.0112 N I_3^- to reach the endpoint, calculate the TEL content in grams per gallon.

*11-48. A solution containing vanadium(V) and uranium(VI) can be analyzed by titration with Ti^{3+}, using a potentiometric endpoint technique. The first endpoint corresponds to the reduction of V(V) to V(IV), while the second endpoint corresponds to the reduction of V(IV) to V(III) *and* the reduction of U(VI) to U(IV). That is, the half-cell potentials for the U(VI)/U(IV) and V(IV)/V(III) couples are too close to be separated. A 5.653-g sample is dissolved in acid and diluted to 250 mL, and a 50-mL aliquot is taken for analysis. It requires 13.71 mL of 0.1506 N Ti^{3+} to reach the first endpoint and 46.12 mL (total) to reach the second endpoint. Calculate percentages of V_2O_5 and U_3O_8 in the original sample.

11-49. The vitamin C (ascorbic acid) content in tomato juice can be determined by oxidation with Br_2:

$$C_6H_8O_6 + Br_2 + 2H_2O \rightleftharpoons C_6H_6O_6 + 2Br^- + 2H_3O^+$$

ascorbic acid   dehydroascorbic acid

A 150-mL sample of tomato juice was acidified and titrated with 0.0107 N $KBrO_3$ containing KBr, using α-naphthoflavone as an indicator to detect excess Br_2. If 14.79 mL was required to reach the endpoint, calculate the vitamin C as milligrams per ounce (1 oz $\cong$ 30 mL), and determine the amount of tomato juice one would have to drink to meet the 30-mg minimum daily requirement.

SPECTROCHEMICAL METHODS

PART FOUR

Principles of Spectrochemical Analysis

12

In Chapter 1, the total analysis process was described, and it was pointed out that the quantitation step may consist of measuring any signal that can be related to the analyte. So far in this text, we have discussed analytical methods based on measurement of mass and volume. Now we turn our attention to the measurement of electromagnetic radiation and its interaction with the sample. Although several old analytical methods had been based on the interaction of *visible* light with a sample and although the descriptive term *colorimetry* is in use, we shall find it generally valuable to consider all wavelengths of the electromagnetic radiation spectrum—not just the visible portion of the spectrum lying between 400 and 700 nanometers (nm), to which the human eye is sensitive. It is also customary to consider how the sample interacts with different wavelengths in a given region of electromagnetic radiation, and this collection of measurement signals as a function of wavelength is called a *spectrum*, leading to the term *spectrochemical analysis*, or *spectroscopy*.

It should be kept in mind that the final analytical procedure may specify measurement at only one wavelength; that is, no spectrum is actually measured. However, the optimum wavelength had most likely been established by measuring a spectrum; so it is not improper to call such a procedure a "spectrochemical" analysis.

12-1 TERMS

Before surveying the different types of spectrochemical analysis, we define several spectrochemical terms that will be used in our discussions.

Electromagnetic Radiation Light can be described in terms of its wavelike properties as well as its particlelike properties. For the former, terms such as *wavelength* and *frequency* will be used. The particlelike properties are expressed in terms of packets of energy called *photons*. These terms are valid throughout the entire energy range and are not limited to what is normally considered to be light (visible, ultraviolet (UV), and infrared).

The wavelike nature of electromagnetic radiation can be visualized as shown in Figure 12-1. It consists of oscillating electric (**E**) and magnetic (**M**) fields, which are at right angles to each other and travel at a constant velocity in a given medium. This velocity in a vacuum is 3×10^{10} cm/sec (the speed of light) and is given the symbol c.

Wavelength   The distance between the peaks (or troughs) of the wave shown in Figure 12-1 is called the *wavelength* and is given the symbol λ. The wavelength may be given in centimeters; but because of the small wavelengths for most types of electromagnetic radiation, units such as nanometers (10^{-9} m) are more convenient. An angstrom (Å) is another common unit; it is 10^{-10} m, or 10^{-8} cm.

Frequency   If an observer sat at some point in the space implied in Figure 12-1 and counted the rate at which wave peaks passed by, this rate would be the *frequency* for the wave; it is given the symbol v. The unit for frequency is hertz (Hz), which is the number of peaks (cycles) per second.

Wavelength and frequency are related as in Equation 12-1:

$$\text{wavelength (cm/peak)} \times \text{frequency (peaks/sec)} = \text{velocity (cm/sec)} \qquad \textbf{(12-1)}$$

$$\lambda \times v = c \text{ (in vacuum and approximately in air)} \qquad \textbf{(12-2)}$$

Thus wavelength and frequency are inversely related to each other. For convenience, another unit is based on this relationship; it is called the *wave number*, which equals $1/\lambda$, is generally expressed in units of reciprocal centimeters (cm^{-1}), and is directly proportional to frequency, and give the symbol v.

Energy   The energy of a photon associated with an electromagnetic wave is directly proportional to its frequency. The proportionality constant (Planck's

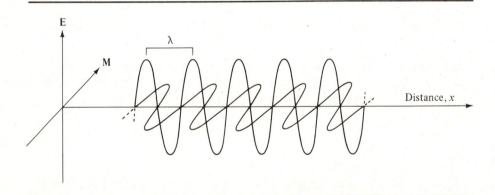

FIGURE 12-1   Schematic representation of an electromagnetic wave. An oscillating electric field (**E** axis) and an oscillating magnetic field (**M** axis) are moving in the x direction with velocity c.

constant) is given the symbol h and equals 6.63×10^{-27} erg-sec:

$$E \propto v$$

$$E = hv \qquad \qquad (12\text{-}3)$$

Energy then is given in ergs per photon. A mole of photons is called an einstein, and its energy would equal Nhv, where N is Avogadro's number. This energy characterization, of course, assumes that all the photons have the same frequency, and they are said to be *monochromatic* ("one color" because they all have the same wavelength and, consequently, the same frequency).

Power   The *radiant power*, P, of a beam of electromagnetic radiation is the energy that impinges on a given area per second. Radiant power and intensity are often used synonymously, but *intensity* is strictly defined as radiant power from a point source per unit solid angle.[1] Both radiant power and intensity are related to the square of the amplitude of a wave such as that shown in Figure 12-1.

Absorption and emission   It was stated earlier that any signal that can be related to the analyte may form the basis for an analytical method. Interactions of electromagnetic radiation with matter may be broadly classified into *absorption processes*, in which electromagnetic radiation from a source is absorbed by the sample and results in a decrease of radiant power (which reaches a detector), and *emission processes*, in which electromagnetic radiation emanates from the sample, resulting in an increase in the radiant power that reaches a detector.

We can amplify the concepts of absorption and emission by considering the processes leading to absorption and emission. For a molecule A—B, the energy as a function of internuclear distance can be schematically represented as in Figure 12-2. The lower curve represents the molecule in its lowest electronic energy state ("ground state"), while the upper curve represents the molecule in an excited electronic state; that is, one of its electrons has been promoted to a higher energy level. If electromagnetic radiation of energy equal to the difference in energy between these states impinges on the sample, some of this energy may be absorbed (A) and cause the molecule to reach this excited state. On the other hand, if the sample contains molecules in an excited state, these molecules may drop back to the ground state and emit electromagnetic radiation equal to this energy difference (E).

In addition to its electronic energy, a molecule may also vibrate with the chemical bond's acting as a restoring force similar to that of a spring, and the molecule may rotate around one of its axes. Thus the molecule will exhibit various vibrational and rotational energy states, as shown in Figure 12-2. Absorption of a photon can lead to a change in one of these states. The energy-level differences are normally in the order $\Delta E_{\text{electronic}} > \Delta E_{\text{vibrational}} > \Delta E_{\text{rotational}}$. Therefore, the frequencies used to observe vibrational and rotational changes are lower than those used to observe electronic state changes.

---

[1] A solid angle is defined as the ratio of an area on a spherical surface to the square of the sphere's radius. Therefore it is dimensionless and is given the units of steradians.

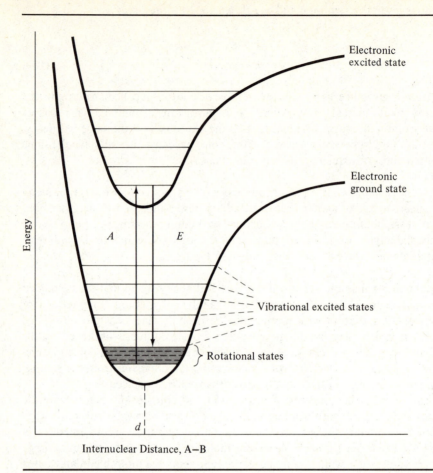

FIGURE 12-2   Energy-level diagram for molecule A—B, with equilibrium bond distance *d*. Vibrational and rotational states are associated with each electronic state. Absorption (*A*) of electromagnetic energy leads to an increase in potential energy in one or more of these states, while emission (*E*) leads to a decrease.

12-2   SURVEY OF SPECTROCHEMICAL METHODS

The electromagnetic spectrum covers many orders of magnitude in frequency or wavelength. In addition to the electronic, vibrational, and rotational states discussed above, changes in nuclear energy levels can be observed at very high energies (γ rays), while changes in nuclear spin states induced by magnetic fields can be observed at very low energies (microwave and radio waves). These are all summarized in Figure 12-3.

Although the underlying principles are similar no matter in what region of the spectrum the analysis falls, it is still useful to classify various spectral regions because

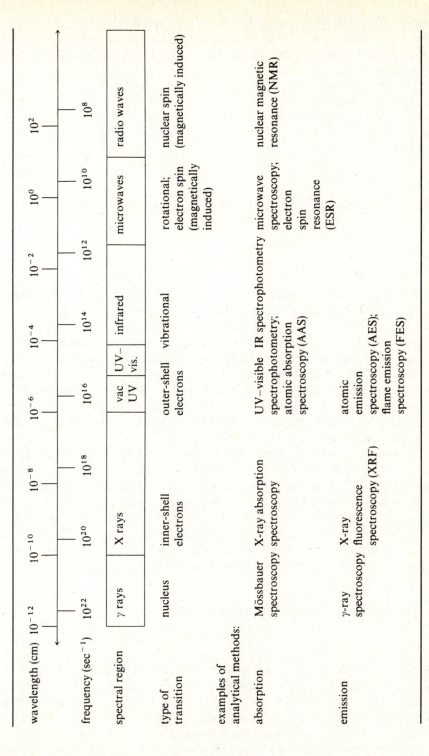

FIGURE 12-3 Electromagnetic spectrum. Spectrochemical methods may all be termed *spectroscopy*, but when a *spectrophotometer* is used for the spectrochemical measurements, the technique is commonly called *spectrophotometry*. Similarly, when a *spectrometer* is used for the measurements, the technique is called *spectrometry* (for example, atomic absorption spectrometry instead of atomic absorption spectroscopy). These various terms with slightly different meanings are discussed in Chapter 13.

wavelength (cm)	10^{-12}	10^{-10}	10^{-8}	10^{-6}	10^{-4}	10^{-2}	10^{0}	10^{2}
frequency (sec^{-1})	10^{22}	10^{20}	10^{18}	10^{16}	10^{14}	10^{12}	10^{10}	10^{8}
spectral region	γ rays	X rays		vac UV / UV–vis.	infrared	microwaves		radio waves
type of transition	nucleus	inner-shell electrons		outer-shell electrons	vibrational	rotational; electron spin (magnetically induced)		nuclear spin (magnetically induced)

examples of analytical methods:

absorption	Mössbauer spectroscopy	X-ray absorption spectroscopy		UV–visible IR spectrophotometry; atomic absorption spectroscopy (AAS)		microwave spectroscopy; electron spin resonance (ESR)		nuclear magnetic resonance (NMR)
emission	γ-ray spectroscopy	X-ray fluorescence spectroscopy (XRF)		atomic emission spectroscopy (AES); flame emission spectroscopy (FES)				

the experimental techniques and chemical information content differ from region to region. Thus the sources for X-rays will be different from sources for UV–visible spectrophotometry or infrared spectrophotometry. Likewise, detectors will be different for each spectral region.

Another factor that varies with spectral region is the probability for emission, which is proportional to v^3. Thus a system in an excited state will relax to its ground state primarily by either energy transfer upon collision with other molecules or emission of a photon whose energy is equal to the energy difference between the excited state and the ground state:

$$\text{absorption:}\quad M + hv \rightarrow M^* \quad \text{(excited state)}$$

$$\text{relaxation:}\quad M^* \rightarrow M + \text{heat} \quad \text{(collisions with neighbors)}$$

or

$$M^* \rightarrow M + hv \quad \text{(emission)}$$

The latter decay mode has a high probability only at high frequency. Therefore, analytical methods based on emission are useful only at frequencies equivalent to visible light or higher, as shown in Figure 12-3.

Although Figure 12-3 shows boundaries for spectral regions, the boundaries are not really well defined. There is considerable overlap: For example, rotational transition energies overlap with vibrational transitions, and vibrational transition energies overlap with electronic transitions. As new instruments are developed, the frequency range that can be measured with a given type of instrument is expanded. The ability of modern infrared spectrophotometers to measure very low frequencies (the far infrared) is a case in point.

In a text of this length it is impractical to discuss each of the analytical techniques listed in Figure 12-3. We shall concern ourselves with UV–visible absorption spectrophotometry and atomic absorption spectroscopy as examples of absorption techniques and, secondly, with flame emission spectroscopy and atomic emission spectroscopy as examples of emission techniques. Other methods, such as infrared spectrophotometry, will be mentioned in comparisons of experimental techniques, but the reader should consult the appropriate chapter-end reading lists for more complete discussions of the various spectrochemical methods of analysis.

12-3 QUANTITATIVE PRINCIPLES OF ABSORPTION: BEER'S LAW

In previous chapters, we saw that gravimetric methods involve calculation of solubilities from solubility products, while volumetric methods involve derivation of titration curves from ionic equilibrium. In parallel fashion, when considering absorption of electromagnetic radiation, we need to know how the radiant power coming from a source is decreased when it interacts with our sample. Basically, we can measure the radiant power by using a detector with or without the sample in the

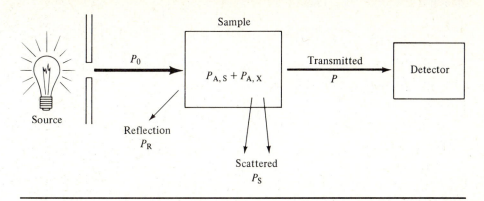

FIGURE 12-4   Interaction of electromagnetic radiation (light) with a sample and its container.

beam of electromagnetic radiation. We ultimately seek the relationship between these measurement signals and the amount of analyte in the sample.

There are several ways in which the radiant power can change when we place a sample in the light beam (for convenience, the word "light" will be used for the longer, more general term "electromagnetic radiation"). These are shown schematically in Figure 12-4. The original radiant power from the source, P_0, must equal the radiant power attributed to all these interactions if we assume that we have accounted for all interactions. Therefore,

$$P_0 = P + P_S + P_R + P_{A,S} + P_{A,X} \tag{12-4}$$

where P_S represents scattering losses; P_R, reflection losses;[2] and $P_{A,S}$, absorption by the solvent. However, the term we should like to know is $P_{A,X}$, absorption by the analyte.

The radiant power reaching the detector, P, is measured; conceivably, P_0 could be measured by removing the sample so that the radiant power reaching the detector would equal P_0. However, there is no simple way to measure $P_{A,S}$, P_R, and P_S. To circumvent this dilemma, a second measurement is taken, not of P_0 but of the radiant power reaching the detector after it passes through a "blank" sample, which we shall label P'. The blank measurement is made under conditions identical with the sample measurement (same sample tube, same solvent, same radiant power from the source), but the blank contains no analyte. $P_{A,X}$ for the blank will equal zero, while all the other terms will remain the same (assuming the presence of the analyte

---

[2] Reflection occurs at a boundary between two materials that exhibit different refractive indexes. When the surface is smooth—for example, a mirror—the angle of incidence equals the angle of reflection. Scattering, on the other hand, involves an interaction of electromagnetic radiation with small particles (dimensions comparable with, or smaller than, the wavelength of the radiation) in its path. This interaction causes radiant power of the same wavelength to be radiated in all directions, with a corresponding decrease of radiant power in the original beam.

does not alter P_S or P_R). Thus

$$P_0 = P' + P_S + P_R + P_{A,S} \tag{12-5}$$

Equations 12-4 and 12-5 can now be combined to eliminate all the terms difficult to measure individually. From Equation 12-4,

$$P_0 - P_{A,S} - P_S - P_R = P + P_{A,X}$$

From Equation 12-5,

$$P_0 - P_{A,S} - P_S - P_R = P'$$

Therefore, $P' = P + P_{A,X}$, and

$$P_{A,X} = P' - P \tag{12-6}$$

Equation 12-6 states that the radiant power absorbed by the analyte is equal to the radiant power reaching the detector from the blank minus the radiant power reaching the detector from the sample. This is equivalent to the *hypothetical* experiment shown in Figure 12-5, in which the radiant power from the source is first passed through the blank and leads to P', which in turn is a source radiant power, P_0', for the analyte particles alone. The analyte absorbs some of the radiant power, while the rest reaches the detector and is measured. Thus

$$P' \equiv P_0' = P_{A,X} + P$$
$$P_{A,X} = P_0' - P \tag{12-7}$$

Clearly, this hypothetical experiment cannot be carried out in the laboratory but is presented to help emphasize how a blank measurement allows the analyst to

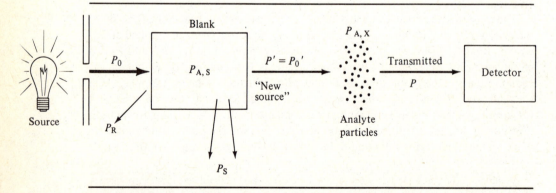

FIGURE 12-5 Hypothetical use of blank to correct for extraneous losses (P_R, P_S, $P_{A,S}$) of radiant power.

measure the amount of radiant power absorbed by the analyte. From this point on, we shall use the term "incident radiant power from the source, P_0," but it should be remembered that, in reality, what is measured in the laboratory is the radiant power transmitted by the blank ($P' \equiv P_0'$) and measured by the detector.

Equation 12-7 gives us a relationship involving the radiant power absorbed by the analyte and the radiant power P transmitted by the sample, which is measured by the detector. But what is the relationship between these terms and the *amount* of analyte present in the sample?

Figure 12-6 shows schematically what occurs. An incident radiant power P_0 (really P_0', the radiant power transmitted by the blank) enters the solution, and absorption by the analyte occurs at various points. Some of the light exits (P) and is measured by the detector. Over 200 yr ago (1729), Pierre Bouguer found that the fraction of light absorbed in each segment of path length (Δx) in the cell is directly proportional to its thickness. Some years later (1768), the relationship was restated by Johann Heinrich Lambert (after whom it is generally called "Lambert's law"), as given in Equation 12-8.

$$\text{fraction absorbed} = \frac{P_{in} - P_{out}}{P_{in}} = \frac{\Delta P}{P_{in}} \propto \Delta x = k'\Delta x \tag{12-8}$$

Lambert's law is not very useful in determining the amount of analyte present in a sample. Over a century ago, A. Beer found that the fraction of light absorbed was also directly proportional to the concentration of absorbing species present, C. This, too, is clear from Figure 12-6. As the number of solute particles increases, the amount of absorption will increase in the same way that increasing the path length will increase the amount of absorption. Mathematically, Beer's law can be expressed

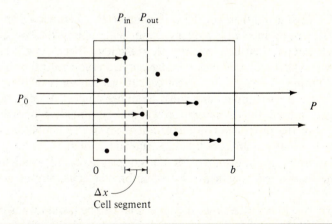

FIGURE 12-6   Absorption of incident radiant power (P_0) in a cell (path length $= b$) containing analyte particles. Light not absorbed exits from the cell as transmitted radiant power (P).

in Equation 12-9:

$$\text{fraction absorbed} = \frac{P_{\text{in}} - P_{\text{out}}}{P_{\text{in}}} = \frac{\Delta P}{P_{\text{in}}} \propto C = k''C \tag{12-9}$$

The two laws can be combined as shown in Equation 12-10:

$$\frac{\Delta P}{P_{\text{in}}} = kC\Delta x \tag{12-10}$$

However, it must be kept in mind that P_{in} is not constant as light moves through the sample but will continually decrease. To derive a mathematical relationship that describes how P changes as the light beam moves through the solution, calculus is used (see Appendix 1-H), with the result of Equation 12-11:

$$\log \frac{P_0}{P} = \varepsilon bC \tag{12-11}$$

Equation 12-11 relates the measurement signals P_0/P (the radiant power transmitted by the blank divided by the radiant power transmitted by the sample) to the cell's path length, b, and to the concentration of analyte in the sample solution. Note that C is *not* proportional to P_0/P but is proportional to the logarithm of P_0/P.

Modern spectrophotometers are designed so that $\log(P_0/P)$ can be read directly, and this term is called the *absorbance*, A. The absorbance *is* directly proportional to concentration

$$\log \frac{P_0}{P} \equiv A = \varepsilon bC \tag{12-12}$$

Equation 12-12 is a mathematical expression of the Lambert–Beer laws, and because of its importance and frequent use in determining the concentration of analyte, its name is often conveniently shortened to "Beer's law." The proportionality constant ε is called the *molar absorptivity* and represents the absorbance of a 1 $M(F)$ solution of analyte in a 1-cm cell. It has the units of liters per mole-centimeter since a logarithmic term must be dimensionless. When other concentration units are used, the concentration will still be directly proportional to the absorbance; but the proportionality constant is given the symbol a to stand for absorptivity (normally, grams per liter concentration units).

The fraction of light transmitted by the sample, T, is P/P_0; and this too can normally be read directly with a spectrophotometer, generally as a percentage ($100P/P_0$).

It is instructive to note the range of values that can be taken by absorbance and transmittance. If the analyte absorbs no light, then $P = P_0$, $T = P/P_0 = 1$ (100%), and $A = \log P_0/P = \log 1 = 0$. On the other hand, if the analyte absorbs all the light so that none reaches the detector, then $P = 0$, $T = P/P_0 = 0$ (0%), and $A = \log P_0/P = \log \infty = \infty$. Thus transmittance ranges from 0 to 100%, while absorbance ranges from 0 to ∞. Table 12-1 summarizes the terms and symbols. Some typical problems are presented in Examples 12-1 to 12-3.

TABLE 12-1. Terms and Symbols Used in Absorption Spectrophotometry

Term	Symbol	Definition
radiant power (incident)	P_0	energy of radiation per second reaching detector from blank
radiant power (transmitted)	P	energy of radiation per second reaching detector from sample
transmittance	T	$\dfrac{P}{P_0}$
absorbance	A	$\log \dfrac{P_0}{P} = \log \dfrac{1}{T} = -\log T$
path length (cell length)	b	length of light path through cell (b normally given in cm)
molar absorptivity	ε	A/bC (C given in mol/L, or mmol/mL; ε, in L/mol-cm, or mL/mmol-cm)
absorptivity	a	A/bC (C given in g/L or other concentration units; a, in L/g-cm for C in g/L)

EXAMPLE 12-1

Trace amounts of iron in drinking water can be determined by forming an intensely colored 1:1 complex of Fe(II) with o-phenanthroline. If a standard solution containing 5.36×10^{-6} F Fe exhibits a transmittance of 76% in a 2.00-cm cell at 510 nm, what are the absorbance of this solution and the molar absorptivity for the Fe(II) o-phenanthroline complex?

Solution

$$T = \frac{P}{P_0} = 0.76$$

$$A = \log \frac{P_0}{P} = -\log T = -\log 0.76 = 0.119$$

$$= \varepsilon b C$$

$$0.119 = \varepsilon \times 2.00 \text{ cm} \times 5.36 \times 10^{-6} \text{ mol/L}$$

$$\varepsilon = 1.11 \times 10^4 \text{ L/mol-cm}$$

(*Note:* The value of the wavelength is not needed for calculation, but all measurements would be carried out at this wavelength.)

EXAMPLE 12-2

A 20-mL aliquot of drinking water containing Fe(III) and Fe(II) is treated with hydroquinone to reduce all dissolved iron to Fe(II), o-phenanthroline is added, and the aliquot is diluted to 50.0 mL. The absorbance in a 2.00-cm cell is 0.085 at 510 nm. Calculate the iron content of the water as parts per million

Solution

$$A = \varepsilon bC \qquad \varepsilon = 1.11 \times 10^4 \text{ L/mol-cm (from Example 12-1)}$$

$$C = \frac{0.085}{2 \text{ cm} \times 1.11 \times 10^4 \text{ L/mol-cm}} = 3.83 \times 10^{-6} \ F$$

The water sample was diluted from 20.0 to 50.0 mL. Therefore, concentration in the original sample is

$$3.83 \times 10^{-6} \frac{\text{mmol}}{\cancel{\text{mL}}} \times \frac{50.0 \ \cancel{\text{mL}}}{20.0 \text{ mL}} = 9.57 \times 10^{-6} \ F \text{ (mmol/mL)}$$

$$9.57 \times 10^{-6} \frac{\cancel{\text{mmol}}}{\cancel{\text{mL}}} \times 55.85 \frac{\text{mg}}{\cancel{\text{mmol}}} \times 1000 \frac{\cancel{\text{mL}}}{\text{L}} = 0.53 \text{ mg/L (ppm)}$$

(two significant figures because 0.085 (absorbance) contains only two significant figures).

EXAMPLE 12-3

What is the percentage transmittance of the solution in Example 12-2?

Solution

$$A = 0.085 = \log \frac{P_0}{P}$$

$$\frac{P_0}{P} = 10^{0.085} = 1.22$$

$$T = \frac{P}{P_0} = \frac{1}{1.22} = 0.82$$

$$\% \ T = 0.82 \times 100 \ (\%) = 82\%$$

12-4   ANALYTICAL USE OF BEER'S LAW

The molar absorptivity, ε, depends not only on the analyte but also on the wavelength of light measured. The spectral response of the analyte is normally measured and displayed as an *absorption spectrum*, as shown in Figure 12-7. The absorbance due to the analyte, x, at any wavelength, λ, will be given by Beer's law:

$$A_{x,\lambda} = \varepsilon_{x,\lambda} b C_x \qquad \text{(12-13)}$$

Therefore, to calculate the analyte concentration, it is necessary to measure the absorbance at some particular wavelength and to know or measure the molar absorptivity at that wavelength. Because ε varies with experimental conditions (detailed composition of the solution and the measuring instrument itself), most often it is measured by using a *standard solution* of the analyte. This is shown in Example 12-4. Frequently a *calibration-curve* method is used, in which the absorbances of several standard solutions are measured and plotted as a function of concentration. The absorbance of the sample containing analyte is then measured at the same wavelength, and the concentration of analyte is read directly from the calibration curve (Example 12-5). The slope of the calibration line is $\varepsilon_{x,\lambda} b$. It sometimes happens, when cylindrical sample cells are used, that an accurate value of b is not known so that a literature value of ε would not be helpful. However, the calibration method still allows calculation of concentration, provided that the same or matched cell is used for both the standard solutions and the sample solution.

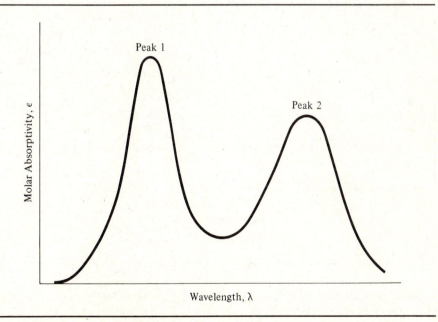

FIGURE 12-7   Absorption spectrum of an analyte, exhibiting two absorption peaks.

EXAMPLE 12-4

A 10.0-ml aliquot of a standard manganese solution containing 62.5 mg/L of Mn is oxidized to MnO_4^- and diluted to 50.0 mL in a volumetric flask. The absorbance measured in a 1.00-cm cell at 525 nm is 0.343.

An 850-mg steel sample containing manganese is dissolved in acid and diluted to 250 mL. A 50.0-mL aliquot is treated with KIO_4 to oxidize manganese to MnO_4^- and diluted to 100 mL. If the absorbance at 525 nm is 0.468 in a 1.00-cm cell, calculate the weight percent of Mn in the steel.

Solution

Determine ε from standard solution

$$A = \varepsilon b C$$

$$C_{std} = \frac{62.5 \text{ mg/L}}{54.9 \text{ mg/mmol}} = 1.14 \times 10^{-3} \ F$$

$$C = C_{std} \times \frac{10.0 \text{ mL}}{50.0 \text{ mL}} = 0.228 \times 10^{-3} \ F \qquad \text{(dilution factor)}$$

$$\varepsilon = \frac{A}{bC} = \frac{0.343}{1 \text{ cm} \times 0.288 \times 10^{-3} \text{ mol/L}} = 1.19 \times 10^3 \text{ L/mol-cm}$$

Determine weight percent of Mn in steel

$$A = \varepsilon b C$$

$$C = \frac{A}{\varepsilon b} = \frac{0.468}{1.19 \times 10^3 \text{ L/mol-cm} \times 1 \text{ cm}} = 3.93 \times 10^{-4} \ F$$

$$C_{aliquot} = 3.93 \times 10^{-4} \ F \times \frac{100 \text{ mL}}{50 \text{ mL}} = 7.86 \times 10^{-4} \ F \ (\text{mmol/mL})$$

$$\text{Mn in sample} = 7.86 \times 10^{-4} \ \frac{\text{mmol}}{\text{mL}} \times 250 \text{ mL} = 0.197 \text{ mmol}$$

$$0.197 \text{ mmol} \times 54.9 \ \frac{\text{mg}}{\text{mmol}} = 10.8 \text{ mg Mn}$$

$$\% \text{ Mn} = \frac{10.8 \text{ mg Mn}}{850 \text{ mg steel}} \times 100 \ (\%) = 1.27\%$$

What wavelength should be used to measure $A_{x,\lambda}$? With Figure 12-7, we could, in principle, pick any wavelength and calculate C_x from Equation 12-13. However, for the greatest sensitivity, we should use the wavelength with the maximum ε value. Recall from Chapter 1 (Equation 1-1) that the proportionality constant relating signal to analyte was the sensitivity factor. In the case of absorption spectrophotometry, the signal is absorbance, and therefore the sensitivity factor is εb. In

EXAMPLE 12-5

The amount of manganese in steel can also be determined by using a calibration curve, as shown in Figure 12-8. A 790-mg steel sample is dissolved in acid and diluted to 250 mL. A 50.0-mL aliquot is treated with KIO_4 to oxidize manganese to MnO_4^- and diluted to 100 mL. If the absorbance at 525 nm is 0.267 in a 1.00-cm cell, calculate the weight percent of Mn in the steel.

Solution

From the calibration curve given in Figure 12-8, the final solution contains 12.2 mg of Mn per liter. The concentration of Mn in the original solution of steel is

$$C_{aliquot} = 12.2 \frac{\text{mg Mn}}{\text{L}} \times \frac{100 \text{ mL}}{50 \text{ mL}} = 24.4 \text{ mg/L}$$

$$\text{Mn in total sample} = 24.4 \frac{\text{mg}}{\text{L}} \times 0.25 \text{ L} = 6.1 \text{ mg}$$

$$\% \text{ Mn} = \frac{6.1 \text{ mg Mn}}{790 \text{ mg steel}} \times 100 \, (\%) = 0.77\%$$

(*Note:* A larger graph than the one shown in Figure 12-8 is needed to read manganese concentration to three significant figures.)

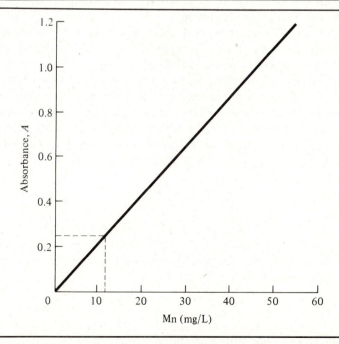

FIGURE 12-8 Calibration curve for spectrophotometric determination of manganese as MnO_4^-.

other words, we can increase the sensitivity by choosing a wavelength where ε is large and/or by using a long path length. Occasionally long path lengths are used (measurement of gas samples) but most spectrophotometers accept only cells of 1 to 10 cm. Thus for highest sensitivity, it is usually recommended that an absorption peak (Figure 12-7) be used for measurement of absorbance. Note that in general spectrophotometric determinations require two measurements: In principle, one measures ε, while the other measures A to allow calculation of C_x.

Reasons for picking an absorption-peak maximum for measurement will be discussed later, but one fairly obvious factor involves uncertainty in the wavelength used. Spectrophotometers have inherent uncertainty just as balances and burets do. When a wavelength selector reads 525 nm, a narrow band of wavelengths centered at 525 nm will supposedly impinge upon the sample. However, it may turn out that, with a reading of 525 nm, the actual central wavelength is 530 nm. The molar absorptivity of the analyte at 530 nm should be used instead of the one at 525 nm assumed from the wavelength selector. Even if the spectrophotometer had no error in its wavelength selector, there is still uncertainty in setting and reading the selector. The operator may think he or she is always selecting 525 nm, but one day the setting may be actually 527 nm, while the next it may be actually 524 nm. The error introduced will depend on how rapidly ε changes with wavelength: on the slope of the spectral curve at this wavelength. When measurement is carried out at the absorption maximum, the slope is essentially zero; and over a narrow wavelength range, little error is introduced by a small error in wavelength. The same argument could be made for measurement at an absorption minimum, but of course the sensitivity would be considerably lower.

One other method frequently used to calculate analyte concentration from absorbance measurements is called *standard addition*. All the constituents in the sample may not be known; so a standard solution with the same chemical composition cannot be prepared. To eliminate any error arising from ε's being different in the standard solution from ε in the sample solution, an amount of standard is *added* to the sample. The two measurements are now absorbance of the sample alone and absorbance of the sample containing analyte plus a known amount of analyte. If a sample of volume V_x containing a concentration C_x of analyte x is diluted to V_t, application of Beer's law gives

$$A_{\lambda,x} = \varepsilon_{\lambda,x} b \frac{C_x V_x}{V_t} \qquad (12\text{-}14)$$

Note that $C_x V_x$ is the millimoles of x and $C_x V_x / V_t$ is the concentration of x in the final sample solution. Now another sample is taken, which also contains a volume V_s of a standard solution of analyte x with a concentration C_s before dilution to the final volume V_t. Application of Beer's law gives

$$A_{\lambda,x+s} = \varepsilon_{\lambda,x} b \left(\frac{C_x V_x}{V_t} + \frac{C_s V_s}{V_t} \right) \qquad (12\text{-}15)$$

Equations 12-14 and 12-15 can be combined to give

$$A_{\lambda,x+s} = \frac{A_{\lambda,x} \cancel{V_t}}{C_x V_x} \left(\frac{C_x V_x}{\cancel{V_t}} + \frac{C_s V_s}{\cancel{V_t}} \right)$$

(note that total volume cancels out) and rearranged to solve for C_x:

$$C_x = \frac{C_s V_s}{V_x} \frac{A_{\lambda,x}}{A_{\lambda,x+s} - A_{\lambda,x}} \qquad (12\text{-}16)$$

Equation 12-16 shows that C_x is still directly proportional to the absorbance of the original solution but, in addition, is directly proportional to the concentration of the analyte added in the standard addition and depends inversely on the absorbance difference between the two solutions. It is important to have reasonable differences in absorbance, or else uncertainty in the denominator term will lead to a large uncertainty in the analyte concentration. A typical calculation is given in Example 12-6.

It was assumed in the derivation of Equation 12-16 that the sample solution and the sample-plus-standard-addition solution had the same final volume, V_t. On occasion, a standard addition is made directly to the sample solution so that the final total volume is $(V_t + V_s)$ for the standard addition measurement. In this procedure, Equation 12-15 becomes

$$A_{\lambda,x+s} = \varepsilon_{\lambda,x} b \left(\frac{C_x V_x}{V_t + V_s} + \frac{C_s V_s}{V_t + V_s} \right) \qquad (12\text{-}17)$$

Equations 12-14 and 12-17 can be combined and rearranged to give

$$C_x = \frac{C_s V_s}{V_x} \frac{A_{\lambda,x}}{A_{\lambda,x+s}[(V_t + V_s)/V_t] - A_{\lambda,x}} \qquad (12\text{-}18)$$

Equation 12-18 is the same as Equation 12-16, *except* that the absorbance with the standard addition, $A_{\lambda,x+s}$, has been *corrected* for the volume increase, namely,

---

EXAMPLE 12-6

A 2.00-mL urine specimen is analyzed for phosphate by adding reagents to develop a color, diluting to 100 mL, and measuring at 700 nm. The absorbance is 0.375. To a second 2.00-mL urine specimen is added 5.00 mL of a standard phosphate solution containing 45.0 mg of phosphate per liter. After color development and dilution to 100 mL, the absorbance is 0.506 at 700 nm. Calculate the concentration of phosphate in the original urine specimen as milligrams per liter.

Solution

From Equation 12-16, we find

$$C_x = \frac{45.0 \text{ mg/L} \times 5.00 \text{ mL}}{2.00 \text{ mL}} \times \frac{0.375}{0.506 - 0.375} = 322 \text{ mg/L}$$

(*Note:* C_x will have the same units as C_s, and the cell path length is not needed as long as both measurements are carried out in the same cell or cells of equal path length.)

$(V_t + V_s)/V_t$. Some analysts correct merely the absorbance reading for the volume change and continue to use Equation 12-16. Also note that, if the standard addition volume, V_s, is small compared to V_t, $(V_t + V_s)/V_t \cong 1$ and Equation 12-18 becomes Equation 12-16.

One last comment on standard addition techniques: Equation 12-15 is a straight line form for absorbance against V_s; if a series of standard additions is made (with the same total volume or corrected for any volume change, as described in the previous paragraph) and the absorbance is plotted against V_s, the x-axis intercept $(V_{s,A\,=\,0})$ can be used to find C_x from Equation 12-19,

$$C_x V_x = -C_s V_{s,A\,=\,0} \tag{12-19}$$

and is shown in Figure 12-9. The left side reflects the millimoles of analyte in the original sample, while the right side reflects the millimoles of standard equivalent to the amount of analyte in the sample. This procedure uses data from several runs to

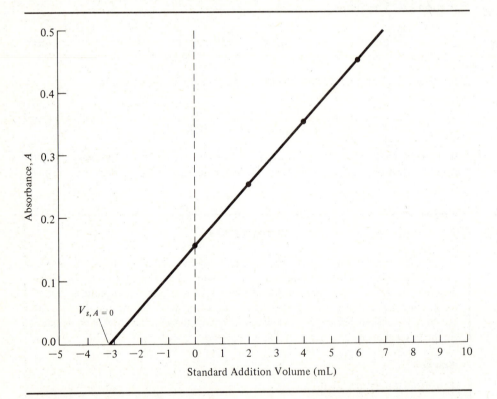

FIGURE 12-9   Standard addition technique. A series of standard additions is made to the sample, and the absorbance (corrected for volume change if necessary) is plotted. The intercept, $V_{s,A=0}$, is determined by the method of least squares (or by "eye"); it is −3.2 mL in this example. The amount of analyte present in the sample is then calculated from Equation 12-19.

obtain the "best" value of C_x—for example, a least-squares fit to the line may be employed.

Standard addition techniques are illustrated in several problems at the ends of Chapters 12 to 14, are described in several experimental procedures in the laboratory manual, and will also be encountered for some electroanalytical methods.

ANALYSIS OF MIXTURES

So far it has been assumed that only analyte x absorbs at the wavelength selected. Suppose that the sample contains two analytes, x and y, which have the spectral curves shown in Figure 12-10. The total absorbance at any wavelength will be the sum of absorbance due to x and absorbance due to y, that is,

$$A_\lambda = A_{\lambda,x} + A_{\lambda,y} = \varepsilon_{\lambda,x} b C_x + \varepsilon_{\lambda,y} b C_y$$

It is possible to use standard solutions containing only x or y to measure $\varepsilon_{\lambda,x}$ and $\varepsilon_{\lambda,y}$, but a measurement of A_λ still leaves one equation and two unknowns, C_x and C_y.

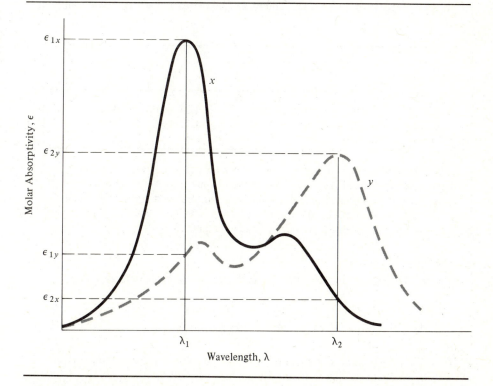

FIGURE 12-10 Absorption spectra for two analytes, x (solid curve) and y (broken curve). Analysis of a mixture containing x and y is accomplished by measuring the absorbance at λ_1 and λ_2.

To solve for both analytes, we need a second equation, which can be provided by measuring at a second wavelength. This gives us a set of simultaneous equations (where 1 refers to the first wavelength and 2 refers to the second wavelength):

$$A_1 = \varepsilon_{1x}bC_x + \varepsilon_{1y}bC_y \tag{12-20}$$

$$A_2 = \varepsilon_{2x}bC_x + \varepsilon_{2y}bC_y \tag{12-21}$$

The molar absorptivities, ε_{1x}, ε_{2x}, ε_{1y}, and ε_{2y}, are measured with standard solutions of x and y and the cell path length, b, is known. Thus Equations 12-20 and 12-21 can be solved for C_x and C_y if absorbance measurements are carried out at the *two* wavelengths 1 and 2. Example 12-7 shows a typical calculation.

EXAMPLE 12-7

Molar absorptivity data for cobalt and nickel complexes with 2,3-quinoxalinedithiol are

	510 nm	656 nm
ε_{Co}	3.64×10^4 L/mol $-$ cm	1.24×10^3 L/mol $-$ cm
ε_{Ni}	5.52×10^3	1.75×10^4

A 250-mg soil sample taken near an industrial metal-processing plant is dissolved, masking agents are added to avoid interferences from complexes other than Co or Ni, color is developed with addition of 2,3-quinoxalinedithiol, and the sample is diluted to 100 mL. The absorbance at 510 nm is 0.517 and at 656 nm is 0.405 in a 1.00-cm cell. Calculate weight percent Co and percentage of Ni in the soil.

Solution

If we assume A is due solely to absorption by the Co and Ni complexes,

$$A_{510} = 3.64 \times 10^4 C_{Co} + 5.52 \times 10^3 C_{Ni} = 0.517$$

$$A_{656} = 1.24 \times 10^3 C_{Co} + 1.75 \times 10^4 C_{Ni} = 0.405$$

These are two simultaneous equations that can be solved by the method in Appendix 1-B. Alternatively, algebraic substitution may be used:

$$C_{Ni} = \frac{0.517 - 3.64 \times 10^4 C_{Co}}{5.52 \times 10^3}$$

This idea can be extended to any number of components, n, by measurement at n wavelengths. However, an algebraic method (as in Example 12-7) is tedious for more than two components, and it is far simpler to use matrix algebra, as illustrated in Appendix 1-B. In this approach, the analyst needs only to enter all molar absorptivity values, the path length (alternatively, εb values can be determined with standard solutions without separation into ε and b), and the absorbances at n wavelengths. A computer will quickly generate the concentrations of all components. It can also verify that all equations are truly independent (for example, $x + 2y + 4z = 10$ and $2x + 4y + 8z = 20$ are not independent because the second one was generated from the first by multiplying by 2) and assign uncertainties to the component values arising from the calculation. Absorption spectrophotometry is

---

EXAMPLE 12-7 (cont.)

$$1.24 \times 10^3 C_{Co} + 1.75 \times 10^4 \left(\frac{0.517 - 3.64 \times 10^4 C_{Co}}{5.52 \times 10^3} \right) = 0.405$$

$$1.24 \times 10^3 C_{Co} + 1.64 - 1.15 \times 10^5 C_{Co} = 0.405$$

$$C_{Co} = \frac{0.405 - 1.64}{-1.14 \times 10^5} = 1.08 \times 10^{-5} \ F$$

$$C_{Ni} = \frac{0.517 - 3.64 \times 10^4 C_{Co}}{5.52 \times 10^3} = 2.24 \times 10^{-5} \ F$$

$$\text{amount Co in soil} = 1.08 \times 10^{-5} \ \frac{\text{mmol}}{\cancel{\text{mL}}} \times 100 \ \cancel{\text{mL}} = 1.08 \times 10^{-3} \ \text{mmol}$$

$$1.08 \times 10^{-3} \ \cancel{\text{mmol}} \times 58.9 \ \frac{\text{mg}}{\cancel{\text{mmol}}} = 0.0636 \ \text{mg}$$

$$\frac{0.0636 \ \text{mg Co}}{250 \ \text{mg soil}} \times 100 \ (\%) = 0.025\% \ \text{Co}$$

$$\text{amount Ni in soil} = 2.24 \times 10^{-5} \ \frac{\text{mmol}}{\cancel{\text{mL}}} \times 100 \ \cancel{\text{mL}} = 2.24 \times 10^{-3} \ \text{mmol}$$

$$2.24 \times 10^{-3} \ \cancel{\text{mmol}} \times 58.7 \ \frac{\text{mg}}{\cancel{\text{mmol}}} = 0.131 \ \text{mg}$$

$$\frac{0.131 \ \text{mg Ni}}{250 \ \text{mg soil}} \times 100 \ (\%) = 0.053\% \ \text{Ni}$$

not well suited to handle many components (the broad peaks will usually lead to large uncertainties with more than three or four components), but systems handling 50 or more simultaneous equations are available for some analytical methods, such as mass spectrometry.

DEVIATIONS FROM BEER'S LAW

The Lambert part of the Lambert–Beer law—absorbance is directly proportional to path length—has no known exceptions. However, when one takes real solutions of varying concentration and measures the absorbance with a real spectrophotometer, it frequently appears that Beer's law is not obeyed over wide ranges of concentration. This is one reason why the calibration-curve method, in which several standard solutions or standard additions are measured, is generally preferred. Frequently, a plot such as that in Figure 12-11 will show curvature at high concentrations, and the solution is said to deviate from Beer's law. What factors can cause deviation?

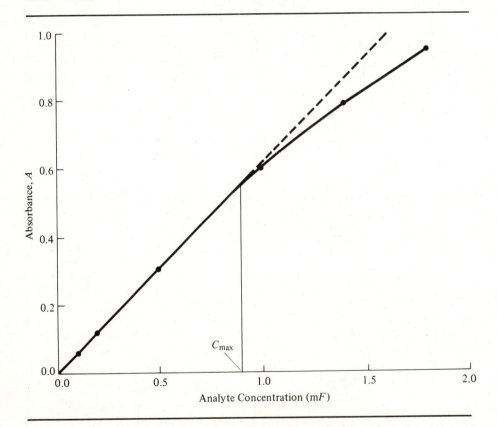

FIGURE 12-11 Spectrophotometric calibration curve. Beer's law holds up to C_{max}, but deviation from a straight line occurs at higher concentrations.

First of all, Beer's law has an inherent limitation: It assumes that there are no concentration-dependent interactions present that might make ε a function of concentration. This is strictly true only in very dilute solutions ($< 10^{-2}$ F) because ε is a function of refractive index, which changes significantly in more concentrated solution, and because interactions among the solute particles can alter ε. That is, Beer's law applies to infinitely dilute solutions in the same way that activity coefficients approach the value of 1. However, this type of deviation is not often a problem because absorbances are usually measured in dilute solution (see Examples 12-4 to 12-6).

Apparent deviations from Beer's law are much more prevalent. It has also been assumed that a *single* wavelength is employed in the absorbance measurement; Beer's law is true for *monochromatic* light. Real spectrophotometers must utilize a band of wavelengths (bandpass width) centering around some nominal wavelength. If ε varies with wavelength across this band of wavelengths, Beer's law is not obeyed. This is then another reason for measuring at an absorption peak, to minimize change in ε across the band of wavelengths used.

Finally, there are apparent deviations from Beer's law caused by chemical reactions involving the analyte. Absorption is due to a particular species in solution, and changes in species will lead to changes in absorbance. For example, the spectrum of a conjugate acid is not necessarily the same as its conjugate base. Standard solutions prepared to measure adherence to Beer's law (calibration curve) often consider only formal concentration and not the concentration of particular species. If the actual absorbing-species concentration is directly proportional to the formal concentration, Beer's law will still be obeyed (but ε calculated from the formal concentration will be incorrect). For example, suppose that absorbance measurements are carried out for solutions containing Fe(III) in a chloride medium and that absorption is due to both Fe^{3+} and $Fe(Cl)_4^-$. Then

$$Fe^{3+} + 4Cl^- \rightleftharpoons FeCl_4^-$$

$$K_f = \frac{[FeCl_4^-]}{[Fe^{3+}][Cl^-]^4}$$

$$\begin{aligned} A &= \varepsilon_{FeCl_4^-} b[FeCl_4^-] + \varepsilon_{Fe^{3+}} b[Fe^{3+}] \\ &= \varepsilon_{FeCl_4^-} bK_f[Fe^{3+}][Cl^-]^4 + \varepsilon_{Fe^{3+}} b[Fe^{3+}] \\ &= (\varepsilon_{FeCl_4^-} K_f[Cl^-]^4 + \varepsilon_{Fe^{3+}})b[Fe^{3+}] \end{aligned}$$

(12-22)

In terms of the formal concentration of Fe(III), Equation 12-22 becomes Equation 12-23,

$$\begin{aligned} f_{Fe} &= [Fe^{3+}] + [FeCl_4^-] \\ &= [Fe^{3+}] + K_f[Fe^{3+}][Cl^-]^4 = (1 + K_f[Cl^-]^4)[Fe^{3+}] \end{aligned}$$

$$[Fe^{3+}] = \frac{f_{Fe}}{1 + K_f[Cl^-]^4}$$

$$A = \left(\frac{\varepsilon_{FeCl_4^-} K_f[Cl^-]^4 + \varepsilon_{Fe^{3+}}}{1 + K_f[Cl^-]^4} \right) b f_{Fe}$$

(12-23)

Equation 12-23 has the same form as Beer's law except that, with formal concentration of iron used, the apparent molar absorptivity is the expression within the large parentheses. As long as the chloride concentration remains constant, Beer's law will be obeyed.

However, if the actual absorbing-species concentration is not directly proportional to the formal concentration, Beer's law is not obeyed. For example, if the absorbance of the air pollutant NO_2 is measured at a wavelength at which only NO_2 absorbs,

$$2NO_2 \rightleftharpoons N_2O_4$$

$$K = \frac{[N_2O_4]}{[NO_2]^2}$$

$$A = \varepsilon_{NO_2} b [NO_2] \qquad \varepsilon_{N_2O_4} = 0$$

$$f_{NO_x} = [NO_2] + 2[N_2O_4] = [NO_2] + 2K[NO_2]^2$$

$$= \frac{A}{\varepsilon_{NO_2} b} + \frac{2KA^2}{\varepsilon_{NO_2}^2 b^2}$$

$$\varepsilon_{NO_2} b f_{NO_x} = A + \frac{2KA^2}{\varepsilon_{NO_2} b} \tag{12-24}$$

The left side of Equation 12-24 has the form of Beer's law, but the right side is now quadratic with respect to absorbance. The absorbance does not rise linearly with formal concentration of nitrogen oxide. As concentration increases, the absorbance will not increase proportionally, and the plot will appear as in Figure 12-11.

The NO_2/N_2O_4 equilibrium involves dimerization in the gas phase. An example of dimerization in an aqueous medium involves formation of $Cr_2O_7^{2-}$ from CrO_4^{2-} in an acidic solution:

$$2CrO_4^{2-} + 2H_3O^+ \rightleftharpoons Cr_2O_7^{2-} + 3H_2O \tag{12-25}$$

Again, a Beer's law plot appears to deviate from a straight line unless the concentrations of the actual absorbing species are used. These examples are really *apparent* deviations from Beer's law because a formal concentration is used instead of the true concentration of absorbing species.

SUPPLEMENTARY READING

ARTICLES DESCRIBING ANALYTICAL METHODS OR APPLICATIONS

Bader, M., "A Systematic Approach to Standard Addition Methods in Instrumental Analysis," *Journal of Chemical Education*, 57:703, 1980.

REFERENCE BOOKS

Meehan, E.J., "Fundamentals of Spectrophotometry," in *Treatise on Analytical Chemistry*, 2nd ed., I.M. Kolthoff and P.J. Elving, eds., pt. 1, vol. 7, ch. 2. New York: John Wiley & Sons, Inc., 1981.

Olsen, E.D., *Modern Optical Methods of Analysis*. New York: McGraw-Hill Book Company, 1975.

Winefordner, J.D., ed., *Spectrochemical Methods of Analysis*. New York: John Wiley & Sons, Inc., 1971.

PROBLEMS

12-1. Complete the following table:

	λ (cm)	v (sec^{-1})	$\bar{v}$ (cm^{-1})
*a.	4.5×10^{-5}		
b.		3×10^{13}	
c.			2×10^{5}
d.		2×10^{10}	
e.	3×10^{-11}		
f.			1×10^{8}

12-2. To what region of the electromagnetic spectrum (that is, visible, UV, X-ray, and so on) does each of the following belong?
  a.   450 nm
  b.   3000 nm
  c. 10^{16} sec^{-1}
  d.   2 cm
  e. 10^{8} sec^{-1}
  f.   1000 cm^{-1}

12-3. To what region of the electromagnetic spectrum (that is, visible, UV, X-ray, and so on) does each of the following belong?
  a. 6×10^{-9} erg/photon
  b. 1×10^{-13} erg/photon
  c. 2.4×10^{5} J/mol (10^{7} ergs = 1 J)
  d. 1×10^{5} kcal/mol (4.18 J = 1 cal)
  e.   46 kcal/mol

*12-4. One compound being considered for utilization of solar energy is HI. The bond energy in HI is 71 kcal/mol. What wavelength and frequency of light are sufficiently energetic to break the HI bond?

12-5. Chlorophyll, the green pigment in plants, absorbs red light around 700 nm. What energy does this correspond to in ergs per photon and in kilocalories per mole?

12-6. One possibility for utilization of solar energy is to break a chemical bond with light energy, to produce a fuel. For example, H_2O could be split into H_2 and O_2, giving us hydrogen fuel to replace gasoline and/or natural gas. However, the

---

*Answers to problems marked with an asterisk will be found at the back of the book.

O—H bond is so strong that it requires light of 40,000 cm^{-1} (wave numbers) to break it.

    a.   In what region of the electromagnetic spectrum is this light?

    b.   Calculate the O—H bond energy in kilocalories per mole. Conversion factors and constants that may be useful are 10^7 ergs/J, 4.18 J/cal, 10^3 cal/kcal.

12-7. *a. Calculate the wavelength in nanometers of a methyl vibrational absorption at 2985 cm^{-1}.

    b.   Calculate the energy of the photon that causes this transition.

12-8. A certain aqueous solution transmits 32.0% of incident red light ($\lambda = 614$ nm). If 1.000 mL of this solution is mixed with 3.000 mL H_2O, what percentage of light will be transmitted?

*12-9. A substance absorbs at 550 nm and at 4×10^{-4} cm. What type of energy transition most likely accounts for each of these absorption processes?

12-10. A substance that has been excited emits electromagnetic radiation at 1.5×10^{18} sec^{-1} and at 6000 Å. What type of energy transition most likely accounts for each of these emission processes?

12-11. A solution exhibits 62% transmittance in a 1.00-cm cell. What will be its percent transmittance in a 2.00-cm cell?

12-12. Complete the following table:

	[x] (F)	A	T (%)	ε (L/mol-cm)	b (cm)
*a.			34	1 540	2.00
b.		0.613		250	10.0
c.	2.5×10^{-2}	0.167			1.00
d.	3.8×10^{-6}		92	4 765	
e.	2.1×10^{-2}			15	1.00

12-13. Complete the following table:

	[x] (F)	A	T (%)	ε (L/mol-cm)	b (cm)
*a.	4×10^{-4}	0.632			1.00
b.	6×10^{-4}			1 250	2.00
c.			75	690	1.00
d.	3×10^{-5}		62	1 387	
e.		0.318		20 400	2.00

12-14. Calculate the molar absorptivity of $K_2Cr_2O_7$ at 455 nm, given that 36.5 mg dissolved in 500 mL exhibits 12% transmittance at 455 nm in a 2-cm cell.

*12-15. The molar absorptivity of $KMnO_4$ is 2.2×10^3 at 525 nm. What is its absorptivity [in (cm-g/L)$^{-1}$]? What is its absorptivity [in (cm-ppm)$^{-1}$]?

12-16. Based on the information given in Problem 12-15, how many ppm $KMnO_4$ are in a solution if its absorbance is 0.238 at 525 nm in a 2.00-cm cell?

12-17. Calculate the molar absorptivity for each of the following substances at the wavelength specified (1-cm cell):

	Substance	Concentration	λ (nm)	Absorbance
*a.	$Fe_2(SO_4)_3$	0.25 g/L	320	1.25
b.	$KMnO_4$	22.0 mg/L	525	0.306
c.	CH_3CHO	0.51 μg/mL	180	0.115
d.	$HC_2H_3O_2$	1.47 mg/mL	208	0.782

*12-18. A simple colorimeter is used to compare a standard Fe o-phenanthroline complex solution (3.3×10^{-5} F) with a sample solution. If the light-path length for the standard is 2.5 cm and the sample-solution color matches that of the standard in intensity when its path length is 8.3 cm, what is the Fe concentration in the sample solution?

12-19. A simple colorimeter can be used to compare the color intensity of two solutions. If solution A, containing 2.2×10^{-3} F Co(II)–EDTA complex, and solution B, containing 8.3×10^{-2} F Ni(II)–NH$_3$ complex, have equal color intensity, what conclusion may be drawn regarding their molar absorptivities? What assumptions must be made to reach this conclusion?

12-20. The molar absorptivity of acetone (FW $= 58$) in hexane solvent is 900 at 188 nm. Calculate the maximum concentration (grams per liter) that could be used in a 5-cm cell so that absorbance will not exceed 0.9.

12-21. a.   The molar absorptivity of a colored substance is 1450 at 550 nm. Calculate the absorbance of a 0.6 mF solution in a 5-mm cell if measured at 550 nm.
  b.   What is the absorbance at 650 nm?

12-22. For many spectrophotometers, the lowest relative uncertainty is achieved when the absorbance is about 0.4. Calculate the optimum concentration (milligrams per liter) for the following metal ions, assuming a 1.00-cm cell:
  *a.   Ni^{2+} (as dimethylglyoxime complex, $\varepsilon = 3.5 \times 10^3$ at 350 nm)
  b.   Co^{2+} (as EDTA complex, $\varepsilon = 230$ at 580 nm)
  c.   Cu^{2+} (as cuproine complex, $\varepsilon = 6.4 \times 10^3$ at 546 nm)

12-23. An alloy sample weighing 487 mg and containing Ni is dissolved in HNO$_3$ and diluted to 100 mL. Part of the sample is transferred to a 1-cm cell, and the absorbance is 0.317 at 390 nm. Then in a 50-mL aliquot of the sample 436 mg of Ni(NO$_3$)$_2 \cdot 6H_2O$ is dissolved, and the absorbance is found to be 0.485 at 390 nm. Calculate the percentage of Ni in the alloy. Assume no volume change with the addition of nickel nitrate.

12-24. A 100-μL sample of a nickel plating solution containing KCN is diluted to 100 mL. Exactly 20 mL is removed and used to rinse and fill a 1-cm cell. The absorbance at 268 nm is 0.602. To the remaining 80 mL is added 10.00 mL of 2.00×10^{-4} F Ni^{2+}. After mixing, the solution is transferred to a cell, and the absorbance is now 0.734 at 268 nm. Calculate grams per liter of Ni(NO$_3$)$_2 \cdot 6H_2O$ in the plating solution.

*12-25. The Fe(II) complex with o-phenanthroline has a molar absorptivity of 1.00×10^4 at 510 nm. If an absorbance of 0.01 is the lowest detectable signal, what concentration (parts per million) can be detected in a 10-cm cell?

12-26. The absorbance of dye solutions is measured on a low-cost spectrophotometer (A) and on a high-quality (expensive) spectrophotometer (B). Account for the results shown below.

[dye] (F)	A	B
2×10^{-5}	0.22	0.24
4×10^{-5}	0.43	0.48
6×10^{-5}	0.63	0.72

12-27. An analyst neglected to use a blank cell when taking an absorbance measurement; that is, the analyst used the light beam itself to measure P_0 and obtained an absorbance reading of 0.38 for the sample. If the blank cell actually transmits only 90% of the radiant power, what was the determinate error (per cent)?

*12-28. A substance may exist in a monomer form (M) and a dimer form (D). Only the monomer form absorbs at 650 nm and has a molar absorptivity of 1200 L/mol-cm. If a solution having a formal concentration of 5×10^{-5} F has an absorbance of 0.120 in a 5.0-cm cell, calculate the equilibrium constant for the reaction.

$$2M \rightleftharpoons D$$

12-29. A weak organic acid in its dissociated (A^-) form absorbs at 360 nm ($\varepsilon = 2.5 \times 10^3$), but the undissociated form (HA) does not. If a 2.0×10^{-3} F solution of the acid in water has an absorbance of 0.50 in a 1.00-cm cell, calculate K_a for the acid.

*12-30. A weak-base dye solution exhibits an absorbance of 0.00 at pH 1, 0.235 at pH 9, and 0.362 at pH 12 and above. What is K_b for the dye?

12-31. A weak-acid dye solution exhibits an absorbance of 0.375 below pH 2, 0.472 at pH 5, and 0.862 above pH 10. What is K_a for the dye?

*12-32. A mixture of Ni^{2+} and Co^{2+} is analyzed by forming the salicylaldoxime complexes in a chloroform solvent. The molar absorptivities at 500 nm are $\varepsilon_{Ni} = 10$ and $\varepsilon_{Co} = 1000$, while at 400 nm both molar absorptivities are 5.00×10^3. If a solution in a 1-cm cell exhibits an absorbance of 0.091 at 500 nm and 0.615 at 400 nm, calculate the Ni^{2+} and Co^{2+} concentrations.

12-33. Rare-earth ions exhibit sharp but not very intense absorption peaks (low ε). The molar absorptivities for Dy^{3+} and Eu^{3+} are given below. If a solution containing these rare earths has an absorbance of 0.206 at 394 nm and 0.201 at 908 nm in a 5-cm cell, calculate the concentrations of Dy^{3+} and Eu^{3+} in the solution.

λ (nm)	ε_{Dy}	ε_{Eu}
394	0.50	3.06
908	2.46	0.00

12-34. Compounds X and Y and a sample containing X and Y have the following spectral information. Plot the spectra for X, Y, and the mixture, and recommend the best wavelengths for analysis. Calculate the concentration of X and Y in the sample.

λ (nm)	ε_X	ε_Y	A_{sample} (1-cm cell)
400	40	35	0.021
420	200	120	0.087
440	350	160	0.134
460	540	200	0.189
480	680	180	0.212
500	500	150	0.162
520	280	220	0.080
540	220	410	0.196
560	210	290	0.151

Spectrophotometry $\quad$ 13

13-1 INSTRUMENTATION

In Chapter 12, the basic principles of relating absorption of light to an analyte's concentration were developed. Now we must devise ways in which the measurements can be taken in the laboratory. From Beer's law, it can be seen that the radiant power transmitted by a blank and the radiant power transmitted by a sample must be measured at a particular wavelength while the radiant-power output of the source remains constant. Thus it is necessary to have (1) a source providing a stable output of radiant power, (2) a selector to allow choice of a particular wavelength, (3) a sample compartment, and (4) a radiant-power detector, which converts this optical signal to some type of readout. A block diagram of these basic components for a spectrophotometric system is shown in Figure 13-1.

We now examine the components to see how the requirements for UV–visible spectrophotometry may be met.

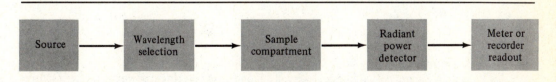

FIGURE 13-1 Block diagram of components of a spectrophotometer.

13-2   SOURCES

Sunlight or room light can be utilized as a radiation source for rough measurements, but such a source does not meet the requirement of high output stability. Also, the source should have its radiant-power peak in the wavelength region of interest. For UV–visible measurements, this range is 200 to 800 nm. As can be seen from Figure 13-2, the xenon arc source meets this requirement quite well although below 300 nm its output drops rapidly. Actually, the xenon arc matches the sun's spectrum closely and is frequently used in solar-energy research.

A tungsten filament lamp, also shown in Figure 13-2, is not ideal, with only 15% of its output in the visible region and little in the UV region. However, it is inexpensive and can provide a reliable, highly stable source. Therefore, it finds widespread use for measurements in the visible region.

The radiant power of a source depends on the operating voltage, V, as given in Equation 13-1,

$$P = kV^x \qquad\qquad\qquad \textbf{(13-1)}$$

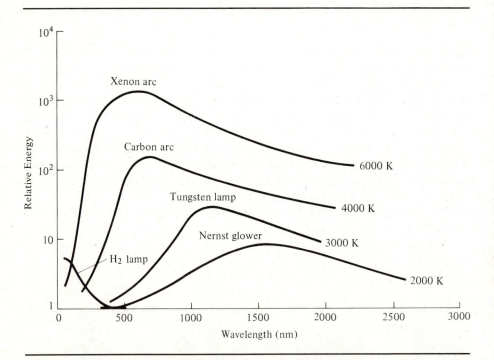

FIGURE 13-2   Sources of radiant power for spectrophotometry. A comparison of the relative energy of various sources will depend on the electrical power of the lamp source. H_2 lamps generally have lower electrical power than other sources do and consequently will exhibit lower relative radiant power than that shown.

where k is a constant and x is an exponent that depends on the type of source. For tungsten filament sources, x lies between 3 and 4. Because of this power dependence, the voltage stability must be high; for example, if $x = 4$, then a 1% fluctuation in operating voltage will give a 4% fluctuation in radiant-power output. A source stable to $\pm 0.2\%$ must possess a regulated voltage supply of $\pm 0.05\%$, or ± 0.003 V for a 6-V system.

For measurements in the UV region, electric-discharge sources are used in which excitation of the gaseous molecules is caused by the passage of electrons through the gas. The hydrogen lamp is common and has an output rising rapidly below 375 nm, as shown in Figure 13-2. There are both high-voltage lamps (~ 2500 V ac) and the more prevalent low-voltage lamps (~ 40 V dc). Radiant-power output continues down to 165 nm, but the quartz source window itself absorbs below 200 nm. The use of deuterium in place of hydrogen results in radiant powers several times greater than those from the hydrogen lamp, but the spectral range is basically the same.

Although the tungsten filament lamp has its peak power in the near infrared, its considerable amount of visible radiation makes it of little use as a source for infrared spectrophotometers. The tungsten filament is operated at a temperature of about 3000 K, while typical sources for infrared spectrophotometers such as the Nernst glower (see Figure 13-2) and the Globar operate at temperatures of < 2000 K. The lower temperature gives less visible light output and allows measurement to about 40,000 nm (40 μm).

13-3 WAVELENGTH SELECTION

Some sources of electromagnetic radiation provide only certain wavelengths. A laser is an obvious example. The helium–neon laser has an emission wavelength of 632.8 nm and is quite useful as a source in Raman spectrometers. Another example of a *line source*, as opposed to a continuum source like the tungsten filament, is the mercury arc. It emits several lines, of which the 435.8-nm line is frequently used in chemical studies. In this case, some type of wavelength selection is needed if one wishes to look at the interaction of one of the mercury arc spectral lines with a sample. In Chapter 14, another type of source will be discussed, which provides radiant power at discrete wavelengths. This source is a hollow cathode tube used in atomic absorption spectroscopy. However, even in this case, several lines are usually present, and some type of wavelength selection is required.

Line sources such as helium–neon lasers are not useful for absorption spectrophotometry because we should like to choose an appropriate wavelength at which the analyte absorbs. We may also wish to measure a mixture of two components at two wavelengths to determine the concentration of each. Tunable dye lasers have been developed and provide some amount of wavelength selection by "tuning" the source itself. In general, though, molecular absorption spectrophotometry is carried out with a source whose output is continuous over a reasonable wavelength region, and wavelength selection is a necessary component of the spectrophotometer.

Before proceeding to methods for wavelength selection, we should mention that absorption measurements can be made with a continuum source without wavelength selection. The color intensity of the sample solution compared with a standard can be measured by eye, using a *colorimeter* with daylight as the source. This method of measuring light absorption is not highly accurate and is not useful when the sample contains more than one colored species. However, there still exist colorimeters that are useful for some analytical applications. With a Duboscq colorimeter, the analyst determines the concentration of an absorbing species by preparing a standard and varying the path length (or path-length ratio) until standard and unknown appear by eye to have the same absorbance. The calculations are as follows:

$$A_x = \varepsilon b_x C_x$$

$$A_s = \varepsilon b_s C_s$$

When absorbances are matched, $A_x = A_s$; therefore,

$$\varepsilon b_x C_x = \varepsilon b_s C_s$$

The path-length ratio, $R = b_s/b_x = C_x/C_s$, and

$$C_x = C_s R \qquad\qquad \text{(13-2)}$$

Analyses accurate to within $\pm 5\%$ can be accomplished with such a system.

One relatively simple way to select a particular wavelength region is the use of a filter. Instruments with filters for wavelength selection are called *photometers*. There are two types of filter: absorption and interference. An absorption filter for a narrow band of wavelengths may actually contain two glass filters: One filter (A in Figure 13-3) absorbs strongly above a certain wavelength, while the second (B in Figure 13-3) absorbs strongly below a certain wavelength. The resultant light transmitted by the filter is curve C in Figure 13-3. Clearly, it would not be possible to pass only one particular wavelength, as in a laser; so the nominal wavelength of the filter is given by the maximum transmittance. A second important characteristic is the bandpass, which is the width of curve C at the half height shown in Figure 13-3. As the bandpass width is decreased, the transmittance of the filter decreases, and an absorption filter with a bandpass of 30 nm may typically transmit only 20% at its maximum point.

Interference filters consist of a thin dielectric film, such as MgF_2, sandwiched between two semitransparent metal films deposited on glass plates. When a beam of light strikes the filter, some light passes through the first plate, and a fraction of this is reflected by the second plate. Like the situation leading to X-ray diffraction, the light will cause constructive (amplitudes will add) interference with the incoming beam (if it is the proper wavelength) or cause destructive (amplitudes will subtract) interference (if it is not). The proper wavelength is given by Equation 13-3,

$$\lambda = \frac{2nb}{m} \qquad\qquad \text{(13-3)}$$

where n is the refractive index of the dielectric material, b is its thickness, and m is an

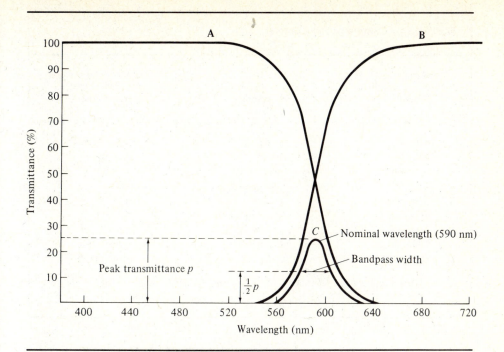

FIGURE 13-3 Wavelength selection with glass filters.

integer called the *order*. Thus the wavelength that will be transmitted is directly proportional to the thickness of the dielectric film, and interference filters are available with nominal wavelengths anywhere from the UV to well into the infrared region. Interference filters have an advantage over absorption filters in that they can exhibit higher transmittances at their nominal wavelength (typically 50%) and narrower bandpass widths (10 to 20 nm).

When picking a filter for a measurement of absorption, we must realize that the color of the filter will be the complementary color of the solution to be measured. Consider first a colorless solution: All wavelengths are transmitted, but the eye perceives this as white light because for every wavelength in the spectrum there is a complementary wavelength that results in white light. If a solution appears yellow, it means that yellow light is not being absorbed but that the complement to yellow (namely, blue) is being absorbed. Thus, to measure absorbance of a yellow solution, the analyst should employ a blue filter. These complementary relationships are set out in Table 13-1.

Wavelength selection in a *spectrophotometer* is accomplished by a *mono-chromator*, consisting of an *entrance slit*, a *dispersing element*, and an *exit slit*. After the incoming beam has been sharply defined by the entrance slit, it is dispersed into its component wavelengths by a prism or a grating dispersing element. Then a small portion of the spectrum is selected with the exit slit. The slit size controls the amount

TABLE 13-1   Complementary Colors

Wavelength region removed by absorption (nm)	Color absorbed	Complementary color of the residual light, as seen by the eye
400–450	violet	yellow-green
450–480	blue	yellow
480–490	green-blue	orange
490–500	blue-green	red
500–560	green	purple
560–580	yellow-green	violet
580–600	yellow	blue
600–650	orange	green-blue
650–750	red	blue-green

SOURCE: Data from W.J. Blaedel and V.W. Meloche, *Elementary Quantitative Analysis*, 2nd ed., p. 508 (New York: Harper & Row, Publishers, Inc., 1963).

of measured radiant power and controls the bandpass. A high-quality spectro-photometer will allow bandpass widths of <1 nm, while less expensive spectro-photometers may have fixed band widths of several nanometers.

Prisms are familiar to everyone. As shown in Figure 13-4, light is refracted as it enters one face and then a second time as it exits through a second face. The change in angle as light moves from one medium to another is illustrated in Figure 13-5 and is given by Snell's law:

$$n = \frac{\sin a}{\sin b} = \frac{v_1}{v_2} \qquad\qquad (13\text{-}4)$$

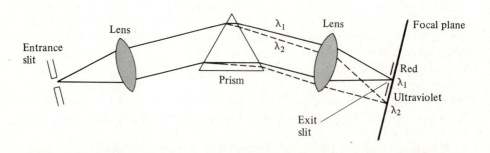

FIGURE 13-4   Diagram of a prism monochromator. Wavelength λ_1 is refracted at the proper angle to reach the exit slit. All other wavelengths, such as λ_2, will not reach the slit.

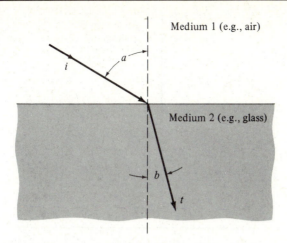

FIGURE 13-5 Refraction of electromagnetic radiation; i = incident beam, a = angle of incidence, b = angle of refraction, and t = transmitted beam.

where n is the index of refraction, v_1 is the velocity of light in medium 1, and v_2 is the velocity of light in medium 2. That the light is refracted as it enters or leaves the prism would not of itself disperse the light into its component wavelengths. However, the velocity of light in a medium depends on n. If medium 1 is a vacuum, the velocity of light is a constant, c, and therefore,

$$n_2 = \frac{c}{v_2} \tag{13-5}$$

in which n_2 is the refractive index of medium 2 with respect to vacuum. The critical point for the operation of a prism as a dispersing element is that v_2 is a function of wavelength and, therefore, so are n_2 and the refraction angle. Refractive-index values are normally quoted for the sodium D line (589.3 nm), and materials with the highest refractive index are best for dispersing the light (for example, $n = 2.42$ for diamond and may range from 1.5 to 1.9 for various types of glass). A material cannot be used as a prism in any wavelength region in which it absorbs; for example, glass absorbs <350 nm. Quartz is normally employed in prisms for UV measurements even though it is not quite so effective as glass in dispersing the light. In the infrared region, both glass and quartz absorb, and prisms made of NaCl or other alkali halides are generally used.

Gratings are the second means of dispersion. UV, visible, and infrared radiation can be dispersed with either a *transmission grating* or a *reflection grating*. Reflection gratings are common in spectrophotometers, and one is shown schematically in Figure 13-6. Basically, a grating is produced by forming a series of parallel grooves on the smooth surface of a material; for a transmission grating, glass may be used,

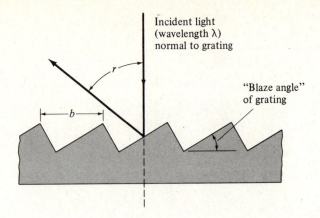

FIGURE 13-6 Diagram of a reflection grating used with normal-incidence light. The angle of reflection is *r*, and constructive interference will occur when Equation 13-6 is fulfilled.

each groove acting as a radiation source. Constructive interference for each wavelength of light will occur at particular angles to the grating.

Reflection gratings can be produced by ruling a metal surface with a diamond point or by depositing—by evaporation—a thin film of a metal (for example, Al) onto a ruled surface. The latter method may result in a *replica grating*, and it is a much less expensive method than producing an original ruled surface. One original, expensive, master grating can be a mold for plastic replicas. The art has been developed to a high degree.

For either transmission or reflection gratings, the spacing between grooves must be comparable to the wavelength of light. For example, for light striking a reflection grating at normal incidence, the angle of reflection *r* that gives constructive interference is

$$\sin r = \frac{m\lambda}{b} \tag{13-6}$$

where *b* is the ruling in centimeters per line when λ is in centimeters. The term *m* is the order and is an integer. This points out one disadvantage of gratings illustrated in Example 13-1. At the same angle, light will be observed corresponding to $m = 1$ (first order), $m = 2$ (second order), and so on. Filters can be used to eliminate unwanted orders, and special methods for ruling the surface emphasize the amount of light due to lower-order reflections.

If *b* becomes much larger than λ, the angle of reflection approaches 0° for all λ; therefore, there will be no dispersion. On the other hand, the sine function can never be greater than 1, and a grating that has *b* smaller than λ will act like a mirror and again will provide no dispersion. For use in the visible, rulings of 5000 to 10,000 lines per centimeter are typical.

EXAMPLE 13-1

A spectrophotometer employs a reflection grating with 1000 grooves per millimeter. At what angle will the first-order diffraction of 700-nm light occur? What other wavelengths will be observed at this same angle?

Solution

$$\sin r = \frac{m\lambda}{b}$$

For the first order,

$$\sin r = \frac{1 \times 700 \times 10^{-7} \text{ cm}}{10^{-4} \text{ cm/groove}} = 0.7$$

$$r = 44.4°$$

For other orders,

$$\lambda = \frac{b \sin r}{m} = \frac{10^{-4} \times 0.7}{2} = 3.5 \times 10^{-5} \text{ cm} = 350 \text{ nm}$$

m	λ (nm)
2	350
3	233
4	175

Of these only the 350-nm light from order 2 will be a problem if glass windows and cells are used because glass absorbs strongly below 350 nm.

Some spectrophotometers employ two dispersing elements, which can be two prisms, two gratings, or a prism and a grating. Double monochromators provide greater dispersion and, therefore, for a given slit size, more spectral resolution (smaller bandpass). In addition, a double monochromator reduces the amount of stray light (light reflecting off various components of the spectrophotometer and reaching the exit slit), which can cause an error in absorbance measurement.

13-4   SAMPLE COMPARTMENT

Recall from Chapter 12 that it was necessary to measure the transmitted radiant power from a blank to compensate for absorption by the solvent and sample container, reflection losses, and scattering losses. The compensation will be accurate only when these terms are exactly the same for both the blank solution and the sample solution. If the analyst uses the same cell (or *cuvette*) for both the blank measurement and the sample measurement, these terms should be identical. However, this approach is inconvenient when several samples are to be analyzed and an appreciable delay between the two measurements may occur. A change in radiant power from the source, perhaps due to voltage fluctuation in the power supply, would introduce a determinate error. More often, the analyst may use *matched cuvettes*, which have the same absorption and reflection characteristics as well as the same path length, so that one may be for the blank solution while the other is for the sample solution. Remember that, if an absorbance reading accurate to $\pm 0.5\%$ is sought, the path-length difference for 1-cm cells must be <0.005 cm! Even matched cuvettes will cause errors if they are not kept immaculately clean and free of fingerprints. Windows of cells must not be touched while samples are being transferred to the cuvette because even a minute amount of grease can significantly change the absorption and reflection characteristics of a cell. Matched cells should be checked periodically and recalibrated if their characteristics have changed.

The cuvette holder must also be designed carefully so that positioning of the cuvettes in the light path will always be the same. The cuvette wall should be exactly perpendicular ($90°$) to the light path to minimize reflection losses; for any change in this angle will alter the reflection losses.

Positioning and cell matching is especially difficult in cylindrical cells. Cylindrical cells are frequently used because they are less expensive than are flat-wall cells (all cuvettes have an inclination to break when dropped or handled improperly). For reasonable accuracy, one must insert the cylindrical cuvette into the sample compartment in exactly the same position each time; and cylindrical cuvettes will generally have an etched mark, which is to be lined up with a mark on the sample compartment. Even so, measurements with cylindrical cuvettes will not meet the same level of accuracy as will measurements with flat-wall cells, but these cuvettes are still quite useful when errors of 1 or 2% are acceptable.

Cuvettes for UV–visible spectrophotometry frequently have path lengths of 1 cm although cells of 0.1 to 10-cm path length can be purchased and will fit sample compartments for many spectrophotometers. For measurements in the UV region, quartz or fused-silica cuvettes can be employed down to 200 nm. Glass, quartz, or fused silica cannot be used in the infrared region because they absorb; so cells with windows made of NaCl are frequently used for infrared measurements. Clearly, aqueous solutions or even organic solvents containing appreciable amounts of water cannot be measured in NaCl cells!

Although some absorption by the solvent can be tolerated because it is compensated for by the measurement with the blank, the solvent must be essentially transparent in the wavelength region under investigation. Otherwise, it would absorb so much of the radiant power of the source that little light would reach the detector from the sample or the blank measurement. Molar absorptivity values near a small absorption peak are typically >10, and values for absorption peaks of analytical interest are usually 1000 to 50,000. The pure solvent will have a molar concentration of perhaps 10 F ($H_2O = 55\ F$, and ethanol $= 17\ F$), and therefore absorbance of solvent in a 1-cm cell would be >100 for even a small molar absorptivity value:

$$A = \varepsilon bC > 10 \times 1 \times 10 = 100$$

This means that the fraction of light transmitted would be 10^{-100}! Many solvents, including H_2O, are sufficiently transparent to be used in the UV–visible region of the spectrum, but no common solvents are transparent throughout the infrared region. Often pure liquid samples are measured, but in light of the discussion above, it is necessary to use extremely narrow path lengths. Molar absorptivities for infrared peaks are quite small; but even so, path lengths of 0.01 cm or less are often required.

13-5 DETECTORS

The human eye is employed as a detector in a colorimeter. Although the eye cannot measure absorbances directly with high accuracy, it is reasonably capable of matching solutions with equal overall absorbance. However, the eye is sensitive only to light of 400 to 700 nm wavelength, and even in this range, it is more sensitive to yellow light (~ 600 nm) than it is to other wavelengths. The eye cannot sort out and look at one color in the presence of others. Thus it is severely limited, and other methods for detecting radiant power are needed.

Radiation detectors generally convert a light signal to an electrical signal by a *transducer*. A major requirement is that the resulting signal, S, should be directly proportional to the original radiant-power signal, P:

$$S \propto P \quad \text{or} \quad S = kP$$

The proportionality constant, k, represents the sensitivity of the detector, which we should like to be high so that low light intensities can be measured. In addition, even when no light is supposedly striking the detector, some background electrical signal is generally observed, which is called the dark current, d:

$$S = kP + d \tag{13-7}$$

Photometers and spectrophotometers usually have a means of compensating for dark current (adjusting signal so that $S = 0$ when $P = 0$; therefore, $S = kP$). Unfortunately, it is not simple to measure k because a calibrated light source would be needed and any losses between source and detector must be accounted for. In addition, k varies with wavelength. However, we are not really interested in measuring P but in measuring absorbance, which is $\log (P_0/P)$. This ratio is determined simply without a knowledge of k, as shown in Equation 13-8:

$$\text{blank:} \quad S_0 = kP_0$$

$$\text{sample:} \quad S = kP$$

Therefore,

$$A = \log \frac{P_0}{P} = \log \frac{S_0/k}{S/k} = \log \frac{S_0}{S} \tag{13-8}$$

Several detectors are available that convert photon signals into electrical signals; they are called *photoelectric detectors*, or *photoelectric transducers*.

A comparatively simple, inexpensive, and rugged photoelectric detector is the *barrier-layer cell*. It consists of an iron electrode coated with a semiconductor, typically selenium. On top of the semiconductor, a thin metal film, such as Ag, is deposited, which acts as the second electrode for this photocell (as shown in Figure 13-7). Photons strike the selenium–silver interface and, if sufficiently energetic (<700-nm wavelength), will promote an electron from the valence band of selenium to its conduction band. The resulting mobile electron is collected by the silver electrode. Thus a potential difference is established between the iron electrode and the silver electrode. The resistance of the system is kept quite low, allowing a current to pass directly proportional to the intensity of the incident radiation. Therefore, no external power supply is required.

The disadvantages of the barrier-layer cell are that it is similar to the human eye in its wavelength response, having a maximum response at 550 nm and little response below 250 nm or above 750 nm. Its electrical output may be several microamperes, which can be read directly; but because of the low internal resistance and slow response, a low signal cannot be amplified—which limits its usefulness to reasonably high light intensities.

A more sensitive photoelectric detector is the *phototube*. The sensitivity is not due to the current produced by the direct conversion of photons to electrons but occurs because the output can be amplified. Thus the output of the phototube is sent to a dc

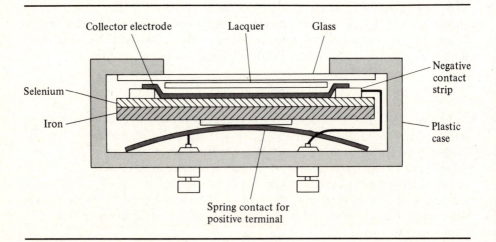

FIGURE 13-7   Diagram of a barrier-layer photocell.

amplifier, which magnifies the signal many times, and the output from the amplifier is then read on a meter.

The phototube looks much like a vacuum tube for a TV or radio. Inside the evacuated chamber is a light-sensitive cathode consisting of layers of silver–silver oxide–alkali metal. This surface has a low work function (it takes little energy to remove an electron). In this case, an impinging photon must have sufficient energy to remove an electron from the surface. The energy required depends on the method of cathode preparation, especially on which alkali metal is used since the ionization potential for the alkali metals decreases as the atomic number increases. Thus a cesium-coated cathode requires less energetic photons (< 1000 nm) than a sodium-coated cathode (< 500 nm) does.

An applied voltage (~ 90 V) between the photocathode and a wire anode causes the liberated electrons to be accelerated by the electric field and move to the anode, as shown in Figure 13-8. This photocurrent is then amplified to yield a measurable signal even at reasonably low light intensities. The low wavelength sensitivity is

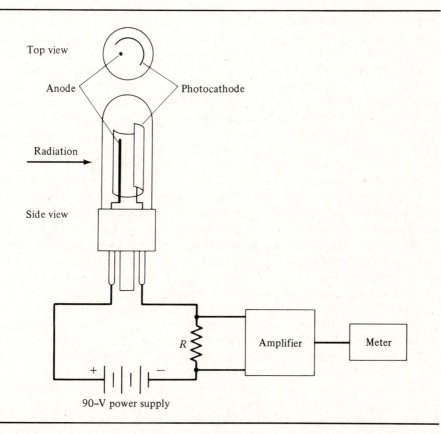

FIGURE 13-8 Schematic diagram of a phototube. Current due to photons striking the photosensitive cathode is measured by the potential drop in R and then amplified.

limited only by the absorptivity of the window; fused-silica windows are available for UV measurements.

The ultimate in sensitivity is achieved in *photomultiplier tubes*. In principle, the conversion of light energy to electrical energy is accomplished in the same way as in a phototube. However, instead of just collecting the electrons that reach the anode, the anode surface consists of a special Be–Cu or Cs–Sb coating that, when an electron strikes it, emits several (two to five) secondary electrons. With an electric field, these secondary electrons are directed to another electrode, and the number of electrons is multiplied again. Typically, the photomultiplier tube utilizes nine of these plates (called *dynodes*) before collecting the electrons at the actual anode. If the average number of secondary electrons for each impinging electron is four from each dynode, then the multiplication factor is $4^9 = 2.6 \times 10^5$. Multiplication factors of >1 million can be accomplished in some photomultiplier tubes. Because each succeeding dynode must be at a higher voltage (~ 90 V) than the previous one to accelerate the electrons, the power supply is typically 900 V, as shown in Figure 13-9. In addition to the millionfold amplification within the photomultiplier tube, external amplification of the signal can also be carried out. The limit of detection is determined by the dark current caused by stray electrons' being emitted. For some applications, especially in the low-sensitivity red region of the spectrum, the photomultiplier tube is cooled to decrease dark current.

In recent years, advances in solid-state electronics have made possible the introduction of optoelectronic image devices (OID) as detectors for spectrochemical measurements. Arrays of photodiodes convert light signals to electrical charge, which is recorded. The entire dispersed light from a prism or grating strikes the array, the photodiode position being associated with a particular wavelength band. In this way, the entire spectrum can be measured at one time instead of one wavelength band at a time, as in the conventional spectrophotometer.

None of the detectors described above is suitable for measurements in the infrared region. Infrared spectrophotometers employ detectors that respond to the heat generated when infrared radiation strikes them. Thus a relatively simple *thermocouple*[1] can be used, one junction of which has a surface blackened to act as an efficient absorber. The extremely small temperature change causes a correspondingly small thermoelectric voltage and results in a small current. The current can be greatly amplified, leading to a measurable signal. A more sensitive detector is a *thermopile*, which consists of a number of thermocouple junctions.

Before leaving the subject of detectors, we should mention the use of photographic detection. Photographic film is sensitive to visible, UV, and high-energy radiation such as X rays. Recording a spectrum on photographic film is accomplished with instruments called *spectrographs*. This method is useful when extremely low levels of emitted radiation must be monitored for long periods of time (such as in X-ray-diffraction or atomic-emission spectroscopy). However, photographic detection is not useful for measuring absorbance since film exposure times may be long and film development slow; film intensity is often determined by eye although

---

[1] A thermocouple is created in a circuit containing two junctions of two dissimilar metals each. When the two junctions are at different temperatures, the thermocouple develops a voltage approximately proportional to the temperature difference between the two junctions.

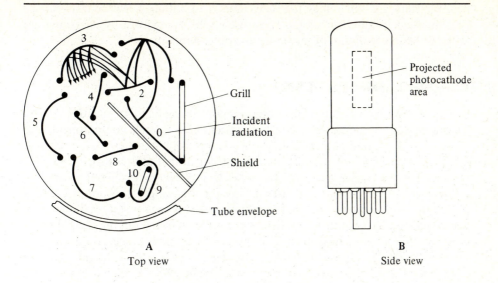

A
Top view

B
Side view

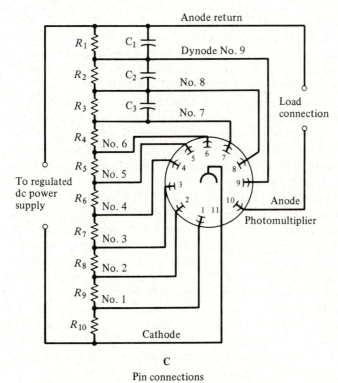

C
Pin connections

FIGURE 13-9   Schematic diagram of a photomultiplier tube.

densitometers are sometimes used. Accuracies to within a few percent can be accomplished in atomic-emission spectroscopy by comparison of film intensity from the sample with that observed with various standards.

13-6 SPECTROPHOTOMETER DESIGN

So far, we have discussed the four basic components required for spectrochemical measurements: (1) source, (2) wavelength selection, (3) sample compartment, and (4) detector. Now we put them together into a spectrophotometer.

First, however, we should note that an instrument that utilizes the entire visible spectrum of the source measures the total "color" of the sample; it is called a *colorimeter*. An instrument that uses a filter for wavelength selection is called a *photometer*; any instrument that measures radiant power (at one or more wavelengths) with a photoelectric detector is a *spectrometer*. On the other hand, a *spectrophotometer* is an instrument that specifically contains a dispersing element: a prism or a grating (or possibly both in a double monochromator instrument).

It should also be noted that there are commercially available many spectrophotometers of varying design and price tag. Prices range from a few hundred dollars to above $20,000. Thus it is not possible to discuss all, or even a major fraction of, the spectrophotometers available. Let us then look at one simple, widely used spectrophotometer: the Spectronic ® 20 manufactured by Bausch & Lomb, Inc. The external appearance of this instrument is pictured in Figure 13-10(**A**); the internal schematic design is shown in Figure 13-10(**B**).

The source in this instrument is a tungsten-filament lamp, and an optical system of lenses and an entrance slit directs the light beam to a replica reflection grating, which disperses the radiation. Wavelength control is accomplished by turning the grating (wavelength cam) so that the desired wavelength region will pass through the exit slit. The slit is fixed; and because dispersion from a grating is constant with wavelength (unlike that from a prism), the bandpass of 20 nm is constant throughout the wavelength region. To set 100% transmission with the blank, the analyst can block part of the light with a light control.

The sample compartment accepts cylindrical cuvettes, and even inexpensive test tubes may be employed. Finally, the light reaches a phototube detector, where it is converted to an electrical signal, amplified, and incorporated into a bridge circuit. A meter, also in the bridge circuit, is graduated in a linear transmittance scale (0 to 100%) and a nonlinear absorbance scale. Normally, a blue-sensitive phototube is used, and the instrument range is 340 to 650 nm. However, the range may be extended to 950 nm when a red-sensitive phototube is substituted. Because higher orders from the grating (see Example 13-1) may interfere, measurements above 660 nm also require a red-colored filter, which passes the long-wavelength red and absorbs any short-wavelength light.

The Spectronic ® 20 has sacrificed some accuracy to achieve a low-cost, simply-operated instrument, which is still capable of yielding absorbance readings accurate to ±1 or 2%.

What kinds of design change are made to achieve higher accuracy and a wider wavelength region? The Beckman DU (more recently designated DU-2) was first

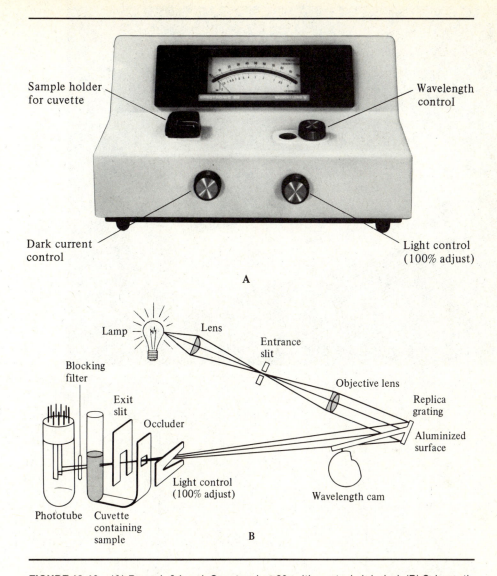

FIGURE 13-10 (**A**) Bausch & Lomb Spectronic ®20, with controls labeled. (**B**) Schematic diagram of the optical system of the Bausch & Lomb Spectronic ®20. (Courtesy of Bausch & Lomb Analytical Systems Division.)

introduced over 30 years ago and is capable of high accuracy; but measurements are relatively slow. By using a hydrogen lamp in addition to a tungsten lamp and a quartz prism, the analyst can select wavelengths down to 190 nm. The upper limit is 1000 nm. The slits are adjustable from 0.01 to 2 mm, allowing a bandpass of 0.01 nm in the UV to 0.5 nm in the visible. A photomultiplier tube is used for measurements below 625 nm.

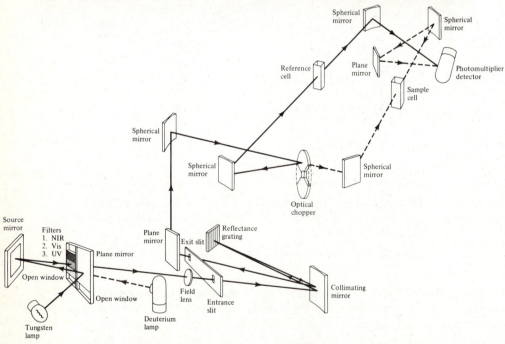

FIGURE 13-11 Perkin-Elmer Lambda 3 microcomputer-controlled UV–visible spectrophotometer: (**A**) photograph and (**B**) schematic of the optical system.

More rapid—but still accurate—measurements can be accomplished with a recording spectrophotometer (available models include those manufactured by Cary, Perkin-Elmer, and Beckman). A continuous change in wavelength requires that a *double-beam instrument* be designed so that light through the blank and through the sample can be monitored continuously. An example of a modern double-beam spectrophotometer is shown in Figure 13-11. This instrument contains a holographic grating monochromator and double-beam operation. It is controlled by a microcomputer, which allows the analyst to carry out repetitive scans of a spectrum, measure absorbance at specified wavelengths, and read concentration directly.

The cost of these instruments is high because of the complicated optics system as well as the recorder needed for recording absorbance *versus* wavelength. The additional cost is often justified because of the tremendous savings in the time required to measure the entire spectral response of new materials or to follow the spectral changes that occur while a chemical reaction takes place.

13-7 INSTRUMENTAL ERRORS IN ABSORPTION MEASUREMENTS

In gravimetric and volumetric analysis, errors involve weight and volume measurements. In absorption measurements, weight and volume measurements are also involved. For example, in the determination of manganese in steel, a sample must be weighed; and after dissolution and formation of MnO_4^-, the sample is diluted to a known volume. The reason for this known-volume requirement is that absorption measurements determine *concentration* (Beer's law), and to find the total *amount* of manganese, we must multiply concentration by volume. In addition to these sources of determinate and indeterminate error, there is uncertainty in the measurement of absorption and the use of Beer's law:

$$C = \frac{A}{\varepsilon b}$$

If a molar absorptivity value is used, then its uncertainty as well as the uncertainty in path length, b, must be included. Usually, a calibration curve is determined, and the slope of the line is εb. Thus there is only one uncertainty that may introduce error, namely, the determination of slope. Fitting the calibration data to a straight line by the method of least squares will also allow determination of a standard deviation for the slope. However, it should be kept in mind that, if several cuvettes are used, uncertainty in path length should be included.

To measure absorbance, the analyst must adjust dark current (0% T) and transmitted light through the blank (100% T) and then measure transmitted light through the sample. Each of the readings involves transmittance; frequently, an overall uncertainty in transmittance results, which is constant for the instrument and operator combination. This value can be determined; and it can be verified that it is constant by taking several replicate measurements of several samples with varying absorbances.

EXAMPLE 13-2

Transmittance measurements on a standard sample give the following values: 36.1, 36.5, 35.9, 36.2, 36.1, and 36.4%. If the standard deviation for the calibration line alone is 0.8%, what is the overall uncertainty in measuring concentration if the absorbance is (a) 0.1, (b) 0.4, and (c) 1.0?

Solution

(a)
$$\bar{T} = 36.2 \qquad \sigma_T = \sqrt{\frac{\Sigma(T - \bar{T})^2}{N - 1}} = 0.2\%$$

$$\sigma_{r,c} = \frac{\sigma_T}{2.303\,T \log T}$$

$$A = 0.1$$

Therefore,

$$T = \frac{1}{10^{0.1}} = 0.79 \qquad \sigma_{r,c} = \frac{0.2}{2.303 \times 0.79 \times \log 0.79} = 1.1\%$$

Note that, if σ_T is given in percent, then $\sigma_{r,c}$ will be in percentage. Also note that $\sigma_{r,c} = -1.1\%$ because $\log 0.79 < 0$. However, standard deviations do not have sign and are really plus or minus since they represent a measure of spread around the central value. If this equation is used to calculate error, it is necessary to recognize that a positive error in T will result in a negative error in C.

$$\text{overall } \sigma_{r,c} = \sqrt{\sigma_{r,c}^2 + \sigma_{r,\text{cal}}^2} = \sqrt{1.1^2 + 0.8^2} = 1.4\%$$

(b)
$$A = 0.4$$

Therefore,

$$T = \frac{1}{10^{0.4}} = 0.40 \qquad \sigma_{r,c} = \frac{0.2}{2.303 \times 0.40 \times \log 0.40} = 0.5\%$$

$$\text{overall } \sigma_{r,c} = \sqrt{\sigma_{r,c}^2 + \sigma_{r,\text{cal}}^2} = \sqrt{0.5^2 + 0.8^2} = 0.9\%$$

(c)
$$A = 1.0$$

Therefore,

$$T = \frac{1}{10^1} = 0.10 \qquad \sigma_{r,c} = \frac{0.2}{2.303 \times 0.10 \times \log 0.10} = 0.9\%$$

$$\text{overall } \sigma_{r,c} = \sqrt{\sigma_{r,c}^2 + \sigma_{c,\text{cal}}^2} = \sqrt{0.9^2 + 0.8^2} = 1.2\%$$

In the case of a constant uncertainty in transmittance, its effect on the measurement of concentration can be calculated with the calculus, as in Appendix 1-J. The result is

$$\text{relative error in } C = \frac{\Delta T}{2.303\, T \log T}$$

$$\frac{\sigma_C}{C} = \sigma_{r,c} = \frac{\sigma_T}{2.303\, T \log T} \tag{13-9}$$

If ΔT is given as a standard deviation, σ_T, then the relative error in C will be a relative standard deviation for the measurement of concentration, $\sigma_{r,c}$.

Equation 13-9 shows that the relative standard deviation (or error if ΔT represents a known error) depends on T and approaches infinity as T approaches 0 or 1. It reaches a minimum when $T = 0.368$ ($A = 0.434$), as shown in Appendix 1-J, and a typical problem is given in Example 13-2.

Sometimes it is found that σ_T is not constant but is a function of T. This occurs in more accurate spectrophotometers, for which reading transmittance has little uncertainty. For example, more expensive spectrophotometers normally have scale-expansion capabilities—equivalent to having transmittance scales of several feet in place of the 5 to 7 in. on direct-reading, low-cost spectrophotometers. Such uncertainties as positioning the cuvette reproducibly in the sample compartment or fluctuations in source intensity during reading have been found to give σ_T as a linear function of T. When all other uncertainties have been made small, it is sometimes possible to see *shot noise* in the electronics, random fluctuations in the output of phototubes and photomultipliers. Here σ_T is proportional to $\sqrt{T^2 + T}$. In this situation, the minimum relative error in C occurs at $T = 0.28$ ($A = 0.55$), not greatly different from the simple case where σ_T is a constant.

13-8 APPLICATIONS

With the principles of spectrophotometry developed in Chapter 12 and the instruments available for measuring absorption, we now briefly examine the kinds of analytical application that involve spectrophotometry.

ABSORBING SPECIES

Figure 12-3 indicated that absorption in the UV–visible region of the electromagnetic spectrum involves electronic energy levels. Therefore, for a species to show an absorption band, it must have a reasonably low-lying excited electronic state. Organic compounds containing only single bonds, such as alkanes, are so stable that the ground state is at a very low energy compared to any excited state. The wavelength required to excite such molecules is <200 nm, a spectral region not ordinarily accessible because nitrogen and oxygen in air absorb in this same high-energy region of the UV. Instruments designed to observe such transitions operate in a vacuum environment, and the name *vacuum ultraviolet* has been applied to this region of the spectrum.

Organic molecules containing double or triple bonds have electronic excited states with energy levels above the ground state corresponding to UV or even visible light. Functional groups that absorb in the UV–visible are called *chromophores*; some examples are given in Table 13-2. The absorption maxima and molar

TABLE 13-2   Electronic Absorption Characteristics for Some Common Chromophores

		Typical values	
Chromophore	Structure	λ_{max} (nm)	ε_{max}
aldehyde	$>\!C\!=\!O$ with H	180 / 290	10 000 / 12
alkene	$>\!C\!=\!C\!<$	190	8 000
alkene, conjugated	$>\!C\!=\!CH\!-\!CH\!=\!C\!<$	220	20 000
alkyne	$-C\!\equiv\!C-$	195	2 000
amine	$-NH_2$	195	2 800
aromatic	(benzene ring)	205 / 255	8 000 / 200
bromide	$-Br$	208	300
carboxyl	$-C(\!=\!O)OH$	210	50
ester	$-C(\!=\!O)OR$	205	50
ketone	$>\!C\!=\!O$	190 / 290	1 000 / 20
nitrate	$-ONO_2$	270	12
nitro	$-NO_2$	210	5 000
nitroso	$-N\!=\!O$	300 / 665	100 / 20
oxime	$-NOH$	190	5 000
thiocyanate	$-S\!-\!C\!\equiv\!N$	250	50
thiol	$-SH$	195	1 400

absorptivity values are only approximate because peaks in this region are broad and shift with slight changes in structure or solvent. Larger changes may occur when more than one chromophore is present, with shifts toward lower energy (higher wavelength), called *bathochromic* shifts, usually observed. In some instances, change in solvent may shift an absorption peak to higher energy (*hypsochromic* shift).

Halogen atoms and other atoms containing nonbonding electron pairs also have excited states that can be reached by absorption of UV light.

Inorganic ions of representative elements such as alkalies and alkaline earths do not absorb in the UV–visible region, but most transition metal ions do. In fact, many of these transitions are in the visible region, giving rise to the familiar colored complexes. The reason for these transitions lies with the *partially* filled d or f levels. Groups (ligands) bonded to the transition metal ion split the d and f levels, as shown in Figure 13-12. If the lower set contains electrons while the upper set is not completely occupied, absorption of light corresponding to the difference in energy levels may occur. However, these d–d transitions have a low probability (they are *forbidden*), and therefore the molar absorptivities are orders of magnitude smaller than they are when probability is high. For example, absorption of light may result in electron density's being transferred from the metal ion to the ligand (or vice versa), and these *charge-transfer* bands frequently exhibit high probability. Consequently, molar absorptivities are generally large for such a process.

Analytical chemists have found that intense absorption charge-transfer bands can be obtained if the metal ions are complexed with particular ligands, thus forming the basis for analytical procedures. (Entire books have been written on reagents for spectrophotometric analysis.) Absorption peaks are again broad for transition-metal ions and their complexes when d levels are involved. However, the $4f$ (lanthanides) and $5f$ (actinides) electrons are screened from the effects of ligands, and ions from these groups often exhibit sharp bands.

Infrared spectrophotometry involves vibrational energy levels (see Figure 12-3), and again organic functional groups exhibit characteristic absorption peaks. In addition, the infrared region is usually rich in peaks because there are so many vibrational modes ($3n - 6$, where n is the number of atoms for any nonlinear molecule). The infrared spectrum can be used as a fingerprint to identify a particular

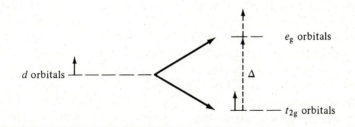

FIGURE 13-12 Splitting of d orbitals in an octahedral field. Absorption of light of energy equal to the energy difference, Δ, will promote an electron from the lower set to the upper set.

molecule. However, the molar absorptivities are much lower than are electronic charge-transfer bands; and the disparity makes quantitative analysis difficult. Infrared spectrophotometry is predominantly applied to qualitative organic analysis. For one thing, inorganic atoms are generally heavier than the C, H, N, and O atoms of organic compounds, and this leads to vibrational energy transitions at low energies (high wavelength), a region not readily accessible. Also, inorganic species are customarily encountered in aqueous media, which cannot be studied in the common NaCl cells used for infrared spectrophotometry. Probably the main reason for the lack of interest in infrared spectrophotometry for studying inorganic species is that other tools, such as UV–visible spectrophotometry, are available and capable of better quantitative accuracy.

To summarize, UV–visible spectrophotometry is capable of reasonably high accuracy ($\pm 1\%$ is typical) for some organic compounds that contain chromophores and for many transition-metal complexes. Molar absorptivities can be $> 10,000$ for charge-transfer bands and some chromophores, allowing analysis of solutions containing $< 10^{-6}$ F. By comparison, infrared spectrophotometry is used primarily for qualitative identification of organic compounds.

INDUSTRIAL APPLICATIONS

Spectrophotometry is ideally suited to routine analyses required for process control, to inspection of incoming materials, and to quality assurance. Spectrophotometers need not be very expensive so that even small companies can afford one. Accuracies of 1 to 2%, routinely achievable in spectrophotometric procedures, are more than adequate for most of these applications. Because samples are usually diluted to some specified volume before absorbance is measured, wide ranges of analyte values can be handled by appropriate choice of this dilution volume, With the use of highly intense absorption bands (high molar absorptivity values), trace amounts of impurities can be successfully analyzed.

There are also convenience features that make spectrophotometry attractive for industrial applications. Operation of an instrument and interpretation of the results can normally be carried out by a lab technician although infrared spectrophotometry is a little more sophisticated and may require an analytical chemist for interpretation. In fact, calculations can become trivial when standard reference samples are used. Consider steel samples to be analyzed for manganese content: Along with the samples for analysis are samples with known manganese content (these can be purchased at reasonable cost and, for work of the highest accuracy, may be obtained with values certified by the National Bureau of Standards); now it does not matter what the dilution factors are, what the molar absorptivity of the absorbing MnO_4^- species is, or how long the path length may be because the absorbance will be directly proportional to concentration and, provided that the same sample weight is used, will be directly proportional to the percentage of Mn in the steel. Thus

$$\text{unknown:} \quad A_x = k(\% \text{ Mn}_x)$$

$$\text{standard:} \quad A_{std} = k(\% \text{ Mn}_{std}) \quad \text{or} \quad k = \frac{A_{std}}{\% \text{ Mn}_{std}}$$

Therefore,

$$\% \ Mn_x = \frac{A_x}{A_{std}} (\% \ Mn_{std}) \qquad\qquad (13\text{-}10)$$

This approach also avoids some problems from interferences. Other constituents in the steel may also absorb a small amount at the wavelength chosen for analysis. If the standard sample contains comparable amounts of these constituents, their effect will be approximately compensated for. The simplicity of this approach, also used in many clinical analyses, is demonstrated in Example 13-3.

EXAMPLE 13-3

A 600-mg steel sample is dissolved in 50 mL of 4 F HNO$_3$, any carbon present is oxidized with ammonium peroxodisulfate, and the sample is diluted to 100 mL in a volumetric flask. A 20.0-mL aliquot is taken, and Mn is oxidized to MnO_4^- with KIO$_4$. The absorbance is measured, using a second 20.0-mL aliquot not containing KIO$_4$ as a blank. The absorbance is 0.452. A reference steel sample containing 0.243% Mn is treated by the same procedure and gives an absorbance reading of 0.479. Calculate the percentage of Mn in the steel sample.

Solution

$$\% \ Mn_x = \frac{A_x}{A_{std}} (\% \ Mn_{std})$$

$$= \frac{0.452}{0.479} \times 0.243 \ (\%) = 0.229\%$$

Finally, UV–visible spectrophotometry can be automated so that a technician can analyze hundreds of samples per day. The cost per determination with such automated equipment is frequently below $1.

It is impossible to describe or even list all spectrophotometric procedures used in industry, but a few examples are given in Table 13-3 and some are commented on here: A sequence of reactions may be necessary to produce an absorbing species. For example, penicillins react with hydroxylamine to give hydroxamic acid. The hydroxamic acid then forms a complex with Fe(III), which is the species measured spectrophotometrically. Sometimes it is necessary to correct for interferences by measuring the absorbance at another wavelength. For example, in the analysis of yeast for ergosterol, the total ergosterols are measured at 281.5 nm while dehydro-ergosterol is measured at 230 nm. The difference is the ergosterol content. Analysis of metals to determine alloy composition is one of the largest areas of application for absorption spectrophotometry. Table 13-3 lists just a handful of examples. For essentially all alloys, procedures have been published for the analysis of major, minor, and trace constituents. Many of these procedures have also been applied to the determination of metal content in environmental samples.

TABLE 13-3   Examples of Some Industrial Applications of Spectrophotometry

Industry	Sample	Analyte	Reagent(s)	λ (nm)
drug	antibiotics	chlortetracycline	chelate with Th	405
	antibiotics	streptomycin	picric acid	525
	antibiotics	penicillins	hydroxylamine, Fe	485, 515, or 622
	hormones	cortisone	phenylhydrazine, H_2SO_4	410
		diethylstilbesterol (DES)	H_2SO_4, Fe(III)	558
food	cod-liver oil	vitamin A	—	325
	flour	Fe	o-phenanthroline	520
	meat	nitrite	α-napthylamine, sulfanilic acid	520
	meat	nitrate	brucine alkaloid	425
	strawberries	anthocyanin	—	500
	yeast	ergosterol	—	difference (281.5 − 230)
fertilizer	fertilizer	total P	molybdovanadate reagent	420
paint	pigments	Ti	H_2O_2	410
glass	rare-earth mixture	Nd, Dy	—	865 (Nd), 915 (Dy)
metal	brass	Al	stilbazochrome	665
	bronze	Be	p-sulfophenylazosalicylic acid	450
	aluminum	V	tungstovanadophosphoric acid	440
	steel	Mn	KIO_4	525
		Ti	tiron	420
		Nb	brompyrogallol red	610

ENVIRONMENTAL APPLICATIONS

The last 10 to 15 years have seen great emphasis placed on detection, control, and removal of toxic substances in our environment. Analysis plays a major role in this effort. The areas of concern include the air we breathe, the water we drink, and the food and other materials with which we come into contact. Environmental chemists use the same kinds of methods as any other analytical chemists, but the requirements and priorities may be quite different. Thus a method that might be quite appropriate for analyzing a mixture of reaction products from an industrial process may be inadequate for the corresponding environmental analysis and vice versa.

Sensitivity takes on new meaning for many environmental analyses: Trace pollutants at the levels of parts per million (ppm), parts per billion (ppb), and even parts per trillion (ppt) may still cause increased occurrences of cancer or impart undesirable taste and odor to drinking-water supplies.

At the same time that sensitivity increases, requirements for selectivity increase. If an analyte is present at the 1-ppb level, major constituents (for example, N_2 and O_2 in an air sample) will be present at levels 10^7 to 10^9 times higher. Any slight interference by these constituents could overwhelm the signal due to analyte. However, the real problem with interferences does not usually arise from major constituents: The analyst knows what they are and can devise methods that show no sensitivity to them or correct for their effect with appropriate blanks. The truly thorny problem comes from interference by other pollutants. A trace contaminant at the 1-ppm level will be 1000 times more prominent than is an analyte at the 1-ppb level. A slight sensitivity of the method to this other contaminant could still yield significant errors. For example,

$$S = kC_A + k'C_i$$

where S is the signal, C_A is the concentration of analyte, and C_i is the concentration of the interference. If the sensitivity k' for the interference is only $10^{-3}k$, but $C_i = 10^3 C_A$, then

$$S = kC_A + (10^{-3}k)(10^3 C_A) = kC_A + kC_A = k(2C_A)$$

In other words, the signal would be equivalent to a sample containing $2C_A$; and if the analysis were left uncorrected, an error of 100% would result. Sometimes the presence of interference is not even realized; so correction is not possible.

Fortunately, while demands for sensitivity and selectivity in environmental applications increase, the demand for accuracy usually decreases. It is important not whether drinking water contains 5 ppb or 20 ppb of cadmium determined to within 0.1 ppb but whether it is below 10 ppb, the maximum allowed level. Errors of 10 to 20% are usually of no consequence although answering such questions as "Is the drinking water in our community better now than it was a decade ago?" may require measurement of small changes in pollutant level. The "Delaney clause" is another case in point: Foods and drugs cannot contain *any* carcinogenic material; so detection is all-important, whereas quantitative level is not. It does not matter whether a substance contains 1 ppb or 1 ppm of carcinogen—it must be rejected in either case.

Spectrophotometry is still one of the basic tools of the environmental chemist even though newer and more sensitive methods are continually being developed. Some examples are shown in Table 13-4.

Developed by P.W. West and G.C. Gaecke in 1956, the spectrophotometric method listed in Table 13-4 for the determination of SO_2 in air is the *reference method*, which means that all other methods must be shown to give equivalent results before they are approved by the EPA. The air sample is bubbled through a solution containing $HgCl_4^{2-}$, which forms disulfitomercurate(II) with SO_2, according to Equation 13-11:

$$HgCl_4^{2-} + 2SO_2 + 6H_2O \rightleftharpoons Hg(SO_3)_2^{2-} + 4Cl^- + 4H_3O^+ \qquad \text{(13-11)}$$

A dye, *p*-rosaniline, is added in the presence of formaldehyde to form a bright-violet complex. The absorbance is measured at 569 nm. The method is sensitive down to 5 ppb, and the only interference is NO_2, which must be removed prior to analysis if its concentration exceeds 2 ppm.

Sometimes an analyte is measured indirectly: It does not participate in the absorbing species but, by prior chemical reaction, produces a stoichiometric

TABLE 13-4   Examples of Environmental Analyses Using Spectrophotometry

Analysis	Reagent(s) and/or Procedure(s)	λ (nm)
air:		
As	Ag diethyldithiocarbaminate	538
H_2S	$FeCl_3$, dimethyl-*p*-phenylenediamine	670
NH_3	K_2HgI_4 (Nessler's reagent)	450
NO, NO_2	sulfanilic acid, N-1-naphthylethylenediamine · 2HCl	550
Pb	dithizone (Pb complex extracted into 1,1, 1-trichloroethane)	510
SO_2	*p*-rosaniline	569
water:		
chloride	release SCN^- from $Hg(SCN)_2$ by Cl^-, Fe(III)	480
Cr(VI)	diphenylcarbazide	540
Cu	reduce to Cu(I), bathocuproine disulfonate	480
Fe	*o*-phenanthroline	510
NO_3^-, NO_2^-	sulfanilamide, N-1-naphthylethylenediamine · 2HCl	520
phenol	alkaline ferricyanide, 4-aminoantipyrine	460, 510
phosphate	ammonium molybdate, reduction to molybdenum blue	700
Si	ammonium molybdate, mask phosphate with oxalic acid	410

amount of a species that does. The determination of chloride in water is one such example. To a water sample containing Cl^- is added $Hg(SCN)_2$, which liberates SCN^- by Equation 13-12,

$$2Cl^- + Hg(SCN)_2 \rightleftharpoons HgCl_2 + 2SCN^- \qquad \text{(13-12)}$$

The liberated thiocyanate ion forms with Fe(III) the red complex used in the Volhard endpoint discussed in Chapter 8. The method thus measures the amount of SCN^-, which equals the amount of Cl^- in the original sample because of the 1:1 stoichiometry in Equation 13-12.

CLINICAL APPLICATIONS

Clinical chemistry is concerned primarily with the analysis of blood, urine, and spinal fluid to gather data relating to the health of a patient. Of these, blood analysis is the most common. There are many parameters that may be determined, and spectrophotometry is ideally suited for nearly all of them; some examples are listed in Table 13-5. Several methods are sometimes available for a particular parameter, such as glucose, and a lab director will choose the one that appears most suited to his or her lab (availability and cost of reagents, instrumentation, and experience of personnel).

One of the methods for glucose listed in Table 13-5 involves the redox properties of sugars. Glucose (and, unfortunately for the accuracy of the method, some other sugars present in blood) reduces copper(II) to Cu_2O according to Equation 13-13 (although the reaction is not completely stoichiometric, and an empirical calibration with known samples must be carried out):

$$C_6H_{12}O_6 + 2Cu^{2+} + 6H_2O \rightleftharpoons Cu_2O + C_6H_{12}O_7 + 4H_3O^+ \qquad \text{(13-13)}$$

TABLE 13-5   Examples of Clinical Analyses Using Spectrophotometry

Parameter	Reagent(s) and/or Procedure(s)	λ (nm)
bilirubin	diazotized sulfanilic acid (Ehrlich's reagent)	540
cholesterol	acetic anhydride–H_2SO_4 (Liebermann–Burchard)	625
glucose	glucose reduces Cu(II) to Cu(I); Cu(I) reduces phosphomolybdic acid (Folin–Wu)	420
glucose	o-toluidine	635
phosphate (inorganic)	Na_2MoO_4, p-methylaminophenol sulfate	700
urea nitrogen	urease, Na_2WO_4, Nessler's reagent	490

This reaction is allowed to proceed for exactly 8 min at 100°C. Then phosphomolybdic acid is added, which is reduced by Cu_2O to give an intensely colored "molybdenum blue."

As an alternative, glucose also forms a specific complex with o-toluidine according to Equation 13-14:

$$CH_2OH{+}CHOH{+}_4CHO + \text{(o-toluidine, NH}_2\text{, CH}_3\text{)} \rightleftharpoons CH_2OH{+}CHOH{+}_4CH + H_2O$$

(13-14)

The development of a diagnostic green color requires 10 min at 100°C.

Clinical labs normally use standard samples and compute analyte concentrations as was shown in Example 13-3. This technique will, at best, give results only as good as the standard values: At one hospital, for example, a restandardization of the laboratory's hemoglobin standard from 13.0 to 13.5 g per 100 ml revised all patients' hemoglobin values upward. The change resulted in a 100-pint-per-month reduction in whole blood used for transfusions because physicians relied on the hemoglobin data for transfusion decisions.

Many clinical labs, especially those associated with large hospitals, carry out blood analyses automatically. This approach decreases the amount of time required by a technician, and hundreds of determinations per hour can be carried out. Autoanalyzers may be "single channel," performing one determination (as shown in Figure 13-13) or "multichannel," performing several different determinations on each sample. The results from this complete battery of tests are plotted on a "blood panel" such as that in Figure 13-14. One advantage of a blood panel is that it is completely objective, whereas a doctor ordering a single test to support a tentative diagnosis might not consider that more than one disorder can produce an abnormal value in a single parameter. The blood panel in Figure 13-14 shows normal ranges (shaded areas) so that an abnormal value in any of the twelve tests can be spotted easily. Also, test results are always printed in the same order so that they form a pattern. Each disorder results in a characteristic pattern and can thus be recognized by the diagnostician. A whole new field of pattern recognition has developed, and computers have been "taught" to recognize patterns whether they are complicated infrared spectra or blood panels. With the multichannel autoanalyzers, it makes little difference in cost whether one determination is ordered or all twelve.

FORENSIC APPLICATIONS

Anyone who watches TV regularly is aware of the role analytical chemistry plays in the field of criminology (in fact, chemical analysis was an important aspect of Sherlock Holmes's ability to solve cases). Today, the crime lab may contain some of

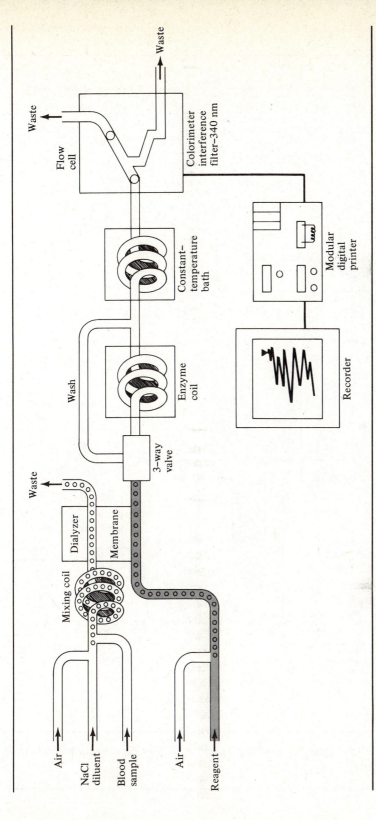

FIGURE 13-13 Typical example of a single-channel autoanalyzer (for glucose analysis by the hexokinase enzyme method). The blood sample is diluted and mixed with air to facilitate mixing in the upper stream. Analyte constituents in the sample diffuse through the membrane and enter the reagent flowing through the lower stream (shown dark). Additional reagents may be added at various points, and the correct amounts are added with the proportioning pump, which controls the rate of flow of the various streams. Color development occurs in the final mixing coil and in the constant-temperature bath, and the solution is brought to the correct temperature for the absorbance measurement. The stream flows through the flow cell, while absorbance is continuously monitored and compared with a reference signal. The absorbance is recorded, each sample exhibiting a discrete peak. Standard samples are run periodically to ensure that the calibration remains valid.

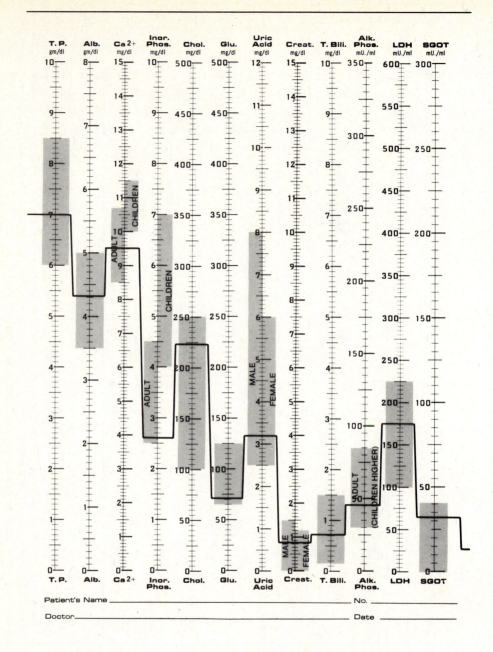

FIGURE 13-14 Blood panel for a patient having all twelve parameters in the normal range (shaded areas).

the most advanced instrumentation, and the field of *forensic chemistry* has greatly developed. The spectrophotometer is one of the chief instruments in the forensic chemist's arsenal.

An example of the use of spectrophotometry in a case involving alleged illicit drugs can be cited: Amphetamine and its derivatives constitute one important class of controlled drugs, and to determine source and trafficking route, it is critical to know which amphetamines are involved. In the case, narcotics agents purchased from a distributor capsules suspected to contain amphetamine or methamphetamine. The capsules contained a pink powder, which was submitted to the forensic laboratory for analysis. UV spectrophotometry was carried out by dissolving the powder in 1 F H_2SO_4; amphetamine and methamphetamine were then ruled out because the peaks at 252, 257, and 263 nm associated with the monosubstituted benzene chromophore were absent. However, peaks at 233 and 287 nm indicated a dioxyamphetamine-like derivative, and several other analytical techniques were used to complete the identification as 3,4-methylenedioxymethamphetamine (MDM). This substance, although chemically similar to 3,4-methylenedioxyamphetamine (MDA), is exempt from Federal control. The rapid analysis averted unjustified prosecution.

Forensic chemists, like other analytical chemists, are concerned with accuracy, sensitivity, and speed of analytical methods. However, they must pay special attention to the legal standing of their methods. It accomplishes nothing if the analytical results cannot be entered as evidence. Frequently, the forensic chemist cannot *prove* guilt; for his evidence is only corroborative, that is, supporting charges. Hair or blood samples may be chemically the same as a suspect's, but since other persons may have the same types of hair and blood, the evidence can only support— not prove—the charges. Therefore, it is important to gather as much such evidence as possible since the probability that a suspect is guilty will increase as more and more evidence piles up. Courts will generally accept a 95% confidence level as proof, but individual juries may vary widely from this level. Finally, the forensic chemist must be able to communicate clearly his results and conclusions to laymen who have never heard of spectrophotometry and other analytical instruments. In such cases, confidence levels may lie more in the tone of voice than in the precision of the results.

SUPPLEMENTARY READING

ARTICLES DESCRIBING ANALYTICAL METHODS OR APPLICATIONS

"The Analytical Chemist and Multielement Chemical Testing in Preventive Medicine," *Analytical Chemistry*, 43:18A (July, 1971).

Eichmeier, L.S., and M.E. Caplis, "The Forensic Chemist—An Analytical Detective," *Analytical Chemistry*, 47:841A (1975).

REFERENCE BOOKS

Henry, R.J., D.C. Cannon, and J.W. Winkelman, eds., *Clinical Chemistry—Principles and Technics*, 2nd ed. New York: Harper & Row, Publishers, Inc., 1974.

IUPAC Commission on Spectrochemical and Other Optical Procedures for Analysis, *Spectrophotometric Data for Colorimetric Analysis*. London: Butterworth Publishers Ltd., 1963.

Meehan, E.J., "Spectroscopic Apparatus and Measurements," in *Treatise on Analytical Chemistry*, 2nd ed., I.M. Kolthoff and P.J. Elving, eds. pt. 1, vol. 7, chap. 3. New York: John Wiley & Sons, Inc., 1981.

Sandell, E.B., and H. Onishi, *Photometric Determination of Traces of Metals*, 4th ed. New York: John Wiley & Sons, Inc., 1978.

PROBLEMS

13-1. Show mathematically that a 1% fluctuation in voltage will lead to a 4% fluctuation in radiant power if in Equation 13-1 the exponent x is 4.

*13-2. A 6-V power supply has a standard deviation of ± 0.01 V. If in Equation 13-1 the exponent x is 4, what uncertainty will be introduced into a transmittance reading of 65%?

13-3. It can be seen from Figure 13-2 that the wavelength for the maximum emission from a lamp source increases as the temperature of the source decreases. Using the data shown in Figure 13-2, plot *frequency* of the maximum as a function of source temperature, and predict the maximum for a source whose temperature is 1400 K.

13-4. What color absorption filter should be used to measure each of the following absorptions?

    *a. MnO_4^-, $\lambda_{max} = 525$ nm

    b. P (heteropoly blue), $\lambda_{max} = 680$ nm

    c. Ti (H_2O_2), $\lambda_{max} = 410$ nm

    d. Glucose (*o*-toluidine), $\lambda_{max} = 635$ nm

13-5. MgF_2 (refractive index $n = 1.38$) is used as the dielectric film in an interference filter. Calculate the wavelength of the first-order light that the filter will pass if the MgF_2 thickness is 300 nm. What other wavelengths will be passed by this interference filter?

*13-6. A spectrophotometer employs a reflection grating (normal incidence) with 7000 lines/cm. At what angle will the first-order diffraction of 1100-nm light occur?

13-7. A reflection grating of 9000 lines/cm is normal to the light beam. What wavelengths for the first two orders of diffraction will be observed at 45°?

*13-8. Suppose that a photomultiplier tube exhibits an average of 3.800 ± 0.002 secondary electrons per dynode and contains nine dynodes. What will be the percent uncertainty in the tube current?

13-9. If the absorbance of the solvent that can be tolerated is 0.8, what is the maximum ε exhibited by CCl_4 (density = 1.595 g/cm^3) at any wavelength at which measurements are to be taken?

13-10. In addition to changing the reflection characteristics, a cell not exactly perpendicular (90°) to the light beam will lead to a longer light path for the sample. If the actual angle in a spectrophotometer is 85°, what error will result because of the increased light path?

---

*Answers to problems marked with an asterisk will be found at the back of the book.

13-11. In what region of the spectrum would you expect to observe electronic transitions of the following compounds?

*a. $CH_3CH_2\overset{\overset{\displaystyle O}{\displaystyle \|}}{C}—OH$

b. [benzene ring with two CH_3 groups in ortho position]

c. $CrCl_3$

d. $CH_2\!\!=\!\!CH—CH\!\!=\!\!CH—CH\!\!=\!\!CH_2$

e. $NiSO_4$

13-12. A Duboscq colorimeter is used to determine the P content in blood serum. A sample of 0.1 mL of blood serum and a sample of 0.1 mL of a standard containing 0.030 mg/mL of P are treated in the same fashion to produce the intensely colored heteropoly blue. If the path-length ratio (b_s/b_x) is 2.2, calculate the P level in the blood sample as milligram percent (milligrams per 100 mL of serum).

*13-13. You have been given the assignment to develop a spectrophotometric method for the determination of Ti in steel (expected range 1 to 2% Ti), using the H_2O_2 complex ($\varepsilon = 720$ at 410 nm). If you plan on 1-cm cells and need to dilute the steel sample to 50 mL prior to measuring the absorbance, what size steel sample do you recommend so that a sample of midrange value will have the optimum absorbance value of 0.43?

13-14. (a) The molar absorptivity of a zinc complex is 5.00×10^3 at 420 nm. A 290-mg brass sample is dissolved in acid, complexing agent is added, and the sample is diluted to 250 mL. The absorbance of the solution in a 0.50-cm cell is 0.752. Calculate the percentage of zinc in the brass. (b) Based on this percent zinc, can you calculate the absorbance at 520 nm?

13-15. A 518-mg sample of steel is dissolved in HNO_3 and diluted to 100 mL in a volumetric flask. Three 20.0-mL aliquots are drawn: One serves as a blank, the second is treated with KIO_4 to oxidize Mn^{2+} to MnO_4^-, and the third has a standard addition of 5.00 mL of 0.124-mg/mL Mn. All aliquots are diluted to 50.0 mL, and the absorbances are measured versus the blank at 525 nm. If the sample gives an absorbance of 0.248 while the sample plus standard addition gives an absorbance of 0.514, calculate the percent Mn in the steel.

*13-16. A standard iron solution contains 350 mg of $Fe(NH_4)_2(SO_4)_2 \cdot 6H_2O$ (FW $= 392$) in 500 mL. If a 2.00-mL aliquot is treated with o-phenanthroline and diluted to 100 mL, calculate the molar absorptivity if the absorbance is 0.361 at 510 nm in a 1.00-cm cell. Calculate absorptivity with concentration given as micrograms of Fe per milliliter.

13-17. A 50-mL sample of drinking water is treated with hydroxylamine to reduce iron to Fe(II), and then the o-phenanthroline complex is formed. The sample is diluted to 100 mL, and the absorbance at 510 nm is 0.043 in a 1-cm cell. Calculate parts per million of Fe (micrograms per milliliter) in the drinking water, using information given in Problem 13-16.

13-18. HNO_3 content in rain water ("acid rain") may be determined by reducing NO_3^- to NO_2^- in a Cd reductor column. The nitrite ion formed undergoes azo dye reactions to form intensely colored species. The absorbance is compared to standards prepared in the same way. A 10-mL rainwater sample is passed through the reductor and collected in a 25-mL volumetric flask. After the azo dye reaction, the solution is diluted to the mark; the absorbance is 0.265 at 535 nm. A 10-mL aliquot of a standard solution containing 1.50 ppm NO_3^- (1.50 μg/mL) treated in the same way gives an absorbance of 0.396. Calculate nitrate content in the rainwater as parts per million (micrograms per milliliter) NO_3^-.

*13-19. A 5.00-mg/L solution of procaine hydrochloride (FW = 273) has an absorbance of 0.385 at 288 nm in a 1.00-cm cell. What is the concentration of a solution exhibiting an absorbance of 0.825, and what is the molar absorptivity of procaine hydrochloride?

13-20. The quinalizarine complex of uranium has a molar absorptivity of 2800 at 610 nm. If a 175-mg sample of uranium ore is dissolved in acid, complexing agent added and pH adjusted, and diluted to 50.0 mL and if the absorbance is 0.268 in a 1.00-cm cell, calculate the percentage of uranium in the ore.

*13-21. Drugs often absorb strongly in the UV. As an example, $\varepsilon_{254} = 16,000$ and $\varepsilon_{267} = 19,000$ for tetracycline, while $\varepsilon_{254} = 16,000$ and $\varepsilon_{267} = 15,000$ for epi-tetracycline, an inactive hydrolysis product. If a mixture exhibits absorbances of 0.402 at 254 nm and 0.432 at 267 nm, what is the concentration of each compound?

13-22. Inorganic phosphorus is spectrophotometrically measured in blood serum by formation of a reduced heteropoly acid (heteropoly blue) and comparison of its absorbance with a standard treated by the same procedure. If a 1.0-mL blood-serum sample gives an absorbance of 0.217, while a 2.00-mL aliquot of a standard containing 91.2 mg KH_2PO_4 per liter gives an absorbance of 0.285, calculate milligram percent of P (milligrams per 100 mL) in the blood sample. (Both aliquots are diluted to the same final volume prior to absorbance measurements.)

13-23. Glucose can be spectrophotometrically measured in blood serum by formation of a complex with o-toluidine and comparison of the absorbance with a standard. If a 0.100-mL blood-serum sample gives an absorbance of 0.228, while a 0.100-mL standard containing 2.08 mg/mL of glucose gives an absorbance of 0.432, calculate the glucose level in the blood sample as milligram percent (milligrams per 100 mL).

13-24. A reference method for determining SO_2 content in polluted air was carried out by scrubbing the air sample (1.5 L/min for 45 min) with 10 mL of $HgCl_4^{2-}$ solution. The color was developed with p-rosaniline and formaldehyde, and the solution was diluted to 25 mL in a volumetric flask. The air sample gave an absorbance value of 0.326. This was compared with a standard prepared with 1.00 mL of 10-μg/mL SO_2, which gave an absorbance reading of 0.121. Calculate SO_2 content in the air as parts per million (micrograms of SO_2 per gram of air), assuming a density of 1.18 g/L for the air sample.

*13-25. The cyclamate concentration in a diet soft drink can be analyzed spectrophotometrically by extracting the cyclamate with ethyl acetate, followed by oxidation of cyclamate ion with NaOCl to produce N, N-dichlorohexylamine, which absorbs at 314 nm. A 100-mL sample yields an absorbance of 0.204. A second 100-mL aliquot spiked with 10 mg of sodium cyclamate gives an absorbance of 0.586. Calculate sodium cyclamate concentration in the soft drink as milligrams per 100 mL.

13-26. Sodium nitrite ($NaNO_2$) is frequently added to meat products, and its presence can be determined spectrophotometrically by azo dye formation with sulfanilic acid and N-1-naphthylethylenediamine. A 20-g sample of meat is extracted and filtered, with the filtrate diluted to 250 mL. A 1.00-mL aliquot is added to 10 mL of the colorimetric reagent; the absorbance at 505 nm is 0.289. A 1.00-mL spike of $NaNO_2$ solution (1500 $\mu g/mL$) is added to the remaining extract and mixed thoroughly. A 1.00-mL aliquot treated as before gives an absorbance of 0.497. Calculate parts per million (micrograms per gram) of $NaNO_2$ in the meat sample, assuming recovery from the meat is 82%.

13-27. A colored complex ion AB_2^{2-} is formed when A^{2+} is mixed with B^{2-}. Various quantities of Na_2B were added to a 2.00×10^{-3} F solution of A^{2+}; it was found that the absorbance became independent of the quantity of this salt at formal concentrations greater than about 0.2. This would indicate that the complex-formation reaction was essentially complete under these conditions. The absorbance in the plateau region was 0.982 in a 1.00-cm cell. When the formal concentration of Na_2B was only 9.00×10^{-3}, the absorbance was 0.670 in a 1.00-cm cell. Calculate the formation constant for the complex.

*13-28. A 1.0×10^{-3} F solution of a dye X shows an absorbance of 0.20 at 450 nm and an absorbance of 0.05 at 620 nm. A 1.0×10^{-4} F solution of dye Y shows 0.00 absorbance at 450 nm and an absorbance of 0.42 at 620 nm. Calculate the concentration of each dye present in a solution that exhibits an absorbance of 0.38 and 0.71 at 450 and 620 nm, respectively. The same cell is used for all measurements.

13-29. Standard solutions of 0.0100 F $KMnO_4$ and 0.0500 F $K_2Cr_2O_7$ are prepared and accurate dilutions of $\frac{1}{100}$, $\frac{3}{100}$, and $\frac{5}{100}$ are made with dilute H_2SO_4. Absorbances are measured in 1.00-cm cells at 440 nm and 545 nm, with the data given below. Calculate molar absorptivities for each at each wavelength and absorptivities, using milligrams per milliliter of the metal.

Aliquot (mL)	A ($KMnO_4$)		A ($K_2Cr_2O_7$)	
	440 nm	545 nm	440 nm	545 nm
1	0.021	0.191	0.170	0.005
3	0.064	0.570	0.512	0.015
5	0.106	0.948	0.849	0.025

13-30. A steel sample weighing 206 mg is dissolved in HNO_3, oxidized with $K_2S_2O_8$ to form MnO_4^- and $Cr_2O_7^{2-}$, and diluted to 250 mL. The absorbance at 440 nm is 0.387, while the absorbance at 545 nm is 0.625. Based on the results to Problem 13-29, calculate the percentages of Mn and Cr in the steel sample.

13-31. The absorption spectra of two new and highly reactive organic ketones are measured. The absorption maxima of smogone are at 360 nm ($\varepsilon = 2600$) and 420 nm ($\varepsilon = 1450$). At these same wavelengths, the molar absorptivities of cancerone are 250 and 3100, respectively. You record a spectrum of a mixture in an appropriate solvent and determine that A at 360 nm is 0.295 and that A at 420 nm is 0.362. Find the concentrations of smogone and cancerone in the solution.

*13-32. In the determination of Fe in H_2O, a 1.00 ± 0.005-mL pipet, a 10.00 ± 0.02-mL pipet, and a 50.00 ± 0.05-mL volumetric flask are used to prepare the colored complex. The calibration line has a slope of 0.145 ± 0.002 absorbance units per part per million of Fe. Transmittance measurements on a standard sample give the following values: 41.3, 41.5, 40.8, 41.9, 41.6, 41.5, and 41.2%. Calculate parts per million of Fe and its uncertainty if the sample gives an absorbance of 0.126.

13-33. Spectrophotometric endpoint signals can be quite sensitive and useful for titration of dilute solutions. For example, samples of $KMnO_4$ may be titrated with ferrous ammonium sulfate while monitoring the MnO_4^- absorbance at 525 nm. If a 5.00-mL aliquot of $KMnO_4$ is titrated with 0.00748 N Fe(II) and gives the absorbance shown below, calculate the normality of the $KMnO_4$ solution:

Fe(II) (mL)	A	Fe(II) (mL)	A	Fe(II) (mL)	A
0.00	>1	3.50	0.45	4.60	0.04
1.00	>1	4.00	0.25	4.80	0.02
2.00	1.00	4.20	0.17	5.00	0.02
3.00	0.65	4.40	0.10	6.00	0.02

13-34. After graduation, you obtain a position as an analyst with the Weaker Steel Co. Show how you would apply the five steps of the total analysis process to the analysis of manganese and chromium in steel, using a spectrophotometric method of analysis.

Atomic Spectroscopy

Chapter 13 is concerned with the absorption of electromagnetic radiation by species in solution. The UV–visible region of the spectrum involves outer electron configuration—the valence electrons—and thus is responsive to chemical bonding. Infrared spectrophotometry involves molecular vibrations; so it too reflects chemical bonding. In this chapter, we shall discuss some analytical methods that are ideally not sensitive to chemical bonding and are, therefore, useful for *elemental analysis*. Primarily, these methods will involve absorption and emission by substances existing in the atomic state—hence the term *atomic spectroscopy*.

The concept of measuring atomic spectra leads to the following advantages:

1. Elemental analysis: Whether the original sample contains Fe(II) or Fe(III), the result will still be percentage of Fe.

2. Fewer interferences: The lines (absorption and emission) are extremely sharp, in contrast to the broad lines normally encountered in absorption spectro-photometry.

Beer's law was developed in Section 12-3 and Appendix 1-H for an absorption process. Because some atomic spectroscopy methods involve emission, it is necessary to see how an emission signal can be related to an analyte concentration.

14-1 PRINCIPLES OF EMISSION

One method by which emission may occur is to excite a species with electromagnetic radiation (absorption occurs); then, as the species relaxes, it emits a photon equal to this energy, as shown in Figure 14-1. Such a process is given the general name of *fluorescence*.

There are other mechanisms by which the excited species can return to the ground state. For example, the species may collide with its neighbors and lose energy

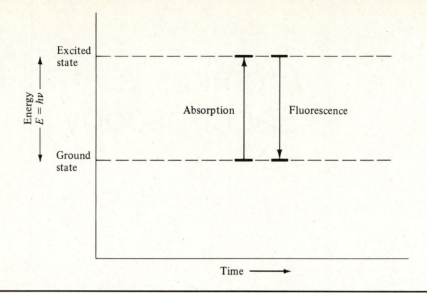

FIGURE 14-1 The process of fluorescence, in which a system absorbs a photon whose energy equals $h\nu$ and then returns to the original energy state with emission of a photon whose energy equals $h\nu$.

by that route. Atomic spectroscopy is carried out in the gaseous phase, which makes this mode of energy transfer less probable. For a given set of conditions (concentration, pressure, and temperature), the amount of fluorescence will be directly proportional to the number of excited species, and this number will be directly proportional to the radiant power absorbed. From Equation 12-7, we see that

$$P_F \propto P_A = P_0 - P$$

$$P_F = k'(P_0 - P) \tag{14-1}$$

where P_F is the radiant power emitted as fluorescence. The proportionality constant, k', will depend on the probability that an excited-state species will fluoresce instead of losing energy by some other mode.

To see how Equation 14-1 can be used to relate analyte concentration to a fluorescence signal, we recall the development of Beer's law given in Appendix 1-H. It is shown that

$$\ln \frac{P}{P_0} = -kbC \tag{14-2}$$

or, in exponential form,

$$\frac{P}{P_0} = e^{-kbC} \tag{14-3}$$

Combining Equations 14-1 and 14-3 we obtain

$$P_F = k'(P_0 - P_0 e^{-kbC}) = k'P_0(1 - e^{-kbC}) \tag{14-4}$$

Equation 14-4 shows a fairly complicated relationship between fluorescence signal, P_F, and concentration, C. However, this relationship can be simplified significantly when the exponential term is small (little absorbance), which is typically true for fluorescence measurements. The exponential can be expanded as

$$e^{-x} = 1 - x$$

with higher terms neglected. For example, with $x = 0.1$, $e^{-0.1} = 0.905$, while $1 - x = 0.900$.

Thus Equation 14-4 becomes

$$P_F = k'P_0(\cancel{1} - \cancel{1} + kbC) = k'P_0 kbC = k''P_0 bC \tag{14-5}$$

According to Equation 14-5, the fluorescence signal will be directly proportional to concentration but will also be proportional to path length b and to the incident radiant power P_0 used to excite the analyte species. The proportionality constant, k'', includes absorptivity and the probability of fluorescent decay. Two fluorescence techniques, *atomic fluorescence spectroscopy* and *X-ray fluorescence spectroscopy*, will be described in Sections 14-5 and 14-6.

There are other methods besides absorption of electromagnetic radiation for exciting an atom in the gaseous state. For example, thermal or electrical energy forms the basis for *atomic emission spectroscopy* (Section 14-2), or thermal energy in the form of flame is used in *flame emission spectroscopy* (Section 14-3). In these techniques, the signal, radiant power emitted, will be directly proportional to the number of excited atoms present. In turn, the number of excited atoms will be directly proportional to the total number of atoms present and will also depend on the thermal or electrical energy exciting them. The proportionality constants then become involved, and the success of these analytical techniques lies in the analyst's ability to run standard samples under the same set of conditions as those for the unknown sample.

14-2 ATOMIC EMISSION SPECTROSCOPY (AES)

When a sample is bombarded by a large amount of thermal or electrical energy (for example, with a dc arc discharge), compounds will dissociate to form free atoms in the gaseous state. The energy is also sufficient to promote atoms to various excited electronic states, and as they drop back to their electronic ground states, they will emit light. About 70 elements can be determined in this way (primarily the metals and metalloids), which makes AES a powerful tool for the detection of impurities in materials. Actually, the energy available is sufficient to excite *all* elements, but for some, the emitted lines lie at such high energies that they are in the vacuum UV region, which is not readily accessible.

The components required for AES include excitation source, dispersing element, and detector, as shown in Figure 14-2 for a typical setup. However, packaging these components into an instrument capable of resolving the hundreds of emitted lines and of measuring their intensities results in price tags in the $100,000 range. Therefore, not many labs own one but will send samples to commercial testing labs for analysis.

Several excitation sources in use include (1) dc arc, (2) ac arc, (3) ac spark, and (4) inductively coupled plasma (ICP). The dc arc employs a dc voltage of 200 to 300 V applied between graphite electrodes. A solid (or, occasionally, a liquid) sample is placed on the lower electrode. The electrodes are momentarily brought together to start the current flowing (5 to 15 A) and then separated to form a gap of about 1 cm. As the current flows, the temperature rises rapidly to 4000 to 8000 K. The high temperature results in a very sensitive technique—sufficiently energetic to excite all detectable elements—but the dc arc is not very reproducible so that quantitative results are not of high accuracy.

The ac arc employs 2000 to 5000 V, which are sufficient to jump the gap (~1 mm) without closing it first (as is done with the dc arc). The ac arc gives better reproducibility since with each cycle (60 Hz) a new arc begins at different points in the sample; a single measurement consists of a large number of such arcs. However, the gap temperature is lower, which means that sensitivity is also lower.

The ac spark is accomplished with a voltage of 10,000 to 40,000 V. In this technique, a high current flows for only a moment (~1 μsec), which gives the atoms in the gas phase temperatures up to 40,000 K; but the electrodes remain relatively cool. The excitation appears to be mainly electrical, from bombardment of the sample with electrons. Only small amounts of sample are volatilized so that sensitivity is low, but this technique is the best for quantitative analysis.

The ICP method of atomizing and exciting samples has developed in the last 10 or 20 yr and offers several advantages over the use of arc and spark sources. A plasma is a gas in which a significant fraction of the atoms or molecules is ionized and, therefore, will interact with magnetic fields. The plasma is produced by

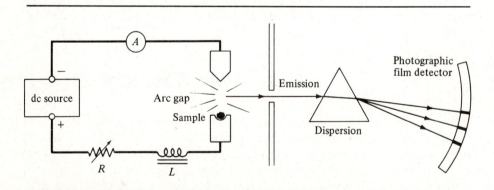

FIGURE 14-2 Atomic emission spectroscopy with a dc arc source.

"seeding" with electrons a carrier gas (Ar) flowing through a tube, to provide some conductivity, and by surrounding the tube with an induction coil, which generates an oscillating magnetic field. The field induces electrons and ions to flow in annular paths (an eddy current) and to be accelerated each half cycle of the high-frequency generator. Resistance to flow causes heating and additional ionization. A sample injected into this gas stream will attain temperatures of 9000 to 10,000 K. The high gas temperatures and reproducibility give this source many advantages: (1) Linear response over several orders of magnitude allows wide ranges of concentration to be handled without dilution, (2) very small amounts of sample can be analyzed (microgram level), (3) no electrodes are used so that contamination does not occur, and (4) the surrounding Ar gas is inert so that combustion and other chemical reactions are not involved. Detection limits, as shown in Table 14-1, for most elements are a few parts per billion, and quantitation to $\pm 5\%$ is attainable.

Emission from a complex mixture may consist of hundreds of lines. Fortunately, atomic spectral lines are exceedingly sharp because, unlike molecules or species in

TABLE 14-1   Detection Limits for Emission Spectroscopy with an ICP Source

Element	Detection limit (ppb)	Element	Detection limit (ppb)
Ag	4	Lu	8
Al	2	Mg	0.7
Au	40	Mn	0.7
Ba	1	Mo	5
Be	5	Na	0.2
Bi	50	Nb	10
Ca	0.07	Ni	6
Cd	2	Pb	8
Ce	7	Pd	7
Co	3	Rh	3
Cr	1	Sc	3
Cu	1	Sr	0.02
Dy	4	Th	3
Er	1	Ti	3
Eu	1	Tm	7
Fe	5	U	30
Ga	14	V	6
Gd	7	W	2
Hf	10	Y	2
Ho	10	Yb	0.09
In	30	Zn	2
La	3	Zr	5

SOURCE: See V.A. Fassel, "Quantitative Elemental Analyses by Plasma Emission Spectroscopy," *Science*, *202*:183 (1978), for a discussion of ICP source.

solution, the electronic levels are not affected by the vibrational or rotational energy present in molecules. However, the emitted lines must be dispersed so that they are sufficiently separated to be resolved at the detector. As with UV–visible spectrophotometers, dispersion is effected by a prism or grating. Gratings are preferred because they give *linear dispersion*; that is, the resolution is constant throughout the wavelength region.

Three kinds of detector are used in AES: photographic detectors, photoelectric detectors, and optoelectronic image devices (OID). As was mentioned earlier, photographic detection is the means in a spectrograph, whereas spectrometers and photometers employ photoelectric detection. Thus there are emission spectrographs and emission spectrometers.

The emission spectrograph is capable of simultaneously detecting hundreds of lines arising from the presence of many elements (up to all 70 that are detectable) in the sample. The photographic detection method also averages out the emission intensity over a period of time—which helps to compensate for the varying nature of the excitation source. On the other hand, development of the film and interpretation of the complicated spectrum (see Figure 14-3) takes considerable time and expertise. Variation in film characteristics normally limits accuracies to the 5 to 10% range, and often results for trace impurities are reported as order-of-magnitude values. A typical report is given in Table 14-2.

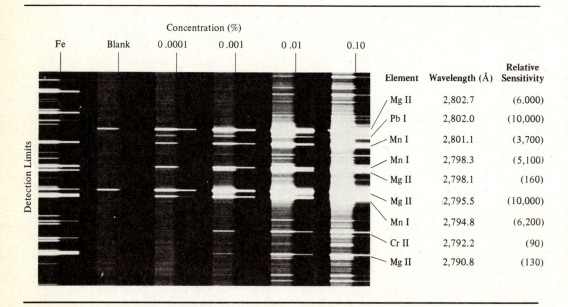

Element	Wavelength (Å)	Relative Sensitivity
Mg II	2,802.7	(6,000)
Pb I	2,802.0	(10,000)
Mn I	2,801.1	(3,700)
Mn I	2,798.3	(5,100)
Mg II	2,798.1	(160)
Mg II	2,795.5	(10,000)
Mn I	2,794.8	(6,200)
Cr II	2,792.2	(90)
Mg II	2,790.8	(130)

FIGURE 14-3 A small portion of emission spectra (279 to 281 nm) used for calibration. Highest-sensitivity lines are readily observed when elemental concentration is 0.0001 or 0.001%. Less sensitive lines require higher concentrations to be detected. The complicated Fe spectrum shows the difficulty encountered when trace elements in steel alloys are to be analyzed. (Courtesy of Fisher Scientific Co., Jarrell-Ash Division.)

TABLE 14-2   Typical Report of Emission Spectrograph Analysis

Element	Fe_2O_3 doped with Cr (%)	Fe_2O_3 doped with V (%)
Fe	69	70
Cr	0.74	ND (<0.0005)
V	ND* (<0.0001)	0.0035
Mn	ND (<0.004)	0.0017
Al	0.0032	0.0039
Cu	0.0043	0.0012
Mo	0.0010	0.0024
Si	0.076	0.030
Ni	ND (<0.001)	ND (<0.001)
Mg	ND (<0.0005)	ND (<0.0005)

*ND stands for not detected with limit given in parentheses.

There are available emission spectrometers that contain a series of photo-multiplier tubes, allowing direct readout of a number of elements simultaneously. However, because of the cost of photomultiplier tubes and space limitations, only a selected group of elements can be analyzed. The instruments are usually custom built, the customer specifying which elements are to be measured. Emission spectrometers are more expensive than spectrographs because of the cost of detection, but quicker results and higher accuracies sometimes justify the added cost. Accuracies of better than $\pm 1\%$ have been obtained. The signal from the photomultiplier tube is integrated over a period of time of about 1 min or less, which helps average out source fluctuations while still yielding rapid results.

Recently, OID's have come on the market that allow simultaneous detection of the entire spectrum; they were described in Section 13-5. The information may be displayed on a cathode-ray tube (TV screen) and stored in a computer for subsequent data handling and printout of analytical results.

The applications of AES are widespread since almost any solid or liquid sample may be analyzed for its elemental content. Lunar rock samples were analyzed by emission spectroscopy at several laboratories to determine the presence of at least 27 elements. New techniques to increase the sensitivity have made emission spectroscopy capable of ultratrace analysis at the parts-per-billion level for environmental analysis. (Table 14-1 shows the detection limits when an ICP source is used.)

In summary, AES is an excellent method for trace analysis and can be used for all metals and metalloids. Small samples (~ 1 mg) can be analyzed generally as received—preliminary separations are unnecessary—in a relatively short time (especially when photoelectric detection is employed). On the negative side, high-quality equipment is expensive, and accuracy better than $\pm 5\%$ is very difficult to achieve. One important point, which can be an advantage or a disadvantage according to the analytical problem, is that AES is limited to elemental analysis.

14-3   FLAME EMISSION SPECTROSCOPY (FES)

Another analytical technique that is also in the classification of emission spectroscopy but uses a flame as an excitation source is FES, also known as *flame photometry*. Probably you have carried out a qualitative flame test by inserting a sample into a Bunsen burner flame and observing the color of light emitted by the sample; making this test into a quantitative analytical technique is the basis of FES. It is traditional to discuss flame emission separately from arc and spark emission spectroscopy because the sample requirements, instrumentation, and information obtained are all different. For example, solid samples are normally used in AES, whereas solutions must be employed in FES.

As an excitation source, the flame has advantages and disadvantages compared to an arc or spark. The lower temperature of the flame produces and excites only a small number of atoms. The most easily excited atoms are the alkali and alkaline-earth metals. Therefore, flame emission is used almost exclusively for these elements although 30 to 40 elements have been determined by flame emission techniques. On the other hand, a flame's characteristics can be more reproducibly controlled than those of an arc or spark so that higher accuracy is achievable.

Higher accuracy also involves homogeneous liquid samples in place of solid samples, which are frequently heterogeneous. However, this means that solid samples must be first dissolved and then diluted to a known volume. Accuracies of 1 to 2% are typical with flame emission, whereas it requires considerable effort to achieve better than 5 to 10% accuracies with an arc or spark method. One common feature is that both FES and AES yield only elemental analysis results.

An instrument for flame emission is shown schematically in Figure 14-4. The liquid sample is aspirated into the flame, and the emitted light is dispersed and detected. Because so few lines are emitted in a flame compared with arc and spark excitation sources, it is often unnecessary to use a high-resolution monochromator. A filter is sufficient, and in this case, by definition, the instrument would be a *flame photometer* (hence flame photometry). However, it is less costly to add flame emission capability to a spectrophotometer than to purchase another instrument dedicated only to flame photometry. Recently, because both techniques use flames, most FES is carried out with spectrometers designed for atomic absorption spectroscopy, which will be discussed in Section 14-4.

There are several types of flame that can be employed in FES. A flame represents the combustion reaction between a fuel and an oxidant, and by varying these components, the analyst can achieve different flame temperatures and other characteristics. Table 14-3 lists the most frequently used flames and the maximum temperatures observed with them. We focus on the flame temperature because it will dictate in large part the sensitivity of the technique for any element; for a sequence of reactions occurs when the liquid sample is aspirated into the flame, as shown schematically in Figure 14-5. Although steps 1 and 2 will occur in any flame, the fraction of sample converted to atoms and excited will vary widely with flame temperature. In general, the higher the flame temperature, the more probable this process becomes, as shown in Table 14-4. Also note how few atoms actually become excited in a flame. Even for alkali metals, less than 1% is typically excited. However, some flames can be so hot that ionization occurs, which acts as an interference. As an

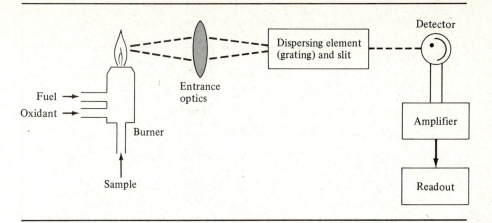

FIGURE 14-4 Schematic design of a flame photometer.

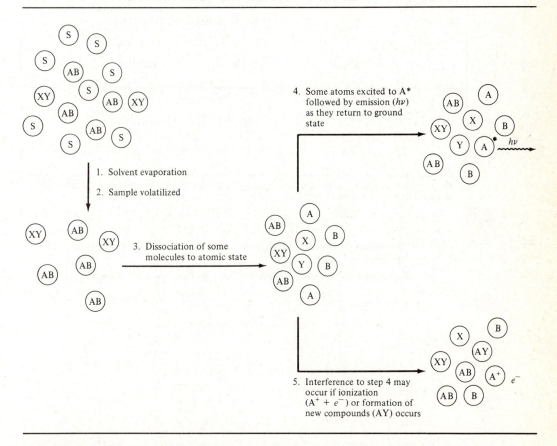

FIGURE 14-5 Sequence of steps as a sample solution containing analyte A (as compound AB) and compound XY in solvent S is aspirated into a flame.

TABLE 14-3 Maximum Temperatures (°C) of Various Flames

Fuel	Oxidant		
	Air	Oxygen	N_2O
propane	1925	2800	—
hydrogen	2100	2780	—
acetylene	2200	3050	2955
cyanogen	2330	4550	—

TABLE 14-4 Percentage of Atoms in the Excited State*

Element	In excited state (%)	
	2000 K	3000 K
Cs	4×10^{-2}	7×10^{-1}
Na	1×10^{-3}	6×10^{-2}
Ca	1×10^{-5}	4×10^{-3}
Zn	7×10^{-13}	6×10^{-8}

*The values in the table can be calculated from the energy difference ($\Delta E = h\nu$) between the excited and ground states and from the Boltzmann distribution:

$$\% \text{ in excited state} \cong 100 \frac{N^*}{N_0} = 100 A e^{-\Delta E/kT}$$

where N^* is the number of excited atoms, N_0 is the number of ground-state atoms, and A is a constant involving statistical probability of the two states. Actually, we really want $100N^*/(N^* + N_0)$; but since so few are excited, $N_0 \cong N^* + N_0$

As an example, the atomic line for Zn is at 214 nm (2.14×10^{-5} cm), and $A = 3$. Therefore,

$$\Delta E = h\nu = \frac{hc}{\lambda} + \frac{6.63 \times 10^{-27} \text{ erg-sec} \cdot 3 \times 10^{10} \text{ cm/sec}}{2.14 \times 10^{-5} \text{ cm}} = 9.29 \times 10^{-12} \text{ erg}$$

$$k = 1.38 \times 10^{-16} \text{ erg/K (Boltzmann's constant)}$$

At 2000 K,

$$\% \text{ in excited state} \cong 300 \exp\left(\frac{-9.29 \times 10^{-12} \text{ erg}}{1.38 \times 10^{-16} \text{ erg K}^{-1} \cdot 2000 \text{ K}}\right) = 7 \times 10^{-13}\%$$

At 3000 K,

$$\% \text{ in excited state} \cong 300 \exp\left(\frac{-9.29 \times 10^{-12} \text{ erg}}{1.38 \times 10^{-16} \text{ erg K}^{-1} \cdot 3000 \text{ K}}\right) = 6 \times 10^{-8}\%$$

Because A is a small number, reasonable estimates can be made by assuming $A = 1$ so that only wavelength is needed for this approximate calculation.

example, cesium is 96% ionized in an oxygen–acetylene flame (3050°C) so that a lower flame temperature would be recommended.

Flames are complicated, consisting of several zones of temperature and of oxidizing nature. Thus the position of the flame with respect to the optical path is an important experimental variable and must be controlled. Normally, standard solutions are run sequentially with the samples under the same flame conditions, and a calibration curve is constructed. This is shown in Example 14-1.

There are several sources of interference that the analyst must consider when devising a flame emission procedure. It would appear that, with the extremely sharp lines observed with atomic spectra, little if any interference should be encountered.

EXAMPLE 14-1

A 45.0-mg sample of a mineral is dissolved in HCl, diluted to 250 mL, and analyzed for calcium by FES. The emission signal is 26 units on the emission scale. Standard calcium solutions gave the following results:

Ca (ppm)	Emission signal (scale units)
5.0	13
10.0	25
15.0	37

Calculate weight percent of Ca in the mineral.

Solution

A calibration curve can be drawn by the method of least squares, or for simplicity, the average emission signal can be calculated as

$$\frac{signal}{ppm} = \frac{13}{5.0} = 2.60 \qquad \frac{25}{10.0} = 2.50 \qquad \frac{37}{15.0} = 2.47$$

$$\frac{average\ signal}{ppm} = 2.52$$

Therefore, the sample contains $\dfrac{26\ \text{units}}{2.52\ \text{units}/ppm} = 10.3$ ppm (mg/L)

$$\% \ Ca = \frac{10.3\ \text{mg}/\cancel{L} \times 0.250\ \cancel{L}}{45.0\ \text{mg}} \times 100\ (\%) = 5.7\%$$

(*Note:* With meter readings of only two significant figures, percent of Ca should be given to only two also.)

Indeed, atomic lines rarely overlap although there are a few cases (for example, an emitted line from Si is 250.690 nm, while an emitted line from V is 250.691 nm). Of more serious concern is the presence of molecular bands, which will be broad in nature and capable of overlapping with the atomic lines being measured. Emission from the flame itself can give rise to molecular bands, such as from C_2, NH, or CN. Molecular bands from other substances in the sample may occur when an exceptionally stable molecular species is produced. For example, when a sample containing calcium is analyzed for sodium, a broad molecular band from CaOH overlaps the Na atomic line. Finally, other constituents in the sample may impede or facilitate the fraction of analyte that is atomized and excited. The emission signal is then a function of the concentration of these other constituents. The dependence is called a *matrix effect*. For example, in an analysis for metal ions of samples that also contain sulfur, phosphorus, or aluminum, formation of metal sulfates, phosphates, or aluminates may occur. These species are so stable that atomization becomes difficult, and a lower emission signal will be observed. Sometimes the use of standard solutions containing the same concentration of interfering substances will compensate for the matrix effects, but frequently their concentration or even their presence is unknown.

It is possible to avoid some matrix effects by the addition of other components to the sample solution. Phosphate plays a major matrix effect in the analysis for calcium by forming an extremely stable calcium phosphate species in the flame. Addition of excess La, Sr, or Mg to the sample forms their phosphate in preference to calcium. In this way, the calcium is released from being bound as the phosphate and can be atomized and excited more efficiently. Such excess reagents are called *releasing agents*. Another approach is to prevent formation of highly stable species by adding a chelate such as EDTA to form the metal complex. This technique is called *protective chelation*.

The solvent itself can play a role in FES. Organic solvents can enhance flame sensitivity (allow a higher proportion of atoms to become excited). This effect is most likely due to the presence in the flame of organic radicals and ions that become excited and then transfer energy to the atoms. However, emission from these species can also act as interference. Finally, physical effects, such as viscosity and surface tension of the sample solution, must be taken into account because these factors affect the rate of aspiration and the size of sample droplets introduced into the flame. Again, it is important to run standards with the same characteristics as the samples. Alternatively, the method of standard addition may be employed.

14-4 ATOMIC ABSORPTION SPECTROSCOPY (AAS)

AAS (or atomic absorption, AA) is atomic spectroscopy involving an absorption instead of an emission process. In a period of a few short years, AAS has become one of the most widely used analytical techniques. In fact, one could say that AAS is to inorganic analysis what gas chromatography is to organic analysis.

Because it measures absorption, an atomic absorption spectrometer will contain the same basic components as an ordinary UV–visible spectrophotometer: namely, source, sample compartment, wavelength selection, and detection. However, the analytical requirements and the physical appearance of the instruments will be considerably different. One important difference to notice is that in AAS a sample must be atomized for the observation of atomic lines, and this is frequently accomplished by a flame. Thus atomic absorption and flame emission spectrometers have this aspect in common, and commercial instruments are usually designed to do both techniques.

Another major difference between AAS and UV–visible spectrophotometry lies in the radiation source. The differing source requirements have to do with line widths: Molecular absorption lines (bands) are inherently broad, whereas atomic absorption lines are extremely sharp. By looking at Figures 14-6 and 14-7, we can see the problem that arises if we try to use a conventional white-light source and monochromator for atomic absorption. In conventional UV–visible spectro-photometry (Figure 14-6), the absorption band is much broader than is the

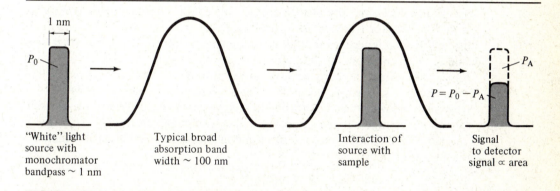

1 nm

P_0

$P = P_0 - P_A$

P_A

"White" light source with monochromator bandpass ~ 1 nm

Typical broad absorption band width ~ 100 nm

Interaction of source with sample

Signal to detector signal ∝ area

FIGURE 14-6   Conventional UV–visible spectrophotometry.

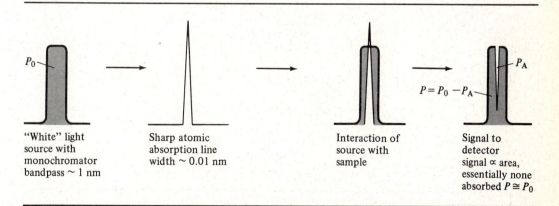

P_0

$P = P_0 - P_A$

P_A

"White" light source with monochromator bandpass ~ 1 nm

Sharp atomic absorption line width ~ 0.01 nm

Interaction of source with sample

Signal to detector signal ∝ area, essentially none absorbed $P \cong P_0$

FIGURE 14-7   Attempt to use conventional UV–visible source for atomic absorption.

bandpass of the radiation source so that all wavelengths included in the source after dispersion will be absorbed equally (particularly at the absorption peak, where little change in absorbance *versus* wavelength occurs). The final signal to the detector will be attenuated from P_0 (blank) to P, and the absorbance is the difference:

$$A = \log P_0 - \log P \qquad\qquad (14\text{-}6)$$

With atomic lines (Figure 14-7), the bandpass of the source would be much broader than the absorption lines so that only a minute fraction of the source radiant power, P_0, would be absorbed. The transmitted radiant power, P, would be almost the same as P_0, which means

$$A = \log P_0 - \log P \cong \log P_0 - \log P_0 = 0$$

For example, if the source were approximated by a rectangle 1 nm wide and the atomic line were a rectangle 0.01 nm wide, only 1% of the incoming radiant power would be absorbed *even if all* the light in that 0.01-nm band were absorbed. Therefore, what is needed for AAS is a source with a bandpass width narrower than that of the absorption line itself. The solution to this problem takes the approach that, if atomic lines are sharp, we ought to use an atomic line source; and the result of this approach is the employment of a *hollow cathode tube*, as shown in Figure 14-8. The cathode is constructed of, or contains, the element to be analyzed. Low-pressure (at or few torr) argon or neon is present in the tube, and application of 100 to 200 V causes a glow discharge, which results in atomic emission from the cathode material via a series of steps:

1. Ionization of filler gas: $Ne + e \rightarrow Ne^+ + 2e$

2. "Sputtering" of metal cathode material: $M(s) \xrightarrow{Ne^+} M(g)$

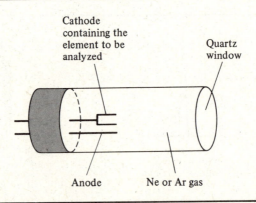

Cathode
containing the
element to be
analyzed

Quartz
window

Anode   Ne or Ar gas

FIGURE 14-8   Schematic diagram of a hollow cathode lamp.

3. Excitation of metal cathode material: $M(g) \xrightarrow{Ne^+} M(g)^*$

4. Emission: $M(g)^* \rightarrow M(g) + h\nu$

The emitted line is similar in width to the absorption line and at exactly the same wavelength, but its width is actually somewhat narrower because the absorption line is more affected by pressure and Doppler broadening.[1]

The hollow cathode tube has a large drawback: The analyst must buy, for a few hundred dollars each, a separate tube for each element to be analyzed. There are, however, some multielement tubes with cathodes containing several elements. Because the source lines are naturally so sharp, there is no real need to use a monochromator for light-source wavelength selection as in a conventional spectrophotometer with a continuum source. However, elements will emit more than one line; and of course, with multielement tubes, it is then necessary to isolate the lines representing each element's atomic emission. Therefore, atomic absorption spectrometers will contain a monochromator for line selection, but the requirements for narrow bandpass are not critical.

AAS differs from UV–visible spectrophotometry in one other aspect. The use of a flame for atomization results in broad-band emitted light's reaching the detector. Through placement of the monochromator *after* the sample compartment (flame), much of this flame emission will be filtered out. However, the signal is still "noisy" and contains a significant contribution from the flame itself at the wavelength of interest. This problem is overcome by *modulating* the signal from the source by means of a *chopper*: Between the source and the sample compartment is placed a rotating circular disk of which alternate quadrants are open to allow light to pass while the rest of the disk blocks the light. As shown in Figure 14-9, the final signal contains a dc component from the flame plus an ac component from the light source. The ac component can be amplified to the point where the dc compartment is negligible. Recent advances in electronics have made it possible to pulse the hollow cathode tube instead of employing the mechanical chopper. An additional advantage of the approach by the pulsed hollow cathode tube is that higher currents may be used and higher sensitivity reached.

Figure 14-9 is a schematic representation of an atomic absorption spectrometer. The detector for an atomic absorption spectrometer is basically the same as for a UV–visible spectrophotometer, photomultiplier tubes generally being used. The output can be calibrated in transmittance or absorbance, and newer instruments can directly read out in analyte concentration.

The burner used to establish the flame—and, in essence, the sample compartment—is also an important design feature for AAS (and for FES). The two types that are in general use are the turbulent flow–total consumption burner and the laminar flow–premix burner; they are shown in Figure 14-10.

In the turbulent flow–total consumption burner the aspirated sample, fuel, and oxidant are all brought together at the base of the burner. High velocity prevents

---

[1]Atoms moving toward a light source "see" a higher-frequency electromagnetic radiation; those moving away from the source see a lower frequency. This is known as the *Doppler effect*, and it is experienced in everyday life as the train-whistle phenomenon: A train-whistle frequency (pitch) is higher when the train approaches an observer than when it is receding.

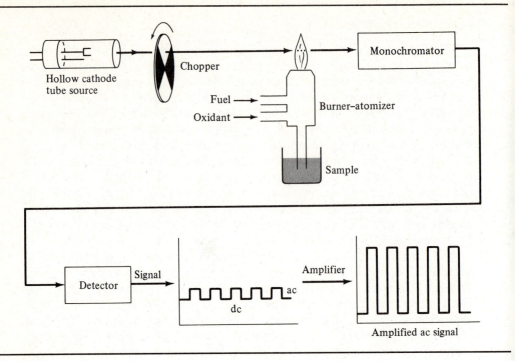

FIGURE 14-9   Diagram of atomic absorption spectrometer.

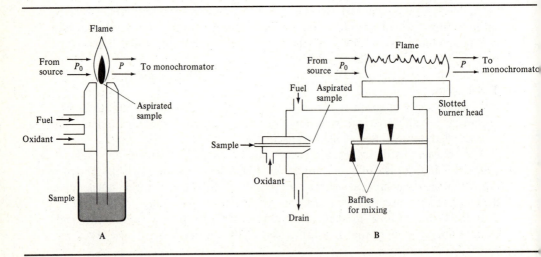

FIGURE 14-10   Burners for AAS: (**A**) turbulent flow–total consumption burner; (**B**) laminar flow–premix burner.

flashback of the flame, which can be a problem when hydrogen fuel is used. Even though the sample is totally consumed, some of the larger droplets can go right through the flame region without atomization. The turbulent flow also makes this burner noisy.

In the laminar flow type, the sample is aspirated into a chamber in which the fuel, oxidant, and analyte solution droplets are mixed. However, larger drops fall to the bottom and are collected. The mixture is then forced out through a long, slotted, burner head. The more homogeneous mixture and relatively slow velocity result in a more reproducible signal. In addition, this is an absorption process and follows Beer's law,

$$A = \varepsilon bC$$

The b term is the path length through the flame and, as can be seen in Figure 14-10, will be considerably greater for the laminar flow burner than for the turbulent flow burner.

In recent years, considerable effort has been spent in developing nonflame atomic absorption techniques. Unlike FES, in which the flame is used as a means of exciting the atoms, atomic absorption measures the interaction of the hollow-cathode-tube source with *ground-state* atoms. Thus any means of atomizing the sample and containing it for some period of time in the optical path would be acceptable; it should preferably *not* excite a significant fraction of the atoms. One widely used approach is that of adding the sample to a graphite furnace, which can be heated up to 3000°C to accomplish atomization. An example of such a furnace is shown in Figure 14-11. It consists of a hollow graphite cylinder a few centimeters long and a few millimeters in cross section. A small amount (1 to 50 μL) of sample is inserted through the sample port, and the solvent is evaporated at a low temperature. Following this, the sample is ashed and then strongly heated to accomplish atomization. The atoms diffuse out the ends of the tube, but their residence time of 2 to 3 sec is sufficient to allow recording of the absorption signal. The signal is recorded as a function of time, and the peak area is proportional to concentration of analyte in the sample. The technique has a very high sensitivity, but severe matrix effects limit accuracy to about ± 5 to 10%.

Like conventional UV–visible absorption spectrophotometry, AAS should ideally follow Beer's law so that

$$A = \varepsilon bC = kC \qquad\qquad\qquad (14\text{-}7)$$

where k is a proportionality constant, including absorptivity and path length through the flame or furnace. However, deviations from linearity occasionally occur when absorbance is plotted against concentration, especially at high absorbance values. Therefore, calibration curves generally determine analyte concentration. Alternatively, the standard addition method can be used by spiking an aliquot of the sample with a known amount of the analyte element. This technique has the advantage of not requiring a run of standards for a calibration curve, and—more important—it will help compensate for any matrix effects. However, the technique assumes linearity between absorbance and concentration in the range used so that

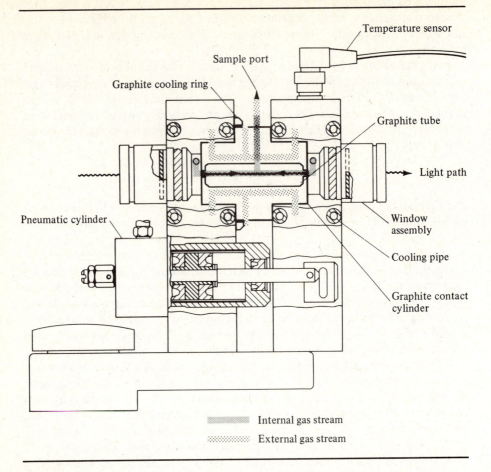

FIGURE 14-11 Cross section of HGA graphite furnace assembly. (Courtesy of *Perkin-Elmer Corp.*)

the proportionality constant k in Equation 14-7 can be eliminated:

$$\text{unknown sample:} \quad A_x = kC_x \quad \text{or} \quad k = \frac{A_x}{C_x} \qquad \textbf{(14-8)}$$

$$\text{sample + standard addition:} \quad A_{x+s} = k(C_x + C_s) \qquad \textbf{(14-9)}$$

(dilute to same volume as sample above). Therefore,

$$A_{x+s} = \frac{A_x}{C_x}(C_x + C_s) \qquad \textbf{(14-10)}$$

Rearrangement of Equation 14-10 leads to

$$C_x = C_s \left(\frac{A_x}{A_{x+s} - A_x} \right) \tag{14-11}$$

A sample calculation using the standard addition method is shown in Example 14-2, and Chapter 12 (especially Section 12-4) should be consulted for a more general approach to standard addition techniques.

Because more than 60 elements can be determined with accuracies of about ± 1 to 2%, AAS finds wide-ranging applications. It is also a relatively sensitive technique, with detection limits for most elements < 100 ppb, as shown in Table 14-5. Although FES is more sensitive for alkali metals (Table 14-5), it cannot compete

---

EXAMPLE 14-2

A 400-mg sample of hair from a young child is digested in HNO_3–$HClO_4$ and diluted to 25 mL. A 10-mL aliquot is diluted to 25 mL and analyzed with a cadmium hollow cathode tube. The absorbance reading is 0.13. To a second 10-mL aliquot is added 1.00 mL of a 0.40-ppm standard cadmium solution, followed by dilution to 25 mL and analysis for cadmium. The absorbance reading is 0.18. Calculate the cadmium content in the hair sample, and compare it with the mean value of 3.3 ppm found for children ($\sigma = 1.29$ ppm).

Solution

The concentration of standard addition, C_s, is

$$0.40 \text{ ppm} \times \frac{1 \text{ mL}}{25 \text{ mL}} = 0.016 \text{ ppm}$$

Therefore, from Equation 14-11, we obtain

$$C_s = 0.016 \left(\frac{0.13}{0.18 - 0.13} \right) = 0.042 \text{ ppm}$$

The original 25-mL solution would contain $0.042 \ (25 \text{ mL}/10 \text{ mL}) = 0.105$ ppm, or 0.105 $\mu g/mL$. The hair contains

$$0.105 \ \mu g/mL \times 25 \text{ mL} = 2.6 \ \mu g$$

$$\text{ppm} (\mu g/g) = \frac{2.6 \ \mu g}{0.4 \text{ g}} = 6.5 \text{ ppm}$$

This value is considerably higher than the mean and even lies above the 95% range of $3.3 \pm 2 \times 1.29 = 0.7$ to 5.9 ppm.

TABLE 14-5   Comparison of Detection Limits for Various Elements by Flame Emission, Atomic Absorption, and Atomic Fluorescence Techniques

Element	Wavelength (nm)*	Detection limit (μg/L; ppb)			
		Flame emission	Atomic absorption, flame	Atomic absorption, nonflame	Atomic fluorescence
Ag	328.1	2	1	0.01	0.1
Al	396.2	3	40	0.1	100
Ca	422.7	0.1	0.5	0.04	20
Cd	326.1, 228.8, 228.8	800	0.06	0.008	0.001
Cr	425.4, 357.9, 357.9	2	2	0.2	5
Fe	372.0, 248.3, 248.3	5	4	1	8
Hg	253.7	10,000	200	2	0.2
K	776.5	0.05	3	4	—
Li	670.8	0.02	1	0.3	0.1
Mg	285.2	5	0.3	0.004	—
Na	589.0	0.5	0.8	0.1	—
Zn	213.8	10,000	1	0.003	0.02

SOURCE: Data from J.D. Winefordner, J.J. Fitzgerald and N. Omenetto, "Review of Multielement Atomic Spectroscopic Methods," *Applied Spectroscopy, 29*:369 (1975), and J.D. Winefordner and R.C. Elser, "Atomic Fluorescence Spectrometry," *Analytical Chemistry, 43(4)*:24A (1971).

*When three wavelengths are given, they represent lines used for flame emission, atomic absorption, and atomic fluorescence, respectively.

with AAS for most analytes. In addition, the development of nonflame techniques has extended the detection limits for many elements to the parts-per-billion range (Table 14-5).

APPLICATIONS

Clearly, any analytical problem involving low concentrations of metal ions in an aqueous solution can be tackled by AAS. Determination of water quality is frequently carried out by this method. Biological fluids are another area of application: For example, zinc can be determined in cow's milk (see the laboratory manual); analysis by AAS of blood and urine for metal content is beginning to replace older spectrophotometric methods. Determination of metal ion concentration in electroplating solutions is easily accomplished with AAS although the solutions must be greatly diluted.

Solid samples can be analyzed by dissolution in a solvent: Alloy analysis, for example, is accomplished by dissolution of a sample in an appropriate acid, as described for a steel sample in the laboratory manual. Hair can be analyzed for metal content by digestion in HNO_3 and then $HClO_4$ to solubilize the metals and eliminate the organic matter. The whole field of hair analysis has developed in recent years because it has been established that metal content varies with the individual, his diet, and his environment. Thus, for forensic purposes, hair analysis is useful for identification purposes although individual hairs vary considerably in composition (so hair bundles should be used). Hair composition also varies with distance from the scalp; and accurate analysis obviously requires scrupulous prior cleaning to eliminate shampoo or hair-spray residues, which frequently contain zinc and other metals. To date, over 25 elements in hair have been determined by AAS, and several other elements have been found by other analytical techniques. Analysis of hair has also been useful for the determination of such pollutants in the environment as lead, cadmium, mercury, and arsenic; a correlation between metal content and learning disability in children has been found for some of these elements. Hair analysis may become a routine diagnostic tool, to complement blood and urine analyses.

Trace-element composition of foods is another area well suited to atomic absorption analysis. Iron, zinc, and copper analyses are well established. Samples are digested in H_2SO_4–H_2O_2, with the whole procedure largely automated so that over 100 sampless per hour can be handled.

Another environmental application involves air pollutants. The nonflame techniques are used both because they are highly sensitive and because the graphite furnace performs the required pretreatment of the sample. Samples are collected by drawing a known amount of air through a filter mounted in a graphite atomizer cup. After the sample is collected, a drop of H_3PO_4 is added and the sample is ashed to eliminate the organic filter and form metal phosphates. The temperature is then increased to atomize the sample and absorbance is measured. Alternatively, there can be used a graphite filter that is part of the atomizer itself. This approach has the advantage that the filter is reusable and that it can be precleaned in the atomizer, resulting in low blank values. Applications of this technique have included the following: Highly toxic beryllium was detected in a room in which a gas-mantle lantern had been lit—especially high levels when the mantle had been lighted for the first time; cadmium and lead have been measured in cigarette puffs; lead concentration in a room reached city-traffic levels when a single Christmas candle having a wire wick stiffener was burning; mercury levels in offices and laboratories have been measured as a function of location and time. These few examples illustrate the tremendous scope of AAS in solving analytical problems.

14-5   ATOMIC FLUORESCENCE SPECTROSCOPY (AF)

We can go one more step in the series of atomic absorption and emission processes described so far. First, a sample is excited thermally or electrically, and atomic emission is observed. Then, a radiation source is used in which atoms are excited electrically and emit light; the emitted light is absorbed by the sample, and atomic absorption is measured. The atoms in the sample have now been excited by the absorption process and will return to the ground state, frequently by emitting light of the same wavelength as that absorbed. This emitted radiation is called *atomic fluorescence*, and the series of steps is shown in Figure 14-12. Measurement of AF represents one of the newer analytical methods.

Fluorescence measurements reported in the literature have usually been obtained with modified atomic absorption spectrometers. As can be seen from Figure 14-12, the only additional feature in principle is a detector (photomultiplier tube) placed at some angle—usually 90°—to the radiation from the source. (The fluorescence will occur at all angles, but at 0° the fluorescence signal would be masked by the transmitted light coming from the source.) The source and the sample system remain the same as those for AAS although a single high-intensity *continuous* source may be used in place of individual hollow cathode tubes.

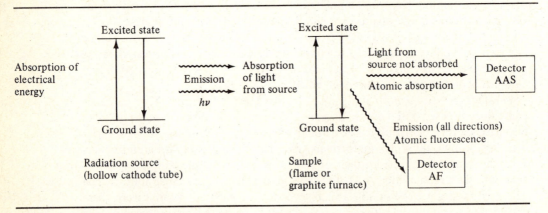

FIGURE 14-12   Comparison of AAS with AF.

The flames used in AAS to atomize the sample can be used in AF, but some flames tend to *quench* emitted radiation. That is, they provide alternative pathways for the excited atoms to return to the ground state. Flames containing CO, CO_2, and N_2 are detrimental to the fluorescence process; H_2–O_2 flames are preferred. Nonflame atomizers are also common.

The applications for AF are similar to those for AAS. As shown in Table 14-5, there are instances where AF exhibits the highest sensitivity of the three related methods. It might seem that, since only atoms that have absorbed radiation and therefore contribute to the absorption signal may fluoresce, AF could never be more sensitive than AAS is. However, it must be remembered that the sensitivity of an absorption method depends on measuring the small *difference* between two signals ($A = \log P_0 - \log P$), whereas fluorescence depends on the direct measurement of emitted radiation. The latter measurement can be made with photomultiplier tubes at very low radiant powers, and the capability gives emission techniques potentially high sensitivities. The high sensitivity, however, will be realized only when a significant fraction of the excited-state species emits light and only provided that this light can be collected efficiently at the detector.

14-6 X-RAY FLUORESCENCE SPECTROSCOPY (XRF)

Although an extensive discussion of XRF is beyond the scope of this text, it would be unfair to leave the topic of atomic spectroscopy without mentioning it.

The wavelengths associated with X-rays are so short (see Figure 12-3) that their energies are capable of exciting inner electrons. Because inner electrons are not involved in chemical bonding, XRF lines for an analyte are not significantly affected by its chemical environment, and like emission spectroscopy, XRF yields elemental analysis information only. (In this case, though, it is unnecessary to atomize the sample first.) As with AF, we need a source that can excite electrons. The excitation is accomplished with X-ray tubes which operate at 30,000 to 50,000 V to provide sufficient electrical energy for X-rays. When the X-rays interact with the sample, they will be capable of removing an inner electron. It is customary to use the older shell designations for electron levels ($1s = K$ shell, $2s$, $2p = L$ shell, and so forth) in discussions of X-ray methods. For example, if the X-ray source excites a K shell electron and an L shell electron drops down to fill the K shell vacancy, an X-ray (called K_α radiation) is emitted and has an energy corresponding to the energy difference between the L and K levels. If, instead, an M shell electron dropped down to fill the K shell, K_β radiation is emitted, as shown in Figure 14-13. On the other hand, if the X-ray source excites an L shell electron, then L_α (M shell to L shell) and L_β (N shell to L shell) radiation may be observed.[2]

---

[2] The designations become more complex as the number of subshells increase. As an example $L_{\beta 2}$ represents one possible $N \to L$ transition while $L_{\beta 1}$ represents one of the allowed $M \to L$ transitions.

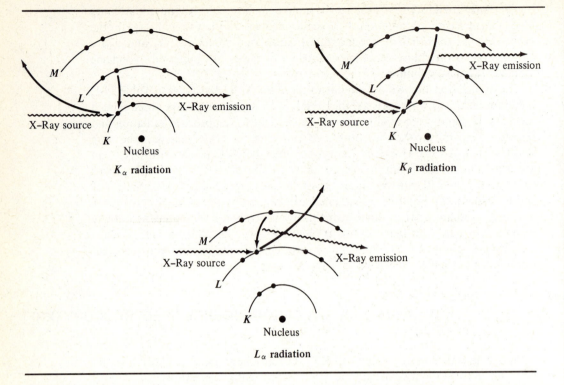

FIGURE 14-13 Examples of XRF emission. It should be noted that not all *L* and *M* shell electrons are at exactly the same energy so that it is possible to observe lines of nearly equal energy, denoted, for example, as K_{α_1} and K_{α_2}.

The components of an XRF spectrometer are shown in Figure 14-14. Basically, they are the same as those for any other spectroscopic technique: source, sample compartment, wavelength selector, and detector.

Requirements for the source limit the technique to about 35 elements that can be analyzed *conveniently*. The energy required to excite an inner electron depends on the number of protons in the nucleus and therefore increases with atomic number. The X-ray tube must contain an element of higher atomic number than any of the elements in the sample. The limit is even more restrictive because much of the emitted radiation from the X-ray tube consists of K_{α} and K_{β} radiation, which represents only the energy *difference* between the K shell and the L or M shell, whereas excitation of the sample requires promoting an electron to infinity (ionization of an inner electron). Therefore, even with a tungsten tube (atomic number 74), the element of highest atomic number that can be analyzed is near cerium (atomic number 58). With L or M spectra or special high-voltage tubes, it is possible to analyze heavier elements, but sensitivities are considerably less. At the light-element limit, the problem is that the X-rays emitted have such long wavelengths (low energy) that they are absorbed by air in the spectrometer. Titanium

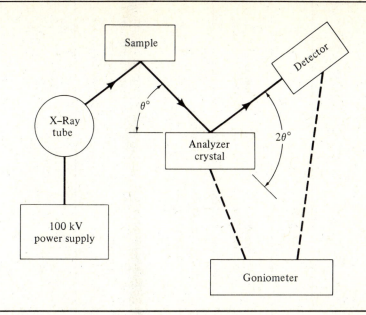

FIGURE 14-14 Typical arrangement of XRF spectrometer components. The goniometer assembly controls the analyzer crystal and detector rotations so that, when the crystal has rotated $\theta°$, the detector rotates $2\theta°$.

(atomic number 22) is the lightest element that can be determined conveniently; if one is willing to accept low sensitivity or to work in a vacuum environment, it is possible to determine elements as light as boron (atomic number 5). Newer surface-analysis X-ray techniques, such as Auger spectroscopy, are better suited for analysis of low-atomic-number elements.

The sample compartment simply holds a liquid or solid sample, and no pretreatment is necessary. It is convenient to press a powdered sample into a pellet before analysis.

Wavelength selection is considerably different from that for UV–visible spectrophotometry. The emitted X-rays are diffracted by an *analyzer crystal* (for example, LiF) and are detected as a function of angle, as shown in Figure 14-14. Bragg's law shows that

$$m\lambda = 2d \sin \theta \tag{14-12}$$

where m is an integer indicating the diffraction order, d is the lattice spacing, and θ is the diffraction angle. For first-order diffraction, $m = 1$, the analyzer crystal has a known lattice spacing d so that the wavelength, λ, is directly proportional to $\sin \theta$.

Detection can be accomplished with photographic film or some type of radiation counter, such as a Geiger counter, a proportional counter, a scintillation counter, or the more recently developed solid-state silicon (doped with Li) detector. The

TABLE 14-6   Comparison of Atomic Spectroscopic Techniques

Technique	Abbreviation	Process measured	Examples of excitation sources	Sample compartment
UV–visible spectrophotometry*	UV–vis	absorption	"white" light (lamp) + monochromator	cuvette
atomic emission spectroscopy	AES	emission	arc, spark, or ICP	electrode gap
flame emission spectroscopy	FES	emission	flame	flame
atomic absorption spectroscopy	AAS	absorption	hollow cathode tube	flame or graphite furnace
atomic fluorescence	AF	emission	hollow cathode tube	flame or graphite furnace
X-ray fluorescence	XRF	emission	X-ray tube	container for liquids or pressed-powder wafers

*UV–visible spectrophotometry is not an atomic spectroscopy, but it has been included for comparison.

frequently used scintillation counter consists of a NaI crystal activated with thallium. When an X-ray is captured by the detector, pulses of light will be emitted, which are then collected by a photomultiplier tube. The output signal is therefore highly amplified, and it is also proportional to the energy of the X-ray absorbed because the number of light pulses produced by each X-ray photon is proportional to its energy. Because of this feature, the scintillation counter can discriminate and count only X-rays of the desired energy.

Accuracy of the XRF method is typically $\pm 5\%$ although, with utmost care and calibration, higher accuracies ($\pm 1\%$) have been reported. Sensitivity is not so high as those for the methods discussed earlier; 1% is typical. The range of detectable elements, low accuracy, and low sensitivity all tend to limit applications, but there are still many situations in which XRF is the method of choice. It requires no pretreatment and can handle both liquid and solid samples. Few lines are observed so that interpretation of the spectrum is relatively easy. Of utmost importance for some applications is the fact that XRF is *nondestructive*. The sample is not dissolved, burned, or destroyed in any way. Analysis of rare objects or of samples needed in court cases is frequently accomplished by XRF.

14-7 SUMMARY

A summing up of atomic spectroscopy is presented in Table 14-6. It is important to realize that all these methods (except the UV–visible spectrophotometry included for comparison) give *elemental* analyses. So they are useful in any analytical problem requiring only determination of the elemental composition of a sample.

SUPPLEMENTARY READING

ARTICLES DESCRIBING ANALYTICAL METHODS OR APPLICATIONS

Birks, L.S., "Pinpointing Airborne Pollutants," *Environmental Science and Technology*, 12:150 (1978).

Dulka, J.J., and T.H. Risby, "Ultratrace Metals in Some Environmental and Biological Systems," *Analytical Chemistry*, 48:640A (1976).

Fassel, V.A., "Simultaneous or Sequential Determination of All Elements at All Concentration Levels—The Renaissance of an Old Approach," *Analytical Chemistry*, 51:1290A (1979).

Fassel, V.A., and R.N. Knisely, "Inductively Coupled Plasmas," *Analytical Chemistry*, 46:1110A, 1155A (1974); 51:1290A (1980).

"Hair: A Diagnostic Tool to Complement Blood Serum and Urine," *Science*, 202:1271 (1978).

Koirtyohann, S.R., "Current Status and Future Needs in Atomic Absorption Instrumentation," *Analytical Chemistry*, 52:736A (1980).

VanLoon, J.C., "Atomic Fluorescence Spectrometry—Present Status and Future Prospects," *Analytical Chemistry*, 53:332A (1981).

Walsh, A., "Stagnant or Pregnant," *Analytical Chemistry*, 46:698A (1974).

REFERENCE BOOKS

Robinson, J.W., *Atomic Absorption Spectroscopy*, 2nd ed. New York: Marcel Dekker, Inc., 1975.

Winefordner, J.D., ed., *Trace Analysis—Spectroscopic Methods for Elements*. New York: John Wiley & Sons, Inc., 1976.

PROBLEMS

14-1. Describe the differences between AAS and conventional absorption spectrophotometry. Include a comparison of source, sample container, and monochromator for the two techniques.

14-2. Why do you think the inert gases (He, Ne, Ar, and so on) are not among the 70 elements that can be analyzed by emission spectroscopy?

14-3. Which do you think would be the better emission spectroscopic technique for analyzing a heterogeneous mineral sample, dc arc or ac arc?

*14-4. A well-water sample is analyzed by FES for potassium. The emission signal is 4.8 units. Standards give the data shown below. Prepare a calibration curve, and determine the K^+ level in the water sample.

K^+ standard (ppm)	Signal (units)
0.2	0.32
0.5	0.80
1	1.4
2	2.7
5	5.7

14-5. In Problem 14-4, how much error would have been introduced if the calibration curve had been assumed to be linear, based on the first two standards?

*14-6. One method for correcting for fluctuations in flame conditions in FES is to use an internal standard. For alkali metal analysis, Li^+ is often used for this purpose. Readings for another element (S_x) are then normalized to the Li^+ signal (S_{Li}); that is, the ratio S_x/S_{Li} is used as the measurement signal. It is assumed that, for example, a 2% decrease in the Li^+ signal would also mean a 2% decrease in the analyte signal. Using the internal standard method, determine a calibration curve for Na^+ analysis, and calculate parts per million of Na^+ in a sample with the data shown below:

	S_{Na}	S_{Li}
standard, 0.2 ppm Na, 500 ppm Li	0.22	48
standard, 0.5 ppm Na, 500 ppm Li	0.53	47
standard, 2.0 ppm Na, 500 ppm Li	2.30	51
standard, 5.0 ppm Na, 500 ppm Li	5.00	46
sample, 500 ppm Li	0.88	48

14-7. A 2.00 mL seawater sample is diluted to 250 mL and analyzed for K^+, using the calibration curve given in Problem 14-4. The emission signal is 3.8 units.

---

*Answers to problems marked with an asterisk will be found at the back of the book.

Calculate parts per million of K^+ in the seawater sample. Why do you think it was wise not to run the seawater sample directly without dilution?

14-8. A 5.0-mL sample of wine is diluted to 250 mL, spiked with 500 parts per million of Li^+, and analyzed by FES, using the standard curve prepared in Problem 14-6. The emission signals are $S_{Na} = 1.6$ and $S_{Li} = 44$. Calculate parts per million of Na^+ in the wine.

14-9. Why is flame instability a more important factor in FES than it is in AAS?

14-10. a. As the chief analyst for the local winery, you have chosen AAS for the determination of copper in brandy, using the standard addition technique. From the following data, calculate the concentration of copper (parts per million) in the brandy:

Sample	Absorbance
pure brandy	0.12
50 mL of brandy + 5 mL of 15-ppm copper soln.	0.28

b. What scientific law did you assume in making this calculation?

*14-11. A 480-mg sample of cat food is digested with HNO_3–$HClO_4$ and diluted to 25 mL in a volumetric flask. Three aliquots of 2.00 mL each are drawn, to which are added 0.1-, 0.2-, and 0.3-mL spikes of 100-ppm Fe standard solution. The absorbance of the original is measured by AAS, and the spiked samples are measured with the values given below. Correct the absorbance values of the spiked samples for dilution, and then plot the absorbance versus parts per million of Fe added. Extrapolate back to zero absorbance (see Chapter 12 for a discussion of standard addition techniques) to find the amount of Fe in the cat-food sample in parts per million (micrograms per gram).

Sample	A
original	0.186
+0.1 mL	0.295
+0.2 mL	0.391
+0.3 mL	0.478

14-12. The technique described in Problem 14-11 was carried out for the determination by AAS of manganese in a 530-mg cat-food sample. From the data below. calculate the amount of manganese present in parts per million micrograms per gram):

Sample	A
original	0.075
+0.1 mL	0.187
+0.2 mL	0.285
+0.3 mL	0.370

*14-13. Magnesium in blood serum can be determined by AAS. A 1.00-mL serum sample is diluted to 100 mL, and its absorbance is 0.125. A standard containing 2.0×10^{-5} F Mg^{2+} gives an absorbance of 0.187. Calculate Mg concentration in the blood as milligram percent (milligrams per 100 mL).

14-14. A 2.00-mL sample of cow's milk was diluted to 50 mL and analyzed for Zn, using AAS. The absorbance was 0.106. A second 2.00-mL aliquot was spiked with 12 μg of Zn before dilution to 50 mL. The absorbance was 0.245. Calculate Zn concentration in the milk as parts per million (micrograms per milliliter).

14-15. For the analysis described in Problem 14-14, the following uncertainties were estimated: 2.00 ± 0.01-mL sample, 12 ± 0.1-μg spike, 0.106 ± 0.002 absorbance for sample, and 0.245 ± 0.004 absorbance for spiked sample. Calculate the uncertainty in parts per million of Zn in the milk sample.

14-16. Small amounts of Cr in steel are to be determined by AAS. A series of standard solutions gives the following absorbance readings: 2 ppm = 0.062, 4 ppm = 0.124, 6 ppm = 0.186, 8 ppm = 0.240, and 10 ppm = 0.300. Four replicate samples (250 mg each), dissolved in acid and diluted to 500 mL, give absorbance readings of 0.265, 0.271, 0.258, and 0.260. What should be the reported value for the percentage of Cr in the steel, and what is the probability that the true value lies below 1.6%, the minimum acceptable value?

*14-17. The determination of zinc in hair can aid in identification for forensic purposes. How many replicate samples should be run to ensure (95% confidence) that the mean value is within 10 ppm of the true value if the AAS technique has a known standard deviation of 14 ppm?

14-18. If the radiant power used to excite a species by AF fluctuates with $\sigma = 1.2\%$, what uncertainty will that introduce into the fluorescence signal?

*14-19. The minimum wavelength of X-rays emitted by an X-ray tube can be calculated by the assumption that all electrical energy of the exciting electron is converted to photon energy in the X-ray. That is,

$$eV = h\nu$$

where e is the charge on the electron (1.6×10^{-19} C). To be consistent with energy in joules, Planck's constant is 6.63×10^{-34} J-sec. Calculate the minimum wavelength for an x-ray tube operated at 50 kV.

14-20. A topaz analyzer crystal is sometimes used for XRF and has a lattice spacing of 1.356 Å. Calculate its useful range (wavelength of X-ray) if 2θ values from 10 to 160° are accessible. Assume only first-order diffraction.

*14-21. A steel sample is to be analyzed for its Mo content by XRF. If the X-ray line is at 0.71 Å, what goniometer setting (2θ) should be used when a LiF analyzer crystal is employed (lattice spacing = 2.014 Å)? Assume first-order diffraction.

14-22. H.G.J. Moseley found that $\nu^{1/2}$ of the emission from the K or L shell is proportional to atomic number. Based on the wavelength given in Problem 14-21 for Mo, where would the corresponding line for Sr be found (in angstroms)? What would its goniometer setting (2θ) be on the assumption that a LiF analyzer crystal is used?

ELECTRO-ANALYTICAL METHODS

PART FIVE

Potentiometry 15

Part Four described analytical techniques based on measurement of the interaction of electromagnetic radiation with a sample. Now we turn our attention to electrical signals that can be related to an analyte. As mentioned previously, analytical chemists exploit any signal that can be measured accurately and reproducibly; electrical signals are no exception. Electrochemical potential, current, and coulombs of charge are three such signals, and they are the basis of electroanalytical methods. In general, the name of the technique reflects the signal used:

Signal	Electroanalytical technique
electrochemical potential	potentiometry
coulombs of charge	coulometry
current as a function of applied voltage	voltammetry

Of course, some techniques have labels that do not fit this simple rule: The classic and still widely-used electroanalytical technique of *polarography* is really a voltammetric method because current as a function of applied voltage is measured. The method will be studied as part of Chapter 17, Voltammetry.

It is important to recognize another classification of analytical techniques because, especially with electroanalytical techniques, some techniques fall into one classification and some into another. One class measures the total *amount* of analyte present, whereas the second measures the *concentration* of analyte present: With gravimetric and volumetric methods, the total amount is measured; with spectrochemical methods, the absorbance signal determines the concentration. Table 15-1 presents this classification scheme and lists various electroanalytical techniques.

Why is the classification significant? For one thing, different methods for handling samples (*quantitative transfer*) will be required in each technique group. When the total amount of analyte is to be measured, the entire sample must be

---

Table 15-1   Classification of Analytical Techniques

---

class I: total amount of analyte:

  gravimetric methods
  volumetric methods
  electroanalytical methods:
    coulometric titration (coulometry)
    controlled-potential electrolysis (coulometry)
  chromatographic methods (absolute amount of analyte present)

class II: concentration of analyte:

  spectrochemical methods
  electroanalytical methods:
    potentiometry
    polarography (voltammetry)
    anodic stripping (voltammetry)
    cyclic voltammetry
  chromatographic methods (relative amount of analyte present)

---

transferred through each step of the process. For example, in the gravimetric determination of iron by precipitation of $Fe(OH)_3 \cdot xH_2O$, it is first necessary to precipitate the iron quantitatively (negligible amount left in solution); then the entire precipitate must be transferred to the filter by a washing out of the vessel used for precipitation, the total amount of wash volume being unimportant; the precipitate is carefully ignited to convert all of it to Fe_2O_3. On the other hand, when iron is determined spectrophotometrically, a reagent (for example, *o*-phenanthroline) is added to form a colored complex, and the solution is diluted to a known volume; this procedure establishes the concentration of analyte. It is unimportant whether or not all the solution is transferred to the cuvette, and usually only an unknown fraction is transferred. However, it is critically important that the cuvette be dry or have been washed out with the solution to be measured so that no *concentration* change will occur when the solution is added to the cuvette for measurement of absorbance. In contrast, it does not matter whether a filter is wet when iron hydroxide precipitate is added. Thus quantitative transfer will mean different things, depending on whether a class I or a class II technique is being carried out.

Even a class I technique can be used to determine concentration of analyte in a sample. For example, at the equivalence point in a volumetric technique,

$$\text{meq analyte} = \text{meq reagent}$$

but if the analyte is contained in a known volume of solution,

$$\text{meq analyte} = NV$$

and therefore,

$$N = \frac{\text{meq}}{V} = nF$$

Likewise, a class II technique can be used to determine total amount of analyte present in a sample. For example, in a spectrophotometric technique, the concentration (millimoles per milliliter) is found from absorbance and the total amount from knowledge of the volume of sample:

$$mmol = CV$$

As a last point, it should be noted that the term "total amount of analyte present" does not preclude the use of aliquots. In this case, a class I technique determines the total amount of analyte present in the *aliquot*; and the dilution factor is then used to calculate the total amount of analyte present in the original sample.

As with other analytical techniques, we shall begin our discussion of each electroanalytical method with theoretical principles. But much of this material was presented in Chapter 11, and you should review that before proceeding.

We now examine various ways by which electrochemical potential measurements (potentiometry) can be used analytically.

15-1 METAL INDICATOR ELECTRODES

Consider the electrochemical cell of Figure 15-1, which can be written in shorthand version as $Pt|Fe^{2+}$ (0.1 F), Fe^{3+} (0.01 F)$\|$ Ag^+ (x F)$|Ag$. With the conventions adopted in Chapter 11, we can state

$$E_{cell} = E_{right} - E_{left} = E_{Ag} - E_{Fe} \qquad (15\text{-}1)$$

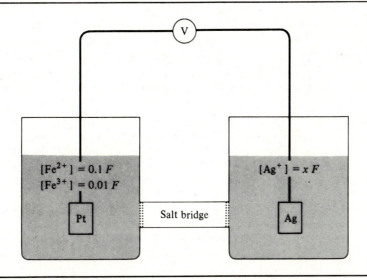

FIGURE 15-1 Electrochemical cell for the measurement of silver ion activity.

The electrochemical cell potential, E_{cell}, represents the driving force for the overall cell reaction when oxidation occurs at the left side (anodic) and reduction at the right (cathodic). That is,

$$Fe^{2+} + Ag^+ \rightleftharpoons Fe^{3+} + Ag \qquad (15\text{-}2)$$

The free energy change for the reaction in Equation 15-2 will be

$$\Delta G = -nE_{cell}F$$

With the Nernst equation, the individual electrode (half-cell) potentials at 25°C can be written as

$$Fe^{3+} + e \rightleftharpoons Fe^{2+}$$

$$E_{Fe} = E_{Fe}^0 - \frac{0.059}{n} \log \frac{[Fe^{2+}]}{[Fe^{3+}]}$$

$$Ag^+ + e \rightleftharpoons Ag$$

$$E_{Ag} = E_{Ag}^0 - \frac{0.059}{n} \log \frac{[Ag]}{[Ag^+]}$$

Therefore,

$$E_{cell} = E_{Ag} - E_{Fe} = E_{Ag}^0 - \frac{0.059}{n} \log \frac{[Ag]}{[Ag^+]} - \left(E_{Fe}^0 - \frac{0.059}{n} \log \frac{[Fe^{2+}]}{[Fe^{3+}]} \right) \qquad (15\text{-}3)$$

Equation 15-3 can be greatly simplified when $[Fe^{2+}]$ and $[Fe^{3+}]$ are known and E^0 and n values are inserted and when we note that $[Ag] = 1$ if a pure Ag metal electrode is used. The simplification is demonstrated in Example 15-1.

Immersing a silver *indicator electrode* in the sample solution allows us to calculate the silver ion activity (assumed to be equal to concentration) when the potential is measured against some known electrode potential. Actually, under these conditions, Equation 15-3 can be written in a simpler form:

$$E_{cell} = +0.799 + 0.059 \log [Ag^+] - E_{Fe} \qquad (15\text{-}4)$$

With the concentration of $[Fe^{2+}]$ and $[Fe^{3+}]$ given in Figure 15-1, $E_{Fe} = +0.771 - 0.059 \log 10 = +0.712$. Thus

$$E_{cell} = +0.799 + 0.059 \log [Ag^+] - 0.712$$

$$= +0.087 + 0.059 \log [Ag^+] = +0.087 - 0.059 \, pAg \qquad (15\text{-}5)$$

or

$$E_{cell} = K - 0.059 \, pAg \qquad (15\text{-}6)$$

EXAMPLE 15-1

The electrochemical potential for the cell shown in Figure 15-1 is $-217\,\text{mV}$. Calculate the silver ion concentration in the sample solution.

Solution

Equation 15-3 is used with the following values:

$$[Fe^{2+}] = 0.1$$

$$[Fe^{3+}] = 0.01$$

$$E^0_{Fe} = +0.771$$

$$E^0_{Ag} = +0.799$$

$$n_{Ag} = n_{Fe} = 1$$

$$E_{cell} = -0.217 = +0.799 - \frac{0.059}{1} \log \frac{1}{[Ag^+]}$$

$$- \left(+0.771 - \frac{0.059}{1} \log \frac{0.1}{0.01} \right)$$

$$0.059 \log \frac{1}{[Ag^+]} = 0.304$$

$$\log [Ag^+] = -5.15$$

$$[Ag^+] = 7.0 \times 10^{-6}\,F$$

(*Note:* The value -5.15 contains two significant figures because -5 determines only the exponent.)

Equation 15-6 shows that there is a straight-line relationship between E_{cell} and pAg with a slope of $-59\,\text{mV}$, as shown in Figure 15-2. Because of the uncertainties in E^0 values (recall that the formal potential must be used in moderately concentrated solutions, and this concentration may be unknown), it is standard procedure to determine the constant in Equation 15-6 by using known solutions (Figure 15-2). In this way, a calibration curve is constructed as is done in Example 15-2. With a calibration curve, it is not necessary to know even the concentration ratio $[Fe^{2+}]/[Fe^{3+}]$ as long as this ratio remains constant throughout the series of measurements.

The problem, then, with this experiment is that any change in the $[Fe^{2+}]/[Fe^{3+}]$ ratio will change the constant, K, in Equation 15-6. If the circuit is closed, the cell

EXAMPLE 15-2

For a cell of the type shown in Figure 15-1, a calibration curve is constructed as given in Figure 15-2. A 150-mg sample of an alloy containing silver is dissolved in HNO_3 and diluted to 250 mL. If the electrochemical cell potential is -72 mV, calculate the percentage of Ag in the alloy.

Solution

Equation 15-5 may be written in millivolts as

$$E_{cell} = K - 59 \text{ pAg}$$

From Figure 15-2, the value of K is $+91$ mV. Therefore,

$$\text{pAg} = \frac{K - E_{cell}}{59} = \frac{91 - (-72)}{59} = 2.76$$

$$[Ag^+] = 1.73 \times 10^{-3} \ F$$

$$\text{mmol Ag} = FV = 1.73 \times 10^{-3} \frac{\text{mmol}}{\text{mL}} \times 250 \ \text{mL} = 0.432 \ \text{mmol}$$

$$\% \ Ag = \frac{0.432 \ \text{mmol} \times 107.9 \ \text{mg/mmol}}{150 \ \text{mg sample}} \times 100 \ (\%) = 31.1\%$$

will run spontaneously right to left ($E_{cell} < 0$) so that some Fe^{3+} will be reduced to Fe^{2+}; this will increase the $[Fe^{2+}]/[Fe^{3+}]$ ratio and increase K. Fortunately, it is possible to measure electrochemical cell potentials with a *potentiometer* without drawing appreciable amounts of current. A potentiometer (Figure 15-3) employs a source of voltage that opposes the cell potential: When no current flows through the detector (balance point, or null), the voltage equals the cell potential. This procedure sounds fine until one realizes that, in order to find the balance point, one has to have the potentiometer unbalanced and observe flow of current. The current can be in either direction, depending on which side of the balance point the analyst begins, and its magnitude will depend on how far off balance the initial reading is. Thus even when no current flows in the final measurement, small amounts of current will have previously flowed through the system and unknown changes in the $[Fe^{2+}]/[Fe^{3+}]$ ratio will have occurred.

In fact, some electrochemical systems can be driven so far from equilibrium while the analyst is trying to balance a potentiometer that the true equilibrium value is not measured even when the null point is reached. For this reason and for convenience, most electrochemical cell measurements are now carried out with a *high-impedance voltmeter*. As the name implies, *voltage* (as opposed to *potential*) is measured; in other words, a current flows while the measurement is made, but the current is so

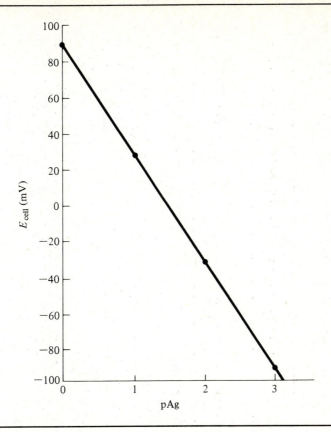

FIGURE 15-2 Cell potential as a function of pAg. For this cell, K is $+91$ mV.

small—typically, 10^{-12} A—that the voltage remains within the experimental error of the true potential for the system. High-impedance voltmeters (for example, a pH meter) are direct reading and may be interfaced to a recorder to provide a continuous record of electrode potential.

The Fe^{3+}/Fe^{2+} couple is not an ideal reference electrode with which to compare the metal indicator electrode. A reference electrode is needed whose constituents have fixed activities even if small amounts of current pass through the cell. This may be accomplished with the *saturated calomel electrode* (SCE), whose half-cell reaction is

$$Hg_2Cl_2 + 2e \rightleftharpoons 2Hg + 2Cl^- \tag{15-7}$$

$$E = E^0_{Hg_2Cl_2} - \frac{0.059}{2} \log \frac{[Cl^-]^2[Hg]^2}{[Hg_2Cl_2]} \tag{15-8}$$

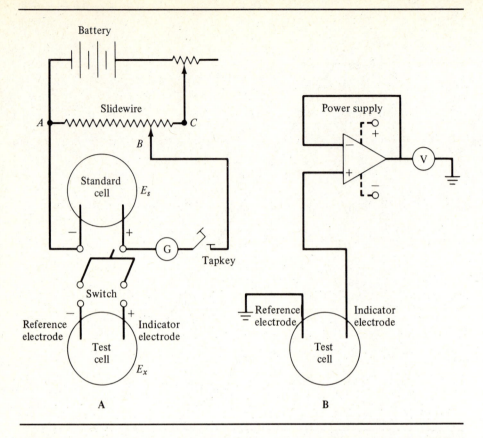

FIGURE 15-3 **(A)** Potential measurement using a classical potentiometer. Potential of cell is determined when point B is adjusted so that no current flows through galvanometer Ⓖ when tapkey is momentarily depressed. Slidewire is calibrated by switching to the standard cell so that $E_x = E_s(AB)_x/(AB)_s$. **(B)** Potential measurement using modern high impedance voltmeter. The input from the cell is fed into the noninverting terminal of an operational amplifier. A direct reading of cell voltage with a high input impedance ($\sim 10^{11}\Omega$) at Ⓥ is accomplished.

A SCE is prepared by mixing together Hg, Hg_2Cl_2, and KCl with some water, there being sufficient KCl to keep the solution saturated. Commercial SCE's hold this mixture in a central tube surrounded by an outer compartment containing saturated KCl. A fiber tip maintains electrical contact with the sample solution, as shown in Figure 11-3. Because pure Hg_2Cl_2 (essentially insoluble) and pure Hg are used, they are in their standard states; so their activities are 1. The chloride activity, although not equal to 1, at a given temperature will remain fixed as long as the solution remains saturated with respect to KCl. Thus Equation 15-8 may be written as (see

Appendix 3-A for $E^0_{Hg_2Cl_2}$)

$$E_{SCE} = +0.268 - \frac{0.059}{2} \log [Cl^-]^2_{satd.\ KCl} \tag{15-9}$$

Example 15-3 shows how this equation can be used in calculations.

EXAMPLE 15-3

(a) Estimate E_{SCE} from the solubility of KCl, which is 4.18 F at 25°C, neglecting activity coefficients.

(b) The actual value of E_{SCE} is $+0.242$ versus NHE. Calculate the activity of chloride ion and its activity coefficient.

Solution

(a) From Equation 15-9 and $[Cl^-] = 4.18$, we can write

$$E_{SCE} = +0.268 - \frac{0.059}{2} \log (4.18)^2$$

$$= +0.268 - 0.037 = +0.231 \text{ V vs. NHE}$$

(b) We see that activity coefficients should not be neglected in such a highly concentrated solution since part (a) was in error by 11 mV. With the measured value of $+0.242$ V vs. NHE,

$$+0.242 = +0.268 - \frac{0.059}{2} \log [Cl^-]^2$$

$$\log [Cl^-] = \frac{0.026}{0.059} = 0.441$$

$$\text{activity} = [Cl^-] = 2.76$$

$$a = fC \text{ (Equation 4-8)}$$

$$\text{activity coefficient} = f = \frac{2.76}{4.18} = 0.66$$

One problem encountered with the SCE is that its potential versus NHE changes significantly with temperature. This is easy to understand from Equation 15-9, where, because the solubility of KCl rises rapidly with temperature, $[Cl^-]$ increases with temperature increase and E_{SCE} decreases. Values at various temperatures are given in Table 15-2. Actually, all electrode potentials will be a function of temperature since free-energy change is a function of temperature. However, when establishing a set of electrode (half-cell) potentials, we had to assign one electrode as

Table 15-2   Potentials of Reference Electrodes at Various Temperatures

	Potential vs. NHE			
Temp. (°C)	SCE	Normal calomel	Decinormal calomel	Ag/AgCl
10	+0.252	+0.284	+0.335	+0.212
20	+0.245	+0.281	+0.334	+0.202
25	+0.242	+0.280	+0.334	+0.197
30	+0.239	+0.279	+0.334	+0.192
50	+0.226	+0.273	+0.332	+0.172
80	+0.206	+0.265	+0.329	+0.142

our reference point. The standard hydrogen electrode was given the value of zero, and this arbitrary reference point is adopted at *all* temperatures. All changes in cell potentials involving the NHE as one electrode are attributed to the other electrode. Therefore, changes in electrode potential with changes in temperature reflect not only any changes in activities of the various species but also any changes in E^0 for the couple in question *plus* any absolute changes in the NHE.

One way of decreasing the significant temperature dependence exhibited by a calomel electrode is to substitute a fixed concentration of KCl in place of a saturated solution. As can be seen from Table 15-2, the temperature dependence is decreased considerably when 1 N or 0.1 N (decinormal) KCl is used. However, now we have the problem mentioned earlier: Drawing current through the reference will change the activity of one or more species involved in the electrode reaction (Equation 15-7) and thereby alter the potential. The lower the concentration, the more sensitive the electrode will be to perturbations caused by current. For that reason, the normal (1 N) calomel electrode is a good compromise because it has a much lower temperature dependence than the SCE does and, at the same time, has a much higher chloride ion activity than the decinormal (0.1 N) calomel electrode does.

Another widely used reference electrode is the silver–silver chloride electrode. Its half-cell reaction is

$$AgCl + e \rightleftharpoons Ag + Cl^- \qquad \textbf{(15-10)}$$

It is constructed (Figure 15-4) in the same fashion as the SCE except that a silver wire is used both as an electrode contact and to establish unit-activity silver. Because it is a saturated type of electrode like the SCE, it exhibits a significant temperature dependence, as shown in Table 15-2.

Another problem encountered with these reference electrodes is the presence in the salt bridge (fiber point or porous plug of some type) of high concentrations of chloride ion, which can diffuse into the sample compartment. In the example discussed at the beginning of this chapter, the measurement of $[Ag^+]$ with a silver

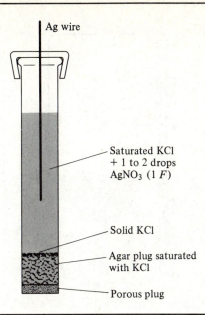

Ag wire

Saturated KCl
+ 1 to 2 drops
$AgNO_3$ (1 F)

Solid KCl

Agar plug saturated
with KCl

Porous plug

FIGURE 15-4   Construction of a Ag/AgCl reference electrode, with a half-reaction of $AgCl + e \rightleftharpoons Ag + Cl^-$.

indicator electrode, the presence of chloride ion could cause precipitation of AgCl and thereby change the silver ion concentration. In such cases, an inert salt bridge (such as a KNO_3 solution) would be needed to separate the sample solution from the reference electrode. Such an arrangement was shown in Figure 11-2.

    With these parameters kept in mind, we should be able to measure the metal ion activity (concentration) in any solution, using a metal indicator electrode. For example, copper ion concentration can be measured as shown in Figure 15-5. The measured potential at 25°C is

$$E_{cell} = E_{Cu} - E_{SCE} = E_{Cu} - (+0.242)$$

$$Cu^{2+} + 2e \rightleftharpoons Cu$$

$$E_{Cu} = E_{Cu}^0 - \frac{0.059}{2}\log\frac{1}{[Cu^{2+}]}$$

$$E_{cell} = +0.337 - 0.242 - \frac{0.059}{2} \quad pCu = +0.095 - 0.0295 \quad pCu \qquad \textbf{(15-11)}$$

In general, for any metal–metal ion couple of the type

$$M^{n+} + ne \rightleftharpoons M$$

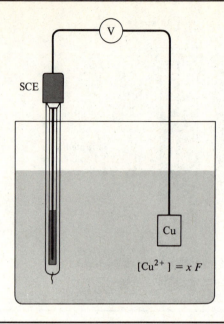

FIGURE 15-5   Electrochemical cell for measuring copper ion activity with SCE reference.

the cell potential will be

$$E_{\text{cell}} = (E_M^0 - E_{\text{ref}}) - \frac{0.059}{n} \ \text{pM} = K - \frac{0.059}{n} \ \text{pM} \qquad \textbf{(15-12)}$$

Equation 15-12 reiterates what was shown in Equation 15-6. A plot of E_{cell} against pM should be a straight line with a slope of $-0.059/n$ and an intercept of K. It is usually better to measure K and the slope with standard solutions because uncertainties in E_M^0 will affect K and discrepancies between concentration and activity will affect both K and the slope.

   In addition, there is one other uncertainty that should be considered (it will be further discussed in Section 15-3), that is *liquid junction potential*, E_j. Whenever two phases are brought into contact with each other, such as a solid electrode with an electrolyte solution, a potential difference arises at the interface. In most electrochemical cells, there is, in addition to the two electrode–electrolyte interfaces, a liquid–liquid interface, and the potential difference at this interface is the liquid junction potential. For example, in Figure 15-5, the reference electrode compartment contains saturated KCl while the analyte solution contains Cu^{2+} and other ions. Whenever the two liquids in contact with each other contain different ions or even the same ions at different concentrations, there will be a liquid junction potential. This potential difference originates in ions' diffusing at different rates. Where a concentration gradient (concentration change with distance) exists, the

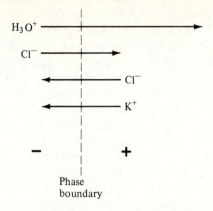

FIGURE 15-6 Development of a liquid junction potential at the phase boundary between HCl and KCl.

faster ions will tend to move more than the slower ions do. For example, H_3O^+ has a much higher diffusion coefficient than does K^+; so, as shown in Figure 15-6, there will be a greater tendency for H_3O^+ to move into the right compartment than there will be for K^+ to move into the left. Therefore, the right compartment will become positive relative to the left, and a positive liquid junction potential is generated.

To include liquid junction potentials in electrochemical cells, we should write

$$E_{cell} = E_{right} - E_{left} + E_j \qquad (15\text{-}13)$$

Calculation of liquid junction potentials can be carried out only for the most simple cases, and usually E_j is unknown for analyte solutions in contact with a reference electrode. Frequently, cells such as the one shown in Figure 15-5 are written schematically as

$$SCE||Cu^{2+}(x)|Cu$$

The double verticals, ||, inform the reader that the liquid junction potential will be ignored. How valid is this assumption? The liquid junction potential can be minimized by a salt bridge connecting the two electrode compartments; the bridge contains a highly concentrated KCl solution. The potential is minimized because the electrical charge carried across the interface when current is drawn will depend on both the ion concentrations and their diffusion coefficients; that is, most of the current is carried across the interface by the most mobile ions (high diffusion coefficients) present at the highest concentration. With a highly concentrated KCl salt bridge, most of the charge will be carried by K^+ and Cl^- ions, which have nearly equal diffusion coefficients (mobilities), so that very little junction potential results. However, the junction potential is not exactly zero (chloride ions are about 4% more mobile than are potassium ions) and will vary somewhat according to the concentration and nature of ions in the analyte solution. It is known that, even with a highly concentrated salt bridge, junction potentials of 1 or 2 mV may exist; without a KCl salt bridge, junction potentials of 10 to 100 mV are common. The bottom line to this liquid junction phenomenon is that the best potentiometric measurements have inherent uncertainties of about ± 1 mV. These circumstances present severe limitations for analytical application, as shown in Example 15-4.

It is shown in Appendix 1-K that the relative uncertainty in concentration does not depend on the absolute potential reading—only on the uncertainty in potential. For $n = 1$, the uncertainty in concentration will be about 4% (at 25°C) for a ± 1-mV uncertainty in potential. It does not matter whether the potential reading is 5 mV or 500 mV. For higher n values, the uncertainty is greater; so the best accuracy one can hope to achieve with direct potentiometry is about 5 to 10%.

It has been assumed throughout the previous discussion that the metal–metal ion couple will reach thermodynamic equilibrium so that Equation 15-12 will be valid. Several metal electrodes exhibit such slow electrode kinetics that equilibrium potentials are not observed. Potential readings for these metal indicator electrodes will not be stable and may be sensitive to other couples in the system, such as O_2/H_2O. Slow electrode kinetics are generally observed when the metal ion is not the simple hydrated species but a more complicated oxygenated species and/or the metal forms tightly adherent oxide coatings that prevent contact with the solution. For this reason, one cannot use a tungsten or molybdenum electrode to measure WO_4^{2-} or MoO_4^{2-} concentrations.

In addition to the limitation imposed by electrode kinetics, the metal indicator electrode must be chemically stable in the solution to be analyzed. Acids react with all metals having an E^0 value <0 versus NHE; for example, a Zn electrode cannot be used in an acid solution to determine $[Zn^{2+}]$. Clearly, alkali metal electrodes could not be used in an aqueous solution at any pH. The most useful metal electrodes are Ag, Hg, Pb, Cd, Cu, and Zn, whereas Pt may be used for measuring solution potentials.

Partly as a result of the inherent low accuracy and of the small number of useful metal–metal ion couples, applications of potentiometry with metal indicator electrodes are quite limited in scope. However, compensating for the fairly low accuracy is the possibility of continuous readout of metal ion concentration, which can be an important consideration for many environmental analyses or for process control. Also, the electrical signal is easy to record and can become part of a feedback loop to control a process.

Another area of application is as an endpoint signal in volumetric titrations. With reasonably sharp endpoints, an uncertainty of 50 or even 100 mV will not introduce a significant error. The electrical signal can be incorporated into a system that automatically titrates a sample and stops at the endpoint. These *autotitrators* are useful when many titrations must be run to justify the high cost compared to a simple buret and voltmeter.

An inert metal electrode (for example, Pt) for the determination of solution potential can also determine concentration and, especially, endpoint signal. The latter was described in Section 11-2 for the titration of Fe(II) with Ce(IV).

EXAMPLE 15-4

The Cu^{2+} concentration in a waste stream from a plating factory is measured with a copper metal indicator electrode. If the potential reading is $+250\,mV$ versus NHE and has an uncertainty of $\pm 1\,mV$, what is the percentage uncertainty in Cu^{2+} concentration?

Solution

First, it should be recognized that, because the signal (electrode potential) is not directly proportional to the analyte concentration, we *cannot* conclude that an uncertainty of 0.4% (1 mV per 250 mV) will introduce an uncertainty of 0.4% in $[Cu^{2+}]$.
    The half-cell reaction is

$$Cu^{2+} + 2e \rightleftharpoons Cu \qquad E^0 = 0.337 \text{ V vs. NHE}$$

$$E_{Cu} = +0.250 = E^0_{Cu} - \frac{0.059}{2} \log \frac{1}{[Cu^{2+}]} = +0.337 + \frac{0.059}{2} \log [Cu^{2+}]$$

$$\log [Cu^{2+}] = \frac{+0.250 - 0.337}{0.059/2} = -2.95$$

$$[Cu^{2+}] = 1.12 \times 10^{-3}\ F$$

Because of the ± 1-mV uncertainty, the electrode potential could just as likely have been $+251$ mV vs. NHE. In that case,

$$\log [Cu^{2+}] = \frac{+0.251 - 0.337}{0.059/2} = -2.92$$

$$[Cu^{2+}] = 1.22 \times 10^{-3}\ F$$

$$\% \text{ uncertainty in } [Cu^{2+}] = \frac{1.22 \times 10^{-3} - 1.12 \times 10^{-3}}{1.12 \times 10^{-3}} \times 100\,(\%) = 8.9\%$$

We could have used $(250 - 1)$ mV equally well, and in that case,

$$[Cu^{2+}] = \frac{+0.249 - 0.337}{0.059/2} = -2.98$$

$$[Cu^{2+}] = 1.04 \times 10^{-3}\ F$$

$$\% \text{ uncertainty in } [Cu^{2+}] = \frac{1.12 \times 10^{-3} - 1.04 \times 10^{-3}}{1.12 \times 10^{-3}} \times 100\,(\%) = 7.1\%$$

The reason that these two calculations give somewhat different answers is that the function is not linear; so the change in concentration with a 1-mV increase is not the same as it is with a 1-mV decrease. What is really needed is the slope of the curve ($[Cu^{2+}]$ vs. E_{Cu}) at $E_{Cu} = 250$ mV. This is shown in Appendix 1-K, with the result that a 1-mV uncertainty in potential generates a 7.8% uncertainty in $[Cu^{2+}]$.

15-2 ELECTRODES OF THE SECOND KIND

So far we have been concerned with simple metal–metal ion couples for which the metal indicator electrode potential measures the metal ion activity directly involved in the half-cell reaction. Such electrodes are said to be *of the first kind*. It is also possible to measure the activity of an ion involved not directly in the metal–metal ion couple but indirectly via some chemical equilibrium (*of the second kind*). We have already seen two such electrodes, the SCE and the silver–silver chloride electrode. They are frequently used as reference electrodes. However, they can also be used to determine chloride ion activity. For example, the silver–silver chloride electrode half-cell reaction is

$$AgCl + e \rightleftharpoons Ag + Cl^- \tag{15-14}$$

$$E_{AgCl} = E_{AgCl}^0 - \frac{0.059}{1} \log [Cl^-] \tag{15-15}$$

As long as the electrode is in equilibrium with pure silver metal (activity = 1) and pure AgCl (activity = 1), the electrode potential will be related to the chloride ion activity (concentration).

It was pointed out in Chapter 11 that *any* half-cell present in a system can be used if equilibrium has been reached because, at equilibrium, all couples must reach the same potential. Therefore, instead of using the reaction given in Equation 15-14, we could just as well use the simple silver–silver ion couple

$$Ag^+ + e \rightleftharpoons Ag \tag{15-16}$$

$$E_{Ag} = E_{Ag}^0 - \frac{0.059}{1} \log \frac{1}{[Ag^+]} \tag{15-17}$$

In the presence of AgCl, the solubility product relates $[Ag^+]$ and $[Cl^-]$:

$$K_{sp} = [Ag^+][Cl^-] \tag{15-18}$$

Therefore, Equation 15-17 can be written in the form

$$E_{Ag} = E_{Ag}^0 - \frac{0.059}{1} \log \frac{[Cl^-]}{K_{sp}}$$

or

$$E_{Ag} = E_{Ag}^0 + 0.059 \log K_{sp} - 0.059 \log [Cl^-] \tag{15-19}$$

According to Equation 15-19, the silver indicator electrode potential measures the chloride ion activity *indirectly* because the solubility product for AgCl relates $[Cl^-]$ to $[Ag^+]$. This, then, is an electrode of the second kind.

EXAMPLE 15-5

The potential for the cell

$$\text{Pt, H}_2 \text{ (1 atm)} | \text{H}_3\text{O}^+ \, (a = 1) \, || \, \text{KCl (0.015 } F) \, | \, \text{AgCl, Ag}$$

is $+0.332$ V. Calculate K_{sp} for AgCl.

Solution

Equation 15-19 can be used to calculate K_{sp} as follows:

$$E_{\text{cell}} = E_{\text{right}} - E_{\text{left}} = E_{\text{Ag}} - E_{\text{H}_2}$$

$$E_{\text{Ag}} = E_{\text{Ag}}^0 + 0.059 \log K_{sp} - 0.059 \log [\text{Cl}^-]$$

$$= +0.799 + 0.059 \log K_{sp} - 0.059 \log 0.015$$

$$E_{\text{H}_2} = E_{\text{H}_2}^0 - \frac{0.059}{2} \log \frac{[\text{H}_2]}{[\text{H}_3\text{O}^+]^2}$$

$$= 0 - \frac{0.059}{2} \log \frac{1}{1^2} = 0 \text{ vs. NHE (this } is \text{ a NHE)}$$

Therefore,

$$E_{\text{cell}} = E_{\text{Ag}}$$

$$+0.332 = +0.799 + 0.059 \log K_{sp} + 0.108$$

$$\log K_{sp} = \frac{-0.575}{0.059} = -9.75$$

$$K_{sp} = 1.8 \times 10^{-10}$$

Because of the involvement of the solubility product, these electrodes are useful for determining solubility product constants, as shown in Example 15-5. With knowledge of the solubility product, the E^0 value for the electrode of the second kind can be calculated (Equation 15-15) as well as the electrode potential for any activity conditions (Example 15-6).

For applications, electrodes of the second kind face the same limitations as do the simple metal indicator electrodes (electrodes of the first kind). Determination of halide ion activities can be carried out by silver–silver halide electrodes; the procedure is useful as a *potentiometric endpoint* signal for the titration of halide ion with $AgNO_3$. Uncertainties of several millivolts will not significantly affect the accuracy of a potentiometric endpoint. Aside from their use in volumetric titrations, electrodes of the second kind are widespread as reference electrodes (SCE and Ag/AgCl). They are much more convenient than NHE and maintain fixed activities for all species in the electrode reaction even when small amounts of current are drawn through the electrode.

EXAMPLE 15-6

(a) The solubility product for AgCl is 1.8×10^{-10}. From this equilibrium constant and $E^0_{Ag} = +0.799$ V vs. NHE, calculate E^0_{AgCl} for Equation 15-14.

(b) With the value of E^0_{AgCl}, estimate the potential for the silver–silver chloride reference electrode.

Solution

(a) The conditions for any E^0 value are that all species *in the equation* are at unit activity. Therefore, for Equation 15-14, we can state that $[Cl^-] = 1$ under E^0 conditions. From the solubility product, we can then calculate $[Ag^+]$:

$$K_{sp} = 1.8 \times 10^{-10} = [Ag^+][Cl^-] = [Ag^+] \times 1$$

$$[Ag^+] = 1.8 \times 10^{-10}$$

Because all couples must be at the same potential at equilibrium, we can use the simple silver–silver ion couple and equate it to E^0_{AgCl}:

$$E^0_{AgCl} = E_{Ag} = E^0_{Ag} - \frac{0.059}{1} \log \frac{1}{[Ag^+]}$$

$$= +0.799 + 0.059 \log 1.8 \times 10^{-10} = +0.224 \text{ V vs. NHE}$$

(*Note:* The actual measured value is $+0.222$ V vs. NHE.)

(b) Equation 15-15 can now be used to calculate E_{AgCl} for any conditions. The silver–silver chloride reference electrode contains saturated KCl. The solubility of KCl at 25°C is about 4.2 F; but as shown in Example 15-3, significant error can be introduced if activity is equated to concentration for such concentrated solutions. Example 15-3 showed that the chloride ion activity in saturated KCl was 2.76. With this value, we can then use Equation 15-15.

$$E_{AgCl} = E^0_{AgCl} - \frac{0.059}{1} \log [Cl^-]$$

$$= +0.224 - 0.059 \log 2.76 = +0.198 \text{ V vs. NHE}$$

(*Note:* The result will be $+0.196$ V vs. NHE if the actual measured value of $+0.222$ V for E^0_{AgCl} is used. The actual measured value for E_{AgCl} reference electrode is $+0.197$ V vs. NHE.) If activity coefficients were ignored, the result would be

$$E_{AgCl} = +0.224 - 0.059 \log 4.2 = +0.187 \text{ V vs. NHE}$$

introducing a 10-mV error compared with the actual measured value.

15-3   ION-SELECTIVE ELECTRODES (ISE)

The analytical applications utilizing potentiometry would be quite limited were it not for the development of *ion-selective electrodes* (ISE's). As the name implies, these electrodes are capable of selectively measuring the activity (concentration) of a particular ionic species. The method can at times be so selective that such electrodes are frequently called *specific-ion electrodes*, a name that implies complete specificity. Although the designation is somewhat overstated, these electrodes *can* be highly selective, as we shall see.

Basically, ISE's make use of the factor that causes so much trouble with metal indicator electrodes, namely, the liquid junction potential. We have seen several times throughout this text that analytical chemists are not above using *any* signal so long as it can be related to an analyte, and the liquid junction potential is no exception.

If we imagine some type of membrane separating one electrolyte compartment from another, as shown in Figure 15-7, then a junction potential will exist because of the different mobilities of the various ions in the two electrolytes. If we consider the case of several ions at activity a_1 on the left side and a_2 on the right side of the membrane, the junction potential will be given by the complicated expression in Equation 15-20,

$$E_j = -\frac{RT}{F}\sum_i \frac{t_i}{Z_i}\ln\frac{a_{i_2}}{a_{i_1}} = +\frac{RT}{F}\sum_i \frac{t_i}{Z_i}\ln\frac{a_{i_1}}{a_{i_2}} \qquad (15\text{-}20)$$

This expression seems to be similar to the activity part of the Nernst equation, but the contribution of each ion (i) must be considered; the junction potential will be the *sum* ($\sum_i$) of all these contributions. There are some other important differences: An ion's contribution to the junction potential depends on its charge, Z_i–not the n value for an electrochemical reaction occurring at an electrode; and finally, as mentioned earlier, the junction potential depends on the mobility of the ions. In Equation 15-20, the mobility has been expressed as the transference number, t_i for each

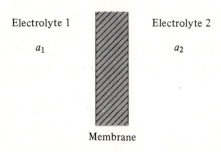

Electrolyte 1   Electrolyte 2

a_1 a_2

Membrane

FIGURE 15-7   Development of liquid junction potential at a membrane.

ion. The transference number is a measure of an ion's ability to carry charge relative to the other ions in the system. Thus the transference number may range from zero (the ion carries none of the charge transferred across the membrane) to one (the ion carries all the charge across the membrane). Example 15-7 shows how this equation can be used to calculate the liquid junction potential for two electrolytes containing only KCl.

EXAMPLE 15-7

Calculate the liquid junction potential across a membrane at 25°C, as in Figure 15-7, in which electrolyte 1 is saturated KCl (4.2 F) and electrolyte 2 is 0.1 F KCl. The transference numbers are $t_K = 0.49$ and $t_{Cl} = 0.51$. Neglect activity coefficients.

Solution

$$E_j = +\frac{RT}{F} \sum \frac{t_i}{Z_i} \ln \frac{a_{i_1}}{a_{i_2}}$$

At 25°C and using base 10 logarithms ($2.303RT/F = 0.059$),

$$E_j = E_j(K^+) + E_j(Cl^-)$$

$$= +\frac{0.059 \times 0.49}{(+1)} \log \frac{4.2}{0.1} + \frac{0.059 \times 0.51}{(-1)} \log \frac{4.2}{0.1}$$

$$= +0.047 - 0.049 = -0.002 \text{ V} = -2 \text{ mV}$$

Equation 15-20 is hopelessly difficult to solve for any real analyte solutions containing several different ions. To make this expression useful for analytical applications, we need a membrane with unique properties. In an ideal ion-selective membrane, $t_i = 1$ for one particular ion, whereas $t_i = 0$ for *all* other ions. With this ideal membrane, Equation 15-20 becomes

$$E_j = +\frac{RT}{FZ} \ln \frac{a_1}{a_2} \tag{15-21}$$

where a_1, a_2, and Z refer to the ion whose transference number is equal to 1. At 25°C and with base 10 logarithms, this equation simplifies to

$$E_j = +\frac{0.059}{Z} \log \frac{a_1}{a_2}$$

Now let us consider an electrochemical cell that incorporates an ion-selective membrane as shown in Figure 15-8. The cell potential will be

$$E_{cell} = E_{right} - E_{left} + E_j \tag{15-22}$$

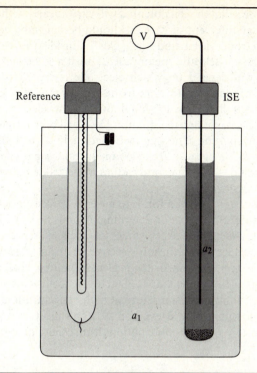

FIGURE 15-8 Electrochemical cell for measuring unknown activity, a_1, with ISE.

If both electrodes are reference electrodes, such as SCE or Ag/AgCl (they need not be the same), then

$$E_{\text{cell}} = E_{\text{ref right}} - E_{\text{ref left}} + E_j$$

$$E_{\text{cell}} = E_{\text{ref right}} - E_{\text{ref left}} + \frac{0.059}{Z} \log \frac{a_1}{a_2} \qquad \textbf{(15-23)}$$

Now we add a fixed activity (concentration) of the selected ion to the electrolyte compartment of $E_{\text{ref right}}$ (a_2) and combine all the constant terms:

$$E_{\text{cell}} = E_{\text{ref right}} - E_{\text{ref left}} - \frac{0.059}{Z} \log a_2 + \frac{0.059}{Z} \log a_1 \qquad E_{\text{cell}} = K + \frac{0.059}{Z} \log a_1$$

For example, if our ion-selective membrane is selective for Cd^{2+},

$$E_{\text{cell}} = K + \frac{0.059}{2} \log [Cd^{2+}] \qquad \textbf{(15-24)}$$

or

$$E_{\text{cell}} = K - \frac{0.059}{2} \text{pCd} \qquad \textbf{(15-25)}$$

Equation 15-25 shows that the cell potential will vary linearly with pCd, with a slope of -29.5 mV and an intercept of K. K, it should be remembered, depends on the two reference electrodes used and the $[Cd^{2+}]$ inside the electrode compartment containing the ion-selective membrane. It must also be noted that the sign of the activity term will depend on which electrode is the "left-hand" one. Some commercial high-impedance voltmeters are designed so that the ISE containing the "magic" membrane will be the left-hand one in the electrochemical cell configuration. In such cases, the sign in Equation 15-24 will be inverted and, of course, K will be the opposite sign too. The sign in Equation 15-24 will also depend on the charge of the selected ion because Z carries an algebraic sign. The slope, as given in Equation 15-24, would $+29.5$ mV for a cadmium electrode but would be -59 mV for a fluoride (F^-) electrode.

The simplest way to use such an ISE is to measure E for some standard solutions and construct a calibration curve or determine K and the slope by the method of least squares. In this way, the exact values of each reference electrode and the activity of the selected ion inside the ion-selective electrode (a_2) are not needed. Any nonideality in the membrane and/or the neglect of activity coefficients can also be handled in this manner. For example, if activity coefficients are neglected, the calibration plot will show some curvature at high concentrations and the slope of the best straight line will be a little less than $59/Z$ mV (absolute value), as shown in Figure 15-9. K also contains any contributions from other liquid junction potentials

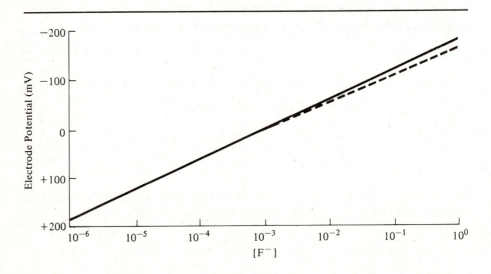

FIGURE 15-9 Calibration curve for the measurement of $[F^-]$ with the fluoride ISE. Solid line: fluoride activity, slope = -59 mV; broken line: fluoride concentration, average slope = -57 mV.

in the electrochemical cell. For example, in Figure 15-8, there will be a small junction potential where the external reference, $E_{ref\ left}$, interfaces with the analyte solution.

Now we explore some of the membrane materials and electrode construction that have led to the development of ISE's.

GLASS ELECTRODE

The first membrane to be developed as an electrode was glass, which is selective for H_3O^+. A glass electrode to measure pH is familiar to almost everyone who has worked in a chemical laboratory; for the phenomenon of glass selectivity was reported as early as 1906 by M. Cremer, and practical glass electrodes have been in use for over 50 years. The electrode is fabricated by sealing a thin glass membrane sensitive to H_3O^+ to a thicker glass tubing. The glass electrode contains a silver–silver chloride reference and an internal solution with a fixed $[H_3O^+]$. Figure 15-10 shows an electrochemical cell for measurement of pH and the shorthand expression for this cell.

Experimentally, it is found that the cell potential follows that predicted by Equation 15-25; namely,

$$E_{cell} = K - 0.059\ \text{pH} \tag{15-26}$$

However, after considerable study, it is now known that the glass electrode is more complicated than the simple diffusional junction potential concept described in the previous subsection. In fact, glass-membrane conductivity is due to migration of Na^+ through the glass, whereas the pH-sensitive potential results from equilibria at the inner and outer surfaces of the glass membrane. As shown in Figure 15-10, there is a potential V_1 at the outer surface and a potential V_2 at the inner surface. The measured cell potential will be

$$E_{cell} = E_{AgCl} - E_{SCE} + E_j = K' + E_m \tag{15-27}$$

the total membrane potential, E_m, being

$$E_m = V_1 + V_2$$

What is the nature of this chemical equilibrium at the glass surface? It is now well established that the surface of the glass contains sites for cations and that Na^+ and H_3O^+ compete for these sites in a way similar to that in an ion-exchange equilibrium (see Chapter 19):

$$H_3O^+_{soln} + Na^+_{glass} \rightleftharpoons Na^+_{soln} + H_3O^+_{glass} \tag{15-28}$$

The equilibrium constant for Equation 15-28 is so high that, on the surface of the glass, essentially all the sites are occupied by hydrogen ions except for the most basic solutions. The presence of water is essential for this surface equilibrium, and a glass electrode that has been dehydrated will not be responsive to pH. A pictorial

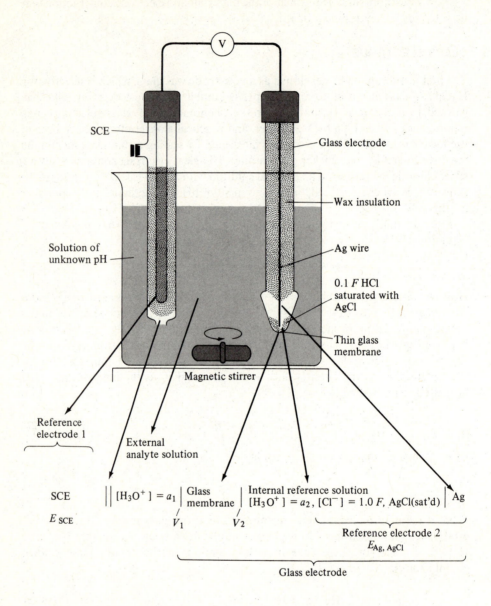

FIGURE 15-10 Construction of electrode system for measuring pH and diagram of potentials developed at the various phase boundaries.

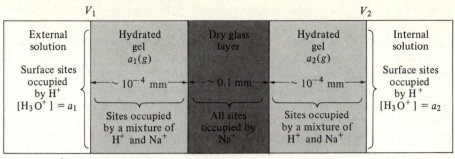

Membrane potential, $E_m = V_1 + V_2$

FIGURE 15-11   Schematic representation of a well-soaked glass membrane.

representation of the glass membrane is shown in Figure 15-11. It turns out that the two surface potentials, V_1 and V_2, follow a Nernstlike expression involving hydrogen ion activity,

$$V_1 = +0.059 \log \frac{a_1}{a_1(g)} \qquad\qquad (15\text{-}29)$$

$$V_2 = +0.059 \log \frac{a_2(g)}{a_2} \qquad\qquad (15\text{-}30)$$

where $a_1(g)$ and $a_2(g)$ are the activities of hydrogen ion in the hydrated gel surfaces (outer and inner, respectively). If the two gel surfaces are identical and all available sites are filled by hydrogen ions, $a_1(g) = a_2(g)$. Thus

$$E_m = V_1 + V_2 = 0.059 \log \frac{a_1 \cdot \cancel{a_2(g)}}{\cancel{a_1(g)} \cdot a_2}$$

$$= 0.059 \log \frac{a_1}{a_2} \qquad\qquad (15\text{-}31)$$

The activity of hydrogen ion in the internal solution (a_2) is a constant so that

$$E_m = K'' + 0.059 \log a_1 \qquad\qquad (15\text{-}32)$$

or

$$E_m = K'' - 0.059 \text{ pH} \qquad\qquad (15\text{-}33)$$

The cell potential will, therefore, follow Equation 15-26, as observed experimentally. Note that K in Equation 15-26 includes K'' from Equation 15-33 as well as the two reference electrode potentials, K' in Equation 15-27.

This development is still not the complete story. For example, if the two reference electrodes are identical (both Ag/AgCl electrodes, for example) and the internal and external solutions are at the same pH, E_{cell} should equal zero. When this experiment is carried out with real glass electrodes, a small potential is observed, which is called the *asymmetry potential*. The asymmetry potential varies from electrode to electrode and, for a given electrode, varies with time. Thus it is necessary to calibrate a pH meter with solutions of known pH. This is accomplished by using buffers containing weak acids with known K_a values. It is important to realize that the absolute measurement of pH then relies on the accuracy of the acid dissociation constants. Because of the uncertainties in K_a values and other uncertainties in measuring pH, absolute pH measurements are no better than about ± 0.01 pH (2.3%). However, changes of pH in a system can be measured with much higher accuracy, and commercial pH meters are available that measure to ± 0.001.

It was mentioned earlier that the surface of the glass membrane will be completely in the hydrogen ion form except in strong base solutions. When the sodium ion concentration in solution is much higher than the hydrogen ion concentration, the glass membrane will show response to $[Na^+]$ and an *alkaline error* results. The measured cell potential will include a contribution from $[Na^+]$, as shown in Equation 15-34:

$$E_{cell} = K + 0.059 \log ([H_3O^+] + K_s[Na^+]) \tag{15-34}$$

where K_s is a *selectivity constant* for the interfering ion. If $K_s = 1$, the membrane responds equally to both ions. Fortunately, glass membranes have been developed for which $K_s < 10^{-10}$. Still, errors can be appreciable when measurements of pH > 11 are attempted, as shown in Example 15-8. The concept of selectivity constants can be applied to other ion-selective electrodes, and manufacturers of these electrodes publish selectivity constants for them. The values are only approximate but are useful for calculating the possibility of interference from other ions in the system. The interferences from all ions must be summed to determine the total amount of interference, as shown in Equation 15-35:

$$E_{cell} = K + \frac{0.059}{Z_A} \log ([A] + \sum_i K_{s,i} [I]_i^{Z_A/Z_{I,i}}) \tag{15-35}$$

where $[A]$ is the concentration of analyte and $[I]_i$ is the concentration of each interfering species. $Z_A/Z_{I,i}$ corrects for a possible charge difference between the analyte and each interference. Normally, the most serious interferences have the same charge as the analyte so that $Z_A/Z_I = 1$.

In addition to the significant alkaline error, glass electrodes will exhibit a small positive error in strong acid solution. Below pH 0 to 1, errors of a few tenths of a pH will be observed. The reasons for this error are not definitely known.

That glass electrodes show response to sodium ions led analytical chemists to develop glass membranes selective to various cations. In other words, while considerable research effort has been spent in developing glasses for pH measurements with low K_s values, other research effort has been directed to finding glass membranes with high K_s values. Failure for one group would be success for the

---

EXAMPLE 15-8

Calculate the error in pH that will be observed in 0.02 F NaOH for a glass electrode whose selectivity constant for sodium ion is 10^{-10}.

Solution

True cell potential

$$[H_3O^+] = \frac{K_w}{[OH^-]} = \frac{10^{-14}}{2 \times 10^{-2}} = 5 \times 10^{-13}$$

$$E_{cell} = K + 0.059 \log 5 \times 10^{-13} = K - 0.726 \text{ V}$$

Measured cell potential

$$E_{cell} = K + 0.059 \log (5 \times 10^{-13} + 10^{-10} \times 2 \times 10^{-2})$$

$$= K + 0.059 \log 2.5 \times 10^{-12} = K - 0.685 \text{ V}$$

Therefore,

$$\text{error in potential} = \cancel{K} - 0.685 - \cancel{K} + 0.726 = +0.041 \text{ V}$$

The potential changes -0.059 V for each pH unit (Equation 15-26) so that

$$\text{error in pH} = \frac{+0.041 \cancel{V}}{-0.059 \cancel{V}/\text{pH}} = -0.69 \text{ pH units}$$

Alternatively, one can calculate the pH error by looking at the logarithmic term. The effective hydrogen ion concentration, including Na^+ interference, is

$$5 \times 10^{-13} + 2 \times 10^{-12} = 2.5 \times 10^{-12}$$

$$\text{pH} = -\log [H_3O^+] = -\log 2.5 \times 10^{-12} = 11.60$$

$$\text{true pH} = -\log 5 \times 10^{-13} = 12.30$$

$$\text{error} = 11.60 - 12.30 = -0.70 \text{ pH units}$$

---

other! It has been found that glasses containing Al_2O_3 or B_2O_3 are responsive to alkali metal ions and that glass membranes responsive to Na^+, K^+, NH_4^+, Rb^+, Cs^+, Li^+, and Ag^+ can be constructed.

FLUORIDE ELECTRODE

During the last 10 to 15 yr, several crystal membranes have been developed that exhibit the required property that the transference number for one ion is 1 and essentially zero for all others. A classic example is the fluoride ion-selective electrode.

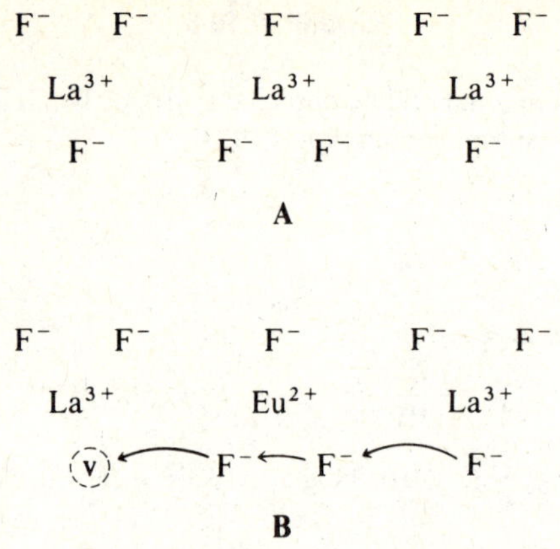

F⁻ F⁻ F⁻ F⁻ F⁻
La3+ La3+ La3+
F⁻ F⁻ F⁻ F⁻
A
F⁻ F⁻ F⁻ F⁻ F⁻
La3+ Eu2+ La3+
v F⁻ F⁻ F⁻
B

FIGURE 15-12 Lanthanum fluoride (LaF_3), as an ion-selective membrane. (**A**) Schematic arrangement of ions in pure LaF_3; (**B**) schematic arrangement of ions in LaF_3, doped with EuF_2, showing possible motion of a fluoride ion by filling a vacant lattice site, (v).

V
SCE
0.1 F KCl, 0.1 F NaF
Ag/AgCl reference
Analyte solution
[F⁻] = x F
LaF3

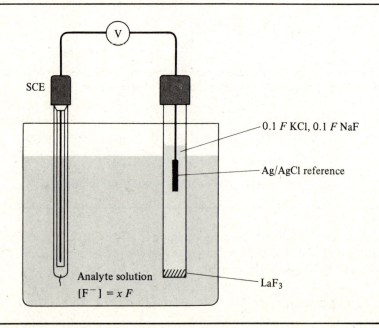

FIGURE 15-13 Electrochemical cell for measurement of fluoride ion concentration.

The membrane employed consists of a LaF_3 crystal doped with EuF_2, as pictured in Figure 15-12. The Eu^{2+} ions substitute for La^{3+}; but since each EuF_2 introduces only two F^- instead of three, there will be one vacant fluoride ion site for each EuF_2 added. The vacant sites allow F^- ions to move (by "watching" the vacancy, one can equally well say that the vacancy moves) and make the material ionically conductive. Ideally, no other ion can move in the crystal membrane so that $t_F = 1$ and $t = 0$ for all other ions, including La^{3+} and Eu^{2+}. Equation 15-24 for the fluoride electrode is

$$E_{cell} = K - 0.059 \log [F^-] = K + 0.059 \, pF \qquad \textbf{(15-36)}$$

The sign change compared to the glass electrode occurred because $Z = -1$ for the fluoride ion.

The construction of the fluoride ISE is shown in Figure 15-13. The internal electrolyte contains a silver-silver chloride reference (only 0.1 F KCl instead of saturated) and a fixed concentration of NaF equal to 0.1 F to establish a_F on one side of the LaF_3 membrane.

This electrode is almost an ideal membrane, but there is one interfering ion (to be an interference, an ion must be similar to F^- in size and charge). Other halide ions are considerably larger than F^- so that they do not interfere. However, OH^- is almost the same size and carries the same -1 charge. The electrode is about 10 times more sensitive to F^- than it is to OH^-. Thus Equation 15-36 can be expanded to include the effect of OH^- ($K_s = 0.1$ for $[OH^-]$):

$$E_{cell} = K - 0.059 \log ([F^-] + 0.1[OH^-]) \qquad \textbf{(15-37)}$$

As shown in Example 15-9, a significant error will be introduced if the fluoride ISE is used in basic solutions, especially when low levels of fluoride are being measured. Generally, fluoride electrodes should be used only in neutral or acidic solution.

For acidic solutions, the analyst must keep in mind that HF, unlike HCl is *not* a strong acid so that the fluoride ion activity will be considerably less than the formal fluoride concentration. The acid dissociation constant, K_a, is 7.2×10^{-4} so that

$$\frac{[F^-]}{[HF]} = \frac{7.2 \times 10^{-4}}{[H_3O^+]}$$

Only in solutions in which pH $> \sim 4$ will F^- be the predominant species. In spite of these difficulties, the fluoride ISE is widely used for the determination of fluoride ion activity. Some of these applications will be described later.

SOLID-STATE ELECTRODES BASED ON SILVER COMPOUNDS

Unfortunately, there are not many compounds such as LaF_3 that can be used to produce an ISE. However, several compounds of silver exhibit conductivity, Ag^+ being the only mobile ion. For example, AgI, AgBr, AgCl, and Ag_2S show this behavior, and a silver ISE can be devised with one of them or a mixture of them. Because Ag_2S is so insoluble, it is preferred. It is important to realize that the use of a silver ISE is different from the use of a silver indicator electrode for measuring

EXAMPLE 15-9

Calculate the error in concentration and in millivolts that would occur if a solution containing $5 \times 10^{-5} F$ F^- at pH 10 was measured with a fluoride ISE.

Solution

The electrode should measure

$$E_{cell} = K - 0.059 \log (5 \times 10^{-5})$$

$$= K + 0.254$$

The interference due to OH^- at pH 10 will be

$$[OH^-] = \frac{K_w}{[H_3O^+]} = \frac{10^{-14}}{10^{-10}} = 10^{-4}$$

From Equation 15-37, we obtain

$$E_{cell} = K - 0.059 \log (5 \times 10^{-5} + 0.1 \times 10^{-4})$$

Therefore,

$$\text{error in concentration} = \frac{0.1 \times 10^{-4}}{5 \times 10^{-5}} = 0.2 \ (20\%)$$

$$E_{cell} = K - 0.059 \log (6 \times 10^{-5})$$

$$= K + 0.249$$

Therefore,

$$\text{error in mV} = K + 249 - (K + 254) = -5 \text{ mV}$$

Note what a large effect a 5-mV error has on the measurement of the concentration.

$[Ag^+]$. The former measures a membrane or junction potential across a Ag_2S pellet and requires two reference electrodes to complete the circuit; the latter measures an electrode potential versus one external reference electrode. The ISE can be used in solutions that are capable of oxidizing silver metal, a versatility that gives it a distinct advantage over the silver indicator electrode.

The Ag_2S-pellet membrane may measure $[S^{2-}]$ as well as $[Ag^+]$. Just as in the case of the silver–silver chloride electrode, the activity of silver ion will be fixed in the presence of an anion that forms an insoluble compound; that is

$$Ag_2S \rightleftharpoons 2Ag^+ + S^{2-}$$

$$[Ag^+] = \sqrt{\frac{K_{sp}}{[S^{2-}]}}$$

(15-38)

Thus the crystalline membrane acts as a *membrane of the second kind*, analogous to an electrode of the second kind.

By the addition of silver halides to the silver sulfide, it is possible to play this game not only with Ag_2S but with other anions. In this way, chloride, bromide, and iodide ion-selective electrodes can be produced; and in a similar fashion, cyanide (CN^-) and thiocyanate (SCN^-) ISE's have been developed. Finally, it has been found that the addition of other sulfide crystals to Ag_2S yields electrodes responsive to the added metal ions. For example, PbS, CdS, and CuS additions form the basis for the lead, cadmium, and copper ISE's, respectively.

With these electrodes, the interference mechanism is different from the selectivity-constant concept discussed earlier. For example, the bromide electrode depends on the membrane AgBr's being in equilibrium with Br^- in the analyte solution. If sufficiently high concentrations of SCN^- are present, an exchange reaction occurs at the surface of the crystal membrane:

$$AgBr + SCN^- \rightleftharpoons AgSCN + Br^- \tag{15-39}$$

Equation 15-39 has an equilibrium constant dependent on the K_{sp} values for AgBr and AgSCN:

$$K = \frac{[Br^-]}{[SCN^-]} = \frac{[Ag^+][Br^-]}{[Ag^+][SCN^-]} = \frac{K_{sp}(AgBr)}{K_{sp}(AgSCN)} \tag{15-40}$$

From K_{sp} values given in Appendix 2-A, Equation 15-40 becomes

$$\frac{[Br^-]}{[SCN^-]} = \frac{5.2 \times 10^{-13}}{1.1 \times 10^{-12}} = 0.47$$

As long as $[Br^-]/[SCN^-] > 0.47$, the membrane surface contains only AgBr, and the electrode responds correctly to $[Br^-]$ in solution. However, if $[SCN^-]$ becomes so great that $[Br^-]/[SCN^-] < 0.47$, then reaction Equation 15-39 proceeds to the right, and the surface of the membrane is converted to AgSCN. The electrode no longer responds to $[Br^-]$. The interfering film can be removed by a scrubbing with a toothbrush to restore the electrode.

LIQUID MEMBRANE ELECTRODES

With the development of liquid membrane electrodes, the field of ISE's has been able to expand despite the limited number of available crystal or glass membranes. Basically, the liquid membrane electrode still consists of an internal reference electrode, an internal solution containing a fixed activity of the ion to be analyzed, and a membrane separating the internal solution from the external analyte solution. Here, however, the membrane consists of an organic liquid (immiscible with water) held on a porous plastic support. The organic liquid (R—H) is an ion-exchange material (see Chapter 19) that can be converted to the form containing a particular metal ion by Equation 15-41,

$$nR{-}H + M^{n+} + nH_2O \rightleftharpoons (R)_n{-} M + nH_3O^+ \tag{15-41}$$

When the membrane has been completely converted to the M^{n+} form, it will then be responsive to $[M^{n+}]$ in the external analyte solution because of the junction potential generated at this membrane.

For example, a calcium ion-selective electrode is pictured in Figure 15-14. The liquid ion exchanger can be a solution of calcium didecylphosphate in di-n-octylphenyl phosphonate. However, newer versions use a polyvinyl chloride–gelled membrane, which looks physically like a solid-state membrane. Just as with the other ISE's, the cell potential for this electrode will be given by Equation 15-24 or 15-25 so that

$$E_{cell} = K + \frac{0.059}{2} \log [Ca^{2+}]$$

or

$$E_{cell} = K - \frac{0.059}{2} \, pCa$$

Several other liquid membrane electrodes have been developed, utilizing the principles described for the calcium ISE. For example, there is a potassium electrode that uses valinomycin in diphenyl ether. Valinomycin, an antibiotic, has a cyclical ether structure whose central opening is just the right size for K^+. Unlike glass membranes, which are always sensitive to Na^+, this liquid membrane electrode has a selectivity constant for Na^+ of 2×10^{-4}.

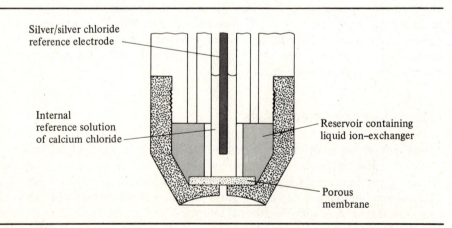

Silver/silver chloride
reference electrode

Internal
reference solution
of calcium chloride

Reservoir containing
liquid ion–exchanger

Porous
membrane

FIGURE 15-14 Schematic diagram of a calcium-selective liquid ion-exchange membrane electrode.

GAS-SENSING ELECTRODES

Analytical chemists have been ingenious in developing selective membranes. In addition to the various ion-selective electrodes, there has also been developed a series of gas-sensitive electrodes. The basic principle underlying these electrodes is that of a gas-permeable membrane, which, for the activity of the gas, allows a relatively rapid equilibrium to be established between the analyte solution and the internal solution. In the internal solution, the gas undergoes a reaction liberating or consuming an ion to be detected by an ISE.

For example, the SO_2 electrode is sensitive to dissolved SO_2 in the analyte solution because SO_2 diffuses through the highly gas-permeable membrane shown in Figure 15-15 until equilibrium is reached. In the internal solution, the SO_2 reacts

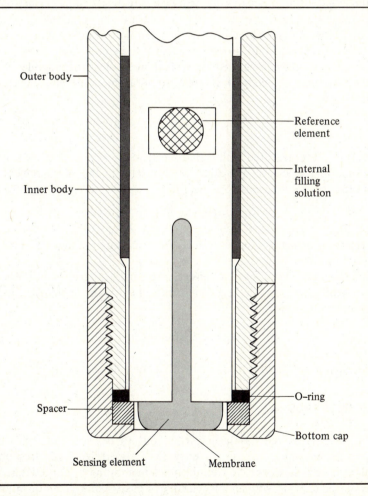

FIGURE 15-15 Construction of the SO_2 electrode, in which the sensing element is a mirror-flat pH electrode. The NH_3 and NO_x electrodes have similar construction. (Courtesy of Orion Research Incorporated.)

with H_2O according to Equation 15-42,

$$SO_2 + 2H_2O \rightleftharpoons HSO_3^- + H_3O^+ \tag{15-42}$$

and the $[H_3O^+]$ is measured with a glass electrode (sensing element) placed immediately behind the gas-permeable membrane. Because $[H_3O^+]$ is directly proportional to $[SO_2]$, the cell potential is given by Equation 15-43:

$$E_{cell} = K + \frac{0.059}{1} \log [SO_2] \tag{15-43}$$

The gas-sensing electrode also contains an internal reference element so that no external reference is required; in other words, it is really a complete electrochemical cell and makes contact with the analyte solution only via the gas-permeable membrane. This allows it to be used in many situations in which a conventional reference electrode or an indicator electrode would be unsuitable.

The same approach can be used for measurement of nitrogen oxides (NO_2 and NO), which, when dissolved in H_2O, are in equilibrium with HNO_2. The HNO_2 is an acid,

$$HNO_2 + H_2O \rightleftharpoons H_3O^+ + NO_2^- \tag{15-44}$$

and the $[H_3O^+]$ is measured with a glass electrode.

The ammonia electrode provides an example of the measurement of the consumption of an ion because NH_3 is a base,

$$NH_3 + H_2O \rightleftharpoons NH_4^+ + OH^- \tag{15-45}$$

A glass electrode detects the decrease in H_3O^+ resulting from the OH^- produced in the internal solution.

Other ISE's may be employed to sense changes in the internal solution caused by an analyte's diffusing through the gas-permeable membrane. A chlorine electrode is based on the equilibrium

$$Cl_2 + 3H_2O \rightleftharpoons 2H_3O^+ + ClO^- + Cl^- \tag{15-46}$$

The chloride ion produced is measured with a chloride ISE.

ENZYME ELECTRODES

In recent years, research has been carried out to develop electrodes that are responsive to enzymes or—via an enzyme-catalyzed reaction—responsive to metabolites, coenzymes, or enzyme inhibitors. A membrane is required that contains an immobilized enzyme, with an ion-selective or gas-sensing electrode immediately behind this membrane, as shown in Figure 15-15. For example, an electrode responsive to urea has been developed, in which urease is immobilized in a thin layer of acrylamide gel held over the surface of a cation electrode by cellophane film. The reaction is

$$NH_2-\overset{\overset{\textstyle O}{\|}}{\underset{\text{urea}}{C}}-NH_2 + 2H_2O \xrightarrow[\text{membrane}]{\text{urease}} NH_3 + NH_4^+ + HCO_3^- \qquad (15\text{-}47)$$

The cation electrode responds to the NH_4^+ produced and is capable of measuring urea concentrations from 5×10^{-5} to 1.6×10^{-1} F. Alternatively, the gas-sensing NH_3 electrode can be used.

An electrode selective for amygdalin is based on the cyanide ISE. An acrylamide gel membrane contains β-glucosidase, which hydrolyzes amygdalin according to Equation 15-48,

$$\underset{\text{amygdalin}}{\underset{\displaystyle OC_{12}H_{21}O_{10}}{C_6H_5\underset{|}{CHCN}}} + 2H_2O \xrightarrow[\text{membrane}]{\beta\text{-glucosidase}} 2C_6H_{12}O_6 + C_6H_5CHO + HCN \quad (15\text{-}48)$$

Using the NH_3 gas-sensing electrode, an electrode responsive to AMP has been reported. A thin layer of the enzyme AMP deaminase in glycerol suspension is held on the sensing surface of the NH_3 electrode by means of a thin dialysis membrane. Enzymatic deamination of AMP occurs to produce a stoichiometric amount of NH_3.

Enzyme electrodes are still in the research and early development stage, but commercial versions may be available in the future.

APPLICATIONS

From the wide variety of ISE's that have been described in the previous subsections, it is easy to see that these electrodes will have far-ranging applications. Table 15-3 lists some of the commercially available ISE's and their properties. Many other electrodes have been produced in research laboratories and reported in the scientific literature.

Before we describe a few examples of applications, we should recall the limitations of these electrodes for quantitative work. As the fundamental relationship (Equation 15-24) between signal and analyte shows, the electrode potential is proportional to the logarithm of the analyte's activity. This mathematical relationship is not very sensitive as to concentration (activity) so that a small error in observing potential will result in a significant error in determining concentration. As was demonstrated in Example 15-4, a 1-mV error in potential would lead to about an 8% error in concentration for $n = 2$ ($Z = 2$ for an ISE). Therefore, high-impedance voltmeters capable of reading potential to ± 0.3 mV (for instance, an expanded-scale pH meter) are preferred. Still, errors of 2 to 5% should be expected under the best conditions. The neglect of activity coefficients can also introduce significant errors, and it is prudent to take all potential readings in solutions of the same ionic strength. This means that the analyte ionic strength must be known so that appropriate standards can be prepared. If the analyte is a very dilute solution, as in the analysis of drinking water, it is possible to add a known amount of inert salt

Table 15-3 Some Commercially Available Ion-Selective Electrodes

Electrode	Model	Type	Concentration range F ppm	$\sim$Slope (mV/decade)	Temperature range (°C)	Interferences
Ammonia (NH_3); (ammonium) (NH_4^+)	95–10	gas-sensing	10^0–10^{-6}; 17,000–0.02	−57	0–50	volatile amines
bromide (Br^-)	94–35	solid-state	10^0–5×10^{-6}; 79,900–0.40	−56	0–80	max level: $S^{2-} \leqq 10^{-7}\,F$; $2 \times 10^{-7}\,I^-$; $8 \times 10^{-8}\,CN^-$; high levels Cl^- and NH_3
cadmium (Cd^{2+})	94–48	solid-state	10^{-1}–10^{-7}; 11,200–0.01	+28	0–80	max level: Ag^+, Hg^{++}, $Cu^{++} \leqq 10^{-7}\,F$; high levels lead and ferric ion interfere
calcium (Ca^{2+})	93–20	plastic	10^0–5×10^{-7}; 40,100–0.02	+28	0–50	max level *(F): $0.2\,Na^+$; $4 \times 10^{-3}\,Hg^{2+}$; H_3O^+; $6 \times 10^{-3}\,Sr^{2+}$; $2 \times 10^{-2}\,Fe^{2+}$; $4 \times 10^{-2}\,Cu^{2+}$; $5 \times 10^{-2}\,Ni^{2+}$; $0.2\,NH_4^+$; $0.3\,Tris^+$; Li^+; $0.4\,K^+$; $0.7\,Ba^{2+}$, $1.1\,Zn^{2+}$, $1.2\,Mg^{2+}$
carbon dioxide (CO_2); carbonate (CO_3^{2-})	95–02	gas-sensing	10^{-2}–10^{-4}; 440–4.4	+56	0–50	volatile weak acids
chloride (Cl^-)	93–17	liquid	10^0–8×10^{-6}; 35,500–0.3	−56	0–50	max level *(F): $6 \times 10^{-6}\,ClO_4^-$; $8 \times 10^{-6}\,I^-$; $3 \times 10^{-5}\,NO_3^-$, SO_4^{2-}; $4 \times 10^{-5}\,Br^-$; $10^{-4}\,OH^-$; $4 \times 10^{-4}\,OAc^-$, $4 \times 10^{-4}\,HCO_3^-$; $7 \times 10^{-4}\,F^-$
	94–17	solid-state	10^0–5×10^{-5}; 35,500–1.8	−56	0–80	max level† (F): $\leqq 10^{-7}\,S^{2-}$; $3 \times 10^{-6}\,Br^-$; $5 \times 10^{-10}\,I^-$; $2 \times 10^{-10}\,CN^-$; NO_3^-, SO_4^{2-}, HCO_3^- do not interfere
	96–17	combination	10^0–5×10^{-5}; 35,500–1.8	−56	0–80	same as 94–17
chlorine (Cl_2)	97–70	solid-state	3×10^{-4}–10^{-7}; 21.27–0.01	+29	0–80	same as for conventional iodimetric titration (strong oxidizing agents)

cupric (Cu^{2+})	94–29	solid-state	sat'd–10^{-8}; sat'd–6.35×10^{-4}	+26	0–80	max level: S^{2-}, Ag^+, Hg^{2+} $\leqq 10^{-7}$ F, high levels chloride, bromide, ferric, and cadmium interfere
cyanide (CN^-)	94–06	solid-state	10^{-2}–10^{-6}; 260–0.03	−58	0–80	max level† (F): $\leqq 10^{-7}$ S^{2-}; 10^{-4} I^-; 5 Br^-; 10^3 Cl^-
fluoride (F^-)	94–09	solid-state	sat'd–10^{-6}; sat'd–0.02	−56	0–80	max level (F): $< 10^{-4}$ OH^-
	96–09	combination	sat'd–10^{-6}; sat'd–0.02	−56	0–80	
fluoroborate (BF_4^-)	93–05	liquid	sat'd–3×10^{-6}; sat'd–0.26	−57	0–50	max level* (F): 2×10^{-2} NO_3^-; 0.2 Br^-, OAc^-, HCO_3^-, F^-, Cl^-, OH^-, SO_4^{2-}
iodide (I^-)	94–53	solid-state	10^0–5×10^{-8}; 127,000–5×10^{-3}	−56	0–80	max level: S^{2-} $\leqq 10^{-7}$ F
lead (Pb^{2+})	94–82	solid-state	10^0–10^{-6}; 20,700–0.2	+25	0–80	max level: Ag^+, Hg^{2+}, Cu^{2+} $\leqq 10^{-7}$ F; high levels cadmium and ferric ion interfere
nitrate (NO_3^-)	93–07	plastic	10^0–7×10^{-6}; 14,000–0.08 as N	−56	0–50	max level* (F): 10^{-7} ClO_4^-; 6×10^{-6} I^-; 6×10^{-5} ClO_3^-; 2×10^{-4} CN^-; 8×10^{-4} Br^-, NO_2^-; 4×10^{-3} HS^-; 2×10^{-2} CO_3^{2-}; 3×10^{-2} HCO_3^-; 4×10^{-2} Cl^-; 0.1 PO_4^{3-}, HPO_4^{2-}, $H_2PO_4^-$; 0.3 OAc^-, 0.7 F^-; 1.0 SO_4^{2-}
nitrogen oxide (NO_x); nitrite (NO_2^-)	95–46	gas-sensing	5×10^{-3}–4×10^{-6}; 230–0.18	+58	0–50	max level* (F): 3×10^{-2} F; CO_2, volatile weak acids interfere
oxygen	97–08	gas-sensing	0–14 ppm	N.A.	0–45	
perchlorate (ClO_4^-)	93–81	liquid	10^0–2×10^{-6}; 99,500–0.2	−58	0–50	max level* (F): 2×10^{-3} I^-; 5×10^{-2} NO_3^-; 4×10^{-2} Br^-
pH (H_3O^+)	91 series; GX series	glass	pH 0–14	+59	0–80 to 0–100	sodium interferes in basic solution

Table 15-3 (cont.)

Electrode	Model	Type	Concentration range F ppm	~Slope (mV/decade)	Temperature range (°C)	Interferences
potassium (K$^+$)	93–19	plastic	10^0–10^{-6}; 39,000–0.4	+56	0–50	max level* (F): 3×10^{-4} Cs$^+$; 6×10^{-3} NH$_4^+$; Tl$^+$; 10^{-2} H$_3$O$^+$; 10^0 Ag$^+$; Tris$^+$; 2.0 Li$^+$, Na$^+$
redox	96–78	combination	not applicable	N.A.	0–80	not applicable
silver/sulfide (Ag$^+$/S^{2-})	94–16	solid-state	Ag$^+$: 10^0–10^{-7}; 107,900–0.01 S^{2-}: 10^0–10^{-7}; 32,100–0.003	+56 −28	0–80	max level: Hg^{2+} < 10^{-7} F
sodium (Na$^+$)	94–11	solid state	sat'd–10^{-6}; sat'd–0.02	+56	0–80	max level* (F): 3×10^{-7} Ag$^+$; 10^{-6} H$_3$O$^+$; 5×10^{-2} Li$^+$; 6×10^{-2} Cs$^+$; 0.1 K$^+$; 0.2 N(C$_2$H$_5$)$_4^+$; 0.5 Tl$^+$
	96–11	combination	sat'd–10^{-6}; sat'd–0.02	+56	0–80	same as 94–11
thiocyanate (SCN$^-$)	94–58	solid-state	10^0–5×10^{-6}; 58,100–0.29	−57	0–80	max level† (F): 0.1 OH$^-$, 3×10^{-6} Br$^-$, 2×10^{-2} Cl$^-$ 1.3 $\times 10^{-4}$ NH$_3$, 10^{-5} S$_2$O$_3^{2-}$; 7×10^{-6} CN$^-$; 10^{-7} I$^-$, S^{2-}
water hardness	93–32	liquid	10^{-2}–6×10^{-6};	+24	0–50	max level* (F) 3×10^{-2} Na$^+$; 3×10^{-5} Cu^{2+}, Zn^{2+}; 6×10^{-5} Fe^{2+}; 10^{-4} Ni^{2+}; 4×10^{-4} Sr^{2+}; 6×10^{-4} Ba^{2+}; 0.1 K$^+$

SOURCE: From *Analytical Methods Guide*, 9th ed. (Cambridge, MA: Orion Research Incorporated, 1978).

*For 10% error in 10^{-3} F concentration of analyte.
†Maximum level for no interference.

to samples and standards to establish the same ionic strength in all solutions. Alternatively, the method of standard addition can be used. Frequently, this method is carried out by measuring the potential of a known volume of analyte solution, followed by the addition of a small volume of a standard more concentrated than the analyte, and then remeasuring the potential. Example 15-10 shows how an expression relating signal (the difference between the two measured potentials) to analyte concentration can be derived.

Now let us turn our attention to a few applications employing ISE's. The fluoride electrode has found widespread use because fluoride is an important constituent in many samples and, until the development of this electrode, was quite difficult to analyze for. The fluoride level in fluoridated water (or milk) can be quickly and easily found with the fluoride ISE. This is a particularly important application because a concentration of 1 ppm is recommended for beneficial effects while a concentration of a few parts per million may lead to mottled teeth. Thus it is necessary to monitor fluoridated water supplies continuously to maintain proper fluoride concentration. Analysis for fluoride in toothpaste is also carried out with the fluoride electrode by simply adding water and comparing with a calibration curve from SnF_2 standards (see the laboratory manual). Fluoride in bone can be determined by dissolving the ashed bone sample in HCl and then buffering to pH 4.7 (recall that the fluoride electrode should be used in neutral or slightly acid solutions to avoid OH^- interference and formation of undissociated HF). Teeth can be analyzed for fluoride content by dissolving the sample in $HClO_4$ and then buffering with sodium citrate. Fluoride present in air and stack gases can be determined by collecting samples on filters, followed by extraction with an acetate buffer and then measurement of potential with the electrode. Similar methods are available for determining fluoride content in minerals, cement, coal, plating baths, vegetation, and even wine.

The calcium ISE is another popular one, especially for use in biochemical studies. Calcium in serum, urine, and saliva can be determined quite simply by using this electrode. Analyses of beer, milk, wine, and various water samples for calcium content have all been reported. As with the fluoride electrode, most of these samples can be analyzed directly after adjustment of pH and ionic strength.

The nitrate ISE is another welcome addition to the analyst's arsenal of instruments because previously analysis for NO_3^- had been quite difficult. Although the nitrate electrode has several interferences (selectivity constants are given in Table 15-4), it is simpler and faster than other methods are and offers comparable accuracy. Analysis of meat for nitrate content is an important application, and nitrate, NO_2^-, content can also be determined by oxidation to nitrate before potential measurement. Nitrate content in various foods, fertilizers, sewage, and water samples has been reported.

The sulfide electrode has been widely used in the paper industry to determine sulfide in wood chips and pulping liquors. Sulfide in cigarette smoke can be determined by bubbling the smoke through a sodium ascorbate solution followed by dilution and potential measurement. Sulfur content in proteins can be determined by titration with $AgNO_3$, using the sulfide electrode for an endpoint signal.

The ammonia gas-sensing electrode has been used to determine ammonia content in various biological samples as well as water samples ranging from fish

EXAMPLE 15-10

Derive an expression relating potential measurements to analyte concentration for the standard addition technique. Assume the sample volume to be 100 mL and that 10 mL of the standard is added.

Solution

Initial measurement

$$E_1 = K + \frac{0.059}{Z} \log X = K + m \log X$$

where X is the concentration of the analyte and m is the actual slope, which may vary somewhat from the theoretical value of $0.059/Z$ because of activity coefficient effects.

Second measurement after standard addition

$$E_2 = K + m \log \left[\frac{100}{110}X + \frac{10}{110}S \right]$$

(reflects the fact that we have diluted the unknown 100 to 110 mL and the standard 10 to 110 mL). The concentration of the standard solution is S.

$$\Delta E = E_2 - E_1$$

$$= K + m \log \left[\frac{100}{110}X + \frac{10}{110}S \right] - K - m \log X$$

$$\frac{\Delta E}{m} = \log \left[\frac{100}{110}X + \frac{10}{110}S \right] - \log X = \log \left[\frac{\frac{100}{110}X + \frac{10}{110}S}{X} \right]$$

$$\text{antilog} \frac{\Delta E}{m} \equiv A = 0.909 + \frac{0.091S}{X}$$

which, after some algebraic maneuvering, becomes

$$X = S \left(\frac{0.091}{A - 0.909} \right)$$

That is, the unknown concentration, X, is directly proportional to the standard concentration, S, and the proportionality constant depends only on ΔE, provided that we always use a 10% volume addition and if we know the slope m. A series of standard solutions is used to determine m, or alternatively, the theoretical value of $0.059/Z$ may be used. Note that the concentration units of X will be the same as those for S so that calculations can be carried out as molar, millimolar, parts per million, grams per liter, and so on.

Table 15-4   Selectivity Constants of Interfering Anions for the Nitrate Ion-Selective Electrode

Interfering anion	Selectivity constant, K_s
ClO_4^-	1×10^3
I^-	20
ClO_3^-	2
Br^-	0.9
S^{2-}	0.57
NO_2^-	6×10^{-2}
CN^-	2×10^{-2}
HCO_3^-	2×10^{-2}
Cl^-	6×10^{-3}
OAc^-	6×10^{-3}
CO_3^{2-}	6×10^{-3}
$S_2O_3^{2-}$	6×10^{-3}
SO_3^{2-}	6×10^{-3}
F^-	9×10^{-4}
SO_4^{2-}	6×10^{-4}
$H_2PO_4^-$	3×10^{-4}
PO_4^{3-}	3×10^{-4}
HPO_4^{2-}	3×10^{-5}

SOURCE: From *Guide to Specific Ion Electrodes and Instrumentation* (Cambridge, MA: Orion Research Incorporated, 1969).

tanks to waste waters. Stack gases can be analyzed for NH_3 by a scrubbing with dilute H_2SO_4 followed by neutralization and potential measurement.

It can be seen from these few examples that ISE's have made an important contribution to chemical analysis; and with the large number of kinds of electrode still in the research stage, the future is bright for this field of potentiometry.

SUPPLEMENTARY READING

ARTICLES DESCRIBING ANALYTICAL METHODS OR APPLICATIONS

Durst, R.A., "Ion-Selective Electrodes in Science, Medicine, and Technology," *American Scientist*, 59:353 (1971).

Frant, M.S., "Detecting Pollutants with Chemical-Sensing Electrodes," *Environmental Science and Technology*, 8:224 (1974).

Rechnitz, G.A., "Ion-Selective Electrodes," *Chemical & Engineering News*, 45:146 (June 12, 1967).

"Two Methods Give Fast Clinical Analyses," *Chemical & Engineering News*, 50:29 (February 7, 1972).

REFERENCE BOOKS

Bates, R.G., *Determination of pH: Theory and Practice*. New York: John Wiley & Sons, Inc., 1964.

Buck, R.P., "Potentiometry: pH Measurements and Ion-Selective Electrodes," in *Physical Methods of Organic Chemistry*, vol. 1, pt. 2A. A. Weissberger, ed., New York: John Wiley & Sons, Inc., 1971.

Durst, R.A., ed., *Ion-Selective Electrodes*, Washington: National Bureau of Standards Special Publication 314, 1969.

PROBLEMS

15-1. Explain the difference between the two types of quantitative transfer: (1) concentration and (2) amount.

15-2. Draw schematic cells for the following cell descriptions. Use a KNO_3 salt bridge to connect the two compartments.

*a. left compartment: Ag wire in 0.01 F $AgNO_3$
right compartment: Pt wire in 0.02 F $FeCl_2$, 0.08 F $FeCl_3$

 b. left compartment: Pb wire in 0.002 F $Pb(NO_3)_2$
right compartment: Pt wire in 0.1 F $FeCl_2$, 10^{-5} F $FeCl_3$

 c. left compartment: Pt wire with H_2 (1 atm) bubbling over it in 0.002 F HCl
right compartment: Ag wire coated with AgCl in 0.002 F KCl

 d. left compartment: Ag wire coated with AgBr in 1 F NaBr
right compartment: Pt wire in 0.001 F NaBr saturated with liquid bromine (Br_2)

*15-3. Write a balanced chemical equation for each of the cells in Problem 15-2.

*15-4. Calculate the cell potential for each of the cells in Problem 15-2.

*15-5. Based on the cell potentials calculated in Problem 15-4, determine the spontaneous direction for each of the cells, and write a balanced chemical equation for the cell in the spontaneous direction.

15-6. Draw schematic cells for the following shorthand cell notations. Assume||indicates a KCl salt bridge.

 a. $Pt|I_2(s), I^-$ (0.02 F)$||Fe^{2+}$ (0.04 F), Fe^{3+} (0.04 F)$|Pt$

 b. Pt, H_2 (0.5 atm)$|H_3O^+$ (0.01 F)$||OH^-$ (0.01 F)$|H_2$ (1 atm), Pt

 c. Ag, AgCl$|Cl^-$ (0.1 F)$||Ti^{3+}$ (0.01 F), TiO^{2+} (0.1 F), H_3O^+ (0.1 F)$|Pt$

 d. Hg, $Hg_2Cl_2|KCl$ (sat'd)$||Cr_2O_7^{2-}$ (0.001 F), Cr^{3+} (0.002 F), H_3O^+ (1 F)$|Pt$

*15-7. Calculate the cell potential for each of the cells in Problem 15-6.

*15-8. Based on the cell potentials calculated in Problem 15-7, determine the spontaneous direction for each of the cells, and write a balanced chemical equation for the cell in the spontaneous direction.

15-9. Write shorthand cell notations for the following descriptions:

*a. left compartment: Cu wire in 0.05 F $CuSO_4$
right compartment: Cd wire in 0.20 F $CdSO_4$

 b. left compartment: Pt wire in 2×10^{-4} F KI, sat'd with I_2
right compartment: Pt wire in 2×10^{-3} F $MnCl_2$, 4×10^{-3} F $KMnO_4$, 0.1 F H_2SO_4

---

*Answers to problems marked with an asterisk will be found at the back of the book.

c. left compartment: Pt wire with H_2 (1 atm) bubbling over it in HCl
(pH = 2.30)
right compartment: Ag wire coated with AgCl in 2×10^{-4} F KCl
d. left compartment: Pb plate coated with $PbSO_4$ in 2 F H_2SO_4
right compartment: PbO_2 plate coated with $PbSO_4$ in 2 F H_2SO_4

*15-10. Calculate the cell potential for each of the cells in Problem 15-9.

*15-11. Based on the cell potentials calculated in Problem 15-10, determine the spontaneous direction for each of the cells, and write a balanced chemical equation for the cell in the spontaneous direction.

15-12. Calculate the unknown concentration for each of the analytes noted as (x F) from the potential data given at 25°C:

*a. $Pt|Fe^{2+}$ (0.05 F), Fe^{3+} (0.25 F)$||Ag^+$ (x F)$|Ag$ $\quad$ $E_{cell} = -0.106$ V
b. $Cd|Cd^{2+}$ (0.01 F)$||I^-$ (x F), I_2 (s)$|Pt$ $\quad$ $E_{cell} = +1.063$ V
c. $Ag, AgBr|Br^-$ (1 F)$||Cu^{2+}$ (x F)$|Cu$ $\quad$ $E_{cell} = +0.183$ V
d. $Ag, AgCl|Cl^-$ (0.1 F)$||Cl^-$ (x F), Cl_2 (1 atm)$|C$ $\quad$ $E_{cell} = +1.400$ V

15-13. Calculate the unknown concentration for each of the analytes noted as (x F) from the potential data given at 25°C:

*a. $SCE||Cu^{2+}$ (x F)$|Cu$ $\quad$ $E_{cell} = -160$ mV
b. $SCE||Pb^{2+}$ (x F)$|Pb$ $\quad$ $E_{cell} = -385$ mV
c. sat'd AgCl ref$||Cd^{2+}$ (x F)$|Cd$ $\quad$ $E_{cell} = -758$ mV
d. sat'd AgCl ref$||Cl^-$ (x F)$|AgCl, Ag$ $\quad$ $E_{cell} = 0$ mV
e. $SCE||Br^-$ (x F)$|AgBr, Ag$ $\quad$ $E_{cell} = -156$ mV

15-14. The following electrodes are proposed as reference electrodes. Calculate their potentials versus NHE with the information given.

*a Ag/AgI in 1 F KI; $K_{sp} = 8 \times 10^{-17}$, assume $f = 1$ for I^-
b. Ag/AgI F KI; $K_{sp} = 8 \times 10^{-17}$, $f = 0.6$ for I^-
c. $Pb/PbSO_4$ in 0.1 F H_2SO_4; $K_{sp} = 1.6 \times 10^{-8}$, $f = 0.3$ for SO_4^{2-}
d. Ag/Ag_2CrO_4 in 0.5 F K_2CrO_4; $K_{sp} = 2 \times 10^{-12}$, assume $f = 1$ for CrO_4^{2-}
e. Hg/Hg_2Br_2 in sat'd KBr (solubility = 50 g/100 mL soln), assume $f = 1$ for Br^-.

15-15. Electrodes of the second kind may employ complexes in place of insoluble compounds. In this case, the formation constant must be used instead of K_{sp}. Calculate potentials of the following electrodes versus NHE with the information given:

*a. $Ag/Ag(CN)_2^-$ (0.1 F), KCN(0.2 F) $\quad$ $\beta_2 = 9.5 \times 10^{12}$
b. $Ag/Ag(NH_3)_2^+$ (0.2 F), NH_3(1 F) $\quad$ $\beta_2 = 1.6 \times 10^7$
c. Cu/Cu–EDTA (0.05 F), EDTA (0.1 F), pH = 7 $\quad$ $K_{MY} = 6.3 \times 10^{18}$

15-16. Calculate K_{sp} for the underlined compounds from the cell potential data given below:

*a. $SCE||Br^-$ (0.05 F)$|\underline{AgBr}, Ag$ $\quad$ $E_{cell} = -0.093$ V
b. $SCE||IO_3^-$ 0.1 F)$|\underline{Pb(IO_3)_2}, Pb$ $\quad$ $E_{cell} = -0.680$ V
c. $SCE||Na_2CO_3$ 0.5 F), pH 8$|\underline{PbCO_3}, Pb$ $\quad$ $E_{cell} = -0.688$ V

15-17. Calculate the ratio of K_{sp}'s for $Fe(OH)_2$ and $Fe(OH)_3$ from the observation that $E_{cell} = -0.669$ V for the following cell: $SCE||pH$ 12$|Fe(OH)_2, Fe(OH)_3$, Pt.

*15-18. The following cell has a voltage of -0.180 V: $Pb|Pb^{2+}$ (0.08 F)$||KI$ (0.100 F)$|PbI_2$, Pb. Calculate K_{sp} for PbI_2.

15-19. A solution was prepared containing 0.2 F NaH_2PO_4 and 0.1 F Na_2HPO_4 and saturated with $Cu(OH)_2$. The potential of a copper electrode was then measured and found to be $+0.201$ V on the hydrogen scale. Calculate K_{sp} for $Cu(OH)_2$.

15-20. The pH of a solution may be determined by measuring the potential of a hydrogen electrode against a reference electrode. The following cell was used:

SCE$\|$H$_3$O$^+$ $(x\,F)$$|H_2$ (1 atm), Pt. Calculate pH of the following solutions from the cell potentials given:

   a. -0.242 V

*b. -0.450 V

   c. -0.867 V

   d. -1.013 V

   e. -0.655 V

15-21. With some potentiometers, the reference electrode is actually the right-hand electrode instead of the indicator electrode. That means, for a pH measurement, Pt, H$_2$ (1 atm)$|$H$_3$O$^+$ $(x\,F)\|$SCE. Calculate pH of the following solutions from the cell potentials given:

   a. $+0.242$ V

*b. $+0.600$ V

   c. $+0.224$ V

   d. $+1.067$ V

   e. $+1.086$ V

15-22. The following cell has a potential of $+0.554$ V: Pt, H$_2$ (1.00 atm)$|$HA (0.15 F), NaA (0.25 F)$\|$SCE. Calculate the dissociation constant of HA.

*15-23. Using the transference number data given in Example 15-7, calculate the liquid junction potential at 25°C across a membrane as shown in Figure 15-7, in which electrolyte 1 is 0.001 F KCl and electrolyte 2 is 0.1 F KCl. Neglect activity coefficients.

15-24. If the KCl electrolyte used in Problem 15-23 is replaced by HCl, in which $t_H = 0.82$ and $t_{Cl} = 0.18$, calculate the new liquid junction potential.

15-25. A high-impedance voltmeter is used in which the glass electrode is the left-hand electrode, while the SCE reference is the right-hand electrode. If a pH 4.00 buffer at 25°C gives a measured cell potential of 256 mV, calculate the pH of the following solutions from the measured cell potentials:

*a. $+487$ mV

  b. $+256$ mV

  c. $+20$ mV

  d. $+808$ mV

15-26. The pH meter and electrodes used in Problem 15-25 are now used to measure the pH of a solution containing 0.35 F HA and 0.05 F A$^-$. The measured cell potential is $+183$ mV. Calculate K_a for HA.

15-27. Suppose a glass membrane has a selectivity constant for Na$^+$ of 2×10^{-10}.

  (a) Calculate the error in pH which will be observed in 0.005 F NaOH.

  (b) Calculate the percentage error in [H$_3$O$^+$].

15-28. Selectivity constants for the nitrate ISE are given in Table 15-4. Calculate the percentage error in nitrate ion concentration that would be observed if the following solutions were measured:

*a. $[NO_3^-] = 1.00 \times 10^{-3}\,F$, $[NO_2^-] = 1.00 \times 10^{-3}\,F$

  b. $[NO_3^-] = 1.00 \times 10^{-3}\,F$, $[Br^-] = 1.00 \times 10^{-3}\,F$.

  c. $[NO_3^-] = 1.00 \times 10^{-2}\,F$, $[Cl^-] = 1.00 \times 10^{-3}\,F$

  d. $[NO_3^-] = 1.00 \times 10^{-4}\,F$, $[PO_4^{3-}] = 2.00 \times 10^{-3}\,F$

*15-29. A 10-mL wine sample is diluted to 100 mL with dilute NH$_3$, and the potential is measured with a sodium ISE. The reading is -52.4 mV. Then 10.00 mL of a standard containing 95 ppm of Na^+ is added; the new potential is -21.6 mV. Calculate sodium content in the wine sample as parts per million of Na. Assume theoretical slope for the electrode.

15-30. A 3.6-g soil sample is treated with $HClO_4$, and HF is distilled from the mixture. The distillate is collected in an ionic strength buffer and diluted to 50 mL. Measurement with a fluoride ISE gives a value of $+95$ mV. Exactly 5.00 mL of a standard containing 1×10^{-3} F F^- is added to the 50-mL sample, and the potential becomes $+68$ mV. Calculate fluoride content in the soil sample as parts per million of F^-. Assume theoretical slope for the electrode.

15-31. If the uncertainty in each of the potential measurements is 0.3 mV in Problem 15-30, calculate the uncertainty in parts per million of F^- (*Hint:* because of the nonlinearity of the relationship, it is simpler to calculate the change that occurs when the signal is $S + \Delta S$.)

15-32. A 2.05-g sample of iodized salt was dissolved in H_2O and diluted to 50 mL. Measurement with an iodide ISE gave a value of $+62.5$ mV. Then 5.00 mL of a standard containing 2.5×10^{-4} F I^- was added; the potential reading was $+42.7$ mV. Standard KI solutions were used to measure the slope m, and the data are 2.5×10^{-5} F: $+95$ mV; 2.5×10^{-4} F: $+39$ mV; 2.5×10^{-3} F: -17 mV. Calculate percentage of KI in the salt sample.

*15-33. Ten milliliters of seawater is diluted to 100 mL, and the potential is measured with a calcium ISE. The reading is $+32$ mV. A standard calcium solution containing 2×10^{-4} F Ca^{2+} gives a reading of $+12$ mV. Calculate calcium content in the seawater as parts per million of Ca. Assume theoretical slope for the electrode.

15-34. a. A volumetric titration of bromide with standard $AgNO_3$ solution is designed, using a potentiometric endpoint. Determine the theoretical potential of a silver metal electrode in contact with the solution at the equivalence point.

b. In practice, a SCE will be used as a reference electrode. What will be the potential of the Ag electrode versus SCE at the equivalence point?

*15-35. If an unknown and uncorrected liquid junction potential introduces an uncertainty of ± 2 mV with a glass electrode and reference, what will be the uncertainty in pH measurements?

15-36. a. The Cu^{2+} concentration is measured using a copper metal electrode potential. If the potential reading is $+300$ mV (hydrogen scale) and has an uncertainty of ± 2 mV, what is the percentage uncertainty in Cu^{2+} concentration?

b. If a calomel electrode is used as the reference, the actual potential reading is $+58$ mV ± 2 mV. What is the percentage uncertainty in Cu^{2+} concentration under these conditions? The calomel potential is $+242$ mV (hydrogen scale).

Coulometry | **16**

In Chapter 15, analytical techniques were divided into two classes: (I) those in which the total amount of analyte present in the sample is determined and (II) those in which the concentration of analyte is determined. In this chapter, two electroanalytical methods that fall into class I will be discussed.

It was pointed out in Section 11-1 that, when 1 mol of an electrochemical reaction takes place, the amount of charge that flows is given by

$$Q = nF$$

If the entire amount of analyte present in the sample, X, undergoes an electrochemical reaction, the amount of charge will be

$$Q(C) = n\left(\frac{eq}{mol}\right)F\left(\frac{C}{eq}\right)X(mol) \tag{16-1}$$

Thus Q, the number of coulombs of charge needed to carry out a reaction, is a signal directly proportional to the amount of analyte, the proportionality constant being nF. Electroanalytical techniques in which electrical charge is measured are called *coulometric methods*, and *coulometric titrations* and *controlled-potential electrolyses* are the two coulometric methods that we shall explore in this chapter.

It was also shown in Section 11-1 that the electrical charge is determined by the current, which is the flow rate of charge (coulombs per second):

$$Q(C) = I\left(\frac{C}{sec}\right)t(sec) \tag{16-2}$$

Equation 16-2 assumes that the current is constant or that I represents an average

521

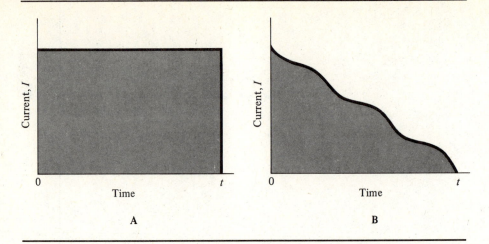

FIGURE 16-1 Integration of current I as a function of time: **(A)** For the constant-current case, area $= Q = It$; **(B)** for the varying-current case, area $= Q = \int_0^t I(t)\, dt$.

current. For situations in which current is varying, the amount of charge would be

$$Q = \int_0^t I(t)\, dt \tag{16-3}$$

In other words, Q would be the area under the curve in Figure 16-1. A coulometric titration involves constant-current conditions (Equation 16-2), whereas a controlled-potential electrolysis involves a current varying with time (Equation 16-3).

It should also be noted that Q/F is an expression of equivalents (or in milliquantities at the equivalence point):

$$\text{meq analyte} = \text{meq electrical charge}$$

$$= \frac{Q(mC)}{F(9.65 \times 10^4\ mC/\text{meq})} \tag{16-4}$$

For a constant-current technique, the amount of analyte present is

$$\text{meq analyte} = \frac{mg}{FW/n} = \frac{I\ (mA)t\ (sec)}{F} = \frac{It}{9.65 \times 10^4} \tag{16-5}$$

16-1 COULOMETRIC TITRATIONS

For Equation 16-1 to be meaningful, it is essential that all measured electrical charge can be attributed to the reaction involving the analyte. Any other reactions occurring would represent interference. When all the current involves reaction with the analyte, the process is said to have 100% current efficiency. The art of developing coulometric titrations lies in designing experimental conditions in which there is 100% current efficiency.

The problem that arises when an electrochemical reaction occurs directly at the electrode can be conceptually considered through a particular reaction. Suppose that you wish to determine Fe^{2+} by oxidizing it to Fe^{3+} at an anode,

$$Fe^{2+} \rightleftharpoons Fe^{3+} + e \qquad \text{(16-6)}$$

At first, there may be sufficient Fe^{2+} present at the electrode so that the process is 100% current efficient. However, as more and more of the Fe^{2+} is oxidized, its concentration in solution decreases to the point where it can no longer supply Fe^{2+} ions to the electrode surface fast enough to give up the required electrons. Under constant-current conditions, the potential of the electrode will increase until some other reaction occurs to satisfy the current. For example, H_2O could be oxidized,

$$6H_2O \rightleftharpoons O_2(g) + 4H_3O^+ + 4e \qquad \text{(16-7)}$$

or if Cl^- were present, it could be oxidized,

$$2Cl^- \rightleftharpoons Cl_2 + 2e \qquad \text{(16-8)}$$

Oxygen does not react rapidly with Fe^{2+} so that Equation 16-7 would represent an interference, and the current efficiency would drop below 100%. Because *more* coulombs of charge would be observed than are needed for Fe^{2+} oxidation, a positive error would result.

On the other hand, in Equation 16-8, the product, Cl_2, reacts with Fe^{2+}:

$$Cl_2 + 2Fe^{2+} \rightleftharpoons 2Fe^{3+} + 2Cl^- \qquad \text{(16-9)}$$

Chlorine produced at the electrode can migrate into the solution and react with Fe^{2+}. Two electrons are transferred to the electrode in Equation 16-8, and two Fe^{3+} ions are produced in Equation 16-9. The net reaction (addition of Equations 16-8 and 16-9) is simply Equation 16-6. Chlorine acts as an electrogenerated intermediate for the reaction. Although all the Fe^{2+} does not react directly at the electrode, all electrons transferred to the electrode ultimately involve oxidation of Fe^{2+}. The net reaction is, therefore, 100% current efficient provided that *all* the chlorine produced oxidizes Fe^{2+} and not some other substance present (for example, Cl_2 reacts slowly even with H_2O). It is important to realize that it does not matter whether the analyte reacts directly at the electrode or in a secondary reaction with the electrogenerated intermediate. At the beginning of the titration, a significant fraction of the analyte may react directly at the electrode; near the end, essentially all will react chemically with the electrogenerated intermediate. The essential point is that all electrical charge ($I \times t$) must lead to reaction of the analyte.

The presence of an intermediate which is produced efficiently at the electrode and which reacts quickly with the analyte is the key to a successful coulometric titration. Also note that the concentration of the intermediate reagent (Cl^-) does not change so long as Fe^{2+} is present to reduce Cl_2 back to Cl^-. Therefore, the problem that arose with direct Fe^{2+} oxidation will not occur with the intermediate reagent.

What will happen after the equivalence point is reached? Current will still flow to oxidize Cl^- to Cl_2, but there will be no Fe^{2+} to react with the Cl_2; the concentration of Cl_2 will increase. This sequence is analogous to the conventional titration of Fe^{2+} with an oxidizing agent (for example, Ce^{4+} was used in Section 11-2). An endpoint signal is needed to alert the analyst that the equivalence point has been reached. Any of the endpoint signals for conventional titrations may also be appropriate for coulometric titrations. In fact, the analogous titration of Fe^{2+} with Ce^{4+} can be carried out coulometrically by adding Ce^{3+} to the analyte solution as an intermediate reagent. The potential of a platinum indicator electrode versus a reference electrode could be used as an endpoint signal (Figure 16-2); so the cell would contain four electrodes: two for the constant-current circuit and two for the potentiometric endpoint system. Alternatively, a redox color indicator (also described in Section 11-4) could be employed. Example 16-1 shows a typical calculation involving a coulometric titration.

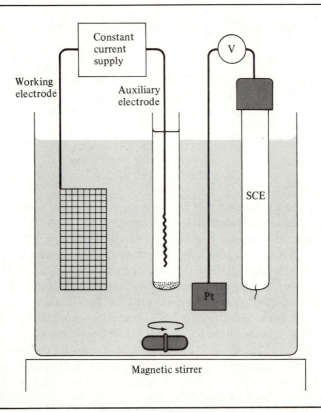

FIGURE 16-2 Cell for a coulometric titration employing a potentiometric endpoint detection system (Pt vs. SCE).

EXAMPLE 16-1

A 107.1-mg alloy sample is dissolved in 100 mL of H_2SO_4 solution containing Fe(II). There is added 20 mL of 0.1 F Ce(III) and the solution is placed in an electrolysis cell for coulometric titrations. A constant current of 48.65 mA is used, and the endpoint is reached in 415.6 sec. Calculate the percentage of Fe in the alloy.

Solution

$$\text{meq Fe} = \text{meq charge}$$

$$\frac{\text{mg Fe}}{55.85/1} = \frac{It}{9.65 \times 10^4} = \frac{48.65 \text{ mA} \times 415.6 \text{ sec}}{9.65 \times 10^4 \text{ mC/meq}}$$

$$\text{mg Fe} = 11.70 \text{ mg}$$

$$\% \text{ Fe} = \frac{11.70 \text{ mg Fe}}{107.1 \text{ mg sample}} \times 100 \, (\%) = 10.93\%$$

A coulometric titration then involves generation of an intermediate at an electrode operated at constant current. The generated intermediate reacts stoichiometrically with the analyte. There is needed an endpoint signal equal to or related to the equivalence point for the reaction. The amount of analyte present is calculated by Equation 16-5.

TITRATION OF AS(III) WITH ELECTROGENERATED I_3^-

Let us consider another example of a widely used coulometric titration, the determination of arsenic with electrogenerated I_3^-. The requirements for a successful intermediate were given above as (1) fast electrode kinetics, yielding 100% current efficiency, and (2) stoichiometric chemical reaction between the electrogenerated intermediate and the analyte. For a redox reaction, the second requirement implies that the electrogenerated intermediate will have a redox potential sufficiently higher (for an oxidation) than the analyte couple has so that the equilibrium constant will be high. This was shown to be a few hundred millivolts in Section 11-5. In Figure 16-3 it can be seen that the Cl_2/Cl^- couple lies nearly 600 mV above the Fe^{3+}/Fe^{2+} couple. In other words, Cl_2 is a much stronger oxidizing agent than Fe^{3+} is and is thermodynamically capable of oxidizing Fe^{2+} essentially to completion (the kinetics of the reaction may be another story). However, when we consider the proposed reaction, Figure 16-3 shows that I_3^- is not more oxidizing than is As(V). In fact, if anything, As(V) should be able to oxidize iodide to iodine. At this point, we should recall that E^0 values as shown in Figure 16-3 refer to standard-state conditions for *all* species in the reaction. For the arsenic couple,

$$H_3AsO_4 + 2H_3O^+ + 2e \rightleftharpoons H_3AsO_3 + 3H_2O \qquad \textbf{(16-10)}$$

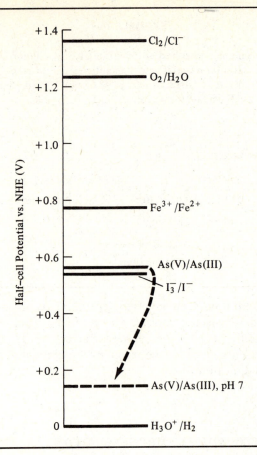

FIGURE 16-3 Diagram of half-cell potentials, showing shift of As(V)/As(III) couple from pH 0 to pH 7.

which means that not only H_3AsO_4 and H_3AsO_3 are at unit activity but also H_3O^+ is. In other words, this is the potential in strongly acidic solution. The reaction shown in Equation 16-10 will be much less favored (lower potential) in neutral or basic solution because of the lower $[H_3O^+]$. We can see how the formal potential of this couple changes with pH as follows:

$$E = E^0 - \frac{0.059}{2} \log \frac{[H_3AsO_3]}{[H_3AsO_4][H_3O^+]^2}$$

$$= +0.559 - \frac{0.059}{2} \log \frac{[H_3AsO_3]}{[H_3AsO_4]} + \frac{0.059}{\cancel{2}} \log [H_3O^+]^{\cancel{2}}$$

With formal potential conditions, $[H_3AsO_3] = [H_3AsO_4] = 1\ F$. Therefore,

$$E_f^0 = +0.559 - 0.059\ \text{pH} \tag{16-11}$$

At pH 7, the formal potential for the As(V)/As(III) couple will be

$$E_f^0 = +0.559 - 0.059 \times 7 = +0.146\ \text{V vs. NHE}$$

At pH 7, the As(V)/As(III) couple lies well below the I_3^-/I^- couple, as shown in Figure 16-3 (dashed line). The I_3^-/I^- couple does not involve H_3O^+ and, therefore, remains at $+0.536$ V vs. NHE at pH 7. Thus a coulometric titration of As(III) can be carried out with electrogenerated I_3^-, provided that a neutral solution is maintained.

To carry out this coulometric titration, the oxidation state of the analyte must be adjusted to As(III). Then the solution is buffered to pH 7 with $NaHCO_3$, and the intermediate reagent, KI, is added. The solution is added to an electrolysis cell, as shown in Figure 16-2, and the constant-current source is activated. At the anode, I_3^- is produced by Equation 16-12,

$$3I^- \rightleftharpoons I_3^- + 2e \tag{16-12}$$

which in turn reacts with As(III) according to Equation 16-13,

$$H_3AsO_3 + I_3^- + 3H_2O \rightleftharpoons H_3AsO_4 + 3I^- + 2H_3O^+ \tag{16-13}$$

The I_3^- is consumed, and the iodide concentration remains constant throughout the titration. After the equivalence point, excess I_3^- is produced, and its yellow color will be observed. For a sharper endpoint, a small amount of a starch solution is added so that the intensely colored, blue, starch–iodine complex will be produced. A typical calculation is given in Example 16-2.

In the example shown it would also be possible to titrate a second aliquot by adding it directly to the solution at the first endpoint in the electrolysis cell. This technique automatically performs a blank correction, provided that the analyst always titrates to the same endpoint signal (the same blue color in this example). Only the first titration would be uncorrected. Instead of wasting the first titration, the analyst is frequently advised to generate the endpoint color before adding the first aliquot.

INSTRUMENTATION

Equipment needed for a coulometric titration is relatively simple. An electrolysis cell was shown in Figure 16-2. A large electrode (10 to 20 cm^2) is used for generation of the intermediate because current efficiency is normally higher at low *current density* (milliamps per square centimeter of electrode area). The auxiliary electrode required to complete the circuit is usually isolated from the analyte solution by placement in a glass tube with a sintered-glass disk to allow transfer of charge. This arrangement helps to prevent any products formed at the auxiliary electrode from reacting with

EXAMPLE 16-2

A 3.547-g sample of an insecticide was dissolved in acid, all arsenic being thereafter reduced to As(III), and diluted to 100.0 mL in a volumetric flask. A 10.00-mL aliquot was added to 150 mL of a solution containing 0.05 F KI and buffered to pH 7. A few drops of starch indicator were added for the endpoint signal. The coulometric titration required 415.7 sec at a constant current of 25.16 mA to reach the endpoint. Calculate the percentage of As_2O_3 in the insecticide.

Solution

$$meq\ As_2O_3 = meq\ charge$$

$$\frac{mg\ As_2O_3}{FW/n} = \frac{It}{9.65 \times 10^4}$$

In the reaction, each As is oxidized from (III) to (V), and there are two arsenic atoms in As_2O_3. Therefore, $n = 4$.

$$\frac{mg\ As_2O_3}{197.8/4} = \frac{25.16\ mA \times 415.7\ sec}{9.65 \times 10^4\ mC/meq}$$

$$mg\ As_2O_3\ in\ aliquot = 5.360\ mg$$

$$mg\ As_2O_3\ in\ sample = 5.360 \times \frac{100}{10} = 53.60\ mg$$

$$\%\ As_2O_3 = \frac{53.60\ mg\ As_2O_3}{3547\ mg\ sample} \times 100\ (\%) = 1.511\%$$

the analyte or the electrogenerated reagent. Whenever air-sensitive reagents are to be generated, an inert gas, such as N_2, is bubbled through the solution during electrolysis. Stirring is needed to mix the electrogenerated reagent produced at the electrode with the analyte solution and is frequently accomplished with a magnetic stirrer.

Electrochemical endpoint signals are often used with coulometric titrations although any of the endpoint signals discussed in Section 7-2 can be used; potentiometric endpoints have been discussed in Section 11-4. Another electro-chemical signal is called an *amperometric endpoint*. As the name implies, a current is measured during the titration. Two indicator electrodes are placed in the analyte solution, and a small voltage is applied. If the analyte and reagent couples react reversibly (fast electrode kinetics in both the anodic and cathodic directions), current will flow when significant concentrations of both oxidized and reduced forms are present. For example, in the titration of Fe^{2+} with electrogenerated Ce^{4+}, the solution contains Fe^{2+} and Ce^{3+} before titration begins. Both of these species can be oxidized at the indicator anode, but neither one can be reduced at the indicator cathode; so no current flows. During the coulometric titration, some of the Fe^{2+} will be oxidized to Fe^{3+} so that Fe^{3+} can be reduced at the indicator cathode while Fe^{2+} can be oxidized at the indicator anode. Indicator current will reach a maximum about halfway through the titration when equal quantities of both species are present. As the equivalence point approaches, the concentration of Fe^{2+} will decrease, and so will the indicator current. At the equivalence point, the concentration of Fe^{2+} is essentially zero. No current flows because, although Fe^{3+} may be reduced at the indicator cathode and Ce^{3+} could be oxidized at the indicator anode, this reaction would require several hundred millivolts of applied voltage. With a small applied voltage (typically 100 mV), no current will flow. After the equivalence point, the solution contains both Ce^{3+} and Ce^{4+}, which can be oxidized at the indicator anode and reduced at the indicator cathode, respectively. The solution contains very little Ce^{4+}, and its concentration limits the current. Thus the indicator current is directly proportional to the amount of excess Ce^{4+} present. The indicator current as a function of time is shown in Figure 16-4. The intersection of the line observed just before the equivalence point (proportional to $[Fe^{2+}]$) with the line just after the equivalence point (proportional to $[Ce^{4+}]$) is the endpoint signal and is close to the true equivalence point.

In addition to the electrolysis cell and endpoint detection system, a coulometric titration requires a high-quality constant-current source and a clock. Constant-current sources accurate to $\pm 0.01\%$ with built-in clocks are available for several hundred dollars. Constant currents of 1 to 200 mA are typically available with commercial instruments. The clock can be the largest source of instrumental error because it must start with activation of the current source and when deactivated, stop with no lag or coasting. A stopclock with a solenoid brake can keep this start-stop error to about 0.01 sec for each switching. However, the current may

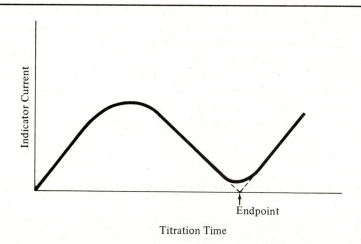

FIGURE 16-4 Coulometric titration employing amperometric endpoint detection system. Titration of Fe(II) with Ce(IV) in 8 F H_2SO_4.

be turned on and off several times near the equivalence point for observation of the endpoint signal so that start-stop errors of about 0.1 sec can be anticipated. In addition, many built-in clocks in commercial coulometric current sources read to only ± 0.1 sec. Thus titration times of > 100 sec are required to keep timing errors $< 0.1\%$. Usually titrations of several hundred seconds are used for high-accuracy work. Under these conditions, the largest error is in observing the endpoint signal.

It should also be noted that, with a current of 10 mA and a titration time of 100 sec, the amount of analyte titrated is only

$$\text{meq} = \frac{I}{9.65 \times 10^4} \, t = \frac{10}{9.65 \times 10^4} \times 100 = 0.0104 \text{ meq}$$

This is about two orders of magnitude less than a typical conventional titration. Coulometric titrations of such small amounts of analyte can be accomplished if the endpoint signal is sufficiently sharp. This is one of the reasons why electrochemical endpoint signals are often recommended because they are more sensitive than color indicators are.

If a commercial constant-current source is not available, titrations accurate to $\pm 0.5\%$ can be accomplished simply by a high-voltage battery (for example, four 45-V B batteries in series to provide a 180-V supply) and a large resistance (say, 10 to 20 kΩ). The current will be given by Ohm's law,

$$I = \frac{V}{R}$$

where V includes the high-voltage supply plus any voltage developed at the electrolysis cell and R includes the large resistance plus resistance of the cell. With a high-voltage supply, the effects of voltage at the cell will be relatively small. For example, if the cell requires 0.5 V, then $V = 180 - 0.5 = 179.5$. Changes in voltage that occur during the titration will probably be only a few tenths of a volt; so the voltage supply should be stable to ± 0.2 to 0.3%. The same argument holds for the total resistance. The cell resistance is typically 10 to 50 Ω, and changes during a titration should be < 10 Ω. Thus R should be stable to $\pm 0.1\%$ with a 10,000-Ω resistance in the circuit. The actual current should be measured during the titration by observation of the voltage drop across a precision resistor, as shown in Figure 16-5.

COULOMETRIC TITRATION COMPARED WITH CONVENTIONAL TITRATION

The designation "coulometric titration" is well chosen for this technique. Although coulombs of charge are measured instead of volume and concentration of a reagent, there are many similarities to the titrations discussed in Part Three. The calculation of milliequivalents involves the product of two factors. In the conventional titration,

$$\text{meq} = NV$$

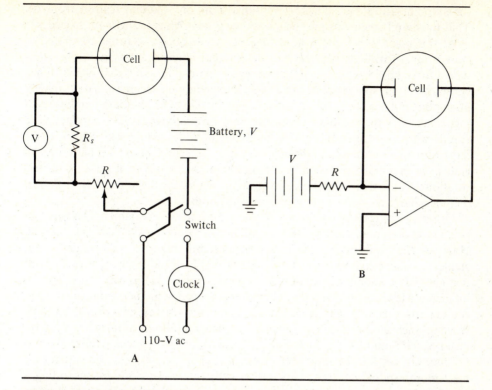

FIGURE 16-5 **(A)** Classical circuit for a coulometric titration. A large battery supply combined with a large resistance R produces an essentially constant current through the cell of $I \cong V/R$. The current is accurately measured on the voltmeter or potentiometer $\widehat{V}$ as a voltage drop, V_s, across the small standard resistance, $I = V_s/R_s$. The circuit is activated and the clock starts when the switch is closed. **(B)** Modern circuit for a constant-current supply, using an operational amplifier. Current, I, through the cell is equal to V/R. A switch and a clock circuit are needed as in part **(A)**.

whereas in the coulometric titration,

$$\text{meq} = \frac{I}{9.65 \times 10^4} t$$

It should be noted that $I/9.65 \times 10^4$ has units of I (C/sec)/9.65 × 10⁴ (C/meq) = meq/sec. Thus the constant-current value is analogous to the titrant normality in a conventional titration; and the titration time is analogous to titrant volume. To measure the titration time, an endpoint signal is required, in the same way that, for a conventional titration, an endpoint signal is required to measure titration volume.

As the equivalence point is approached, it is good procedure to let current flow for a short period of time and then check for the endpoint signal. When no endpoint is evident, another short burst of current is run through the cell in the same way that the endpoint is approached in a conventional titration by adding reagent dropwise. Thus the on-off switch for the current source is analogous to the buret stopcock.

At this point, some of the advantages of coulometric titrations compared to conventional titrations become evident. Reagent concentration has to be determined by standardization (or prepared determinately) for conventional titrations; constant-current source values are usually known to high accuracy ($\pm 0.1\%$ readily attainable) and need be checked only periodically. Time can also be determined with higher accuracy than reading a buret can, and current flow can be controlled more readily with a switch than can volume flow with a buret stopcock. Therefore, the largest source of error in a coulometric titration is usually the determination of the equivalence point, an error source common to both conventional and coulometric titrations. Coulometric titration is inherently one of the most accurate analytical methods and is comparable to gravimetric methods.

Other advantages are apparent from the arsenic titration example. This titration could be carried out in a conventional fashion with a standard I_3^- reagent solution. However, the vapor pressure of iodine is sufficiently high that this reagent is not stable for long periods of time. Standardization of I_3^- solutions must be carried out frequently. In coulometric titration, I_3^- is generated only as it is used and is present for merely a few seconds at the endpoint after the reaction has been completed. Thus coulometric titrations are useful for carrying out reactions involving unstable reagents, such as strong reducing agents that react rapidly with air or even H_2O (for instance, Ti^{3+} and Cr^{2+}). Such oxidizing agents as Ag^{2+}, which oxidize H_2O to produce O_2, can also be successfully generated at an electrode, provided that they react more rapidly with the analyte than with H_2O. Therefore, by generating a reagent at an electrode, the analyst can avoid most of the problems of storage, preparation, and standardization of solutions. The constant-current source is a universal "standard solution," the electron itself being the universal "titrant."

It might appear that coulometric titrations could be applied only to redox reactions because a redox reaction is involved in the generation of the intermediate. However, coulometric titrations have been devised for all four of the reaction types (see Section 3-3) that have been utilized in volumetric methods (Part Three). Examples from each of these reaction types follow.

Precipitation Reactions   In Sections 8-2 and 8-3, the survey of applications involving precipitation titrations focused on reactions of Ag^+ to form an insoluble silver compound. In the conventional precipitation titration, the Ag^+ reagent is added as a solution of $AgNO_3$. In the coulometric titration, the analyst can carry out the same reaction by generating Ag^+ at an electrode, simply substituting a silver metal anode for the platinum electrode shown in Figure 16-2:

$$Ag \rightleftharpoons Ag^+ + e$$

The electrogenerated silver ion can be used to determine Cl^-, Br^-, I^-, S^{2-}, mercaptans, and other anions that form insoluble silver compounds. Endpoint signals discussed in Section 8-2 may be used for the coulometric analog. Potentiometric or amperometric endpoints are common probably because, if the analyst is using two electrodes for generating a reagent, he or she may as well use two more for an endpoint signal.

With a potentiometric endpoint, mixtures of halides can theoretically be analyzed. The most insoluble halide (solubility of AgI < AgBr < AgCl) will precipitate first, and then a potential inflection will occur as the second precipitation begins, as shown in Figure 16-6. In practice, AgI is somewhat of a problem because it tends to adsorb halide ions on its surface (coprecipitation), leading to negative errors. Also, the solubility of AgCl is not sufficiently low to give a very sharp break in potential in an aqueous medium, and H_2O–ethanol mixtures are recommended.

Other precipitation reactions include the generation of Hg_2^{2+} from a mercury anode and the generation of $Fe(CN)_6^{4-}$ by reduction of $Fe(CN)_6^{3-}$ at a platinum cathode. The electrogenerated $Fe(CN)_6^{4-}$ can be used to determine zinc by precipitation of $K_2Zn_3[Fe(CN)_6]_2$, which was mentioned in Section 8-3. These examples of coulometric titration are summarized in Table 16-1.

Acid–base Reactions Electrolysis of water leads to the production of H_3O^+ and OH^- according to Equations 16-14 and 16-15:

$$\text{anode:}\quad 6H_2O \rightleftharpoons 4H_3O^+ + O_2 + 4e \qquad \textbf{(16-14)}$$

$$\text{cathode:}\quad 2H_2O + 2e \rightleftharpoons 2OH^- + H_2 \qquad \textbf{(16-15)}$$

The electrogenerated H_3O^+ and/or OH^- can be used to carry out acid–base titrations coulometrically. Clearly, the two electrodes must be kept separated, or else

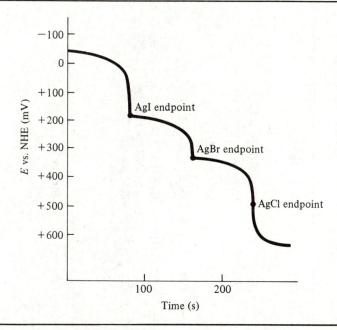

FIGURE 16-6 Coulometric titration of an equimolar mixture of halides with potentiometric endpoint detection (ideal curve).

TABLE 16-1   Examples of Coulometric Titrations Involving the Four Reaction Types

Reaction type	Electrogenerated reagent	Generation reaction	Example analytes	Titration reaction with electrogenerated reagent
precipitation	Ag^+	$Ag \rightleftharpoons Ag^+ + e$	$X^- = Cl^-, Br^-, I^-$; S^{2-} ; RSH (mercaptan)	$Ag^+ + X^- \rightleftharpoons AgX$; $2Ag^+ + S^{2-} \rightleftharpoons Ag_2S$; $Ag^+ + RSH + H_2O \rightleftharpoons AgSR + H_3O^+$
	Hg_2^{2+}	$2Hg \rightleftharpoons Hg_2^{2+} + 2e$	$X^- = Cl^-, Br^-, I^-$	$Hg_2^{2+} + 2X^- \rightleftharpoons Hg_2X_2$
	$Fe(CN)_6^{4-}$	$Fe(CN)_6^{3-} + e \rightleftharpoons Fe(CN)_6^{4-}$	Zn^{2+}	$3Zn^{2+} + 2K^+ + 2Fe(CN)_6^{4-}$ $\rightleftharpoons K_2Zn_3[Fe(CN)_6]_2$
acid–base	H_3O^+	$6H_2O \rightleftharpoons 4H_3O^+ + O_2 + 4e$	OH^- ; amines	$H_3O^+ + OH^- \rightleftharpoons 2H_2O$; $H_3O^+ + RNH_2 \rightleftharpoons RNH_3^+ + H_2O$
	OH^-	$2H_2O + 2e \rightleftharpoons 2OH^- + H_2$	H_3O^+ ; HA (weak acids)	$OH^- + H_3O^+ \rightleftharpoons 2H_2O$; $OH^- + HA \rightleftharpoons H_2O + A^-$
complexation	EDTA (HY^{3-})	$HgNH_3Y^{2-} + NH_4^+ + 2e \rightleftharpoons Hg$ $+ 2NH_3 + HY^{3-}$	$M^{2+} = Ca^{2+}, Cu^{2+}, Pb^{2+}, Zn^{2+}$	$HY^{3-} + M^{2+} \rightleftharpoons MY^{2-} + H_3O^+$
redox	Cl_2	$2Cl^- \rightleftharpoons Cl_2 + 2e$	As(III) ; Tl(I) ; fatty acids	$Cl_2 + As(III) \rightleftharpoons 2Cl^- + As(V)$; $Cl_2 + Tl(I) \rightleftharpoons 2Cl^- + Tl(III)$; $Cl_2 + C{=}C \rightleftharpoons -\underset{Cl}{C}-\underset{Cl}{C}-$
	Br_2	$2Br^- \rightleftharpoons Br_2 + 2e$	U(IV) ; phenol ; oxine ; H_2S ; SO_2	$Br_2 + U(IV) \rightleftharpoons 2Br^- + U(VI)$; $3Br_2 + C_6H_6O \rightleftharpoons C_6H_3OBr_3 + 3HBr$; $2Br_2 + C_9H_7NO \rightleftharpoons C_9H_5NOBr_2 + 2HBr$; $Br_2 + H_2S + 2H_2O \rightleftharpoons S + 2H_3O^+ + 2Br^-$; $Br_2 + SO_2 + 6H_2O \rightleftharpoons 2Br^- + SO_4^{2-} + 4H_3O^+$

Reagent	Half-reaction	Substance determined	Reaction
I_3^-	$3I^- \rightleftharpoons I_3^- + 2e$	Sb(III)	$I_3^- + Sb(III) \rightleftharpoons 3I^- + Sb(V)$
		$S_2O_3^{2-}$	$I_3^- + 2S_2O_3^{2-} \rightleftharpoons 3I^- + S_4O_6^{2-}$
		As(III)	$I_3^- + As(III) \rightleftharpoons 3I^- + As(V)$
Ce^{4+}	$Ce^{3+} \rightleftharpoons Ce^{4+} + e$	Fe^{2+}	$Ce^{4+} + Fe^{2+} \rightleftharpoons Ce^{3+} + Fe^{3+}$
		$Fe(CN)_6^{4-}$	$Ce^{4+} + Fe(CN)_6^{4-} \rightleftharpoons Ce^{3+} + Fe(CN)_6^{3-}$
		Ti(III)	$Ce^{4+} + Ti(III) \rightleftharpoons Ce^{3+} + Ti(IV)$
		U(IV)	$2Ce^{4+} + U(IV) \rightleftharpoons 2Ce^{3+} + U(VI)$
Mn^{3+}	$Mn^{2+} \rightleftharpoons Mn^{3+} + e$	As(III)	$2Mn^{3+} + As(III) \rightleftharpoons 2Mn^{2+} + As(V)$
		$H_2C_2O_4$	$2Mn^{3+} + H_2C_2O_4 + 2H_2O \rightleftharpoons 2Mn^{2+} + 2CO_2 + 2H_3O^+$
Ag^{2+}	$Ag^+ \rightleftharpoons Ag^{2+} + e$	Ce^{3+}	$Ag^{2+} + Ce^{3+} \rightleftharpoons Ag^+ + Ce^{4+}$
		V(IV)	$Ag^{2+} + V(IV) \rightleftharpoons Ag^+ + V(V)$
		$H_2C_2O_4$	$2Ag^{2+} + H_2C_2O_4 + 2H_2O \rightleftharpoons 2Ag^+ + 2CO_2 + 2H_3O^+$
$CuCl_2^-$	$Cu^{2+} + 2Cl^- + e \rightleftharpoons CuCl_2^-$	Br_2	$2CuCl_2^- + Br_2 \rightleftharpoons 2Cu^{2+} + 2Br^- + 4Cl^-$
		$Cr_2O_7^{2-}$	$6CuCl_2^- + Cr_2O_7^{2-} + 14H_3O^+ \rightleftharpoons 6Cu^{2+} + 2Cr^{3+} + 12Cl^- + 21H_2O$
		V(V)	$CuCl_2^- + V(V) \rightleftharpoons Cu^{2+} + 2Cl^- + V(IV)$
Fe^{2+}	$Fe^{3+} + e \rightleftharpoons Fe^{2+}$	Ce^{4+}	$Fe^{2+} + Ce^{4+} \rightleftharpoons Fe^{3+} + Ce^{3+}$
		$Cr_2O_7^{2-}$	$6Fe^{2+} + Cr_2O_7^{2-} + 14H_3O^+ \rightleftharpoons 6Fe^{3+} + 2Cr^{3+} + 21H_2O$
		MnO_4^-	$5Fe^{2+} + MnO_4^- + 8H_3O^+ \rightleftharpoons 5Fe^{3+} + Mn^{2+} + 12H_2O$
Ti^{3+}	$TiO^{2+} + 2H_3O^+ + e \rightleftharpoons Ti^{3+} + 3H_2O$	U(VI)	$2Ti(III) + U(VI) \rightleftharpoons 2Ti(IV) + U(IV)$
		V(V)	$2Ti(III) + V(V) \rightleftharpoons 2Ti(IV) + V(III)$ or $Ti(III) + V(V) \rightleftharpoons Ti(IV) + V(IV)$
		Mo(VI)	$Ti(III) + Mo(VI) \rightleftharpoons Ti(IV) + Mo(V)$

the electrogenerated H_3O^+ and OH^- would neutralize each other instead of the analyte. Because of the high mobility of H_3O^+ and OH^-, the usual isolation chamber shown in Figure 16-2 is not adequate. One technique is to use external generation of the reagents, as shown in Figure 16-7. This method is inconvenient and destroys some of the advantages of coulometric titrations over conventional titrations. However, it should be recalled that essentially all acid and base standard solutions must be standardized, whereas no standardization is required for a coulometric titration.

Alternatively, an innocuous electrode reaction can be carried out in the isolation chamber to avoid the necessity of external generation. For example, if generation of OH^- at the cathode is desired, a silver metal anode can be used in a solution containing some Br^- (or Cl^-). The anode reaction taking place will then be

$$Ag + Br^- \rightleftharpoons AgBr + e$$

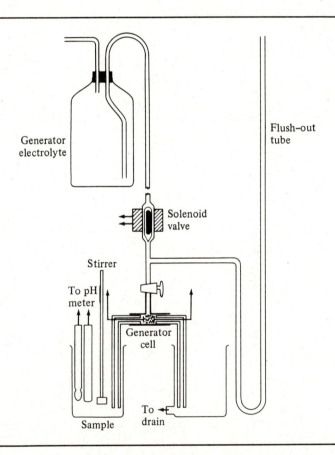

FIGURE 16-7 Apparatus for automatic coulometric titration with externally generated reagents.

The endpoint signal for an acid–base coulometric titration is typically obtained by a glass electrode and a pH meter, but color indicators have also been employed. In addition to the advantage that no standardization of reagents is needed for coulometric titration, the problems associated with CO_2 in base solutions and discussed in Section 9-6 are easily avoided. Carbon dioxide in the analyte solution is removed by bubbling in an inert gas or by boiling before generation of the OH^- coulometric titrant.

Complexation Reactions Generation of a ligand to carry out a complexation titration is not so feasible as is the employment of reagents for other reaction types. However, one example can be mentioned: The complex of ammine mercury with EDTA binds EDTA so tightly that it cannot form a complex with other metal ions in solution; but if the complex is reduced to mercury metal at a cathode, the EDTA is released as shown in Equation 16-16,

$$HgNH_3Y^{2-} + NH_4^+ + 2e \rightleftharpoons Hg + 2NH_3 + HY^{3-} \qquad \text{(16-16)}$$

The electrogenerated HY^{3-} can then be used to titrate Ca^{2+}, Cu^{2+}, Pb^{2+}, or Zn^{2+} It should be noted from Equation 16-16 that it takes two electrons to produce each HY^{3-}, which in turn forms 1:1 complexes with metal ions. Therefore, $n = 2$ for all metal ions titrated with electrogenerated EDTA. Reactions are summarized in Table 16-1.

Redox Reactions Coulometric titrations are especially well suited for carrying out redox titrations. Most of the redox reagents discussed in Chapter 11 can be generated at an anode (oxidizing agent) or a cathode (reducing agent), and several other reagents that are too unstable to be used in conventional titrations may be electrogenerated. These include Br_2, Cl_2, Mn^{3+}, Ag^{2+}, Ti^{3+}, and Cr^{2+}.

Back-titration techniques may also be carried out coulometrically. For example, Br^- and Cu^{2+} may be added to a solution containing an organic analyte such as aniline. Excess bromine is generated at the anode, which reacts with aniline according to Equation 16-17:

$$\text{(16-17)}$$

After reaction is complete, polarity of the electrode is reversed, and Cu^+ is generated by reduction of Cu^{2+}. The electrogenerated Cu^+ reduces excess Br_2:

$$2Cu^+ + Br_2 \rightleftharpoons 2Cu^{2+} + 2Br^-$$

An amperometric endpoint has been used for this coulometric titration. Several examples of coulometric titrations involving redox reactions are given in Table 16-1.

16-2   CONTROLLED-POTENTIAL ELECTROLYSIS

In Section 16-1, it was pointed out that an analyte cannot be completely oxidized (or reduced) at an electrode under constant-current conditions. As the reactant concentration decreases, a point will be reached when insufficient reactant to satisfy the current occurs, and the potential of the electrode will shift to some other electrochemical couple to maintain constant-current conditions. Hence the need for an intermediate for a coulometric titration. However, if, instead of controlling the current, the analyst controlled the potential, he would find it possible to exhaustively oxidize (or reduce) an analyte directly at an electrode.

Let us again consider the oxidation of Fe^{2+}. Electrochemical potentials are shown in Figure 16-8; if the platinum electrode is kept at some point above

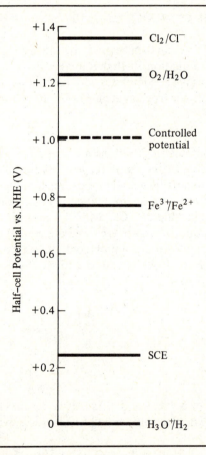

FIGURE 16-8   Diagram of half-cell potentials, showing controlled-potential electrolysis value of $+1.007$ V vs. NHE to oxidize Fe^{2+} to Fe^{3+}.

$+0.77$ V vs. NHE but below $+1.23$ V vs. NHE, only Fe^{2+} may be oxidized because the electrode is not sufficiently anodic to oxidize H_2O or Cl^-. The actual potential required for quantitative oxidation of Fe^{2+} can be calculated with the Nernst expression for the Fe^{3+}/Fe^{2+} couple:

$$Fe^{3+} + e \rightleftharpoons Fe^{2+}$$

$$E = E_{Fe}^0 - \frac{0.059}{1} \log \frac{[Fe^{2+}]}{[Fe^{3+}]} \qquad (16\text{-}18)$$

As can be seen from Equation 16-18, it is not possible to oxidize Fe^{2+} *completely* because, if its activity is reduced to zero, the potential would need to be $+\infty$ because $\log 0$ approaches $-\infty$. If the potential is higher than $+1.23$ or $+1.36$ V vs. NHE, oxidation of H_2O or Cl^- would occur. On the other hand, *quantitative* oxidation of Fe^{2+} can be accomplished if $[Fe^{2+}]/[Fe^{3+}] < 10^{-3}$. For a measure of safety, let us use $[Fe^{2+}]/[Fe^{3+}] = 10^{-4}$. Thus

$$E = +0.771 - 0.059 \log \frac{[Fe^{2+}]}{[Fe^{3+}]}$$

$$E = +0.771 - 0.059 \log 10^{-4} = +0.771 + 0.236 = +1.007 \text{ V vs. NHE}$$

If the potential of the electrode is maintained at $+1.007$ V vs. NHE, only Fe^{2+} may be oxidized, and this oxidation will be quantitative.

However, the current will not be constant. At the beginning of the electrolysis, the concentration of Fe^{2+} at the electrode surface will be high; therefore, the current will be high. As the concentration of Fe^{2+} decreases, the current will also decrease. Finally, when essentially all the Fe^{2+} has been oxidized, the current will be essentially zero. The current as a function of time will have the shape shown in Figure 16-9. The area under the curve is equal to Q, the number of coulombs required to oxidize Fe^{2+}. Therefore, at the end of electrolysis,

$$\text{meq Fe} = \text{meq charge} = \frac{Q \text{ (mC)}}{9.65 \times 10^4 \text{ (mC/meq)}} \qquad (16\text{-}19)$$

A typical calculation is given in Example 16-3.

Sometimes the electrolysis current will not drop completely to zero but will reach a reasonably steady *residual current*. This may arise from some other species' reacting slowly at the electrode, for example, Cl^- in the iron example. Even though the potential is maintained below E^0, that value refers only to all species at unit activity. At $+1.007$ V vs. NHE, we can see that

$$Cl_2 + 2e \rightleftharpoons 2Cl^-$$

$$E = E_{Cl}^0 - \frac{0.059}{2} \log \frac{[Cl^-]^2}{[Cl_2]}$$

$$1.007 = 1.359 - \frac{0.059}{2} \log \frac{[Cl^-]^2}{[Cl_2]}$$

EXAMPLE 16-3

A 497.3-mg sample of an iron ore is dissolved in acid, reduced in a Jones reductor to give Fe^{2+}, and diluted to 1 L. A 10.00-mL aliquot is electrolyzed at $+1.007$ V vs. NHE. The current as a function of time is plotted as in Figure 16-9, with 1 cm^2 of graph paper $= 80$ mA-sec (mC). If the total number of squares (square millimeters) is 5474, calculate the percentage of Fe in the ore.

Solution

From Equation 16-19 and the fact that Fe^{2+} is oxidized to Fe^{3+} in the electrolysis reaction ($n = 1$),

$$\text{meq Fe} = \frac{Q}{9.65 \times 10^4}$$

$$\frac{\text{mg Fe}}{\text{FW (Fe)}/1} = \frac{Q}{9.65 \times 10^4}$$

$$\text{Area} = \frac{5474 \text{ mm}^2}{100 \text{ mm}^2/\text{cm}^2} = 54.74 \text{ cm}^2$$

$$Q = 54.74 \text{ cm}^2 \times 80 \frac{\text{mC}}{\text{cm}^2} = 4379 \text{ mC}$$

$$\frac{\text{mg Fe}}{55.85} = \frac{4379}{9.65 \times 10^4}$$

$$\frac{\text{mg Fe}}{\text{aliquot}} = 2.534 \text{ mg}$$

(Note the very small amount of Fe needed for the analysis.)

$$\frac{\text{mg Fe}}{\text{sample}} = 2.534 \times \frac{1000}{10} = 253.4 \text{ mg}$$

$$\% \text{ Fe} = \frac{253.4 \text{ mg Fe}}{497.3 \text{ mg sample}} \times 100 \, (\%) = 50.96\%$$

If the solution happens to contain $[Cl^-] = 0.1$,

$$\log \frac{(0.1)^2}{[Cl_2]} = \frac{2(1.359 - 1.007)}{0.059} = 11.93$$

$$\frac{(0.1)^2}{[Cl_2]} = 10^{11.93} = 8.6 \times 10^{11}$$

$$[Cl_2] = \frac{10^{-2}}{8.6 \times 10^{11}} = 1.2 \times 10^{-14}$$

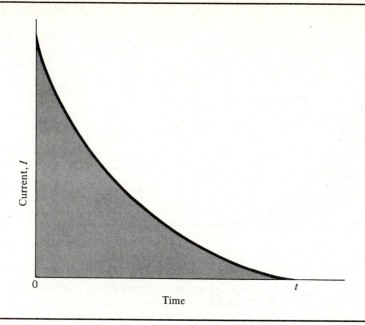

FIGURE 16-9 Controlled-potential electrolysis current I as a function of time. Coulombs of charge Q = area = $\int_0^t I(t)\, dt$.

This value is essentially zero; yet a very small amount of current could flow to produce this activity of $[Cl_2]$ at the electrode surface. In cases where such a background reaction results in significant residual currrents, the area due to analyte can be found as shown in Figure 16-10. The charge from electrolysis of the analyte is the shaded area, and it can be calculated by subtracting from the total area the area due to the background reaction assumed to occur at essentially constant current $(I_r t)$.

INSTRUMENTATION

It can be seen from the iron example that careful control of potential is essential to the success of a controlled-potential electrolysis. If the potential in the iron electrolysis dropped to $+900\ \text{mV}$ vs. NHE,

$$\log \frac{[Fe^{2+}]}{[Fe^{3+}]} = \frac{+0.900 - 0.771}{-0.059} = -2.19$$

$$\frac{[Fe^{2+}]}{[Fe^{3+}]} = 6.5 \times 10^{-3}$$

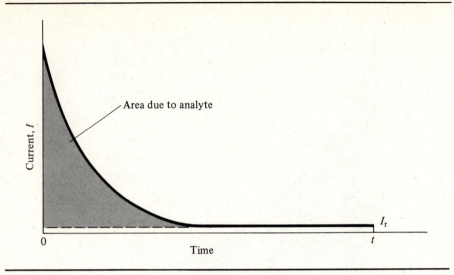

FIGURE 16-10 Controlled-potential electrolysis current I as a function of time, with a residual current I_r. Area due to analyte $= Q = \int_0^t I(t)\, dt - I_r t$.

resulting in a negative error of $>0.6\%$ because a significant fraction of the Fe^{2+} is not oxidized. On the other hand, an increase in potential to >1.2 V vs. NHE would lead to some oxidation of H_2O and/or Cl^-, resulting in a positive error. Simply applying a voltage of $+1.007$ V would not be correct even if the counterelectrode were NHE. The main reason is that some of the applied voltage would represent an ohmic voltage drop in the solution because of its resistance. It is called an "ohmic drop" because it is given by Ohm's law,

$$I = \frac{V}{R} \quad \text{or} \quad V = IR$$

For example, if the solution resistance is $10\ \Omega$ and the initial current is 15 mA,

$$V = 15\ \text{mA} \times 10\ \Omega = 150\ \text{mV}$$

The potential of the anode would be only $1.007 - 0.150 = +0.857$ V vs. NHE. The applied voltage could be increased to compensate for this IR loss; but then, as the current dropped during the electrolysis, the potential of the anode would become too high. What is needed is a system to control the potential of the anode (the *working electrode*) at all times at $+1.007$ V vs. NHE. The control is accomplished with a three-electrode technique.

A three-electrode cell is shown schematically in Figure 16-11. A voltage from a variable power supply is applied between the working electrode (anode in the iron

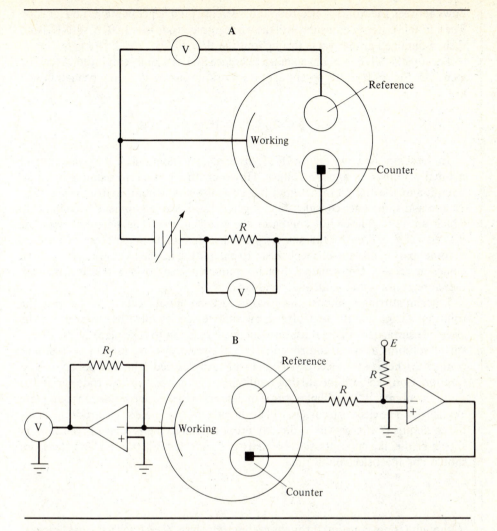

FIGURE 16-11 (**A**) Classical arrangement for a three-electrode cell. The current flowing between the working electrode and the counterelectrode is measured as the voltage drop across the standard resistor R. The variable power supply is adjusted to maintain the desired potential on the working electrode versus the reference. (**B**) Modern circuit for a three-electrode potentiostat, in which the working electrode is kept at potential E versus the reference electrode. Current at the working electrode is measured as a voltage at voltmeter $\textcircled{v}$, with $I = V/R_f$.

example) and a counterelectrode (kept in an isolation chamber to prevent undesirable side reactions with the analyte). The potential of the working electrode is monitored by measuring against a reference electrode with a high-impedance voltmeter. This is effectively a *potential* measurement because essentially no current flows between the working electrode and reference (and hence contains no IR drop). The variable power is adjusted until the working electrode is $+1.007$ V vs. NHE and then maintained at that value throughout the electrolysis.

Because the NHE is inconvenient, a reference SCE is usually employed. As can be seen from Figure 16-8, the potential of the working electrode should be maintained at

$$E = +1.007 - 0.242 = +0.765 \text{ V vs. SCE}$$

In the setup shown in Figure 16-11, the analyst could monitor the potential and manually adjust the applied voltage. However, this procedure requires constant attention by the analyst throughout the electrolysis, which normally requires 15 to 30 min and sometimes considerably longer. Commercial equipment is available in which an electronic circuit compares the working electrode potential with the desired value chosen by the analyst. Any discrepancy between these two values becomes part of a feedback mechanism to change the applied voltage until the two values are equal. Unfortunately, such equipment costs over $1000 dollars and, therefore, is not found in all analytical labs.

Efficient stirring is important for controlled-potential electrolysis. Because the solution is exhaustively electrolyzed, all analyte species must be swept up to the electrode surface to enable the electrochemical reaction to take place. With a well-stirred solution, a typical controlled-potential electrolysis may take 15 to 30 min to complete, whereas the electrolysis time can easily exceed 1 hr in a poorly stirred solution. Coupled with the stirring requirement is the ratio of electrode surface to solution volume. It too will play a role in determining the time needed for complete electrolysis. The electrical current will be directly proportional to electrode area, but the coulombs of charge will be directly proportional to the volume of solution. As shown below, the concentration of analyte, C, affects both Q and I and therefore should not affect electrolysis time:

$$Q = It$$

(Although I is not constant throughout the electrolysis, $Q = It$ for *any* small interval of time during which I is essentially constant. Functionally, then, Q equals the product It.)

$$I \propto CA$$

Therefore,

$$Q \propto CAt$$

In terms of the amount of analyte present,

$$Q \propto CV$$

Therefore,

$$\cancel{\mathscr{C}}At \propto \cancel{\mathscr{C}}V$$

$$t \propto \frac{V}{A}$$

or

$$\text{speed of electrolysis} = \frac{1}{t} \propto \frac{A \text{ (area of electrode)}}{V \text{ (volume of solution)}}$$

With the electrolysis cell and instrumentation shown schematically in Figure 16-11, we can now proceed to carry out a controlled-potential electrolysis. The next question is "How can we measure the number of coulombs of charge?" One method is to measure current as a function of time and determine the area under the curve, as shown in Figure 16-9 (or 16-10). The current can be simply measured by recording the voltage drop V across a standard resistor R in the circuit, as indicated in Figure 16-11 (recall that $I = V/R$, which is Ohm's law).

There are several ways to calculate the area under the curve. One method, which was used in Example 16-3, is to count squares of graph paper. Another is to cut out a known area of graph paper and weigh it to find milligrams per square centimeter for the paper used (it may vary from sheet to sheet). Then the area for the electrolysis (shaded area in Figure 16-9) is cut out and weighed. From these two weighings, the area is determined. A third method employs a planimeter. This mechanical device can be traced around any area; when the original starting point is reached, a dial reading gives the enclosed area. A known area must be circumscribed to calibrate the dial reading, and this standard area must be equivalent to a standard amount of charge (millicoulombs per square centimeter) determined from the graphing scale.

Instead of measuring current as a function of time, the analyst can directly measure coulombs of charge by some type of *coulometer* or *integrator*. The use of a chemical coulometer involves carrying out a chemical reaction in series with the analyte reaction. For example, a second electrolysis cell can be added to the circuit shown in Figure 16-9, in which silver is plated on the cathode. The cathode is weighed before and after electrolysis to determine the weight of silver deposited. Although the number of coulombs can be calculated, this value can be bypassed, and the number of milliequivalents of analyte is found from the milliequivalents of silver, as shown in Example 16-4. Even with the high equivalent weight of silver (107.9), it would be necessary to have at least 1 meq of analyte present to keep weighing uncertainty to $\pm 0.1\%$ with the standard laboratory balance.

Another chemical coulometer involves the electrolysis of water to give hydrogen plus oxygen; the total volume of gas produced is measured. At STP conditions, the milliequivalent volume of H_2 is $22.4/2 = 11.2$ mL, while the milliequivalent volume of O_2 is $22.4/4 = 5.60$ mL. Therefore, the H_2–O_2 coulometer yields 16.8 mL/meq after the measured volume is corrected to standard conditions. Both the silver and H_2–O_2 coulometers lose many of the advantages of electroanalytical techniques because a mass or volume measurement must be made. One way of utilizing only electrical signals with a silver coulometer is to use electrical readout, which is accomplished by plating silver on an inert electrode, such as gold. After completion

EXAMPLE 16-4

A 965-mg sample of an insecticide is digested in 150 mL of 1 F H_2SO_4, arsenic being brought into solution as As(III). The solution is transferred to an electrolysis cell, and controlled-potential electrolysis is carried out at $+1.0$ V vs. SCE. A silver coulometer is placed in series with the electrolysis cell. From the weight changes of the coulometer cathode, calculate the percentage of As_2O_3 in the insecticide.

Solution

wt of cathode after electrolysis: 22.1613 g

wt of cathode before electrolysis: <u>22.0947 g</u>

wt change: 0.0666 g

$$\text{meq As} = \text{meq Ag}$$

$$\frac{\text{mg As}_2O_3}{\text{FW (As}_2O_3)/n_{As}} = \frac{\text{mg Ag}}{\text{FW (Ag)}/n_{Ag}} = \frac{66.6}{107.9/1}$$

In the reaction, As(III) is oxidized to As(V). Therefore, $n = 2$ for each arsenic atom, and since As_2O_3 contains two arsenic atoms, $n = 4$ for As_2O_3.

$$\frac{\text{mg As}_2O_3}{197.8/4} = \frac{66.6}{107.9/1}$$

$$\text{mg As}_2O_3 = 30.5 \text{ mg}$$

$$\% \text{ As}_2O_3 = \frac{30.5 \text{ mg As}_2O_3}{965 \text{ mg sample}} \times 100 \text{ (\%)} = 3.16\%$$

of the controlled-potential electrolysis, the silver coulometer is electrolyzed in the opposite direction at a constant current I. When all the silver has been stripped from the gold electrode, the voltage climbs and this provides an endpoint signal. The coulombs of charge $Q = It$, where t is the time required to strip the silver. This technique is easily accurate to about $\pm 2\%$ even in the millicoulomb range (10^{-5} meq).

Mechanical and electronic integrators are now readily available although they add significantly to the cost of equipment. Recorders with built-in ball-and-disk integrators provide a readout of the area under the curve. They are accurate to about 1 or 2% and can be used for very small values of charge. Probably the most elegant method for determining coulombs of charge for a controlled-potential electrolysis makes use of an electronic integrator. Available on commercial instruments, such integrators can yield values of charge with accuracies better than $\pm 0.5\%$.

If the analyst knows from experience that the controlled-potential electrolysis represents a single electrochemical step and gives the current–time curves theoretically expected, a complete electrolysis is not needed to determine the number of coulombs. It has been found from theory and experiment that well-behaved controlled-potential electrolyses obey Equation 16-20,

$$I = I_0 e^{-kt} \qquad \text{(16-20)}$$

where I_0 is the initial current and k is a constant that depends on electrode area and solution volume, stirring, and the diffusion coefficients for the reacting species. In logarithmic form,

$$\ln I = \ln I_0 - kt$$

which is a straight-line form, with a slope of $-k$ and an intercept of $\ln I_0$ when $\ln I$ is plotted against t.

The amount of charge is given by

$$Q = \int_0^t I \, dt = \int_0^t I_0 e^{-kt} \, dt$$

The value of this integral is

$$Q = \frac{I_0}{k} (1 - e^{-kt})$$

For complete electrolysis, t is long so that e^{-kt} is much less than 1. Therefore,

$$Q = \frac{I_0}{k} \qquad \text{(16-21)}$$

From at least two measurements of current as a function of time, the values of I_0 and k can be obtained from the plot shown in Figure 16-12(B). The amount of charge

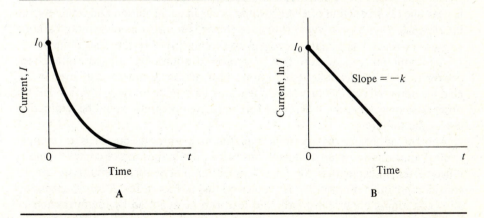

FIGURE 16-12   **(A)** Ideal current–time curve for controlled-potential electrolysis; **(B)** ln I–time line for ideal current–time curve.

is then calculated from Equation 16-21. Although this method is not of high accuracy (typically, $\pm 5\%$), the value of Q can be found after a few minutes of electrolysis instead of the delay for complete electrolysis. This method is especially useful for slow electrolyses, which might take hours to complete. However, it should be kept in mind that slow electrolyses are frequently the result of complicated electrode reactions, in which case Equation 16-20 may not necessarily be obeyed.

APPLICATIONS

Controlled-potential electrolysis is capable of determining small amounts of analytes (0.01 to 1 meq) with a reasonably high degree of accuracy ($\pm 0.5\%$). With care, accuracies of $\pm 0.1\%$, comparable to gravimetric methods, can be attained. In the typical cell, a sample solution of 50 to 200 mL is needed to enable the large working electrode to be totally immersed. Smaller volumes can be handled, especially when a mercury pool cathode is used as the working electrode.

One of the most useful areas of application is the cathodic reduction of metal ions to the metallic state. The working electrode may be platinum or a mercury pool. The possibility of determining several metal ions in the same solution can be seen from Figure 16-13. If the cathode is controlled at about ± 0.25 V vs. NHE ($+0.008$ V vs. SCE), shown as dashed line (1) in Figure 16-13, the potential is sufficiently cathodic to reduce Cu^{2+} to Cu^0 but is not sufficiently cathodic to reduce any of the other metal ions shown. After the current decays to zero and the number of coulombs corresponding to the copper content has been determined, the potential is decreased to dashed line (2) in Figure 16-13. Now the potential is low enough (~ 0 V vs. NHE) to reduce Bi(III), as $BiCl_4^-$, to Bi^0. A controlled-potential electrolysis at this potential determines the bismuth content. Following this electrolysis, the potential is dropped to dashed line (3) in Figure 16-13 to determine lead, and finally to dashed line (4) to determine zinc. This particular group of metals is frequently found in various copper-based alloys, such as brass.

A few details must be noted. As can be seen in Figure 16-13, the hydrogen couple is near line (2). Reduction of H_3O^+ would occur in acid solution and represents an interference. One way to avoid this interference is to use a mercury pool cathode because mercury has a high hydrogen overvoltage; that is, the electrokinetics of H_3O^+ reduction are so slow that a significant additional applied voltage (approximately 1 V) is necessary to carry out this reaction. In other words, even in 1 F acid the potential of a mercury electrode would have to be below -1 V vs. NHE for observation of reduction of H_3O^+. More negative potentials would be required in less acidic solutions.

Another approach that permits a platinum working electrode is to employ neutral solutions and keep the metal ions in solution in the form of complex ions. It must be remembered that the formal potentials under these conditions can be considerably different from the E^0 values quoted in Figure 16-13. For example, an alloy containing copper, bismuth, and lead can be analyzed by the formation of tartrate complexes with the adjustment of the pH to 6. Copper is reduced at the platinum electrode at -0.30 V vs. SCE, followed by bismuth at -0.40 V and lead at -0.60 V. The formation of complexes requires more negative potentials to bring

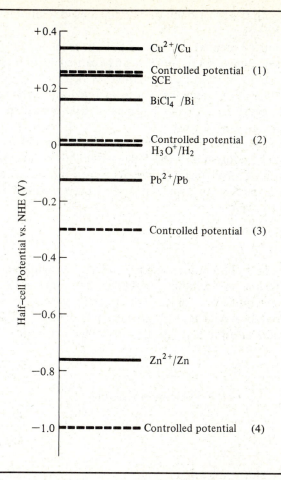

FIGURE 16-13 Diagram of half-cell potentials, showing multistep controlled-potential electrolysis. At (1), Cu^{2+} is reduced to Cu; at (2), $BICl_4$, to BI; at (3), Pb^{2+}, to Pb; and at (4), Zn^{2+}, to Zn.

about reduction as expected. This type of procedure was mentioned in Section 6-5 as electrogravimetry, in which the electrode is weighed after each electrolysis to determine the amount of each analyte present. The coulometric method does not require removing the electrode after each electrolysis, drying, and weighing; so it is considerably faster. In addition, much smaller amounts of material can be determined when coulombs are measured instead of mass. And, finally, it is far easier to use a mercury pool cathode in the coulometric technique. On the other hand, unlike the coulometric method, the electrogravimetric method does not require 100% current efficiency.

We shall encounter the controlled-potential electrolysis of metal ions in Chapter 17 as part of the anodic stripping technique. In this application, controlled-potential electrolysis concentrates trace amounts of metal ions in solution by reducing them to the metal at an electrode.

Not all controlled-potential electrolyses need be reduction of metal ions to the metallic state. Reduction of UO_2^{2+} to U^{4+} can be carried out in H_2SO_4 at a mercury pool cathode (-0.6 V vs. SCE) as a method for determining uranium. Samples containing 7 to 75 mg of uranium have been analyzed with an accuracy of $\pm 0.1\%$. In a citrate medium at pH 4.5, quantities as small as 0.0075 mg of uranium have been determined with about -3% error.

In the previous discussion, it was mentioned that Fe^{2+} could be oxidized to Fe^{3+} and As(III) could be oxidized to As(V). Other oxidations include the determination of halide ions with a silver metal anode. As shown in Figure 16-14, a potential controlled above the AgI/Ag couple at dashed line (1) will allow the oxidation of silver to AgI to occur:

$$Ag + I^- \rightleftharpoons AgI + e$$

This potential is not sufficiently anodic to produce any significant amount of free Ag^+ or to cause the corresponding reactions with Br^- or Cl^-. More anodic potentials can determine Br^- or Cl^-. Some mixtures can be handled, but the difference between the AgCl and AgBr potentials is not large—which means that it is difficult to produce AgBr quantitatively without the formation of AgCl. Also, AgI

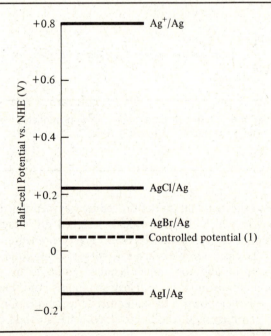

FIGURE 16-14 Diagram of half-cell potentials, showing possibility of controlled-potential electrolysis of I^- in the presence of other halides by the formation of AgI at 1.

tends to coprecipitate halide ions; the coprecipitation can lead to negative errors. Actually, iodide can be determined more easily by oxidation to I_2 at $+0.7$ V vs. SCE. This is a useful method for determining the amount of iodide in "iodized" salt.

Several organic electrochemical reactions have been carried out by controlled-potential electrolysis. Frequently, these reactions are used not for analytical purposes but for determining the n value for the reaction. With this knowledge, the electrochemical reaction may be deduced. For example, picric acid is reduced at a mercury pool cathode in a reaction in which $n = 18$ (the earliest papers reported $n = 17$, which was an error). The reaction product is concluded to be triaminophenol, as shown in Equation 16-22,

$$O_2N\text{—}\overset{\displaystyle OH}{\underset{\displaystyle NO_2}{\bigodot}}\text{—}NO_2 + 18H_3O^+ + 18e \rightleftharpoons H_2N\text{—}\overset{\displaystyle OH}{\underset{\displaystyle NH_2}{\bigodot}}\text{—}NH_2 + 24H_2O \quad \textbf{(16-22)}$$

picric acid   triaminophenol

This method of determining n values (Equation 16-23),

$$\text{meq analyte} = \text{meq charge}$$

$$\frac{mg}{FW/n} = \frac{Q}{9.65 \times 10^4} \qquad \textbf{(16-23)}$$

$$n = \frac{Q \cdot FW}{9.54 \times 10^4 \cdot mg}$$

is especially useful for studying new electrochemical reactions taking place in nonaqueous solvents.

Because the method electrolyzes the entire sample, it is also useful for synthesizing new species or for producing known amounts of unstable reaction products. The electron itself is the reagent for carrying out these reactions: No chemical reagents with their reaction products contaminate the solution. Thus controlled-potential electrolysis finds application not only as an analytical technique but also as a tool for studying electrochemical reactions and synthesizing novel chemical species.

SUPPLEMENTARY READING

Bard, A.J., and L.R. Faulkner, *Electrochemical Methods, Fundamentals and Applications.* New York: John Wiley & Sons, Inc., 1980.

E. Bishop, "Coulometric Analysis," in *Comprehensive Analytical Chemistry*, vol. 2D. C.L. Wilson and D.W. Wilson, eds., New York: Elsevier North-Holland, Inc., 1975.

Lingane, J.J., *Electroanalytical Chemistry*, 2nd ed. New York: Interscience Publishers, Inc., 1954.

Rechnitz, G.A., *Controlled-Potential Analysis.* New York: Macmillan, Inc., 1963.

PROBLEMS

16-1. List the advantages and disadvantages a coulometric titration has over a conventional volumetric titration.

16-2. What are the coulometric titration terms analogous to the conventional volumetric titration terms given below?
   a. normality of standard solution
   b. volume of standard solution
   c. endpoint signal
   d. buret stopcock
   e. volume uncertainty
   f. concentration uncertainty

16-3. Compare the overall relative uncertainty in a volumetric titration with a coulometric titration, given the following estimates (both titrations have about equal milliequivalents): volumetric titration, 20 ± 0.02 mL each volume reading, ± 0.04 mL endpoint, 0.02000 ± 0.00002 N reagent; coulometric titration, 50 ± 0.02 mA, 800 ± 0.2 sec, ± 1-sec endpoint.

*16-4. A constant-current supply is constructed with a 90-V battery and a resistance of 8000 Ω. If cell voltage variation during a coulometric titration is ± 0.3 V and resistance variation is ± 20 Ω, calculate the uncertainty in the constant-current value.

16-5. A 3.67-g sample of an insecticide was decomposed in acid, any As(V) reduced to As(III), and diluted to 250 mL in a volumetric flask. A 5-mL aliquot was added to 125 mL of 0.05 F KI buffered to pH 7. A coulometric titration was carried out with electrogenerated I_3^-, which oxidized arsenic to As(V). It required 287 sec at 24.25 mA to reach the starch–iodine endpoint signal. Calculate the percentage of As_2O_3 in the insecticide.

16-6. A mineral sample (465 mg) containing stibnite (Sb_2S_3) is decomposed, dissolved in acid, and diluted to 100 mL. A 5-mL aliquot is added to 150 mL of an electrolyte containing 2 F HCl and 0.2 F KBr. Electrogenerated bromine oxidizes Sb(III) to Sb(V), and the coulometric titration requires 256.7 sec at 48.50 mA to reach the amperometric endpoint signal. Calculate the percentage of stibnite in the sample.

*16-7. The Al in a 55.3-mg sample was precipitated with 8-hydroxyquinoline (Q). The precipitate (AlQ_3) was filtered, washed, and redissolved, and the Q was brominated with electrogenerated bromine (see equation below). With a current of 150 mA, the coulometric titration endpoint was reached in 647 sec. Calculate the percentage of Al_2O_3 in the sample.

---

*Answers to problems marked with an asterisk will be found at the back of the book.

16-8. Water containing H_2S can be analyzed by a coulometric titration in which H_2S is oxidized by electrogenerated I_3^- according to the equation

$$H_2S + I_3^- + 2H_2O \rightleftharpoons S + 2H_3O^+ + 3I^-$$

To a 100-mL water sample is added 2 g KI and some starch indicator. If the coulometric titration requires 552 sec at 73.1 mA, determine the H_2S content in the water sample as parts per million (milligrams per liter).

*16-9. The sulfur in an organic compound was determined by burning a 47.1-mg sample in a stream of oxygen and collecting the resulting SO_2 in a neutral solution of H_2O_2, which converted the SO_2 to H_2SO_4:

$$SO_2 + H_2O_2 \rightleftharpoons H_2SO_4$$

The sulfuric acid was titrated coulometrically by generation of hydroxide ions from water at a platinum electrode. With a current of 48.0 mA, 10.20 min were required to reach a phenolphthalein endpoint. Calculate the percentage of sulfur in the sample.

16-10. A 100-mL water sample containing trace amounts of Fe(III) is transferred to an electrolysis cell, and 80 mL of H_2SO_4 plus 20 mL of 0.5 F Ti(IV) is added. A coulometric titration is carried out at 5.67 mA with electrogenerated Ti(III), which reduces iron to Fe(II). An amperometric endpoint signal is used with data given below. Determine the iron level in the water as parts per million of Fe.

time(sec)	$I(\mu A)$	time(sec)	$I(\mu A)$	time(sec)	$I(\mu A)$
110	20	125	8	140	8
115	16	130	5	145	14
120	12	135	4	150	20

16-11. Phenol is easily brominated according to the equation

$$C_6H_5OH + 3Br_2 \rightleftharpoons C_6H_2Br_3OH + 3HBr$$

which forms the basis for a coulometric titration with electrogenerated bromine. A 200-mL sample of polluted H_2O is acidified, and 5 g KBr is added. The coulometric titration requires 362 sec at 28.6 mA to reach the amperometric endpoint. Calculate the phenol content in the sample as parts per million (milligrams per liter).

*16-12. A 965-mg sample of uranium ore is dissolved in acid and diluted to 100 mL. A 5.00-mL aliquot is added to 8 F H_2SO_4 containing Ti(IV). A coulometric titration is carried out at 25.67 mA with electrogenerated Ti(III), which reduces U(VI) to U(IV) in the presence of Fe(II) catalyst. An amperometric endpoint is used with the data given below. Calculate the percentage of U_3O_8 in the ore sample.

time(sec)	μA	time(sec)	μA
240	0	255	4
245	0	260	11
250	0	265	18
		270	25

16-13. A 4.5-g sample of Al_2O_3 catalyst containing Tl is fused in alkali and then dissolved. The resulting solution is diluted to 100 mL. A 10-mL aliquot is added to 1 F H_2SO_4 and electrolyzed at $+1.34$ V vs. SCE to oxidize Tl(I) to Tl_2O_3. The electrolysis data are tabulated below. Plot the data and determine the area under the curve by counting squares or weighing the paper. Calculate the percentage of Tl in the catalyst.

Time(min)	I(mA)	Time(min)	I(mA)	Time(min)	I(mA)
0	4.80	30	1.90	70	0.52
5	3.90	40	1.50	80	0.30
10	3.30	50	1.15	87	0.0
20	2.50	60	0.78		

16-14. Mercaptans (RSH) react rapidly with Ag^+ to form insoluble RSAg, and their presence in petroleum fuels can be determined by coulometric titration with electrogenerated Ag^+ and a Ag anode. An ammoniacal electrolyte is pretitrated to some arbitrary amperometric current value, and then 1 mL of fuel-oil sample is added. If a titration at 4.57 mA requires 86.5 sec to reach the amperometric endpoint signal, calculate RSH concentration in the oil as parts per million of CH_3SH (micrograms per gram), assuming the oil density is 0.75 g/mL.

16-15. Copper content in an outlet line from a plating plant is determined by generating EDTA from $HgNH_3Y^{2-}$. If a 50-mL sample required 252 sec at 21.67 mA to reach the endpoint signal produced with a copper ion-selective electrode, calculate the copper content as parts per million of Cu.

*16-16. A 35.7-mg sample of a purified organic base is dissolved in 150 mL of alcohol–water solvent, and H_3O^+ is generated externally at 49.3 mA and delivered to the solution. If 475 sec are required to reach the methyl orange endpoint, calculate the equivalent weight (EW) of the organic base.

16-17. If, in Problem 16-16, the equivalence point is actually at pH 5.0 but the methyl orange endpoint is at pH 4.0, calculate the endpoint error and the corrected equivalent weight. Assume no volume change with addition of the electrogenerated reagent.

16-18. A 5.00-mL sample of liquid bleach (containing NaOCl) is diluted to 250 mL in a volumetric flask. Then a 5.00-mL aliquot of the diluted bleach is transferred to an electrolysis cell containing a KI electrolyte. A 10.00-mL aliquot of $Na_2S_2O_3$ solution is added to reduce I_3^- produced by the reaction

$$OCl^- + 3I^- + 2H_3O^+ \rightleftharpoons Cl^- + I_3^- + 3H_2O$$

and provide an excess. The excess $Na_2S_2O_3$ is oxidized by electrogenerated I_3^-, and 162.5 sec at 49.57 mA is required to reach the starch–iodine endpoint signal. A 5.00-mL aliquot of $Na_2S_2O_3$ (and no bleach) requires 247.5 sec at 49.57 mA to reach the starch–iodine endpoint signal. Calculate the percentage weight per volume of NaOCl in the bleach sample.

16-19. A 288-mg organic sample containing aniline is dissolved in alcohol and diluted to 250 mL in a volumetric flask. A 10-mL aliquot is added to an electrolysis cell containing 1 F HCl, 0.1 F NaBr, and 0.02 F $CuSO_4$. Electrogenerated Br_2 is produced at 15.67 mA for 250.1 sec, and the bromination reaction shown in Equation 16-17 is allowed to proceed. Polarity is reversed, and a back-titration with electrogenerated Cu(I) is carried out to reduce the excess Br_2. If the back-titration requires 96.8 sec at 15.67 mA, calculate the percentage of aniline in the sample.

*16-20. A 3.91-g sample of iodized salt is dissolved in H_2O and electrolyzed at $+0.72$ V vs. SCE to oxidize I^- to I_2. A charge of 257 mC passes before the current reaches its residual current value of 14 μA. Electrolysis time is 35 min. Calculate the percentage of KI in the salt sample.

16-21. A sample of brass weighing 345 mg is dissolved in acid and diluted to 500 mL. A 10-mL aliquot is transferred to an electrolysis cell containing a tartrate electrolyte at pH 6. The solution is electrolyzed at -0.3 V vs. SCE to reduce copper(II), which requires 15.86 C. Then electrolysis at -0.4 V vs. SCE is carried out to reduce bismuth(III), which requires 475 mC. Finally, electrolysis at -0.6 V vs. SCE is carried out to reduce lead(II), which requires 876 mC. Calculate the percentage of Cu, Bi, and Pb in the brass sample.

*16-22. Methyl iodide, CH_3I, may be reduced at -2 V vs. SCE to yield methane(CH_4). A 681-mg sample is dissolved in a dioxane–water solvent, and controlled-potential electrolysis is carried out with a Ag coulometer. If the Pt electrode in the coulometer gains 35.8 mg, calculate the percentage of CH_3I in the sample.

16-23. A 16-mg sample of 2,3-dibromobutane is reduced, using controlled-potential electrolysis. If 14.17 C of charge is required, determine the n value for the reaction. Postulate an electrochemical reaction consistent with this n value.

16-24. A sample of quinone is reduced, using controlled-potential electrolysis. If 45 min is required to reach the residual current level of 28 μA and the total charge recorded is 35.72 C, calculate the residual current correction, corrected charge, and the relative determinate error (percent) that would be introduced if the correction were not made.

16-25. A 98.7-mg ore sample containing iron is dissolved in acid, and Fe(III) is reduced to Fe(II), using controlled-potential electrolysis. The initial current is 75 mA, and 5 min later the current is 48 mA. Calculate the percentage of Fe_2O_3 in the ore.

Voltammetry | **17**

Now we take up electroanalytical techniques in which current is measured as a function of voltage (potential). The general term for these techniques is *voltammetry*, and because voltammetric methods measure analyte *concentration*, they fall into the class II described in Chapter 15. The most widely used voltammetric technique has been given the name *polarography*; developed by J. Heyrovsky over 50 yr ago, its importance to chemistry is attested to by his having received the Nobel prize in 1959 for his work. Much of this chapter will be devoted to a discussion of polarography.

17-1 CURRENT–POTENTIAL CURVES

To understand any voltammetric technique, we must consider what a current–potential curve will look like for the sample. In other words, we apply an increasing voltage at our working electrode and observe the flowing current as a function of that voltage. Suppose that we immerse a platinum electrode in a solution of KCl and record the current as a function of voltage. To avoid problems associated with the counterelectrode and the IR drop in the solution, we could use a three-electrode setup, as shown in Figure 16-11 for controlled-potential electrolysis. However, the currents in polarography are typically so small that a voltage applied directly between the working electrode and a saturated calomel reference electrode (SCE) can be assumed to represent the potential of the working electrode versus SCE. In any case, we observe curve A in Figure 17-1 for the KCl solution. At low voltage values, nothing can be oxidized or reduced at the platinum electrode, and therefore no current flows. However, as the platinum electrode is driven more and more negative relative to the SCE, a point is reached at which reduction occurs, and a cathodic current is observed. Something is being reduced at the platinum electrode. Platinum metal and chloride ion cannot be reduced; so that something must be H_2O or K^+. Even if K^+ were reduced to K^0, the potassium would immediately react chemically with H_2O to produce H_2. Thus the reduction product is H_2.

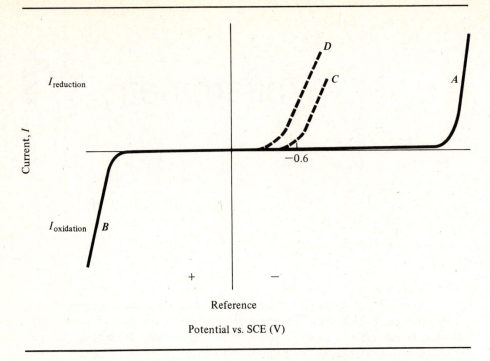

FIGURE 17-1 Current–potential curve for solution of KCl (solid curve *AB*) and for KCl containing CdCl$_2$ (broken curves *C* and *D*). Note that negative potentials are plotted toward the right and reduction currents have been given positive values.

Suppose that we reverse the polarity and drive the platinum working electrode in a positive direction with respect to the SCE. Again, no current flows until an electrochemical reaction can occur. At a sufficiently high potential, curve *B* in Figure 17-1, the electrode becomes so anodic that H$_2$O, Cl$^-$, or the platinum electrode itself can be oxidized. For example, curve *B* may be the result of Equation 17-1,

$$2Cl^- \rightleftharpoons Cl_2 + 2e \qquad \text{(17-1)}$$

Now let us add a small amount of CdCl$_2$ to the KCl solution and record the current–potential curve again (curve *C* in Figure 17-1). The only new species in the system is Cd^{2+}, which can be reduced to Cd^0 when the electrode is near, or more negative than, E^0 for the Cd^{2+}/Cd couple. The E^0 value for this couple is -0.40 V vs. NHE (-0.64 V vs. SCE). Thus, as the potential becomes more negative than E^0 is for the couple, a current flows—which was not observed for the solution containing only KCl. It should not be surprising that, if we add more CdCl$_2$ to the solution more current will flow (curve *D* in Figure 17-1). However, it will still require an applied voltage of approximately E^0 to cause current to flow. This, then, is the basis for all voltammetric techniques. Current will flow when the potential at the working electrode is sufficiently cathodic for reduction to occur or sufficiently anodic for oxidation to occur. The potential at which current begins to flow is related to E^0 for the couple and, therefore, is a *qualitative* analytical signal. The magnitude of the current is related to the concentration of the substance reacting at the electrode and, therefore, is a *quantitative* analytical signal.

17-2 POLAROGRAPHY

Although the concept described above was recognized over 100 yr ago, before a useful analytical technique could be developed, it was necessary to design conditions such that the observed current could be accurately related to the concentration of analyte in solution. The first step in achieving this objective is to use a very small working electrode compared to the counterelectrode (the reference in a traditional two-electrode polarographic setup). In this way, the working electrode will become *polarized*; that is, the current will no longer follow Ohm's law but will depend on the rate at which electroactive species can reach the working electrode surface— hence the name "polarography." In other words, for our example of $CdCl_2$ in the KCl solution, the current due to reduction of Cd^{2+} will depend on the rate at which Cd^{2+} ions can reach the electrode surface. The electrode is said to be "concentration polarized."

What are the mechanisms by which a cadmium ion out in the solution may reach the electrode surface? There are three basic mechanisms, or modes of migration:

1. **Convection** Migration due to stirring, density gradients, or temperature gradients in the solution.

2. **Electromigration** Migration due to cations' moving toward the cathode and anions' moving toward the anode. That is, when current flows, the charge is carried through the solution by ions according to their transference number (see Section 15-3).

3. **Diffusion** Migration due to a concentration gradient. That is, a substance will spontaneously migrate from a region of high concentration to a region of low concentration.

With all three of these modes of migration at work, the situation is hopelessly complex, making it impossible to relate current to concentration.

However, with careful control, the analyst can eliminate convection by *not* stirring the solution (unlike the conditions for the coulometric electroanalytical techniques discussed in Chapter 16) and by using low concentrations of analyte so that electrochemical reactions at the electrode will not lead to significant density or temperature changes.

Electromigration cannot be eliminated because, when current flows, charge must be carried through the solution. However, as we saw in Section 15-3, the current will be carried by the species with the highest transference numbers, and this depends on mobility and concentration. If a large concentration of an inert salt is present, such as KCl in our example, essentially all the charge is carried by this salt (K^+ and Cl^-) and none by Cd^{2+}; so electromigration as a means of transporting Cd^{2+} to the electrode can, therefore, be essentially eliminated. The inert salt is called a *supporting electrolyte*, and it should be present at a concentration 50 to 100 times higher than that of the analyte. Typical experimental conditions for a polarographic analysis would be 10^{-3} F analyte and 0.1 F supporting electrolyte.

With the first two modes effectively removed, the Cd^{2+} ion in our example reaches the electrode surface only via diffusion. The observed current is thus said to be *diffusion controlled*. The driving force for diffusion is a concentration gradient— the change in concentration as a function of distance. When the experiment is begun, $[Cd^{2+}]$ is the same at all points, and no diffusion will take place [Figure 17-2(A)]. However, when the potential nears E^0, some of the Cd^{2+} at the electrode surface will be reduced, and $[Cd^{2+}]$ at the electrode surface (C_0) will be lower than $[Cd^{2+}]$ is in the bulk of the solution (C_b). This situation is shown in Figure 17-2(B). Now there is a concentration gradient at the electrode given by

$$\text{concentration gradient} = \frac{\Delta C}{\Delta x} = \frac{C_b - C_0}{\delta} \tag{17-2}$$

where δ is called the "diffusion layer thickness." As the potential becomes more negative, a higher percentage of the Cd^{2+} ions at the electrode surface will be reduced, until finally the electrode is so negative that essentially all Cd^{2+} is reduced. Complete reduction leads to the highest concentration gradient and thereby the highest observed current, I_d,

$$I \propto \text{concentration gradient} = \frac{C_b - C_0}{\delta} \tag{17-3}$$

$$I_d \propto \frac{C_b - 0}{\delta} = \frac{C_b}{\delta} \tag{17-4}$$

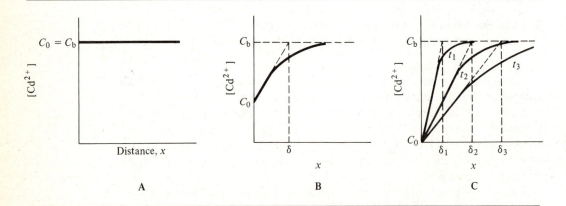

FIGURE 17-2 Concentration as a function of distance x from the electrode surface. (**A**) Potential not sufficient to cause reduction, I = 0. (**B**) Potential sufficient to reduce $[Cd^{2+}]$ at electrode surface to C_0. (**C**) Potential sufficient to reduce all Cd^{2+} at electrode surface so that $C_0 = 0$. Diffusion layer thickness δ increases with time.

shown in Figure 17-2(**C**). Since this is the maximum current that can flow, the current reaches a plateau and does not increase as the potential is made more negative. The current–potential curve will show a "wave" due to Cd^{2+} reduction, as illustrated in Figure 17-3. The plateau current, I_d, is limited by diffusion (I_d is called the "diffusion current") and represents the situation in which every Cd^{2+} reaching the electrode surface is reduced to Cd^0.

Up to this point the analysis proceeds smoothly, but one serious complication now appears: If we observe the plateau current at a fixed negative potential, the current will decrease with time: $t_1 < t_2 < t_3$ in Figure 17-2(**C**). The diffusion layer thickness increases, and thus the concentration gradient decreases. The diffusion current is still proportional to the bulk concentration at potentials sufficiently negative that C_0 approaches 0, but it also depends on other variables, including time:

$$I \propto \frac{nCD^{\frac{1}{2}}A}{t^{\frac{1}{2}}} \qquad \textbf{(17-5)}$$

where D is the diffusion coefficient for the diffusing species (Cd^{2+} in our example), A

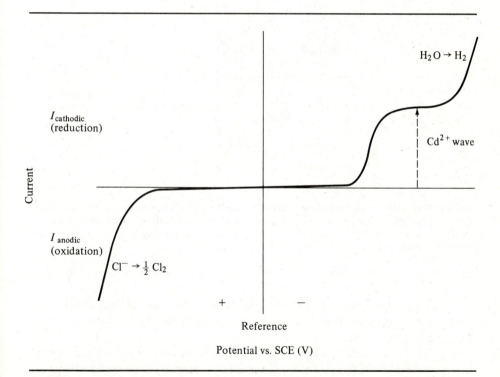

FIGURE 17-3 Current–potential curve for KCl containing $CdCl_2$, showing diffusion-controlled reduction wave for Cd^{2+}.

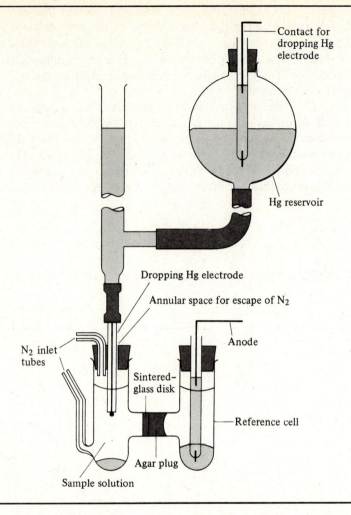

Contact for
dropping Hg
electrode

Hg reservoir

Dropping Hg electrode

Annular space for escape of N$_2$

Anode

N$_2$ inlet
tubes

Sintered-
glass disk

Reference cell

Agar plug

Sample solution

FIGURE 17-4   DME assembly and H cell for polarographic measurements. [Redrawn, with permission, from a paper by J.J. Lingane and H.A. Laitinen, "Cell and Dropping Electrode for Polarographic Analysis," *Industrial and Engineering Chemistry* (*Analytical Edition*) 11:504 (1939).]

is the area of the electrode, and t is the time.[1] The time dependence is very troublesome, especially when the voltage is scanned (changed continuously while the current is recorded).

A significant advance was Heyrovsky's study of a new electrode—the dropping mercury electrode (DME). As shown in Figure 17-4, mercury is allowed to drop into

---

[1]The derivation of the mathematical relationship shown in Equation 17-5 is beyond the scope of this text. The interested reader should consult the supplementary reading.

the solution from a capillary tube. The rate of dropping (typically 2 to 6 sec) can be controlled by the size of the capillary tube and the height of the mercury column. How does this electrode solve the time-dependency problem for the diffusion current? First, it establishes a reproducible time base. Every time a drop falls, as shown in Figure 17-5(**D**), the solution is stirred and the system returns to its initial conditions ($t = 0$). The current that flows at the moment just before the drop falls [Figure 17-5(**C**)] will be at a maximum for $t = $ drop time. This gives us a value to use for t in Equation 17-5. Second, the area is no longer constant, but it too is a function of time. With mercury flowing at a constant rate—(milligrams per second) then the area as a function of time is given by Equation 17-6, derived as follows:

Volume of Hg Drop

$$V = \tfrac{4}{3}\pi r^3$$

$$V(\text{mL}) = \frac{m\,(\text{mg/sec})\;t\,(\text{sec})}{\rho\,(\text{mg/mL})} \qquad \rho = \text{Hg density}$$

Therefore,

$$\frac{mt}{\rho} = \tfrac{4}{3}\pi r^3$$

$$r^3 = \frac{3mt}{4\pi\rho}$$

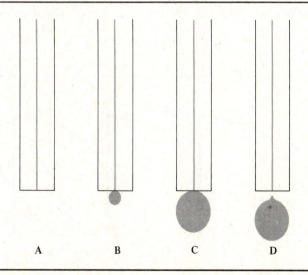

FIGURE 17-5 Growth of mercury drop at the DME: (**A**) New electrode begins to form, $I \cong 0$; (**B**) growth of drop, $I < I_{\text{max}}$; (**C**) maximum growth of drop, $I = I_{\text{max}}$; (**D**) drop dislodges, I decreases to 0, and system returns to (**A**).

Area of Hg Drop

$$A = 4\pi r^2 = 4\pi \left(\frac{3mt}{4\pi\rho}\right)^{\frac{2}{3}}$$ (17-6)

When Equation 17-6 is inserted into Equation 17-5, it is seen that the diffusion current now increases with time instead of decreasing,

$$I \propto \frac{nCD^{\frac{1}{2}}A}{t^{\frac{1}{2}}} \propto \frac{nCD^{\frac{1}{2}}m^{\frac{2}{3}}t^{\frac{2}{3}}}{t^{\frac{1}{2}}} = nCD^{\frac{1}{2}}m^{\frac{2}{3}}t^{\frac{1}{6}}$$ (17-7)

The current that flows as a function of time is shown in Figure 17-6. The current increases proportionally to $t^{\frac{1}{6}}$ as the drop grows, and then it suddenly falls to essentially zero when the drop falls and a new one begins. Most recorders do not respond quickly enough to show the drop to zero but will give a small oscillation, as Figure 17-6 (dashed curve) shows. The maximum diffusion current, $I_{d,\max}$, will be the value found from Equation 17-7, with t = drop time. When all the constants are assembled, including a correction for the curved electrode surface, Equation 17-7 takes the form known as the *Ilkovic equation* (Equation 17-8):

$$I_{d,\max}(\mu A) = 708nD^{\frac{1}{2}}m^{\frac{2}{3}}t^{\frac{1}{6}}C\,(mF)$$ (17-8)

This equation can be used to calculate the expected diffusion current when all the parameters are known, as in Example 17-1. For a particular capillary and pressure head of mercury, $m^{\frac{2}{3}}t^{\frac{1}{6}}$ should remain constant; and for a particular species, n and $D^{\frac{1}{2}}$ are also constant. Therefore, for analytical purposes, Equation 17-8 may be

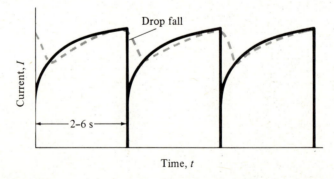

FIGURE 17-6 Current growth during three successive drops of a DME. The solid line could be observed with an oscilloscope, whereas the gray line will be observed with a conventional recorder.

EXAMPLE 17-1

(a) Calculate the diffusion current that would be expected from the reduction of $2.00 \times 10^{-3}\,F$ Pb^{2+}. The diffusion coefficient for Pb^{2+} is $1.01 \times 10^{-5}\,cm^2/sec$, and the mercury drop characteristics are $m = 1.90$ mg/sec and $t = 3.47$ sec/drop.

(b) An unknown solution containing lead gives a diffusion current of $11.7\,\mu A$ with the same drop characteristics as in (a). What is the $[Pb^{2+}]$ in this solution?

Solution

(a) From the Ilkovic Equation 17-8, we see that

$$I_{d,\text{max}} = 708\, nD^{\frac{1}{2}}m^{\frac{2}{3}}t^{\frac{1}{6}}C$$

For

$$Pb^{2+} + 2e \rightleftharpoons Pb$$

$$I_{d,\text{max}} = 708 \times 2(1.01 \times 10^{-5})^{\frac{1}{2}}(1.90)^{\frac{2}{3}}(3.47)^{\frac{1}{6}}2.00$$

$$= 17.0\,\mu A$$

Note that, when concentration is given in mF, the current will be in microamps. These are convenient units for polarographic analysis. Compare this calculated value with the actual polarogram shown in Figure 17-7.

(b) Now that we know that a $2\,mF$ solution gives a diffusion current of $17.0\,\mu A$ from part (a), Equation 17-9 can be used. From the standard solution,

$$I_{d,\text{max}} = kC$$

$$k = \frac{I_{d,\text{max}}}{C} = \frac{17.0}{2} = 8.50\,\mu A/mF$$

For the unknown solution,

$$C = \frac{I_{d,\text{max}}}{k} = \frac{11.7\,\mu A}{8.50\,\mu A/mF} = 1.38\,mF\ (1.38 \times 10^{-3}\,F)$$

written as

$$I_{d,\text{max}} = kC \tag{17-9}$$

In other words the signal, I, is directly proportional to concentration. Instead of measuring $m^{\frac{2}{3}}t^{\frac{1}{6}}$ and trying to obtain an accurate value for $D^{\frac{1}{2}}$, which may not be available, the analyst customarily uses standard solutions to determine k ($\mu A/mF$) in Equation 17-9. For example, polarographic waves for two concentrations of Pb^{2+} reduction are given in Figure 17-7. The plateau current I_d—from this point on,

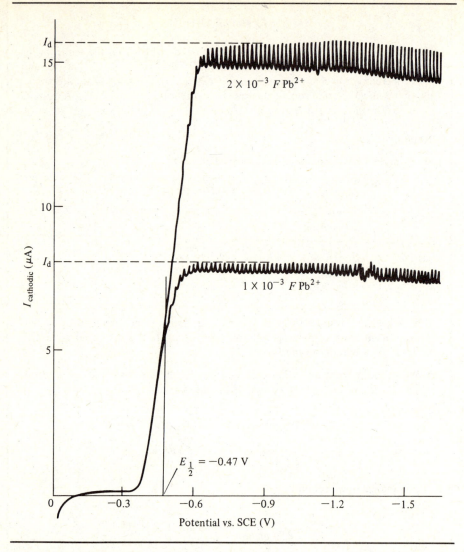

FIGURE 17-7 Polarograms for $1.00 \times 10^{-3}\,F\,Pb^{2+}$ and $2.00 \times 10^{-3}\,F\,Pb^{2+}$. Note that the half-wave potential remains nearly constant at $-0.47\,V$ vs. SCE.

we shall always assume that the maximum value during the life of the Hg drop is being measured and drop the distinction $I_{d,\text{max}}$—is directly proportional to concentration, whereas the potential required for reduction remains the same.

Simply put, polarography is voltammetry with a DME. This electrode has several advantages summarized in Table 17-1. The reproducible time base and the high hydrogen overvoltage mentioned in previous chapters give mercury a decided edge over other possible electrode materials. The available "window" for observing the

---

TABLE 17-1   Advantages and Disadvantages of the DME

---

advantages:

1. New electrode surface every few seconds, which reestablishes initial conditions: $t = 0$ at beginning of each drop. This allows diffusion equations to be solved for current as function of concentration.

2. New electrode surface every few seconds, which prevents buildup of impurities or electrochemical reaction products at the electrode.

3. Most metal products (reduction of metal ions to M^0) dissolve in Hg electrode, which maintains a relatively pure Hg electrode surface. (In contrast, metal deposition at solid electrodes leads to area changes as well as electrochemical effects.)

4. High hydrogen overvoltage allows observation of reduction waves considerably more negative than reduction of H_3O^+ or H_2O.

disadvantages:

1. Hg easily oxidized, which prevents observation of polarographic waves more positive than Hg oxidation (~ 0 V vs. SCE).

2. DME cumbersome to set up and use.

3. Continual Hg dropping costly.

4. Hg highly toxic and DME fragile—a risky combination.

---

cathodic region extends to -1 V vs. SCE in acidic solution and to < -2 V vs. SCE in neutral solution. The latter value is so negative that reduction of alkali metal ions in an aqueous medium can be observed at a DME. (All these advantages were recognized by Heyrovsky as early as 1922.)

On the other hand, mercury has disadvantages, also listed in Table 17-1. The most serious is that electrochemical reactions occurring at potentials more positive than mercury oxidation cannot be studied. This limited anodic window depends to a small extent on the anions present. In a high concentration of chloride ion, the mercury is essentially in the same environment as the SCE. Therefore, oxidation occurs at 0 V vs. SCE. A somewhat higher positive potential is required to oxidize Hg in a SO_4^{2-} or ClO_4^- medium, but oxidation will still occur at $+0.2$ to $+0.4$ V vs. SCE. In an iodide medium, oxidation of mercury is even easier than it is in a chloride medium ($Hg_2I_2 + 2e \rightleftharpoons Hg + 2I^-$; $E^0 \cong -0.3$ V vs. SCE) so that the anodic window is cut off on the negative side of the SCE.

The DME in Figure 17-4 needs a little further discussion. It should be stressed that the capillary tube must have a fine bore (~ 0.05 mm) to provide reasonable drop times of 2 to 6 sec; it is, therefore, easily plugged with impurities. The mercury must be scrupulously clean, and the DME should *always* have a positive head of mercury pressure to prevent solution from backing up into the capillary. The analyst can store a DME overnight by first washing the outside of the capillary tube while mercury is still flowing; then the DME can be left in air with a small mercury pressure to prevent air bubbles from entering the electrode system and to prevent dust particles from lodging in the capillary. With care, a DME capillary tube should give months of reproducible results. That the capillary is becoming plugged can sometimes be deduced from the drop characteristics: Visual inspection to ensure that drop formation and dropping is smooth (as in Figure 17-5) is one way; also, any deviations from the smooth fluctuations in current during the life of the drop (Figures 17-6 and 17-7) should be noted.

INSTRUMENTATION

At this point, let us examine how a polarographic analysis can be carried out in the laboratory. The setup consists of a DME, a cell for containing the sample, and a system for scanning the voltage (potential) while observing the current.

Instruments designed for controlled-potential electrolysis (see Section 16-2) can be used for polarographic measurements, provided that the selected potential can be varied slowly with time. This flexibility is typically available in commercial instruments. An $x-y$ recorder is used, the potential of the DME versus SCE being measured on the x-axis while the current flowing at each potential is recorded on the y-axis. Such equipment is fairly expensive; therefore, many polarographic measurements are taken with less sophisticated instruments.

With the very small diffusion currents normally encountered (0.01 to 100 μA), only a few millivolts of IR drop in an aqueous solution will occur. If a large calomel reference is used as the counterelectrode, the small diffusion currents will not produce significant polarization in the SCE. Therefore, the applied *voltage* in this two-electrode system will very nearly equal the *potential* of the DME versus SCE. It should be noted that polarographic measurements taken in solutions of high resistance, such as organic solvents, may experience large IR drops, and the three-electrode system is preferred.

Two-electrode systems can be constructed easily in the laboratory especially if the analyst is willing to plot the *polarogram* (current–potential curve) point by point. Two or three 1.5-V dry cells or a small stable power supply furnishes the applied voltage, and a potential divider (R_1) allows selection of a fraction of the total available voltage, as shown in Figure 17-8. Automatic scanning of the applied voltage can be accomplished by a motor-driven 10-turn potentiometer as R_1 and by recording the current ($V_R = IR_2$) on a strip-chart recorder (Figure 17-8).

A polarographic cell (called an "H cell" because of its shape) for the analyte solution and SCE counterelectrode (two-electrode system) was shown in Figure 17-4. A KCl salt bridge (in agar to decrease diffusion between the analyte solution and reference electrode) provides electrical contact between the two electrodes. Most polarographic measurements involve reduction at the DME, and it is, therefore, the cathode. The counterelectrode will then be the anode, as indicated in Figure 17-4.

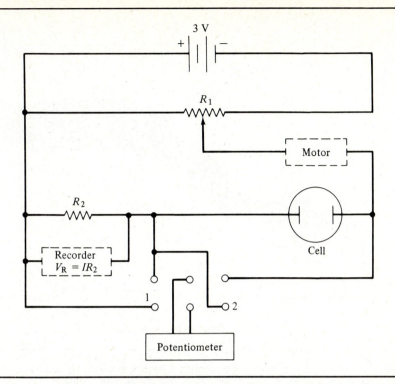

FIGURE 17-8 A simple circuit for polarographic measurements. [From J.J. Lingane, *Analytical Chemistry*, 21:45 (1949). With permission of the American Chemical Society.]

If a polarogram of an air-saturated solution of KCl is recorded, curve *A* of Figure 17-9 will be obtained. The explanation for the two waves in that curve is that O_2 is a reducible species like Cd^{2+}: The wave observed around -0.1 V vs. SCE is the reduction of O_2 to peroxide; the broad second wave, around -0.9 V vs. SCE, represents reduction to H_2O:

$$O_2 + 2H_3O^+ + 2e \xrightleftharpoons{-0.1\ V} H_2O_2 + 2H_2O \qquad \textbf{(17-10)}$$

$$H_2O_2 + 2H_3O^+ + 2e \xrightleftharpoons{-0.9\ V} 4H_2O \qquad \textbf{(17-11)}$$

The solubility of O_2 in water happens to be about 0.2 mF, which is right in the concentration range probed by polarography. Therefore, diffusion currents due to O_2 reduction are comparable to metal ion reduction and must be eliminated before polarographic analysis. As N_2 inlets are provided for the sample cell compartment (see Figure 17-4), the elimination of such diffusion currents is easily accomplished by bubbling N_2 (H_2O saturated to prevent evaporation) into the solution for 5 to 10 min before recording the polarogram. After the deaeration step, the second N_2 inlet provides a N_2 cover for the solution while the polarogram is recorded.

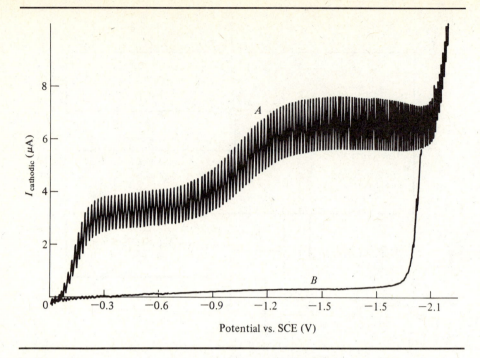

FIGURE 17-9 Curve *A* represents the polarographic reduction of O_2 dissolved in 0.1 *F* KCl supporting electrolyte; *B*, residual current after deaeration with N_2.

Bubbling N_2 into the solution cannot be continued while the polarogram is being recorded because this would introduce convection as a means of transport in addition to diffusion; the currents would be higher, erratic and nonreproducible. After the KCl is deaerated, its polarogram becomes curve *B* of Figure 17-9. Essentially no current flows until the electrode is so negative (-2 V vs. SCE) that reduction of K^+ occurs. The small current observed prior to K^+ reduction is called the *residual current*. Some of the residual current is due to trace impurities and/or oxygen in the solution. However, even with ultrapure solutions, a current is observed, which is called the *charging current*. Each mercury drop must be brought to the potential corresponding to the applied voltage at that point in the polarogram. This requires that some electrical charge must exist at the mercury–electrolyte interface. As the applied voltage increases (DME more negative), the amount of charge required also increases, and thus the charging current increases. It can be seen from *B* in Figure 17-9 that the charging current changes sign around -0.4 V vs. SCE: At more positive potentials, a negative current flows (anodic), whereas at more negative potentials, a cathodic current flows. (The potential at which no current is required for charging the mercury drop is called the *point-of-zero charge*. At this point, the mercury drop is neutral with respect to the solution surrounding it.) The charging current places a lower limit on the concentrations that can be studied by conventional polarographic techniques. Diffusion currents comparable with, or less than, the charging current will have a large uncertainty that occurs around 10^{-5} F. Therefore, most polarographic measurements are carried out for analyte concentrations of 10^{-4} to 10^{-2} F. The high end is limited by the requirements for a supporting electrolyte (50 to 100 times more concentrated than the analyte) and diffusion control (large concentration changes at the electrode will lead to convection effects). The two-electrode system also requires that only low currents be measured so that the applied voltage is a good approximation to the potential of the DME.

The role of the supporting electrolyte in eliminating electromigration of the analyte can be seen in Figure 17-10. When no supporting electrolyte is present, Cd^{2+}

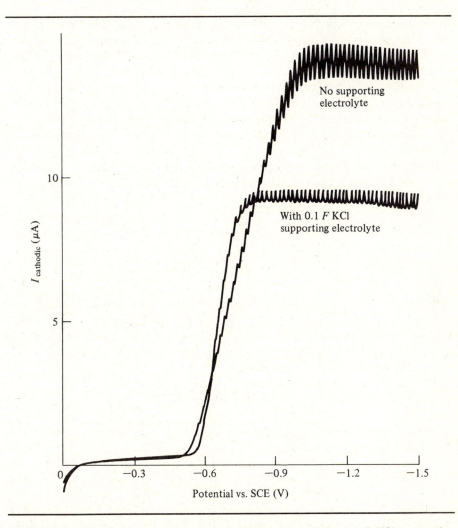

FIGURE 17-10   Reduction of 0.001 F CdBr$_2$ with and without 0.1 F KCl supporting electrolyte.

is transported to the electrode surface by both diffusion and electromigration. For the 0.001 F $CdBr_2$ solution containing 0.1 F KCl as a supporting electrolyte, diffusion becomes the only mode of transport. The plateau current is now smaller, as expected. One might argue that the higher current wave offers more sensitivity; but although that is true, the effect is fairly small and fails to compensate for the added difficulty that, when electromigration current is present, the total current will depend not only on the analyte concentration but on the concentration of all other species in the solution.

HALF-WAVE POTENTIALS

In the discussion of current–potential curves, it was mentioned that the potential at which current flows depends on E^0 for the electrochemical couple and provides a qualitative signal. However, a study of Figure 17-7, depicting polarograms for Pb^{2+} reduction, shows that the point at which current starts to flow (that is, current greater than the residual current) is not well defined. At the same time, the voltage at which the plateau current is reached is also not well defined. However, the voltage at which the current is half the limiting value is a well-defined value that does not vary with concentration, as shown in Figure 17-7. This point is called the *half-wave potential*.

The relationship of the half-wave potential to E^0 becomes clear if we consider the Nernst equation for the electrochemical reaction. In Figure 17-7, the wave represents reduction of Pb^{2+},

$$Pb^{2+} + 2e \rightleftharpoons Pb \tag{17-12}$$

$$E = E^0_{Pb} - \frac{0.059}{2} \log \frac{[Pb]_0}{[Pb^{2+}]_0} \tag{17-13}$$

where $[Pb]_0$ and $[Pb^{2+}]_0$ refer to activities at the electrode surface. The lead metal product dissolves in the mercury drop, and therefore its activity will be proportional to its concentration in Hg at the surface. In Equation 17-3, it was stated that the current is proportional to the concentration difference between the electrode surface and the bulk solution value. Therefore,

$$I = k([Pb^{2+}]_b - [Pb^{2+}]_0) \tag{17-14}$$

When the current reaches the diffusion plateau, essentially every Pb^{2+} is reduced at the electrode surface so that

$$I_d = k([Pb^{2+}]_b - 0) = k[Pb^{2+}]_b \tag{17-15}$$

We can also state that, when current flows, Pb^{2+} will be reduced to Pb so that the lead metal activity at the electrode surface will be directly proportional to the current,

$$I = k'[Pb]_0 \tag{17-16}$$

Equations 17-14 to 17-16 can be combined to give the expressions for $[Pb]_0$ and $[Pb^{2+}]_0$ needed for Equation 17-13

$$[Pb]_0 = \frac{I}{k'} \tag{17-17}$$

$$[Pb^{2+}]_0 = \frac{I_d - I}{k} \tag{17-18}$$

Therefore,

$$E = E^0_{Pb} - \frac{0.059}{2} \log \frac{I/k'}{(I_d - I)/k} \tag{17-19}$$

$$= E^0_{Pb} - \frac{0.059}{2} \log \frac{k}{k'} - \frac{0.059}{2} \log \frac{I}{I_d - I} \tag{17-20}$$

Equation 17-20 can be used to calculate the current at any potential provided that E^0, k, k', and I_d are known. Of particular interest is the potential at which $I = I_d/2$:

$$E_{\frac{1}{2}} = E^0_{Pb} - \frac{0.059}{2} \log \frac{k}{k'} - \frac{0.059}{2} \log \frac{I_d/2}{I_d - I_d/2}$$

$$= E^0_{Pb} - \frac{0.059}{2} \log \frac{k}{k'} - \frac{0.059}{2} \log 1$$

Therefore,

$$E_{\frac{1}{2}} = E^0_{Pb} - \frac{0.059}{2} \log \frac{k}{k'} \tag{17-21}$$

The proportionality constants, k and k', depend on the diffusion and activity coefficients for Pb^{2+} in H_2O and Pb in Hg. Because numerical values are often reasonably close, $k/k' \cong 1$, and therefore, as a rough approximation (± 50 mV),

$$E_{\frac{1}{2}} \cong E^0_{Pb} \tag{17-22}$$

Combination of Equation 17-21 with Equation 17-20 gives

$$E = E_{\frac{1}{2}} - \frac{0.059}{n} \log \frac{I}{I_d - I} \tag{17-23}$$

If $\log [I/(I_d - I)]$ versus E is plotted, a straight line should be obtained, as shown in Figure 17-11, from which values for $E_{\frac{1}{2}}$ and n can be found.

For polarographic analysis, a table of $E_{\frac{1}{2}}$ values is analogous to a table of E^0 values. Some $E_{\frac{1}{2}}$ values are given in Appendix 3-B. They are usually quoted on the

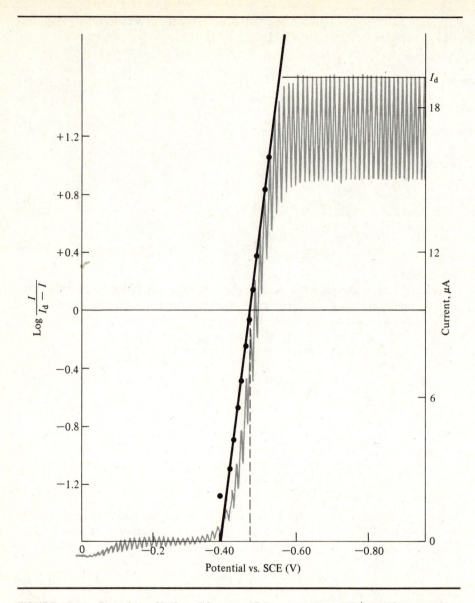

FIGURE 17-11 Plot of log $[I/(I_d - I)]$ versus E for reduction of Tl$^+$ in KCl supporting electrolyte. The actual polarographic wave is shown in the background, with current scale at right. The half-wave potential (-0.47 V vs. SCE) is found from the potential at which log $[I/(I_d - I)] = 0$.

SCE scale because this is the traditional reference electrode. After adjustment to the NHE scale, it can be seen that $E_{\frac{1}{2}}$ values for simple metal ion reductions are reasonably close to the E^0 value. However, the log (k/k') term is generally not negligible so that it is prudent to use $E_{\frac{1}{2}}$ values instead of E^0 values for predicting polarographic waves.

POLAROGRAPHIC ANALYSIS OF MIXTURES

When an analyte solution contains more than one reducible species, diffusion currents resulting from each of them may be observed. If the potential on the DME is sufficiently negative so that all the species will be reduced, the diffusion current will be just the sum of all the individual diffusion currents. But at a potential only cathodic enough to reduce the most easily reduced species in the solution, the diffusion current will be due to that one species. A polarographic wave will be observed, which can be attributed to the most easily reduced species. As the potential is swept more negative, a second species will be reduced and contributes its diffusion current to the diffusion current of the first species. A second polarographic wave is then observed.

Clearly, if two species are reduced at exactly the same potential (their half-wave potentials are equal), only one polarographic wave will be observed with a diffusion current equal to the sum of the two individual diffusion currents. How much separation in half-wave potentials is needed to observe two distinct waves? To answer this question, we must decide what criterion for separation is suitable and then use Equation 17-23. So let us calculate how much potential range it requires to move from the "foot" of the wave to the plateau. In a mathematical sense, there is some diffusion current at any potential; but to be more practical, we define the foot of the wave to be at $I = 0.01 I_d$, in other words, 1% up the wave. At the same time, the wave never mathematically achieves the absolute top of the wave; that is, the concentration of reducible species at the electrode surface is never exactly zero unless E approaches $-\infty$. When $I = 0.99 I_d$ (99% up the wave), we are for all practical purposes at the top. Now with Equation 17-23, the width of a polarographic wave can be calculated:

$$E_{\text{foot}} = E_{\frac{1}{2}} - \frac{0.059}{n} \log \frac{0.01 I_d}{I_d - 0.01 I_d}$$

$$= E_{\frac{1}{2}} + \frac{0.118}{n}$$

$$E_{\text{top}} = E_{\frac{1}{2}} - \frac{0.059}{n} \log \frac{0.99 I_d}{I_d - 0.99 I_d}$$

$$= E_{\frac{1}{2}} - \frac{0.118}{n}$$

$$\text{width} = E_{\text{foot}} - E_{\text{top}} = E_{\frac{1}{2}} + \frac{0.118}{n} - E_{\frac{1}{2}} + \frac{0.118}{n} = \frac{0.236}{n} \text{ V}$$

For $n = 1$, it takes 236 mV to move from the foot to the top of the wave. If we consider two waves to be separated for analytical purposes when the top of the first wave coincides with the foot of the second wave, then

$$E_{1,\text{top}} = E_{\frac{1}{2},1} - \frac{0.118}{n_1}$$

$$E_{2,\text{foot}} = E_{\frac{1}{2},2} + \frac{0.118}{n_2}$$

$$E_{1,\text{top}} = E_{2,\text{foot}}$$

$$E_{\frac{1}{2},1} - \frac{0.118}{n_1} = E_{\frac{1}{2},2} + \frac{0.118}{n_2}$$

$$E_{\frac{1}{2},1} - E_{\frac{1}{2},2} = \Delta E_{\frac{1}{2}} = \frac{0.236}{n} \text{ V} \quad (\text{if } n_1 = n_2) \tag{17-24}$$

In other words, the difference in half-wave potentials must be at least $(0.236/n)$ V. In round numbers, the minimum separation is

n	$\Delta E_{1/2}$ (mV)
1	240
2	120
3	80

    With a cathodic window of about 2 V in a neutral or basic medium, it is conceivable that perhaps eight or more individual polarographic waves could be observed. This would require very fortunate spacing of $E_{\frac{1}{2}}$ values. Although the analyst has some control of $E_{\frac{1}{2}}$ with choice of the supporting electrolyte (see the later subsection on the polarography of complexes), he does not have complete freedom to vary $E_{\frac{1}{2}}$ for one substance without some effect on the other components. A more realistic picture of a polarographic analysis of mixtures is shown in Figure 17-12. In this example, the solution contains four reducible analytes: Cu^{2+}, Cd^{2+}, Ni^{2+} and Zn^{2+}. The supporting electrolyte is $2 F \, NH_3 – 2 F \, NH_4Cl$, which yields a slightly basic medium and forms ammine complexes with the metal ions. The polarogram consists of five peaks, not four, because copper can be reduced to Cu^+, with $E_{\frac{1}{2}} = -0.25$ V vs. SCE, and to Cu^0, with $E_{\frac{1}{2}} = -0.5$ V vs. SCE. This is more than enough $\Delta E_{\frac{1}{2}}$ to make the two individual waves apparent. Cadmium is reduced with $E_{\frac{1}{2}} = -0.8$ V vs. SCE, and the amount of cadmium can be determined by measuring from the top of the second copper wave to the top of the cadmium wave. For the analysis of mixtures, this simple additivity feature often makes

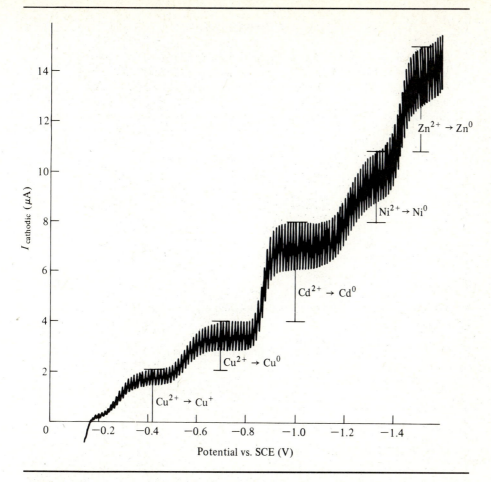

FIGURE 17-12 Polarogram of a mixture containing Cu^{2+}, Cd^{2+}, Ni^{2+}, and Zn^{2+} in 2 F NH_3–2 F NH_4Cl supporting electrolyte (see Example 17-2). Added as a maximum suppressor is 1 mL of 0.2% Triton X–100.

polarography preferred to electroanalytical techniques that do not possess additivity. Typical calculations with a polarogram such as Figure 17-12 are shown in Example 17-2.

POLAROGRAPHIC MAXIMA

So far, we have considered only ideal polarographic waves. In practice, quantitative analysis can be made more difficult by the occurrence of phenomena that distort the

EXAMPLE 17-2

An alloy sample containing Cu, Cd, Ni, and Zn and weighing 395 mg is dissolved in acid and diluted to 250 mL. A 10.0-mL aliquot is placed in a 100-mL volumetric flask, and 50 mL of a $2F$ NH_3–$2F$ NH_4Cl solution plus 1 mL of a 0.2% Triton X-100 solution are added. The sample is diluted to volume with H_2O, and after deaeration, the polarogram shown in Figure 17-12 is recorded. Standard solutions give the following sensitivity values:

$$Cu, 7.2\ \mu A/mF \text{ (total of two waves)}$$

$$Cd, 7.6\ \mu A/mF$$

$$Ni, 6.6\ \mu A/mF$$

$$Zn, 7.8\ \mu A/mF$$

Calculate the alloy composition.

Solution

The diffusion currents are read off the polarogram (top of each plateau) and are as follows:

$$Cu, 3.9\ \mu A \text{ (total of two waves)}$$

$$Cd, 3.9\ \mu A$$

$$Ni, 2.9\ \mu A$$

$$Zn, 4.2\ \mu A$$

The concentration of each species can now be calculated:

$$[Cu^{2+}] = \frac{3.9\ \mu A}{7.2\ \mu A/mF} = 0.54\ mF$$

$$[Cd^{2+}] = \frac{3.9\ \mu A}{7.6\ \mu A/mF} = 0.51\ mF$$

$$[Ni^{2+}] = \frac{2.9\ \mu A}{6.6\ \mu A/mF} = 0.44\ mF$$

$$[Zn^{2+}] = \frac{4.2\ \mu A}{7.8\ \mu A/mF} = 0.54\ mF$$

EXAMPLE 17–2   (cont.)

The concentration of each species in the original solution is

$$[Cu^{2+}] = 0.54 \text{ mF} \times \frac{100}{10} = 5.4 \text{ mF}$$

$$[Cd^{2+}] = 0.51 \text{ mF} \times \frac{100}{10} = 5.1 \text{ mF}$$

$$[Ni^{2+}] = 0.44 \text{ mF} \times \frac{100}{10} = 4.4 \text{ mF}$$

$$[Zn^{2+}] = 0.54 \text{ mF} \times \frac{100}{10} = 5.4 \text{ mF}$$

The amount of each species in the sample is

$$\text{mg Cu} = 5.4 \frac{\text{mmol}}{\cancel{L}} \times 0.25 \cancel{L} \times 63.5 \frac{\text{mg}}{\text{mmol}} = 86 \text{ mg}$$

$$\text{mg Cd} = 5.1 \frac{\text{mmol}}{\cancel{L}} \times 0.25 \cancel{L} \times 112.4 \frac{\text{mg}}{\text{mmol}} = 143 \text{ mg}$$

$$\text{mg Ni} = 4.4 \frac{\text{mmol}}{\cancel{L}} \times 0.25 \cancel{L} \times 58.7 \frac{\text{mg}}{\text{mmol}} = 65 \text{ mg}$$

$$\text{mg Zn} = 5.4 \frac{\text{mmol}}{\cancel{L}} \times 0.25 \cancel{L} \times 65.4 \frac{\text{mg}}{\text{mmol}} = \frac{88 \text{ mg}}{382 \text{ mg total accounted for}}$$

The percentage composition is

$$Cu = \frac{86 \text{ mg Cu}}{395 \text{ mg sample}} \times 100 \text{ (\%)} = 22\%$$

$$Cd = \frac{143 \text{ mg Cd}}{395 \text{ mg sample}} \times 100 \text{ (\%)} = 36\%$$

$$Ni = \frac{65 \text{ mg Ni}}{395 \text{ mg sample}} \times 100 \text{ (\%)} = 16\%$$

$$Zn = \frac{88 \text{ mg Zn}}{395 \text{ mg sample}} \times 100 \text{ (\%)} = 22\%$$

(*Note:*   Only two significant figures are used in reporting the results because, in a complex mixture, it is difficult to measure the diffusion current of each component in the presence of others. Although accuracies of ± 1 or 2% can be obtained with single waves, it is more realistic to expect accuracies of ± 3 to 10% for mixtures.)

waves. One of these is the polarographic maximum. In curve *A* of Figure 17-13, a polarogram of another complicated mixture is shown. The copper wave rises to a maximum current before dropping back to the expected diffusion current plateau. A similar effect can be seen for Al^{3+} reduction. In this case, it is difficult to determine the diffusion current because reduction of the K^+ supporting electrolyte begins before the maximum has passed.

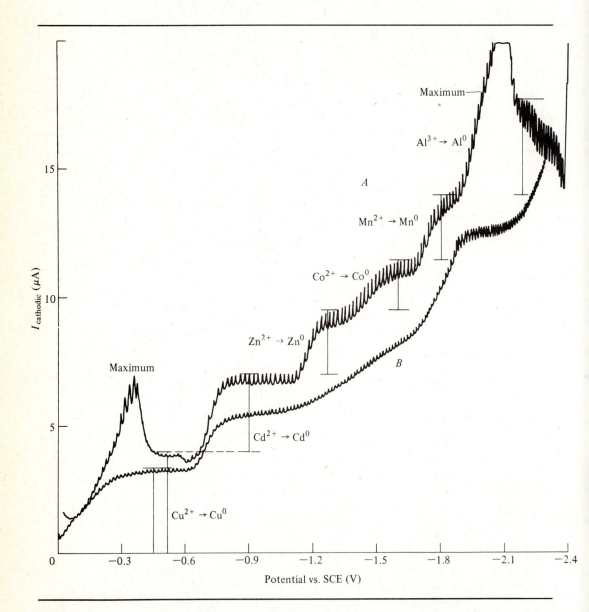

FIGURE 17-13 Polarogram of a mixture containing Cu^{2+}, Cd^{2+}, Zn^{2+}, Co^{2+}, Mn^{2+}, and Al^{3+}: curve *A*, with no maximum suppressor, showing polarographic maxima; *B*, with 0.01% gelatin added.

Several explanations have been proposed for polarographic maxima, and evidence for more than one cause exists. Stirring effects at the mercury surface can add convection to diffusion as transport mode and thereby increase the observed currents. Why these effects are seen only on the rising portion of the wave is not clearly understood. In any case, it is known that many polarographic maxima can be eliminated by the addition of surface-active agents, which adsorb on the mercury drop. Many years ago, gelatin and dyes such as methyl red were found to be effective, but the synthetic surfactant Triton X-100 is now the maximum suppressor of choice. The effect of gelatin can be seen in curve B of Figure 17-13. The addition of 0.01% gelatin completely eliminated the copper wave maximum, but its effect on the other components is quite deleterious. The waves for Co^{2+}, Mn^{2+} and Al^{3+} are so distorted that only from the current flow can the analyst see that reduction is occurring. Fortunately, Triton X-100 (used for Figure 17-12) is an effective maximum suppressor that does not distort waves as much as gelatin does. However, the moral of this story is that the polarogram of a new sample should be first taken *without* a maximum suppressor. Then, if maxima are a problem, a small amount of suppressor (typically 0.002%) can be added and a second polarogram recorded. In this way, no information will be lost, as would be the case if the analyst first ran the polarogram for curve B in Figure 17-13.

IRREVERSIBLE WAVES

In the development of Equation 17-23, it was assumed that the Nernst equation is obeyed at the surface of the mercury drop. That is, when the potential at the DME is sufficiently negative to cause reduction, the electrochemical reaction is fast and maintains thermodynamic equilibrium at all times. Such electrochemical reactions are said to be *reversible*. If the kinetics of the electrochemical reaction are so slow that thermodynamic equilibrium cannot be maintained, the reaction is said to be *irreversible*.

When the reaction rate is slow, less current will flow at a given potential than if it were reversible. However, electrochemical reaction rates depend on potential, and by application of sufficient additional voltage, the full diffusion current will ultimately be attained. This means that the width of an irreversible wave will be greater than that of a reversible wave (that is, $> 0.236/n$ V), and a plot of log $[I/(I_d - I)]$ against E will *not* give the theoretical slope. Such a plot is useful in determining the reversibility of a polarographic wave.

The effect of an irreversible wave on a polarographic analysis can be seen in Figure 17-12. The nickel wave is broader than theoretically predicted for $n = 2$, which makes it difficult to measure its diffusion current before the onset of the zinc wave even though the $\Delta E_{\frac{1}{2}}$ is greater than that required by Equation 17-24 ($\Delta E_{\frac{1}{2}} = -1.20 - (-1.41) = 0.21$ V $> 0.236/2$ V). Some waves are so irreversible that the full diffusion current is not attained throughout the entire voltage scan. This behavior not only hinders quantitative analysis for this component but can interfere with measurement of any waves occurring at more negative potentials. Another example of an irreversible wave can be seen in curve A of Figure 17-9, the reduction of O_2 to H_2O at -0.9 V vs. SCE. The width of this wave is about 600 mV, far greater than the theoretical value for a reversible wave.

POLAROGRAPHY OF COMPLEXES

It has already been mentioned that half-wave potentials can be shifted by the choice of supporting electrolyte. The shift is readily seen by an examination of Appendix 3-B. The basic concept is that complexation between the metal ion and the supporting electrolyte anion will shift $E_{\frac{1}{2}}$ to more negative potentials. Why the shift will be in this direction can be seen from the Nernst equation for a simple metal ion couple, say, Pb^{2+}:

$$Pb^{2+} + 2e \rightleftharpoons Pb$$

$$E = E^0_{Pb} - \frac{0.059}{n} \log \frac{[Pb]}{[Pb^{2+}]} \tag{17-25}$$

The potential becomes more negative as the activity of Pb^{2+} decreases. When a complex forms, the formation constant is

$$Pb^{2+} + xL \rightleftharpoons PbL_x \quad \text{(for simplicity, charge value on L neglected)}$$

$$K = \beta_x = \frac{[PbL_x]}{[Pb^{2+}][L]^x}$$

or

$$[Pb^{2+}] = \frac{[PbL_x]}{K[L]^x} \tag{17-26}$$

Thus the activity of Pb^{2+} decreases as K, and the ligand concentration increases. The effect on $(E_{\frac{1}{2}})_c$, the half-wave potential for the complex, can be seen by combining Equations 17-25 and 17-26:

$$E = E^0_{Pb} - \frac{0.059}{n} \log \frac{[Pb]}{[PbL_x]} - \frac{0.059}{n} \log K - \frac{0.059x}{n} \log [L] \tag{17-27}$$

The values for $[Pb]$ and $[PbL_x]$ can now be evaluated in terms of the current, assuming that the complex is reversibly reduced:

$$E = E^0_{Pb} - \frac{0.059}{n} \log \frac{k}{k'} - \frac{0.059}{n} \log \frac{I}{I_d - I}$$
$$- \frac{0.059}{n} \log K - \frac{0.059x}{n} \log [L] \tag{17-28}$$

From Equation 17-21 (and for simplicity, it is assumed that k remains the same for the complex as for the simple ion), we see that

$$E = E_{\frac{1}{2}} - \frac{0.059}{n} \log \frac{I}{I_d - I} - \frac{0.059}{n} \log K - \frac{0.059x}{n} \log [L]$$

where $E_{\frac{1}{2}}$ is the half-wave potential for the simple metal ion couple. When $I = I_d/2$ for the complex, $E = (E_{\frac{1}{2}})_c$:

$$(E_{\frac{1}{2}})_c = E_{\frac{1}{2}} - \frac{0.059}{n} \log K - \frac{0.059x}{n} \log [L] \qquad \text{(17-29)}$$

The half-wave potential shift is

$$(E_{\frac{1}{2}})_c - E_{\frac{1}{2}} = \Delta E_{\frac{1}{2}} = -\frac{0.059}{n} \log K - \frac{0.059x}{n} \log [L] \qquad \text{(17-30)}$$

Thus, by judicious choice of complexing agent and its concentration, the half-wave potential can be shifted almost at will—at least in the negative direction. However, it should be kept in mind that the ligand may also complex the other components in the mixtures to the same extent and that no increase in separation will result. Usually though, K values will differ by orders of magnitude, and waves will be shifted relative to each other in a mixture.

A plot of $\Delta E_{\frac{1}{2}}$ versus log [L] for a reversible wave will be a straight line, from which x can be determined from the slope and K can be determined from the intercept.

ANODIC POLAROGRAPHIC WAVES

It has been tacitly assumed throughout the previous discussion that reduction occurs at the DME and a cathodic current is measured. Most—but not all—polarographic analysis involves reduction at potentials negative with respect to the SCE. This type of electrochemical reaction involves one quadrant of the current–potential plane. Reductions at potentials positive with respect to the SCE may also be observed, but the wave may be obscured by the oxidation of mercury. The first reduction wave of Cu^{2+} shown in Figure 17-13 is an example. The actual wave occurs at potentials more positive than mercury oxidation, and only the diffusion current plateau can be seen.

The other two quadrants of the current–potential plane involve oxidations. Just as the current drops below zero when mercury is oxidized, an oxidizable species will given an anodic wave. Ion–ion couples such as Cu^{2+}/Cu^{+} or Fe^{3+}/Fe^{2+} can yield reduction waves when the higher oxidation state is present or oxidation waves when the lower oxidation state is present. For a reversible process, the half-wave potentials for the cathodic and anodic waves will be equal. When a mixture of oxidation states is present, a wave containing both an anodic current plateau and a cathodic current plateau will be observed, as shown in Figure 17-14. For such a mixture, the half-wave potential refers to the point halfway between the anodic diffusion current and the cathodic diffusion current. For a reversible system, the half-wave potential for the mixture of oxidation states will be the same as for the pure cathodic or pure anodic wave, as indicated on Figure 17-14.

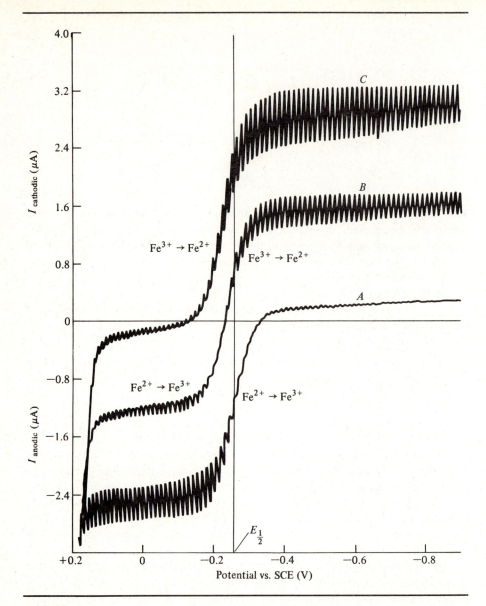

FIGURE 17-14 Polarograms of iron in a 0.5 F sodium citrate medium. Curve A is for Fe^{2+}, showing only anodic wave; B, mixture of Fe^{2+} and Fe^{3+}, showing both anodic and cathodic currents; and C, Fe^{3+}, showing only cathodic wave. Note the nearly constant half-wave potential at $-0.26\,V$ vs. SCE.

Some organic compounds can be oxidized at the DME yielding anodic waves. One reversible couple is the quinone–hydroquinone couple. The reaction is

$$+ 2H_3O^+ + 2e \rightleftharpoons \quad + 2H_2O \qquad \text{(17-31)}$$

quinone   hydroquinone

As Equation 17-31 indicates, the reaction is dependent upon H_3O^+, and therefore $E_{\frac{1}{2}}$ varies with pH. The reduction becomes easier at high H_3O^+, and $E_{\frac{1}{2}}$ will be more positive. At pH 7, the half-wave potential is about $+0.05$ V vs. SCE. Because the reaction is reversible, a solution of hydroquinone gives an anodic wave with the same half-wave potential of $+0.05$ V vs. SCE at pH 7.

INORGANIC APPLICATIONS

It should be clear after the previous discussion that polarography can be a useful technique for the determination of metal content in a sample. Analytical applications include determination of alloy compositions as well as trace metal content in waste water, foods, paint, glass, and biological specimens.

One area especially well suited for the polarographic method is the analysis of copper-based alloys containing various other metals. However, since copper is easily reduced (see Figures 17-12 and 17-13), it could obscure succeeding waves. In other words, if the large copper waves are to be kept on scale, current sensitivity may be too low for the accurate determination of minor constituents. One way to circumvent this problem is to remove most or all of the copper in the alloy prior to polarographic analysis. The removal can be accomplished by dissolving the alloy in acid and then depositing copper on an electrode with controlled-potential electrolysis at -0.35 V vs. SCE. Part of the remaining solution is then made basic, and a polarogram is run for the determination of lead, which is reduced, with $E_{\frac{1}{2}} = -0.76$ V vs. SCE. To another aliquot of the sample is added NH_4Cl, and a polarogram is taken of tin plus lead reduction. The amount of lead is subtracted (I_d for Pb in $NH_4Cl = 1.04 I_d$ for Pb in NaOH) to yield the tin content. Finally a third aliquot is used to determine nickel and zinc. It is recommended to reduce Pb and Sn with controlled-potential electrolysis at -0.70 V vs. SCE to avoid any interference from these constituents. For some alloys, this step is unnecessary. In NH_3–NH_4Cl medium, the zinc and nickel waves are well separated (see Figure 17-12). The half-wave potentials are -1.06 V vs. SCE for Ni and -1.33 V vs. SCE for Zn. Results from several copper-base alloys are given in Table 17-2. It can be seen that, for constituents present at $>0.1\%$, results agree with National Bureau of Standards values to within $\pm 4\%$ and in most cases to within $\pm 2\%$.

One example in which preelectrolysis is not necessary is the analysis of Monel metal. This alloy consists of approximately 30% Cu and 70% Ni, and the waves are

TABLE 17-2   Polarographic Analysis of Copper-Base Alloys

Sample	National Bureau of Standards (%)				Polarographic analysis (%)			
	Pb	Sn	Ni	Zn	Pb	Sn	Ni	Zn
phosphor bronze 63a	8.92	9.76	0.32	0.61	9.04	9.84	0.32	0.60
cast bronze 52b	0.01	8.00	0.72	2.96	0.02	7.82	0.70	2.96
ounce metal 124a	4.86	4.81	0.001	5.25	4.84	5.00	—	5.29
brass 37c	0.97	0.96	0.58	27.22	0.92	0.97	0.59	27.8

well separated in NH_3–NH_4Cl medium (Figure 17-12). Results are accurate to $\pm 2\%$.

Procedures for the polarographic determination of copper, nickel, cobalt, manganese, chromium, molybdenum, vanadium, and tungsten in steel and other ferrous alloys have been reported. One problem is interference from iron. Iron(III) is reduced to iron(II) at positive potentials versus SCE, and its large diffusion current would be present throughout the polarographic range. In the determination of Cu and Ni, iron may be precipitated as $Fe(OH)_3 \cdot xH_2O$ with NH_3. Precipitation of iron with $BaCO_3$ is satisfactory for samples containing small amounts of Co and Mn. Various supporting electrolytes have been used for the various constituents, depending on the alloy being analyzed. One interesting method for vanadium is based on an anodic wave in basic solution, resulting from oxidation of V(IV) to V(V) at $E_{\frac{1}{2}} = -0.42$ V vs. SCE.

The chemical and electrochemical characteristics of lead are very favorable for its polarographic determination from a variety of supporting electrolytes. As can be seen from Appendix 3-B, the half-wave potential is about -0.4 V vs. SCE except in strong base or cyanide, where $E_{\frac{1}{2}}$ is about -0.75 V vs. SCE. The determination of lead in copper-based alloys has already been mentioned. Determination of lead in flue dust and slags is accomplished simply by fusing the sample with sodium peroxide and diluting with HCl followed by polarographic analysis. Lead in gasoline can be determined by extracting the lead with HCl (see Section 18-4 for liquid–liquid extraction). The HCl extract containing lead is diluted and analyzed polarographically. Details are given in the laboratory manual. Polarographic determinations of lead in paint, food, plants, and insecticides have been reported. For example, lead and zinc content in paint can be determined by digesting the extracted paint pigment in HCl, followed by dilution and polarographic analysis. Trace amounts of lead in blood can be found by treating a sample with HCl and recording the polarogram directly. The lead wave is at -0.4 V vs. SCE, and none of the organic constituents apparently interferes.

Zinc is another element well suited to polarographic analysis. However, zinc reduction in acidic media occurs near reduction of H_3O^+ and therefore is usually observed in neutral or basic media. Since reduction of copper, cadmium, lead, and nickel precede zinc, these elements may have to be removed prior to zinc analysis if their amounts are considerably higher than the zinc concentrations. Determination in copper-based alloys was mentioned earlier. Determination of trace amounts of zinc in plants and biological materials is accomplished by ashing the material below 500°C, followed by dissolution in 1 F HCl. Zinc, nickel, cadmium, and lead are extracted into chloroform as the dithizone complexes. These elements are transferred back to an aqueous HCl medium by a second liquid–liquid extraction. Finally, a polarographic analysis is carried out in a supporting electrolyte containing ammonium acetate and potassium thiocyanate. Waves due to reduction of Pb, Cd, Ni, and Zn (in that order) are observed.

Glass can be analyzed polarographically to determine metal content. With the high hydrogen overvoltage in basic solution, it is even possible to determine barium ($E_{\frac{1}{2}} = -1.9$ V vs. SCE), sodium ($E_{\frac{1}{2}} = -2.1$ V vs. SCE), potassium ($E_{\frac{1}{2}} = -2.1$ V vs. SCE), and calcium ($E_{\frac{1}{2}} = -2.1$ V vs. SCE). Clearly, the last three constitute one wave, but a second sample can be treated with H_3PO_4 to precipitate calcium phosphate. The second polarogram will yield Na plus K, and the difference between the first and second polarograms will yield the Ca amount. The supporting electrolyte for this determination cannot contain any alkali metal ion, and tetramethylammonium hydroxide is recommended. Other metal constituents in glass, such as zinc, lead, and aluminum, can also be determined polarographically.

ORGANIC APPLICATIONS

It might be presumed that polarography is strictly a technique for the determination of inorganic species, but this is far from the truth. Many organic compounds are reducible at the DME, and a few others exhibit anodic waves. However, it is true that polarography of organic species can be more difficult to carry out than can inorganic polarography. For example, many organic compounds are not soluble in H_2O, and some nonaqueous solvent is required. The nonaqueous solvent must still be capable of dissolving an adequate concentration of an ionic substance that plays the role of the supporting electrolyte. Thus the choice of solvent for organic polarography can be difficult and critical to the success of the analysis.

Another troublesome factor in organic polarography is that most waves are irreversible. This means, as shown in Figure 17-12 for reduction of Ni^{2+}, that the wave will be broad and may not reach its diffusion current plateau before some subsequent process occurs. The reason for irreversibility is the complicated reaction pathway for many organic redox reactions. Unlike the reduction of a metal ion from one oxidation state to another (say, Fe^{3+} to Fe^{2+}), in which the species simply accepts or donates an electron, the organic electrochemical reaction may involve breaking of chemical bonds along with the electron transfer. Such a reaction is often kinetically slow at the electrode, leading to irreversibility. There are, however, some reversible organic electrochemical reactions, such as the reduction of a quinone to a hydroquinone given in Equation 17-31. Because it is reversible, the anodic wave for oxidation of hydroquinone to quinone can also be measured.

One aspect of organic polarography that can be either advantageous or disadvantageous is that most reactions involve hydronium ion and are, therefore, pH dependent. The half-wave potential will be a function of pH, just as $E_{\frac{1}{2}}$ varies with ligand concentration for reduction of metal complexes. Usually, the reduction reaction consumes H_3O^+ so that $E_{\frac{1}{2}}$ will become more positive at higher acid concentration (lower pH). For example, the half-wave potential for the quinone case given in Equation 17-31, using Equation 17-21, would be

$$E_{\frac{1}{2}} = E_Q^0 - \frac{0.059}{2} \log \frac{k}{k'} + \frac{0.059}{z} \log [H_3O^+]^z$$

$$\simeq E_Q^0 + 0.059 \log [H_3O^+] = E_Q^0 - 0.059 \text{ pH} \qquad \textbf{(17-32)}$$

For most organic electrochemical reactions, the number of hydronium ions equals the number of electrons so that the term -0.059 pH will be present. Equation 17-32 may not always be obeyed for irreversible waves, but pH dependence will be observed. The pH dependence can be an advantage when the half-wave potential must be shifted because of an interference such as mercury oxidation. On the other hand, it may turn out that the reaction proceeds only in a pH region that is not accessible—for example, $E_{\frac{1}{2}} < -1$ V vs. SCE in acid solution—beyond hydrogen ion reduction.

In spite of these difficulties, several organic functional groups can be reduced at the DME; these are summarized in Table 17-3. It can be seen that simple saturated hydrocarbons or olefins are not on the list as well as many common groups such as alcohols and amines. However, the carbon–carbon double bond can be reduced when it is conjugated with another double bond, such as in an aromatic ring, Φ. Carbonyls in the form of aldehydes or quinones are reducible, and even ketone reduction has been observed at very negative potentials ($E_{\frac{1}{2}} = -2.46$ V vs. SCE for acetone). Of more use to analysis is the ability to reduce many organic compounds containing nitrogen, sulfur, or halogens.

Amines and acids are not directly reducible, but they may be determined polarographically because they generally catalyze the hydrogen wave; that is, they decrease the hydrogen overvoltage. Acids containing a double bond conjugated with the carboxyl group are reducible and thus quite useful for analyzing samples containing maleic or fumaric acids.

One valuable aspect of organic polarography is that many important biochemical compounds are electroactive. For example, vitamin C (ascorbic acid) is oxidizable and exhibits an anodic wave that has been used for its determination in fruit juices. The wave lies close to the region of mercury oxidation, especially in acidic solution ($E_{\frac{1}{2}} = +0.25$ V vs. SCE at pH 1.5). As predicted from Equation 17-32, the half-wave shifts approximately 60 mV for each pH unit even though the wave is not completely reversible. For analysis of fruit juice, a pH of 6 to 8 is recommended, in which case the half-wave potential is about -0.04 V vs. SCE.

Several vitamin B factors can be determined polarographically. For example, riboflavin exhibits a wave at -0.35 V vs. SCE in 0.1 F KCl, and thiamin gives a wave at -1.25 V vs. SCE in the same medium. Niacin is reduced at -1.56 V vs. SCE in a borate buffer (pH 8.7).

Ketose sugars (fructose and sorbose), unlike other sugars, give well-defined waves, which can be utilized in the determination of fructose in honey, candy, or even urine. The presence of other sugars does not interfere.

TABLE 17-3   Organic Functional Groups Reducible at the DME

Functional group*	$E_{\frac{1}{2}}$ vs. SCE (pH = 7)[†]
$\Phi-C=C-$	-2.3
$\Phi-C\equiv C-$	-2.3
$>C=CH-CH=C<$	-2.6
$\Phi-CHO$	$-1.1, -1.4$
$>C=O$	-2.2
$>C=CH-CH=O$	$-1.1, -1.3$
$O=\langle\bigcirc\rangle=O$	-0.1
$>C=C-COOH$	-1.3
$-NO_2$	$-0.9, -1.5$
$-N=N-$	-0.4
$-OOH$	-0.3 (pH = 1)
$>CCl_2$	-2.0
$-CCl_3$	$-1.7, -2.0$
$>CBr_2$	$-1.5, -1.8$
$-CBr_3$	$-0.6, -1.5$
$>CI_2$	$-1.1, -1.5$
$-CI_3$	$-0.5, -1.1, -1.5$

SOURCE: Data from *Treatise on Analytical Chemistry*, pt. 1, vol. 4 I.M. Kolthoff and P.J. Elving, eds., (New York: Interscience Publishers, Inc., 1963).

*Φ is an abbreviation for a phenyl group, $\langle\bigcirc\rangle-$.
[†]When more than one $E_{\frac{1}{2}}$ value is listed, the group exhibits more than one reduction wave; that is, it is reduced in a series of steps.

Many other organic and biochemical compounds of interest can be determined polarographically because they give catalytic hydrogen waves. These compounds include steroids, cysteine, alkaloids, and amino acids.

In conclusion, it may be said that, although polarography is especially well suited for inorganic analysis, there are hundreds of possible applications for the determination of organic analytes.

DERIVATIVE POLAROGRAPHY

In recent years, many techniques have been developed to make polarography more sensitive or more convenient. As one example, it is now possible, with modern commercial instruments, to record the current just before the drop falls. In other words, the electronic circuit "knows" when the drop is about to fall, "samples" the current value (I_{max} in Figure 17-6) at this point, and sends it to the recorder, which holds it until it is time to sample the next drop. Such a polarogram has been recorded with a *sampled dc* signal and will not exhibit all the current oscillations of a conventional polarogram (as the previous figures in this chapter have been recorded). A sampled dc polarogram is shown in Figure 17-15. Clearly, it is necessary for the electronic circuit to know the exact drop time. Natural dropping varies from solution to solution and even varies with the applied voltage so that it is not constant during the recording of a polarogram. To circumvent this problem, mechanical droppers are available; they "knock" the capillary at a prescribed time to force the drop to fall. In this way, a 2-sec drop time, as an example, can be maintained from electrode to electrode, day to day, and throughout the applied voltage range.

It was stated earlier that polarography is useful for solutions typically 10^{-4} to 10^{-3} F in analyte concentration. Several methods have been developed to extend

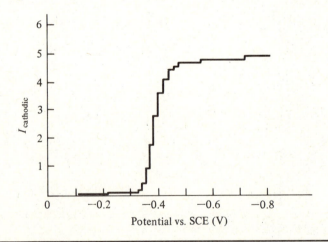

FIGURE 17-15 Polarogram of Pb^{2+} reduction, using sampled dc technique.

the sensitivity to lower concentrations. It has also been mentioned that analysis of mixtures can be difficult when a more easily reduced species is present at much higher concentration than the more difficultly reduced species is. The problem arises because the sensitivity (microamps per millimeter of chart paper) must be decreased to keep the high diffusion current of the first species on scale. Instruments normally have some ability to compensate for high currents, which will allow the operator to set the diffusion current plateau of the first peak at zero, but this compensation is limited to perhaps a ratio of 10:1 for I_d (first peak)/I_d (second peak). In addition, the top of the diffusion current plateau of the first peak will be "noisy" (it will show random variations), which will make it difficult to establish the baseline needed to measure the second reduction wave.

Both problems (low sensitivity and measurement of small waves in the presence of large diffusion currents from prior waves) can be circumvented by using *derivative polarography*. There are several instrumental methods for accomplishing the taking of the derivative of a polarographic wave, but in principle, they all yield the same type of information. (As you may know from the calculus, the derivative is the slope of the curve at each point.) Consider the single polarographic wave shown in Figure 17-15 for the reduction of Pb^{2+}. Until the potential is sufficiently negative for reduction to occur, the slope is essentially zero although the residual (charging) current will give the slope a small constant value. At the half-wave potential, the current is changing fastest, and therefore the slope reaches a maximum value. When the top of the wave is reached, the current is essentially constant, and so the slope is again nearly zero. The derivative polarogram is shown in Figure 17-16(A). The peak value can be shown to be directly proportional to concentration. The advantages of this technique can be seen in Figure 17-16(B). In this example, a large first wave is followed by a small second wave. When the slope returns to nearly zero after the first wave (at CS), the sensitivity can now be increased many times in order to observe the

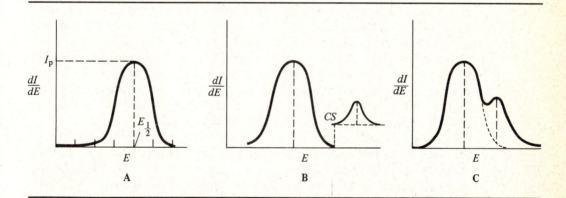

FIGURE 17-16 Derivative polarography. (**A**) Reduction of Pb^{2+}, showing peak current at $E_{\frac{1}{2}} = -0.4\,V$ vs. SCE. (**B**) Measurement of small wave following large wave by change of sensitivity at CS. (**C**) Resolution of two polarographic waves whose $E_{\frac{1}{2}}$ values are too close to be separated by conventional polarography.

small second wave. In addition to being able to measure small succeeding peaks, the derivative technique is more sensitive than conventional polarography is. Concentrations of $<10^{-6}$ F can often be measured. However, the peak slope value is not so reproducible as is the diffusion current value so that accuracy of derivative polarography is not so high as that of conventional polarography.

One other advantage of derivative polarography is seen in Figure 17-16(**C**). In this case, two waves are too close together to measure the diffusion current; that is, the slope does not return to zero before the second wave begins. The derivative clearly shows the existence of two peaks. The analyte concentrations can be determined from the peak values although the second peak must be corrected for contributions from the first peak [Figure 17-16(**C**)]. Again, accuracy suffers; but in such situations, conventional polarography may not be at all possible.

Some of the methods used to obtain derivative-type polarograms involve pulsing an additional voltage signal on top of the linearly increasing (in a negative direction) dc voltage associated with the conventional polarogram. These pulsing techniques (*differential pulsed polarography*) are capable of detecting analytes at concentrations down to 10^{-8} F and of resolving two steps whose half-wave potentials differ by only 0.05 V.

17-3   ANODIC STRIPPING

A recently developed voltammetric technique capable of detecting trace amounts of analyte (as low as 10^{-9} F) has been given the name *anodic stripping*. Actually, it involves two analytical techniques: controlled-potential electrolysis and anodic voltammetric scan.

The basic concept of anodic stripping is that of concentrating the analyte by depositing it as the metal on the electrode (or in it, in the case of a liquid Hg electrode). The deposition is accomplished by electrolyzing the solution with stirring for a prescribed amount of time at some potential more negative than the reduction potential for any of the analytes. Following this treatment, a voltammetric scan (unstirred solution) is taken in the anodic (toward more positive potentials) direction. When the potential becomes sufficiently positive to oxidize one of the analyte metals, an anodic current flows. However, the electrode is rapidly depleted of the analyte so that a peak current is observed instead of a polarographic plateau. This two-step process is depicted schematically in Figure 17-17, where the conventional polarographic scan is shown for comparison.

Obviously, a DME cannot be used for this technique because the voltammetric scan must be carried out with the same electrode as the one utilized for the entire controlled-potential electrolysis. Nevertheless, mercury is the preferred electrode material because of its high hydrogen overvoltage. One technique is to use a hanging-drop mercury electrode. As an alternative, a mercury pool electrode is sometimes used. The hanging drop has the advantage of being so small that a large concentration is effected in a tiny mercury drop by cathodic reduction of metal ions present in a large volume of solution. Hanging-drop assemblies are commercially available in which an accurately known drop of mercury can be formed by extruding mercury from a fine tube. The mercury drop will cling to the end of the tube

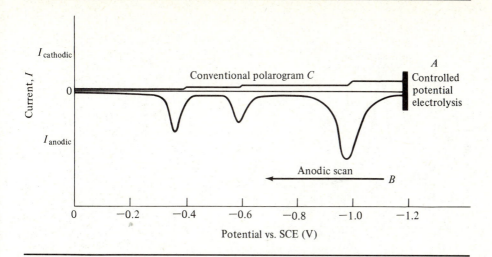

FIGURE 17-17 Anodic stripping voltammetry. The solution is electrolyzed at controlled potential (*A*) and then scanned in the anodic direction (*B*). A conventional polarogram is shown for comparison (*C*).

throughout the process even while the solution is stirred during the controlled-potential electrolysis. When the analysis is completed, the drop can be knocked off and a new one formed.

There are several variables that must be controlled if reproducible results are to be obtained. The controlled-potential electrolysis step is not exhaustive, as described in Section 16-2. Exhaustive electrolysis would require an inordinate amount of time because of the very small ratio of electrode area to solution volume. An electrolysis time of a few minutes is typical, but longer times are sometimes necessary for extremely low concentrations ($< 10^{-7}$ F). Since only a fraction of the metal ions is reduced, it is important that the electrolysis time and stirring conditions be kept constant. After the controlled-potential electrolysis, a short period of time without stirring is recommended to allow convection to decay. During this time, diffusion of the deposited metals into the mercury drop will also continue. Therefore, this waiting period (15 to 30 sec) must also be controlled carefully. The anodic sweep rate (millivolts per second) is another experimental variable that will affect peak currents. When all variables are carefully controlled, Equation 17-33 will be valid:

$$I_p = kC \tag{17-33}$$

where I_p is the current and C is the analyte concentration in the original solution. Because of the many variables, quantitation is usually accomplished with the standard addition method, which has been described several times previously in this text. Peak currents may be only a few nanoamps (10^{-9} A) for 10^{-8} F solutions; and,

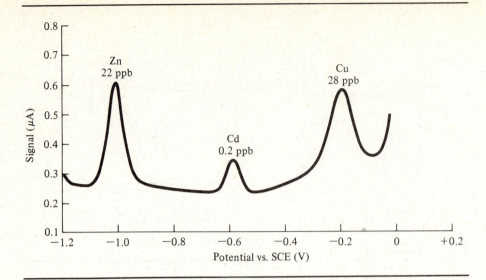

FIGURE 17-18 Differential pulse anodic stripping voltammetry. Tap water sample in ammonia buffer at pH 8.4, showing presence of zinc, cadmium, and copper.

of course, the mercury, solvents, and cell must be scrupulously clean so that residual currents arising from impurities do not mask these small signals.

Pulsing techniques mentioned in the previous section can also be applied to anodic stripping to improve sensitivity. Metal ion impurities in tap water at the level of parts per billion have been determined, as shown in Figure 17-18: Controlled-potential electrolysis was carried out at -1.2 V vs. SCE (in order to observe zinc, which has $E_{\frac{1}{2}} = -1.0$ V vs. SCE). The peak potentials for Zn, Cd, and Cu were -1.0, -0.6, and -0.2 V vs. SCE. Note that the scan recording has been rotated 180° from the one in Figure 17-17: Anodic currents are now in the "up" direction, and positive potentials are to the right of zero. A similar procedure has been reported for the determination of lead in blood.

17-4 CYCLIC VOLTAMMETRY

Several other voltammetric techniques have been developed, primarily to use solid electrodes in place of the DME. Mercury has one big disadvantage in that electrochemical reactions occurring at positive potentials versus SCE cannot be studied. Platinum is an ideal electrode to use for electrochemical reactions taking place in this potential region. Several other electrodes, such as those made of carbon, gold, and even semiconductors, have also been studied.

The main problem in using solid electrodes for voltammetry is the troublesome time dependence shown in Equation 17-5. When a potential is reached at which reduction (or oxidation) occurs, the concentration of reactant at the electrode surface decreases, as predicted from the Nernst equation. This will cause an increasing concentration gradient, which leads to an increase in current until the concentration at the electrode reaches zero. However, the diffusion region thickness [δ in Figure 17-2(**B**)] begins to increase—which acts to lower the concentration gradient. Thus the current will increase up to a peak value and then fall off as the time dependence takes over. The peak current is observed just past $E_{\frac{1}{2}}$ (28 mV/n), as shown in Figure 17-19.

If the scan is reversed (Figure 17-19), the electrochemical *products* present have an opportunity to react. Thus, if the initial scan involved reduction, such as Fe^{3+} to Fe^{2+}, then during the reverse scan an anodic peak due to oxidation of Fe^{2+} to Fe^{3+} can be seen. The same diffusion conditions apply so that the peak will be beyond $E_{\frac{1}{2}}$ by 28 mV/n. The difference in peak potentials will therefore be 56 mV/n. This analytical method is called *cyclic voltammetry*.

The peak current is proportional to concentration, which makes the method useful for analysis. However, for mixtures, the complicated dependence of current on potential means that a second peak current will in general depend on the concentration of the first analyte and the peak separation as well as on the concentration of the second analyte. This makes calculating the concentration of the second analyte more difficult.

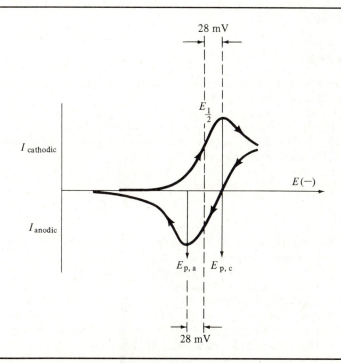

FIGURE 17-19 Cyclic voltammetry. The initial scan shows reduction of Fe^{3+} to Fe^{2+}, whereas the reverse scan shows oxidation to Fe^{3+} of Fe^{2+} produced during initial scan. Peak potentials lie beyond $E_{\frac{1}{2}}$ by 28 mV, and peak separation is 56 mV.

Cyclic voltammetry is used primarily to study new electrochemical reactions. The potentials for oxidation and reduction steps are readily determined as well as the n values for the reactions. As might be anticipated, the peak separation of 56 mV/n is valid only if the reaction is reversible, and therefore measurement of peak separation yields information on reversibility and electrochemical kinetics.

Several other voltammetric techniques are available to the electrochemist for the study of electrochemical reactions. Some, like cyclic voltammetry, use unstirred solutions so that diffusion is the sole means of transport. Others employ controlled stirring, such as with a rotating-disk electrode. Again, these techniques are used more for the study of electrochemical reactions than for routine analysis.

SUPPLEMENTARY READING

Articles Describing Analytical Methods or Applications

Adams, R.N., "Probing Brain Chemistry with Electroanalytical Techniques," *Analytical Chemistry*, 48:1126A (1976).

Copeland, T.R., and R.K. Skogerboe, "Anodic Stripping Voltammetry", *Analytical Chemistry*, 46:1257A (1974).

Koryta, J., "Discovery of Polarography," *Journal of Chemical Education*, 49:183 (1972).

Stock, J.T., "Automated Anodic Stripping Voltammetry," *Journal of Chemical Education* 57:A125 (1980).

Reference Books

Bard, A.J., and L.R. Faulkner, *Electrochemical Methods, Fundamentals and Applications.* New York: John Wiley & Sons, Inc., 1980.

Heyrovsky, J., and P. Zuman, *Practical Polarography.* New York: Academic Press, Inc., 1968.

Kolthoff, I.M., and J.J. Lingane, *Polarography*, Interscience Publishers, Inc., 1952.

Willard, H.H., L.L. Merritt, Jr., and J.A. Dean, *Instrumental Methods of Analysis*, 5th ed., New York: D. Van Nostrand Co., 1974.

PROBLEMS

*17-1. For the Ilkovic equation, the flow rate of mercury is frequently found experimentally as milligrams per drop. If mercury flow is 15 mg per drop and the drop time is 4.0 sec per drop, calculate the drop factor, $m^{\frac{2}{3}}t^{\frac{1}{6}}$.

17-2. For the DME described in Problem 17-1, calculate the expected diffusion current for the reduction of 5.0×10^{-4} F Zn^{2+}, which has a diffusion coefficient of 0.72×10^{-5} cm^2/sec.

*17-3. The diffusion coefficient (D) for Tl^+ is 2.00×10^{-5} cm^2/sec, whereas D for Cd^{2+} is 0.72×10^{-5} cm^2/sec. If a 1.00×10^{-3} F solution of Cd^{2+} gives a diffusion current of 8.15 μA, predict the diffusion current for 1.00×10^{-3} F Tl^+ under the same conditions.

*Answers to problems marked with an asterisk will be found at the back of the book.

17-4. A test of diffusion control as opposed to kinetic control (rate of electrochemical reaction) is to measure the limiting current for a polarographic wave as a function of pressure head on the DME. Keeping in mind that the drop rate (drop per second) is approximately proportional to the pressure head while the drop size (milligrams per drop) is reasonably constant, deduce the expected relationship between pressure head and observed current if the wave is diffusion controlled (follows the Ilkovic equation).

17-5. Lead(II) is reduced polarographically with a half-wave potential of -0.40 V vs. SCE. A 0.353-g sample of an ore containing lead is dissolved in acid and diluted to 100 mL. The diffusion current is 8.53 μA. Standard solutions give the following results:

Pb^{2+} (mF)	$I_d (\mu A)$
0.5	3.92
1.0	7.90
2.0	15.85

a. Sketch the polarogram from $+0.5$ to -2.0 V vs. SCE.
*b. Calculate the percentage of Pb_3O_4 in the ore.

17-6. Cadmium(II) is reduced polarographically with a half-wave potential of -0.605 V vs. SCE. A 0.850-g sample of an ore containing cadmium is dissolved in acid and diluted to 250 mL. The diffusion current is 9.61 μA. Standard solutions give the following results:

Cd^{2+} (mF)	$I_d (\mu A)$
0.5	4.12
1.0	8.26
2.0	16.56

a. Sketch the polarogram from $+0.5$ to -2.0 V vs. SCE.
b. Calculate the percentage of Cd in the ore.

17-7. Zinc(II) is reduced polarographically with a half-wave potential of -1.1 V vs. SCE. A 650-mg sample of brass is dissolved and diluted to 50 mL. A 10-mL aliquot buffered with NH_3/NH_4Cl and diluted to 100 mL gives a diffusion current of 5.13 μA. To a second 10-mL aliquot 10 mL of 1.60×10^{-2} F $ZnSO_4$ is added and then buffered and diluted to 100 mL. The diffusion current is now 18.25 μA.

a. Sketch the first polarogram from 0 to -2.0 V vs. SCE.
b. Calculate the percentage of Zn in the brass sample.

17-8. Thallium(I) is reduced polarographically with a half-wave potential of -0.46 V vs. SCE. A 100-mL solution containing 0.15 F KCl is saturated with TlCl and exhibits a diffusion current of 8.25 μA. Standard thallium solutions give the following results:

$Tl(I)$ (mF)	$I_d (\mu A)$
0.5	2.90
1.0	5.78
2.0	11.62

a. Sketch the polarogram from 0 to -2.0 V vs. SCE.
*b. Calculate K_{sp} for TlCl.

17-9. A 764-mg nickel ore sample is dissolved in acid and diluted to 50 mL. A 10-mL aliquot is added to 50 mL NH_3/NH_4Cl buffer plus Triton X-100 (as a maximum suppressor) and diluted to 100 mL. A polarogram is recorded, and the nickel wave at -1.1 V vs. SCE has a diffusion current of 8.25 μA. A second 10-mL aliquot of the dissolved ore sample is added to 50 mL NH_3/NH_4Cl buffer plus Triton X-100 and spiked with 10 mL of 0.0135 F Ni(II) before dilution to 100 mL. The nickel wave is now 17.92 μA. Calculate the percentage NiO in the ore sample.

*17-10. Determination of riboflavin in vitamin formulations can be carried out polarographically because it is more easily reduced ($E_{\frac{1}{2}} = -0.47$ V vs. SCE at pH 7) than are other vitamin B factors. A 10-mL aliquot of a liquid diet supplement was added to a 0.1 F phosphate supporting electrolyte at pH 7.2 and diluted to 50 mL. The diffusion current was 0.28 μA. A standard 40-ppm riboflavin solution in the same electrolyte gave a diffusion current of 0.45 μA. Calculate the riboflavin content in the diet supplement as milligrams per 100 mL.

17-11. A 20-mL aliquot of gasoline is extracted with 20 mL of HCl to remove lead. The HCl solution is diluted to 100 mL in a volumetric flask. A polarogram is run, and the diffusion current is 0.82 μA at -0.6 V vs. SCE. A standard containing $5.0 \times 10^{-4} F$ Pb^{2+} gives a diffusion current of 3.75 μA. Calculate tetraethyllead (TEL) content as grams of TEL per gallon. The formula for TEL is $(C_2H_5)_4Pb$; there are 3.79 L/gal.

17-12. Vitamin C (L-ascorbic acid) gives an anodic wave at $+0.1$ V vs. SCE that can be used for analysis. A 20-mL sample of orange juice stabilized with the addition of a small amount of oxalic acid is filtered, buffered to pH 8, and diluted to 50 mL. The anodic wave has a limiting current of 8.17 μA with a drop factor ($m^{\frac{2}{3}}t^{\frac{1}{6}}$) value of 2.65. The vitamin C wave has a diffusion current constant $I_d/Cm^{\frac{2}{3}}t^{\frac{1}{6}} = 3.00$. Calculate the vitamin C content as milligrams per 100 mL.

*17-13. An analyst can determine sulfate concentration by adding excess Pb^{2+} to form $PbSO_4$ and polarographically determining the amount of excess Pb^{2+}. The analyst prepares a standard by adding 10.00 mL of 0.0100 F $Pb(NO_3)_2$ to 20 mL of ethanol (to decrease K_{sp} for $PbSO_4$), adding two drops of Triton X-100, and diluting to 100 mL with 0.1 F NaCl. The diffusion current is 7.87 μA. A sample trial is prepared in the same way, but in addition, a 2.00-mL aliquot of seawater is added before dilution to 100 mL. The diffusion current is now 4.35 μA. Determine the sulfate (SO_4^{2-}) concentration in seawater as parts per million (milligrams per liter).

17-14. A 2.50-g sample of plant tissue is ashed at 500°C, and the residue is dissolved in 1 F HCl. Zinc is removed by extraction with dithizone and then leached back into an aqueous medium with HCl. The sample is diluted to 100 mL in a volumetric flask. Part of the sample is analyzed polarographically with a diffusion current for Zn(II) of 0.58 μA. Then a 5.00-mL aliquot of 1.2 $\times$ $10^{-3} F$ Zn(II) is added to a 50-mL aliquot of the remaining sample solution. The diffusion current for the zinc wave is now 1.35 μA. Calculate the zinc content in the plant tissue as parts per million (micrograms per gram).

17-15. A 400-mg sample containing iron is dissolved and treated to make 100 mL of a Fe(II) solution. A 20.0-mL portion of this is polarographed, yielding a diffusion current of 42.0 μA. To the remaining portion of the initial solution 5.00 mL of a $1.00 \times 10^{-2} F$ ferrous ammonium sulfate solution is added. A 20.0-mL aliquot portion of the resulting solution yields a diffusion current of 58.5 μA. Calculate the percentage of Fe in the original sample.

*17-16. A polarographic wave gives the data tabulated below. Plot the data according to Equation 17-23; and determine whether the reduction appears reversible (that is, gives a straight line with a theoretically consistent slope), find the n value for the wave, and graphically determine the half-wave potential $(E_{\frac{1}{2}})$.

E (V vs. SCE)	μA	E (V vs. SCE)	μA
−0.37	0.8	−0.42	11.7
−0.38	1.4	−0.43	13.3
−0.39	3.2	−0.44	15.4
−0.40	5.4	−0.45	16.2
−0.41	8.5	−0.60	$17.2 = I_d$

17-17. A polarographic wave gives the data tabulated below. Plot the data according to Equation 17-23; and determine whether the reduction appears reversible (that is, gives a straight line with a theoretically consistent slope), find the n value for the wave, and graphically determine the half-wave potential $(E_{\frac{1}{2}})$.

E (V vs. SCE)	μA	E (V vs. SCE)	μA
−0.56	0.62	−0.66	7.68
−0.58	1.12	−0.68	8.64
−0.60	2.76	−0.70	9.60
−0.62	4.80	−0.80	$10.24 = I_d$
−0.64	6.24		

17-18. The overall formation constant for $Cd(NH_3)_4^{2+}$ is 1.3×10^7. Predict $E_{\frac{1}{2}}$ for Cd(II) reduction in 0.5 F NH_3, based on $E_{\frac{1}{2}} = -0.60$ V vs. SCE for the simple Cd^{2+} reduction to Cd metal.

17-19. $E_{\frac{1}{2}}$ for Cu(II) reduction to Cu metal in a noncomplexing HNO_3 supporting electrolyte is −0.01 V vs. SCE, whereas, in 0.05 F EDTA at pH 10, $E_{\frac{1}{2}}$ shifts to −0.51 V vs. SCE. Calculate the formation constant for the Cu–EDTA complex.

*17-20. A 50-mL water sample is buffered to pH 8.4 and diluted to 100 mL. Then differential pulse anodic stripping is carried out as described in the text. Using Figure 17-18 as a reference sample, from the peak values given below, determine metal ion contamination as parts per billion in the water sample:

E vs. SCE(V)	Signal (μA)
−1.0	0.82
−0.6	0.0
−0.2	1.65

17-21. A sample of pottery being considered for import is leached with 4% acetic acid for 24 hr. A 50-mL aliquot is transferred to an electrolysis cell, and after controlled-potential electrolysis at −1.0 V vs. SCE, an anodic scan is recorded with the differential pulse method. A 2.5-μA peak due to Pb is observed at −0.4 V vs. SCE. Then 10-μL aliquots of a standard containing 12 mF $Pb(NO_3)_2$ are added; the peak values are given below. Plot the peak current as a function of Pb standard added, and extrapolate to zero current to determine the amount of Pb present in the pottery. Does the pottery meet the FDA standard of <7 ppm?

μL added	I_p (μA)
0	2.5
10	3.7
20	4.9
30	6.0

ANALYTICAL SEPARATIONS

PART SIX

Principles of Analytical Separations

18

Chapters 5 to 17 dealt with signals that could be used in the quantitation step of the total analysis process, and it was generally assumed that the analyte had been separated from any substance that would interfere with that quantitation signal. We now turn our attention to the methods used to accomplish step 3—separation—of the total analysis process.

Quite simply, the separation of an analyte from an interference requires a physical phase separation: That is, the analyte ends up in one physical phase (solid, liquid, or gas), while the interference ends up in another physical phase (solid, liquid, or gas). The analyte and interference are originally in the same phase (initial phase for the separation process), and then either the analyte or the interference is transferred to a new phase (final phase for the separation process). It does not matter to the development of analytical separations whether the analyte or the interference is transferred although practical laboratory procedures may depend on the transfer.

There are only three different physical phases so that all analytical separations can be classified into one of nine possible phase transfers. These are listed in Table 18-1, with examples of analytical techniques for each.

Some of these combinations are more useful for analysis than others are. For example, much of analytical chemistry is carried out in solution; so the initial phase is a liquid. But this does not mean that the original sample must be a liquid. Frequently, the sample will be solid, but the first step in the analysis will be the dissolution of a weighed amount of sample to form a solution containing the analyte as well as any possible interferences. This solution is the initial phase for the separation step in the analysis.

One simple separation method is the precipitation of the analyte or the interference from the initial phase. This represents a liquid–solid separation. There are other examples of liquid–solid separations that will be discussed in Chapter 19, Chromatography.

TABLE 18-1   Phase Separations

Initial phase	Final phase	Analytical technique
liquid	solid	precipitation ion-exchange chromatography (IEC) liquid–solid chromatography (LSC) thin-layer chromatography (TLC)* high-performance liquid chromatography (HPLC)[†]
liquid	liquid	liquid–liquid extraction (solvent extraction) liquid chromatography (LLC) high-performance liquid chromatography (HPLC)[†] thin-layer chromatography (TLC)* paper chromatography (PC)
liquid	gas	distillation
gas	solid	adsorption absorption gas–solid chromatography (GSC)
gas	liquid	absorption gas–liquid chromatography (GLC)
gas	gas	gaseous diffusion (isotope separation)
solid	solid	magnetic separation
solid	liquid	leaching
solid	gas	sublimation

*TLC may be either a liquid–solid or a liquid–liquid separation, depending on experimental conditions (see Chapter 19).
[†]HPLC represents a high-pressure method for carrying out chromatographic separations that can be applied to several liquid–solid or liquid–liquid separations.

The final phase may also be a liquid, leading to liquid–liquid separations. The distribution of a substance between two immiscible liquid phases is an important analytical tool, known as "solvent extraction" or "liquid–liquid extraction." The principles of solvent extraction discussed later in this chapter help to form the basis of all chromatographic methods.

If the initial liquid phase is heated, it may be possible to volatilize the analyte or the interference, the volatilization constituting a distillation method of separation. We shall see later that it is possible to separate two volatile substances by fractional distillation.

There are times when the initial phase is a gas. When the gas is passed through a solid, it is frequently possible to *absorb* or *adsorb* the analyte or the interference. For example, when an air sample is passed through a tube containing solid KOH, any CO_2 will be absorbed by the KOH. A solution of KOH could also be used; then the separation would actually be a gas–liquid type. Absorption of CO_2 was an essential part of the carbon–hydrogen analysis discussed in Section 5-1. You may know the technique of collecting toxic pollutants in air by adsorption on activated charcoal, as in a gas mask. Such adsorption processes, as gas–solid separation, can be used as part of an analytical method.

One of the most widely used analytical separation methods involving a gaseous initial phase is gas–liquid chromatography (abbreviated GLC or just gas chromatography, GC). The basis for GLC separations is the selective solubility of volatile components in the final liquid phase. It is also possible to accomplish a chromatographic separation with gas–solid chromatography (GSC), but this technique is not nearly so widely used as is GLC.

Unlike liquids, all gases are mutually soluble; so, strictly speaking, we cannot take an initial gaseous phase and separate out a second gaseous phase. However, we can separate two substances in the gaseous phase by making use of their different rates of diffusion. Thus $^{235}_{92}U$ can be separated from $^{238}_{92}U$ by gaseous diffusion of the volatile UF_6. This type of separation is rarely used for analytical work.

Solid initial phases are not so easy to manipulate as fluid phases are so that there are few solid separation processes of analytical importance. However, in the processing of ores, magnetic separation is widely used, in which magnetic solid particles are separated from nonmagnetic particles. *Leaching* of analytes (or of an interference) from a solid sample by contact with a liquid is more common: For example, with a toluene–chloroform solvent, nicotine can be leached from tobacco prior to titration.

As in the case of liquid–gas separations by distillation, it is also occasionally possible by heating to volatilize a component from a solid sample. The direct transfer from solid to gas is called *sublimation*. One important example is the volatilization of mercury from solid samples; the volatilized mercury can be collected on gold and then quantified.

Thus all nine combinations may be described, but only a few lead to important analytical methods. A few of them will be probed in this chapter; Chapter 19 will be devoted to the many chromatographic separations listed in Table 18-1.

The separation step can be such an important part of the total analysis that, instead of named according to the quantitation step as is usually done (gravimetric, volumetric, spectrochemical, or electroanalytical), the analysis may be named according to the separation step (see Table 1-1). Thus you may see, for example, articles on the determination of pollutants in automobile exhaust gases by gas–liquid chromatography. The actual quantitation in such an analysis will depend on the detection system used, but this is just incidental to the separation step, which is critical to the success of the analysis.

18-1 EQUILIBRIUM CONSTANT

One unifying principle is applicable to all separation processes. The separation process will have an equilibrium constant of the form

$$A(1) \rightleftharpoons A(2) \tag{18-1}$$

$$K = \frac{[A(2)]}{[A(1)]} \tag{18-2}$$

in which substance A is transferred from the initial phase (1) to the final phase (2). This expression of a physical separation process is analogous to a chemical reaction. Substance A in the initial phase is the "reactant," whereas in the final phase it is the "product" for the separation process. Just as in any chemical reaction, the equilibrium constant equals the product (final) activity divided by the reactant (initial) activity. It is important to remember that *activities* are involved because, with two different phases (1 and 2), the standard state for A may be different for the two phases. However, we shall frequently develop approaches in which the activities will be approximated by concentrations.

18-2 PRECIPITATION

Precipitation as a means of separating an analyte from the rest of the sample was discussed at length in Chapter 5. After separation, the quantitation could involve weighing the precipitate in a gravimetric method. However, the precipitation step could be used only for separation, with quantitation carried out by some other method: For example, calcium may be precipitated as calcium oxalate, followed by titration of the oxalate with $KMnO_4$, which is then a volumetric method of analysis.

With calcium oxalate as an example, the precipitation separation equilibrium would be

$$Ca^{2+} + C_2O_4^{2-} \rightleftharpoons CaC_2O_4(s)$$

$$K = \frac{[CaC_2O_4]}{[Ca^{2+}][C_2O_4^{2-}]} = \frac{1}{[Ca^{2+}][C_2O_4^{2-}]} \tag{18-3}$$

The standard state for solid CaC_2O_4 is the pure CaC_2O_4; and therefore, whenever the system is in equilibrium with pure CaC_2O_4, its activity will be 1, as indicated in Equation 18-3.

Traditionally, the precipitation equilibrium constant is not quoted but rather the solubility product, which is the equilibrium constant for the reverse reaction,

$$K_{sp} = [Ca^{2+}][C_2O_4^{2-}] = \frac{1}{K} \tag{18-4}$$

It was also demonstrated in Section 5-2 that separation of two insoluble components in a solution could sometimes be accomplished by control of the precipitant concentration (Examples 5-4 to 5-7). That is, if sufficient difference in solubility product existed between the two components, a concentration of precipitant could be calculated at which the more insoluble one was quantitatively precipitated while the more soluble one remained in solution.

One can also envisage a separation process in which a "clean" separation as described above would not be possible. The separation would require several precipitation steps, and each one would represent an equilibrium *stage*. This concept of employing several equilibrium stages to accomplish a complete separation will be of great use in other separation methods, particularly chromatographic separations.

For precipitation, the method is not very useful because one precipitation is laborious and several would be impractical. Also, nonequilibrium effects, such as coprecipitation, would prevent attainment of true equilibrium conditions at each stage. This would make the process less efficient so that 10 precipitations might represent only three or four true equilibrium stages. In fact, double or even triple precipitations are sometimes required to eliminate the nonequilibrium effects (coprecipitation) for separations that theoretically should be possible with one precipitation step.

Needed for fast, effective separations are methods by which the analyst can accomplish many equilibrium stages without going through each stage one at a time and the remainder of this book is devoted to such methods.

18-3   DISTILLATION

One simple way to separate an analyte or an interference from a liquid solution is to transfer it to the gaseous state by *distillation*. Many students may be quite familiar with distillation from laboratory work in organic chemistry. However, we shall examine the process of distillation from an analytical viewpoint, as an introduction to other separation processes.

For separation by distillation to be successful, the substance must exhibit some volatility, which is normally expressed as its *vapor pressure* in the liquid system. The vapor pressure of a pure substance, P^0, is determined by only one variable: the temperature. Thus the vapor pressure of pure water at 25°C is 24 torr; at 100°C, the vapor pressure is 760 torr (1 atm), the normal boiling point. When a solute is added to the water, the vapor pressure of water is decreased. Thus the temperature must be higher than 100°C to reach a vapor pressure of 760 torr. The solute has decreased the vapor pressure and elevated the boiling point. The vapor pressure of water in the solution will depend on both the temperature *and* the activity of solute. For an *ideal solution* (no interaction between solvent and solute particles), the activity of solvent can be expressed as its mole fraction, χ. The vapor pressure at a given temperature will be directly proportional to its mole fraction,

$$P \propto \chi$$

$$P = k\chi \tag{18-5}$$

when $\chi = 1$, the solvent is pure, and $P = P^0$. Therefore,

$$P = P^0\chi \tag{18-6}$$

Equation 18-6 is an expression of *Raoult's law*.

If the solute also has a vapor pressure and forms an ideal solution, its vapor pressure will also be given by Equation 18-6, where P^0 will be the vapor pressure of the pure solute and χ its mole fraction in the solution. The *total* vapor pressure of the solution will be the sum of the vapor pressures of the constituents. It is not necessary to identify one as the solvent and the others as solutes. For an ideal solution containing two components, A and B, Raoult's law can be written as

$$P_{total} = P_A^0 \chi_A + P_B^0 \chi_B \tag{18-7}$$

Because the system contains only two components, it follows that

$$\chi_A + \chi_B = 1$$
$$\chi_B = 1 - \chi_A \tag{18-8}$$

Therefore

$$P_{total} = P_A^0 \chi_A + P_B^0 (1 - \chi_A) \tag{18-9}$$

A plot of vapor pressure for each of the components and the total vapor pressure as a function of composition (χ_A) at a given temperature will resemble Figure 18-1. The vapor pressure of each component will be a straight line (Equation 18-6), and the sum of two straight lines is also a straight line.

If we consider distillation as a separation process,

$$A(l) \rightleftharpoons A(g)$$

$$K = \frac{[A(g)]}{[A(l)]} \tag{18-10}$$

As usual, the equilibrium constant involves the activity in the final phase divided by the activity in the initial phase. The standard state for A(g) will be chosen as 1 atm

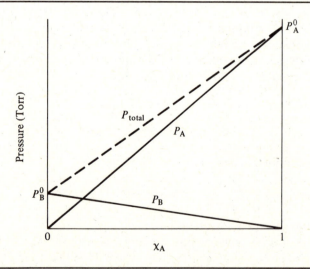

FIGURE 18-1 Vapor pressure–composition diagram for an ideal solution obeying Raoult's law.

pressure so that $[A(g)] = P_A$. The standard state for $[A(l)]$ will be chosen as pure A, where $\chi_A = 1$. Therefore, for an ideal solution, $[A(l)] = \chi_A$, and the equilibrium constant becomes

$$K = \frac{P_A}{\chi_A}$$

$$P_A = K\chi_A \qquad\qquad (18\text{-}11)$$

For an ideal solution, we recognize from Equation 18-6 that the equilibrium constant for the distillation process is just the vapor pressure of pure A:

$$K = P_A^0$$

When the solution is nonideal, the relationship between vapor pressure and composition will be more complex. However, for very dilute solutions, the vapor pressure of a dilute solute will often follow Equation 18-11, but the constant, K, is not equal to P_A^0. The more general relation shown in Equation 18-11 is known as Henry's law, and K is the Henry's law constant for the substance in the solution considered.

For distillation, we really need to know how the boiling point of the solution varies with composition. A diagram of boiling point against composition is constructed as in Figure 18-2 (dark curve). For a solution obeying Raoult's law, it is

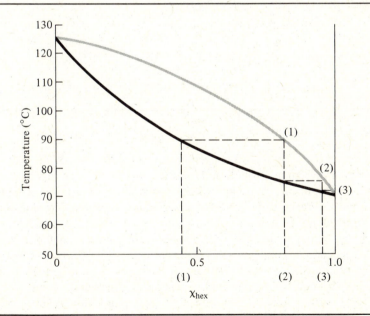

FIGURE 18-2 Boiling point–composition curve for mixtures of *n*-hexane and *n*-octane. The solid curve shows boiling point as a function of liquid composition. The lighter curve shows the corresponding composition of vapor (horizontal tie lines).

simple to find the composition that will boil at a given temperature [Example 18-1(b)]. However, it is more difficult to calculate the boiling point of a given composition—which is the more likely problem. One needs to know how the vapor pressure varies with temperature, and this involves the heat of vaporization for each of the components. Therefore boiling point–composition diagrams are convenient and save computation time. It should be noted, though, that they given information for only one *total* pressure: 760 torr.

EXAMPLE 18-1

(a) Calculate the vapor pressure of a solution containing 20 mol % *n*-hexane and 80 mol % *n*-octane at 90°C, assuming Raoult's law is obeyed. The vapor pressures for the pure components at 90°C are the following: *n*-hexane, $P^0 = 1390$ torr; *n*-octane, $P^0 = 253$ torr.

(b) Calculate the composition of *n*-hexane/*n*-octane solution that boils at 90°C.

Solution

(a)
$$P_{total} = P_{hex} + P_{oct} = P^0_{hex}\chi_{hex} + P^0_{oct}\chi_{oct}$$

$$= 1390 \times 0.20 + 253 \times 0.80 = 480 \text{ torr}$$

(b)
$$P_{total} = 760 = P^0_{hex}\chi_{hex} + P^0_{oct}\chi_{oct}$$

$$760 = 1390\chi_{hex} + 253(1 - \chi_{hex})$$

$$\chi_{hex} = \frac{760 - 253}{1390 - 253} = 0.446 \ (44.6 \text{ mol } \% \ n\text{-hexane})$$

$$\chi_{oct} = 1 - \chi_{hex} = 1 - 0.446 = 0.554 \ (55.4 \text{ mol } \% \ n\text{-octane})$$

Suppose we boil the solution given in Example 18-1(b). What will be the composition of the vapor? With Raoult's law we can calculate the vapor pressure of each of the components, and this will be its partial pressure in the gas phase (Example 18-2). Note how the vapor composition is enriched in *n*-hexane, the more volatile constituent. The vapor composition for each solution composition can also be shown on the boiling point–composition diagram, the lighter curve in Figure 18-2. The distillation has certainly given us some vapor that contains a higher percentage of *n*-hexane than the original solution does, but it is by no means pure. However, the process can be repeated by condensing the vapor (composition 1 in Figure 18-2) to form a new liquid solution (2) of this composition. When solution 2 is boiled, the boiling point will be lower (around 75°C), and the vapor composition will be at point 2 on the lighter vapor-composition curve (around 96 mol% *n*-hexane). To improve the separation, a third distillation can be carried out as illustrated on Figure 18-2.

EXAMPLE 18-2

For the solution given in Example 18-1(b), calculate the vapor composition.

Solution

$$P_{total} = P_{hex} + P_{oct} = P^0_{hex}\chi_{hex} + P^0_{oct}\chi_{oct}$$

Therefore,

$$P_{hex} = P^0_{hex}\chi_{hex} = 1390 \times 0.446 = 620 \text{ torr}$$

$$P_{oct} = P^0_{oct}\chi_{oct} = 253 \times 0.554 = 140 \text{ torr}$$

$$\overline{\phantom{P_{oct} = P^0_{oct}\chi_{oct} = 253 \times 0} 760 \text{ torr total}}$$

Vapor composition:

$$\chi_{hex} = \frac{620}{760} = 0.816 \text{ (81.6 mol \% } n\text{-hexane)}$$

$$\chi_{oct} = \frac{140}{760} = 0.184 \text{ (18.4 mol \% } n\text{-octane)}$$

The ability to separate two components by distillation depends on their difference in vapor pressure. The mole fraction of component A in the vapor ($\chi_{A,v}$) is

$$\chi_{A,v} = \frac{P_A}{P_{total}} = \frac{\chi_{A,1}P^0_A}{P_{total}} \tag{18-12}$$

For component B, we can state

$$\chi_{B,v} = \frac{P_B}{P_{total}} = \frac{\chi_{B,1}P^0_B}{P_{total}} \tag{18-13}$$

Equations 18-12 and 18-13 can be combined to find the ratio $\chi_{A,v}/\chi_{B,v}$:

$$\frac{\chi_{A,v}}{\chi_{B,v}} = \frac{\chi_{A,1}P^0_A/P_{total}}{\chi_{B,1}P^0_B/P_{total}} = \frac{\chi_{A,1}P^0_A}{\chi_{B,1}P^0_B} \tag{18-14}$$

The ratio P^0_A/P^0_B is the relative volatility of A with respect to B and is given the symbol α. For a two-component mixture we can write

$$\frac{\chi_{A,v}}{\chi_{B,v}} = \frac{\chi_{A,v}}{1 - \chi_{A,v}} = \alpha\frac{\chi_{A,1}}{\chi_{B,1}} = \alpha\frac{\chi_{A,1}}{1 - \chi_{A,1}} \tag{18-15}$$

One method for carrying out a multistep distillation process is to use a bubble-cap distillation tower (Figure 18-3), commonly found in petroleum refineries. The first distillation (stage 1) occurs in the still pot, and the vapor is condensed on the first plate. This condensate is distilled (stage 2), and the vapor is condensed on the second plate. The temperature of each plate will be progressively lower as the composition becomes more and more enriched in the more volatile component.

Equation 18-15 can be applied to each plate if we recall that the liquid composition for each stage will be the same as the vapor composition for the previous stage. Thus on plate 1 (stage 2),

$$\frac{\chi_{A,v_2}}{1 - \chi_{A,v_2}} = \alpha \frac{\chi_{A,1_2}}{1 - \chi_{A,1_2}} = \alpha \frac{\chi_{A,v_1}}{1 - \chi_{A,v_1}} = \alpha^2 \frac{\chi_{A,1_1}}{1 - \chi_{A,1_1}} \tag{18-16}$$

So the vapor composition can always be related to the original known liquid composition in the still pot. On plate $n - 1$ (n stages),

$$\frac{\chi_{A,v_n}}{1 - \chi_{A,v_n}} = \alpha^n \left(\frac{\chi_{A,1_1}}{1 - \chi_{A,1_1}} \right) \tag{18-17}$$

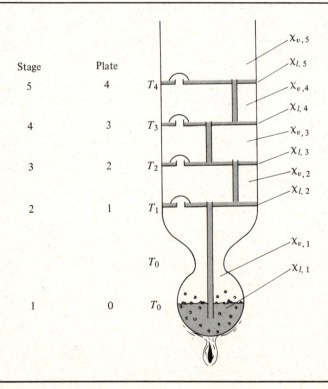

Stage	Plate
5	4	T_4
4	3	T_3
3	2	T_2
2	1	T_1
		T_0
1	0	T_0

$X_{v,5}$
$X_{l,5}$
$X_{v,4}$
$X_{l,4}$
$X_{v,3}$
$X_{l,3}$
$X_{v,2}$
$X_{l,2}$
$X_{v,1}$
$X_{l,1}$

FIGURE 18-3   Operation of a bubble-cap distillation column.

The left side of Equation 18-17 is the final composition, whereas the term in parenthesis on the right side is the original composition. The separation factor, α, enters to a power equal to the number of theoretical stages. With a knowledge of the separation factor, the number of stages, and the initial conditions, we can calculate the final conditions, as shown in Example 18-3.

This textbook is not intended to be a study of distillation, but the terms and concepts developed here will be useful as we look at other separation processes. In general, the final separation conditions will depend on some separation factor raised to a power equal to the number of *theoretical stages* or *theoretical plates* (the terms are used synonomously although there is a slight difference between n and $n - 1$ in the bubble-cap-tower example).

EXAMPLE 18-3

A 40-ft bubble-cap tower is used to separate *p*-xylene from *o*-xylene. If the still pot contains 25 mol % *p*-xylene and the final condensate contains 99.9 mol % *p*-xylene, calculate the number of theoretical stages the column contains. Vapor pressures at 138°C are *o*-xylene = 637 torr and *p*-xylene = 760 torr.

Solution

Equation 18-17 is used to calculate n:

$$\frac{0.999}{1 - 0.999} = \alpha^n \frac{0.25}{1 - 0.25}$$

$$\alpha^n = \frac{0.999/0.001}{0.25/0.75} = 2997$$

$$\alpha = \frac{P^0_{p\text{-xylene}}}{P^0_{o\text{-xylene}}} = \frac{760}{637} = 1.19$$

Therefore,

$$n \log \alpha = \log 2997$$

$$n = \frac{\log 2997}{\log 1.19} = \frac{3.48}{0.077} \cong 45$$

It is also possible to develop laboratory apparatus that will accomplish many theoretical stages in a fashion less defined than that of a bubble-cap tower containing actual physical plates. For example, distillation can be carried out in the laboratory with fractionation columns in which some type of baffle or packing allows vapors to condense and redistill several times as they move up the column. By measurement of the initial and final compositions, Equation 18-17 can be used to determine n, the equivalent number of theoretical stages. In fact, this approach is necessary for the bubble-cap tower too because no distillation column is completely ideal. That is, a tower containing 50 actual plates might contain only 40 equivalent theoretical plates.

The idea of using columns equivalent to many theoretical plates is the basis for chromatographic separations. One way of expressing a column's efficiency is to state the column height required for one theoretical plate. This is called the *height equivalent to a theoretical plate* (HETP). The column described in Example 18-3 had an HETP value of

$$\text{HETP} = \frac{\text{column height}}{\text{theoretical plates}} = \frac{40 \text{ ft}}{45 \text{ plates}} = 0.89 \text{ ft/plate} = 27 \text{ cm/plate}$$

The reason why chromatographic techniques are such potent separation methods is that HETP values are several orders of magnitude smaller than values for real plates in a distillation column; for example, 0.02 cm would be a typical value. Thus a chromatographic column several feet long may contain the equivalent of thousands of theoretical plates.

18-4 LIQUID–LIQUID EXTRACTION (SOLVENT EXTRACTION)

Now we consider another separation process that has proved to be of great value to analysis. When two immiscible liquids (having low mutual solubility) are in contact with each other, a solute species will be distributed between them. For example, many organic solvents have low mutual solubility with water; and a solute species A will be distributed between water and the organic phase according to Equation 18-18:

$$\text{A}(aq) \rightleftharpoons \text{A}(org) \qquad\qquad\qquad \textbf{(18-18)}$$

Equation 18-18 represents a separation process that will have an equilibrium constant as shown in Equation 18-19,

$$K_D = \frac{[\text{A}(org)]}{[\text{A}(aq)]} \qquad\qquad\qquad \textbf{(18-19)}$$

The equilibrium constant, K_D, is called the *distribution coefficient*, or *partition coefficient*, and like any equilibrium constant, it equals final phase activity divided by initial phase activity.

Another way of viewing Equation 18-18 is that solute A has been extracted from an aqueous solvent into an organic solvent. There is no reason why the reverse reaction could not be considered—the extraction of A from the organic solvent to the aqueous solvent. The distribution coefficient for the reverse reaction would equal K_D^{-1}. Thus it is important to know which direction is being considered; to avoid confusion, we shall always assume in an organic–aqueous system that the direction of extraction is from aqueous to organic, as in Equation 18-18.

It is also important to note that charge neutrality must be maintained for each of the phases. That is, cation or anion solutes cannot be transferred to another solvent without some charge compensation mechanism. Generally, this is accomplished by the formation of a neutral extractable species, A. Because A is a solute, we can define its standard state as the ideal 1 m solution so that, in general, activity will equal concentration. The requirement that this is valid only in very dilute solution is not so stringent for a neutral species as it is for ionic solutes. Activity coefficients for neutral species are nearly 1 for all reasonable concentrations.

The distribution coefficient is dimensionless so that any concentration units may be used in Equation 18-19, provided that the same units are used for both phases. Molarity is commonly used, but grams per liter or parts per million could be used just as easily.

Suppose we begin an extraction process with x_0 mmol of A in water and contact it with an organic solvent. This will be accomplished in a separatory funnel, as shown in Figure 18-4. The separatory funnel is stoppered and then shaken vigorously to establish equilibrium. After the two solvents separate, with the denser liquid's forming the lower layer, the two phases can be physically separated by drawing off the bottom layer through the stopcock.

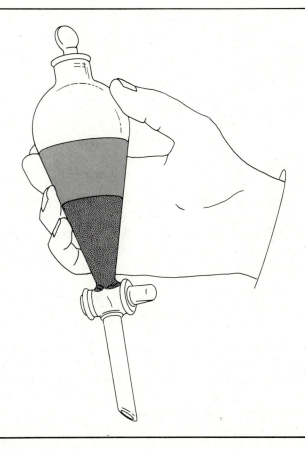

FIGURE 18-4 A separatory funnel, containing two immiscible liquids that have been shaken together and allowed to separate.

At equilibrium, the aqueous layer will now contain x_1 mmol of A. The organic phase must then contain $x_0 - x_1$ mmol of A. To use Equation 18-19, we must know the concentration of A in each of the phases, and this means that we must know the volume of solvent. That is,

$$[A(org)] = \frac{x_0 - x_1}{V_{org}}$$

$$[A(aq)] = \frac{x_1}{V_{aq}}$$

$$K_D = \frac{(x_0 - x_1)/V_{org}}{x_1/V_{aq}}$$

where V_{org} and V_{aq} are, respectively, the volumes of the organic and aqueous phases in milliliters. The volume terms can be combined with the distribution coefficient to give Equation 18-20:

$$K_D \frac{V_{org}}{V_{aq}} = \frac{x_0 - x_1}{x_1} = \frac{x_0}{x_1} - 1 \tag{18-20}$$

Rearrangement of Equation 18-20 leads to

$$\frac{x_0}{x_1} = 1 + K_D \frac{V_{org}}{V_{aq}}$$

$$x_1 = x_0 \left(\frac{1}{1 + K_D V_{org}/V_{aq}} \right) \tag{18-21}$$

According to Equation 18-21, the amount of A remaining in the aqueous phase after the extraction is directly proportional to the initial amount with a proportionality constant of $(1 + K_D V_{org}/V_{aq})^{-1}$. The term $K_D V_{org}/V_{aq}$ will appear regularly as we look at various separation processes; it is called the *capacity factor*, k'. It too is dimensionless and represents the ratio of the amount of A in the organic phase to the amount of A in the aqueous phase. This can be seen as follows:

$$k' = K_D \frac{V_{org}}{V_{aq}} = \frac{[A(org)](mmol/mL)}{[A(aq)](mmol/mL)} \cdot \frac{V_{org}(mL)}{V_{aq}(mL)} = \frac{(mmol\ A)_{org}}{(mmol\ A)_{aq}} \tag{18-22}$$

It may turn out that one extraction is not sufficient to remove A quantitatively from the aqueous phase into the organic phase. To remove more A from the aqueous phase, the analyst may contact the aqueous phase with a fresh volume of the organic solvent. The results of the second extraction will also be determined by Equation 18-21 for the second extraction; that is,

$$x_2 = x_1 \left(\frac{1}{1 + K_D V_{org}/V_{aq}} \right) \tag{18-23}$$

If the new volume of organic phase is the same as that used for the first extraction, then Equations 18-21 and 18-23 can be combined to give

$$x_2 = x_0 \left(\frac{1}{1 + K_D V_{org}/V_{aq}} \right)^2 \qquad \text{(18-24)}$$

For n extractions (theoretical stages), Equation 18-24 can be generalized to

$$x_n = x_0 \left(\frac{1}{1 + K_D V_{org}/V_{aq}} \right)^n \qquad \text{(18-25)}$$

Now we can see the similarity between solvent extraction (Equation 18-25) and distillation (Equation 18-17). The final conditions (x_n) are directly proportional to the initial conditions (x_0) multiplied by a separation factor raised to the nth power, where n is the number of theoretical stages.

Often the analyst may not know how much A is present initially (x_0) but is interested in extracting a certain percentage of it (for instance, at least 99%). The fraction of A remaining in the aqueous phase after n extractions can be found by rearranging Equation 18-25,

$$\frac{x_n}{x_0} = \text{fraction remaining} = \left(\frac{1}{1 + K_D V_{org}/V_{aq}} \right)^n \qquad \text{(18-26)}$$

or as a percentage,

$$\frac{100 x_n}{x_0} = \% \text{ remaining} = 100 \left(\frac{1}{1 + K_D V_{org}/V_{aq}} \right)^n \qquad \text{(18-27)}$$

Uses of Equations 18-25 and 18-27 are shown in Examples 18-4 to 18-6. It should be noted that any units for x_0 may be used, but x_n will then have the same units.

EXAMPLE 18-4

Iodine may be extracted from an aqueous solution into various organic solvents. The distribution coefficient for extraction by CCl_4 is 85. If 50 mL of an aqueous solution containing 2.00×10^{-2} mmol of I_2 is contacted with 30 mL of CCl_4, calculate the amount of I_2 in the aqueous phase and the amount in the CCl_4.

Solution

$$x_1 = x_0 \left(\frac{1}{1 + K_D V_{org}/V_{aq}} \right)^1$$

$$= 2.00 \times 10^{-2} \left(\frac{1}{1 + 85 \cdot 30/50} \right)$$

$$= 3.85 \times 10^{-4} \text{ mmol remaining in the aqueous phase}$$

$$\text{iodine in } CCl_4 = x_0 - x_1 = 2.00 \times 10^{-2} - 3.85 \times 10^{-4} = 1.96 \times 10^{-2} \text{ mmol}$$

EXAMPLE 18-5

(a) Compound X has a distribution coefficient of 2.7 between water and benzene. If a 50-mL solution contains 4.5 mg of X, calculate the volume of benzene needed for a single extraction to remove 99.0% of X.

(b) How many extractions using 50 mL of benzene would be needed to extract 99.0% of X?

Solution

(a) If 99.0% is extracted, then the percentage remaining in the aqueous phase is 1.0%:

$$1.0 = 100 \left(\frac{1}{1 + 2.7 V_{org}/50} \right)^1$$

$$1.0 + 0.054 V_{org} = 100$$

$$V_{org} = \frac{99}{0.054} = 1.83 \times 10^3 \text{ mL}$$

(b)
$$1.0 = 100 \left(\frac{1}{1 + 2.7 \cdot 50/50} \right)^n$$

$$\frac{1.0}{100} = \left(\frac{1}{3.7} \right)^n$$

$$\log 0.01 = n \log 0.270$$

$$n = \frac{-2}{-0.568} = 3.52$$

Since n must be an integer, it will take four extractions to remove at least 99% of X from the aqueous solution. Note how much more efficient several extractions of small volume are, compared with a single extraction. Four extractions with 50 mL each used a total of 200 mL to remove more than 99% (with $n = 4$, the percentage extracted is 99.5%), whereas a single extraction required 1830 mL to remove 99%.

Equation 18-25 was developed with x as millimoles of solute, but moles, grams, pounds, and so forth, could be used instead. In fact, it is not even necessary to use *amounts* of solute; *concentrations* are also acceptable since Equation 18-25 can be converted to Equation 18-28 by dividing both sides by V_{aq},

$$\frac{x_n}{V_{aq}} = \frac{x_0}{V_{aq}} \left(\frac{1}{1 + K_D V_{org}/V_{aq}} \right)^n$$

$$[A(aq)]_n = [A(aq)]_0 \left(\frac{1}{1 + K_D V_{org}/V_{aq}} \right)^n \qquad \textbf{(18-28)}$$

EXAMPLE 18-6

(a) LSD may be removed from an aqueous solution by contact with ether. The distribution coefficient is about 10. If you had an aqueous solution containing 2.5 mg of LSD, how much could be extracted with an equal volume of ether?

(b) What percentage of the LSD could be extracted by dividing the ether volume into four equal portions and extracting four times instead of once, as in part (a)?

Solution

(a)
$$x_1 = 2.5\left(\frac{1}{1 + 10 V_{org}/V_{aq}}\right)^1$$

$$= 2.5\left(\frac{1}{11}\right)^1 = 0.23 \text{ mg remaining}$$

Therefore,

$$\text{amount extracted} = 2.5 - 0.23 = 2.27 \simeq 2.3 \text{ mg (91\%)}$$

(*Note:* Units for x_1 are the same as units for x_0—milligrams.)

(b)
$$V_{org} = \frac{V_{aq}}{4}$$

$$\% \text{ remaining} = 100\left(\frac{1}{1 + 10 V_{aq}/4 V_{aq}}\right)^4$$

$$\% \text{ remaining in aqueous phase} = 0.67\%$$

Therefore,

$$\% \text{ extracted} = 100 - 0.67 = 99.33\%$$

Effects of chemical equilibria on liquid–liquid extraction

The distribution coefficient, K_D, refers to the partitioning of a single species between two solvents. In many chemical systems, however, an analyte may exist in *more than one* form, and the amount of each will depend on the chemical equilibria involved. As a relatively simple example, consider the distribution of a weak acid between water and an organic solvent. In water, the weak acid may exist in either the protonated form, HA, or the dissociated form, A^-, related by the acid dissociation constant, K_a

$$HA + H_2O \rightleftharpoons H_3O^+ + A^- \tag{18-29}$$

$$K_a = \frac{[H_3O^+][A^-]}{[HA]} \tag{18-30}$$

The distribution coefficient, on the other hand, refers to the distribution of the neutral HA species between the two solvents

$$K_D = \frac{[\text{HA}]_{org}}{[\text{HA}]_{aq}} \qquad (18\text{-}31)$$

When an analyst *observes* the distribution of the weak acid between the two solvents, he or she will be measuring the total concentration of acid in both forms in each of the solvents. In other words, a *distribution ratio, D,* is measured, involving formal concentrations. For the moment, let us assume that Equation 18-29 is the only important chemical equilibrium; then the distribution ratio is

$$D = \frac{(f_{\text{HA}})_{org}}{(f_{\text{HA}})_{aq}} = \frac{[\text{HA}]_{org}}{[\text{HA}]_{aq} + [\text{A}^-]_{aq}} \qquad (18\text{-}32)$$

Equation 18-32 can be simplified by incorporating Equation 18-30,

$$D = \frac{[\text{HA}]_{org}}{[\text{HA}]_{aq} + \dfrac{K_a}{[\text{H}_3\text{O}^+]}[\text{HA}]_{aq}} = \frac{[\text{HA}]_{org}}{[\text{HA}]_{aq}(1 + K_a/[\text{H}_3\text{O}^+])} \qquad (18\text{-}33)$$

Now the true distribution coefficient, Equation 18-31, can be used to arrive at

$$D = K_D \frac{1}{1 + K_a/[\text{H}_3\text{O}^+]} \qquad (18\text{-}34)$$

Although K_D is not a function of K_a or pH, the observed distribution ratio is. In strongly acidic solutions, $[\text{H}_3\text{O}^+] \gg K_a$, and $D \cong K_D$. Chemically this is easy to see. In strongly acidic solutions, essentially all the acid exists as undissociated HA, and only the distribution coefficient is of importance. On the other hand, in basic solution, $K_a \gg [\text{H}_3\text{O}^+]$, and $D \to 0$. In other words, essentially all the acid in the aqueous phase is dissociated as A^- so that there is little HA form to distribute between the two phases.

This additional chemical equilibrium gives the analyst more flexibility in designing separation methods. Two weak acids whose K_D values are quite similar may still be separated, provided that their K_a values are significantly different. Calculations for such a situation are carried out in Example 18-7, and D as a function of pH for this example is shown in Figure 18-5.

Control of the pH variable must be considered, especially in multistage extractions. If pH is maintained constant with the use of a buffer system, D may be used in place of K_D in Equation 18-25. However, in an unbuffered system, the pH will change with each extraction by removal of some HA and cause a shift to the left in Equation 18-29 (LeChâtelier principle). It must be kept in mind that buffers themselves are weak-acid (or weak-base) systems, which may distribute between the two solvents.

EXAMPLE 18-7

Two weak acids, HA and HB, have K_a values of 1×10^{-6} and 3×10^{-4}, respectively. Their distribution coefficients between water and ether are 12.2 for HA and 8.3 for HB. Design a separation procedure for the removal of HA ($\sim 99\%$) from an aqueous solution, while retaining HB ($\sim 99\%$) in the aqueous solution.

Solution

We see that, since K_a for HA is smaller than K_a is for HB, it may be possible to find conditions in which HA will be mostly undissociated and able to be extracted while HB is mostly dissociated and will remain in the aqueous phase. It would be impossible to perform the reverse extraction of HB, leaving HA in the aqueous phase.

With the design goal that 99% of the HB remains in the aqueous phase, Equation 18-34 can be used to determine the minimum pH:

$$D = 0.01 = 8.3 \frac{1}{1 + 3 \times 10^{-4}/[H_3O^+]}$$

$$0.01 = \frac{8.3 [H_3O^+]}{[H_3O^+] + 3 \times 10^{-4}}$$

$$[H_3O^+] = \frac{3 \times 10^{-6}}{8.3 - 0.01} = 3.6 \times 10^{-7} \qquad pH = 6.44$$

At any lower pH (higher $[H_3O^+]$), the distribution ratio will be higher. At pH 6.44, the distribution of HA will be

$$D = 12.2 \frac{1}{1 + 1 \times 10^{-6}/3.6 \times 10^{-7}}$$

$$= 3.23$$

This is not sufficient to remove 99% HA in one extraction; but Equation 18-26, with D in place of K_D and assuming equal volumes of the two phases, shows

$$\text{fraction remaining} = 0.01 = \left(\frac{1}{1 + 3.23}\right)^n = (0.236)^n$$

$$n = \frac{\log 0.01}{\log 0.236} = \frac{-2.00}{-0.626} = 3.2$$

Since $n > 3$, a minimum of four extractions will be necessary. This, of course, no longer meets the HB requirement because 1% will be extracted at *each* stage for a total of $\sim 4\%$. However, it can be shown that, by increasing the pH slightly to 6.6, the extraction of HA will be $\sim 99\%$ in four extractions and extraction of HB will be $< 3\%$.

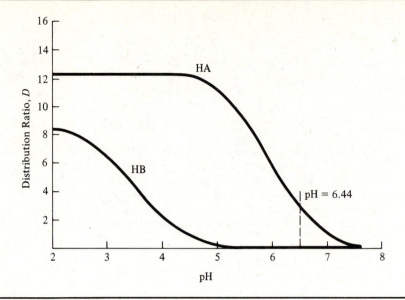

FIGURE 18-5 Distribution ratio as a function of pH for HA ($K_a = 1 \times 10^{-6}$; $K_D = 12.2$) and for HB ($K_a = 3 \times 10^{-4}$; $K_D = 8.3$). See Example 18-7.

Chemical equilibria may also be occurring in the organic solvent. In a nonpolar organic solvent, such as benzene, an organic acid may dimerize by forming species of the type

$$RC \overset{O \;\cdots\; HO}{\underset{OH \;\cdots\; O}{\big\langle}} CR$$

A chemical equilibrium can be written for the dimerization reaction

$$2HA \rightleftharpoons (HA)_2 \qquad\qquad\qquad \text{(18-35)}$$

$$K_{di} = \frac{[(HA)_2]_{org}}{[HA]_{org}^2} \qquad\qquad\qquad \text{(18-36)}$$

The observed distribution will now depend on both acid dissociation in the aqueous phase and dimerization in the organic phase

$$D = \frac{(f_{HA})_{org}}{(f_{HA})_{aq}} = \frac{[HA]_{org} + 2[(HA)_2]_{org}}{[HA]_{aq} + [A^-]_{aq}} \qquad\qquad \text{(18-37)}$$

The coefficient of 2 for the $[(HA)_2]_{org}$ term is necessary because each mole of $(HA)_2$ will contribute 2 mol of organic acid to the total formal concentration in the organic solvent. The chemical equilibrium constants K_a and K_{di} can be incorporated into

Equation 18-37 to give

$$D = \frac{[HA]_{org} + 2K_{di}[HA]^2_{org}}{[HA]_{aq}(1 + K_a/[H_3O^+])}$$

$$D = K_D\left(\frac{1 + 2K_{di}[HA]_{org}}{1 + K_a/[H_3O^+]}\right) \tag{18-38}$$

Equation 18-38 contains the complication that D depends on the concentration of HA in the organic phase because the dimerization reaction is nonlinear (K_{di} depends on $[HA]_{org}$ to the second power). Mathematically, however, the equation is still solvable because the system consists of four unknowns at a given pH: $[HA]_{org}$, $[(HA)_2]_{org}$, $[HA]_{aq}$, and $[A^-]_{aq}$; and we have four equations available to us. They are

1. Mass balance (because of possible volume differences between the aqueous and organic phases, millimoles must be used in place of formal concentration):

   mmol HA total $= [HA]_{aq}V_{aq} + [A^-]_{aq}V_{aq} + [HA]_{org}V_{org} + 2[(HA)_2]_{org}V_{org}$

2. $$K_D = \frac{[HA]_{org}}{[HA]_{aq}}$$

3. $$K_a = \frac{[H_3O^+][A^-]_{aq}}{[HA]_{aq}}$$

4. $$K_{di} = \frac{[(HA)_2]_{org}}{[HA]^2_{org}}$$

Besides acid dissociation, several other types of chemical equilibrium are important to solvent extraction. Many metal ions can be extracted from aqueous solutions through the formation of neutral complexes. For example, Fe(III) can be extracted with ether by the formation of $HFeCl_4$,

$$Fe^{3+} + H_3O^+ + 4Cl^- \rightleftharpoons HFeCl_4 + H_2O \tag{18-39}$$

The formation constant for this reaction is

$$K_f = \frac{[HFeCl_4]_{aq}}{[Fe^{3+}][H_3O^+][Cl^-]^4} \tag{18-40}$$

The neutral $HFeCl_4$ can be partitioned between ether and water:

$$K_D = \frac{[HFeCl_4]_{org}}{[HFeCl_4]_{aq}} \tag{18-41}$$

If Fe^{3+} and $HFeCl_4$ are the only predominant species, the observed distribution ratio will be

$$D = \frac{(f_{Fe})_{org}}{(f_{Fe})_{aq}} = \frac{[HFeCl_4]_{org}}{[HFeCl_4]_{aq} + [Fe^{3+}]_{aq}} \tag{18-42}$$

Equation 18-42 can be put into a more useful form by combination with Equations 18-40 and 18-41:

$$D = \frac{[HFeCl_4]_{org}}{[HFeCl_4]_{aq} + [HFeCl_4]_{aq}/K_f[H_3O^+][Cl^-]^4}$$

$$D = K_D\left(\frac{1}{1 + 1/K_f[H_3O^+][Cl^-]^4}\right) \tag{18-43}$$

Equation 18-43 shows that, if $K_f[H_3O^+][Cl^-]^4 \gg 1$, then $D = K_D$, but for all other conditions $D < K_D$. In other words, K_D is the largest distribution ratio that can be observed and will occur when essentially all the iron is present as $HFeCl_4$ in the aqueous phase. But in reality, there may be several iron species present so that the observed distribution ratio will be a more complicated function of $[H_3O^+]$, $[Cl^-]$, and K_f values.

Equations 18-43 shows the great potential in finding appropriate conditions for analytical separations. Metal ions not only will have different K_D values but also, in general, will have different K_f values. In fact, they may form different types of complexes that exhibit different dependences on pH and ligand concentration.

Once the metal has been extracted from the aqueous phase, it may be desirable to leach it from the organic phase for the quantitation step. The leaching can be accomplished by application of Equation 18-43. An aqueous solution that contains little Cl^- and/or is at a higher pH may have a low D value because $K_f[H_3O^+][Cl^-]^4 \ll 1$. Alternatively, an aqueous solution containing a complexing agent that forms a strong ionic complex will effectively remove $[Fe^{3+}]$ and shift Equation 18-39 to the left, resulting in a low observed distribution ratio; in other words, the iron will be leached from the organic solvent back into the new aqueous solvent. This second distribution process also presents possibilities for analytical separation: Of two metals capable of extraction during the first distribution step, perhaps one can be leached out while the other remains in the organic phase.

APPLICATIONS

When considering the distribution of a species between an aqueous and an organic phase, we ought to recognize that the solvation properties will be quite different. Water has a great tendency to dissolve ionic substances, whereas nonpolar organic solutes show high solubility in nonpolar organic solvents—the old saying that "like dissolves like." Thus liquid–liquid extraction will be useful only for substances showing intermediate properties (for example, polar but nonionic) or for substances having both ionic and nonionic forms (for example, the weak-acid case discussed previously). One application area is the extraction of organic solutes from an aqueous medium in which they have reasonable solubility into an organic solvent.

The extraction of species occurring in an organic medium by an aqueous phase is also known. The analysis of gasoline for tetraethyllead (TEL) content can be accomplished by contacting the gasoline phase with an HCl solution. The lead forms an ionic chlorocomplex in the aqueous phase and is readily removed from the gasoline. Details are given in the laboratory manual.

A much larger area of application is the separation of inorganic species in an aqueous phase by extraction with an organic solvent. In many instances, the analyst may choose precipitation or solvent extraction as a separation method. Solvent extraction can be considerably faster because precipitation and filtration are inherently slow and laborious. For this reason, the extraction procedure may be preferred. And the conditions under which a metal ion will form an extractable neutral complex are quite specific so that separations with selective extractions are often feasible. The most frequently used complexes are chlorides, nitrates, and chelates.

We have already discussed how iron(III), as the $HFeCl_4$ species, can be extracted into ether from an HCl solution. Many other metal ions form extractable metal chlorides, which are listed in Table 18-2. The usefulness of the procedure, also shown

---

TABLE 18-2 Extraction of Metal Chlorides from HCl Solution

Metal	Aqueous phase (HCl)	Organic phase	Extraction (%)
easy extraction, >90%:			
Sb(V)	6.5–8.5 F	isopropyl ether	99.5
Ga(III)	6 F	ethyl ether	97
Au(III)	6 F	ethyl ether	95
Fe(III)	6 F	ethyl ether	99
Mo(VI)	5 F	amyl acetate	99
Sc(III)	8 F	tributyl phosphate	98
Tl(III)	6 F	ethyl ether	90–95
moderate extraction, 40–90%:			
Sb(V)	6 F	ethyl ether	81
As(III)	6 F	ethyl ether	68
Ge(IV)	6 F	ethyl ether	40–60
Mo(VI)	6 F	ethyl ether	80–90
difficult extraction, <40%:			
Sb(III)	6 F	ethyl ether	6
As(V)	6 F	ethyl ether	2–4
Co(II)	4.5 F	2-octanol	9
In(III)	8 F	ethyl ether	3
Hg(II)	6 F	ethyl ether	0.2
Te(IV)	6 F	ethyl ether	34
Sn(IV)	6 F	ethyl ether	17
Sn(II)	6 F	ethyl ether	15–30
V(V)	7.8 F	isopropyl ether	22

SOURCE: Data from G.H. Morrison and H. Freiser, *Solvent Extraction in Analytical Chemistry* (New York: John Wiley & Sons, Inc., 1957).

in Table 18-2, is that many other metals are *not* extracted under these conditions. The distribution equilibria are quite complicated, as mentioned earlier in connection with the iron complex. The experimentally observed distribution ratio for Fe depends not only on HCl concentration but also on iron concentration, making it difficult to remove the last traces of iron from the aqueous phase. There is actually a maximum distribution ratio when the HCl concentration is about 6 *F*, as shown in Figure 18-6. The simplistic approach used to develop Equation 18-43 would not predict such behavior. At such high ionic strengths, activity coefficients cannot be neglected, and the existence of other chlorocomplexes must be considered. In addition, HCl itself can be extracted by ether—which changes the properties of the ether solvent. In any case, essentially all the iron(III) can be removed from a 6 *F* HCl solution with just one ether extraction. This is useful in the analysis of steel samples for trace metal content because the high concentration of iron originally present may interfere with the quantitation steps.

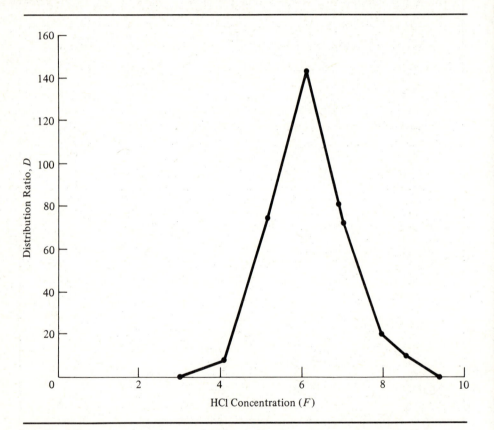

FIGURE 18-6 Extraction of Fe(III) from aqueous HCl solutions with ether. Over 99% of the iron can be removed in one extraction if equal volumes of solvents are used at the optimum concentration near 6 *F* HCl. (*Note:* Above 8 *F* HCl, the H_2O–ether two-phase system is unstable so that data points do not reflect true equilibrium conditions.)

A few metal ions form neutral nitrate complexes, which can be extracted by such organic solvents as ethyl ether, tributyl phosphate, and methyl isobutyl ketone (MIBK). Uranium(VI) is a prime example: It exists in water as UO_2^{2+} and forms $UO_2(NO_3)_2$ in solutions containing high concentrations of NO_3^-. Recommended experimental conditions in the aqueous phase for extraction by ethyl ether are 1.5 F HNO_3 saturated with NH_4NO_3. Extraction of nitrates has been especially useful for the determination of actinide elements such as Am(VI), Np(VI), and Pu(VI) in addition to uranium. Other elements that extract well as nitrate complexes include Ce(IV), Au(III), Sc(III), Pa(IV), Th(IV), and Fe(III).

Chelates offer a rich choice of possible neutral complexes to be extracted from an aqueous medium. In addition to solvents such as ether, the complexes can often be extracted by nonpolar organic solvents, such as chloroform, carbon tetrachloride, or benzene. For example, nickel dimethylglyoxime (described in Chapter 6 in connection with a gravimetric method for the determination of nickel) can be extracted with chloroform. The distribution coefficient is about 300, and quantitative extraction can be accomplished at pH 4.5. Quantitation can be carried out spectrophotometrically by measurement of the absorbance of the red complex in the chloroform solvent.

Another organic chelating agent described previously for gravimetric applications is 8-hydroxyquinoline (oxine). It too forms metal chelates soluble in chloroform. Because the metal ion displaces protons on the chelating agent, the equilibrium will be pH dependent. By controlling the pH, the analyst can accomplish separations when the metals have different formation constants for their chelate complexes.

One other widely used chelating agent in solvent extraction is dithizone (diphenylcarbazone). With a dipositive metal ion, the formation reaction is

$$M^{2+} + 2 \text{ (dithizone)} \rightleftharpoons \text{(complex)} + 2H_2O$$

dithizone

(18-44)

$$\text{(dithizone-metal chelate)} + 2H_3O^+$$

although there is evidence that some metal ions may bond to two nitrogen atoms instead of one nitrogen and one sulfur, as shown in Equation 18-44. Dithizone and its metal chelates are soluble in chloroform and carbon tetrachloride—which makes extraction possible. However, as Equation 18-44 shows by the H_3O^+ released, the formation equilibrium is pH dependent (as with the 8-hydroxyquinoline complexes). Many of the dithizone complexes are intensely colored in the organic solvent so that spectrophotometric determination can be utilized as the quantitation step.

Another type of inorganic compound that exhibits high solubility in organic solvents is the class of heteropoly acids. These are large molecules, such as phosphotungstic acid $(H_3PW_{12}O_{40} \cdot xH_2O)$, in which a central heteroatom (P in the example) is surrounded by MO_6 octahedra (M = W in the example). Tungsten and molybdenum are the primary metals forming the MO_6 groups, but many elements can act as heteroatoms (for example, P, Si, As, or Ge). The heteropoly acids can be extracted with ether or, more efficiently, with solvents such as tributyl phosphate.

Solvent extraction can be used to separate elements whose chemistry is extremely similar, such as actinides, rare earths, or members of the same periodic-table group. For example, niobium can be quantitatively extracted from 11 F HCl with a methylene chloride solvent containing 8% tribenzylamine. Tantalum, which occurs with niobium and usually reacts very similarly, does not extract under these conditions. Another example is the separation of hafnium from zirconium, for which a method has been developed that uses extraction of the thiocyanate complex with ethyl ether: Both hafnium and zirconium partition between the two phases, but there is sufficient difference in their distribution coefficients to effect separation by a multistage extraction procedure. This procedure is different from merely contacting an aqueous solution several times with an organic solvent to remove an analyte quantitatively. The organic layers also contain the interference and must be contacted with fresh aqueous phase to remove the interference. So now we consider multistage extractions and how they may be used to carry out separations.

18-5 MULTISTAGE EXTRACTIONS

Suppose that two substances, A and B, have different distribution coefficients between an aqueous phase and an organic phase but that the coefficients are not sufficiently different to allow quantitative separation in one theoretical stage. As long as there is *some* difference in K_D values, separation of A from B should be possible, but it will require a multistage extraction process.

To visualize a multistage extraction, first consider only the distribution of A, and begin with A in the aqueous phase. The extraction process will then be

$$A(aq) \rightleftharpoons A(org) \tag{18-45}$$

$$K_D = \frac{[A]_{org}}{[A]_{aq}} \tag{18-46}$$

After the first extraction with a volume of organic phase, the aqueous phase moves on to be extracted with a second volume of organic phase. This is similar to the process described earlier. The difference is that the first organic phase at this second step will also be contacted with a volume of fresh aqueous phase containing no A. As shown in Figure 18-7, we have a series of stationary containers containing organic solvent and a series of mobile containers containing aqueous phase, only the first of which originally contains analyte A.

Instead of focusing on the distinction between aqueous and organic, we can generalize Equations 18-45 and 18-46 to identify mobile and stationary phases, as in Equations 18-47 and 18-48. This concept will be quite useful when discussing chromatography in Chapter 19. Note that K is still the ratio of final phase to initial phase.

$$A \text{ (mobile)} \rightleftharpoons A \text{ (stationary)} \qquad \textbf{(18-47)}$$

$$K_D = \frac{[A]_s}{[A]_m} \qquad \textbf{(18-48)}$$

After each stage, we can calculate the distribution of A between the mobile and stationary phases for each vessel. The vessel could be a separatory funnel although it would work conveniently only if the mobile phase had the higher density so that it could be drawn off and transferred to the next vessel.

The concentration of A in the original aqueous phase (farthest advanced to the right in Figure 18-7) after n stages will be given by Equation 18-25; but what about the concentration of A in each of the other vessels? For simplicity, let us consider the case in which $V_{\text{stationary}} = V_{\text{mobile}}$ and $K_D = 1$. Then after two transfers (three equilibrium stages), the fraction of A in the aqueous (mobile) phase in vessel 2

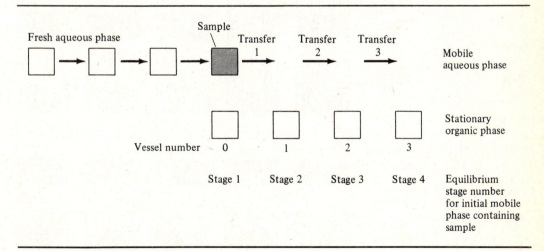

FIGURE 18-7 Multistage extraction. The sample containing A is initially present in the mobile aqueous phase at the first stage and is equilibrated with the stationary organic phase (vessel 0); then the mobile phase is transferred to stage 2 (vessel 1). At the first transfer, fresh mobile phase is introduced into vessel 0.

would be

$$f = \frac{[A]_{m,3}}{[A]_0} = \left(\frac{1}{1 + K_D V_s / V_m}\right)^3$$

$$= (\tfrac{1}{2})^3 = \tfrac{1}{8} = 2^{-n}$$

Because we have assumed that $K_D = 1$, the amount of A will be divided equally between the two phases in each vessel. Therefore,

$$\frac{[A]_{s,3}}{[A]_0} = f = \tfrac{1}{8} = 2^{-n}$$

Figure 18-8 shows the distribution of A between the mobile and stationary phases in each vessel. Note the difference in counting transfers and equilibrium stages. After four transfers, there have been five equilibrium stages so that the fraction of A in the mobile phase will be $2^{-5} = \frac{1}{32}$.

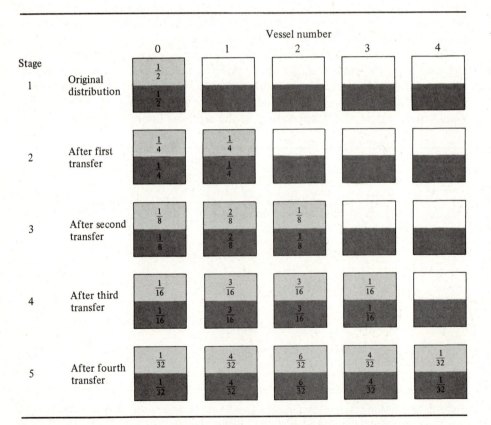

FIGURE 18-8 Multistage extraction: distribution of analyte A after each transfer for $K_D = 1$, $V_m = V_s$.

Counting transfers is mathematically convenient when the total amount of A (or fraction of A) present in a *vessel* is considered. From the example above, after four transfers, there will be $\frac{1}{32}$ of A in the mobile phase and $\frac{1}{32}$ of A in the stationary phase so that the total amount of A in the fourth vessel is $\frac{1}{32} + \frac{1}{32} = \frac{1}{16} = 2^{-m}$, where m is the number of transfers. For a large number of transfers, $m \cong n$, and the distinction can be dropped. The fraction of A in each vessel for zero to four transfers is given in Table 18-3. You may recognize that the numbers are terms in the binomial expansion of $(p + q)^m$,

$$(p + q)^m = p^m + mp^{m-1}q + \frac{m(m - 1)}{2!} p^{m-2}q^2 + \frac{m(m - 1)(m - 2)}{3!} p^{m-3}q^3 + \cdots q^m \quad \textbf{(18-49)}$$

Each term on the right side is the fraction of A in each *vessel*. The term p represents the fraction of A in the stationary phase, and q represents the fraction of A in the mobile phase for any vessel. It is also true that p is the fraction of A in the total stationary phase for the entire system and that q is the fraction of A in the total mobile phase. Therefore,

$$p = \frac{\text{mmol A}_s}{\text{mmol A}_{\text{total}}} = \frac{[A]_s V_s}{[A]_s V_s + [A]_m V_m} \quad \textbf{(18-50)}$$

$$q = \frac{\text{mmol A}_m}{\text{mmol A}_{\text{total}}} = \frac{[A]_m V_m}{[A]_m V_m + [A]_s V_s} \quad \textbf{(18-51)}$$

If numerator and denominator of Equation 18-50 are divided by $[A]_s V_s$ and of Equation 18-51 by $[A]_m V_m$ and if we recall that $K_D = [A]_s/[A]_m$, we obtain

$$p = \frac{1}{1 + [A]_m V_m/[A]_s V_s} = \frac{1}{1 + (K_D V_s/V_m)^{-1}} = \frac{1}{1 + (k')^{-1}} \quad \textbf{(18-52)}$$

$$q = \frac{1}{1 + [A]_s V_s/[A]_m V_m} = \frac{1}{1 + K_D V_s/V_m} = \frac{1}{1 + k'} \quad \textbf{(18-53)}$$

TABLE 18-3   Fraction of A in Each Vessel for a Multistage Extraction Process

Stages, n	Transfers, m	Vessel					Multiplier, 2^{-m}
		0	1	2	3	4	
1	0	1	0	0	0	0	2^0
2	1	1	1	0	0	0	2^{-1}
3	2	1	2	1	0	0	2^{-2}
4	3	1	3	3	1	0	2^{-3}
5	4	1	4	6	4	1	2^{-4}

where k' is the capacity factor introduced earlier. Note that all A must be accounted for by p and q so that

$$p + q = 1 \qquad\qquad (18\text{-}54)$$

Now with Equations 18-49, 18-52, and 18-53, it is possible to calculate the amount of A in every phase of every vessel after m transfers for *any* values of K_D, V_s and V_m, as demonstrated in Example 18-8.

Example 18-8 shows that A is "smeared out" among all four vessels but that the maximum amount is found in vessel 1, which contains 44.4% of A. To see how this process can be used for separation, let us compare these results with those for analyte B, which we shall assume has a K_D value of 1.0 instead of 2.5. The results are summarized in Table 18-4. Analyte B is also spread out among all four vessels, but

---

EXAMPLE 18-8

Determine the distribution of A after three transfers for the case where $V_s = 4$ mL, $V_m = 5$ mL, and $K_D = 2.5$.

Solution

Capacity factor:

$$k' = \frac{K_D V_s}{V_m} = \frac{2.5 \times 4}{5} = 2.0$$

$$p = \frac{1}{1 + (k')^{-1}} = \frac{1}{1 + 2^{-1}} = 0.67$$

$$q = \frac{1}{1 + k'} = \frac{1}{1 + 2} = 0.33$$

or

$$q = 1 - p = 0.33$$

From Equation 18-49,

$$\text{vessel } 0 = p^m = (0.67)^3 = 0.296 \ (29.6\%)$$

$$K_D = \frac{[A]_s}{[A]_m} = 2.5$$

TABLE 18-4   Distribution of A and B after three transfers*

Vessel	A (%)	B (%)
0	29.6	8.8
1	44.4	32.9
2	22.2	41.2
3	3.7	17.2
	99.9[†]	100.1[†]

*$V_s = 4$ mL, $V_m = 5$ mL, $K_D(A) = 2.5$, and $K_D(B) = 1.0$.
[†]Does not equal 100.0% because of rounding error.

Example 18.8 (cont.)

Therefore,

$$[A]_s = 2.5[A]_m$$

$$[A]_s + [A]_m = [A]_{total} = 29.6\%$$

$$2.5[A]_m + [A]_m = 29.6\%$$

$$[A]_m = \frac{29.6}{3.5} = 8.5\%$$

$$[A]_s = 21.1\%$$

$$\text{vessel } 1 = mp^{m-1}q = 3(0.67)^2(0.33) = 0.444 \ (44.4\%)$$

$$[A]_m = \frac{44.4}{3.5} = 12.7\%$$

$$[A]_s = 31.7\%$$

$$\text{vessel } 2 = \frac{m(m-1)}{2!}p^{m-2}q^2 = \frac{3 \cdot 2}{2}(0.67)(0.33)^2 = 0.222 \ (22.2\%)$$

$$[A]_m = \frac{22.2}{3.5} = 6.3\%$$

$$[A]_s = 15.9\%$$

$$\text{vessel } 3 = q^m = (0.33)^3 = 0.037 \ (3.7\%)$$

$$[A]_m = \frac{3.7}{3.5} = 1.1\%$$

$$[A]_s = 2.6\%$$

the maximum amount is found in vessel 2, which contains 41.2%. Because analyte B has a lower K_D value, it tends to stay in the mobile phase more than analyte A does and, therefore, travels more rapidly from vessel to vessel. Clearly three transfers are not sufficient to obtain a quantitative separation, but if 100 transfers were carried out, the fraction in each vessel would be as shown in Figure 18-9.

Now it is clear that separation of A and B has been accomplished. If the vessels from 20 to 38 are combined, they will contain 95% of analyte A, while vessels from 40 to 60 will contain 95% of analyte B. The separation factor, α, for this process is

$$\alpha = \frac{K_D(A)}{K_D(B)} = \frac{2.5}{1.0} = 2.5$$

For a separation factor >2.5, fewer transfers would be needed to accomplish the same amount of separation. On the other hand, for small separation factors, many more transfers would be required. The value of chromatography is that the equivalent of thousands of transfers may be carried out conveniently.

Can the process described for multistage extractions be accomplished in the laboratory? Yes, it can; for there is an apparatus (called "Craig" after its designer) that allows small volumes of one solvent to move with respect to the other. The solvent of

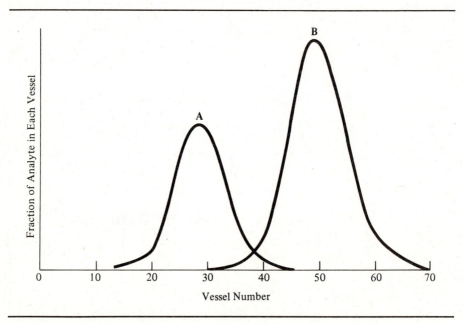

FIGURE 18-9 Distribution of analytes A ($K_D = 2.5$) and B ($K_D = 1.0$) after 100 transfers with $V_s = V_m$. Note that A has moved more slowly than B has, so that nearly quantitative separation has occurred.

lower density is the mobile phase, and hundreds of tiny vessels can be mounted together to provide the many transfers necessary for separation.

The analyst can also hold one solvent stationary by adsorbing it on a porous support material and then flowing the mobile solvent through it. The stationary phase need not be absolutely fixed. The mobile phase may move in one direction, while the stationary phase moves in the opposite direction; and such a process is called *countercurrent extraction*. The number of transfers in such a continuous system cannot be physically observed, but the separation results will be *equivalent* to *m* theoretical transfers. The continuous system will be said to contain *n* theoretical stages (or theoretical plates, reminiscent of fractional distillation).

With these concepts, we can now proceed to chromatography as a separation process.

SUPPLEMENTARY READING

Morrison, G.H., and H. Freiser, *Solvent Extraction in Analytical Chemistry*. New York: John Wiley & Sons, Inc., 1957.

Pecsok, R.L., L.D. Shields, T. Cairns, and I.G. McWilliam, *Modern Methods of Chemical Analysis*, 2nd ed. New York: John Wiley & Sons, Inc., 1976.

Sekine, T., and Y. Hasegawa, *Solvent Extraction Chemistry*. New York: Marcel Dekker, Inc., 1977.

PROBLEMS

18-1. Calculate the relative volatility for the following pairs of compounds:

    *a.   *n*-Pentane ($P^0 = 611$ torr) and *n*-heptane ($P^0 = 58$ torr) at 30°C

    b.   *n*-Heptane ($P^0 = 11$ torr) and *n*-octane ($P^0 = 3$ torr) at 0°C

    c.   *n*-Heptane ($P^0 = 1047$ torr) and *n*-octane ($P^0 = 482$ torr) at 110°C

18-2. Using the data in Problem 18-1, calculate the vapor pressure of a mixture containing 30 mol % *n*-pentane and 70 mol % *n*-heptane at 30°C.

18-3. *a.   Using the data in Problem 18-1, determine the composition of a mixture of *n*-heptane and *n*-octane that boils at 110°C.

    b.   What is its vapor composition?

18-4. Using the data in Problem 18-1, determine the final composition, after five distillation stages, for a mixture originally containing 50 mol % *n*-heptane and 50 mol % *n*-octane. Assume that the relative volatility remains constant at its value at 110°C throughout the distillation column.

18-5. For the situation in Problem 18-4, how many distillation stages would be required to reach 99.9% *n*-heptane?

*18-6.** Calculate the total pressure of the system *and* the mole fraction of A in the vapor of a mixture of A and B in which $\chi_A = 0.20$ at a temperature where the vapor pressure of pure A is 700 torr and that of pure B is 300 torr.

---

*Answers to problems marked with an asterisk will be found at the back of the book.

18-7. Various types of distillation column are available for laboratory separations. From the information below, calculate HETP for each:
*a.   Vigreux, 20 cm long, $n = 2$
b.   Metal helices, 15 cm long, $n = 12$
c.   Concentric tube, 30 cm long, $n = 55$

18-8. The distribution coefficient of iodine between CCl_4 and water is 85. After extraction with two 50-mL portions of CCl_4, how many millimoles of I_2 remain in 100 mL of an aqueous solution that was originally 1.00×10^{-3} F?

18-9. As you may remember from previous chapters, I_2 forms I_3^- in aqueous solutions containing I^-. The equation and its equilibrium constant are

$$I_2 + I^- \rightleftharpoons I_3^- \qquad K_f = 7 \times 10^2$$

Based on the distribution coefficient of 85 for I_2 distributed between H_2O and CCl_4, calculate the observed distribution ratio for 10^{-4} F iodine between H_2O containing the following concentrations of I^- and CCl_4:
*a.   0.02 F
b.   0.1 F
c.   2 F

18-10. Problem 18-9 is somewhat complicated but still solvable when the $[I^-]$ is comparable to $[I_2]$. Calculate the observed distribution ratio for 2×10^{-3} F iodine between water containing 4×10^{-3} $F\, I^-$ initially and an equal volume of CCl_4.

18-11. A substance A has a distribution coefficient of 3.7 between water and CCl_4. Calculate the fraction removed (percentage) from 100 mL of the aqueous phase with 100 mL *total* CCl_4 by
*a.   1 extraction with 100 mL of CCl_4
b.   2 extractions with 50 mL of CCl_4
c.   4 extractions with 25 mL of CCl_4
d.   10 extractions with 10 mL of CCl_4

18-12. *a.   The distribution coefficient for X between CCl_4 and H_2O is 19. If 50 mL of an aqueous solution of 0.250 F X is contacted with 100 mL of CCl_4, calculate the percentage of X remaining in the water.
*b.   If the aqueous solution is contacted with two 50-mL portions of CCl_4 instead of one 100-mL portion, calculate the percentage of X remaining in the water. Assume no volume changes during the process.

18-13. a.   Compound Y has a distribution coefficient of 4.0 between water and chloroform. If a 75-mL aqueous solution contains 35 μg of X, calculate the volume of chloroform needed for a single extraction to remove 99.0% of compound Y.
b.   How many extractions of 75 mL of chloroform each would be needed to extract 99.0% of compound Y?

*18-14. A 250-mg sample of an organic compound (FW = 84) is dissolved in 200 mL of H_2O. The aqueous phase is contacted with 50 mL of benzene and allowed to reach equilibrium. Analysis of the aqueous phase now shows that it contains only 63 mg of the organic compound. Calculate K_D.

18-15. The observed distribution ratio for the extraction of UO_2^{2+} (as the nitrate complex) from 8 F HNO_3 into ether is 1.86. What will be the fraction remaining (percentage) in the aqueous phase after three extractions with an equal volume of ether?

18-16. The observed distribution ratio for the extraction of In(III) (as the bromide complex) from 2 F HBr into ether is 5.76. How many extractions with 50 mL of ether will be required to remove $>99\%$ of the In(III) from 100 mL of 2 F HBr?

*18-17. An organic acid ($K_a = 2 \times 10^{-6}$) has a distribution coefficient of 18.0 between water and methyl isobutyl ketone. If 100 mL of an aqueous solution buffered at pH 6.00 is extracted three times with 50 mL of methyl isobutyl ketone, calculate the percentage remaining in the aqueous phase.

18-18. A 150-mL aqueous solution of an organic acid ($K_a = 3 \times 10^{-5}$) is buffered to pH 4.5 and extracted three times with 100 mL of methyl isobutyl ketone. If the distribution coefficient, K_D, is 6.8, determine the percentage of organic acid remaining in the aqueous phase.

18-19. Two weak acids, HA and HB, have K_a values of 4×10^{-7} and 8×10^{-4}, respectively. Their distribution coefficients between water and ether are 5.6 for HA and 4.6 for HB. Design a separation procedure for the removal of HA (98%) from an aqueous solution, while retaining HB in the aqueous phase with no more than 1% loss with each extraction.

18-20. The observed distribution ratio for acetic acid (HOAc) between an aqueous phase buffered at pH 5 and ether is given below as a function of initial formal concentration of HOAc.

   *a. If dissociation in the aqueous phase ($K_a = 1.8 \times 10^{-5}$) were the only important chemical equilibrium, calculate the apparent K_D for each of the formal concentrations.

   b. Why is K_D not constant?

Initial f_{HOAc}	D
0.28	0.167
0.80	0.174
2.90	0.204

18-21. It can be seen from Equation 18-38 that dimerization in the organic phase will become negligible as the concentration approaches zero. Using the data in Problem 18-20, plot D versus f_{HOAc}, and determine D and K_D when $f_{HOAc} = 0$. With the value of K_D now known, calculate K_{di}, assuming equal volumes of solutes for the data in Problem 18-20.

18-22. **a.** An aqueous solution containing 1.5×10^{-4} F of an organic dye has an absorbance of 0.256 at 450 nm. A 50-mL portion of this solution is contacted with 50 mL of benzene. After this extraction step, the absorbance reading at 450 nm for the aqueous solution is 0.029. The absorbance of the benzene solution is now 0.365, If all the absorbance measurements were made in 2.00-cm cells, calculate the molar absorptivity for the dye in water and in benzene and its distribution coefficient between water and benzene.

   b. Calculate the absorbance at 450 nm in a 2.00-cm cell for the 50-mL aqueous phase after a second contact with benzene—this time with 100 mL of benzene.

*18-23. Metal ions may form a series of complexes leading to a neutral species such as MX_2. By using the concept of α developed for weak acids (Chapter 9) and complexes (Chapter 10), show how the observed distribution ratio is related to K_D and α_2 (for a neutral MX_2).

18-24. Cd(II) forms iodide complexes with $K_1 = 125$ and $K_2 = 21$. If the distribution coefficient for CdI_2 distributed between water and ether is 100, calculate the observed distribution ratio with a 0.1 F I^- solution. (See Problem 18-23 for an approach to this problem.)

18-25. The observed distribution ratio for the extraction of Hg(II) as the chloride complex from 0.12 F HCl into ethyl acetate is 4.0. The formation constants are $K_1 = 5.5 \times 10^6$ and $K_2 = 3.0 \times 10^6$. Calculate K_D.

18-26. The observed distribution ratio for Sn(II) as the bromide complex between aqueous HBr and ether is 0.47 and 1 F HBr and increases to 1.8 in 2 F HBr as a greater percentage of Sn(II) forms neutral $SnBr_2$. However, in 6 F HBr, the distribution ratio is only 0.56. How can you account for this decrease in D at high HBr concentration?

*18-27. Determine the distribution of an analyte A after four transfers for the case when $V_s = 3$ mL, $V_m = 6$ mL, and $K_D = 6.0$.

Chromatography | **19**

Many separation processes are available, as indicated in Chapter 18, but in today's analytical laboratory, chromatographic methods are the most widely used. In the 1980 edition of *Official Methods of Analysis*, 27% of the newly listed methods were chromatographic, making chromatography the fastest growing area of analysis. The reason for popularity lies in the fact that the analyst simply allows his sample (dissolved in a mobile phase) to flow through a column containing a stationary second phase; separations corresponding to thousands of equilibrium stages occur. Chromatographic methods can be classified according to types of mobile and stationary phases, as shown in Figure 19-1.

Credit is usually given to Michael Tswett for scientifically establishing the basic principles of chromatographic separations. In 1903, he began publishing papers describing the separation of plant pigments into colored bands on a calcium carbonate column; and in 1906, he coined the term "chromatography" from the Greek words *chroma*, "color," and *graphein*, "to write." Tswett's plant extracts were dissolved in petroleum ether (the mobile phase) and placed at the top of a column containing firmly packed calcium carbonate powder. The samples were transported through the column by addition of pure petroleum ether, the various plant pigments moving through the column at different rates and resulting in colored bands, as shown in Figure 19-2. According to Tswett, the more strongly adsorbed pigments displaced the more weakly adsorbed ones. In other words, strongly adsorbed components moved slowly through the column (for example, C in Figure 19-2), while weakly adsorbed components moved rapidly (for example, A in Figure 19-2). The chromatographic method used by Tswett was liquid–solid chromatography because liquid petroleum ether was the mobile phase and powdered calcium carbonate was the stationary phase.

Figure 19-2 shows two methods of observing the separated bands. Tswett added sufficient petroleum ether to the column to cause separation, and at some specified

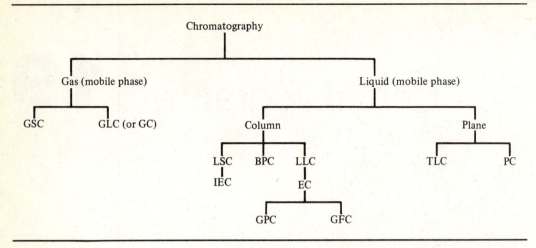

FIGURE 19-1 Alphabet soup of chromatographic methods: GSC, gas–solid chromatography; GLC (often GC), gas–liquid chromatography; LSC, liquid–solid chromatography (adsorption chromatography); IEC, ion-exchange chromatography (a special case of LSC); BPC, bonded-phase chromatography (a gray area between LSC and LLC); LLC, liquid–liquid chromatography (partition chromatography); EC, exclusion chromatography (a special case of LLC); GPC, gel permeation chromatography (a type of EC); GFC, gel filtration chromatography (a type of EC); TLC, thin-layer chromatography (a special case of LSC or LLC); PC, paper chromatography (a special case of LLC).

time he removed the packing; then the individual constituents were extracted from their "zones" and identified. The concentration profile at this specified time is a function of *distance* through the column, as shown in Figure 19-2(**A**) as a distance profile. The second method involves adding sufficient mobile phase to transport the constituents through the entire column—a method shown in Figure 19-2(**B**) as an elution time profile; the mobile phase eluting agent is the *eluent*. The observer monitors the concentration profile as a function of *time* and notes the time required for a particular component to reach the column exit; the mobile phase exiting the column is called the *eluate*. Essentially all modern chromatographic techniques except plane chromatography employ elution time profile analysis. Because an eluent is used in both methods, they are classified as elution chromatography.

The technique had been forgotten for many years although biochemists reintroduced it in 1931. The possibility of using a gaseous mobile phase and a liquid stationary phase was mentioned in the 1940s, but it was not until 1952 that A.J.P. Martin and A.T. James described what is now known as gas–liquid chromatography (GLC, but commonly called simply *gas chromatography* and abbreviated GC). The value of the technique was quickly recognized so that within a few years organic chemists around the world were separating mixtures by GLC. Martin and R.L.M. Synge were awarded the Nobel prize in chemistry in 1952 for their theoretical development of chromatographic separation (partition chromatography). The basic concept of this theoretical development is that a distribution

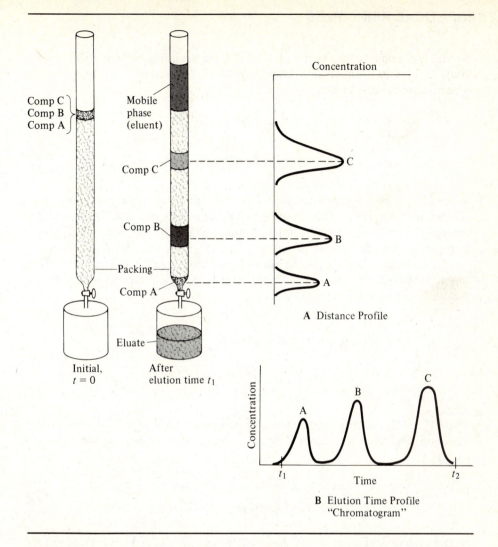

FIGURE 19-2 Chromatographic separation as carried out by Michael Tswett. Separation of components *A*, *B*, and *C* can be observed from either (**A**) a distance profile or (**B**) an elution time profile. At t_1, component *A* is just starting to emerge from the column, and its concentration begins to rise in the elution time profile. When t_2 is reached, component *C* has completely exited the column, and a "chromatogram" of the three components is recorded.

equilibrium occurs at each point on the column, leading to a separation of the components similar to that in liquid–liquid extraction and fractional distillation. Terms introduced for those methods, such as theoretical plates and the distribution coefficient, will be useful in developing the theoretical principles of chromatography.

19-1   THEORETICAL PRINCIPLES

As the mobile phase containing some analyte species A is passed through the column containing the stationary phase, the analyte distributes itself between the two phases according to the distribution coefficient

$$A \text{ (mobile)} \rightleftharpoons A \text{ (stationary)}$$

$$K_D = \frac{[A]_s}{[A]_m} \tag{19-1}$$

where $[A]_s$ is the concentration of A in the stationary phase at this point in the column and $[A]_m$ is the concentration of A in the mobile phase. A can move down through the column only when it is in the mobile phase. In other words, if K_D is very large, analyte A will be predominantly in the stationary phase and will elute very slowly. To see how fast an analyte will elute, we need to know what fraction of the time it will be in the mobile phase. This fraction is simply the fraction of analyte A in the mobile phase at any time. Therefore,

$$f = \text{fraction of time in mobile phase} = \frac{\text{mmol A in mobile phase}}{\text{total mmol of A in column}}$$

$$f = \frac{[A]_m V_m}{[A]_m V_m + [A]_s V_s} \tag{19-2}$$

where V_m and V_s are the volumes of mobile phase and stationary phase, respectively. Equation 19-2 can be simplified by dividing numerator and denominator by $[A]_m V_m$:

$$f = \frac{1}{1 + [A]_s V_s / [A]_m V_m} = \frac{1}{1 + K_D V_s / V_m} \tag{19-3}$$

It can be seen from Equation 19-3 that, even though $[A]_s$ and $[A]_m$ vary through the column, the ratio $[A]_s/[A]_m$ always equals K_D provided that equilibrium is attained. The term $K_D V_s / V_m$ has already been introduced as the capacity factor, k', so that

$$f = \frac{1}{1 + k'} \tag{19-4}$$

The velocity of A through the column u_A, will equal the velocity of mobile phase, u, during the time A is in the mobile phase and zero when it is in the stationary phase. Therefore,

$$u_A = uf = u\left(\frac{1}{1 + k'}\right) \tag{19-5}$$

The term $1/(1 + k')$ must always be ≤ 1 because $k' \geq 0$; so the velocity of A can never be greater than the velocity of the mobile phase (eluent).

As shown in Figure 19-2, analyte A will eventually elute from the column at some time t_R, called the *retention time*. From the velocity of A and the length of the column, the retention time can be calculated to be

$$t_R = \frac{\text{column length (\text{cm})}}{\text{velocity of A (\text{cm}/\text{sec})}} = \frac{L}{u_A} = \frac{L}{u}(1 + k') \qquad \text{(19-6)}$$

The mobile phase will flow through the column at velocity u so that the time it spends in the column, t_m, is

$$t_m = \frac{L}{u} \qquad \text{(19-7)}$$

Equations 19-6 and 19-7 can be combined to give

$$t_R = t_m(1 + k') \qquad \text{(19-8)}$$

The retention time of any substance that spends no time in the stationary phase $(K_D = 0$, and therefore $k' = 0)$ will also be t_m. This factor allows us to measure t_m easily by observing the rate at which a substance not retained by the column is eluted—such as air in a gas chromatographic column.

The retention time depends on the flow rate (F) of mobile phase through the column, which is an experimental variable that sometimes varies from experiment to experiment. On the other hand, the *volume* of mobile phase required to elute the analyte will not depend on the flow rate. The retention volume V_R will be

$$V_R \text{ (mL)} = t_R \text{ (\text{sec})} \, F \left(\frac{\text{mL}}{\text{sec}}\right) \qquad \text{(19-9)}$$

At the same time, the volume of mobile phase, V_m, present in the column

$$V_m \text{ (mL)} = t_m \text{ (\text{sec})} \, F \left(\frac{\text{mL}}{\text{sec}}\right) \qquad \text{(19-10)}$$

V_m is also known as the void, or interstitial, volume in the column because any column volume not occupied by stationary phase will be occupied by mobile phase. Equations 19-8 to 19-10 can now be combined to give

$$t_R = t_m(1 + k')$$

$$\frac{V_R}{F} = \frac{V_m}{F}(1 + k')$$

$$V_R = V_m(1 + k') \qquad \text{(19-11)}$$

If we recall the definition of the capacity factor, k', we may also write Equation 19-11 as

$$V_R = V_m + V_m k' = V_m + \frac{V_m K_D V_s}{V_m}$$

$$= V_m + K_D V_s \qquad (19\text{-}12)$$

The retention volume depends on the volume of mobile phase in the column, the volume of stationary phase, and the distribution coefficient. Of even more fundamental importance is the difference $V_R - V_m$. This is the volume of eluting agent over and above the volume needed to fill the void space in the column; it is known as the adjusted retention volume, V_R',

$$V_R' = V_R - V_m = K_D V_s \qquad (19\text{-}13)$$

The adjusted retention volume is directly proportional to the equilibrium constant for the separation process. The proportionality constant is the volume of stationary phase in the column that will be a constant for any given column. If the volume of stationary phase is known, the adjusted retention volume can be divided by V_s to yield the specific retention volume, V_g,

$$V_g = \frac{V_R'}{V_s} = K_D \qquad (19\text{-}14)$$

In one sense, V_g is dimensionless because V_R' and V_s are both volumes; but in another sense, V_g is the retention *volume* to elute an analyte per unit volume or amount of stationary phase. In fact, the weight of stationary phase may be used in place of volume V_s for quoting specific retention volume, in which case $V_g = K_D/\rho$, where ρ is the density of the stationary phase. The important feature to remember is that the specific retention volume is essentially the same as an equilibrium constant for the separation process.

The adjusted retention volume can also be calculated from retention time and the flow rate F:

$$V_R' = V_R - V_m = t_R F - t_m F = (t_R - t_m)F = t_R' F \qquad (19\text{-}15)$$

The term $t_R - t_m$ is an adjusted retention time, t_R'.

Frequently the value of V_s is not known, but the retention time or retention volume of a substance can be compared with some standard substance. The ratio is known as the *relative retention*, α,

$$\alpha = \frac{V_R'}{V_{R,\text{std}}'} = \frac{(t_R - t_m)F}{(t_{R,\text{std}} - t_m)F} = \frac{t_R'}{t_{R,\text{std}}'} = \frac{K_D V_s}{K_{D,\text{std}} V_s} = \frac{K_D}{K_{D,\text{std}}} \qquad (19\text{-}16)$$

Thus α is the ratio of distribution coefficients for the analyte and standard and can be calculated from retention time or retention volume data alone.

At this point let us consider a *chromatogram* in terms of the equations developed above. Figure 19-3(**A**) shows the elution of two analytes, A and B, from a chromatographic column. It also shows the volume of mobile phase by noting the volume required for some nonretained substance to reach the exit. The area under the curve for A represents the integral

$$\text{area} = \int_{V_1}^{V_2} [A]\, dV = \text{mmol A} \qquad (19\text{-}17)$$

This is the product of concentration and volume and therefore is the amount of A present in the sample.

In Figure 19-3(**B**), the mobile phase (eluent) is passed through the column at a constant flow rate, and the retention time is measured. In place of actual analyte concentration along the y-axis, a detector signal as a function of time is plotted. As long as the detector signal, S, is proportional to analyte concentration—and from Equation 19-9, we know that $t_R \propto V_R$—it follows that

$$\text{area} = \int_{t_1}^{t_2} S\, dt \propto \text{mmol A} = R_A\ \text{mmol A} \qquad (19\text{-}18)$$

where R_A is the signal response for analyte A (square centimeters per millimole of A). Calculation of various parameters is shown in Example 19-1. Note that a good

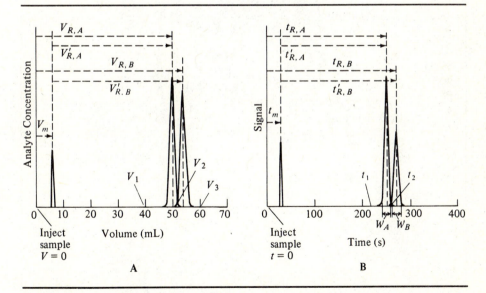

FIGURE 19-3 Chromatograms for mixture containing analytes A and B: (**A**) concentration as a function of eluent volume; (**B**) measurement signal ($\propto C$) as a function of time. (See Example 19-1.)

EXAMPLE 19-1

Using the chromatogram in Figure 19-3 and $V_s = 10$ mL, calculate the quantities listed below.

Solution

Quantity	Calculation
t_R	250 sec (A), 270 sec (B) (graph)
t_m	30 sec (graph)
V_R	50 mL (A), 54 mL (B) (graph)
V_m	6 mL (graph)
F	$F = \dfrac{V_R}{t_R} = \dfrac{50}{250} = 0.2$ mL/sec
V_R'	$V_R' = V_R - V_m = 50 - 6 = 44$ mL (A)
	$V_R' = V_R - V_m = 54 - 6 = 48$ mL (B)
t_R'	$t_R' = t_R - t_m = 250 - 30 = 220$ sec (A)
	$t_R' = t_R - t_m = 270 - 30 = 240$ sec (B)
α (A as standard)	$\alpha = \dfrac{V_R'}{V_{R,\,std}'} = \dfrac{48}{44} = 1.09$
	or
	$\alpha = \dfrac{t_R'}{t_{R,\,std}'} = \dfrac{240}{220} = 1.09$
α (B as standard)	$\alpha = \dfrac{V_R'}{V_{R,\,std}'} = \dfrac{44}{48} = 0.917$
k'	$k' = \dfrac{V_R}{V_m} - 1 = \dfrac{50}{6} - 1 = 7.33$ (A)
	or
	$k' = \dfrac{t_R}{t_m} - 1 = \dfrac{250}{30} - 1 = 7.33$ (A)
	$k' = \dfrac{V_R}{V_m} - 1 = \dfrac{54}{6} - 1 = 8.00$ (B)
K_D	$K_D = \dfrac{V_R - V_m}{V_s} = \dfrac{50 - 6}{10} = 4.4$ (A)
	$K_D = \dfrac{V_R - V_m}{V_s} = \dfrac{54 - 6}{10} = 4.8$ (B)

separation of A and B has been accomplished even though their K_D values differ by less than 10% (4.4 and 4.8). Also note that analyte B, with the higher K_D value, spends more time on the column (stationary phase) and, therefore, has a longer retention time than does analyte A.

If the chromatographic separation shown in Figure 19-3 were ideal, all the molecules of A would elute at t_R (250 sec), and a spike in signal would be recorded. However, in the real world, some molecules of A elute earlier than t_R and some later, with a distribution as shown in Figure 19-3. This distribution is very similar to the one in Figure 18-9 for multistage liquid–liquid extraction involving a large number of transfers or theoretical plates.

How many theoretical plates are associated with the chromatographic column for the chromatogram in Figure 19-3? To answer this question, we must recognize that the binomial expansion distribution shown in Equation 18-49 is the same distribution found for the normal error curve. In other words, the chromatographic peak will be Gaussian in shape. The breadth of the curve is associated with the standard deviation σ. It was also mentioned in Chapter 2 that the breadth increases as $\sqrt{n}$, where n is the number of random choices or, in the case of chromatography, the number of equilibrium transfers—the number of theoretical plates. On the other hand, the retention time will increase directly with n. Thus

$$\sigma \propto \sqrt{n}$$

$$t_R \propto n$$

Therefore,

$$\frac{t_R}{\sigma} \propto \frac{n}{\sqrt{n}} = \sqrt{n}$$

or

$$n \propto \left(\frac{t_R}{\sigma}\right)^2$$

The simple theory shows that the proportionality constant is 1 so that

$$n = \left(\frac{t_R}{\sigma}\right)^2 \qquad \text{(19-19)}$$

Normally the width W is measured experimentally, as shown in Figure 19-3, and $W \cong 4\sigma$. Therefore,

$$n = \left(\frac{t_R}{W/4}\right)^2 = \left(\frac{4t_R}{W}\right)^2 \qquad \text{(19-20)}$$

Example 19-2 shows how the number of theoretical plates can be calculated from the observed peaks in a chromatogram. Example 19-2 also shows how the *height*

EXAMPLE 19-2

(a) Calculate the number of theoretical plates for the chromatographic column, given the chromatogram shown in Figure 19-3.
(b) If the column is 120 cm long, calculate the HETP for this column.

Solution

(a) From Equation 19-20 and the data in Figure 19-3, we see that, for peak A,

$$n = \left(\frac{4t_R}{W}\right)^2$$

$$= \left(\frac{4 \times 250}{15}\right)^2 = 4.44 \times 10^3 \text{ theoretical plates}$$

for peak B,

$$n = \left(\frac{4 \times 270}{17}\right)^2 = 4.04 \times 10^3 \text{ theoretical plates}$$

(*Note:* Because of nonideality differences among substances, the number of theoretical plates will not be exactly the same for each of the peaks. This column is approximately equivalent to 4.2×10^3 theoretical plates.)

(b) $$\text{HETP} = H = \frac{L}{n} = \frac{120}{4.2 \times 10^3} = 0.029 \text{ cm/theoretical plate}$$

equivalent to a theoretical plate (HETP) can be calculated from n and the column length L. The value of 0.029 cm per theoretical plate means that each equilibrium transfer requires only 0.029 cm of column length on the average. In terms of liquid–liquid extraction, the separatory funnel is only 0.029 cm high. This small dimension also requires that only small samples be used so that the sample is not larger than this 0.029-cm "container."

The number of theoretical plates may be calculated with retention volume, V_R, in place of retention time, but, of course, the units of W would also have to be volume to be consistent.

It is interesting to realize how many theoretical plates a chromatographic column contains, but the bottom line is the ability to separate two substances. That is, how well resolved are the peaks associated with analyte A and analyte B? In Figure 19-3, we see that A and B are reasonably well separated, peak A nearly reaching the baseline (signal for pure eluent) before peak B begins. Clearly, the resolution of two peaks depends on how far apart they are (difference in t_R or V_R values) and how wide the peaks are (W values). From the first peak, it will require $W_A/2$ of the time (or volume) to reach the baseline if we neglect the tail of the normal error curve. The second peak will begin $W_B/2$ before the second peak if we again neglect the tail. Thus the two peaks should be resolved if

$$\Delta t_R = t_{R,B} - t_{R,A} \geq \frac{W_A}{2} + \frac{W_B}{2} = \frac{W_A + W_B}{2}$$

or

$$R \equiv \frac{\Delta t_R}{(W_A + W_B)/2} \geq 1 \qquad\qquad \text{(19-21)}$$

The quantity $2\,\Delta t_R/(W_A + W_B)$ is defined as the resolution R. Because of the tails on the peaks, a value of 1.5 is required to give "baseline resolution." You may recall, from the normal error curve, that $\mu \pm 2\sigma$ includes 95% of the area so that only 2.5% of the area lies above $\mu + 2\sigma$ and 2.5% below $\mu - 2\sigma$. This, for a chromatographic peak, would be $t_R + W/2$ and $t_R - W/2$. Therefore, when the resolution value is 1.0, there will be 2.5% of A in peak B and 2.5% of B in peak A. When the resolution is 1.5, the overlap is $<0.2\%$. The resolution of peaks A and B in Figure 19-3 is calculated in Example 19-3.

EXAMPLE 19-3

Calculate the resolution for the peaks in Figure 19-3.

Solution

$$R = \frac{\Delta t_R}{(W_A + W_B)/2} = \frac{270 - 250}{(15 + 17)/2} = 1.25$$

This resolution is reasonable for quantitative separation ($<2\%$ of A will be in peak B, and $<2\%$ of B will be in peak A) but not quite for baseline resolution.

One method of increasing resolution is simply to use a longer column. If we assume that the HETP remains the same (0.029 cm in the example), we find the number of theoretical plates to be directly proportional to column length. Example 19-4 shows how this changes the retention time, peak width, and resolution.

If the chromatographic column were truly ideal, it would contain an infinite number of theoretical plates, and the value of W would approach zero (signal spike instead of signal peak with finite width). Although the column in our example contains over 4000 theoretical plates, it requires a very small but finite length of column to achieve one theoretical transfer (one equilibrium stage). What factors play a role in determining the value of H, the HETP?

We have already seen that n and t_R are directly proportional to the column length, but certainly the analyte does not go through the column in a straight line. The

EXAMPLE 19-4

Calculate the resolution for the separation of A and B if a 200-cm column were used in place of one of 120 cm (see Example 19-2).

Solution

From Equation 19-6, we see that the retention time is directly proportional to column length L, provided that the flow rate and capacity factor do not change with the longer column. Therefore,

$$t_{R,A} = 250\,\frac{200}{120} = 417 \text{ sec}$$

$$t_{R,B} = 270\,\frac{200}{120} = 450 \text{ sec}$$

The number of theoretical plates will also be directly proportional to column length,

$$n = 4.2 \times 10^3\,\frac{200}{120} = 7.0 \times 10^3$$

The width of the peaks can now be calculated from Equation 19-20:

$$n = \left(\frac{4t_R}{W}\right)^2 \quad \text{or} \quad W = \sqrt{\frac{16t_R^2}{n}}$$

$$W_A = \sqrt{\frac{16(417)^2}{7.0 \times 10^3}} = 20 \text{ sec}$$

$$W_B = \sqrt{\frac{16(450)^2}{7.0 \times 10^3}} = 22 \text{ sec}$$

The resolution is now

$$R = \frac{2\,\Delta t_R}{W_A + W_B} = \frac{2(450 - 417)}{20 + 22} = 1.57$$

This is better than baseline resolution (1.5).

actual path length through the column for one analyte molecule may be different from that for another. Thus they will exit at different times, giving rise to band spreading (W) and an increase in H. This contribution from nonequal paths is also known as *eddy diffusion*.

A second contribution to H can be seen by consideration of Figure 19-2. The concentration profile as a function of distance will have a maximum at the center of the zone as it moves through the column. However, diffusion occurs whenever there is a concentration gradient, which was the basis for the voltammetric techniques discussed in Chapter 17. The analyte will tend to diffuse both forward and backward from the high concentration peak—a tendency that results in band spreading and an increase in H. This contribution from diffusion along the column is called *longitudinal diffusion*.

Finally, it must be remembered that the distribution process, like a chemical reaction, requires a finite amount of time to reach equilibrium. As the mobile phase front containing the analyte moves through the column, less than the equilibrium amount may transfer to the stationary phase, and $[A]_s/[A]_m < K_D$. On the other hand, at the rear of the zone, where the analyte is being transferred back to the mobile phase, any nonequilibrium will result in $[A]_s/[A]_m > K_D$. Because the velocity is proportional to $1/K_D$, this relation means that the front will move faster than predicted from K_D while the rear of the zone will move more slowly than predicted from K_D. Thus the band will be spread out, and H will be increased. Contributions to H from nonequilibrium effects are also called *mass transfer* effects since the distribution reaction represents a transfer of material from one phase to another.

These three contributions to H have been identified for gas chromatography and expressed mathematically by some Dutch petroleum chemists as Equation 19-22, known as the Van Deemter equation:

$$H = \text{contributions from nonequal paths} + \text{longitudinal diffusion}$$
$$+ \text{nonequilibrium}$$

$$H = A + \frac{B}{u} + Cu \qquad\qquad \textbf{(19-22)}$$

where u is the mobile phase (carrier) velocity through the column. The units of u would normally be centimeters per second, as indicated in Equation 19-6, but it is more convenient to use flow rate (milliliters per second), which is directly proportional to velocity. The dimensions of B and C must be consistent with the units employed for u. The contribution A from nonequal paths or eddy diffusion does not depend on carrier velocity and, for a well-packed column, will be small. For capillary columns it may usually be neglected.

Band spreading resulting from diffusion, B/u, is inversely proportional to carrier velocity because, the slower the velocity through the column, the more time is available for diffusion to take place. In other words, if the analyte is eluted very quickly, there will be little time for diffusion to take place. To minimize this term, the analyst should utilize high flow rates.

Countering the diffusion term, the nonequilibrium term, Cu, is directly proportional to the carrier velocity. That is, when the analyte moves rapidly through the column, insufficient time is available for reaching equilibrium, and band spreading will result.

The three contributions to H for a GC process are shown graphically in Figure 19-4, with the total as the solid curve. It can be seen that there will be an optimum flow velocity that exhibits the minimum value of H. Clearly, the lowest value of H will yield the highest number of theoretical plates since $n = L/H$ (see Example 19-2) and will therefore provide the best resolution for a given separation.

Appendix 1-L shows how calculus can be used to derive Equations 19-23 for optimum velocity and 19-24 for the minimum value of H:

$$u_{opt} = \sqrt{\frac{B}{C}} \tag{19-23}$$

$$H_{min} = A + 2\sqrt{BC} \tag{19-24}$$

Example 19-5 shows how the Van Deemter equation can be used to determine the

---

EXAMPLE 19-5

A 150-cm chromatographic column yields the following experimental results for the elution of an analyte. Determine the optimum flow rate and the number of theoretical plates for the column when operated under optimum conditions.

Run	u (cm/sec)	t_R(sec)	W(sec)
1	1.0	500	108
2	2.5	200	34
3	5.0	100	18

Solution

The Van Deemter equation contains three constants, A, B, and C, and therefore a minimum of three points is required to determine them. The values of H are first calculated by determining the number of theoretical plates:

run 1:
$$n = \left(\frac{4t_R}{W}\right)^2 = \left(\frac{4 \times 500}{108}\right)^2 = 343$$

$$H = \frac{L}{n} = \frac{150}{343} = 0.437 \text{ cm}$$

run 2:
$$n = \left(\frac{4t_R}{W}\right)^2 = \left(\frac{4 \times 200}{34}\right)^2 = 554$$

$$H = \frac{L}{n} = \frac{150}{554} = 0.271 \text{ cm}$$

run 3:
$$n = \left(\frac{4t_R}{W}\right)^2 = \left(\frac{4 \times 100}{18}\right)^2 = 494$$

$$H = \frac{L}{n} = \frac{150}{494} = 0.303 \text{ cm}$$

optimum flow rate for a column. Unfortunately, the optimum flow rate may turn out to be inconveniently low (especially true for liquid chromatography), leading to long retention times. As can be seen from Figure 19-4, the rise in H after the optimum point is not rapid so that flow velocities of $2u_{opt}$ or higher may be more practical.

Optimization of flow rate offers another means of improving resolution. But in addition to the column length mentioned earlier and optimization of flow rate, what other methods are available for improving resolution? Temperature is an experimental variable that may play an important role in determining resolution.

These values of H and the corresponding values of u can be used to set up three simultaneous equations:

$$H = A + \frac{B}{u} + Cu$$

$$0.437 = A + \frac{B}{1.0} + 1.0C$$

$$0.271 = A + \frac{B}{2.5} + 2.5C$$

$$0.303 = A + \frac{B}{5.0} + 5.0C$$

The set of equations can be solved for A, B, and C by algebraic substitution or the matrix approach described in Appendix 1-B. The solution is

$$A = 0.0075 \text{ cm}$$

$$B = 0.386 \text{ cm}^2/\text{sec}$$

$$C = 0.044 \text{ sec}$$

The optimum velocity is

$$u_{opt} = \sqrt{\frac{B}{C}} = 3.0 \text{ cm/sec}$$

At the optimum velocity, the value of H is

$$H_{min} = A + 2\sqrt{BC} = 0.268 \text{ cm}$$

At the optimum velocity, the number of theoretical plates is given by

$$n = \frac{L}{H} = \frac{150}{0.268} = 559 \text{ theoretical plates}$$

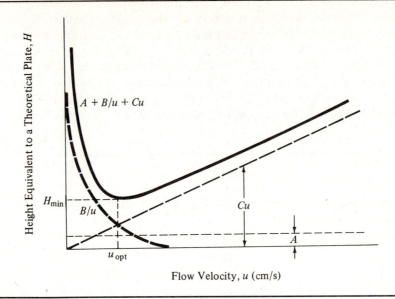

FIGURE 19-4   Van Deemter plot for GC process, showing the relative contributions of each term as a function of mobile phase velocity.

The difference in retention times depends on K_D, which, like any equilibrium constant, is a function of temperature. For GC, the relative retention, α, becomes greater as the temperature is decreased (at high temperatures all analytes will behave similarly), and therefore, resolution will be increased by operating the column at a *lower* temperature. On the other hand, in LC, the resolution depends little on column temperature. In fact, there are instances where the resolution increases slightly with increasing temperature. Most LC separations are carried out at room temperature for convenience. Other methods for increasing α include a change in mobile and/or stationary phase since these will determine K_D.

It was mentioned earlier that small samples are required in chromatography because each theoretical plate is so small. Ideally, the sample should take up zero volume, and the smaller the sample, the closer to ideality it becomes. Therefore, resolution may be improved by *decreasing* the sample size.

The A term in the Van Deemter equation can be decreased by ensuring uniform, tight packing of the column and using a smaller particle size for the packing. Diffusion and nonequilibrium effects are also affected by packing, and especially affected is the stationary phase film thickness in the case of a liquid stationary phase held on a solid support. These methods for improving resolution are summarized in Table 19-1. Note that group 2 methods can be carried out *without* changing the column and, therefore, are considerably more convenient for trying before group 1 methods.

The same three effects leading to band spreading (H) apply to LC, but because diffusion rates in liquids are $> 10^3$ times lower than they are in gases, the relative

TABLE 19-1   Methods for Improving Resolution in a
Chromatographic Separation

1. *based on column construction:*

   (a) column length increased
   (b) more uniform packing and smaller particles
   (c) a different mobile phase (LC)
   (d) a different stationary phase
   (e) column diameter decreased (GC)

2. *based on column operation:*

   (a) flow rate optimization
   (b) sample size decreased
   (c) column temperature decreased (GC)

importance of the terms changes significantly. In turn, this leads to a markedly different Van Deemter plot. Longitudinal diffusion, for example, is negligible for LC because of the low diffusion rates. However, mass transfer effects are now the most important contributors to H. In addition to mass transfer in the stationary phase and in the liquid mobile phase, there is also a mass transfer effect involving transfer of the analyte from the free-moving mobile phase to a stagnant mobile phase trapped in the pores of the column packing. A typical Van Deemter plot for LC is shown in Figure 19-5, the several contributions to H as broken lines. Each of the

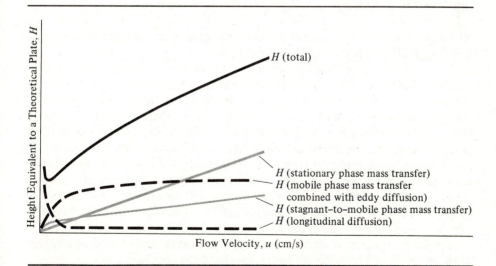

FIGURE 19-5   Van Deemter plot for LC, showing various contributions to H.

mass transfer terms depends on the square of the diameter of the column packing. Therefore, it is critically important to use the smallest particles possible to achieve small values of H, that is, to achieve a large number of theoretical plates in the chromatographic column. Also, the optimum point (minimum value of H) normally occurs at very low flow rates, and in practice, LC is carried out at flow rates considerably above the optimum.

19-2   GAS CHROMATOGRAPHY (GC)

Now let us turn our attention to some particular chromatographic methods. GLC as an analytical method took off like a rocket in the 1950s, and thousands of scientific articles have been written during its development. At present more than 200,000 gas chromatographs are in operation around the world. As the name implies, the sample is carried through the column in a gaseous mobile phase, whereas the stationary phase in the column is a liquid. To prevent the liquid from moving, a solid support material, which is a fine powder packed into the column, holds it. Capillary columns, in which the liquid stationary phase is held on the column wall and no packing is present, have also become popular in recent years.

The reason for this great popularity is that quantitative separation of similar organic compounds—such as o-, m-, and p-xylenes—which is impossible by fractional distillation, is relatively simple with a gas chromatograph. Also, very complicated mixtures, such as automobile exhaust gases containing over 100 components, can be separated by this technique. Before describing a few of these applications, we shall briefly examine GC instrumentation and the information obtained from a chromatogram.

INSTRUMENTATION

Although there are many types of gas chromatograph available, the basic features common to all are diagrammed in Figure 19-6.

The first component shown is a high-pressure tank of carrier gas. Several gases have been used for GC, including helium, nitrogen, hydrogen, and argon, but helium is the most common carrier gas, especially in the United States, where helium is relatively inexpensive. Hydrogen, although used in the past—particularly in Europe, where helium is expensive—may react with easily reduced analytes and poses a severe safety problem. The carrier gas should not interact with the sample and merely carries it through the column. In other words, the distribution coefficient does not depend on the nature of the carrier gas, only on the nature of the liquid stationary phase. However, the detector signal will depend on the difference in properties between the carrier gas and the analytes. For many detectors, helium is preferred over nitrogen and argon.

A pressure regulator is needed to control the pressure inside the column (gauge pressures of 10 to 50 pounds per square inch) and the flow rate.

Next a means of introducing samples into the column must be available. So that the sample forms a sharp band at the top of the column, its volume must be as small

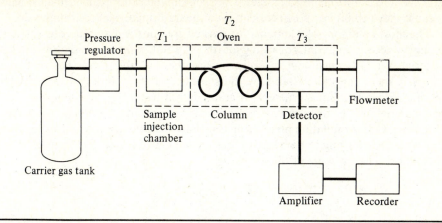

FIGURE 19-6 Essential features of a gas chromatograph, in which the temperatures of the injection chamber, column, and detector are controlled separately.

as possible, and it must be placed on the column in the shortest possible time interval. Typically a syringe is used to inject the sample directly on the column or into a heated sample chamber, which quickly volatilizes a liquid sample. Liquid samples are injected from a syringe through a high-temperature silicone polymer septum with a swift, smooth motion, which requires some experience to achieve. It should be noted that, although the sample is carried through the column in the vapor phase by the carrier gas, the actual sample being analyzed may be a gas, liquid, or solid. The only requirement is that the sample has sufficient vapor pressure to be volatilized in the sample chamber. The sample chamber may be at a temperature somewhat higher than the column to facilitate this volatilization process.

Liquid samples of 0.1 to 10 μL are routinely handled in GC, but smaller samples (10^{-3} to 10^{-2} μL) must be used in the unpacked capillary-tube columns. It is sometimes necessary to split a small sample (for example, 5% of 0.1 μL equals 0.005 μL) to obtain such tiny volumes accurately. Gaseous samples, on the other hand, can be several milliliters in volume since they are already in the gaseous phase and will not expand greatly when placed on the column as liquid samples will. It was mentioned earlier that the ideal sample should have zero volume. To determine the retention time for an analyte at zero volume, the analyst can run the analyte at several different volumes and extrapolate a plot of t_R versus V_{sample} to $V_{sample} = 0$.

The third component in Figure 19-6 is the column itself, placed in an oven to control the column temperature. The most common columns for analysis are metal (or occasionally glass) tubes 1 to 20 m long and 3 to 10 mm in diameter. They are bent or coiled to fit in the oven chamber. The column is packed with a powdered, solid, support material on which the stationary liquid phase is held. Alternatively, the stationary liquid phase may be in the form of a film on the tube wall itself; there is no packing. These open capillary tubes may be up to a mile in length and may be equivalent to $> 10^5$ theoretical plates. The open capillary tubes can be made longer because they exhibit less pressure drop than the traditional packed columns do.

The most common solid support for packed columns is diatomaceous earth, a spongy siliceous material consisting of the skeletal residues of diatoms. This material, a main constituent in firebrick, has the desirable properties of being relatively inert but still wettable by the stationary liquid phase and has a high specific surface area (square meters per gram). It is also stable to high temperatures. The material has the disadvantages of a fragile surface, and it contains numerous impurities. One formulation involves mixing diatomaceous earth with clay, baking at 900°C, and crushing to powder. It is designated Chromosorb-P (P stands for its pink color). The structure of the surface is

$$
\begin{array}{ccc}
\text{OH} & & \text{OH} \\
| & & | \\
\text{—Si} & \text{—O—Si} & —
\end{array}
$$

Several Chromosorb materials are available, with varying specific surface areas and chemical reactivity. They contain intricate networks of fine pores, which require nearly 1% by weight of the liquid phase to form a monolayer coverage of the surface. Coatings of 10 to 20% are commonly used, in which the liquid-film thickness may approach 100 nm.

In GSC, the solid support is the stationary phase. Adsorption effects play a critical role and often lead to poorly defined chromatographic peaks (*band tailing*), especially with polar analytes such as water. Porous styrenedivinylbenzene polymers (Porapaks and Chromosorb Century series) have been developed for use in GSC. The porous polymer beads have low affinity for water and other polar molecules; the low affinity results in fast elution and well-defined symmetrical peaks. These stationary phases are extensively used for the analysis of permanent gases (for instance, O_2, N_2, CO_2, and CH_4) as well as for small polar compounds, such as glycols, acids, and amines.

In GLC, the major problem in coating the support material with the stationary liquid phase is to ensure a uniform coating. The liquid is usually dissolved in a low-boiling solvent to facilitate mixing with the solid support. The solvent is evaporated, and the column is filled by pouring the solid support (which still appears "dry" because the liquid film is so thin) into the column. Shaking, tapping, or vibrating is usually necessary to give tight, uniform packing of the column. The column is bent to fit the oven after it is filled.

From a chemical standpoint, the most important variable in GC is the material for the stationary liquid phase. Ideally, the carrier gas and solid support do not interact with the analyte, whereas the distribution coefficient, K_D, is determined entirely by the nature of the liquid phase and the temperature. Clearly, for the liquid phase to be stationary, it cannot exhibit any significant vapor pressure. This will limit the temperature at which the column may be operated. Distribution coefficients for the analytes in the sample must be intermediate: If K_D is very large, the retention time will be inconveniently long; and if K_D's are very small, retention time for all analytes will be nearly the same as t_m (the air peak), and no separation will be possible. The distribution coefficient is a function of the volatility of the analyte dissolved in the stationary liquid phase. This volatility will depend on the interaction between analyte molecules and the molecules of the stationary liquid phase. Fortunately, there are many possible liquid phases so that the analyst can usually find one that will yield an efficient separation. Some examples are given in Table 19-2. Nonpolar liquid phases tend to show less interaction with the analytes so that retention times will be in the order of their boiling points. Polar liquid phases show strong interactions with some analytes so that prediction of retention time is difficult.

As a general rule, the analyst will use a stationary phase similar in polarity to the components in the sample. For example, squalane, a saturated hydrocarbon itself, is useful for the separation of hydrocarbon mixtures. Carbowax, a polyethylene glycol, is useful for the separation of alcohols. When a sample contains components with

TABLE 19-2   Popular Stationary Liquid Phases and Their Use

Common name	Chemical composition	Maximum Temperature (°C)	Polarity*	Typical use
Squalane[†]	$C_{30}H_{62}$	150	NP	hydrocarbons
OV-1	polymethyl siloxane	350	NP	general purpose, nonpolar compounds
SE-30	polydimethyl siloxane	350	NP	hydrocarbons, separation by boiling point
Apiezon L	high-boiling petroleum hydrocarbon	300	NP	hydrocarbons, ethers, esters
Dexsil-300 GC	polycarborane siloxane	500	NP	compounds with low vapor pressure
DEG adipate	diethyleneglycol adipate (polyester)	200	I	esters, fatty acids, pesticides
—	dinonyl phthalate	150	I	general purpose, ketones, ethers, sulfur compounds
Carbowax 20M	polyethylene glycol	250	P	alcohols, aromatics, esters
QF-1	polytrifluoropropylmethyl siloxane	250	P	amino acids, steroids, nitrogen compounds
XE-30	polycyanomethyl siloxane	275	P	alkaloids, drugs, halogen compounds

*NP, nonpolar; I, intermediate polarity; P, polar.
[†]Squalane is obtained by reducing squalene, a component of shark-liver oil.

similar boiling points but differing polarity, the component with polarity like that of the stationary phase will be retarded on the column and will elute with a longer retention time. For example, if a mixture of nonpolar paraffin hydrocarbons and more polar aromatic hydrocarbons are separated on a Carbowax column, the aromatics will be retarded relative to the paraffins. The reverse would be true if squalane were used as the stationary phase.

High-temperature operation is required in order to analyze mixtures exhibiting low vapor pressures, and many liquid stationary phases have been developed with high boiling points and decomposition temperatures. One outstanding example is the polycarborane siloxanes (Dexsil), which can be used up to 500°C. Another approach is the bonding of the liquid phase directly to the solid support, yielding a stationary phase with essentially no vapor pressure. Carbowax bonded to the surface of a porous silica bead has been marketed as the Durapak series. On the other hand, resolution in GC improves as the column temperature is decreased, and some consideration should be given to the minimum temperature of operation, which is fixed by the melting point of the liquid stationary phase or its viscosity. Most stationary phases can be operated at room temperature, and some (fluorosilicone, phenyl silicone, or β, β'-oxypropionitrile) are operable at 0°C. However, one popular stationary phase, methyl silicone (OV–1), should be operated above 100°C.

Thousands of liquid stationary phases have been reported, but a mere handful (Table 19-2) accounts for most GC separations. But for thorny separation problems, some exotic stationary phases have proved useful. For example, the addition of $AgNO_3$ to a polyethylene glycol produces a stationary phase that interacts strongly with double bonds. Isobutene and butene-1 have nearly identical retention times on most columns. On the $AgNO_3$-containing column, however, butene-1 has a retention time nearly double that of isobutene. Stationary phases consisting of liquid crystals are sensitive to an analyte molecule's shape, and optically active stationary phases are capable of separating racemic mixtures.

Up to this point, we have injected a sample onto a column containing a stationary liquid phase and have carried it through (eluted) with the carrier gas. As the various components of the sample emerge from the column exit, they must be detected. The demands on the detector are severe: One of the components to be identified in a 0.1-μL sample might be present at the 1% level so that 0.001 μL (weighing perhaps 10^{-6} g) must be detected quickly at the moment it emerges from the column. Yet detectors today are capable of measuring amounts of analyte several orders of magnitude smaller. There are several types of detector that have been used in GC, but the most common are those based on (1) thermal conductivity, (2) flame ionization, and (3) electron capture. We shall briefly examine each of these detector systems.

The *thermal conductivity detector* is probably the most widely used because it is simple, inexpensive, reasonably rugged, and nondestructive to the sample. The last advantage is important if the analyst attempts to collect the material as it elutes from the column and to use other analytical methods to identify the components.

The operation of the detector is based on the thermal balance of a heated filament. If electrical power is converted to heat in a resistant filament (as it is in a light bulb), the temperature will continue to climb until heat power loss from the filament equals the electrical input. The filament may lose heat by radiation to a cooler surface and by conduction to the molecules coming into contact with it, the ability of a colliding molecule to carry off heat depending on its thermal conductivity. Hydrogen and helium have high thermal conductivities and therefore will be more efficient at "cooling" a heated filament than other gases will. Schematically, the thermal conductivity detector is shown in Figure 19-7. It contains two filaments: one exposed only to carrier gas, while the other is exposed to the gas as it exits from the column. When the exiting gas is only carrier gas, the two filaments can be balanced; instead of a direct measurement of filament temperature, the filament resistance, which is a function of temperature, is measured. Now when an analyte elutes, its thermal conductivity is lowered, and the filament in the exit line becomes hotter than does the other filament. Its resistance increases, and this imbalance between filament resistances is measured and a signal recorded. The detector signal must be calibrated for each compound because the relationship between an analyte's properties (thermal conductivity) and the signal observed is quite complicated. Known samples are used to determine the amount of signal per mole (or per gram) of analyte; the calibration will be later discussed in more detail.

The *flame ionization detector* is the second commonly used detector for GC. The carrier gas flows into a burner assembly after it leaves the column, is mixed with H_2

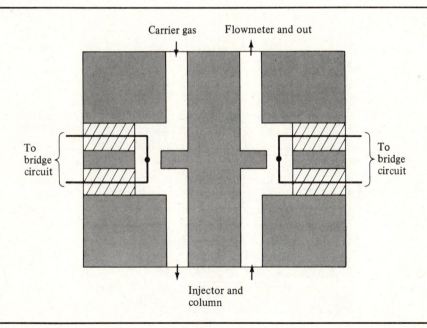

FIGURE 19-7 Thermal conductivity detector.

and air (or O_2), and is ignited as shown in Figure 19-8. The H_2–O_2 flame is nearly colorless; but when the carrier gas contains an organic compound that has eluted from the column, the flame burns yellow. Although a simple detector based on the color change can be constructed, the ionization properties of the flame can be used to produce a more sensitive detector. When organic compounds are burned in a flame, they tend to produce some ions in the combustion process. Electrodes placed near the flame (Figure 19-8) will detect the presence of ions by the small current that flows through the electrical circuit. Even though the currents are only $\sim 10^{-12}$ A, they can be readily measured. The *increase* in ionization current observed from pure carrier gas to carrier gas containing eluted analyte constitutes the recorded signal. The ion current is roughly proportional to the number of carbon atoms. Thus octane (C_8H_{18}) would give twice as much signal as butane (C_4H_{10}) for samples containing equal moles of each. The detector is sensitive neither to highly oxidized carbon (such as carbonyl groups) nor to inorganic compounds, including H_2O, air, and the like.

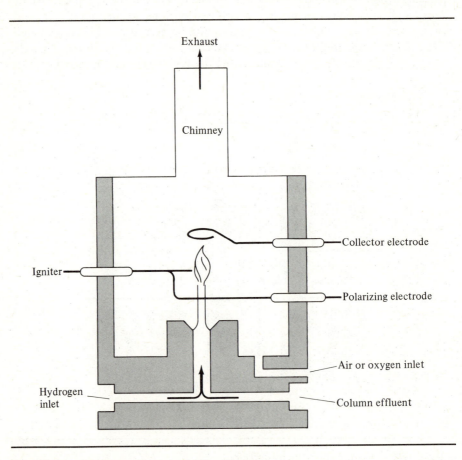

FIGURE 19-8   Hydrogen flame ionization detector.

The third commonly used detector employs *electron capture*. It too detects ions in the gas exiting from the GC column by electrodes in the detector compartment. In this case, the ions are produced by a radioactive source such as ^{3}H or ^{63}Ni, which emits β rays. For example, nitrogen carrier gas can be ionized,

$$\text{ionization: } N_2 + \beta \rightarrow N_2^+ + 2e$$

and these ions will establish a "standing current" in the detector circuit. When an analyte that captures electrons is eluted, it will change this standing current, especially if the compound contains a highly electronegative element X, which tends to capture free electrons and increase the amount of ion recombination:

$$\text{electron capture: } \quad X + e \rightarrow X^-$$
$$\text{ion recombination: } \quad X^- + N_2^+ \rightarrow X + N_2$$

Thus the current will *decrease*, and this decrease constitutes the signal recorded. The detector is especially sensitive to compounds containing halogens, lead, phosphorus, nitro groups, silicon, and polynuclear aromatics. The detector shows little sensitivity to compounds without these elements or functional groups. The great usefulness for environmental analysis of the electron capture detector has emerged in recent years. The list above includes the most *unwanted* species in the environment (insecticides, vinyl chloride, PCB's, fluorocarbons, and the like), and detection at lower and lower levels has been required to meet federal regulations.

GC INFORMATION

A typical gas chromatogram will look like Figure 19-3(**B**), in which detector signal is plotted along the y-axis and time is plotted along the x-axis. When a constituent from the sample elutes and is detected by the detector, a signal is recorded. The time at which the peak signal is observed is the retention time, t_R. The retention time is a function of K_D and, therefore, gives *qualitative* information concerning the sample, in the same way that the half-wave potential in polarography helps to identify a constituent in a sample.

Unfortunately, a table of retention times is not normally available. There are so many variables that affect the observed retention time (flow rate, temperature, sample size, stationary phase, V_s, t_m, and average pressure in the column are the primary factors) that reporting of t_R for one set of conditions would not be helpful to other analysts—it being difficult enough to reproduce all these variables from one day to the next in the same lab. Reporting specific retention volume, V_g, can be of use to other analysts but requires careful measurement of all variables. One of the problems is that gases are compressible so that volume depends on pressure. This means that a measurement of flow rate (in milliliters per second) under atmospheric-pressure conditions at the column exit will not equal the measurement of the flow rate at the top of the column, which may be at a considerably higher pressure. The apparent retention volume found from Equation 19-9 must be corrected for such a pressure change in the column.

One method for avoiding all these correction factors is to measure some reference compound under the same conditions and report relative retention (Equation 19-16). Unfortunately, no universal reference substance (analogous to the NHE in electrochemistry) has been adopted. However, one useful approach is the comparison of relative retention with the normal paraffin hydrocarbons. In this *Kovats index*, each paraffin is given the index value of 100z, where z is the number of carbons ($CH_4 = 100$, $C_2H_6 = 200$, $C_3H_8 = 300$, and so on). A plot of $\log(t_R - t_m)$ versus index value is a straight line, as shown in Figure 19-9, and an analyte having a Kovats index value of 540 would be equivalent to a hypothetical normal paraffin containing 5.4 carbon atoms.

The *quantitative* information from a chromatogram is obtained from the detector signal as a function of time. The area under the peak is proportional to the amount of analyte present, as discussed in regard to Figure 19-3(**B**). The amount can be expressed in mole units or weight units, but, of course, the proportionality constant will be different.

$$S = \text{area under peak} = RA$$

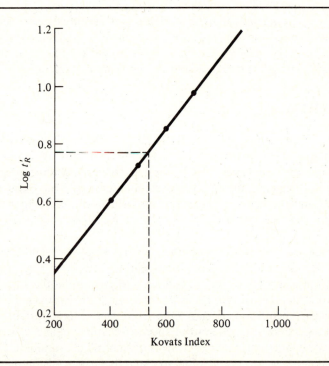

FIGURE 19-9 Kovats index. The logarithm of the adjusted retention time ($t'_R = t_R - t_m$) for a series of normal paraffin hydrocarbons is plotted versus 100 times the carbon number. An unknown compound with $t'_R = 5.89$ min gives an index value of 540, equivalent to a hypothetical normal paraffin of 5.4 carbon atoms.

where R = area per mole when A is in moles and R = area per gram when A is in grams (millimoles, micromoles, milligrams, or micrograms may be preferable units). A standard sample must be run to determine R, which means that the peak identity must be known. Sometimes the molar response (or weight response) can be assumed to be about the same for chemically similar components in a mixture. If so, the percentage composition can be calculated from the fraction of the total peak area that each component represents, as illustrated in Example 19-6.

EXAMPLE 19-6

A mixture containing only benzene (C_6H_6) and bromobenzene (C_6H_5Br) gives peaks of 9.50 and 4.78 cm² respectively, in a gas chromatogram. Assuming the molar response to be the same, calculate mole percent and weight percent of benzene and bromobenzene in the mixture.

Solution

	Area (cm²)	% total area (mol %)
benzene	9.50	$(9.50/14.28) \times 100\ (\%) = 66.5$
bromobenzene	4.78	$(4.78/14.28) \times 100\ (\%) = 33.5$
total	14.28	

With 100 mol of mixture, 66.5 mol would be benzene, and 33.5 mol would be bromobenzene. The weight of this mixture would be

	Moles × FW (g/mol) =	g	Weight %
benzene	66.5 × 78	= 5,187	$(5,187/10,447) \times 100\ (\%) = 49.7$
bromobenzene	33.5 × 157	= 5,260	$(5,260/10,447) \times 100\ (\%) = 50.3$
	total wt =	10,447	

How can the area under the peak be found? This problem was mentioned in Section 16-2 in regard to controlled-potential electrolysis. Many gas chromatographs are now equipped with an electronic digital integrator. If an integrator is not available, two approximate methods work quite well and give results within a few percent of the true area. They are shown in Figure 19-10. In Figure 19-10(**A**), the peak is approximated by a triangle, and in Figure 19-10(**B**), the peak is approximated by a rectangle. In both methods, it is assumed that the peak area neglected is equal to the geometric area lying outside the peak. The rectangle method of Figure 19-10(**B**) appears to be more reproducible and is generally used in preference to

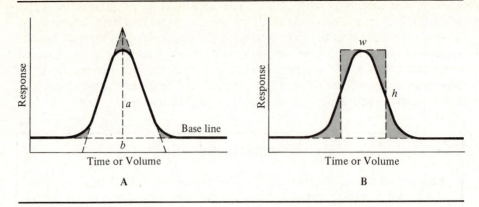

FIGURE 19-10   Measurement of peak areas: **(A)** area of peak $\cong$ area of triangle $= \frac{1}{2}ab$; **(B)** area of peak $\cong$ area of rectangle $= hw$.

the triangulation method. Clearly, both methods require a reasonable width to give a reliable area value. Although the digital integrator can give area values reproducible to better than $\pm 1\%$, the approximate-area methods are usually reproducible to ± 3 to 4%.

When the peak width is very small, the analyst may resort to measuring peak height only. The peak height will be directly proportional to peak area (and thereby proportional to the amount of analyte) if all variables are controlled closely. However, a slight change in flow rate, sample injection rate, or temperature can make a significant change in peak height even though the area remains nearly constant; thus measurement of peak height only may lead to greater errors.

APPLICATIONS

From the discussion above, we can conclude that nearly all mixtures of organic compounds may be separated by GC. Successful separations include not only routine mixtures encountered as part of a synthesis reaction (unreacted starting materials, products, and products from side reactions) but also many complex, difficult-to-separate mixtures encountered in natural products or in the environment. Besides its resolution ability, GC can be quite sensitive: In some cases, picograms (10^{-12} g) of material can be detected.

A GC separation of a complicated mixture of environmental concern is shown in Figure 19-11. Samples from the exhaust gas of an automobile are being analyzed, and over 100 components have been identified. As you might imagine, it is very difficult to *identify* all the peaks. The analyst may know, from GC, that a mixture contains at least 20 components and, from standard samples, may have identified 12 of them, but what are the other 8? It is now possible to purchase GC–mass

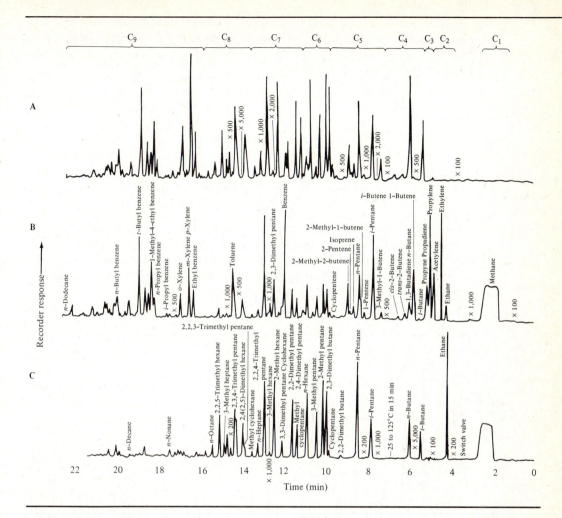

FIGURE 19-11 Chromatograms showing fuel and exhaust hydrocarbon separations: (**A**) Indolene 30 fuel, complete sample; (**B**) exhaust gas from modified engine, 15 to 30 mph, complete sample; (**C**) same exhaust gas as (**B**), unsaturated hydrocarbons removed. [From O.J. McEwen, "Automobile Exhaust Hydrocarbon Analysis by Gas Chromatography," *Analytical Chemistry*, 38:1047 (1966).]

spectrometry equipment in which the eluted gas components are analyzed by mass spectrometry. This is one situation in which a nondestructive GC detector is required. In mass spectrometry, the molecule is ionized and broken into fragments. From the "cracking pattern" of the fragment ions, it is possible to deduce the molecular weight and bonding arrangement. In fact, the cracking pattern is something like a "fingerprint" of the molecule. The information from the GC and mass spectrometer is collected by a computer system, which, in most cases, will give the analyst the identity of each of the GC peaks.

Mixtures extremely difficult to separate by other techniques are often relatively simple by GC because the column contains so many theoretical plates. When benzene is chlorinated in the presence of UV light, an addition of chlorine takes place as given by Equation 19-25:

$$\text{benzene} + 3Cl_2 \xrightleftharpoons{UV} \text{benzenehexachloride} \qquad (19\text{-}25)$$

It may appear that there is only one possible product (isomer), but actually there are several isomers of benzenehexachloride. These highly chlorinated compounds are quite toxic, and one isomer is the active ingredient in the insecticide Lindane. The isomers are very similar chemically, but their K_D values are sufficiently different that separation by GC is possible, as in Figure 19-12.

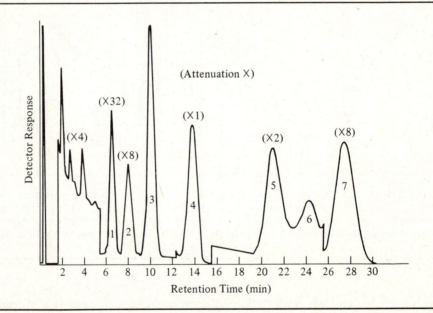

FIGURE 19-12 Chromatogram of standard benzenehexachloride sample (with attenuation indicated in parentheses at peaks): (1) α-benzenehexachloride; (2) γ-heptachlorocyclohexane; (3) γ-benzenehexachloride (Lindane); (4) ε-heptachlorocyclohexane; (5) β-benzenehexachloride; (6) ε-benzenehexachloride; (7) δ-benzenehexachloride. [From A. Davis and H.M. Joseph, "Quantitative Determination of Benzenehexachlorides by Gas Chromatography," *Analytical Chemistry*, 39:1016 (1967).]

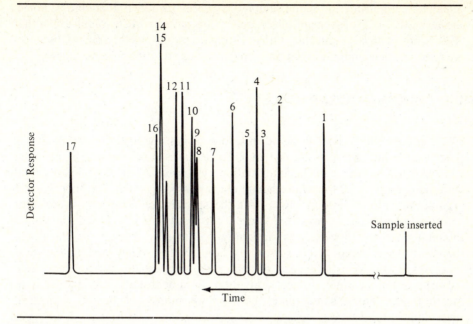

FIGURE 19-13 Separation of octane isomers on a capillary column: glass column, 600 ft, 0.010-in. bore; stationary phase, squalane; film thickness, ~0.3 μ; eluent gas, nitrogen; inlet pressure, 23 psi; temperature, 50°C; efficiency, ~300,000 plates. [From D.H. Desty and A. Goldup, in R.P.W. Scott, ed., *Gas Chromatography 1960*, p. 162 (Washington: Butterworths, 1960).]

Another example of a separation of isomers is presented in Figure 19-13. In this case, a mixture of 17 isomers of octane is nearly completely resolved (numbers 14 and 15 have the same retention time under the conditions used so that only 15 of the isomers have been resolved) by a capillary column 600 ft long, containing about 300,000 theoretical plates!

GC has not been widely used for inorganic mixtures because most inorganic compounds do not have sufficient vapor pressure. However, some complexes are sufficiently volatile to be analyzed by GC. One of the most widely used ligands is hexafluoroacetylacetone,

$$CF_3-\overset{\overset{\displaystyle O}{\|}}{C}-CH_2-\overset{\overset{\displaystyle O}{\|}}{C}-CF_3$$

Not only does this ligand form volatile complexes with many metals, but the presence of the halogen atoms allows detection at very low levels with the electron capture detector.

The list of applications is nearly endless. Even complicated mixtures encountered in the biochemical field have been analyzed by GC. Hundreds of components in human breath, urine, and blood samples have been identified. It should be noted, however, that many biochemical compounds decompose when volatilized so that GC is not always practicable. Other chromatographic methods have been developed for such compounds, and we now turn our attention to some of them.

19-3   LIQUID CHROMATOGRAPHY (LC)

When the mobile phase is a liquid instead of a gas, the analytical technique is known as liquid chromatography (LC). To be more exact, there is liquid–solid chromatography (LSC), in which the stationary phase is a solid, and liquid–liquid chromatography (LLC), in which the stationary phase is a liquid. The initial work of Tswett was LSC. However, except for the special technique of ion-exchange chromatography (IEC), most LC carried out today is LLC. The distinction is becoming blurred, however, as stationary phases may actually be chemically bonded to a solid support. The reason why LSC is not so widely used for analytical purposes is that K_D as a function of concentration is usually not constant when adsorption on the solid surface is involved. A nonconstant K_D value leads to nonideal peak shapes (band tailing) and to a lower number of theoretical plates in the column.

One significant difference between LC and GC is the high diffusion rates of gases compared with liquids. This means that flow rate must be very slow to prevent large mass-transfer contributions (*Cu* term in the Van Deemter equation) in LC. The optimum flow rate is usually too slow to be practical, and the columns are not operated under optimum conditions.

Another significant difference between GC and LC is that the mobile phase interacts with the sample in the latter technique and represents another critical experimental variable. The mobile liquid phase and the stationary liquid phase also must not show significant mutual solubility. Traditionally, water is made the stationary phase by formation of an aqueous film on a silica gel solid support, which is then used to pack the column. Silica gel can adsorb up to 50% water before it becomes too "moist" for use in a chromatographic column. The sample is carried through the column in a nonpolar solvent, which acts as the liquid mobile phase.

Alternatively, a *reversed-phase* LC can be carried out by coating the solid support with a nonpolar solvent and using a polar mobile phase. For example, powdered rubber coated with benzene as the stationary liquid phase can be used, with the sample carried through the column with an aqueous mobile phase. However, the greatest use of reversed-phase LC at present is with bonded phases (Section 19-6).

19-4   ION-EXCHANGE CHROMATOGRAPHY (IEC)

A special case of LSC is IEC. The solid stationary phase consists of a material containing ions that can be replaced by ions present in the liquid mobile phase. Because ions are involved, the mobile phase is normally an aqueous solution. What materials possess these ion-exchange properties? It turns out that some naturally occurring minerals, such as clays and zeolites, are quite effective ion exchangers. They consist of anionic aluminosilicate networks, R—, which are charge balanced by alkali ions such as Na$^+$. When this mineral, R—Na, is in contact with an aqueous solution containing Ca^{2+}, an equilibrium is established (Equation 19-26):

$$2R-Na + Ca^{2+} \rightleftharpoons (2R)-Ca + 2Na^+ \qquad (19\text{-}26)$$

Because calcium is $2+$, it is necessary to use two cation sites on the mineral to maintain charge neutrality. Equation 19-26 is the basis for water softeners. Hard water contains dipositive ions, such as Ca^{2+} and Mg^{2+}, which are replaced by Na$^+$ when the water is passed through the ion exchanger. The equilibrium lies far to the right so that the water is essentially free from the troublesome $2+$ ions. However, after the ion exchanger has been converted predominantly to the $2+$ ion form, it will no longer be effective. At this point, it must be regenerated. Regeneration is accomplished by washing the ion exchanger with a *concentrated* NaCl solution, which reverses Equation 19-26 by virtue of the LeChâtelier principle.

Although mineral ion exchangers are useful for softening water, their properties vary somewhat from sample to sample, a variation not conducive to analytical use. For analysis, synthetic ion exchangers are preferred and can be purchased with properties specifically tailored to particular analytical problems. In fact, synthetic *ion-exchange resins* are available that exchange anions instead of cations.

The ion-exchange resin is produced by a polymerization reaction in the same way that plastics and other polymers are produced. For example, styrene can be polymerized to form polystyrene

$$(19\text{-}27)$$

styrene   polystyrene

If the phenyl group is sulfonated, the polymer will be

The sulfonic acid is a strong acid and will readily exchange H$^+$ with other cations in an analogy to Equation 19-26. The material containing only long straight chains, as shown above, does not possess good physical properties for an ion-exchange resin. Adding about 8% divinylbenzene (DVB) to provide *cross linking* is an improvement:

divinylbenzene

As can be visualized from the structure above, divinylbenzene can be polymerized at both double bonds to provide a polymer with a three-dimensional structure. The cross linking makes the polymer mechanically stronger and less porous, lowers solubility and swelling in water, and changes the equilibrium constant for the exchange reaction.

The chemical properties can be adjusted by changing the strength of the acidic group present. If an amine group is used instead of sulfonic acid, the resin will be an anion exchanger. Some examples are given in Table 19-3. Also included in Table 19-3 is the *exchange capacity* for the resin. The exchange capacity is the total number of replaceable ions per unit volume, expressed as milliequivalents per milliliter. For weak-acid or -base types, the exchange capacity will be a function of pH.

The analyst can measure exchange capacity by simply taking a sample of resin in the H^+ form and placing it in a solution containing a reasonably high concentration of Na^+. The following exchange reaction takes place:

$$R—H + Na^+ + H_2O \rightleftharpoons R—Na + H_3O^+ \qquad \text{(19-28)}$$

TABLE 19-3   Typical Ion-Exchange Resins

Type	Composition*	Examples of commercial names	Exchange capacity (meq/mL)
strong cation	polystyrene–DVB with sulfonic acid group	Dowex 50 Amberlite IR-200 Ionac C-242	1.7–2.0
	porous polymer beads, sulfonic acid	Aminex A-7[†]	
	polymer-coated glass beads, sulfonic acid	Zipax SCX[†]	
Weak cation	methacrylic acid (carboxylic acid group)	Amberlite IRC-50	3.5–4.0
strong anion	polystyrene–DVB with quaternary ammonium group porous polymer beads with quaternary ammonium group	Dowex 1 Amberlite IRA-900 Aminex A-28[†]	1.0–1.4
weak anion	polystyrene–DVB with polyamine porous silica gel, siloxane, bonded with primary amine	Dowex 3 Amberlite IRA-93 Micropak–NH$_2$[†]	1.2–2.0

*DVB stands for divinylbenzene used for cross linking.
[†]Ion-exchange materials suitable for HPLC-IEC.

The resulting H_3O^+ is titrated with base. Typical values are 1 to 5 meq/mL, and this value will be stated on the bottle. However, this capacity value is for the dry resin, and it should be realized that, when the resin is placed in water, it will swell to a larger volume. The reason for this swelling involves osmotic pressure because the ionic strength inside the resin beads (5 meq/mL = 5 N) is usually considerably greater than is the ionic strength in the surrounding solution. The forces are so strong that addition of water to a column tightly packed with dry resin beads can break the column. Therefore, the resin should be wetted before it is packed into the column.

The distribution coefficient for the exchange reaction is still "products over reactants," like any other equilibrium constant, but in this case, a charged species is involved. For every ionic charge that transfers from the mobile phase to the stationary phase, one must move in the reverse direction. The equilibrium constant represents a competition and, for Equation 19-28, may be written (neglecting the activity of H_2O)

$$K_D = \frac{[R—Na][H_3O^+]}{[R—H][Na^+]} \tag{19-29}$$

K_D is also called "selectivity coefficient" for the ion-exchange reaction. With ions involved and high ionic strengths on the resin, activity coefficients play an important role. Recalling Equation 4-8, we can rewrite Equation 19-29 in terms of *concentrations* and activity coefficients, f,

$$K_D = \frac{(R—Na)(H_3O^+)}{(R—H)(Na^+)} \cdot \frac{f_{R-Na}f_{H_3O^+}}{f_{R-H}f_{Na^+}} \tag{19-30}$$

where the terms in parentheses are concentrations. Unfortunately, the activity coefficients are usually unknown and will be incorporated into K_D, resulting in an equilibrium constant which is a function of concentration.

What factors determine K_D? If we consider Equation 19-28, in which H^+ on a cation exchanger is being replaced by some other metal ion, K_D is increased under the following conditions:

1. Charge on M^{n+} increases.

2. Hydrated ionic radius of M^{n+} decreases.

3. Greater cross linking of resin (all K_D values approach 1 as cross linking decreases).

4. Adsorption effects are present, as with some high-molecular-weight organic ions.

Factor 1 shows that water softeners work well because the ions that make water "hard" are 2+, whereas the resin (or natural ion-exchange mineral) contains 1+ ions. Factor 2 becomes important when ions of similar chemistry, such as the rare earths or actinides, are to be separated. As K_D increases, the ion will elute from the column more slowly so that, for a series of similar metal ions, the ions will elute in reverse order of ionic radius (largest first).

Suppose that a sample containing an analyte A^+ is placed at the top of an ion-exchange column and some eluent as the mobile phase is continuously added. How much volume of mobile phase is required before analyte A^+ is eluted? IEC is, in principle, no different from any of the other chromatographic methods, and Equation 19-11 should be applicable:

$$V_R = V_m(1 + k')$$

where V_R is the retention (or elution) volume, V_m is the volume of mobile phase contained in the column (void volume), and k' is the capacity factor, which is

$$k' = \frac{(\text{mmol } A^+)_s}{(\text{mmol } A^+)_m} \tag{19-31}$$

In IEC, k' is also known as D, the distribution ratio. However, in ion exchange, a competition is involved so that k' will depend on the nature and concentration of the competing ion (for example, H_3O^+ in Equation 19-28). Thus the retention volume will depend on the concentration of ions in the eluent.

Of even greater concern is the ability to separate two analytes, A^+ and B^+. Suppose that we consider their competition with H_3O^+; then

$$K_{D,A} = \frac{[R—A][H_3O^+]}{[R—H][A^+]} \tag{19-32}$$

$$K_{D,B} = \frac{[R—B][H_3O^+]}{[R—H][B^+]} \tag{19-33}$$

A separation factor α can be defined as

$$\alpha = \frac{K_{D,A}}{K_{D,B}} = \frac{[R—A][\cancel{H_3O}]^+/[\cancel{R—H}][A^+]}{[R—B][\cancel{H_3O}]^+/[\cancel{R—H}][B^+]} \tag{19-34}$$

$$\alpha = \frac{[R—A][B^+]}{[A^+][R—B]} \tag{19-35}$$

Thus the separation factor does *not* depend on the concentration of other ions even though the retention volume will. The separation factor depends on the two separable ions' competing for sites on the ion-exchange resin. However, activity coefficients may complicate this simplistic picture.

QUANTITATION

Suppose that two analytes are separated on an ion-exchange column and elute at two different retention volumes. How can we determine the amount of each analyte present in the original sample? Detectors useful for GC or LC, will not generally be useful for IEC. Traditionally, the eluate is collected as a series of small samples, and the presence of analyte is determined by some other analytical technique. For example, if the analyte absorbs light, a spectrophotometric method can be used; or if the analyte is reducible, a polarographic method is possible. Where it is known that the analyte will elute, say, between 100 and 150 mL of elution volume, this entire volume can be collected and titrated in a volumetric method. This technique has been quite useful in the separation and determination of metal ions by ion exchange for separation and titration with EDTA for quantitation. However, for IEC to be completely on a par with GC and LC, some sort of universal detector is required that will provide a signal proportional to concentration as analytes elute from the column. This approach is taken in a recent technique known as *ion chromatography*.

ION CHROMATOGRAPHY

Since 1975, the technique of ion chromatography has developed rapidly: It incorporates a universal detector, such as a conductivity measurement, at the exit of the column; but to prevent interference from competing ions in the eluent, a second column is added, called a *suppressor column*. The concept can be seen more clearly through an example. Suppose that a mixture of anions is to be analyzed. The mixture is separated on an analytical column containing an anion-exchange resin in the HCO_3^- form. The exchange reaction is

$$R—HCO_3 + X^- + Na^+ \rightleftharpoons R—X + HCO_3^- + Na^+$$

Because the resin is an anion exchanger, Na^+ is not involved but is shown for electroneutrality. The resulting $NaHCO_3$ solution exits from the column and enters the suppressor column, which is a cation exchanger in the H^+ form. The following exchange takes place:

$$R—H + Na^+ + HCO_3^- \rightleftharpoons R - Na + H_2CO_3$$

As the solution elutes from the suppressor column, it contains nonionic carbonic acid (CO_2 dissolved in H_2O), which shows little ionic conductivity. The signal is essentially zero. However, when X^- elutes from the analytical column, the eluate contains NaX and, after passing through the suppressor column, contains HX (for instance, HCl when the anion is Cl^-). The conductivity will be quite high, provided that HX is a reasonably strong acid (stronger than carbonic acid). Figure 19-14 shows an example of an ion chromatogram.

ION-EXCHANGE APPLICATIONS

It should be clear that ion exchange is most useful for separation of inorganic ions. Mixtures of metallic ions can be separated even when the metal ions are chemically very similar. Separations of Zr from Hf, Nb from Ta, the rare earths, and the actinides are all examples. These mixtures may be separated with cation exchangers; or by formation of anionic complexes, the separation may be carried out with an anion-exchange resin.

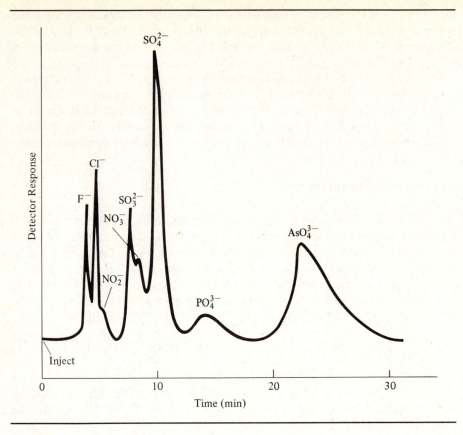

FIGURE 19-14   Separation of anions by ion chromatography.

Sometimes a complexing agent is used to elute a metal ion from a cation-exchange column. For example, a mixture of 3 + rare-earth ions, L^{3+}, can be placed on a cation-exchange column and elution carried out with a citrate solution. The formation of citrate complexes will decrease $[L^{3+}]$ in the mobile phase to a value that depends on the formation constant for the complex. This not only aids in the elution but increases the separation factors because, in addition to varying K_D values, each lanthanide citrate complex will have its own formation constant.

Minute amounts of radioactive actinides can be separated by ion exchange, and radioactive counting can be used as the detector. The eluate is collected *dropwise* and placed in a counter. A typical ion-exchange chromatogram for actinides is shown in Figure 19-15.

Although ion-exchange methods are considered to be of greatest use for separation of inorganic ions, it should not be overlooked that some biochemical systems can be ionic in nature. Amino acids can be cationic or anionic, depending on pH. This property is extremely useful for separation procedures. A mixture of amino

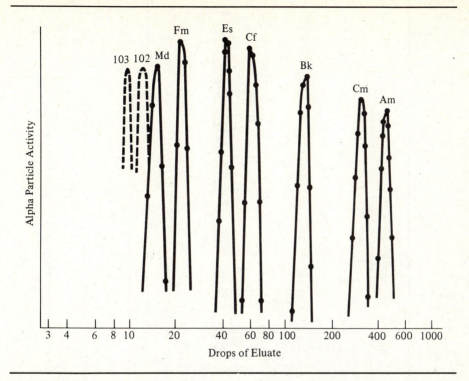

FIGURE 19-15   Ion-exchange chromatographic separation of tripositive actinide ions from Dowex 50 ion-exchange resin with ammonium α-hydroxyisobutyrate. Predicted positions for elements 102 and 103 are indicated by broken lines. [From Glenn T. Seaborg, *Man-made Transuranium Elements* (Englewood Cliffs, N.J.: Prentice-Hall, Inc., 1963).]

acids can be placed on a cation-exchange column (for example, Dowex 50) by use of a low pH. With increasing pH, the amino acids will elute, depending on their isoelectric point (the pH at which the species is neutral). Figure 19-16 is an impressive example.

An ion-exchange system containing both a cation-exchange resin and an anion-exchange resin is useful for providing pure water for the laboratory. Any cations present in the water supply are replaced by H_3O^+, and any anions are replaced by OH^-. Because of the requirement of electroneutrality, H_3O^+ and OH^- will be produced in equivalent quantities and will neutralize each other,

$$H_3O^+ + OH^- \rightleftharpoons 2H_2O$$

The product is called "deionized water" and is used in place of distilled water in many labs. The ion-exchange system will not, however, remove any nonionic impurities.

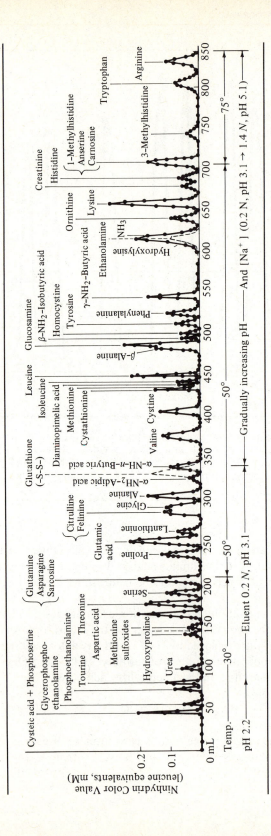

FIGURE 19-16 Separation of amino acids and related compounds on Dowex 50–X4 column, 0.9 by 150 cm. [After Stanford Moore and W.H. Stein, "Procedures for the Chromatographic Determination of Amino Acids on Four Per Cent Cross-Linked Sulfonated Polystyrene Resins," *Journal of Biological Chemistry*, 211:893 (1954).]

19-5   EXCLUSION CHROMATOGRAPHY (EC)

A special case of LLC used in the separation of analytes by size is EC. As shown in Figure 19-1, EC has been subdivided into two classifications, *gel permeation chromatography* (GPC) and *gel filtration chromatography* (GFC). The distinction between the two is more historical than scientific, and the IUPAC recommends that the name "gel filtration" be dropped: Columns containing such gels as Sephadex, which swell in aqueous media and are used for the separation of biochemical compounds, were considered to be GFC; on the other hand, columns employing more rigid gels, such as polystyrene or silica, used in conjunction with organic solvents for the separation of organic compounds, were considered to be GPC. A single theory can be used for both, and thus a single name appears appropriate.

One unusual feature of GPC is that the same solvent is both the mobile and the stationary phase. Solvent that fills the pores of the gel is stationary, whereas solvent that fills the interstitial volume of the column is mobile. For example, consider a soft Sephadex gel often used in the separation of biochemical compounds in an aqueous environment. The gels are cross-linked polymers containing polar groups that can absorb water (or other polar solvents). Thus water held on the gel constitutes the stationary phase, and water used to elute the sample constitutes the mobile phase. The absorption process causes a swelling of the gel bead to many times its original volume, creating pores of varying size. A very simple picture is shown in Figure 19-17. The pores contain a volume, V_s, of solvent; the volume of solvent outside the beads is V_m. A very large molecule (A in Figure 19-17) cannot enter the pores (is excluded from the pores) and therefore is carried along only in the mobile phase, and

$$\text{large molecule:} \quad V_R = V_m$$

On the other hand, a very small molecule (C in Figure 19-17) must traverse both V_m and V_s because the entire pore volume is accessible to it. Thus

$$\text{small molecule:} \quad V_R = V_m + V_s$$

For intermediate-size molecules (B in Figure 19-17), a fraction of pore volume is accessible. Therefore,

$$V_R = V_m + KV_s \tag{19-36}$$

where K is some fraction ($0 \leq K \leq 1$). In other words, $K = 0$ for a large molecule, and $K = 1$ for a very small molecule. What is considered small and large depends on the type of gel used. Sephadex, which is produced from polysaccharide dextrans, is available in several types. Some are useful for separating molecules in the molecular weight range of 100 to 1000, whereas others can be used for molecules of molecular weight $> 100,000$.

The value of K will depend on molecular weight for molecules of intermediate size (fractionation range). Thus, if a mixture of species of high molecular weight is passed through a GPC column, the components of higher molecular weight will elute first because they have smaller K values. In the fractionation range, the

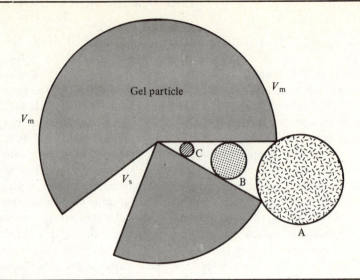

FIGURE 19-17 Schematic representation of gel particle for GPC, showing pores containing stationary phase, V_s. Analyte A in mobile phase V_m is too large to enter pores, analyte B must traverse a fraction of the pore volume, and analyte C is so small that all pore volume is accessible to it.

retention (elution) volume varies linearly with the logarithm of the molecular weight:

$$V_R = a - b \log \text{MW}$$

where a and b are constants dependent on the gel used. An example is shown in Figure 19-18. Sucrose is so small that $V_R \cong V_m + V_s \cong 240 \, \text{mL}$, while blue dextran is so large that $V_R \cong V_m \cong 80 \, \text{mL}$. Therefore, for this example, $V_s \cong 240 - 80 = 160 \, \text{mL}$. Once the line is established in the linear fractionation region (MW $\cong 2 \times 10^4 - 2 \times 10^5$ in Figure 19-18), the molecular weight of an unknown analyte can be determined from its retention volume. As can be seen from this figure, EC is an extremely useful technique for the study of large molecules of biochemical interest. The fact that there is no change in solvent when the analyte transfers from the mobile phase to the stationary phase is particularly useful when dealing with sensitive biochemical compounds, such as proteins.

For separation of organic analytes, gels of polystyrenedivinylbenzene (Styragels) or porous silica (Porasils) are commonly employed. Styragels can be used in most organic solvents and are available in different pore sizes, which allow separations of analytes from MW = 200 to > 50×10^6. Columns of polystyrenedivinylbenzene gels are particularly useful for determining molecular weight distributions of polymers dissolved in a solvent such as tetrahydrofuran. Porasils are rigid gels so that their pore size does not change when solvent is added. Separations from

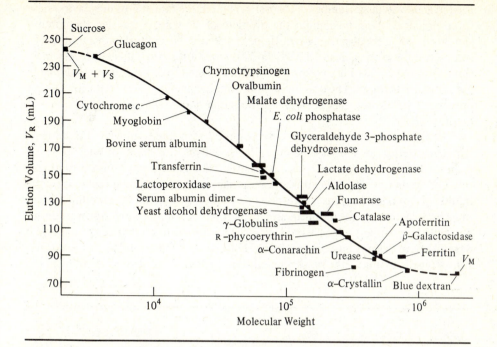

FIGURE 19-18   Relation between elution volume and molecular weight for proteins on a Sephadex G-200 column, 2.5 by 50 cm, at pH 7.5. [After P. Andrews, "The Gel-Filtration Behaviour of Proteins Related to Their Molecular Weights over a Wide Range," *Biochemical Journal*, 96:595 (1965).]

MW <200 to $>10^6$ are possible by choice of the appropriate pore size. In fact, separation of heptane (MW = 100) from hexane (MW = 86) with silica gel beads has been reported.

It should be noted that EC separations depend on the size of the analyte molecules; yet analysts often present their results based on molecular weight, as shown in Figure 19-18. Size is directly related to molecular weight only if the shape and the atomic weights of the constituent atoms are kept constant. For accurate measure of molecular weight, a calibration chart must be made with reference compounds having the same shape (spherical, rodlike, and so on) and the same approximate elemental composition.

19-6   HIGH-PERFORMANCE LIQUID CHROMATOGRAPHY (HPLC)

Figure 19-2 showed a chromatographic column that was used for LC. The analyst may just pour the sample and eluent (mobile phase) onto the column and allow gravity to carry the mobile phase through the column. However, if long columns are used to achieve a large number of theoretical plates, the rate of flow may be too slow to be practical. Also, to combat the problem of slow mass transfer, the column should contain a small-particle-size, tightly packed, solid support with a thin film of stationary phase on each small particle. This situation again will lead to extremely slow flow rates unless a high pressure is applied at the top of the column to drive the mobile phase through. The development of suitable packings and operation at high pressures ($\sim$5000 psi) has been given the name *high-performance liquid chromatography* (HPLC).

HPLC can be applied to any of the column LC techniques shown in Figure 19-1. The basic setup, as shown in Figure 19-19, will be the same for each of these techniques, with the appropriate column inserted. In many ways it is quite similar to a GC system in which the tank of carrier gas as eluent has been replaced by the liquid-solvent reservoir plus a high-pressure pumping system. This pumping system must be able to pump the mobile phase through the column at high pressure without any pulses or surges, which would distort the chromatographic peaks observed. Several pump designs have been used in HPLC instruments, but the reciprocating-piston type is currently the most popular. On the backward stroke, eluent from the

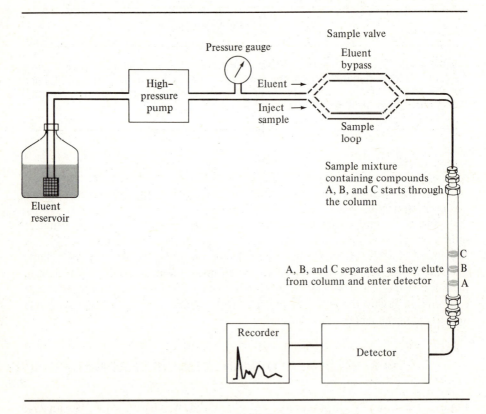

FIGURE 19-19 Schematic arrangement of HPLC components.

solvent reservoir is sucked in, while on the forward stroke, it is pushed onto the column. Although this type of pump is inexpensive, it produces flow pulses that must be damped to produce a relatively smooth flow to the detector. For example, multiple pistons may be employed, which are operated out of phase with each other to help smear out the pressure surges.

Sample Injection   It is necessary to inject the sample onto the column while the system is at high pressure, and the injection is accomplished by a sample valve system. The sample is injected into a sample loop with a typical volume of 10 to 20 μL. While the sample is injected, the mobile phase at high pressure moves through the sample valve directly to the column. Then the valve is rotated (so that the mobile phase passes through the sample loop), sweeping the sample into the column.

Columns   The design of columns will depend to some extent on the type of chromatographic separation planned, but typical analytical columns are precision-bore stainless-steel tubes of 3 to 4 mm diameter and 10 to 30 cm long (columns for EC are usually wider and up to 100 cm long). Because normal roughness is greater than the particle size of the packing, the tubes need to be polished on the inside. Glass-lined or Teflon-lined columns have been used on occasion.

The feature of column packing that has made HPLC one of the fastest growing analytical areas is the small and tightly controlled particle size. Porous silica gel beads of 5 to 10 $\pm$ 1 or 2 μm diameter are generally used. LSC or IEC can be carried out with the dry beads, or the beads may be coated with a liquid stationary phase. Porous beads may be used for EC.

With the high pressures used, a liquid stationary phase could be literally carried out of the column by the mobile phase; one way of avoiding this problem is to bond the stationary phase (bonded phase) chemically to the silica gel support. The porous silica gel beads can be esterified to form silicate esters or highly stable silicone polymers. Usually, the surface retains its polar nature, but if the entire surface is converted to hydrophobic groups, a nonpolar surface results, and reversed-phase chromatography can be carried out.

Detectors   Another problem encountered with LC is the lack of a universal detector, such as the thermal conductivity detector used in GC. Recently, with the development of HPLC, detectors employing a UV source have been generally adopted. It was pointed out in Section 13-8 that many organic compounds absorb in the UV, whereas some solvents that contain no chromophores will not. In many instruments, a fixed wavelength (for instance, the 254-nm line of a mercury lamp) is used; in others, wavelength selection from 195 to 750 nm is possible. A phototube detects the light transmitted, and any change from the value observed with pure mobile phase in the detector represents absorption by an eluted component. This signal is recorded.

Other detectors that have been utilized for HPLC include refractive index and flame ionization detectors, refractive index gaining in popularity in new commercial instruments. Electrochemical detectors are available in which a potential may be set independently on two electrodes, and electrochemical oxidation or reduction of the analyte is measured. Picogram (10^{-12} g) quantities of electroactive substances, such as phenols, catecholamines, and some vitamins can be detected. This type of detector also offers some specificity because many substances are not electroactive and would give no signal; clearly, this type of detector can be used only when the analyst knows the nature of the sample or should be supplementary to a general-purpose detector so that components in the sample are not overlooked.

APPLICATIONS

In LC the sample is dissolved in a solvent and separation is carried out at or near room temperature. The method is particularly useful for biochemical compounds that would decompose if vaporization were attempted. For example, lipids in fat-soluble vitamins can be separated by HPLC as well as peptides in various protein samples. Urine samples have been analyzed by HPLC, more than 100 aromatic acids having been identified by reversed-phase LC.

Another example is demonstrated in Figure 19-20, in which a mixture of isocyanates is separated by HPLC. Isocyanates are used in the manufacture of polyurethane foams, coatings, and elastomers; because of their toxicity, workplace atmospheres must be monitored at all times. Sensitivity is important because some

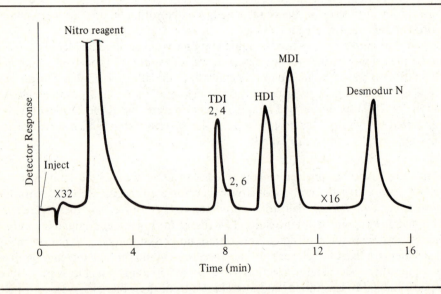

FIGURE 19-20 Separation of isocyanates by HPLC. [Figure reprinted with permission from K.L. Dunlap, R.L. Sandridge, and Jurgen Keller, "Determination of Isocyanates in Working Atmospheres by High Speed Liquid Chromatography," *Analytical Chemistry,* 48:497–499 (1976). Copyright, 1976 by the American Chemical Society.]

isocyanates have low threshhold limits: For example, the allowed limit for toluene diisocyanate (TDI) is 20 ppb per 8-hr day. The HPLC method is sensitive to 0.2 ppb.

Pentachlorophenol (PCP) is another highly toxic, persistent chemical frequently introduced into the water environment in effluents from pesticide, wood-pulp, and paper manufacturing. An HPLC method has been developed that will detect PCP at <1 ppm in water, the allowed limit.

Identification of toxic or decomposition products in consumer products is another area well suited to HPLC techniques. Aflatoxins, highly toxic substances produced by a fungus on grain and peanuts, can be dangerous to health if their concentration exceeds 20 ppb. A rapid HPLC determination of various aflatoxins in peanut butter at the 1- to 5 ~ ppb level (a few nanograms of analyte) is shown in Figure 19-21. HPLC offers a rapid method of analyzing aspirin tablets for the presence of salicylic acid, a decomposition product that forms after long periods of storage. Salicylic acid is responsible for some toxic side effects associated with aspirin, and the FDA regulates the amount of salicylic acid that may be present.

The value of HPLC techniques to IEC and EC has rejuvenated these analytical methods, expanded their range of applications, and increased their speed and resolving power. Figure 19-22 shows the separation of several amino acids on 8-μm strong cation-exchange resin packing. Compare this chromatogram produced in 1 hr with Figure 19-16, in which each point represents a separate quantitation. In a similar application, complicated mixtures of mono-, di-, and triphosphate nucleotides may be separated on a strong anion-exchange resin (10-μm particles).

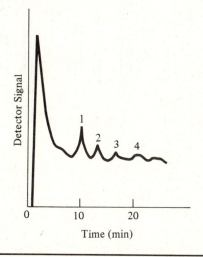

FIGURE 19-21 Chromatogram of peanut-butter sample containing aflatoxins: peak 1, aflatoxin B$_1$ (5 ppb); peak 2, aflatoxin G$_1$ (1 ppb); peak 3, aflatoxin B$_2$ (3 ppb); peak 4, aflatoxin G$_2$ (1 ppb). [From H.M. McNair, "Basic Considerations in HPLC," *American Laboratory*, 12:33 (May, 1980).]

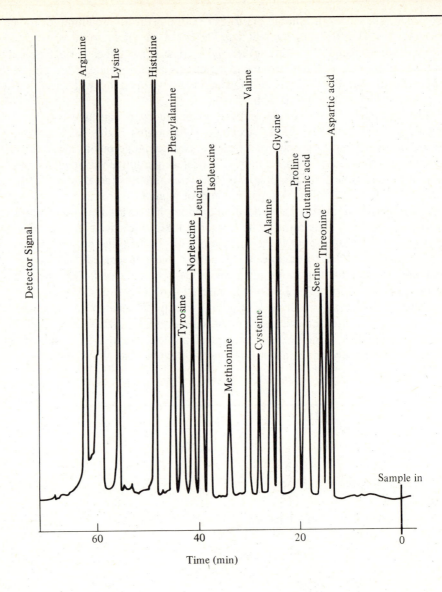

FIGURE 19-22 HPLC of amino acid mixture by IEC. The 35-cm column was packed with 8-μm spheres of a strong cation resin. The eluent was gradually raised from pH 2.2 to 11.5 at a pressure of 600 psi. Detection was carried out by measuring fluorescent emission at 479 nm after forming a fluorescent derivative of the amino acid and exciting at 365 nm. [From J.H. Knox, ed., *High Performance Liquid Chromatography*, p. 57 (Edinburgh: Edinburgh University Press, 1979).]

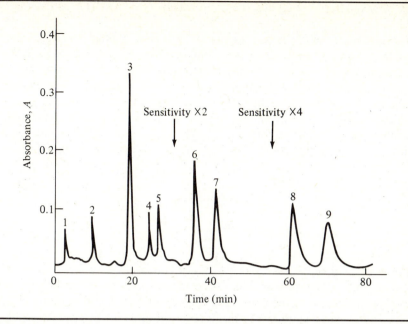

FIGURE 19-23 HPLC of polynuclear aromatic hydrocarbons (PAH's), using bonded-phase reversed-phase chromatography. The 25-cm column was packed with Zorbax-ODS. The eluent was 65% methanol to 35% H_2O (volume to volume) at 60°C and 1250 psi. Detection was accomplished with UV photometry at 254 nm. The peaks are (1) solvent peak, (2) naphthalene, (3) anthracene, (4) fluoranthene, (5) pyrene, (6) triphenylene, (7) benzo [a] anthracene, (8) perylene, and (9) benzo [a] pyrene.

It was mentioned earlier in this section that bonded-phase chromatography (BPC) employs stationary phases chemically bonded to the solid support particles, and either normal-phase (polar stationary phase with nonpolar mobile phase) or reversed-phase (nonpolar stationary phase with polar mobile phase) chromatography may be carried out. An example is shown in Figure 19-23, in which a mixture of highly carcinogenic polynuclear aromatic hydrocarbons (PAH's) has been separated on a bonded-phase packing of the HPLC column. The stationary phase (Zorbax-ODS) consists of silica gel to which has been bonded octadecyl groups ($C_{18}H_{37}$) to give the surface a nonpolar character. The polar eluent was 65% methanol and 35% water.

19-7   PLANE CHROMATOGRAPHY (PC, TLC)

All chromatographic techniques described so far employ columns, and they represent a three-dimensional process. It is also possible to carry out LC essentially on a plane or, in other words, by a two-dimensional process. Two techniques are *paper chromatography* (PC) and the more widely used *thin-layer chromatography* (TLC).

In PC, shown schematically in Figure 19-24, a piece of filter paper performs the role of the column. Cellulose has a great affinity for water, and the water held on the cellulose through hydrogen bonding constitutes the stationary phase. A minute amount of sample dissolved in a solvent is spotted near one end of the strip of paper, and the paper is placed in a tightly covered jar. Inside the jar is a container with an organic solvent in contact with the end of the paper. The organic solvent constitutes the mobile phase and, by capillary action, moves up the paper. The analyte, A, is carried by the organic solvent (m) but will be in equilibrium with the water stationary phase (s):

$$A(m) \rightleftharpoons A(s)$$

$$K_D = \frac{[A]_s}{[A]_m}$$

The analyte will move less rapidly than the organic solvent will because it spends part of the time in the stationary phase. In plane chromatography, a distance profile is measured at some specified time—as opposed to a retention time for the material to elute. At this specified time, the paper is removed, and the distances the solute and

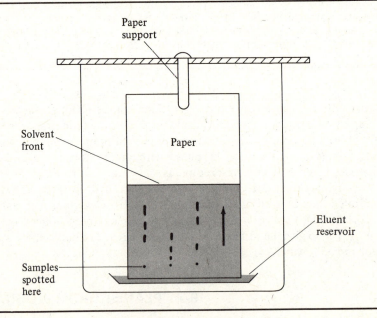

FIGURE 19-24 Basic apparatus for PC with the conventional ascending-solvent technique, in which solvent moves upward by capillary action.

solvent have moved are noted. A retardation factor, R_f, is calculated,

$$R_f = \frac{\text{distance solute moved}}{\text{distance solvent moved}}$$

R_f values are analogous to K_D or t_R values and will, in general, be different for each analyte. It should be noted that R_f values range from 0 ($K_D \rightarrow \infty$) to 1 ($K_D = 0$).

PC is simple to carry out and can be used for very small samples, but quantitation is difficult. The final product is a piece of paper containing spots associated with each of the analytes if separation has been achieved. The spots may be detected qualitatively if the analytes are colored, but frequently the spots are invisible to the naked eye. Detection of spots can sometimes be accomplished with UV or infrared radiation, Especially useful are organic compounds that fluoresce when irradiated with UV light. Counting techniques can be used for radioactive samples, Other techniques developed for plane chromatography include spraying the paper with some reagent or exposing it to some vapor that reacts with the analytes to make them visible. For example, amines and amino acids form a blue color when a ninhydrin reagent spray is used.

Even after a spot is identified, it hardly represents a quantitative analysis. Most plane chromatography is for identification purposes and not usually for quantitative measurement. However, once the spot has been identified, the analyte can be leached from the paper by cutting out the spot and treating it with a solvent. The resulting solution could be quantitated by any appropriate technique, such as UV spectrophotometry.

TLC is based on the same concept as that for PC, except that a substrate (for instance, a glass plate) is coated with a finely divided adsorbent, such as Al_2O_3 or silica gel. The thin layer of adsorbent is produced by forming a slurry of the material with a solvent (water), coating the substrate, and allowing the solvent to evaporate. If drying is carried out near room temperature, the adsorbent will hold water as the paper in PC does. The separation process will then be LLC. However, if the drying is carried out at a temperature high enough to drive off adsorbed water, the separation process will be LSC. The art of successful TLC lies in finding the optimum conditions of drying and choosing the mobile-phase solvent to carry out the required separation.

TLC can be more reproducible than PC can because the properties of paper vary so much from sample to sample. Qualitative information is expressed as R_f values, and quantitation still faces the same problems as mentioned above for PC. Organic analytes can often be made visible through exposure of the plate to I_2 vapor. The iodine vapor does not react with the organic compounds but condenses preferentially in the organic analyte spot. The spots appear orange against a yellow background. The spots must be marked promptly because the color does not persist. Another detection method is to spray the plate with H_2SO_4 and warm gently. The organic analytes will char and turn dark.

Plane chromatography has found widespread use in forensic chemistry. Minute samples of evidence can be analyzed for drug identification as one typical example. Another is the analysis of ink in suspected forgery cases. A tiny amount of ink can be taken from the document and spotted on paper or a thin-layer plate. The R_f values and colors for the ink components are compared with reference ink values or with a sample from the suspect's pen. Plant extracts and other biochemical samples can also be identified with plane chromatography.

SUPPLEMENTARY READING

Articles Describing Analytical Methods or Applications

Carroll, M.A., E.R. White, and J.E. Zarembo, "Over-the-Counter Drug Analyses with HPLC," *Analytical Chemistry,* 53:1111A (1981).

Gosink, T.A., "GC in Environmental Analysis," *Environmental Science and Technology,* 9:630 (1975).

Halász, I., "Columns for Reversed Phase Liquid Chromatography," *Analytical Chemistry,* 52: 1393A (1980).

Lynch, M., and E. Weiner, "HPLC: High Performance Liquid Chromatography," *Environmental Science and Technology,* 13:666 (1979).

Mulik, J.D., and E. Sawicki, "Ion Chromatography," *Environmental Science and Technology,* 13:804 (1979).

Novotny, M., "Contemporary Capillary Gas Chromatography," *Analytical Chemistry,* 50:16A (1978).

Pensenstadler, D.F., and M.A. Fulmer, "Pure Steam Ahead," *Analytical Chemistry,* 53:859A (1981).

Reference Books

Altgelt, K.H., and L. Segal, eds., *Gel Permeation Chromatography.* New York: Marcel Dekker, Inc., 1971.

Giddings, J.C., *Dynamics of Chromatography,* pt. 1, New York: Marcel Dekker, Inc., 1965.

Grob, R.L., ed., *Modern Practice of Gas Chromatography.* New York: John Wiley & Sons, Inc., 1977.

Snyder, L., and J.J. Kirkland, *Introduction to Modern Liquid Chromatography,* 2nd ed. New York: John Wiley & Sons, Inc., 1979.

PROBLEMS

19-1.   What are the basic principles of the three types of GC detector in common use? What are their advantages and disadvantages?

*19-2.   A GC column exhibits 2400 theoretical plates. If the retention times for benzene and toluene are 12 and 14 min, respectively, will baseline resolution be achieved?

19-3.   What increase in resolution will be achieved by doubling the length of a chromatographic column?

19-4.   a.   A mixture of 2-methyl hexane and 3-methyl hexane is separated on a GC column exhibiting 2500 theoretical plates. Calculate the resolution if the

---

*Answers to problems with an asterisk will be found at the back of the book.

retention times are 15 min for 2-methyl hexane and 16 min for 3-methyl hexane.

b. Is this sufficient resolution for quantitative separation?

c. Give three methods for increasing the resolution *without* changing the columns.

19-5. A substance C, which is not retained in the stationary phase, exits from a chromatography column 5.0 min after injection. Another substance D, which is retained, exits in 20.0 min. The flow rate of mobile phase is 25 mL/min.

*a. How long does D spend in the mobile phase?

*b. How long does D spend in the stationary phase?

*c. What is the value of k' for D?

*d. What is the value of K_D for C?

*e. What is the retention volume, V_R, of D?

*f. What is V_m?

19-6. The following data are obtained from a GC analysis: $t_R = 10.0$ min, $W = 0.30$ min, $t_{air} = 0.50$ min, $V_s = 4.0$ mL, and $F = 40$ mL/min. Calculate:

a. k' (capacity factor)

b. V_m (mobile phase volume)

c. K_D (distribution coefficient)

d. V_R (retention volume)

e. n (number of theoretical plates)

19-7. A mixture containing only toluene (C_7H_8) and chlorobenzene (C_6H_5Cl) gives peaks of 8.75 and 1.13 cm^2, respectively, in a gas chromatogram.

*a. Assuming that the molar response is the same, calculate mole percent and weight percent of toluene and of chlorobenzene in the mixture.

b. What detector could be used to increase the sensitivity toward chlorobenzene?

19-8. GC analysis of a mixture of organic liquids gave the following peak areas (cm^2): pentane = 2.7, hexane = 1.6, pentanol = 1.8 and benzene = 0.5.

a. Calculate the mole percent composition of the mixture, assuming the molar response of the detector (area per mole) to be the same for each component.

b. Calculate the weight percent composition of the mixture, assuming the molar response of the detector (area per mole) to be the same for each component.

c. Calculate the weight percent composition, given that R's (weight per unit area) of the detector for each of these compounds are 0.80, 0.98, 0.63, and 1.00, respectively.

19-9. Retention times for several compounds separated by GC are given below.

a. Calculate the relative retention of 1-pentene with respect to *n*-pentane.

b. Calculate the relative retention of *n*-hexane with respect to *n*-pentane.

substance	t_R (sec)
air	75
n-pentane	190
n-hexane	330
1-pentene	302

*19-10. With the data given in Problem 19-9, determine the Kovats index of 1-pentene.

19-11. In GC, columns can be prepared in which the random-path term in the Van Deemter equation is negligible. If such a column (2 m long) has an optimum velocity of 1.2 cm/sec and exhibits 9000 theoretical plates at this optimum velocity, determine the Van Deemter equation for the column.

19-12. The number of theoretical plates at various flow velocities for a GC column 200 cm long is determined from the retention time and peak width for an analyte. The data are given below.

a. Plot H as a function of velocity, and determine the Van Deemter equation for the column.

b. Determine the optimum flow velocity, minimum H value, and the number of theoretical plates that will be achieved at this optimum velocity.

u (cm/sec)	n (plates)
50	870
30	1000
5	980

*19-13. A nonretained substance is eluted from an HPLC column in 0.6 min at a flow rate of 1.8 mL/min. What is the void volume in the column?

19-14. For the HPLC column described in Problem 19-13, morphine is eluted in 4.7 min. What is its capacity factor?

*19-15. Trace amounts of phenol in water can be detected by HPLC. If a 50-ng phenol standard gives a peak area of 140 units and a 20-μL water sample gives a peak area due to phenol of 26 units, calculate phenol content as parts per million (micrograms per milliliter).

19-16. In HPLC, the detector is frequently UV absorption at the 254-nm Hg line. To prevent significant bandbroadening due to the finite volume in the detector (that is, the ideal detector would have no volume), the detector volume should be no more than 10 μL. However, to maintain sensitivity, the optical path length should be as long as possible (recall Beer's law).

a. What must be the diameter of a detector tube that has a 10-μL volume and a 1-cm path length?

b. With a 1-cm path length, what is the sensitivity (nanograms of analyte in the detector) for an analyte with $FW = 100$ and $\varepsilon_{254} = 10^4$ if an absorbance signal of 0.0001 A can be detected?

19-17. Compare GC with IEC as a separation technique. Include phases involved, eluents, quantitation techniques, and typical applications.

19-18. A particular strong cation-exchange resin has an exchange capacity of 1.8 meq/mL. If a column contains 33 mL of the resin, how many milligrams of Ni^{2+} could it hold?

*19-19. A 200-mL sample of hard water is passed through a cation-exchange column in the H^+ form. The column is eluted with water, and the eluate is titrated with 0.0558 N NaOH. If 22.3 mL is required to reach the phenolphthalein endpoint, calculate the water hardness as parts per million (milligrams per liter) of $CaCO_3$.

19-20. The elution volume for $CoCl_4^{2-}$ from an anion-exchange resin is 220 mL, using 6 F HCl. If the total column volume is 40 mL while the resin volume is 32 mL, calculate the capacity factor (distribution ratio).

19-21. A 247-mg sample containing Ni and Co is dissolved in 50 mL of 8 F HCl to form anionic chlorocomplexes. The sample is passed through an anion-exchange column that retains Co(II) but allows Ni(II) to pass through. The eluate is collected, neutralized, and titrated with 0.0217 F EDTA. To reach the bromopyrogallol endpoint signal, 27.6 mL is required. Then 150 mL of H_2O is passed through the column to elute Co(II). The eluate is collected, neutralized, and titrated with 0.0217 F EDTA. To reach the xylenol orange endpoint signal, 18.7 mL is required. Calculate percentages of Ni and Co in the sample.

19-22. A cation-exchange resin (exchange capacity = 2 meq/mL) is being considered for the separation of Dy^{3+} and Gd^{3+}. Some preliminary equilibrium studies are carried out, in which 10 mL of the resin is shaken with 30 mL of 0.04 F solution of one of the metal ions in 0.2 F HNO_3. After equilibration, the aqueous phase is analyzed for the metal ion, with the results given below. Calculate K_D values and the separation factor.

Metal ion	Equilibrium conc (F)
Gd^{3+}	0.00025
Dy^{3+}	0.00047

19-23. For a gel chromatographic column, the elution volume for ovalbumin (MW = 4.0×10^4) is 180 mL; for γ-globulin (MW = 1.5×10^5), it is 100 mL. What is the molecular weight of an enzyme that elutes in 90 mL?

19-24. A column made from 2.30 g of dry Sephadex G-100 has an upper exclusion limit of 100,000 MW. Blue dextran (MW = 2×10^6) elutes in 15.0 mL, and sorbitol (MW = 182) elutes in 38.0 mL. If chymotrypsinogen elutes in 22.3 mL, how much of the interior volume is available to chymotrypsinogen molecules?

19-25. Tinkerbellase is an enzyme that catalyzes the invention of correct answers on final examinations. On a G-100 column from which ferritin (MW = 7.3×10^5) and blue dextran (MW = 2×10^6) elute in 87.5 mL and glycerol (MW = 92) elutes in 263.0 mL, tinkerbellase elutes in 152.0 mL. α-Conarachin (MW = 3.0×10^5) elutes in 150.0 mL and R-phycoerythrin (MW = 2.6×10^5) elutes in 155.0 mL.
a. Find the percentage of interior volume available to tinkerbellase.
b. Estimate the molecular weight of tinkerbellase.

19-26. What qualitative and quantitative signals are obtained from plane chromatographic techniques?

19-27. A series of dyes is separated by TLC, with the data given below. Calculate the R_f value for each dye.

Substance	Distance moved (cm)
solvent	6.6
sudan IV	0.0
Bismarck brown	1.6
rhodamine B	3.8
fast green FCF	5.6

*19-28. Using the data in Problem 19-27, predict the distance rhodamine B will move if the same solvent is used and moves 11.5 cm.

19-29. It is your responsibility to set up a quick, inexpensive, analytical test for purity for your company's product, benzoic acid. The product will be rejected if it is less than 99.9% pure. How would you approach this problem?

19-30. Your first job after graduation is in the Xylene Division of Absurdprofits Oil Company. Show how you would apply the total analysis process for the analysis of o-, m-, and p-xylene mixtures.

19-31. Your second job after leaving school (the job at Absurdprofits didn't work out so well) is with Cheaphighs Drug Co., Ltd. Show how you would apply the five steps of the total analysis process for the analysis of APC tablets for acetylsalicylic acid (aspirin), phenacetin, and caffeine content.

APPENDIXES

APPENDIX 1-A   Method of Least Squares

To determine the best fit to a straight line from experimental data, the method of least squares minimizes the square of the deviations, that is, the distance measured parallel to the y-axis from the experimental point (x_i, y_i) to the straight line. This distance, Δy_i, was shown in Figure 2-8. The equation of a straight line is

$$y = mx + b$$

where m is the slope of the line and b is the y intercept. Thus the difference between the experimental point (x_i, y_i) and the value of y for the straight line at x_i is

$$y_i - y = \Delta y_i = y_i - mx_i - b$$

The square of this distance is

$$\Delta y_i^2 = (y_i - mx_i - b)^2$$
$$= y_i^2 - 2mx_i y_i - 2by_i + m^2 x_i^2 + 2bmx_i + b^2$$

The sum of all Δy_i^2 values for N experimental points will be

$$\sum_i^N (\Delta y_i)^2 = \sum_i^N (y_i)^2 - 2m \sum_i^N (x_i y_i) - 2b \sum_i^N y_i + m^2 \sum_i^N (x_i)^2 + 2bm \sum_i^N x_i + Nb^2$$

To find the minimum value, the partial derivative of $\sum_i^N (\Delta y_i)^2$ is taken with respect to each of the variables, m (best slope) and b (best intercept), and set equal to zero. This results in two equations with two unknowns, m and b. The equations are

695

solved for m and b with algebraic substitutions or determinants or by the method in Appendix 1-B.

$$\frac{\partial \sum\limits_{i}^{N} (\Delta y_i)^2}{\partial b} = 0 - 0 - 2 \sum_{i}^{N} y_i + 0 + 2m \sum_{i}^{N} x_i + 2Nb = 0$$

$$\frac{\partial \sum\limits_{i}^{N} (\Delta y_i)^2}{\partial m} = 0 - 2 \sum_{i}^{N} (x_i y_i) - 0 + 2m \sum_{i}^{N} (x_i)^2 + 2b \sum_{i}^{N} x_i = 0$$

After rearrangement, these equations are

$$(N)b + \left(\sum_{i}^{N} x_i \right)m = \sum_{i}^{N} y_i$$

$$\left(\sum_{i}^{N} x_i \right)b + \left(\sum_{i}^{N} x_i^2 \right)m = \sum_{i}^{N} x_i y_i$$

The solution of this set of equations for b and m is

$$b = \frac{\sum\limits_{i}^{N} y_i \sum\limits_{i}^{N} x_i^2 - \sum\limits_{i}^{N} x_i \sum\limits_{i}^{N} x_i y_i}{N \sum\limits_{i}^{N} x_i^2 - \left(\sum\limits_{i}^{N} x_i \right)^2}$$

$$m = \frac{N \sum\limits_{i}^{N} x_i y_i - \sum\limits_{i}^{N} x_i \sum\limits_{i}^{N} y_i}{N \sum\limits_{i}^{N} x_i^2 - \left(\sum\limits_{i}^{N} x_i \right)^2}$$

These equations may look difficult, but with a calculator, m and b can be determined with little effort. It is necessary to compute only four sums (x_i, y_i, $x_i y_i$, and x_i^2). Note that the sum of $x_i^2 \neq$ (sum of x_i)^2.

Equations can also be derived that yield information about how well the data fit the determined straight line and the standard deviation in m and in b.

There are many variations on this theme of least squares. If a straight-line fit going through the origin is desired, that is,

$$y = mx$$

there is only one variable; and by the method described above, it can be shown that

$$m = \frac{\sum\limits_{i}^{N} x_i y_i}{\sum\limits_{i}^{N} x_i^2}$$

Note that this is not the same as $\sum\limits_{i}^{N} y_i \Big/ \sum\limits_{i}^{N} x_i$.

The method of least squares can be used to find best fits to other equations such as

$$y = a + bx + cx^2 + dx^3 + \cdots$$

or

$$y = a + bx_1 + cx_2 + dx_3 + \cdots$$

With four variables to be determined (a, b, c, d) it is necessary to derive four simultaneous equations by taking the partial derivative with respect to each of them and setting it equal to zero.

Another variation is to give each of the experimental points a "weighting" when the investigator believes some points are more reliable than others.

APPENDIX 1-B   Solving simultaneous equations with matrices

When a set of simultaneous equations is developed of the type

$$a_1 w + b_1 x + c_1 y + \cdots = k_1$$
$$a_2 w + b_2 x + c_2 y + \cdots = k_2$$
$$a_3 w + b_3 x + c_3 y + \cdots = k_3$$
$$\vdots \qquad \vdots \qquad \vdots \qquad\qquad \vdots$$

the coefficients form a matrix of the type

$$\begin{vmatrix} a_1 & b_1 & c_1 \cdots \\ a_2 & b_2 & c_2 \cdots \\ a_3 & b_3 & c_3 \cdots \\ \vdots & \vdots & \vdots \end{vmatrix} = C$$

the unknowns form a matrix of the type

$$\begin{vmatrix} w \\ x \\ y \\ \vdots \end{vmatrix} = X$$

and the constants form a matrix of the type

$$\begin{vmatrix} k_1 \\ k_2 \\ k_3 \\ \vdots \end{vmatrix} = K$$

Therefore,

$$X \cdot C = K$$

To solve for X, we need C^{-1} (the inverse of C), which has the property $C \cdot C^{-1} = 1$:

$$X \cdot C \cdot C^{-1} = X \cdot 1 = X = K \cdot C^{-1}$$

Finding the inverse to a matrix is a laborious job by hand, but it can be done in a flash by a desk-top calculator.

The following program will solve any set of simultaneous equations (up to 10 variables as written, but it could be expanded, depending on the capability of the calculator available), the only time-consuming aspect being the necessity to enter all the coefficients and constants. This program shows only the basic calculation and, in practice, would normally have a test that the determinant is not zero (giving erroneous results); the program would calculate a value of uncertainty in each of the unknowns.

```
10   OPTION BASE 1
20   DIM C(10,10),K(10),C1(10,10),X(10)            !Set up matrices as in text
30   INPUT "ENTER # OF EQUATIONS",N
40   REDIM C(N,N),K(N),C1(N,N),X(N)                !Redimension for N equations
50   FOR Eq=1 TO N
60   FOR Coeff=1 TO N
70   INPUT "ENTER COEFFICIENTS OF EQ.",C(Eq,Coeff)   !Enter coeff. for eq.
80   NEXT Coeff                                   !Repeat for each coeff. in eq.
90   PRINT "ENTER COEFF OF NEXT EQ"
100  NEXT Eq                                      !Repeat for each equation
110  FOR Eq=1 TO N
120  INPUT "ENTER CONSTANT FOR EQ.",K(Eq)          !Enter constant term for eq.
130  NEXT Eq                                      !Repeat for each equation
140  MAT C1=INV(C)                                !Invert matrix C; C1 is inverse
150  MAT X=C1*K                       !Multiply inverse matrix by constant matrix
160  FOR Unknown=1 TO N
170  PRINT "UNKNOWN #";Unknown;"=";X(Unknown)      !Tabulate unknown value
180  NEXT Unknown                                 !Repeat for each unknown
190  END
```

APPENDIX 1-C   Iterative approach

Computer program written in BASIC language for solving equations of the type

$$K = \frac{(x)(x)}{(F - x)}$$

for which $K < 1$ so that x is small compared to F.

```
10   ! SOLVING EQUATIONS OF TYPE: K=X^2/(F-X)
20   ! REWRITE EQUATION AS: K(F-X)=X^2
30   ! X VALUES ON LEFT SIDE = X1
40   ! X VALUES ON RIGHT SIDE = X2
50   INPUT "WHAT IS VALUE OF K",K
```

```
60    INPUT "WHAT IS VALUE OF F",F
70    LET X1=0                          !FIRST APPROXIMATION
80    PRINT "DETERMINATION OF X FOR K=";K, "F=";F
90    LET D=F-X1
100   LET X2=(K*(F-X1))^.5             !SOLVE FOR X ON RIGHT SIDE
110   PRINT X1,X2
120   IF ABS(X2-X1)<(X2/1000 THEN 150  !ITERATION STOPS WHEN DIFF.<0.1%
130   LET X1=X2
140   GOTO 90                          !REPEAT CALCULATION USING X2 AS NEW X1 VALUE
150   PRINT "FINAL ANSWER: X=";X2, "(F-X)=";F-X2
160   END
```

```
                          EXAMPLES

DETERMINATION OF X FOR K= .0001          F= .025
0                       1.58113883020E-03
1.58113883020E-03       1.53032222658E-03
1.53032222658E-03       1.53198165056E-03
1.53198165056E-03       1.53192749020E-03
FINAL ANSWER: X= 1.53192749020E-03       (F-X)= 2.34680725098E-02


DETERMINATION OF X FOR K= .001           F= .01
0                       3.16227766022E-03
3.16227766022E-03       2.61490388733E-03
2.61490388733E-03       2.71755333234E-03
2.71755333234E-03       2.69860087228E-03
2.69860087228E-03       2.70211012506E-03
2.70211012506E-03       2.70146069302E-03
FINAL ANSWER: X= 2.70146069302E-03       (F-X)= .007298539307


DETERMINATION OF X FOR K= .001           F= .005
0                       2.23606797752E-03
2.23606797752E-03       1.66250775121E-03
1.66250775121E-03       1.82688046932E-03
1.82688046932E-03       1.78132521763E-03
1.78132521763E-03       1.79406654912E-03
1.79406654912E-03       1.79051206395E-03
1.79051206395E-03       1.79150437792E-03
FINAL ANSWER: X= 1.79150437792E-03       (F-X)= 3.20849562208E-03
```

APPENDIX 1-D Optimization of excess common ion to minimize solubility losses

For the case discussed in Section 5-2, the following equilibria determined the solubility of AgCl:

$$AgCl(s) \rightleftharpoons Ag^+ + Cl^- \qquad K_{sp} = [Ag^+][Cl^-]$$

$$AgCl(s) + Cl^- \rightleftharpoons AgCl_2^- \qquad K_1 = \frac{[AgCl_2^-]}{[Cl]}$$

$$AgCl_2^- + Cl^- \rightleftharpoons AgCl_3^{2-} \qquad K_2 = \frac{[AgCl_3^{2-}]}{[AgCl_2^-][Cl^-]}$$

The total solubility of AgCl is

$$\text{solubility} = [Ag^+] + [AgCl_2^-] + [AgCl_3^{2-}]$$

$$= \frac{K_{sp}}{[Cl^-]} + K_1[Cl^-] + K_1 K_2[Cl^-]^2$$

To determine the chloride concentration at which solubility is a minimum (the optimum chloride concentration), the derivative is taken and set equal to zero:

$$\frac{d(\text{solubility})}{d[Cl^-]} = \frac{-K_{sp}}{[Cl^-]^2} + K_1 + 2K_1 K_2[Cl^-] = 0$$

or

$$K_{sp} = K_1[Cl^-]^2 + 2K_1 K_2[Cl^-]^3$$

Third-order equations may be solved exactly although it is usually easier to use a simple computer program to run through values of $[Cl^-]$ until the right side of the equation equals the left side. This equation will be valid for any MX compound forming complexes MX_2^- and MX_3^{2-}. For the constants given in Chapter 5 for the AgCl case, the optimum value is

$$[Cl^-]_{optimum} = 0.003$$

for which the solubility is

$$\text{solubility} = \frac{1.8 \times 10^{-10}}{0.003} + 2.0 \times 10^{-5}(0.003) + 2.0 \times 10^{-5} \times 1(0.003)^2$$

$$= 1.2 \times 10^{-7}$$

It can be seen that the third term is negligible at this concentration of chloride, and if this term can be neglected, the derivative is

$$\frac{d(\text{solubility})}{d[Cl^-]} = -\frac{K_{sp}}{[Cl^-]^2} + K_1 = 0$$

$$[Cl^-]^2 = \frac{K_{sp}}{K_1} = 9.0 \times 10^{-6}$$

$$[Cl^-]_{optimum} = 0.003$$

The relationship

$$[X^-]_{optimum} = \sqrt{\frac{K_{sp}}{K_1}}$$

will hold whenever MX_2^- is the predominant complex formed when MX dissolves in excess X^-.

For the case of a compound of the type MX_2 forming a complex MX_3^-, the optimum concentration of X^- will be

$$[X^-]_{optimum} = \sqrt[3]{\frac{2K_{sp}}{K_1}}$$

For $PbCl_2$, $K_{sp} = 1.6 \times 10^{-5}$ and $K_1 = 4.2 \times 10^{-2}$. Therefore,

$$[Cl^-]_{optimum} = \sqrt[3]{\frac{2 \times 1.6 \times 10^{-5}}{4.2 \times 10^{-2}}} = 0.09$$

Minimum solubility of $PbCl_2 = K_{sp}/[Cl^-]^2 + K_1[Cl^-] = 5.8 \times 10^{-3}\ F$

APPENDIX 1-E Program for a weak-acid–strong-base titration curve

```
10    !   DERIVE ACID-BASE TITRATION CURVE USING SUB-ROUTINES FOR EACH REGION
20    !
30    INPUT "ENTER ACID CONC., VOL. IN mL, Ka",Fa,Va,Ka      !ENTER DATA
40    INPUT "ENTER INITIAL VOLUME INCLUDING ACID",Vi         !ENTER DATA
50    INPUT "ENTER CONCENTRATION OF BASE",Fb                 !ENTER DATA
60    LET Kw=1E-14                          !ENTER WATER DISSOCIATION CONSTANT
70    DIM Table(100,1)                      !SET UP TABLE FOR RESULTS
80    LET Eqpt=Fa*Va/Fb                     !CALCULATE EQUIV. PT.
90    LET Vb=0                              !START AT 0 VOLUME OF BASE
100   FOR Pt=0 TO 100                       !CALCULATE 101 POINTS
110   IF Vb=0 THEN GOSUB Initial            !CHOOSE APPROPRIATE SUBROUTINE
120   IF (Vb<Eqpt) AND (Vb>0) THEN GOSUB Buffer    ! "
130   IF Vb=Eqpt THEN GOSUB Equiv                  ! "
140   IF Vb>Eqpt THEN GOSUB Excess                 ! "
150   LET Ph=-LGT(H)                        !DEFINE pH
160   LET Table(Pt,0)=Vb                    !ENTER VOLUME OF BASE IN COLUMN 0
170   LET Table(Pt,1)=Ph                    !ENTER pH IN COLUMN 1
180   LET Vb=Vb+Eqpt/50                     !ADD INCREMENT OF BASE
190   NEXT Pt                               !CALCULATE NEXT POINT
200   DEG                 !  STEPS 200-470 SET UP GRAPH FOR TITRATION CURVE
210   PLOTTER IS 13,"GRAPHICS"
220   GRAPHICS
230   LOCATE 15,115,15,95
240   SCALE 0,2*Eqpt,0,14
250   AXES Eqpt/10,1,0,0,5,2,5
260   FRAME
270   CSIZE 3
280   LDIR 0
290   LORG 6
300   FOR Xposition=0 TO 2*Eqpt STEP Eqpt
310   MOVE Xposition,-.2
320   LABEL USING "DDD.DD";Xposition
```

```
330   NEXT Xposition
340   MOVE Eqpt,-1
350   CSIZE 4
360   LABEL USING "K";"VOLUME OF BASE, mL"
370   CSIZE 3
380   LORG 8
390   FOR Yposition=0 TO 14 STEP 2
400   MOVE 0,Yposition
410   LABEL USING "DD.DX";Yposition
420   NEXT Yposition
430   MOVE -Eqpt/5,7
440   CSIZE 4
450   LDIR 90
460   LORG 4
470   LABEL USING "K";"pH"
480   FOR Pt=0 TO 100         ! STEPS 480-540 PLOT POINTS ON GRAPH
490   CSIZE 3
500   LORG 5
510   MOVE Table(Pt,0),Table(Pt,1)
520   LDIR 0
530   LABEL USING "K";"."
540   NEXT Pt
550   DUMP GRAPHICS               ! PRINT TITRATION CURVE
560   PRINTER IS 0       ! STEPS 560-670 PRINT TABLE OF RESULTS
570   PRINT
580   PRINT "FOR TITRATION OF";Va;"mL OF";Fa;"F WEAK ACID with Ka OF";Ka
590   PRINT "INITIAL VOLUME IS";Vi;"mL.  TITRANT IS";Fb;"F STRONG BASE."
600   FIXED 3
610   PRINT
620   PRINT "EQUIVALENCE-POINT VOLUME IS";Eqpt;"mL"
630   PRINT
640   PRINT "VOLUME OF BASE,mL";TAB(22),"pH"
650   FOR Pt=0 TO 100 STEP 5         !PRINT EVERY 5th POINT TO SAVE SPACE
660   PRINT Table(Pt,0);TAB(20),Table(Pt,1)
670   NEXT Pt
680   STANDARD
690   GOTO 960        ! FINISHED
700   !    SUB-ROUTINES FOLLOW
710   !
720 Initial: !
730   LET H1=0
740   LET H=(Ka*(Fa*Va/Vi-H1))^.5
750   IF ABS(H-H1)<H/100 THEN 780
760   LET H1=H
770   GOTO 740                      !ITERATIVE ROUTINE UNTIL 1% ACCURACY REACHED
780   RETURN
790 Buffer:  !
800   LET H=Ka*(Va*Fa-Vb*Fb)/(Vb*Fb)
810   RETURN
820 Equiv:  !
830   LET Kh=Kw/Ka
840   LET Oh1=0
850   LET D=Va*Fa/(Vi+Vb)-Oh1
860   LET Oh=(Kh*D)^.5
870   IF ABS(Oh-Oh1)<Oh/100 THEN 900
880   LET Oh1=Oh
890   GOTO 850                 ! ITERATIVE ROUTINE UNTIL 1% ACCURACY REACHED
900   LET H=Kw/Oh
910   RETURN
920 Excess:  !
930   LET Oh=Fb*(Vb-Eqpt)/(Vi+Vb)
940   LET H=Kw/Oh
950   RETURN
960   END
```

EXAMPLE 1—Appendix 1-E

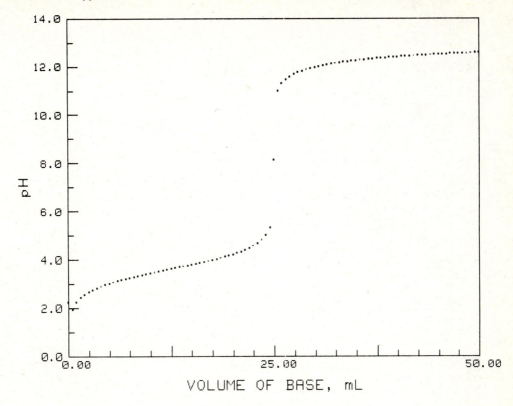

FOR TITRATION OF 50 mL OF .1 F WEAK ACID with Ka OF .000177
INITIAL VOLUME IS 50 mL.   TITRANT IS .2 F STRONG BASE.

EQUIVALENCE-POINT VOLUME IS 25.000 mL

VOLUME OF BASE,mL	pH
0.000	2.385
2.500	2.798
5.000	3.150
7.500	3.384
10.000	3.576
12.500	3.752
15.000	3.928
17.500	4.120
20.000	4.354
22.500	4.706
25.000	8.288
27.500	11.810
30.000	12.097
32.500	12.260
35.000	12.372
37.500	12.456
40.000	12.523
42.500	12.578
45.000	12.624
47.500	12.664
50.000	12.699

EXAMPLE 2—Appendix 1-E

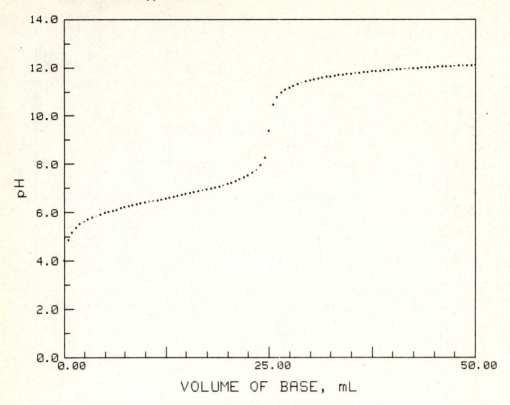

VOLUME OF BASE, mL

FOR TITRATION OF 50 mL OF .05 F WEAK ACID with Ka OF .0000002
INITIAL VOLUME IS 100 mL.   TITRANT IS .1 F STRONG BASE.

EQUIVALENCE-POINT VOLUME IS 25.000 mL

VOLUME OF BASE, mL	pH
0.000	4.151
2.500	5.745
5.000	6.097
7.500	6.331
10.000	6.523
12.500	6.699
15.000	6.875
17.500	7.067
20.000	7.301
22.500	7.653
25.000	9.500
27.500	11.292
30.000	11.585
32.500	11.753
35.000	11.870
37.500	11.959
40.000	12.030
42.500	12.089
45.000	12.140
47.500	12.183
50.000	12.222

EXAMPLE 3—Appendix 1-E

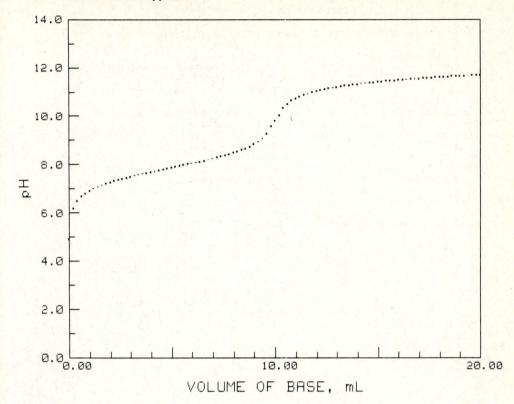

VOLUME OF BASE, mL

FOR TITRATION OF 40 mL OF .02 F WEAK ACID with Ka OF .00000001
INITIAL VOLUME IS 100 mL. TITRANT IS .08 F STRONG BASE.

EQUIVALENCE-POINT VOLUME IS 10.000 mL

VOLUME OF BASE, mL	pH
0.000	5.049
1.000	7.046
2.000	7.398
3.000	7.632
4.000	7.824
5.000	8.000
6.000	8.176
7.000	8.368
8.000	8.602
9.000	8.954
10.000	9.928
11.000	10.858
12.000	11.155
13.000	11.327
14.000	11.448
15.000	11.541
16.000	11.617
17.000	11.680
18.000	11.734
19.000	11.782
20.000	11.824

APPENDIX 1-F   Program for a weak-acid–strong-base titration curve for exact solution

```
10     !    DERIVE ACID-BASE TITRATION CURVE USING EXACT SOLUTION IN EQ. 9-17
20     !
30     INPUT "ENTER ACID CONC., VOL. IN mL, Ka",Fa,Va,Ka       ! ENTER DATA
40     INPUT "ENTER TOTAL VOLUME",Vt                ! CONSTANT VOLUME CALCULATION
50     INPUT "ENTER CONCENTRATION OF BASE",Fb       ! ENTER DATA
60     LET Kw=1E-14                    ! ENTER WATER DISSOCIATION CONSTANT
70     OPTION BASE 1
80     DIM Table(100,2)               ! SET UP TABLE FOR RESULTS
90     LET Eqpt=Fa*Va/Fb              ! CALCULATE EQUIV. PT.
100    LET Phmin=-LGT(Fa+1E-7)   ! LOWEST pH = COMPLETE DISSOC. + WATER DISSOC.
110    LET Phmax=-LGT(Kw/Fb)     ! HIGHEST pH = pH of BASE TITRANT
120    LET N=1                       ! START WITH FIRST POINT
130    FOR Ph=Phmin TO Phmax STEP (Phmax-Phmin)/100 ! pH INCREMENT FOR TITRATION
140    LET H=10^(-Ph)                 ! CALCULATE [H] FROM pH
150    !
160    !    STEP 180 CALCULATES [Na] FROM GIVEN pH USING EQ. 9-17
170    !
180    LET Na=(Ka*Fa*Va*H/Vt+Kw*H+Ka*Kw-H^3-Ka*H^2)/(H^2+Ka*H)
190    LET Vb=Na*Vt/Fb               ! CALCULATE VOLUME OF BASE FROM [Na]
200    IF Vb<0 THEN 250              ! SKIP POINTS WHERE VOL < 0
210    IF Vb>2*Eqpt THEN 260        ! STOP WHEN VOL. OF BASE > TWICE EQ. PT. VOL
220    LET Table(N,1)=Vb            ! ENTER VOLUME OF BASE IN COLUMN 1
230    LET Table(N,2)=Ph            ! ENTER pH IN COLUMN 2
240    LET N=N+1                    ! READY FOR NEXT ENTRY
250    NEXT Ph                      ! INCREASE pH WITH NEXT INCREMENT
260    DEG              ! STEPS 260-530 SET UP GRAPH FOR TITRATION CURVE
270    PLOTTER IS 13,"GRAPHICS"
280    GRAPHICS
290    LOCATE 15,115,15,95
300    SCALE 0,2*Eqpt,0,14
310    AXES Eqpt/10,1,0,0,5,2,5
320    FRAME
330    CSIZE 3
340    LDIR 0
350    LORG 6
360    FOR Xposition=0 TO 2*Eqpt STEP Eqpt
370    MOVE Xposition,-.2
380    LABEL USING "DDD.DD";Xposition
390    NEXT Xposition
400    MOVE Eqpt,-1
410    CSIZE 4
420    LABEL USING "K";"VOLUME OF BASE, mL"
430    CSIZE 3
440    LORG 8
450    FOR Yposition=0 TO 14 STEP 2
460    MOVE 0,Yposition
470    LABEL USING "DD.DX";Yposition
480    NEXT Yposition
490    MOVE -Eqpt/5,7
500    CSIZE 4
510    LDIR 90
520    LORG 4
530    LABEL USING "K";"pH"
540    FOR Pt=1 TO N                ! STEPS 540-600 PLOT POINTS ON GRAPH
550    CSIZE 3
560    LORG 5
570    MOVE Table(Pt,1),Table(Pt,2)
580    LDIR 0
590    LABEL USING "K";"."
600    NEXT Pt
610    DUMP GRAPHICS                      ! PRINT TITRATION CURVE
620    PRINTER IS 0                       ! STEPS 620-730 PRINT TABLE OF RESULTS
630    PRINT
```

```
640   PRINT "FOR TITRATION OF";Va;"mL OF";Fa;"F WEAK ACID with Ka OF";Ka
650   PRINT "TOTAL VOLUME IS";Vt;"mL.   TITRANT IS";Fb;"F STRONG BASE."
660   FIXED 3
670   PRINT
680   PRINT "EQUIVALENCE-POINT VOLUME IS";Eqpt;"mL"
690   PRINT
700   PRINT "VOLUME OF BASE,mL";TAB(22),"pH"
710   FOR Pt=1 TO N STEP 5
720   PRINT Table(Pt,1);TAB(20),Table(Pt,2)
730   NEXT Pt
740   STANDARD
750   END
```

EXAMPLE 1—Appendix 1-F

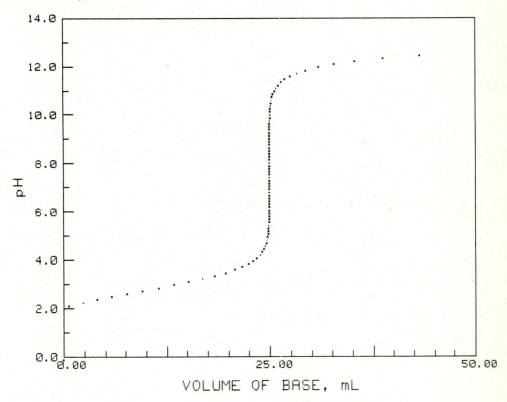

```
FOR TITRATION OF 50 mL OF .1 F WEAK ACID with Ka OF .001
TOTAL VOLUME IS 100 mL.   TITRANT IS .2 F STRONG BASE.

EQUIVALENCE-POINT VOLUME IS 25.000 mL

VOLUME OF BASE,mL      pH
 .686                 2.230           25.000           7.766
 9.581                2.845           25.001           8.381
 18.392               3.460           25.005           8.996
 23.019               4.075           25.020           9.611
 24.490               4.690           25.084           10.226
 24.874               5.305           25.347           10.841
 24.969               5.920           26.428           11.456
 24.993               6.535           30.887           12.071
 24.998               7.151           49.263           12.686
```

EXAMPLE 2—Appendix 1-F

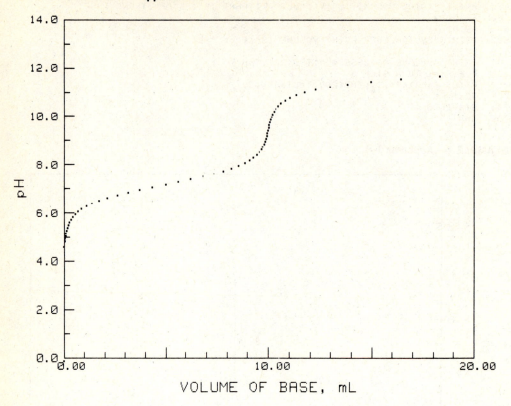

VOLUME OF BASE, mL

FOR TITRATION OF 40 mL OF .02 F WEAK ACID with Ka OF .00000005
TOTAL VOLUME IS 110 mL.   TITRANT IS .08 F STRONG BASE.

EQUIVALENCE-POINT VOLUME IS 10.000 mL

VOLUME OF BASE,mL	pH
.000	4.724
.088	5.284
.336	5.844
1.126	6.405
3.156	6.965
6.262	7.525
8.590	8.085
9.573	8.646
9.899	9.206
10.046	9.766
10.282	10.326
11.056	10.886
13.844	11.447

EXAMPLE 3—Appendix 1-F

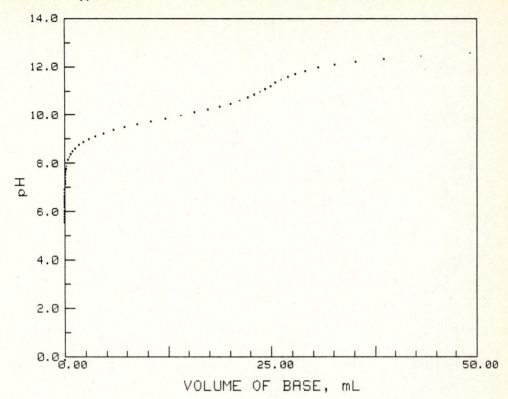

FOR TITRATION OF 50 mL OF .1 F WEAK ACID with Ka OF .0000000001
TOTAL VOLUME IS 100 mL.   TITRANT IS .2 F STRONG BASE.

EQUIVALENCE-POINT VOLUME IS 25.000 mL

VOLUME OF BASE, mL	pH
.000	5.674
.005	6.289
.020	6.904
.083	7.520
.337	8.135
1.333	8.750
4.712	9.365
12.256	9.980
20.130	10.595
24.358	11.210
27.972	11.825
38.679	12.440

EXAMPLE 4—Appendix 1-F

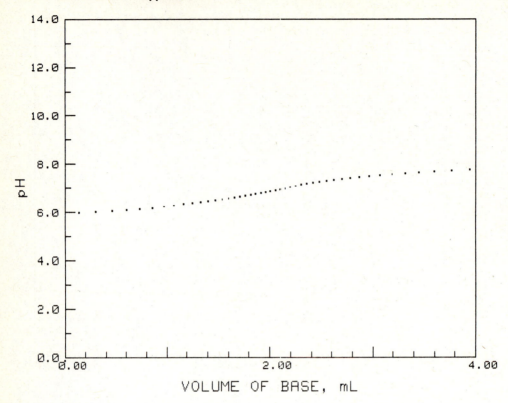

FOR TITRATION OF 200 mL OF .000001 F WEAK ACID with Ka OF .1
TOTAL VOLUME IS 250 mL.   TITRANT IS .0001 F STRONG BASE.

EQUIVALENCE-POINT VOLUME IS 2.000 mL

VOLUME OF BASE,mL	pH		
.138	6.120		
.862	6.322	2.153	7.131
1.336	6.524	2.422	7.333
1.664	6.726	2.783	7.535
1.917	6.929	3.318	7.737

APPENDIX 1-G   Program for a redox titration curve

```
10    PRINT "REDOX TITRATION CURVE FOR ANALYTE AND TITRANT OF EQUAL N VALUE"
20    INPUT "ENTER NORMALITY, VOLUME, Eo OF ANALYTE",Na,Va,Eoa
30    INPUT "ENTER NORMALITY, Eo OF TITRANT",Nt,Eot
40    INPUT " ENTER N VALUE",N
50    DIM Table(100,1)
60    Vt=0
70    Veqpt=Na*Va/Nt
80    FOR Pt=0 TO 100
90    IF Vt=0 THEN 150
100   IF Vt<Veqpt THEN GOSUB Buffer
110   IF Vt=Veqpt THEN GOSUB Equiv
120   IF Vt>Veqpt THEN GOSUB Excess
```

```
130   LET Table(Pt,0)=Vt
140   LET Table(Pt,1)=E
150   LET Vt=Vt+Veqpt/50
160   NEXT Pt
170   PRINTER IS 0
180   PRINT PAGE
190   PRINT "FOR TITRATION OF";Va;"mL OF";Na;"N ANALYTE WITH Eo OF";Eoa;"VOLTS vs
  NHE"
200   PRINT "TITRANT IS";Nt;"N WITH Eo OF";Eot;"VOLTS vs NHE.  N VALUE IS";N
210   FIXED 3
220   PRINT
230   PRINT "EQUIVALENCE-POINT VOLUME IS";Veqpt;"mL.  EQUIV. POT.=";Table(50,1);"
 V vs NHE"
240   PRINT
250   PRINT "VOLUME OF TITRANT (mL)";TAB(27),"E"
260   FOR Pt=5 TO 100 STEP 5
270   PRINT Table(Pt,0);TAB(25),Table(Pt,1)
280   NEXT Pt
290   STANDARD
300   PRINTER IS 16
310   DEG
320   PLOTTER IS 13,"GRAPHICS"
330   GRAPHICS
340   LOCATE 15,115,15,95
350   Emin=0
360   Emax=2
370   IF (Eot<0) OR (Eoa<0) THEN Emin=-1
380   IF (Eot<1) AND (Eoa<.8) THEN Emax=1
390   SCALE 0,2*Veqpt,Emin,Emax
400   AXES Veqpt/10,.1,0,Emin,5,5,5
410   FRAME
420   CSIZE 3
430   LDIR 0
440   LORG 6
450   FOR Xposition=0 TO 2*Veqpt STEP Veqpt
460   MOVE Xposition,Emin-.02
470   LABEL USING "DDD.DD";Xposition
480   NEXT Xposition
490   MOVE Veqpt,Emin-.2
500   CSIZE 4
510   LABEL USING "K";"VOLUME OF TITRANT (mL)"
520   CSIZE 3
530   LORG 8
540   FOR Yposition=Emin TO Emax
550   MOVE 0,Yposition
560   LABEL USING "MD.DX";Yposition
570   NEXT Yposition
580   MOVE -Veqpt/5,(Emin+Emax)/2
590   CSIZE 4
600   LDIR 90
610   LORG 4
620   LABEL USING "K";"POTENTIAL (V vs NHE)"
630   FOR Pt=1 TO 100
640   CSIZE 3
650   LORG 5
660   MOVE Table(Pt,0),Table(Pt,1)
670   LDIR 0
680   LABEL USING "K";"."
690   NEXT Pt
700   DUMP GRAPHICS
710   GOTO 850
720 Buffer:    !
730   LET Ratio=(Veqpt-Vt)/Vt
740   IF Eoa>Eot THEN Ratio=1/Ratio
750   LET E=Eoa-.059/N*LGT(Ratio)
760   RETURN
```

```
770 Equiv:    !
780   LET E=(Eoa+Eot)/2
790   RETURN
800 Excess:   !
810   LET Ratio=Veqpt/(Vt-Veqpt)
820   IF Eoa>Eot THEN Ratio=1/Ratio
830   LET E=Eot-.059/N*LGT(Ratio)
840   RETURN
850   END
```

EXAMPLE 1—Appendix 1-G

FOR TITRATION OF 50 mL OF .05 N ANALYTE WITH Eo OF .771 VOLTS vs NHE
TITRANT IS .1 N WITH Eo OF 1.7 VOLTS vs NHE.   N VALUE IS 1

EQUIVALENCE-POINT VOLUME IS 25.000 mL.   EQUIV. POT.= 1.236 V vs NHE

VOLUME OF TITRANT (mL)	E		
2.500	.715	27.500	1.641
5.000	.735	30.000	1.659
7.500	.749	32.500	1.669
10.000	.761	35.000	1.677
12.500	.771	37.500	1.682
15.000	.781	40.000	1.687
17.500	.793	42.500	1.691
20.000	.807	45.000	1.694
22.500	.827	47.500	1.697
25.000	1.236	50.000	1.700

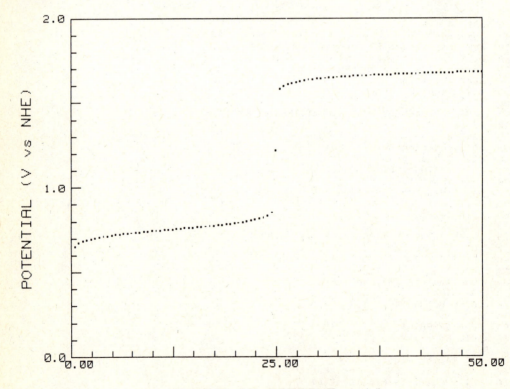

EXAMPLE 2—Appendix 1-G

```
FOR TITRATION OF 50 mL OF .05 N ANALYTE WITH Eo OF 1.2 VOLTS vs NHE
TITRANT IS .1 N WITH Eo OF 1.6 VOLTS vs NHE.   N VALUE IS 1

EQUIVALENCE-POINT VOLUME IS 25.000 mL.   EQUIV. POT.= 1.400 V vs NHE

VOLUME OF TITRANT (mL)       E
   2.500                  1.144
   5.000                  1.164
   7.500                  1.178
  10.000                  1.190
  12.500                  1.200
  15.000                  1.210
  17.500                  1.222
  20.000                  1.236
  22.500                  1.256
  25.000                  1.400
  27.500                  1.541
  30.000                  1.559
  32.500                  1.569
  35.000                  1.577
  37.500                  1.582
  40.000                  1.587
  42.500                  1.591
  45.000                  1.594
  47.500                  1.597
  50.000                  1.600
```

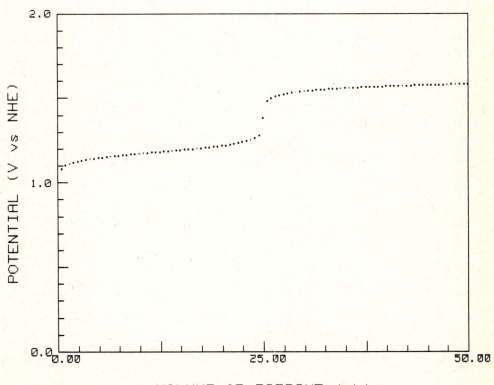

EXAMPLE 3—Appendix 1-G

```
FOR TITRATION OF 50 mL OF .05 N ANALYTE WITH Eo OF .5 VOLTS vs NHE
TITRANT IS .1 N WITH Eo OF-.2 VOLTS vs NHE.   N VALUE IS 1

EQUIVALENCE-POINT VOLUME IS 25.000 mL.   EQUIV. POT.= .150 V vs NHE

VOLUME OF TITRANT (mL)      E
  2.500                    .556
  5.000                    .536
  7.500                    .522
 10.000                    .510
 12.500                    .500
 15.000                    .490
 17.500                    .478
 20.000                    .464
 22.500                    .444
 25.000                    .150
 27.500                   -.141
 30.000                   -.159
 32.500                   -.169
 35.000                   -.177
 37.500                   -.182
 40.000                   -.187
 42.500                   -.191
 45.000                   -.194
 47.500                   -.197
 50.000                   -.200
```

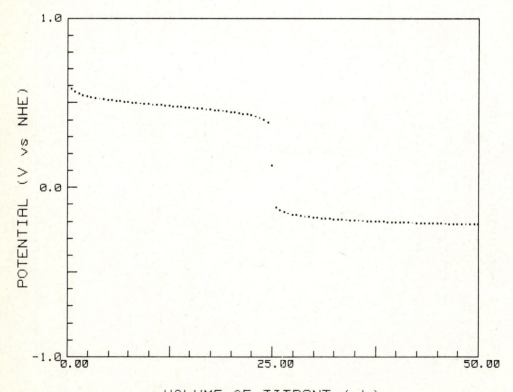

APPENDIX 1-H Derivation of Lambert–Beer law

Figure 12-6 showed that the incident radiant power P_0 was decreased to P as it moved through the solution from 0 to b. If we look at a small segment dx of cell length, the radiant power will change by dP. Thus Equation 12-10 in differential form is

$$\frac{dP}{P} = -kC\,dx$$

The minus sign has been inserted because the radiant power *decreases* as we move through the solution from 0 to b. This differential equation assumes that the concentration does not change with distance (the solution is homogeneous). Integration of the equation can be carried out with the limits stated above:

$$\int_{P_0}^{P} \frac{dP}{P} = -kC\int_{0}^{b} dx \qquad \ln\frac{P}{P_0} = -kCb$$

or

$$\ln\frac{P_0}{P} = kCb$$

It is convenient to use base-10 logarithms and to give the new proportionality constant a special symbol, ε, when concentration is given in moles per liter and path length is given in centimeters.

$$2.303\log\frac{P_0}{P} = kCb \qquad \log\frac{P_0}{P} = \frac{k}{2.303}Cb = \varepsilon Cb$$

The term $\log(P_0/P)$ is called the *absorbance* and is given the symbol A. The final form, then, is

$$A = \varepsilon Cb$$

APPENDIX 1-J Calculation of transmittance, giving minimum error in concentration

In terms of transmittance, Beer's law may be written

$$A = \log\frac{P_0}{P} = -\log T = \varepsilon bC$$

To see how C changes with changes in T, the expression is differentiated:

$$-\log T = \frac{-\ln T}{2.303} = \varepsilon bC \qquad -\frac{1}{2.303\,T}dT = \varepsilon b\,dC$$

or

$$dC = -\frac{dT}{2.303\varepsilon bT}$$

The relative change in C will be

$$\frac{dC}{C} = \frac{-dT/2.303\varepsilon bT}{-(\log T)/\varepsilon b} = +\frac{dT}{2.303T\log T}$$

Thus a small error in transmittance ΔT will cause a relative error in concentration $\Delta C/C$ equal to

$$\text{error } E = \frac{\Delta C}{C} = \frac{\Delta T}{2.303T\log T} = \frac{\Delta T}{T\ln T}$$

$\Delta C/C$ approaches infinity as T approaches 0 or 1.

We can calculate the value of T that will give a minimum error in concentration for a given value of ΔT by taking the derivative and setting it equal to 0:

$$\frac{dE}{dT} = -\frac{\Delta T}{(\ln T)^2 T^2} - \frac{\Delta T}{(\ln T) T^2} = 0$$

Therefore,

$$\ln T = -1$$

$$T = 0.368 \ (36.8\%)$$

$$A = -\log 0.368 = 0.434$$

APPENDIX 1-K   Calculation of uncertainty in concentration from uncertainty in electrode potential

Referring to Example 15-4, we saw, for the copper electrode, that

$$E_{Cu} = E_{Cu}^0 - \frac{0.059}{n} \log \frac{1}{[Cu^{2+}]}$$

We wish to know how $[Cu^{2+}]$ changes with E_{Cu}. Therefore, the equation should be differentiated (first written in natural logarithms):

$$E_{Cu} = E_{Cu}^0 - \frac{RT}{nF}\ln\frac{1}{[Cu^{2+}]} = E^0 + \frac{RT}{nF}\ln[Cu^{2+}]$$

$$dE_{Cu} = 0 + \frac{RT}{nF}\frac{1}{[Cu^{2+}]}d[Cu^{2+}]$$

$$\frac{d[Cu^{2+}]}{[Cu^{2+}]} = \frac{dE}{RT/nF}$$

For a finite change

$$\frac{\Delta[Cu^{2+}]}{[Cu^{2+}]} = \frac{\Delta E}{RT/nF}$$

The left side is the relative uncertainty in $[Cu^{2+}]$ for an uncertainty in $E_{Cu} = \Delta E$. Percentage uncertainty in $[Cu^{2+}]$ for a 1-mV uncertainty in E_{Cu} at 25°C is given by

$$\% \text{ uncertainty in } [Cu^{2+}] = \frac{\Delta E}{RT/nF} \times 100 \, (\%)$$

$$= \frac{0.001}{8.31 \times 298/2 \times 9.65 \times 10^4} \times 100 \, (\%) = 7.8\%$$

It should be noted that the percent uncertainty in concentration depends only on ΔE, not on the absolute potential, the temperature, and the n value for the reaction. At 25°C, *all* cell reactions with $n = 1$ will have an uncertainty of 3.9% for a ± 1-mV uncertainty in electrode potential; 7.8% for $n = 2$; and 11.7% for $n = 3$.

APPENDIX 1-L Derivation of optimum velocity for a chromatographic column obeying the Van Deemter equation

The Van Deemter equation is

$$H = A + \frac{B}{u} + Cu$$

and the minimum value of H will be observed when the slope, dH/du, is zero. Therefore,

$$\frac{dH}{du} = 0 - \frac{B}{u^2} + C = 0$$

The velocity when $dH/du = 0$ will be the optimum value shown in Figure 19-4:

$$u_{opt}^2 = \frac{B}{C}$$

$$u_{opt} = \sqrt{\frac{B}{C}}$$

The minimum value of H will then be

$$H_{min} = A + \frac{B}{\sqrt{B/C}} + C\sqrt{\frac{B}{C}}$$

$$= A + 2\sqrt{BC}$$

APPENDIX TWO

APPENDIX 2-A   Solubility products*

Substance	Activity (concentration) expression	K_{sp}	pK_{sp}[†]
aluminum hydroxide	$[Al^{3+}][OH^-]^3$	2×10^{-32}	31.70
barium arsenate	$[Ba^{2+}]^3[AsO_4^{3-}]^2$	7.7×10^{-51}	50.11
barium carbonate	$[Ba^{2+}][CO_3^{2-}]$	5.1×10^{-9}	8.29
barium chromate	$[Ba^{2+}][CrO_4^{2-}]$	1.2×10^{-10}	9.92
barium oxalate	$[Ba^{2+}][C_2O_4^{2-}]$	2.3×10^{-8}	7.64
barium sulfate	$[Ba^{2+}][SO_4^{2-}]$	1.3×10^{-10}	9.89
beryllium hydroxide	$[Be^{2+}][OH^-]^2$	7×10^{-22}	21.15
bismuth oxide hydroxide	$[BiO^+][OH^-]$	4×10^{-10}	9.40
bismuth iodide	$[Bi^{3+}][I^-]^3$	8.1×10^{-19}	18.09
cadmium carbonate	$[Cd^{2+}][CO_3^{2-}]$	2.5×10^{-14}	13.60
cadmium hydroxide	$[Cd^{2+}][OH^-]^2$	5.9×10^{-15}	14.23
cadmium oxalate	$[Cd^{2+}][C_2O_4^{2-}]$	9×10^{-8}	7.05
cadmium sulfide	$[Cd^{2+}][S^{2-}]$	2×10^{-28}	27.70
calcium carbonate	$[Ca^{2+}][CO_3^{2-}]$	4.8×10^{-9}	8.32
calcium fluoride	$[Ca^{2+}][F^-]^2$	4.9×10^{-11}	10.31
calcium oxalate	$[Ca^{2+}][C_2O_4^{2-}]$	2.3×10^{-9}	8.64
calcium phosphate	$[Ca^{2+}]^3[PO_4^{3-}]^2$	2.0×10^{-29}	28.70
calcium sulfate	$[Ca^{2+}][SO_4^{2-}]$	1.2×10^{-6}	5.92
chromium(II) hydroxide	$[Cr^{2+}][OH^-]^2$	1.0×10^{-17}	17.00
chromium(III) hydroxide	$[Cr^{3+}][OH^-]^3$	6×10^{-31}	30.22
copper(I) bromide	$[Cu^+][Br^-]$	5.2×10^{-9}	8.28
copper(I) chloride	$[Cu^+][Cl^-]$	1.2×10^{-6}	5.92
copper(II) hydroxide	$[Cu^{2+}][OH^-]^2$	2.2×10^{-20}	19.66
copper(I) iodide	$[Cu^+][I^-]$	1.1×10^{-12}	11.96
copper(II) sulfide	$[Cu^{2+}][S^{2-}]$	6×10^{-36}	35.22
copper(I) thiocyanate	$[Cu^+][SCN^-]$	4.8×10^{-15}	14.32
iron(II) hydroxide	$[Fe^{2+}][OH^-]^2$	8×10^{-16}	15.10
iron(III) hydroxide	$[Fe^{3+}][OH^-]^3$	4×10^{-38}	37.40
iron(II) sulfide	$[Fe^{2+}][S^{2-}]$	6×10^{-18}	17.22
lanthanum iodate	$[La^{3+}][IO_3^-]^3$	6.2×10^{-12}	11.21
lead bromide	$[Pb^{2+}][Br^-]^2$	3.9×10^{-5}	4.41
lead carbonate	$[Pb^{2+}][CO_3^{2-}]$	3.3×10^{-14}	13.48
lead chloride	$[Pb^{2+}][Cl^-]^2$	1.6×10^{-5}	4.80
lead chromate	$[Pb^{2+}][CrO_4^{2-}]$	1.8×10^{-14}	13.74
lead hydroxide	$[Pb^{2+}][OH^-]^2$	1.2×10^{-15}	14.92
lead iodide	$[Pb^{2+}][I^-]^2$	7.1×10^{-9}	8.15

*Values are reported near room temperature from various sources.
[†]All pK_{sp} values are given to two significant figures for ease in reading the table.

APPENDIX 2-A (cont.)

Substance	Activity (concentration) expression	K_{sp}	$pK_{sp}^{\dagger}$
lead sulfate	$[Pb^{2+}][SO_4^{2-}]$	1.6×10^{-8}	7.80
lead sulfide	$[Pb^{2+}][S^{2-}]$	1×10^{-28}	28.00
magnesium ammonium phosphate	$[Mg^{2+}][NH_4^+][PO_4^{3-}]$	3×10^{-13}	12.52
magnesium carbonate	$[Mg^{2+}][CO_3^{2-}]$	1×10^{-5}	5.00
magnesium fluoride	$[Mg^{2+}][F^-]^2$	6.5×10^{-9}	8.19
magnesium hydroxide	$[Mg^{2+}][OH^-]^2$	1.8×10^{-11}	10.74
magnesium oxalate	$[Mg^{2+}][C_2O_4^{2-}]$	8.6×10^{-5}	4.07
manganese(II) hydroxide	$[Mn^{2+}][OH^-]^2$	1.9×10^{-13}	12.72
manganese(II) sulfide	$[Mn^{2+}][S^{2-}]$	3×10^{-13}	12.52
mercury(I) bromide	$[Hg_2^{2+}][Br^-]^2$	5.8×10^{-23}	22.24
mercury(I) chloride	$[Hg_2^{2+}][Cl^-]^2$	1.3×10^{-18}	17.89
mercury(I) iodide	$[Hg_2^{2+}][I^-]^2$	4.5×10^{-29}	28.35
mercury(II) sulfide	$[Hg^{2+}][S^{2-}]$	4×10^{-53}	52.40
nickel(II) sulfide	$[Ni^{2+}][S^{2-}]$	3×10^{-19}	18.52
silver arsenate	$[Ag^+]^3[AsO_4^{3-}]$	1×10^{-22}	22.00
silver bromide	$[Ag^+][Br^-]$	5.2×10^{-13}	12.28
silver chloride	$[Ag^+][Cl^-]$	1.8×10^{-10}	9.74
silver chromate	$[Ag^+]^2[CrO_4^{2-}]$	1.2×10^{-12}	11.92
silver iodate	$[Ag^+][IO_3^-]$	3.0×10^{-8}	7.52
silver iodide	$[Ag^+][I^-]$	8.3×10^{-17}	16.08
silver sulfide	$[Ag^+]^2[S^{2-}]$	6×10^{-50}	49.22
silver thiocyanate	$[Ag^+][SCN^-]$	1.1×10^{-12}	11.96
strontium carbonate	$[Sr^{2+}][CO_3^{2-}]$	1.1×10^{-10}	9.96
strontium chromate	$[Sr^{2+}][CrO_4^{2-}]$	3.6×10^{-5}	4.44
strontium oxalate	$[Sr^{2+}][C_2O_4^{2-}]$	1.6×10^{-7}	6.80
strontium sulfate	$[Sr^{2+}][SO_4^{2-}]$	3.2×10^{-7}	6.49
thallium(I) bromide	$[Tl^+][Br^-]$	3.4×10^{-6}	5.47
thallium(I) chloride	$[Tl^+][Cl^-]$	1.7×10^{-4}	3.77
thallium(I) iodide	$[Tl^+][I^-]$	6.5×10^{-8}	7.19
thallium(I) sulfide	$[Tl^+]^2[S^{2-}]$	5×10^{-21}	20.30
tin(II) sulfide	$[Sn^{2+}][S^{2-}]$	1×10^{-25}	25.00
zinc carbonate	$[Zn^{2+}][CO_3^{2-}]$	1.4×10^{-11}	10.85
zinc ferrocyanide	$[Zn^{2+}]^2[Fe(CN)_6^{4-}]$	4.1×10^{-16}	15.39
zinc hydroxide	$[Zn^{2+}][OH^-]^2$	1.2×10^{-17}	16.92
zinc oxalate	$[Zn^{2+}][C_2O_4^{2-}]$	2.8×10^{-8}	7.55
zinc sulfide	$[Zn^{2+}][S^{2-}]$	2×10^{-24}	23.70

APPENDIX 2-B   Acid and base dissociation constants*

System name	Conjugate acid/Conjugate base	K_a	$pK_a^\dagger$	K_b	$pK_b^\dagger$
acetic acid	CH_3COOH/CH_3COO^-	1.75×10^{-5}	4.76	5.71×10^{-10}	9.24
ammonia	NH_4^+/NH_3	5.56×10^{-10}	9.26	1.80×10^{-5}	4.74
aniline	$C_6H_5NH_3^+/C_6H_5NH_2$	2.54×10^{-5}	4.60	3.94×10^{-10}	9.40
arsenic acid	$H_3AsO_4/H_2AsO_4^-$ (K_1)	6.0×10^{-3}	2.22	1.7×10^{-12}	11.78
	$H_2AsO_4^-/HAsO_4^{2-}$ (K_2)	1.05×10^{-7}	6.98	9.52×10^{-8}	7.02
	$HAsO_4^{2-}/AsO_4^{3-}$ (K_3)	3.0×10^{-12}	11.52	3.3×10^{-3}	2.48
arsenious acid	$H_3AsO_3/H_2AsO_3^-$	6.0×10^{-10}	9.22	1.7×10^{-5}	4.78
L-ascorbic acid (vitamin C)	$HC_6H_7O_6/C_6H_7O_6^-$	9×10^{-5}	4.05	1.1×10^{-10}	9.95
benzoic acid	$C_6H_5COOH/C_6H_5COO^-$	6.14×10^{-5}	4.21	1.63×10^{-10}	9.79
boric acid	$H_3BO_3/H_2BO_3^-$	5.83×10^{-10}	9.23	1.71×10^{-5}	4.77
bromoacetic acid	$BrCH_2COOH/BrCH_2COO^-$	1.25×10^{-3}	2.90	8.00×10^{-12}	11.10
1-butanoic acid	$CH_3CH_2CH_2COOH/$ $CH_3CH_2CH_2COO^-$	1.52×10^{-5}	4.82	6.56×10^{-10}	9.18
carbonic	$H_2CO_3(CO_2)/HCO_3^-$ (K_1)	4.45×10^{-7}	6.35	2.25×10^{-8}	7.65
	HCO_3^-/CO_3^{2-} (K_2)	4.7×10^{-11}	10.33	2.1×10^{-4}	3.67
chloroacetic acid	$ClCH_2COOH/ClCH_2COO^-$	1.36×10^{-3}	2.87	7.34×10^{-12}	11.13
citric acid	$H_3C_6H_5O_7/H_2C_6H_5O_7^-$ (K_1)	7.45×10^{-4}	3.13	1.34×10^{-11}	10.87
	$H_2C_6H_5O_7^-/HC_6H_5O_7^{2-}$ (K_2)	1.73×10^{-5}	4.76	5.77×10^{-10}	9.24
	$HC_6H_5O_7^{2-}/C_6H_5O_7^{3-}$ (K_3)	3.98×10^{-7}	6.40	2.51×10^{-8}	7.60
dimethylamine	$(CH_3)_2NH_2^+/(CH_3)_2NH$	1.7×10^{-11}	10.77	5.9×10^{-4}	3.23
ethanolamine	$HOC_2H_4NH_3^+/HOC_2H_4NH_2$	3.14×10^{-10}	9.50	3.18×10^{-5}	4.50
ethylamine	$CH_3CH_2NH_3^+/CH_3CH_2NH_2$	2.34×10^{-11}	10.63	4.28×10^{-4}	3.37
ethylenediamine	$NH_3C_2H_4NH_3^{2+}/$ $NH_2C_2H_4NH_3^+$ (K_2)	1.4×10^{-7}	6.85	7.1×10^{-8}	7.15
	$NH_2C_2H_4NH_3^+/$ $NH_2C_2H_4NH_2$ (K_1)	1.2×10^{-10}	9.93	8.5×10^{-5}	4.07
EDTA‡	H_4Y/H_3Y^- (K_1)	1.0×10^{-2}	2.00	1.0×10^{-12}	12.00
	H_3Y^-/H_2Y^{2-} (K_2)	2.16×10^{-3}	2.67	4.63×10^{-12}	11.33
	H_2Y^{2-}/HY^{3-} (K_3)	6.92×10^{-7}	6.16	1.45×10^{-8}	7.84
	HY^{3-}/Y^{4-} (K_4)	5.50×10^{-11}	10.26	1.82×10^{-4}	3.74

*Equilibrium constants for proton transfer reactions in aqueous solution near room temperature are compiled from various sources.

$$\text{reaction of conjugate acid:} \quad HA + H_2O \rightleftharpoons H_3O^+ + A^- \quad K_a = \frac{[H_3O^+][A^-]}{[HA]}$$

$$\text{reaction of conjugate base:} \quad A^- + H_2O \rightleftharpoons HA + OH^- \quad K_b = \frac{[HA][OH^-]}{[A^-]}$$

(*Note:* Conjugate acid need not be a neutral species, as shown above, but analogous reactions for HA^-/A^{2-} and HA^+/A (and so on) systems can be written.
†All pK values are given to two significant figures for ease in reading the table.
‡See Chapter 10 for chemical structure.

APPENDIX 2-B (cont.)

System name	Conjugate acid/Conjugate base	K_a	$pK_a^\dagger$	K_b	$pK_b^\dagger$
fluoroacetic acid	FCH_2COOH/FCH_2COO^-	2.59×10^{-3}	2.59	3.85×10^{-12}	11.41
formic acid	$HCOOH/HCOO^-$	1.77×10^{-4}	3.75	5.65×10^{-11}	10.25
glycolic acid	$HOCH_2COOH/HOCH_2COO^-$	1.48×10^{-4}	3.83	6.78×10^{-11}	10.17
hydrazine	$H_2NNH_3^+/H_2NNH_2$	7.8×10^{-9}	8.11	1.3×10^{-6}	5.89
hydrogen cyanide	HCN/CN^-	6.2×10^{-10}	9.21	1.6×10^{-5}	4.79
hydrogen fluoride	HF/F^-	7.2×10^{-4}	3.14	1.4×10^{-11}	10.86
hydrogen sulfide	H_2S/HS^- (K_1)	5.7×10^{-8}	7.24	1.8×10^{-7}	6.76
	HS^-/S^{2-} (K_2)	1.2×10^{-15}	14.9	8.3×10^0	-0.9
hydroxylamine	$HONH_3^+/HONH_2$	1.1×10^{-6}	5.96	9.1×10^{-9}	8.04
hypochlorous acid	$HClO/ClO^-$	3.0×10^{-8}	7.52	3.3×10^{-7}	6.48
iodic acid	HIO_3/IO_3^-	1.7×10^{-1}	0.78	6.0×10^{-14}	13.22
lactic acid	$CH_3CHOHCOOH/$ $CH_3CHOHCOO^-$	1.37×10^{-4}	3.86	7.28×10^{-11}	10.14
maleic acid	$H_2C_4H_2O_4/HC_4H_2O_4^-$ (K_1)	1.20×10^{-2}	1.92	8.34×10^{-13}	12.08
	$HC_4H_2O_4^-/C_4H_2O_4^{2-}$ (K_2)	5.96×10^{-7}	6.22	1.68×10^{-8}	7.78
malonic acid	$HOOCCH_2COOH/$ $HOOCCH_2COO^-$ (K_1)	1.40×10^{-3}	2.86	7.16×10^{-12}	11.14
	$HOOCCH_2COO^-/$ $OOCCH_2COO^{2-}$ (K_2)	2.01×10^{-6}	5.70	4.97×10^{-9}	8.30
methylamine	$CH_3NH_3^+/CH_3NH_2$	2.1×10^{-11}	10.68	4.8×10^{-4}	3.32
nitrilotriacetic acid[‡]	H_3Y/H_2Y^- (K_1)	2.2×10^{-2}	1.65	4.5×10^{-13}	12.35
	H_2Y^-/HY^{2-} (K_2)	1.1×10^{-3}	2.95	8.9×10^{-12}	11.05
	HY^{2-}/Y^{3-} (K_3)	5.2×10^{-11}	10.28	1.9×10^{-4}	3.72
oxalic acid	$HOOCCOOH/HOOCCOO^-$ (K_1)	5.36×10^{-2}	1.27	1.87×10^{-13}	12.73
	$HOOCCOO^-/OOCCOO^{2-}$ (K_2)	5.42×10^{-5}	4.27	1.85×10^{-10}	9.73
phenol	$C_6H_5OH/C_6H_5O^-$	1.00×10^{-10}	10.00	1.00×10^{-4}	4.00

APPENDIX 2-B (cont.)

System name	Conjugate acid/Conjugate base	K_a	$pK_a^\dagger$	K_b	$pK_b^\dagger$
phosphoric acid	$H_3PO_4/H_2PO_4^-$ (K_1)	7.1×10^{-3}	2.15	1.4×10^{-12}	11.85
	$H_2PO_4^-/HPO_4^{2-}$ (K_2)	6.3×10^{-8}	7.20	1.6×10^{-7}	6.80
	HPO_4^{2-}/PO_4^{3-} (K_3)	4.2×10^{-13}	12.38	2.4×10^{-2}	1.62
phosphorous acid	$H_3PO_3/H_2PO_3^-$ (K_1)	1.00×10^{-2}	2.00	1.00×10^{-12}	12.00
	$H_2PO_3^-/HPO_3^{2-}$ (K_2)	2.6×10^{-7}	6.59	3.8×10^{-8}	7.41
o-phthalic acid	$H_2C_8H_4O_4/HC_8H_4O_4^-$ (K_1)	1.12×10^{-3}	2.95	8.91×10^{-12}	11.05
	$HC_8H_4O_4^-/C_8H_4O_4^{2-}$ (K_2)	3.91×10^{-6}	5.41	2.56×10^{-9}	8.59
picric acid	$(NO_2)_3C_6H_2OH/$ $(NO_2)_3C_6H_2O^-$	5.1×10^{-1}	0.29	1.9×10^{-14}	13.71
piperidine	$C_5H_{10}NH_2^+/C_5H_{10}NH$	7.6×10^{-12}	11.12	1.3×10^{-3}	2.88
propanoic acid	$CH_3CH_2COOH/$ $CH_3CH_2COO^-$	1.34×10^{-5}	4.87	7.48×10^{-10}	9.13
pyridine	$C_5H_5NH^+/C_5H_5N$	6.0×10^{-6}	5.22	1.7×10^{-9}	8.78
salicylic acid	$C_6H_4OHCOOH/$ $C_6H_4OHCOO^-$	1.05×10^{-3}	2.98	9.5×10^{-12}	11.02
succinic acid	$H_2C_4H_4O_4/HC_4H_4O_4^-$ (K_1)	6.21×10^{-5}	4.21	1.61×10^{-10}	9.79
	$HC_4H_4O_4^-/C_4H_4O_4^{2-}$ (K_2)	2.32×10^{-6}	5.64	4.32×10^{-9}	8.36
sulfamic acid	$H_2NSO_3H/H_2NSO_3^-$	1.03×10^{-1}	0.99	9.73×10^{-14}	13.01
sulfuric acid	H_2SO_4/HSO_4^-	strong			
	HSO_4^-/SO_4^{2-} (K_2)	1.20×10^{-2}	1.92	8.34×10^{-13}	12.08
sulfurous acid	H_2SO_3/HSO_3^- (K_1)	1.72×10^{-2}	1.76	5.81×10^{-13}	12.24
	HSO_3^-/SO_3^{2-} (K_2)	6.43×10^{-8}	7.19	1.56×10^{-7}	6.81
tartaric acid	$H_2C_4H_4O_6/HC_4H_4O_6^-$ (K_1)	9.20×10^{-4}	3.04	1.09×10^{-11}	10.96
	$HC_4H_4O_6^-/C_4H_4O_6^{2-}$ (K_2)	4.31×10^{-5}	4.37	2.32×10^{-10}	9.63
trichloroacetic acid	Cl_3CCOOH/Cl_3CCOO^-	1.29×10^{-1}	0.89	7.8×10^{-14}	13.11
trimethylamine	$(CH_3)_3NH^+/(CH_3)_3N$	1.60×10^{-10}	9.80	6.25×10^{-5}	4.20
tris(hydroxymethyl)- aminomethane	$(HOCH_2)_3CNH_3^+/$ $(HOCH_2)_3CNH_2$	8.41×10^{-9}	8.08	1.19×10^{-6}	5.92

APPENDIX 2-C   Formation constants of metal ion complexes*

Metal ion	Acetate log K_1	Acetate log K_2	Ammonia log K_1	Ammonia log K_2	Ammonia log K_3	Ammonia log K_4	Bromide log K_1	Bromide log K_2	Bromide log K_3	Chloride log K_1	Chloride log K_2	Chloride log K_3
Ag^+	0.4	−0.2	3.2	3.8			4.2	3.0	1.8	2.8	2.2	0.0
Al^{3+}	0.4											
Ba^{2+}										−0.1		
Bi^{3+}							4.3	1.3	0.3	2.4	2.0	1.4
Ca^{2+}	0.5		−0.2	−0.6	−0.8	−1.1						
Cd^{2+}	1.3	1.0	2.6	2.1	1.4	0.9	2.2	0.8	−0.2	1.5	0.4	0.4
Ce^{3+}	1.7	1.0								0.1		
Co^{2+}			2.1	1.6	1.0	0.8				−2.4		
Cu^+			5.9	4.9							4.9(β)	
Cu^{2+}	1.8	1.1	4.3	3.7	3.0	2.3	0.3			0.1	−0.6	
Fe^{2+}										0.4	0.0	
Fe^{3+}						3.7(β)	−0.3	−0.5		1.5	0.6	−1.0
Hg^{2+}		8.4(β)	8.8	8.7	1.0	0.8	9.0	8.3	1.4	5.3	7.5	1.1
Mg^{2+}	0.5		0.2	0.1	−0.3	−1.0						
Mn^{2+}			0.8	0.5						0.0		
Ni^{2+}	0.7	0.6	2.8	2.2	1.7	1.2						
Pb^{2+}	2.7	1.5					1.2	0.7	0.9	1.1	1.2	−0.4
Sn^{2+}							0.7	0.4	0.2	1.1	0.6	0.0
Sr^{2+}	0.4											
Zn^{2+}	1.0		2.4	2.4	2.5	2.1	0.2	−0.3	−0.6	−0.2	−0.4	−0.4

*Stepwise formation constants are given except when noted (β), in which case the overall formation constant for the complex is given. See Chapter 4 for definition of stepwise and overall formation constants. Values are given to one significant figure (except for EDTA formation constants) for ease in reading the table.

APPENDIX 2-C Formation constants of metal ion complexes

Metal ion	Cyanide				EDTA*	Iodide				NTA†	Oxalate		Thiocyanate			
	$\log K_1$	$\log K_2$	$\log K_3$	$\log K_4$	$\log K_{MY}$	$\log K_1$	$\log K_2$	$\log K_3$	$\log K_4$	$\log K_{MY}$	$\log K_1$	$\log K_2$	$\log K_1$	$\log K_2$	$\log K_3$	$\log K_4$
Ag^+		$19.9(\beta)$			7.32			$13.9(\beta)$	-0.2		0.0				$8.2(\beta)$	1.1	0.7
Al^{3+}					16.13								$13(\beta)$				
Ba^{2+}					7.76					6.4	2.3				1.1		
Bi^{3+}									$19.4(\beta_6)$					1.1	1.1		
Ca^{2+}					10.70					8.2	3.0						
Cd^{2+}	5.5	5.1	4.6	3.6	16.46	2.4	1.6	1.0	1.1	9.5	3.5	1.8	1.0	0.7	0.6	1.0
Ce^{3+}											6.5	4.0				
Co^{2+}					16.31					10.6	4.7	2.4	2.3	0.7	-0.7	0.0
Cu^+		$24(\beta)$	4.6	1.7			$8.8(\beta)$							$12.1(\beta)$		
Cu^{2+}					18.80					12.7	6.2	1.8				
Fe^{2+}				$24(\beta_6)$	14.33					8.8	4.7	3.0	1.0			
Fe^{3+}				$31(\beta_6)$	25.1					15.9	9.4	6.8	2.1	1.3		
Hg^{2+}	18.0	16.7	3.8	3.0	21.80	12.9	10.9	3.8	2.2					$17.5(\beta)$		
Mg^{2+}					8.69					7.0	3.4	1.0				
Mn^{2+}					13.79					7.4	3.9	1.9				
Ni^{2+}					18.62					11.5	5.3	2.3	1.2	0.5	0.2	
Pb^{2+}				$22(\beta)$	18.04	1.3	1.5	0.6	0.5	11.5		$6.5(\beta)$	1.1	1.4		
Sn^{2+}																
Sr^{2+}					8.63					5.0	2.5					
Zn^{2+}			$17.5(\beta)$	2.7	16.50	-1.3				10.5	4.9	2.7	1.6			

SOURCE: From L. Meites, ed., *Handbook of Analytical Chemistry* (New York: McGraw-Hill Book Company, 1963).

*EDTA formation constants are for the reaction

$$M^{n+} + Y^{4-} \rightleftharpoons MY^{n-4}$$

See Chapter 10 for discussion of EDTA complexes.
†NTA formation constants are for the reaction

$$M^{n+} + Y^{3-} \rightleftharpoons MY^{n-3}$$

APPENDIX 3-A   Selected Standard and Formal Half-Cell Potentials

Element	Half-reaction	(Formal potential conditions)	E^0 (V)*	E_f^0 (V)*
aluminum				
	$Al^{3+} + 3e \rightleftharpoons Al$		−1.66	
antimony				
	$Sb_2O_3 + 6H_3O^+ + 6e \rightleftharpoons 2Sb + 9H_2O$		+0.152	
	$Sb_2O_5 + 6H_3O^+ + 4e \rightleftharpoons 2SbO^+ + 9H_2O$		+0.581	
	$Sb(V) \rightleftharpoons Sb(III)$	6 F HCl		+0.82
arsenic				
	$As_2O_3 + 6H_3O^+ + 6e \rightleftharpoons 2As + 9H_2O$		+0.234	
	$H_3AsO_4 + 2H_3O^+ + 2e \rightleftharpoons H_3AsO_3 + 3H_2O$		+0.559	
barium				
	$Ba^{2+} + 2e \rightleftharpoons Ba$		−2.90	
beryllium				
	$Be^{2+} + 2e \rightleftharpoons Be$		−1.85	
bismuth				
	$BiO^+ + 2H_3O^+ + 3e \rightleftharpoons Bi + 3H_2O$		+0.32	
	$BiOCl + 2H_3O^+ + 3e \rightleftharpoons Bi + 3H_2O + Cl^-$		+0.16	
bromine				
	$Br_2(aq) + 2e \rightleftharpoons 2Br^-$		+1.087	
	$Br_2(l) + 2e \rightleftharpoons 2Br^-$		+1.065	
	$Br_3^- + 2e \rightleftharpoons 3Br^-$		+1.05	
	$2BrO_3^- + 12H_3O^+ + 10e \rightleftharpoons Br_2 + 18H_2O$		+1.52	
cadmium				
	$Cd^{2+} + 2e \rightleftharpoons Cd$		−0.403	
	$Cd(CN)_4^{2-} + 2e \rightleftharpoons Cd + 4CN^-$		−1.09	
	$Cd(NH_3)_4^{2+} + 2e \rightleftharpoons Cd + 4NH_3$		−0.61	
calcium				
	$Ca^{2+} + 2e \rightleftharpoons Ca$		−2.87	
carbon				
	$2CO_2 + 2H_3O^+ + 2e \rightleftharpoons H_2C_2O_4 + 2H_2O$		−0.49	
cerium				
	$Ce(IV) + e \rightleftharpoons Ce(III)$	1 F HClO_4		+1.70
		1 F HCl		+1.28
		1 F HNO_3		+1.60
		1 F H_2SO_4		+1.44

*Potentials are given in volts versus the NHE, which is assigned a value of 0.000 V. Signs conform to the Stockholm convention, as discussed in Chapter 11. Formal potentials are measured cell potentials for solutions containing 1 F concentration of the redox couple species in the designated medium. The values refer to a temperature of 25°C.

Element	Half-reaction	(Formal potential conditions)	E^0 (V)*	E_f^0 (V)*
chlorine				
	$Cl_2 + 2e \rightleftharpoons 2Cl^-$		+1.359	
	$2HOCl + 2H_3O^+ + 2e \rightleftharpoons Cl_2 + 4H_2O$		+1.63	
	$ClO_4^- + 2H_3O^+ + 2e \rightleftharpoons ClO_3^- + 3H_2O$		+1.19	
	$2ClO_3^- + 12H_3O^+ + 10e \rightleftharpoons Cl_2 + 18H_2O$		+1.47	
chromium				
	$Cr^{2+} + 2e \rightleftharpoons Cr$		−0.56	
	$Cr(III) + e \rightleftharpoons Cr(II)$		−0.41	
		0.1–0.5 F H_2SO_4		−0.37
	$Cr_2O_7^{2-} + 14H_3O^+ + 6e \rightleftharpoons 2Cr^{3+} + 21H_2O$		+1.33	
		1 F HCl		+1.00
		1 F HClO$_4$		+1.025
cobalt				
	$Co^{2+} + 2e \rightleftharpoons Co$		−0.28	
	$Co(III) + e \rightleftharpoons Co(II)$		+1.842	
		8 F H_2SO_4		+1.82
	$Co(NH_3)_6^{3+} + e \rightleftharpoons Co(NH_3)_6^{2+}$		+0.1	
copper				
	$Cu^{2+} + 2e \rightleftharpoons Cu$		+0.337	
	$Cu^{2+} + e \rightleftharpoons Cu^+$		+0.153	
	$2Cu^{2+} + 2I^- + 2e \rightleftharpoons Cu_2I_2$		+0.86	
	$Cu(CN)_2^- + e \rightleftharpoons Cu + 2CN^-$		−0.43	
	$CuCl_2^- + e \rightleftharpoons Cu + 2Cl^-$		+0.177	
	$Cu(EDTA)^{2-} + 2e \rightleftharpoons Cu + EDTA^{4-}$	0.1 F EDTA, pH 4–5		+0.13
fluorine				
	$F_2 + 2e \rightleftharpoons 2F^-$		+2.87	
gold				
	$Au^{3+} + 2e \rightleftharpoons Au^+$		+1.41	
	$Au^{3+} + 3e \rightleftharpoons Au$		+1.50	
	$Au(CN)_2^- + e \rightleftharpoons Au + 2CN^-$		−0.60	
hydrogen				
	$2H_3O^+ + 2e \rightleftharpoons H_2 + 2H_2O$	NHE	0.000	
	$2H_2O + 2e \rightleftharpoons H_2 + 2OH^-$		−0.828	
indium				
	$In^{3+} + 3e \rightleftharpoons In$		−0.33	
iodine				
	$I_2(s) + 2e \rightleftharpoons 2I^-$		+0.5345	
	$I_3^- + 2e \rightleftharpoons 3I^-$		+0.536	
	$2ICl_2^- + 2e \rightleftharpoons I_2 + 4Cl^-$		+1.06	
	$2IO_3^- + 12H_3O^+ + 10e \rightleftharpoons I_2 + 18H_2O$		+1.19	
	$H_5IO_6 + H_3O^+ + 2e \rightleftharpoons IO_3^- + 4H_2O$		+1.16	

Element	Half-reaction	(Formal potential conditions)	E^0 (V)*	E_f^0 (V)*
iron				
	$Fe^{2+} + 2e \rightleftharpoons Fe$		-0.440	
	$Fe(III) + e \rightleftharpoons Fe(II)$		$+0.771$	
		$1\ F\ HClO_4$		$+0.735$
		$1\ F\ HCl$		$+0.70$
		$10\ F\ HCl$		$+0.53$
		$1\ F\ H_2SO_4$		$+0.68$
		$2\ F\ H_3PO_4$		$+0.46$
	$Fe(CN)_6^{3-} + e \rightleftharpoons Fe(CN)_6^{4-}$		$+0.356$	
lead				
	$Pb^{2+} + 2e \rightleftharpoons Pb$		-0.126	
	$PbSO_4 + 2e \rightleftharpoons Pb + SO_4^{2-}$		-0.356	
	$PbO_2 + SO_4^{2-} + 4H_3O^+ + 2e \rightleftharpoons PbSO_4 + 6H_2O$		$+1.685$	
	$PbO_2 + H_2O + 2e \rightleftharpoons PbO + 2OH^-$		$+0.28$	
lithium				
	$Li^+ + e \rightleftharpoons Li$		-3.045	
magnesium				
	$Mg^{2+} + 2e \rightleftharpoons Mg$		-2.37	
manganese				
	$Mn^{2+} + 2e \rightleftharpoons Mn$		-1.19	
	$Mn(III) + e \rightleftharpoons Mn(II)$	$7.5\ F\ H_2SO_4$		$+1.5$
	$MnO_2 + 4H_3O^+ + 2e \rightleftharpoons Mn^{2+} + 6H_2O$		$+1.23$	
	$MnO_4^- + 8H_3O^+ + 5e \rightleftharpoons Mn^{2+} + 12H_2O$		$+1.51$	
	$MnO_4^- + 4H_3O^+ + 3e \rightleftharpoons MnO_2 + 6H_2O$		$+1.695$	
	$MnO_4^- + e \rightleftharpoons MnO_4^{2-}$		$+0.564$	
mercury				
	$Hg_2^{2+} + 2e \rightleftharpoons Hg$		$+0.792$	
	$2Hg^{2+} + 2e \rightleftharpoons Hg_2^{2+}$		$+0.907$	
	$Hg^{2+} + 2e \rightleftharpoons Hg$		$+0.854$	
	$Hg_2Br_2 + 2e \rightleftharpoons 2Hg + 2Br^-$		$+0.139$	
	$Hg_2Cl_2 + 2e \rightleftharpoons 2Hg + 2Cl^-$		$+0.268$	
		sat'd KCl, SCE		$+0.242$
	$Hg_2I_2 + 2e \rightleftharpoons 2Hg + 2I^-$		-0.040	
molybdenum				
	$Mo(VI) + e \rightleftharpoons Mo(V)$		$+0.48$	
		$2\ F\ HCl$		$+0.53$
	$Mo(IV) + e \rightleftharpoons Mo(III)$	$4.5\ F\ H_2SO_4$		$+0.1$
	$Mo(V) + 2e \rightleftharpoons Mo(III)$ (green)	$2\ F\ HCl$		-0.25
	$Mo(V) + 2e \rightleftharpoons Mo(III)$ (red)	$2\ F\ HCl$		$+0.11$
nickel				
	$Ni^{2+} + 2e \rightleftharpoons Ni$		-0.23	
	$Ni(OH)_2 + 2e \rightleftharpoons Ni + 2OH^-$		-0.72	

Element	Half-reaction	(Formal potential conditions)	E^0 (V)*	E_f^0 (V)*
nitrogen				
	$NO_3^- + 4H_3O^+ + 3e \rightleftharpoons NO + 6H_2O$		+0.96	
	$NO_3^- + 3H_3O^+ + 2e \rightleftharpoons HNO_2 + 4H_2O$	1 F HNO_3		+0.92
	$2NO_3^- + 4H_3O^+ + 2e \rightleftharpoons N_2O_4 + 6H_2O$		+0.80	
	$HNO_2 + H_3O^+ + e \rightleftharpoons NO + 2H_2O$		+0.99	
	$N_2 + 4H_3O^+ + 2e \rightleftharpoons 2NH_3OH^+ + 2H_2O$		−1.87	
	$NO_2 + H_3O^+ + e \rightleftharpoons HNO_2 + H_2O$		+1.07	
oxygen				
	$O_3 + 2H_3O^+ + 2e \rightleftharpoons O_2 + 3H_2O$		+2.07	
	$O_2 + 4H_3O^+ + 4e \rightleftharpoons 6H_2O$		+1.229	
	$O_2 + 2H_3O^+ + 2e \rightleftharpoons H_2O_2 + 2H_2O$		+0.682	
	$H_2O_2 + 2H_3O^+ + 2e \rightleftharpoons 4H_2O$		+1.77	
palladium				
	$Pd(II) + 2e \rightleftharpoons Pd$	4 F $HClO_4$		+0.987
phosphorus				
	$H_3PO_3 + 2H_3O^+ + 2e \rightleftharpoons H_3PO_2 + 3H_2O$		−0.50	
	$H_3PO_4 + 2H_3O^+ + 2e \rightleftharpoons H_3PO_3 + 3H_2O$		−0.276	
platinum				
	$Pt^{2+} + 2e \rightleftharpoons Pt$		+1.2	
	$PtCl_4^{2-} + 2e \rightleftharpoons Pt + 4Cl^-$		+0.73	
	$PtCl_6^{2-} + 2e \rightleftharpoons PtCl_4^{2-} + 2Cl^-$		+0.68	
potassium				
	$K^+ + e \rightleftharpoons K$		−2.925	
selenium				
	$Se + 2H_3O^+ + 2e \rightleftharpoons H_2Se + 2H_2O$		−0.40	
	$H_2SeO_3 + 4H_3O^+ + 4e \rightleftharpoons Se + 7H_2O$		+0.74	
	$SeO_4^{2-} + 4H_3O^+ + 2e \rightleftharpoons H_2SeO_3 + 5H_2O$		+1.15	
silver				
	$Ag^+ + e \rightleftharpoons Ag$		+0.799	
		1 F $HClO_4$		+0.792
	$Ag(II) + e \rightleftharpoons Ag(I)$	4 F HNO_3		+1.93
	$AgBr + e \rightleftharpoons Ag + Br^-$		+0.073	
	$AgCl + e \rightleftharpoons Ag + Cl^-$		+0.222	
		sat'd KCl		+0.197
	$AgI + e \rightleftharpoons Ag + I^-$		−0.152	
	$Ag_2O + H_2O + 2e \rightleftharpoons 2Ag + 2OH^-$		+0.342	
sodium				
	$Na^+ + e \rightleftharpoons Na$		−2.713	
strontium				
	$Sr^{2+} + 2e \rightleftharpoons Sr$		−2.89	

Element	Half-reaction	(Formal potential conditions)	E^0 (V)*	E_f^0 (V)*
sulfur				
	$S + 2H_3O^+ + 2e \rightleftharpoons H_2S + 2H_2O$		+0.141	
	$2SO_3^{2-} + 2H_2O + 2e \rightleftharpoons S_2O_4^{2-} + 4OH^-$		-1.12	
	$S_4O_6^{2-} + 2e \rightleftharpoons 2S_2O_3^{2-}$		+0.09	
	$SO_4^{2-} + 4H_3O^+ + 2e \rightleftharpoons SO_2 + 6H_2O$		+0.17	
		$1\,F\,H_2SO_4$		+0.07
	$S_2O_8^{2-} + 2e \rightleftharpoons 2SO_4^{2-}$		+2.01	
	$SO_3^{2-} + 3H_2O + 4e \rightleftharpoons S + 6OH^-$		-0.66	
thallium				
	$Tl^+ + e \rightleftharpoons Tl$		-0.336	
	$Tl(III) + 2e \rightleftharpoons Tl(I)$		+1.28	
		$1\,F\,HCl$		+0.78
tin				
	$Sn^{2+} + 2e \rightleftharpoons Sn$		-0.140	
	$SnCl_4^{2-} + 2e \rightleftharpoons Sn + 4Cl^-$	$1\,F\,HCl$		-0.19
	$Sn(IV) + 2e \rightleftharpoons Sn(II)$		+0.154	
		$1\,F\,HCl$		+0.14
titanium				
	$Ti^{2+} + 2e \rightleftharpoons Ti$		-1.63	
	$Ti^{3+} + e \rightleftharpoons Ti^{2+}$		-0.37	
	$TiO^{2+} + 2H_3O^+ + e \rightleftharpoons Ti^{3+} + 3H_2O$		+0.10	
tungsten				
	$W(V) + 2e \rightleftharpoons W(III)$ (green)	$12\,F\,HCl$		+0.1
	$W(V) + 2e \rightleftharpoons W(III)$ (red)	$12\,F\,HCl$		-0.2
	$W(V) + e \rightleftharpoons W(IV)$	$12\,F\,HCl$		-0.3
	$2WO_3 + 2H_3O^+ + 2e \rightleftharpoons W_2O_5 + 3H_2O$		-0.03	
	$W(VI) + e \rightleftharpoons W(V)$	$12\,F\,HCl$		+0.26
uranium				
	$U(IV) + e \rightleftharpoons U(III)$		-0.61	
		$1\,F\,HCl$		-0.64
	$UO_2^{2+} + 4H_3O^+ + 2e \rightleftharpoons U^{4+} + 6H_2O$		+0.334	
	$UO_2^{2+} + 4H_3O^+ + 3e \rightleftharpoons U^{3+} + 6H_2O$		+0.019	
	$UO_2^{2+} + e \rightleftharpoons UO_2^+$		+0.05	
	$UO_2^+ + 4H_3O^+ + e \rightleftharpoons U^{4+} + 6H_2O$		+0.62	
vanadium				
	$V^{2+} + 2e \rightleftharpoons V$		-1.18	
	$V^{3+} + e \rightleftharpoons V^{2+}$		-0.255	
	$VO^{2+} + 2H_3O^+ + e \rightleftharpoons V^{3+} + 3H_2O$		+0.337	
		$1\,F\,H_2SO_4$		+0.360
	$VO_2^+ + 2H_3O^+ + e \rightleftharpoons VO^{2+} + 3H_2O$		+0.999	
zinc				
	$Zn^{2+} + 2e \rightleftharpoons Zn$		-0.763	
	$Zn(NH_3)_4^{2+} + 2e \rightleftharpoons Zn + 4NH_3$		-1.04	

APPENDIX 3-B   Selected polarographic half-wave potentials*

Supporting electrolyte

Metal ion	1 F HCl	0.1 F [Cl⁻]	NH₃/NH₄Cl, pH 10	Citrate, pH 4	1 F HCl
Al(III)	NR†	−1.75			
Bi(III)	−0.09			−0.19	
Cd(II)	−0.64	−0.60	−0.81	−0.59	−1.18
Co(II)		−1.20	−1.29	NR	−1.3
Cr(III)		−0.61, −0.85, −1.47	−1.43, −1.71		−1.38
Cu(II)	+0.04, −0.22	+0.04	−0.24, −0.51	−0.03	NR
Fe(II)		−1.3	−0.34 (A), −1.49	−0.05 (A)	
Fe(III)	>0	>0	ppt‡	+0.05	
Mo(VI)	−0.26, −0.63		−1.71		
Ni(II)		−1.1	−1.10	NR	−1.36
Pb(II)	−0.44	−0.40	ppt	−0.43	−0.72
Sn(II)	−0.1 (A), −0.47			−0.21 (A)	
Sn(IV)	−0.1, −0.47		>0.3, −0.52		
Tl(I)	−0.48	−0.46	−0.48	−0.45	>0
Zn(II)			−1.35	−1.04	NR

Supporting electrolyte

Metal ion	Tartrate, pH 4	EDTA, pH 7	1 F NaOH	1 F KSCN	HOAc/OAc⁻, pH 5	7.3 F H₃PO₄
Al(III)	−1.79 (pH 2)		NR		NR	NR
Bi(III)	−0.14	−0.66	−0.6	>0	−0.25	
Cd(II)	−0.59	−1.27	−0.78	−0.65	−0.65	−0.15
Co(II)	−1.58	NR	−1.46	−1.08	−1.19	−1.20
Cr(III)				−1.05	−1.2	−1.02
Cu(II)	−0.03	−0.31	−0.41	>0, −0.54	−0.07	−0.09
Fe(II)	−0.12 (A)		−0.9 (A)	−1.5	NR	NR
Fe(III)	−0.12, −1.52	−0.15	ppt	>0	ppt	+0.06
Mo(VI)			NR		−0.6, −1.1, −1.2	0.0, −0.49
Ni(II)	−1.13	NR		−0.68	−1.1	−1.18
Pb(II)	−0.42	−1.37	−0.76	−0.44	−0.50	−0.53
Sn(II)	−0.18 (A), −0.49	−1.26	−0.73 (A), −1.22	−0.46	−0.16 (A), −0.62	−0.58
Sn(IV)			NR	−0.50	−1.1	−0.65
Tl(I)	−0.45	−0.50	−0.48	−0.52	−0.47	−0.63
Zn(II)	−1.04	NR	−1.53	−1.06	−1.1	−1.13

*All potentials are quoted as $E_{\frac{1}{2}}$, V vs. SCE at 25°C. Potentials noted (A) are anodic waves. When more than one potential is given, the species exhibits more than one wave.

†NR indicates that no reduction is observed in the supporting electrolyte.

‡The abbreviation ppt indicates that the metal ion precipitates in the supporting electrolyte.

Answers to Selected Problems

CHAPTER 2

2-1 a. 16.5 ppm

2-2 a. 16.3 ppm

2-3 a. 16.8 ppm

2-4 a. 5.7 ppm (from mean)

2-5 a. 6.6 ppm

2-6 a. 43 ppm^2

2-7 a.

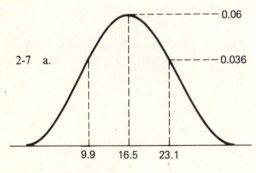

2-8 a. 9.6–23.4 ppm

2-9 a. 12.5–20.5 ppm

2-10 a. 5.6–27.4 ppm

2-14 a. Constant error. 1.2%

2-16 a. 5; b. 2

2-18 11

2-21 a. $Q_{expt} = 0.80$; retain

2-22 Buy A

2-25 a. Ca, $\pm0.4\%$; Mg, $\pm0.5\%$

2-28 a. 7.92 ± 0.21 g/cm^3, rounded to
7.9 g/cm^3

2-29 0.11 ppm

CHAPTER 3

3-1 b. 323.45

3-2 b. 0.464 mmol

3-3 b. 0.989 mmol

3-4 b. 67.3 mg

3-5 b. Acid–base

3-8 743 mg

3-10 a. 3.99 g I$_2$ ($\sim$4 g) dissolved in 500 mL
ethanol (neglect volume change)

3-11 a. 7.58 g NaCl dissolved in 150 mL
H$_2$O (neglect volume change)

3-12 a. 12.4 F

3-13 a. 1.39 mL added to H$_2$O, diluted to
500 mL

3-15 a. 1.47 g added to H$_2$O, diluted to
100 mL

3-16 a. 1.56 g

3-17 a. 1.48×10^3 mg

3-18 a. 43.8 mmol

3-20		0.109 ppm
3-23	a.	50 mL
3-24	b.	2
3-25	b.	165.60
3-26	b.	0.2 N
3-27	b.	2.07×10^3 mg
3-29		0.531 N
3-32		15.1% N
3-34		2.16×10^5 mC
3-36		8.0 mL
3-38		26.7%
3-39		40 nickels, 125 pennies

CHAPTER 4

4-3		2.9×10^{21} torr^{-1}
4-5	a.	$HC_2H_3O_2 + H_2O \rightleftharpoons H_3O^+$ $+ C_2H_3O_2^-$; acid dissociation; $$K_a = \frac{[H_3O^+][C_2H_3O_2^-]}{[HC_2H_3O_2]}$$
4-9	a.	$BaSO_4 \rightleftharpoons Ba^{2+} + SO_4^{2-}$; $K_{sp} = [Ba^{2+}][SO_4^{2-}]$
4-10	a.	1.3×10^{-10}
4-11		5.8×10^{-15}
4-13		6×10^{-2} F
4-18		0.65 M, 0.66 m
4-20	a.	0.991 ($\sim$0.99)
4-21	b.	7.00
4-22	b.	3.4×10^{-10}
4-23		1.60×10^{-13} F
4-27		1.38×10^{-4}
4-31		7.6×10^3
4-33	a.	5.0×10^{-10} F
4-35	a.	1.1×10^{-19} F
4-37	a.	$6Fe^{2+} + Cr_2O_7^{2-} + 14H_3O^+ \rightleftharpoons$ $6Fe^{3+} + 2Cr^{3+} + 21H_2O$; $$K = \frac{[Fe^{3+}]^6[Cr^{3+}]^2}{[Fe^{2+}]^6[Cr_2O_7^{2-}][H_3O^+]^{14}}$$
4-39	a.	$(HCHO)_{ether} \rightleftharpoons (HCHO)_{aq}$; $K_D = [HCHO]_{aq}/[HCHO]_{ether}$
	b.	5.0×10^{-4} F

4-40	a.	0.21 atm; b. 0.184 atm; c. $+12\%$
4-43	a.	$[H_2O] = 0.06$ atm, $[H_2] = 0.040$ atm, $[O_2] = 0.020$ atm
4-44	a.	$[H_2] = [CO_2] = 0.044$ atm, $[H_2O]$ $= [CO] = 0.056$ atm

CHAPTER 5

5-1	a.	530 mg/L; therefore, reject
5-2		C_3H_8O
5-7	a.	2.66 mg/L
5-8	a.	5.2×10^{-13}
5-9	a.	8.6×10^{-4} mg/L
5-10		0.19% loss
5-14		0.06 F
5-17		3.6×10^{-5} F in H_2O; 5.9×10^{-5} F in ionic strength 0.1
5-19		8.6 mg
5-22		7.2–8.6
5-26		4.6 mg/L
5-29		$[OH^-]_{solubility}$, solubility $= 5.8 \times 10^{-6}$ F
5-30		2.8 F
5-32	a.	0.1110
5-35	a.	65.4% $CaCO_3$, 34.6% $SrCO_3$

CHAPTER 6

6-7	solubility $= 0.81$ mg/100 mL H_2O, 2.1 mg/100 mL 0.2 HCl
6-8	0.0724 N
6-11	190 mg
6-14	644 mg
6-16	1.68%
6-18	16.9%
6-20	2.33%
6-22	13.2% P_2O_5, 23.4% superphosphate
6-24	3.840%
6-27	$G_f = 0.01445$; 692 mg
6-29	19.5 mg K/100 mL serum
6-33	72.3% Cu, 4.67% Pb
6-35	Fe_2O_3
6-39	2.54% Al

CHAPTER 7

7-1 a. 169.9; f. 67.00

7-2 a. 97.18; e. 31.60

7-3 a. 0.1020 N

7-6 0.0934 N

7-9 -0.11%

7-13 0.1266 N

7-15 8.384 mg/mL

7-22 -0.06 mL

7-26 54.10%

7-28 31.6%

CHAPTER 8

8-1 a. 8.386 mg/mL

8-4 a. $V = 50$ mL, pCa $= 3.52$

8-6 19.4 mg/L

8-9 0.455%

8-12 16.0 ppm

8-15 0.04%

8-18 4.1×10^{-8} F (too small to be detected by color change)

8-22 $x = 3.43$

8-24 a.

Vol.	[Cl$^-$]	pCl	[Ag$^+$]	pAg
0	0.20	0.70	0	—
10	0.15	0.84	1.2×10^{-9}	8.91
25	0.08	1.10	2.2×10^{-9}	8.65
40	0.029	1.54	6.3×10^{-9}	8.20
48	5.4×10^{-3}	2.27	3.3×10^{-8}	7.48
49	2.7×10^{-3}	2.57	6.7×10^{-8}	7.17
50	1.3×10^{-5}	4.87	1.3×10^{-5}	4.87
51	6.8×10^{-8}	7.17	2.6×10^{-3}	2.58
52	3.4×10^{-8}	7.47	5.3×10^{-3}	2.28
60	7.2×10^{-9}	8.14	0.025	1.60
100	1.8×10^{-9}	8.74	0.10	1.00

CHAPTER 9

9-1 a. $[H_3O^+] = 3 \times 10^{-3}$, pH $= 2.5$

9-3 a. $[H_3O^+] = 1.9 \times 10^{-3}$, pH $= 2.73$

9-4 a. $[H_3O^+] = 1.05 \times 10^{-11}$, pH $= 10.98$

9-5 Iteration needed, pH $= 3.66$

9-7 a. $[H_3O^+] = 2.4 \times 10^{-9}$, pH $= 8.62$

9-10 a. 6.468 mg/mL; e. 5.278 mg/mL

9-11 b. $\triangle$pH $= 5.27$

9-12 a. $[H_3O^+] = 5.8 \times 10^{-6}$, pH $= 5.23$

9-15 0.36 mol/pH

9-18 a.

Vol.	pH
0	2.38
10	3.58
20	4.35
24	5.13
25	8.29
26	11.42
30	12.10
50	12.70

9-19 a.

Vol.	pH
0	10.97
10	9.43
20	8.65
24	7.88
25	5.37
26	2.88
30	2.20
50	1.60

9-23 a. $\triangle$pH $= 2.68$; OK

9-25 EW $= 52.4$

9-27 a. H_2CO_3

9-29 b. $\alpha(H_2Ox) = 0.651$, $\alpha(HOx) = 0.349$, $\alpha(Ox) = 1.9 \times 10^{-4}$

9-33 At pH 3, $\alpha(H_3Cit) = 0.57$, $\alpha(H_2Cit) = 0.42$, $\alpha(HCit) = 7.3 \times 10^{-3}$, $\alpha(Cit) = 2.9 \times 10^{-6}$

9-36 1.95%

9-41 $+0.2\%$

9-45 741 ppm

CHAPTER 10

10-1 a. $n = 4$, EW $= 54.15$; e. $n = 8$, EW $= 20.68$

10-2 a. $Hg^{2+} + SCN^- \rightleftharpoons HgSCN^+$; $K_1 = [HgSCN^+]/[Hg^{2+}][SCN^-]$ $HgSCN^+ + SCN^- \rightleftharpoons Hg(SCN)_2$; $K_2 = [Hg(SCN)_2]/[HgSCN^+][SCN^-]$ $Hg^{2+} + 2SCN^- \rightleftharpoons Hg(SCN)_2$; $\beta_2 = K_1K_2 = [Hg(SCN)_2]/[Hg^{2+}][SCN^-]^2$

10-6 a. 1.176 mg/mL

10-7 0.0624 F

10-12 a. 4.9×10^{10}

10-14 a. $\alpha(H_2Y) = 0.95$

10-16 a. 6.35

10-17

Vol.	pCa
0	1.70
10	1.88
25	2.18
40	2.65
48	3.39
50	6.12
52	8.84
60	9.54
100	10.24

10-19 a. $\triangle pM = 2.7$, OK

10-21 97.8%

10-25 358 mg/100 mL

10-28 34.1% Mg, 65.9% Zn

10-30 $CaCO_3 = 507$ ppm, $MgCO_3 = 121$ ppm

CHAPTER 11

11-1 a. 1.93×10^5 C

11-2 a. $UO_2^{2+} + 2Ti^{3+} \rightleftharpoons U^{4+} + 2TiO^{2+}$

11-3 a. oxidant, UO_2^{2+}; reductant, Ti^{3+}

11-4 a. $n = 2$, $Q = 1.93 \times 10^5$ C

11-5 a. -1.004×10^5 J/mol

11-6 a. $E = +0.52 - \dfrac{0.059}{2} \times \log[U^{4+}]$
$\times [TiO^{2+}]^2/[UO_2^{2+}][Ti^{3+}]^2$

11-7 a. 4×10^{17}

11-9 $+1.26$ V vs. NHE

11-11 a. $+0.730$ V vs. NHE

11-12 a. $+0.488$ V vs. SCE

11-15 a. $\triangle E^\circ = +0.73$ V, OK

11-16 $E = 0 = E^\circ_{AgBr} - E^\circ_{Ag} - 0.059 \log K_{sp}$,
$K_{sp} = 5.0 \times 10^{-13}$

11-18 a. $+0.722$ V vs. NHE

11-20 a.

Vol.	E vs. NHE
10	-0.291
25	-0.255
40	-0.219
48	-0.174
49	-0.155
50	$+0.722$
51	$+1.600$
52	$+1.618$
60	$+1.659$
100	$+1.700$

11-22 $+0.4\%$

11-25 21.62%

11-28 0.1007 N

11-32 4.01%

11-35 0.59 ppm

11-40 79.2%

11-43 63.9%

11-48 16.6% V_2O_5, 34.9% U_3O_8

CHAPTER 12

12-1 a. $v = 6.7 \times 10^{14}$ sec^{-1},
$\bar{v} = 2.2 \times 10^4$ cm^{-1}

12-4 $\lambda = 4032$ Å, $v = 7.44 \times 10^{14}$ sec^{-1}
(visible region)

12-7 a. 3350 nm

12-9 550 nm, outer electron transition; 4×10^{-4} cm, vibrational transition

12-12 a. $[x] = 1.52 \times 10^{-4}$ F, $A = 0.469$

12-13 a. $T = 23.3\%$, $\varepsilon = 1580$ L/mol-cm

12-15 1.39×10^{-2} (cm-ppm)$^{-1}$

12-17 a. 2.0×10^3 L/mol-cm

12-18 9.9×10^{-6} F

12-22 a. 6.7 mg/L

12-25 ~ 0.006 ppm

12-28 3.8×10^4

12-30 5.4×10^{-6}

12-32 $[Co] = 9.1 \times 10^{-5}$ F,
$[Ni] = 3.2 \times 10^{-5}$ F

CHAPTER 13

13-2 $\sim 1\%$

13-4 a. Green

13-6 $50.4°$

13-8 0.51%

13-11 a. UV

13-13 ~ 100 mg

13-16 0.181 μg/mL-cm^{-1}

13-19 $C = 10.7$ mg/L $= 3.93 \times 10^{-5}$ F,
$\varepsilon = 2.1 \times 10^4$ L/mol-cm

13-21 $[tet] = 1.38 \times 10^{-5}$ F,
$[epi] = 1.13 \times 10^{-5}$ F

13-25   5.34 mg/100 mL

13-28 $[X] = 1.9 \times 10^{-3}\,F$,

        $[Y] = 1.5 \times 10^{-4}\,F$

13-32 0.869 ± 0.019 ppm

CHAPTER 14

14-4   3.9 ppm

14-6   0.80 ppm

14-11 4.2×10^2 ppm

14-13   3.25 mg/100 mL

14-17   8

14-19   0.249 Å

14-21   20.3°

CHAPTER 15

15-2   a.

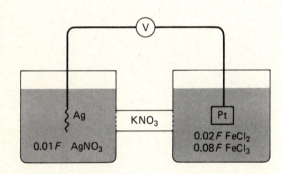

15-3   a. $Ag + Fe^{3+} \rightleftharpoons Ag^+ + Fe^{2+}$

15-4   a. $+0.126$ V

15-5   a. Same direction as in 15-3a

15-7   a. $+0.136$ V

15-8   a. $2I^- + 2Fe^{3+} \xrightarrow{\text{spontaneous}} I_2(s) + 2Fe^{2+}$

15-9   a. $Cu|CuSO_4(0.05\ F)\|CdSO_4(0.20\ F)|Cd$

15-10 a. -0.722 V

15-11 a. $Cu^{2+} + Cd \xrightarrow{\text{spontaneous}} Cu + Cd^{2+}$

15-12 a. $[Ag^+] = 0.0265\ F$

15-13 a. $[Cu^{2+}] = 2.3 \times 10^{-9}\ F$

15-14 a. -151 mV vs. NHE

15-15 a. $+57$ mV vs. NHE

15-16 a. 4.8×10^{-13}

15-18 6.3×10^{-9}

15-20 b. 3.53

15-21 b. 6.07

15-23 $+0.0024$ V

15-25 a. 7.92

15-28 a. $+6\%$

15-29   35.8 ppm

15-33   385 ppm

15-35 ± 0.034 pH units

CHAPTER 16

16-4 11.25 ± 0.05 mA

16-7   7.73%

16-9   10.36%

16-12   19.5%

16-16   EW = 147

16-20   0.0100%

16-22   3.46%

CHAPTER 17

17-1 $3.04\ \text{mg}^{2/3}\text{-sec}^{-1/2}$

17-3   6.79 μA

17-5   b. 6.99%

17-8   b. 2.1×10^{-4}

17-10   12.4 mg/100 mL

17-13 2.15×10^3 ppm

17-16 $n = 1.90(\sim 2$; appears reversible),

        $E_{1/2} = -0.411$ V vs. SCE

17-20   0.10 ppm Zn, no Cd, 0.34 ppm Cu

CHAPTER 18

18-1   a. 10.5

18-3   a. 49% heptane, 51% octane

18-6 $P_{total} = 380$ torr, $\chi_{A,v} = 0.37$

18-7   a. 10 cm/plate

18-9   a.   5.7

18-11   a.   79%

CHAPTER 19

18-12   a.   2.6%;   b. 0.25%

18-14   11.9

18-17   1.6%

18-20   a. [HOAc]   "K_D"

0.28	0.47
0.80	0.49
2.90	0.57

18-23 $D = K_D\alpha_2$ (generalized, $D = K_D\alpha_n$ for neutral MX_n complex)

18-27

Vessel	$[A]_m(\%)$	$[A]_s(\%)$
0	4.5	27.1
1	6.0	36.2
2	3.0	18.1
3	0.7	4.0
4	0.06	0.34

19-2 $R = 1.89$, yes

19-5   a.   5 min;   b. 15 min;   c. 3;   d. 0;
        e.   500 mL;   f. 125 mL

19-7   a.   88.6% toluene, 11.4% chlorobenzene

19-10   585

19-13   1.1 mL

19-15   0.46 ppm

19-19   311 ppm

19-28   6.7 cm

John Wiley & Sons, Inc. **19.11** Reprinted with permission from O. J. McEwen, *Analytical Chemistry,* 38 (1966). Copyright 1982 American Chemical Society. **19.12** Reprinted with permission from A. Davis and H. M. Joseph, *Analytical Chemistry*, 39 (1967). Copyright 1982 American Chemical Society. **19.13** Reprinted with permission from D. H. Desty and A. Goldup, *Analytical Chemistry*, 50 (1978). Copyright 1982 American Chemical Society. **19.14** Reprinted with permission from Roy Wetzel, *Environmental Science and Technology*, 13 (1979). Copyright 1982 American Chemical Society. **19.15** Glenn T. Seaborg, *Man-Made Transuranium Elements*, © 1963, p. 49. Reprinted by permission of Prentice-Hall, Inc., Englewood Cliffs, N.J. **19.20** Reprinted with permission from K. L. Dunlap, R. L. Sandridge and Jurgen Keller, *Analytical Chemistry,* 48 (1976). Copyright 1982 American Chemical Society. **19.22, 19.23** From *High Performance Liquid Chromatography*, ed. J. H. Knox, 1979, Edinburgh University Press.

INDEX

C

Cadmium
determination of, 461, 594
polarographic reduction, 571, 576
Caffeine
detection of, **134, 135**
Calcium
determination of, 66, 159–62, 298, 303–305, 308, 351, 513, **77–81, 87, 88**
separation from magnesium, 126
Calcium carbonate, 176
primary standard, 194, 303
Calcium oxalate, 174, 175
Calcium sulfate, **33**
Calibration
buret, **22, 23**
pipet, **19, 20**
Calibration curve, 389
Calmagite, 300
Calomel electrode, *see* saturated calomel electrode
Capacity factor, (k'), 614, 642
Capillary column, 657, 658, 669
Carbon
determination of, 178, 202
Carbon arc lamp, 406
Carbonate error, 269
Carbonate-bicarbonate mixture
determination of, 274, **70**
Carbon dioxide
determination of, 271
Carbon-hydrogen analysis
microdetermination of, 114–17
Carbon monoxide
determination of, 370
Carbon tetrachloride
hazard, **49**
Carbowax
stationary liquid phase, 659
Carrier gas, 654
Cathode, 169
mercury, 172, 549–51
Cathodic depolarizer, 171
Cation
determination of, 302
Cation exchange resin, 671, 672
Cell, *see* electrochemical cell
Central value $(\bar{X})$, 16, 38
Cerium (IV), 356
electrogenerated, 529
primary standard, 356
standard solution, 194, 356
Characteristic, 19
Charge balance, 232
Charge-transfer absorption, 427
Charring, **36**
Chelate, 281, 627
Chemical coulometer, 545
Chemical equation, 59
Chemicals

chemically pure (CP) grade, **38**
hazardous, **49**
incompatible, **50**
reagent grade, **38**
specifications for, **38**
technical grade, **39**
USP grade, **38**
Chloride
determination of, 131, 134, 135, 140, 209–22, 432, **53–55, 63–65**
Chlorine
determination of, 179, 532, 550
electrogenerated, 523
Chlorophenol red
acid-base indicator, 251
Cholesterol
determination of, 433
Chopping, 457, 458
Chromatogram, 641, 645, 663–66
Chromatographic peak
area, 666
band tailing, 659
height, 666
shape, 647
width (W), 647–56
Chromatographic separation
improving resolution, 655
Chromatography, 639–92
elution time profile, 641
gas–solid (GSC), 604
liquid–solid (LSC), 604
methods, 640
theoretical principles, 642–56
Chromium
determination of, **99, 100, 112, 113**
Chromium (II)
titrant, 366
Chromium (VI)
determination of, 156, 201, 432
Chromophore, 426
Chromosorb-P, 658
Cleaning solution, **13**
"Clear point"
endpoint signal, 196, 212
Cobalt
determination of, 164, 396
Colorimeter, 408, 415, 420
Columns, *see* gas chromatography *and* high performance liquid chromatography
Common ion effect, 119–26
Complexation titration 281–311
applications, 302–308
endpoint signal, 296–302
Complementary colors, 410
Concentration, 60–63
Concentration gradient, 560
Conditional formation constant (K'), 287, 294–96
Conductometric titration
endpoint signal, 196
Confidence interval, 31–35, 38

Formula weights*

AgBr	187.77	$HC_7H_5O_2$ (benzoic acid)	122.12
AgCl	143.32	$H_2C_2O_4 \cdot 2H_2O$ (oxalic acid)	126.07
Ag_2CrO_4	331.73	HCl	36.46
AgI	234.77	$HClO_4$	100.46
$AgNO_3$	169.87	H_5IO_6	227.94
AgSCN	165.95	HNO_3	63.01
$Al(C_9H_6NO)_3$ (oxinate)	459.44	H_2O	18.02
Al_2O_3	101.96	H_2O_2	34.01
$Al_2(SO_4)_3$	342.14	H_3PO_4	98.00
As_2O_3	197.84	H_2S	34.08
As_2S_3	246.02	H_2SO_4	98.07
$BaCl_2 \cdot 2H_2O$	244.27	Hg_2Br_2	560.99
$BaSO_4$	233.39	Hg_2Cl_2	472.09
Bi_2O_3	465.96	$HgCl_2$	271.50
C_2H_2 (acetylene)	26.04	HgO	216.59
C_2H_5NS (thioacetamide)	75.13	HgS	232.65
$C_6H_{12}O_6$ (glucose)	180.16	I_2	253.81
C_6H_5OH (phenol)	94.11	$KB(C_6H_5)_4$	358.33
$C_4H_8N_2O_2$ (dimethylglyoxime)	116.12	KBr	119.00
CO	28.01	$KBrO_3$	167.00
CO_2	44.01	KCN	65.12
CON_2H_4 (urea)	60.06	KCl	74.55
$CaCO_3$	100.09	$KClO_3$	122.55
CaC_2O_4	128.10	$KClO_4$	138.55
$CaCl_2$	110.99	$K_2Cr_2O_7$	294.18
CaF_2	78.08	$K_3Fe(CN)_6$	329.25
CaO	56.08	$K_4Fe(CN)_6$	368.35
$Ca_3(PO_4)_2$	310.18	$KHC_8H_4O_4$ (phthalate)	204.22
$CaSO_4$	136.14	$KH(IO_3)_2$	389.91
$CaSO_4 \cdot 2H_2O$	172.17	K_2HPO_4	174.18
$Cd(OH)_2$	146.42	KH_2PO_4	136.09
$Ce(NH_4)_2(NO_3)_6$	548.23	KI	166.00
Cr_2O_3	151.99	KIO_3	214.00
CuO	79.55	KIO_4	230.00
CuS	95.61	$KMnO_4$	158.03
$CuSO_4 \cdot 5H_2O$	249.68	KNO_3	101.10
$Fe(C_9H_6NO)_3$ (oxinate)	488.30	K_2O	94.20
$Fe(NH_4)_2(SO_4)_2 \cdot 6H_2O$	392.13	KOH	56.11
Fe_2O_3	159.69	KSCN	97.18
$Fe(OH)_2$	89.86	K_2SO_4	174.25
$Fe(OH)_3$	106.87	$La(IO_3)_3$	663.61
$FeSO_4$	151.90	LiCl	42.39
$Fe_2(SO_4)_3$	399.87	LiI	133.85
$HCHO_2$ (formic acid)	46.03	$Mg_2As_2O_7$	310.45
$HC_2H_3O_2$ (acetic acid)	60.05	$Mg(C_9H_6NO)_2$ (oxinate)	312.61
$H_2C_4H_4O_6$ (tartaric acid)	150.09	$MgCO_3$	84.31
$HC_6H_7O_6$ (ascorbic acid)	176.13	$MgNH_4AsO_4$	181.26